To Sabine

Helmut Lütkepohl

Introduction to Multiple Time Series Analysis

Second Edition

With 34 Figures

Springer-Verlag

Berlin Heidelberg New York
London Paris Tokyo
Hong Kong Barcelona
Budapest

Professor Dr. Helmut Lütkepohl
Institute of Statistics and Econometrics
Department of Economics
Humboldt-University of Berlin
Spandauer Straße 1
10178 Berlin, FRG

ISBN 3-540-56940-5 Springer-Verlag Berlin Heidelberg New York Tokyo
ISBN 0-387-56940-5 Springer-Verlag New York Berlin Heidelberg Tokyo

ISBN 3-540-53194-7 1. Aufl. Springer-Verlag Berlin Heidelberg New York Tokyo
ISBN 0-387-53194-7 1st Ed. Springer-Verlag New York Berlin Heidelberg Tokyo

Library of Congress Cataloging-in-Publication Data
Lütkepohl, Helmut.
Introduction to multiple time series analysis / Helmut Lütkepohl. - - 2nd ed.
p. cm. Includes bibliographical references and indexes.
ISBN 3-540-56940-5 (Berlin). - - ISBN 0-387-56940-5 (New York)
1. Time-series analysis. I. Title.
QA280.L87 1993
519.5'5- - dc20

Printing: Druckhaus Beltz, Hemsbach
Bookbinding: J. Schäffer GmbH u. Co. KG, Grünstadt
2142/7130-54321 - Printed on acid-free paper

Preface to the Second Edition

Since the book sold unexpectedly well a second printing had to be prepared less than two years after it was first published. In order to ensure a continuous supply there was only very little time to make changes. Therefore it was decided to reprint the book in its orginal form except for very minor changes such as the correction of some printing errors. In particular the references are not updated. For marketing purposes the publisher felt that it was still suitable to use the lable "Second Edition" for the present version of the book.

Berlin, May 1993 Helmut Lütkepohl

Preface to the First Edition

A few years ago, when I started teaching a course on multiple time series analysis for graduate students in business and economics at the University of Kiel, Germany, a suitable textbook was not available. Therefore I prepared lecture notes from which the present text evolved. There are a number of books on time series analysis that contain portions on multiple time series models and methods. However, none of them covers the full range of models and methods that are now commonly used in the economics and econometrics literature. The present text also neglects some tools which are currently in widespread use. It goes beyond other time series books, however. Unlike advanced books on the topic such as Hannan & Deistler (1988) this text is written at a level which should be accessible to graduate students in business and economics. The issues and models covered in detail reflect my personal interests. I admit that some important models and methods are omitted just because I havn't used them much in the past. Especially the omission of spectral methods may perhaps be regreted by some instructors.

Although multiple time series analysis is applied in many disciplines I have prepared the text with economics and business students in mind. The examples and exercises are chosen accordingly. Despite this orientation I hope that the book will also serve multiple time series courses in other fields. It contains enough material for a one semester course on multiple time series analysis. It may also be combined with univariate time series books such as Schlittgen & Streitberg (1984) and Pankratz (1983) or with texts like Fuller (1976) to form the basis of a one or two semester course on univariate and multivariate time series analysis. Alternatively, it is also possible to select some of the chapters or sections for a special topic of a graduate level econometrics course.

Chapters 1–4 contain an introduction to the vector autoregressive methodology. Chapters 1–9 may be used as an introduction to vector autoregressive and mixed autoregressive moving average models. Chapter 10 briefly reviews econometric dynamic simultaneous equations models; Chapter 11 considers the currently popular cointegration topic; in Chapter 12 models with systematically varying coefficients are treated, and state space models are discussed in Chapter 13. In a course or special topic on vector autoregressive models all or some of Chapters 11–13 may be treated right after Chapter 5. It is also possible to cover Chapters 1–5 and 11 first and then proceed with Chapters 6–9 and, if time permits, conclude a course with Chapters 12 and 13.

The students participating in my multiple time series course typically have knowledge of matrix algebra. They also have been introduced to mathematical statistics, for instance, based on textbooks like Mood, Graybill & Boes (1974), Hogg & Craig (1978), or Rohatgi (1976). Moreover, many of them have a working knowledge of the Box-Jenkins approach and other univariate time series techniques. Although, in principle, it may be possible to use the present text without any prior knowledge of univariate time series analysis if the instructor provides the required motivation, it is clearly an advantage to have some time series background. Also, a previous introduction to econometrics will be helpful. Matrix algebra and an introductory mathematical statistics course plus the multiple regression model are necessary prerequisites.

Since this is meant to be an introductory exposition I am not striving for utmost generality. For instance, quite often I use the normality assumption although the considered results hold under more general conditions. The emphasis is on explaining the underlying ideas and not on generality. In Chapters 2–5 a number of results are proven to illustrate some of the techniques that are often used in the multiple time series arena. Most proofs may be skipped without loss of continuity. Therefore the beginning and the end of a proof are usually clearly marked. Many results are summarized in propositions for easy reference. A listing of the propositions is provided at the end of the book.

The appendices contain a collection of useful results on matrix algebra, multivariate normal distributions, asymptotic theory and so on. Some results and topics were added in response to questions and comments by students. It is not necessary to know all of these results before going through the chapters of the book. It may be useful, however, to have them nearby because many of them are used and referred to in the main body of the text. Therefore they are included in the appendix.

Exercises are given at the end of each chapter with the exception of Chapter 1. Some of the problems may be too difficult for students without a good formal training. Some are just included to avoid details of proofs given in the text. In most chapters empirical exercises are provided in addition to algebraic problems. Solving the empirical problems requires the use of a computer. Matrix oriented software such as GAUSS, MATLAB, or SAS will be most helpful. A menu driven program based on GAUSS has been developed which can do most of the examples and exercises.

Many persons have contributed directly or indirectly to this book and I am very grateful to all of them. Many students have commented on my lecture notes. Thereby they have helped to improve the presentation and to correct errors. A number of colleagues have commented on portions of the manuscript and have been available for discussions on the topics covered. These comments and discussions have been very helpful for my own understanding of the subject and have resulted in improvements to the manuscript.

Although the persons who have contributed to the project in some way or other are too numerous to be listed here I wish to express my special gratitude to some of them. Special thanks go to Theo Dykstra who read and commented on a large part of the manuscript during his visit in Kiel in the summer of 1990. Hans-Eggert Reimers read the entire manuscript, suggested many improvements, and pointed out numerous errors. Wolfgang Schneider helped with the example in Chapter 13 and also commented on parts of the manuscript. Bernd Theilen prepared the final versions of most figures, Knut Haase and Holger Claessen performed the computations for the examples. I deeply appreciate the help of all these collaborators. Last but not least I owe an obligation to Mrs. Milda Tauchert who performed the difficult technical typing job. She skillfully typed and retyped numerous versions of the manuscript. Of course, I assume full responsibility for any remaining errors and I welcome any comments by readers.

Kiel, August 1990 Helmut Lütkepohl

Table of Contents

Chapter 1. Introduction 1

1.1 Objectives of Analyzing Multiple Time Series 1
1.2 Some Basics 2
1.3 Vector Autoregressive Processes 3
1.4 Outline of the Following Chapters 5

Part I. Finite Order Vector Autoregressive Processes 7

Chapter 2. Stable Vector Autoregressive Processes 9

2.1 Basic Assumptions and Properties of VAR Processes 9
2.1.1 Stable VAR(p) Processes 9
2.1.2 The Moving Average Representation of a VAR Process . 13
2.1.3 Stationary Processes 19
2.1.4 Computation of Autocovariances and Autocorrelations of Stable VAR Processes 21
2.1.4a Autocovariances of a VAR(1) Process 21
2.1.4b Autocovariances of a Stable VAR(p) Process 23
2.1.4c Autocorrelations of a Stable VAR(p) Process ... 25

2.2 Forecasting 27
2.2.1 The Loss Function 27
2.2.2 Point Forecasts 28
2.2.2a Conditional Expectation 28
2.2.2b Linear Minimum MSE Predictor 30
2.2.3 Interval Forecasts and Forecast Regions 33

2.3 Structural Analysis with VAR Models 35
2.3.1 Granger-Causality and Instantaneous Causality 35
2.3.1a Definitions of Causality 35
2.3.1b Characterization of Granger-Causality 37
2.3.1c Characterization of Instantaneous Causality 40
2.3.1d Interpretation and Critique of Instantaneous and Granger-Causality 41

2.3.2 Impulse Response Analysis ... 43
2.3.2a Responses to Forecast Errors ... 43
2.3.2b Responses to Orthogonal Impulses ... 48
2.3.2c Critique of Impulse Response Analysis ... 55
2.3.3 Forecast Error Variance Decomposition ... 56
2.3.4 Remarks on the Interpretation of VAR Models ... 58
2.4 Exercises ... 59

Chapter 3. Estimation of Vector Autoregressive Processes ... 62

3.1 Introduction ... 62
3.2 Multivariate Least Squares Estimation ... 62
3.2.1 The Estimator ... 62
3.2.2 Asymptotic Properties of the Least Squares Estimator ... 65
3.2.3 An Example ... 70
3.2.4 Small Sample Properties of the LS Estimator ... 73
3.3 Least Squares Estimation with Mean-Adjusted Data and Yule-Walker Estimation ... 75
3.3.1 Estimation when the Process Mean Is Known ... 75
3.3.2 Estimation of the Process Mean ... 76
3.3.3 Estimation with Unknown Process Mean ... 78
3.3.4 The Yule-Walker Estimator ... 78
3.3.5 An Example ... 79
3.4 Maximum Likelihood Estimation ... 80
3.4.1 The Likelihood Function ... 80
3.4.2 The ML Estimators ... 81
3.4.3 Properties of the ML Estimators ... 82
3.5 Forecasting with Estimated Processes ... 85
3.5.1 General Assumptions and Results ... 85
3.5.2 The Approximate MSE Matrix ... 87
3.5.3 An Example ... 89
3.5.4 A Small Sample Investigation ... 91
3.6 Testing for Granger-Causality and Instantaneous Causality ... 93
3.6.1 A Wald Test for Granger-Causality ... 93
3.6.2 An Example ... 94
3.6.3 Testing for Instantaneous Causality ... 95
3.7 The Asymptotic Distributions of Impulse Responses and Forecast Error Variance Decompositions ... 97
3.7.1 The Main Results ... 97
3.7.2 Proof of Proposition 3.6 ... 103

3.7.3 An Example ... 105
3.7.4 Investigating the Distributions of the Impulse Responses by Simulation Techniques ... 113

3.8 Exercises ... 114
3.8.1 Algebraic Problems ... 114
3.8.2 Numerical Problems ... 116

Chapter 4. VAR Order Selection and Checking the Model Adequacy ... 118

4.1 Introduction ... 118
4.2 A Sequence of Tests for Determining the VAR Order ... 119
4.2.1 The Impact of the Fitted VAR Order on the Forecast MSE ... 119
4.2.2 The Likelihood Ratio Test Statistic ... 121
4.2.3 A Testing Scheme for VAR Order Determination ... 125
4.2.4 An Example ... 126

4.3 Criteria for VAR Order Selection ... 128
4.3.1 Minimizing the Forecast MSE ... 128
4.3.2 Consistent Order Selection ... 130
4.3.3 Comparison of Order Selection Criteria ... 132
4.3.4 Some Small Sample Simulation Results ... 135

4.4 Checking the Whiteness of the Residuals ... 138
4.4.1 The Asymptotic Distributions of the Autocovariances and Autocorrelations of a White Noise Process ... 139
4.4.2 The Asymptotic Distributions of the Residual Autocovariances and Autocorrelations of an Estimated VAR Process ... 142
4.4.2a Theoretical Results ... 142
4.4.2b An Illustrative Example ... 148
4.4.3 Portmanteau Tests ... 150

4.5 Testing for Nonnormality ... 152
4.5.1 Tests for Nonnormality of a Vector White Noise Process ... 152
4.5.2 Tests for Nonnormality of a VAR Process ... 155

4.6 Tests for Structural Change ... 159
4.6.1 A Test Statistic Based on one Forecast Period ... 159
4.6.2 A Test Based on Several Forecast Periods ... 161
4.6.3 An Example ... 163

4.7 Exercises ... 164
4.7.1 Algebraic Problems ... 164
4.7.2 Numerical Problems ... 165

Chapter 5. VAR Processes with Parameter Constraints 167

5.1 Introduction 167
5.2 Linear Constraints 168
5.2.1 The Model and the Constraints 168
5.2.2 LS, GLS, and EGLS Estimation 169
5.2.2a Asymptotic Properties 169
5.2.2b Comparison of LS and Restricted EGLS Estimators 172
5.2.3 Maximum Likelihood Estimation 173
5.2.4 Constraints for Individual Equations 174
5.2.5 Restrictions on the White Noise Covariance Matrix 175
5.2.6 Forecasting 177
5.2.7 Impulse Response Analysis and Forecast Error Variance Decomposition 178
5.2.8 Specification of Subset VAR Models 179
5.2.8a Elimination of Complete Matrices 180
5.2.8b Top-down Strategy 180
5.2.8c Bottom-up Strategy 182
5.2.8d Monte Carlo Comparison of Strategies for Subset VAR Modeling 183
5.2.9 Model Checking 186
5.2.9a Residual Autocovariances and Autocorrelations . 186
5.2.9b Portmanteau Tests 188
5.2.9c Other Checks of Restricted Models 188
5.2.10 An Example 189

5.3 VAR Processes with Nonlinear Parameter Restrictions 192
5.3.1 Some Types of Nonlinear Constraints 192
5.3.2 Reduced Rank VAR Models 195
5.3.3 Multivariate LS Estimation of Reduced Rank VAR Models 197
5.3.4 Asymptotic Properties of Reduced Rank LS Estimators . 198
5.3.5 Specification and Checking of Reduced Rank VAR Models . 201
5.3.6 An Illustrative Example 202

5.4 Bayesian Estimation 206
5.4.1 Basic Terms and Notations 206
5.4.2 Normal Priors for the Parameters of a Gaussian VAR Process 206
5.4.3 The Minnesota or Litterman Priors 208
5.4.4 Practical Considerations 209
5.4.5 An Example 210
5.4.6 Classical versus Bayesian Interpretation of $\bar{\alpha}$ in Forecasting and Structural Analyses 211

5.5 Exercises 212
5.5.1 Algebraic Exercises 212
5.5.2 Numerical Problems 213

Part II. Infinite Order Vector Autoregressive Processes 215

Chapter 6. Vector Autoregressive Moving Average Processes 217

6.1 Introduction 217
6.2 Finite Order Moving Average Processes 217
6.3 VARMA Processes 220
6.3.1 The Pure MA and Pure VAR Representations of a VARMA Process 220
6.3.2 A VAR(1) Representation of a VARMA Process 223
6.4 The Autocovariances and Autocorrelations of a VARMA(p, q) Process 226
6.5 Forecasting VARMA Processes 228
6.6 Transforming and Aggregating VARMA Processes 230
6.6.1 Linear Transformations of VARMA Processes 231
6.6.2 Aggregation of VARMA Processes 235
6.7 Interpretation of VARMA Models 236
6.7.1 Granger-Causality 236
6.7.2 Impulse Response Analysis 238
6.8 Exercises 239

Chapter 7. Estimation of VARMA Models 241

7.1 The Identification Problem 241
7.1.1 Nonuniqueness of VARMA Representations 241
7.1.2 Final Equations Form and Echelon Form 246
7.1.3 Illustrations 249
7.2 The Gaussian Likelihood Function 252
7.2.1 The Likelihood Function of an MA(1) Process 252
7.2.2 The MA(q) Case 254
7.2.3 The VARMA(1, 1) Case 255
7.2.4 The General VARMA(p, q) Case 257
7.3 Computation of the ML Estimates 259
7.3.1 The Normal Equations 260
7.3.2 Optimization Algorithms 262
7.3.3 The Information Matrix 264
7.3.4 Preliminary Estimation 265
7.3.5 An Illustration 268

7.4 Asymptotic Properties of the ML Estimators 271
7.4.1 Theoretical Results 271
7.4.2 A Real Data Example 277
7.5 Forecasting Estimated VARMA Processes 278
7.6 Estimated Impulse Responses 281
7.7 Exercises 282

Chapter 8. Specification and Checking the Adequacy of VARMA Models 284

8.1 Introduction 284
8.2 Specification of the Final Equations Form 285
8.2.1 A Specification Procedure 285
8.2.2 An Example 287
8.3 Specification of Echelon Forms 289
8.3.1 A Procedure for Small Systems 289
8.3.2 A Full Search Procedure Based on Linear Least Squares Computations 291
8.3.2a The Procedure 291
8.3.2b An Example 292
8.3.3 Hannan-Kavalieris Procedure 294
8.3.4 Poskitt's Procedure 295
8.4 Remarks on other Specification Strategies for VARMA Models 297
8.5 Model Checking 298
8.5.1 LM Tests 298
8.5.2 Residual Autocorrelations and Portmanteau Tests 300
8.5.3 Prediction Tests for Structural Change 301
8.6 Critique of VARMA Model Fitting 302
8.7 Exercises 302

Chapter 9. Fitting Finite Order VAR Models to Infinite Order Processes 305

9.1 Background 305
9.2 Multivariate Least Squares Estimation 305
9.3 Forecasting 309
9.3.1 Theoretical Results 309
9.3.2 An Example 311
9.4 Impulse Response Analysis and Forecast Error Variance Decompositions 313
9.4.1 Asymptotic Theory 313
9.4.2 An Example 316
9.5 Exercises 317

Part III. Systems with Exogenous Variables and Nonstationary Processes 321

Chapter 10. Systems of Dynamic Simultaneous Equations 323

10.1 Background 323
10.2 Systems with Exogenous Variables 324
10.2.1 Types of Variables 324
10.2.2 Structural Form, Reduced Form, Final Form 325
10.2.3 Models with Rational Expectations 328
10.3 Estimation 331
10.4 Remarks on Model Specification and Model Checking 333
10.5 Forecasting 334
10.5.1 Unconditional and Conditional Forecasts 334
10.5.2 Forecasting Estimated Dynamic SEMs 337
10.6 Multiplier Analysis 338
10.7 Optimal Control 339
10.8 Concluding Remarks on Dynamic SEMs 342
10.9 Exercises 343

Chapter 11. Nonstationary Systems with Integrated and Cointegrated Variables 346

11.1 Introduction 346
11.1.1 Integrated Processes 346
11.1.2 Cointegrated Processes 351
11.2 Estimation of Integrated and Cointegrated VAR(p) Processes 355
11.2.1 ML Estimation of a Gaussian Cointegrated VAR(p) Process 356
11.2.1a The ML Estimators and their Properties 356
11.2.1b An Example 360
11.2.1c Discussion of the Proof of Proposition 11.2 363
11.2.2 Other Estimation Methods for Cointegrated Systems 368
11.2.2a Unconstrained LS Estimation 369
11.2.2b A Two-Stage Procedure 370
11.2.3 Bayesian Estimation of Integrated Systems 372
11.2.3a Generalities 372
11.2.3b The Minnesota or Litterman Prior 373
11.2.3c An Example 374
11.3 Forecasting and Structural Analysis 375
11.3.1 Forecasting Integrated and Cointegrated Systems 375
11.3.2 Testing for Granger-Causality 378

11.3.2a The Noncausality Restrictions 378
11.3.2b A Wald Test for Linear Constraints 378
11.3.3 Impulse Response Analysis 379
11.3.3a Theoretical Considerations 379
11.3.3b An Example 380

11.4 Model Selection and Model Checking 382
11.4.1 VAR Order Selection 382
11.4.2 Testing for the Rank of Cointegration 384
11.4.3 Prediction Tests for Structural Change 387

11.5 Exercises 388
11.5.1 Algebraic Exercises 388
11.5.2 Numerical Exercises 390

Chapter 12. Periodic VAR Processes and Intervention Models 391

12.1 Introduction 391
12.2 The VAR(p) Model with Time Varying Coefficients 392
12.2.1 General Properties 393
12.2.2 ML Estimation 394

12.3 Periodic Processes 396
12.3.1 A VAR Representation with Time Invariant Coefficients 397
12.3.2 ML Estimation and Testing for Varying Parameters .. 399
12.3.2a All Coefficients Time Varying 400
12.3.2b All Coefficients Time Invariant 401
12.3.2c Time Invariant White Noise 401
12.3.2d Time Invariant Covariance Structure 402
12.3.2e LR Tests 402
12.3.2f Testing a Model with Time Varying White Noise only Against one with all Coefficients Time Varying 403
12.3.2g Testing a Time Invariant Model Against one with Time Varying White Noise 404
12.3.3 An Example 406
12.3.4 Bibliographical Notes and Extensions 408

12.4 Intervention Models 408
12.4.1 Interventions in the Intercept Model 408
12.4.2 A Discrete Change in the Mean 410
12.4.3 An Illustrative Example 411
12.4.4 Extensions and References 412

12.5 Exercises 413

Chapter 13. State Space Models ... 415

13.1 Background ... 415
13.2 State Space Models ... 416
13.2.1 The General Linear State Space Model ... 416
13.2.1a A Finite Order VAR Process ... 418
13.2.1b A VARMA(p, q) Process ... 419
13.2.1c The VARX Model ... 420
13.2.1d Systematic Sampling and Aggregation ... 420
13.2.1e Structural Time Series Models ... 421
13.2.1f Factor Analytic Models ... 422
13.2.1g VARX Models with Systematically Varying Coefficients ... 424
13.2.1h Random Coefficient VARX Models ... 424
13.2.2 Nonlinear State Space Models ... 426
13.3 The Kalman Filter ... 428
13.3.1 The Kalman Filter Recursions ... 429
13.3.1a Assumptions for the State Space Model ... 429
13.3.1b The Recursions ... 429
13.3.1c Computational Aspects and Extensions ... 431
13.3.2 Proof of the Kalman Filter Recursions ... 432
13.4 Maximum Likelihood Estimation of State Space Models ... 434
13.4.1 The Log-Likelihood Function ... 434
13.4.2 The Identification Problem ... 435
13.4.3 Maximization of the Log-Likelihood Function ... 436
13.4.3a The Gradient of the Log-Likelihood ... 437
13.4.3b The Information Matrix ... 437
13.4.3c Discussion of the Scoring Algorithm ... 437
13.4.4 Asymptotic Properties of the ML Estimators ... 438
13.5 A Real Data Example ... 439
13.6 Exercises ... 444

Appendices ... 447

Appendix A. Vectors and Matrices ... 449

A.1 Basic Definitions ... 449
A.2 Basic Matrix Operations ... 450
A.3 The Determinant ... 451
A.4 The Inverse, the Adjoint, and Generalized Inverses ... 453
A.4.1 Inverse and Adjoint of a Square Matrix ... 453
A.4.2 Generalized Inverses ... 454

A.5 The Rank ... 455
A.6 Eigenvalues and -vectors – Characteristic Values and Vectors . 455
A.7 The Trace ... 457
A.8 Some Special Matrices and Vectors ... 457
A.8.1 Idempotent and Nilpotent Matrices ... 457
A.8.2 Orthogonal Matrices and Vectors ... 458
A.8.3 Definite Matrices and Quadratic Forms ... 458
A.9 Decomposition and Diagonalization of Matrices ... 459
A.9.1 The Jordan Canonical Form ... 459
A.9.2 Decomposition of Symmetric Matrices ... 460
A.9.3 The Choleski Decomposition of a Positive Definite Matrix ... 461
A.10 Partitioned Matrices ... 462
A.11 The Kronecker Product ... 463
A.12 The vec and vech Operators and Related Matrices ... 464
A.12.1 The Operators ... 464
A.12.2 The Elimination, Duplication, and Commutation Matrices ... 465
A.13 Vector and Matrix Differentiation ... 467
A.14 Optimization of Vector Functions ... 474
A.15 Problems ... 478

Appendix B. Multivariate Normal and Related Distributions ... 480

B.1 Multivariate Normal Distributions ... 480
B.2 Related Distributions ... 481

Appendix C. Convergence of Sequences of Random Variables and Asymptotic Distributions ... 484

C.1 Concepts of Stochastic Convergence ... 484
C.2 Asymptotic Properties of Estimators and Test Statistics ... 487
C.3 Infinite Sums of Random Variables ... 488
C.4 Maximum Likelihood Estimation ... 491
C.5 Likelihood Ratio, Lagrange Multiplier, and Wald Tests ... 492

Appendix D. Evaluating Properties of Estimators and Test Statistics by Simulation and Resampling Techniques ... 495

D.1 Simulating a Multiple Time Series with VAR Generation Process 495
D.2 Evaluating Distributions of Functions of Multiple Time Series by Simulation ... 496
D.3 Evaluating Distributions of Functions of Multiple Time Series by Resampling ... 497

Appendix E. Data Used for Examples and Exercises 498

References 509

List of Propositions and Definitions 518

Index of Notation 521

Author Index 527

Subject Index 531

Chapter 1. Introduction

1.1 Objectives of Analyzing Multiple Time Series

In making choices between alternative courses of action, decision makers at all structural levels often need predictions of economic variables. If time series observations are available for a variable of interest and the data from the past contain information about the future development of a variable, it is plausible to use as forecast some function of the data collected in the past. For instance, in forecasting the monthly unemployment rate, from past experience a forecaster may know that in some country or area a high unemployment rate in one month tends to be followed by a high rate in the next month. In other words, the rate changes only gradually. Assuming that the tendency prevails in future periods, forecasts can be based on current and past data.

Formally this may be expressed as follows. Let y_t denote the value of the variable of interest in period t. Then a forecast for period $T + h$, made at the end of period T, may have the form

$$\hat{y}_{T+h} = f(y_T, y_{T-1}, \ldots), \tag{1.1.1}$$

where $f(\cdot)$ denotes some suitable function of the past observations $y_T, y_{T-1}, \ldots$. For the moment it is left open how many past observations enter into the forecast. One major goal of univariate time series analysis is to specify sensible forms of the functions $f(\cdot)$. In many applications linear functions have been used so that, for example,

$$\hat{y}_{T+h} = \nu + \alpha_1 y_T + \alpha_2 y_{T-1} + \cdots .$$

In dealing with economic variables often the value of one variable is not only related to its predecessors in time but, in addition, it depends on past values of other variables. For instance, household consumption expenditures may depend on variables such as income, interest rates, and investment expenditures. If all these variables are related to the consumption expenditures it makes sense to use their possible additional information content in forecasting consumption expenditures. In other words, denoting the related variables by $y_{1t}, y_{2t}, \ldots, y_{Kt}$, the forecast of $y_{1,T+h}$ at the end of period T may be of the form

$$\hat{y}_{1,T+h} = f_1(y_{1,T}, y_{2,T}, \ldots, y_{K,T}, y_{1,T-1}, y_{2,T-1}, \ldots, y_{K,T-1}, y_{1,T-2}, \ldots).$$

Similarly, a forecast for the second variable may be based on past values of all variables in the system. More generally, a forecast of the k-th variable may be expressed as

$$\hat{y}_{k,T+h} = f_k(y_{1,T}, \dots, y_{K,T}, y_{1,T-1}, \dots, y_{K,T-1}, \dots). \tag{1.1.2}$$

A set of time series y_{kt}, $k = 1, \dots, K$, $t = 1, \dots, T$, is called a *multiple time series* and the previous formula expresses the forecast $\hat{y}_{k,T+h}$ as a function of a multiple time series. In analogy with the univariate case, it is one major objective of multiple time series analysis to determine suitable functions $f_1, \dots, f_K$ that may be used to obtain forecasts with good properties for the variables of the system.

It is also often of interest to learn about the dynamic interrelationships between a number of variables. For instance, in a system consisting of investment, income, and consumption one may want to know about the likely impact of an impulse in income. What will be the present and future implications of such an event for consumption and for investment? Under what conditions can the effect of such an impulse be isolated and trace through the system? Alternatively, given a particular subject matter theory, is it consistent with the relations implied by a multiple time series model which is developed with the help of statistical tools? These and other questions regarding the structure of the relationships between the variables involved are occasionally investigated in the context of multiple time series analysis. Thus, obtaining insight into the dynamic structure of a system is a further objective of multiple time series analysis.

1.2 Some Basics

In the following chapters we will regard the values a particular economic variable has assumed in a specific period as realizations of random variables. A time series will be assumed to be generated by a stochastic process. Although the reader is assumed to be familiar with these terms it may be useful to briefly review some of the basic definitions and expressions at this point in order to make the underlying concepts precise.

Let $(\Omega, \mathcal{M}, \Pr)$ be a *probability space*, where Ω is the set of all elementary events (sample space), $\mathcal{M}$ is a sigma-algebra of events or subsets of Ω and Pr is a probability measure defined on $\mathcal{M}$. A *random variable* y is a real valued function defined on Ω such that for each real number c, $A_c = \{\omega \in \Omega | y(\omega) \leq c\} \in \mathcal{M}$. In other words, A_c is an event for which the probability is defined in terms of Pr. The function $F\colon \mathbb{R} \to [0, 1]$ defined by $F(c) = \Pr(A_c)$ is the distribution function of y.

A K-dimensional *random vector* or a K-dimensional *vector of random variables* is a function y from Ω into the K-dimensional Euclidean space $\mathbb{R}^K$, that is, y maps $\omega \in \Omega$ on $y(\omega) = (y_1(\omega), \dots, y_K(\omega))'$ such that for each $c = (c_1, \dots, c_K)' \in \mathbb{R}^K$,

$$A_c = \{\omega | y_1(\omega) \leq c_1, \dots, y_K(\omega) \leq c_K\} \in \mathcal{M}.$$

The function $F\colon \mathbb{R}^K \to [0, 1]$ defined by $F(c) = \Pr(A_c)$ is the joint *distribution function* of y.

Suppose Z is some index set with at most countably many elements like, for instance, the set of all integers or all positive integers. A (discrete) *stochastic process* is a real valued function

$$y: Z \times \Omega \to \mathbb{R}$$

such that for each fixed $t \in Z$, $y(t, \omega)$ is a random variable. The random variable corresponding to a fixed t is usually denoted by y_t in the following. The underlying probability space will usually not even be mentioned. In that case it is understood that all the members y_t of a stochastic process are defined on the same probability space. Usually the stochastic process will also be denoted by y_t if the meaning of the symbol is clear from the context.

A stochastic process may be described by the joint distribution functions of all finite subcollections of y_t's, $t \in S \subset Z$. In practice the complete system of distributions will often be unknown. Therefore, in the following chapters we will often be concerned with the first and second moments of the distributions. In other words, we will be concerned with the means $E(y_t) = \mu_t$, the variances $E(y_t - \mu_t)^2$ and the covariances $E(y_t - \mu_t)(y_s - \mu_s)$.

A K-dimensional *vector stochastic process* or *multivariate stochastic process* is a function

$$y: Z \times \Omega \to \mathbb{R}^K,$$

where, for each fixed $t \in Z$, $y(t, \omega)$ is a K-dimensional random vector. Again we usually use the symbol y_t for the random vector corresponding to a fixed $t \in Z$. For simplicity we often denote the complete process by y_t. The particular meaning of the symbol should be clear from the context. With respect to the stochastic characteristics the same applies as for univariate processes. That is, the stochastic characteristics are summarized in the joint distribution functions of all finite subcollections of random vectors y_t. In practice interest will often focus on the first and second moments of all random variables involved.

A realization of a (vector) stochastic process is a sequence (of vectors) $y_t(\omega)$, $t \in Z$, for a fixed ω. In other words a realization of a stochastic process is a function $Z \to \mathbb{R}^K$ where $t \to y_t(\omega)$. A (multiple) time series is regarded as such a realization or possibly a finite part of such a realization, that is, it consists, for instance, of values (vectors) $y_1(\omega), \ldots, y_T(\omega)$. The underlying stochastic process is said to have *generated* the (multiple) time series or it is called the *generating* or *generation process* of the time series. A time series $y_1(\omega), \ldots, y_T(\omega)$ will usually be denoted by $y_1, \ldots, y_T$ or simply by y_t just like the underlying stochastic process, if no confusion is possible. The number of observations T is called the *sample size* or *time series length*. With this terminology available we may now return to the problem of specifying forecast functions.

1.3 Vector Autoregressive Processes

Since linear functions are relatively easy to deal with it makes sense to begin with forecasts that are linear functions of past observations. Let us consider a univariate time series y_t and a forecast $h = 1$ period into the future. If $f(\cdot)$ in (1.1.1) is a linear function we have

$$\hat{y}_{T+1} = \nu + \alpha_1 y_T + \alpha_2 y_{T-1} + \cdots.$$

Assuming that only a finite number p, say, of past y values are used in the prediction formula we get

$$\hat{y}_{T+1} = \nu + \alpha_1 y_T + \alpha_2 y_{T-1} + \cdots + \alpha_p y_{T-p+1}. \tag{1.3.1}$$

Of course, the true value y_{T+1} will usually not be exactly equal to the forecast $\hat{y}_{T+1}$. Let us denote the forecast error by $u_{T+1} = y_{T+1} - \hat{y}_{T+1}$ so that

$$y_{T+1} = \hat{y}_{T+1} + u_{T+1} = \nu + \alpha_1 y_T + \cdots + \alpha_p y_{T-p+1} + u_{T+1}. \tag{1.3.2}$$

Now, assuming that our numbers are realizations of random variables and that the same data generation law prevails in each period T, (1.3.2) has the form of an *autoregressive process*

$$y_t = \nu + \alpha_1 y_{t-1} + \cdots + \alpha_p y_{t-p} + u_t, \tag{1.3.3}$$

where the quantities $y_t, y_{t-1}, \ldots, y_{t-p}$, and u_t are now random variables. To actually get an autoregressive (AR) process we assume that the forecast errors u_t of different periods are uncorrelated, that is, u_t and u_s are uncorrelated for $s \neq t$. In other words, we assume that all useful information in the y_t is used in the forecasts so that there are no systematic forecast errors.

If a multiple time series is considered an obvious extension of (1.3.1) would be

$$\hat{y}_{k,T+1} = \nu_k + \alpha_{k1,1} y_{1,T} + \alpha_{k2,1} y_{2,T} + \cdots + \alpha_{kK,1} y_{K,T} + \cdots + \alpha_{k1,p} y_{1,T-p+1} + \cdots + \alpha_{kK,p} y_{K,T-p+1}, \quad k = 1, \ldots, K. \tag{1.3.4}$$

To simplify the notation let $y_t := (y_{1t}, \ldots, y_{Kt})'$, $\hat{y}_t := (\hat{y}_{1t}, \ldots, \hat{y}_{Kt})'$, $\nu := (\nu_1, \ldots, \nu_K)'$ and

$$A_i := \begin{bmatrix} \alpha_{11,i} & \cdots & \alpha_{1K,i} \\ \vdots & \ddots & \vdots \\ \alpha_{K1,i} & \cdots & \alpha_{KK,i} \end{bmatrix}.$$

Then (1.3.4) can be written compactly as

$$\hat{y}_{T+1} = \nu + A_1 y_T + \cdots + A_p y_{T-p+1}. \tag{1.3.5}$$

If the y_t are regarded as random vectors this predictor is just the optimal forecast obtained from a vector autoregressive model of the form

$$y_t = \nu + A_1 y_{t-1} + \cdots + A_p y_{t-p} + u_t, \tag{1.3.6}$$

where the $u_t = (u_{1t}, \ldots, u_{Kt})'$ form a sequence of independently identically distributed random K-vectors with zero mean vector.

Obviously such a model represents a tremendous simplification compared with the general form (1.1.2). Because of its simple structure it enjoys enormous popularity in applied work. We will study this particular model in the following chapters in some detail.

1.4 Outline of the Following Chapters

In Part I of the book, consisting of the next four chapters, we will investigate some basic properties of vector autoregressive (VAR) processes such as (1.3.6). Forecasts based on these processes are discussed and it is shown how VAR processes may be used for analyzing the dynamic structure of a system of variables. Throughout Chapter 2 it is assumed that the process under study is completely known including its coefficient matrices. In the third chapter it is realized that in practice the coefficients of an assumed VAR process will be unknown and must be estimated from a given multiple time series. The estimation of the coefficients is considered and the consequences of using estimated rather than known processes for forecasting and economic analysis are explored. In Chapter 4 the specification and model checking stages of an analysis are considered. Criteria for determining the order p of a VAR process are given and possibilities for checking the assumptions underlying a VAR analysis are discussed.

In systems with many variables and/or large VAR order p the number of coefficients is quite substantial. As a result the estimation precision will be low if estimation is based on time series of the size typically available in economic applications. In order to improve the estimation precision it is useful to place restrictions from nonsample sources on the parameters and thereby reduce the number of coefficients to be estimated. In Chapter 5 VAR processes with parameter constraints and restricted estimation are discussed. Zero restrictions, nonlinear constraints, and Bayesian estimation are treated.

In Part II of the book it is realized that an upper bound p for the VAR order is often not known with certainty. In such a case one may not want to impose any upper bound and allow for an infinite VAR order. There are two ways to make the estimation problem for the potentially infinite number of parameters tractable. First, it may be assumed that they depend on a finite set of parameters. This assumption leads to vector autoregressive moving average (VARMA) processes. Some properties of these processes, parameter estimation and model specification are discussed in Chapters 6–8. In the second approach for dealing with infinite order VAR processes it is assumed that finite order VAR processes are fitted and that the VAR order goes to infinity with the sample size. This approach and its consequences for the estimators, forecasts, and structural analyses are treated in Chapter 9.

In Part III extensions of the models considered in Parts I and II are studied. In many econometric applications it is assumed that some of the variables are determined outside the system under consideration. In other words, they are exogenous variables. VAR processes with exogenous variables are dealt with in Chapter 10. In the econometrics literature such systems are often called systems of dynamic simultaneous equations. In the time series literature they are sometimes referred to as multivariate transfer function models.

Chapters 11, 12, and 13 are devoted to trending variables and processes with time varying coefficients. In Chapters 2–9 trending variables are excluded by assumption to simplify the theoretical analysis. In practice they are the rule rather than the exception. Therefore Chapter 11 deals with models for certain types of trending variables that have been found useful in practice. In Chapter 12 VAR processes with time varying coefficients are considered. The coefficient variability may be due to a one-time intervention from outside the system or it may result from seasonal variation. Finally, in Chapter 13 so-called state space models are introduced. The models represent a very general class which encompasses all the models previously discussed and includes in addition VAR models with stochastically varying coefficients. A brief review of these and other important models for multiple time series is given. The Kalman filter is presented as an important tool for dealing with state space models.

The reader is assumed to be familiar with vectors and matrices. The rules used in the text are summarized in Appendix A. It may be a good idea to review the basic definitions and rules before starting with Chapter 2. Some results on the multivariate normal and related distributions are listed in Appendix B and stochastic convergence and some asymptotic distribution theory are reviewed in Appendix C. In Appendix D a brief outline is given of the use of simulation techniques in evaluating properties of estimators and test statistics. Although it is not necessary for the reader to be familiar with all the particular rules and propositions listed in the appendices it is implicitly assumed in the following chapters that the reader has knowledge of the basic terms and results. A number of examples are given throughout the book. The data used in these examples and in the exercises are listed in Appendix E.

Part I
Finite Order Vector Autoregressive Processes

In the four chapters of this part finite order, stationary vector autoregressive (VAR) processes and their uses are discussed. Chapter 2 is dedicated to processes with known coefficients. Some of their basic properties are derived, their use for prediction and analysis purposes is considered. Unconstrained estimation is discussed in Chapter 3, model specification and checking the model adequacy is treated in Chapter 4 and estimation with parameter restrictions is the subject of Chapter 5.

Chapter 2. Stable Vector Autoregressive Processes

In this chapter the basic, finite order vector autoregressive (VAR) model will be introduced. Some important properties will be discussed. The main uses of vector autoregressive models are forecasting and structural analysis. These two uses will be considered in Sections 2.2 and 2.3. Throughout this chapter the model of interest is assumed to be known. Although this assumption is unrealistic in practice it helps to see the problems inherent to VAR models without contamination by estimation and specification problems. The latter two aspects of an analysis will be treated in detail in subsequent chapters

2.1 Basic Assumptions and Properties of VAR Processes

2.1.1 Stable VAR(p) Processes

The subject of interest in the following is the VAR(p) model (VAR model of order p)

$$y_t = \nu + A_1 y_{t-1} + \cdots + A_p y_{t-p} + u_t, \qquad t = 0, \pm 1, \pm 2, \ldots, \tag{2.1.1}$$

where $y_t = (y_{1t}, \ldots, y_{Kt})'$ is a $(K \times 1)$ random vector, the A_i are fixed $(K \times K)$ coefficient matrices, $\nu = (\nu_1, \ldots, \nu_K)'$ is a fixed $(K \times 1)$ vector of intercept terms allowing for the possibility of a nonzero mean $E(y_t)$. Finally, $u_t = (u_{1t}, \ldots, u_{Kt})'$ is a K-dimensional *white noise* or *innovation process*, that is, $E(u_t) = 0$, $E(u_t u_t') = \Sigma_u$, and $E(u_t u_s') = 0$ for $s \neq t$. The covariance matrix Σ_u is assumed to be nonsingular if not otherwise stated.

At this stage it may be worth thinking a little more about which process is described by (2.1.1). In order to investigate the implications of the model let us assume for the moment that $p = 1$ and let us consider the VAR(1) model

$$y_t = \nu + A_1 y_{t-1} + u_t. \tag{2.1.2}$$

If this generation mechanism starts at some time $t = 1$, say, we get

$$\begin{aligned}
y_1 &= \nu + A_1 y_0 + u_1, \\
y_2 &= \nu + A_1 y_1 + u_2 = \nu + A_1(\nu + A_1 y_0 + u_1) + u_2 \\
&= (I_K + A_1)\nu + A_1^2 y_0 + A_1 u_1 + u_2, \\
&\ \ \vdots
\end{aligned}$$

$$y_t = (I_K + A_1 + \cdots + A_1^{t-1})\nu + A_1^t y_0 + \sum_{i=0}^{t-1} A_1^i u_{t-i} \qquad (2.1.3)$$

$$\vdots$$

Hence, the vectors $y_1, \ldots, y_t$ are uniquely determined by $y_0, u_1, \ldots, u_t$. Also, the joint distribution of $y_1, \ldots, y_t$ is determined by the joint distribution of y_0, u_1, $\ldots, u_t$.

Although we will sometimes assume that a process is started in a specified period it is often convenient to assume that it has been started in the infinite past. This assumption is in fact made in (2.1.1). What kind of process is consistent with the mechanism (2.1.1) in this case? To investigate this question we consider again the VAR(1) process (2.1.2). From (2.1.3) we have

$$\begin{aligned} y_t &= \nu + A_1 y_{t-1} + u_t \\ &= (I_K + A_1 + \cdots + A_1^j)\nu + A_1^{j+1} y_{t-j-1} + \sum_{i=0}^{j} A_1^i u_{t-i}. \end{aligned}$$

If all eigenvalues of A_1 have modulus less than 1 the sequence A_1^i, $i = 0, 1, \ldots$, is absolutely summable (see Appendix A, Section A.9.1). Hence, the infinite sum

$$\sum_{i=1}^{\infty} A_1^i u_{t-i}$$

exists in mean square (Appendix C, Proposition C.7). Moreover,

$$(I_K + A_1 + \cdots + A_1^j)\nu \xrightarrow[j\to\infty]{} (I_K - A_1)^{-1}\nu$$

(Appendix A, Section A.9.1). Furthermore, A_1^{j+1} converges to zero rapidly as $j \to \infty$ and thus we ignore the term $A_1^{j+1} y_{t-j-1}$ in the limit. Hence, if all eigenvalues of A_1 have modulus less than 1, by saying that y_t is the VAR(1) process (2.1.2) we mean that y_t is the well-defined stochastic process

$$y_t = \mu + \sum_{i=0}^{\infty} A_1^i u_{t-i}, \qquad t = 0, \pm 1, \pm 2, \ldots, \qquad (2.1.4)$$

where

$$\mu := (I_K - A_1)^{-1}\nu.$$

The distributions and joint distributions of the y_t are uniquely determined by the distributions of the u_t process. From Appendix C.3, Proposition C.8, the first and second moments of the y_t process are seen to be

$$E(y_t) = \mu \qquad \text{for all } t \qquad (2.1.5)$$

and

$$\begin{aligned} \Gamma_y(h) := E(y_t - \mu)(y_{t-h} - \mu)' &= \lim_{n\to\infty} \sum_{i=0}^{n} \sum_{j=0}^{n} A_1^i E(u_{t-i} u_{t-h-j}')(A_1^j)' \\ &= \lim \sum_{i=0}^{n} A_1^{h+i} \Sigma_u A_1^{i\prime} = \sum_{i=0}^{\infty} A_1^{h+i} \Sigma_u A_1^{i\prime} \qquad (2.1.6) \end{aligned}$$

because $E(u_t u_s') = 0$ for $s \neq t$ and $E(u_t u_t') = \Sigma_u$ for all t.

Since the condition for the eigenvalues of the matrix A_1 is of importance we call a VAR(1) process *stable* if all eigenvalues of A_1 have modulus less than 1. By Rule 7 of Appendix A.6 the condition is equivalent to

$$\det(I_K - A_1 z) \neq 0 \qquad \text{for } |z| \leq 1. \tag{2.1.7}$$

It is perhaps worth pointing out that the process y_t for $t = 0, \pm 1, \pm 2, \dots$ may also be defined if the *stability condition* (2.1.7) is not satisfied. We will not do so here because we will always assume stability of processes started at $-\infty$.

The previous discussion can be extended easily to VAR(p) processes with $p > 1$ because any VAR(p) process can be written in VAR(1) form. More precisely, if y_t is a VAR(p) as in (2.1.1), a corresponding Kp-dimensional VAR(1)

$$Y_t = \mathbf{v} + \mathbf{A} Y_{t-1} + U_t \tag{2.1.8}$$

can be defined, where

$$\underset{(Kp \times 1)}{Y_t := \begin{bmatrix} y_t \\ y_{t-1} \\ \vdots \\ y_{t-p+1} \end{bmatrix}}, \quad \underset{(Kp \times 1)}{\mathbf{v} := \begin{bmatrix} v \\ 0 \\ \vdots \\ 0 \end{bmatrix}}, \quad \underset{(Kp \times Kp)}{\mathbf{A} := \begin{bmatrix} A_1 & A_2 & \dots & A_{p-1} & A_p \\ I_K & 0 & \dots & 0 & 0 \\ 0 & I_K & & 0 & 0 \\ \vdots & & \ddots & \vdots & \vdots \\ 0 & 0 & \dots & I_K & 0 \end{bmatrix}},$$

$$\underset{(Kp \times 1)}{U_t := \begin{bmatrix} u_t \\ 0 \\ \vdots \\ 0 \end{bmatrix}}.$$

Following the foregoing discussion Y_t is *stable* if

$$\det(I_{Kp} - \mathbf{A}z) \neq 0 \qquad \text{for } |z| \leq 1. \tag{2.1.9}$$

Its mean vector is

$$\boldsymbol{\mu} := E(Y_t) = (I_{Kp} - \mathbf{A})^{-1}\mathbf{v}$$

and the autocovariances are

$$\Gamma_Y(h) = \sum_{i=0}^{\infty} \mathbf{A}^{h+i} \Sigma_U (\mathbf{A}^i)' \tag{2.1.10}$$

where $\Sigma_U := E(U_t U_t')$. Using the $(K \times Kp)$ matrix

$$J := [I_K \quad 0 \dots 0] \tag{2.1.11}$$

the process y_t is obtained as $y_t = J Y_t$. Since Y_t is a well-defined stochastic process the same is true for y_t. Its mean is $E(y_t) = J\boldsymbol{\mu}$ which is constant for all t and the autocovariances $\Gamma_y(h) = J\Gamma_Y(h)J'$ are also time invariant.

It is easy to see that

$$\det(I_{Kp} - \mathbf{A}z) = \det(I_K - A_1 z - \cdots - A_p z^p)$$

(see Problem 2.1). Given the definition of the characteristic polynomial of a matrix we call this polynomial the *reverse characteristic polynomial* of the VAR(p) process. Hence, the process (2.1.1) is *stable* if its reverse characteristic polynomial has no roots in and on the complex unit circle. Formally y_t is stable if

$$\det(I_K - A_1 z - \cdots - A_p z^p) \neq 0 \qquad \text{for } |z| \leq 1. \tag{2.1.12}$$

This condition is called the *stability condition.*

In summary, we say that y_t is a stable VAR(p) process if (2.1.12) holds and

$$y_t = JY_t = J\boldsymbol{\mu} + J \sum_{i=0}^{\infty} \mathbf{A}^i U_{t-i}. \tag{2.1.13}$$

Since the $U_t := (u_t' \quad 0 \ldots 0)'$ involve the white noise process u_t the process y_t is seen to be determined by its white noise or innovation process. Often specific assumptions regarding u_t are made which determine the process y_t by the foregoing convention. An important example is the assumption that u_t is *Gaussian white noise*, that is, $u_t \sim N(0, \Sigma_u)$ for all t and u_t and u_s are independent for $s \neq t$. In that case it can be shown that y_t is a *Gaussian process*, that is, subcollections $y_t, \ldots, y_{t+h}$ have multivariate normal distributions for all t and h.

The condition (2.1.12) provides an easy tool for checking the stability of a VAR process. Consider, for instance, the three-dimensional VAR(1) process

$$y_t = \nu + \begin{bmatrix} .5 & 0 & 0 \\ .1 & .1 & .3 \\ 0 & .2 & .3 \end{bmatrix} y_{t-1} + u_t. \tag{2.1.14}$$

For this process the reverse characteristic polynomial is

$$\det\left(\begin{bmatrix} 1 & 0 & 0 \\ 0 & 1 & 0 \\ 0 & 0 & 1 \end{bmatrix} - \begin{bmatrix} .5 & 0 & 0 \\ .1 & .1 & .3 \\ 0 & .2 & .3 \end{bmatrix} z \right) = \det \begin{bmatrix} 1 - .5z & 0 & 0 \\ -.1z & 1 - .1z & -.3z \\ 0 & -.2z & 1 - .3z \end{bmatrix}$$
$$= (1 - .5z)(1 - .4z - .03z^2).$$

The roots of this polynomial are easily seen to be

$$z_1 = 2, \qquad z_2 = 2.1525, \qquad z_3 = -15.4858.$$

They are obviously all greater than 1 in absolute value. Therefore the process (2.1.14) is stable.

As another example consider the bivariate (two-dimensional) VAR(2) process

$$y_t = \nu + \begin{bmatrix} .5 & .1 \\ .4 & .5 \end{bmatrix} y_{t-1} + \begin{bmatrix} 0 & 0 \\ .25 & 0 \end{bmatrix} y_{t-2} + u_t. \tag{2.1.15}$$

Its reverse characteristic polynomial is

$$\det\left[\begin{bmatrix}1 & 0\\ 0 & 1\end{bmatrix} - \begin{bmatrix}.5 & .1\\ .4 & .5\end{bmatrix}z - \begin{bmatrix}0 & 0\\ .25 & 0\end{bmatrix}z^2\right] = 1 - z + .21z^2 - .025z^3.$$

The roots of this polynomial are

$$z_1 = 1.3, \qquad z_2 = 3.55 + 4.26i, \qquad \text{and} \qquad z_3 = 3.55 - 4.26i.$$

Note that the modulus of z_2 and z_3 is $|z_2| = |z_3| = \sqrt{3.55^2 + 4.26^2} = 5.545$. Thus, the process (2.1.15) satisfies the stability condition (2.1.12) since all roots are outside the unit circle. Although the roots for higher dimensional and higher order processes are often difficult to compute by hand, efficient computer programs exist that do the job.

To understand the implications of the stability assumption it may be helpful to visualize time series generated by stable processes and contrast them with realizations from unstable VAR processes. In Figure 2.1 three pairs of time series generated by three different stable bivariate (2-dimensional) VAR processes are depicted. Although they differ considerably a common feature is that they fluctuate around constant means and their variability (variance) does not change as they wander along. In contrast the pairs of series plotted in Figures 2.2 and 2.3 are generated by unstable, bivariate VAR processes. The time series in Figure 2.2 have a trend and those in Figure 2.3 exhibit quite pronounced seasonal fluctuations. Both shapes are typical of certain instabilities although they are quite common in practice. Hence, the stability assumption excludes many series of practical interest. We shall therefore discuss unstable processes in more detail in Chapter 11. For that analysis understanding the stable case first is helpful.

2.1.2 The Moving Average Representation of a VAR Process

In the previous subsection we have considered the VAR(1) representation

$$Y_t = \mathbf{v} + \mathbf{A}Y_{t-1} + U_t$$

of the VAR(p) process (2.1.1). Under the stability assumption the process Y_t has a representation

$$Y_t = \boldsymbol{\mu} + \sum_{i=0}^{\infty} \mathbf{A}^i U_{t-i}. \tag{2.1.16}$$

This form of the process is called the *moving average* (MA) *representation*, where Y_t is expressed in terms of past and present error or innovation vectors U_t and the mean term $\boldsymbol{\mu}$. This representation can be used to determine the autocovariances of Y_t and the mean and autocovariances of y_t can be obtained as outlined in Section 2.1.1. Moreover, an MA representation of y_t can be found by premultiplying (2.1.16) by the matrix $J := [I_K \quad 0 \dots 0]$ (defined in (2.1.11)),

$$\begin{aligned} y_t = JY_t &= J\boldsymbol{\mu} + \sum_{i=0}^{\infty} J\mathbf{A}^i J'JU_{t-i} \\ &= \mu + \sum_{i=0}^{\infty} \Phi_i u_{t-i}. \end{aligned} \tag{2.1.17}$$

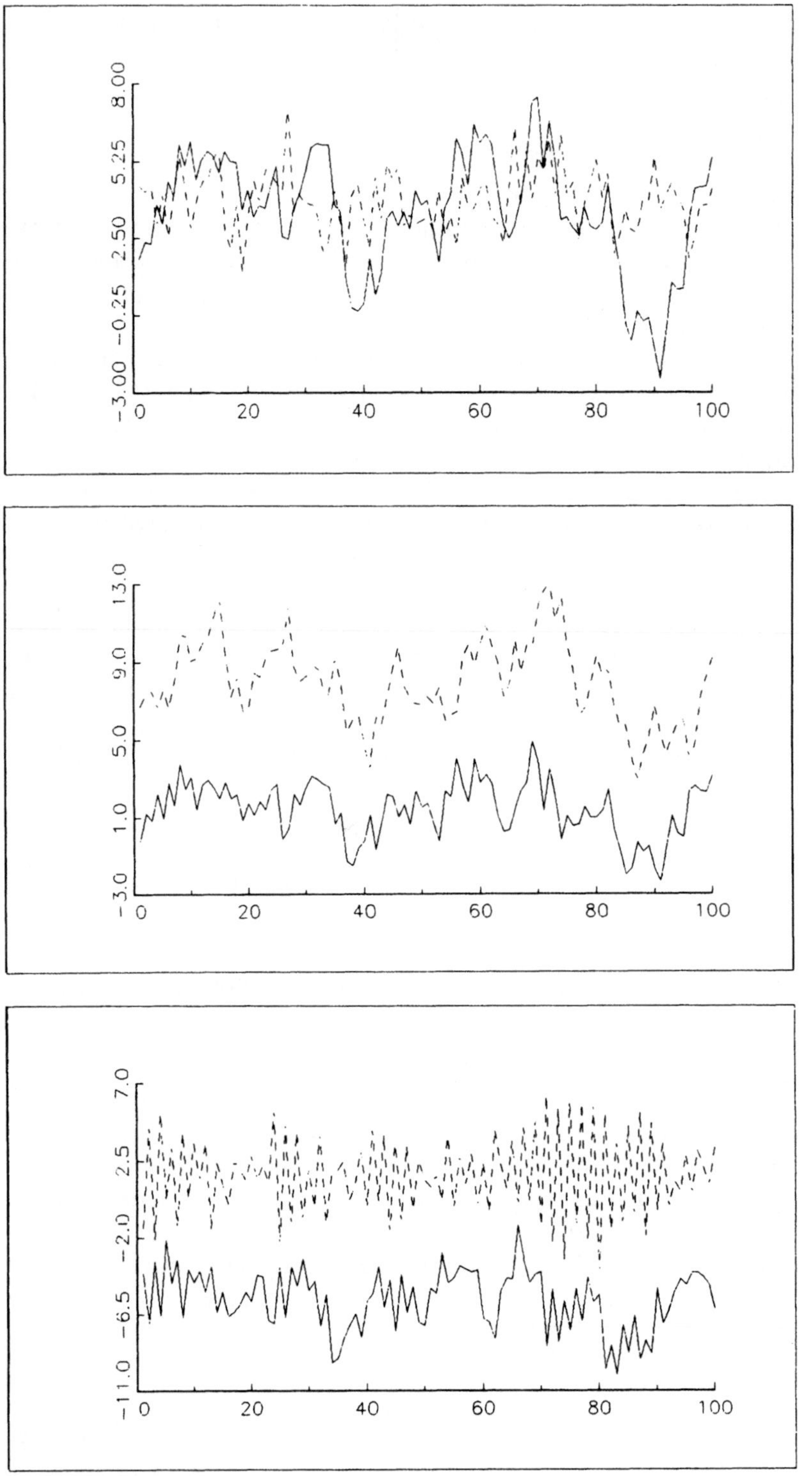

Fig. 2.1. Bivariate time series generated by stable processes.

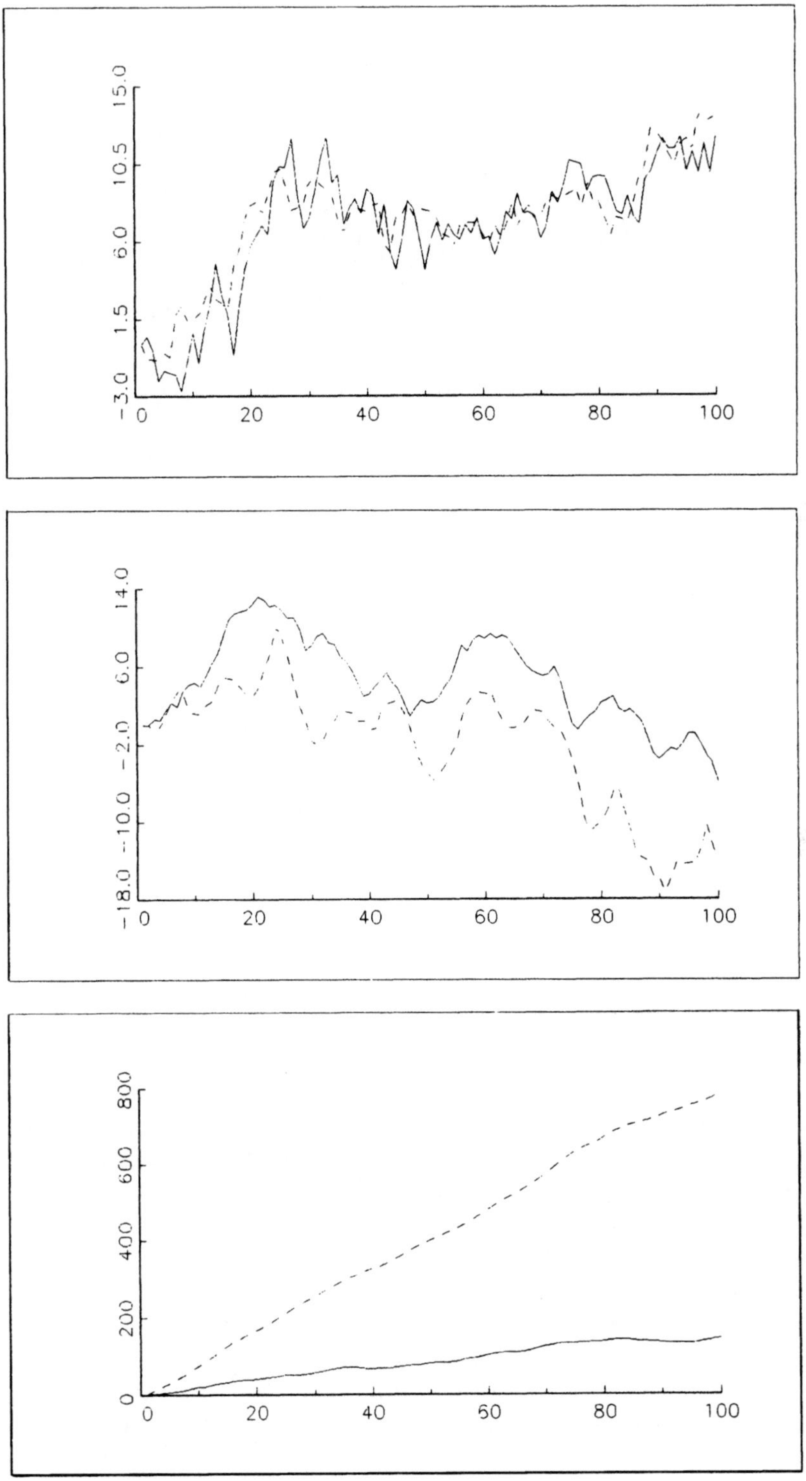

Fig. 2.2. Bivariate time series generated by unstable VAR processes.

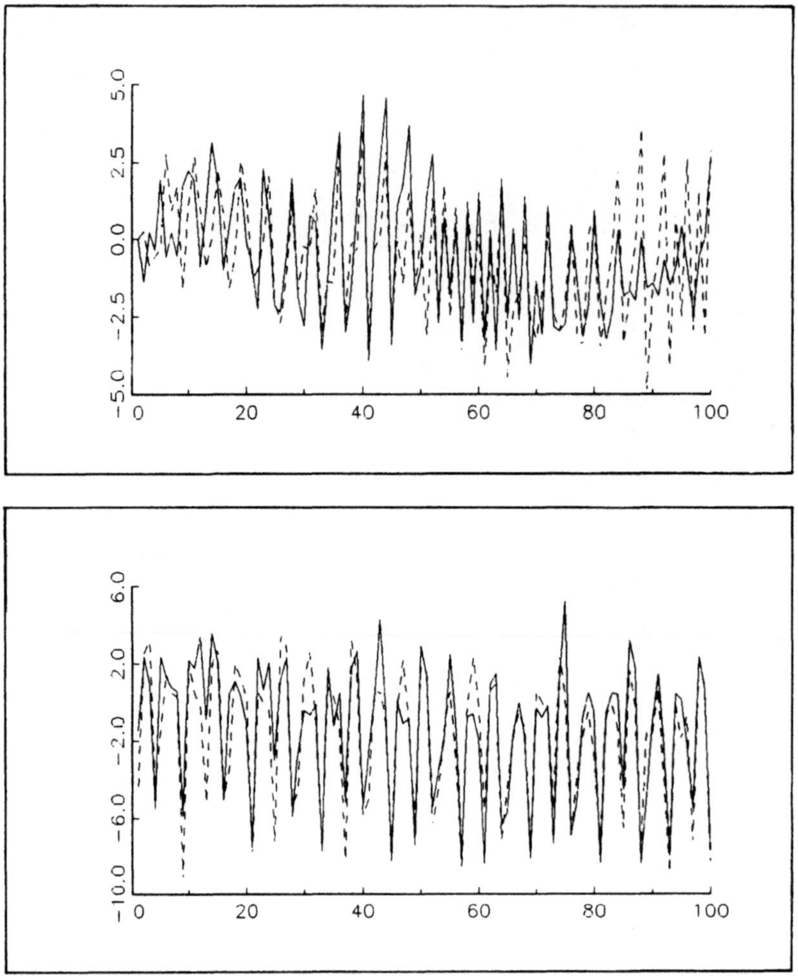

Fig. 2.3. Unstable seasonal time series.

Here $\mu := J\boldsymbol{\mu}$, $\Phi_i := J\mathbf{A}^i J'$ and, due to the special structure of the white noise process U_t, we have $U_t = J'JU_t$ and $JU_t = u_t$. Since the $\mathbf{A}^i$ are absolutely summable the same is true for the Φ_i.

Later we will also consider other MA representations of a stable VAR(p) process. The unique feature of the present representation is that the zero order coefficient matrix $\Phi_0 = I_K$ and the white noise process involved consists of the error terms u_t of the VAR representation (2.1.1). In Section 2.2.2 the u_t will be seen to be the errors of optimal forecasts made in period $t-1$. Therefore, to distinguish the present representation from other MA representations we will sometimes refer to it as the *canonical* or *fundamental* or *prediction error representation.*

Using Proposition C.8 of Appendix C.3 the representation (2.1.17) provides a possibility for determining the mean and autocovariances of y_t:

$$E(y_t) = \mu$$

and

$$\begin{aligned}\Gamma_y(h) &= E(y_t - \mu)(y_{t-h} - \mu)' \\ &= E\left(\sum_{i=0}^{h-1} \Phi_i u_{t-i} + \sum_{i=0}^{\infty} \Phi_{h+i} u_{t-h-i}\right)\left(\sum_{i=0}^{\infty} \Phi_i u_{t-h-i}\right)' \\ &= \sum_{i=0}^{\infty} \Phi_{h+i} \Sigma_u \Phi_i'. \end{aligned} \tag{2.1.18}$$

There is no need to compute the MA coefficient matrices Φ_i via the VAR(1) representation corresponding to y_t as in the foregoing derivation. A more direct way for determining these matrices results from writing the VAR(p) process in *lag operator* notation. The lag operator L is defined such that $Ly_t = y_{t-1}$, that is, it lags (shifts back) the index by one period. Because of this property it is sometimes called *backshift operator*. Using this operator, (2.1.1) can be written as

$$y_t = \nu + (A_1 L + \cdots + A_p L^p) y_t + u_t$$

or

$$A(L) y_t = \nu + u_t, \tag{2.1.19}$$

where

$$A(L) := I_K - A_1 L - \cdots - A_p L^p.$$

Let

$$\Phi(L) := \sum_{i=0}^{\infty} \Phi_i L^i$$

be an operator such that

$$\Phi(L) A(L) = I_K. \tag{2.1.20}$$

Premultiplying (2.1.19) by $\Phi(L)$ gives

$$\begin{aligned} y_t &= \Phi(L)\nu + \Phi(L) u_t \\ &= \left(\sum_{i=0}^{\infty} \Phi_i\right)\nu + \sum_{i=0}^{\infty} \Phi_i u_{t-i}. \end{aligned} \tag{2.1.21}$$

The operator $\Phi(L)$ is the *inverse* of $A(L)$ and is therefore sometimes denoted by $A(L)^{-1}$. Generally we call the operator $A(L)$ *invertible* if $|A(z)| \neq 0$ for $|z| \leq 1$. If this condition is satisfied the coefficient matrices of $\Phi(L) = A(L)^{-1}$ are absolutely summable and hence the process $\Phi(L)u_t = A(L)^{-1}u_t$ is well-defined (see Appendix C.3). The coefficient matrices Φ_i can be obtained from (2.1.20) using the relations

$$\begin{aligned}I_K &= (\Phi_0 + \Phi_1 L + \Phi_2 L^2 + \cdots)(I_K - A_1 L - \cdots - A_p L^p)\\ &= \Phi_0 + (\Phi_1 - \Phi_0 A_1)L + (\Phi_2 - \Phi_1 A_1 - \Phi_0 A_2)L^2 + \cdots\\ &\quad + \left(\Phi_i - \sum_{j=1}^{i} \Phi_{i-j} A_j\right) L^i + \cdots\end{aligned}$$

or

$$\begin{aligned}I_K &= \Phi_0\\ 0 &= \Phi_1 - \Phi_0 A_1\\ 0 &= \Phi_2 - \Phi_1 A_1 - \Phi_0 A_2\\ &\;\;\vdots\\ 0 &= \Phi_i - \sum_{j=1}^{i} \Phi_{i-j} A_j\\ &\;\;\vdots\end{aligned}$$

where $A_j = 0$ for $j > p$. Hence, the Φ_i can be computed recursively using

$$\begin{aligned}\Phi_0 &= I_K,\\ \Phi_i &= \sum_{j=1}^{i} \Phi_{i-j} A_j, \qquad i = 1, 2, \ldots\end{aligned} \tag{2.1.22}$$

The mean μ of y_t can be obtained as follows:

$$\mu = \Phi(1)\nu = A(1)^{-1}\nu = (I_K - A_1 - \cdots - A_p)^{-1}\nu. \tag{2.1.23}$$

For a VAR(1) process the recursions (2.1.22) imply that $\Phi_0 = I_K$, $\Phi_1 = A_1, \ldots,$ $\Phi_i = A_1^i, \ldots$. This result is in line with (2.1.4). For the example VAR(1) process (2.1.14) we get $\Phi_0 = I_3$,

$$\Phi_1 = \begin{bmatrix} .5 & 0 & 0\\ .1 & .1 & .3\\ 0 & .2 & .3 \end{bmatrix}, \qquad \Phi_2 = \begin{bmatrix} .25 & 0 & 0\\ .06 & .07 & .12\\ .02 & .08 & .15 \end{bmatrix},$$

$$\Phi_3 = \begin{bmatrix} .125 & 0 & 0\\ .037 & .031 & .057\\ .018 & .038 & .069 \end{bmatrix} \tag{2.1.24}$$

etc.. For a VAR(2) the recursions (2.1.22) result in

$$\begin{aligned}\Phi_1 &= A_1\\ \Phi_2 &= \Phi_1 A_1 + A_2 = A_1^2 + A_2\\ \Phi_3 &= \Phi_2 A_1 + \Phi_1 A_2 = A_1^3 + A_2 A_1 + A_1 A_2\\ &\;\;\vdots\\ \Phi_i &= \Phi_{i-1} A_1 + \Phi_{i-2} A_2\\ &\;\;\vdots\end{aligned}$$

Thus, for the example VAR(2) process (2.1.15) we get the MA coefficient matrices $\Phi_0 = I_2$,

$$\Phi_1 = \begin{bmatrix} .5 & .1 \\ .4 & .5 \end{bmatrix}, \quad \Phi_2 = \begin{bmatrix} .29 & .1 \\ .65 & .29 \end{bmatrix}, \quad \Phi_3 = \begin{bmatrix} .21 & .079 \\ .566 & .21 \end{bmatrix} \tag{2.1.25}$$

etc.. For both example processes the Φ_i matrices approach zero as $i \to \infty$. This property is a consequence of the stability of the two processes.

It may be worth noting that the MA representation of a stable VAR(p) process is not necessarily of infinite order. That is, the Φ_i may all be zero for i greater than some finite integer q. For instance, for the bivariate VAR(1)

$$y_t = \nu + \begin{bmatrix} 0 & \alpha \\ 0 & 0 \end{bmatrix} y_{t-1} + u_t$$

the MA representation is easily seen to be

$$y_t = \mu + u_t + \begin{bmatrix} 0 & \alpha \\ 0 & 0 \end{bmatrix} u_{t-1}$$

because

$$\begin{bmatrix} 0 & \alpha \\ 0 & 0 \end{bmatrix}^i = 0$$

for $i > 1$.

2.1.3 Stationary Processes

A stochastic process is *stationary* if its first and second moments are time invariant. In other words, a stochastic process y_t is stationary if

$$\text{(i)} \quad E(y_t) = \mu \qquad \text{for all } t \tag{2.1.26a}$$

and

$$\text{(ii)} \quad E[(y_t - \mu)(y_{t-h} - \mu)'] = \Gamma_y(h) = \Gamma_y(-h)' \quad \text{for all } t \text{ and } h = 0, 1, 2, \ldots. \tag{2.1.26b}$$

Condition (2.1.26a) means that all y_t have the same finite mean vector μ and (2.1.26b) requires that the autocovariances of the process do not depend on t but just on the time period h the two vectors y_t and y_{t-h} are apart. Note that, if not otherwise stated, all quantities are assumed finite. For instance, μ is a vector of finite mean terms and $\Gamma_y(h)$ is a matrix of finite covariances. Other definitions of stationarity are often used in the literature. For example, the joint distribution of n consecutive vectors may be assumed time invariant for all n. We shall, however, use the foregoing definition in the following. By that definition the white noise process u_t used in (2.1.1) is an obvious example of a stationary process. Also, from (2.1.18) we know that a stable VAR(p) process is stationary. We state this fact as a proposition.

PROPOSITION 2.1 (*Stationarity Condition*)

A stable VAR(p) process y_t, $t = 0, \pm 1, \pm 2, \ldots$, is stationary. ■

Since stability implies stationarity, the stability condition (2.1.12) is often referred to as *stationarity condition* in the time series literature. The converse of Proposition 2.1 is not true. In other words, an unstable process is not necessarily nonstationary. Because unstable stationary processes are not of interest in the following we will not discuss this possibility here.

At this stage it may be worth thinking about the generality of the VAR(p) processes considered in this and many other chapters. In this context an important result due to Wold (1938) is of interest. He has shown that every stationary process x_t can be written as the sum of two uncorrelated processes z_t and y_t,

$$x_t = z_t + y_t,$$

where z_t is a deterministic process that can be forecast perfectly from its own past and y_t is a process with MA representation

$$y_t = \sum_{i=0}^{\infty} \Phi_i u_{t-i}, \tag{2.1.27}$$

where $\Phi_0 = I_K$, the u_t constitute a white noise process and the infinite sum is defined as a limit in mean square although the Φ_i are not necessarily absolutely summable (Hannan (1970, Chapter III)). The term "deterministic" will be explained more formally in Section 2.2. This result is often called *Wold's Decomposition Theorem*. If we assume that in the system of interest the only deterministic component is the mean term, the theorem states that the system has an MA representation. Suppose the Φ_i are absolutely summable and there exists an operator $A(L)$ with absolutely summable coefficient matrices satisfying $A(L)\Phi(L) = I_K$. Then $\Phi(L)$ is invertible ($A(L) = \Phi(L)^{-1}$) and y_t has a VAR representation of possibly infinite order,

$$y_t = \sum_{i=1}^{\infty} A_i y_{t-i} + u_t, \tag{2.1.28}$$

where

$$A(z) = I_K - \sum_{i=1}^{\infty} A_i z^i = \left(\sum_{i=0}^{\infty} \Phi_i z^i \right)^{-1} \quad \text{for } |z| \leq 1.$$

The A_i can be obtained from the Φ_i by recursions similar to (2.1.22).

The absolute summability of the A_i implies that the VAR coefficient matrices converge to zero rapidly. In other words, under quite general conditions, every stationary, purely nondeterministic process (a process without a deterministic component) can be approximated well by a finite order VAR process. This is a very powerful result which demonstrates the generality of the processes under study. Note that economic variables can rarely be predicted without error. Thus the assumption of having a nondeterministic system except perhaps for a mean term is not a very restrictive one. The crucial and restrictive condition is the stationarity of the system, however. We will talk about nonstationary processes later. For that discussion it is useful to understand the stationary case first.

An important implication of Wold's Decomposition Theorem is worth noting at this point. The theorem implies that any *subprocess* of a purely nondeterministic, stationary process y_t consisting of any subset of the components of y_t also has an MA representation. Suppose, for instance, that interest centers on the first M components of the K-dimensional process y_t, that is, we are interested in $x_t = Fy_t$, where $F = [I_M \quad 0]$ is an $(M \times K)$ matrix. Then $E(x_t) = FE(y_t) = F\mu$ and $\Gamma_x(h) = F\Gamma_y(h)F'$ and thus x_t is stationary. Application of Wold's theorem then implies that x_t has an MA representation.

2.1.4 Computation of Autocovariances and Autocorrelations of Stable VAR Processes

Although the autocovariances of a stationary, stable VAR(p) process can be given in terms of its MA coefficient matrices as in (2.1.18) that formula is unattractive in practice because it involves an infinite sum. For practical purposes it is easier to compute the autocovariances directly from the VAR coefficient matrices. In this section we will develop the relevant formulas.

2.1.4a Autocovariances of a VAR(1) Process

In order to illustrate the computation of the autocovariances when the process coefficients are given suppose that y_t is a stationary, stable VAR(1) process

$$y_t = \nu + A_1 y_{t-1} + u_t$$

with white noise covariance matrix $E(u_t u_t') = \Sigma_u$. Alternatively the process may be written in mean-adjusted form as

$$y_t - \mu = A_1(y_{t-1} - \mu) + u_t, \tag{2.1.29}$$

where $\mu = E(y_t)$, as before. Postmultiplying by $(y_{t-h} - \mu)'$ and taking expectations gives

$$E[(y_t - \mu)(y_{t-h} - \mu)'] = A_1 E[(y_{t-1} - \mu)(y_{t-h} - \mu)'] + E[u_t(y_{t-h} - \mu)'].$$

Thus, for $h = 0$,

$$\Gamma_y(0) = A_1\Gamma_y(-1) + \Sigma_u = A_1\Gamma_y(1)' + \Sigma_u \tag{2.1.30}$$

and for $h > 0$,

$$\Gamma_y(h) = A_1\Gamma_y(h-1). \tag{2.1.31}$$

These equations are usually referred to as *Yule-Walker equations*. If A_1 and the covariance matrix $\Gamma_y(0) = \Sigma_y$ of y_t are known the $\Gamma_y(h)$ can be computed recursively using (2.1.31).

If A_1 and Σ_u are given, $\Gamma_y(0)$ can be determined as follows. For $h = 1$ we get from (2.1.31), $\Gamma_y(1) = A_1\Gamma_y(0)$. Substituting $\Gamma_y(1)$ in (2.1.30) gives

$$\Gamma_y(0) = A_1\Gamma_y(0)A_1' + \Sigma_u$$

or

$$\begin{aligned}\operatorname{vec}\Gamma_y(0) &= \operatorname{vec}(A_1\Gamma_y(0)A_1') + \operatorname{vec}\Sigma_u \\ &= (A_1 \otimes A_1)\operatorname{vec}\Gamma_y(0) + \operatorname{vec}\Sigma_u.\end{aligned}$$

(For the definition of the Kronecker product $\otimes$, the vec operator and the rules used here, see Appendix A). Hence,

$$\operatorname{vec}\Gamma_y(0) = (I_{K^2} - A_1 \otimes A_1)^{-1}\operatorname{vec}\Sigma_u. \tag{2.1.32}$$

Note that the invertibility of $I_{K^2} - A_1 \otimes A_1$ follows from the stability of y_t because the eigenvalues of $A_1 \otimes A_1$ are the products of the eigenvalues of A_1 (see Appendix A). Hence the eigenvalues of $A_1 \otimes A_1$ have modulus less than 1. Consequently, $\det(I_{K^2} - A_1 \otimes A_1) \neq 0$.

Using, for instance,

$$\Sigma_u = \begin{bmatrix} 2.25 & 0 & 0 \\ 0 & 1.0 & .5 \\ 0 & .5 & .74 \end{bmatrix} \tag{2.1.33}$$

we get for the example process (2.1.14),

$$\begin{aligned}\operatorname{vec}\Gamma_y(0) &= (I_9 - A_1 \otimes A_1)^{-1}\operatorname{vec}\Sigma_u \\
&= \begin{bmatrix} .75 & 0 & 0 & 0 & 0 & 0 & 0 & 0 & 0 \\ -.05 & .95 & -.15 & 0 & 0 & 0 & 0 & 0 & 0 \\ 0 & -.10 & .85 & 0 & 0 & 0 & 0 & 0 & 0 \\ -.05 & 0 & 0 & .95 & 0 & 0 & -.15 & 0 & 0 \\ -.01 & -.01 & -.03 & -.01 & .99 & -.03 & -.03 & -.03 & -.09 \\ 0 & -.02 & -.03 & 0 & -.02 & .97 & 0 & -.06 & -.09 \\ 0 & 0 & 0 & -.01 & 0 & 0 & .85 & 0 & 0 \\ 0 & 0 & 0 & -.02 & -.02 & -.06 & -.03 & .97 & -.09 \\ 0 & 0 & 0 & 0 & -.04 & -.06 & 0 & -.06 & .91 \end{bmatrix}^{-1} \begin{bmatrix} 2.25 \\ 0 \\ 0 \\ 0 \\ 1.0 \\ .5 \\ 0 \\ .5 \\ .74 \end{bmatrix} \\
&= \begin{bmatrix} 3.000 \\ .161 \\ .019 \\ .161 \\ 1.172 \\ .674 \\ .019 \\ .674 \\ .954 \end{bmatrix}.\end{aligned}$$

It follows that

$$\Gamma_y(0) = \begin{bmatrix} 3.000 & .161 & .019 \\ .161 & 1.172 & .674 \\ .019 & .674 & .954 \end{bmatrix},$$

$$\Gamma_y(1) = A_1\Gamma_y(0) = \begin{bmatrix} 1.500 & .080 & .009 \\ .322 & .335 & .355 \\ .038 & .437 & .421 \end{bmatrix}, \tag{2.1.34}$$

$$\Gamma_y(2) = A_1\Gamma_y(1) = \begin{bmatrix} .750 & .040 & .005 \\ .194 & .173 & .163 \\ .076 & .198 & .197 \end{bmatrix}.$$

Note that the results are rounded after the computation. A higher precision has been used in intermediate steps.

2.1.4b *Autocovariances of a Stable VAR(p) Process*

For a higher order VAR(p) process

$$y_t - \mu = A_1(y_{t-1} - \mu) + \cdots + A_p(y_{t-p} - \mu) + u_t \tag{2.1.35}$$

the *Yule-Walker equations* are also obtained by postmultiplying with $(y_{t-h} - \mu)'$ and taking expectations. For $h = 0$, using $\Gamma_y(i) = \Gamma_y(-i)'$,

$$\begin{aligned}\Gamma_y(0) &= A_1\Gamma_y(-1) + \cdots + A_p\Gamma_y(-p) + \Sigma_u \\ &= A_1\Gamma_y(1)' + \cdots + A_p\Gamma_y(p)' + \Sigma_u\end{aligned} \tag{2.1.36}$$

and for $h > 0$,

$$\Gamma_y(h) = A_1\Gamma_y(h-1) + \cdots + A_p\Gamma_y(h-p). \tag{2.1.37}$$

These equations may be used to compute the $\Gamma_y(h)$ recursively for $h \geq p$ if $A_1, \ldots, A_p$ and $\Gamma_y(p-1), \ldots, \Gamma_y(0)$ are known.

The initial autocovariance matrices for $|h| < p$ can be determined using the VAR(1) process that corresponds to (2.1.35),

$$Y_t - \boldsymbol{\mu} = \mathbf{A}(Y_{t-1} - \boldsymbol{\mu}) + U_t, \tag{2.1.38}$$

where Y_t, $\mathbf{A}$, and U_t are as in (2.1.8) and $\boldsymbol{\mu} := (\mu', \ldots, \mu')' = E(Y_t)$. Proceeding as in the VAR(1) case gives

$$\Gamma_Y(0) = \mathbf{A}\Gamma_Y(0)\mathbf{A}' + \Sigma_U$$

where $\Sigma_U = E(U_t U_t')$ and

$$\Gamma_Y(0) = E\begin{bmatrix} y_t - \mu \\ \vdots \\ y_{t-p+1} - \mu \end{bmatrix}[(y_t - \mu)', \ldots, (y_{t-p+1} - \mu)']$$

$$= \begin{bmatrix} \Gamma_y(0) & \Gamma_y(1) & \ldots & \Gamma_y(p-1) \\ \Gamma_y(-1) & \Gamma_y(0) & \ldots & \Gamma_y(p-2) \\ \vdots & \vdots & \ddots & \vdots \\ \Gamma_y(-p+1) & \Gamma_y(-p+2) & \ldots & \Gamma_y(0) \end{bmatrix}.$$

Thus, the $\Gamma_y(h)$, $h = -p+1, \ldots, p-1$, are obtained from

$$\operatorname{vec} \Gamma_Y(0) = (I_{(Kp)^2} - \mathbf{A} \otimes \mathbf{A})^{-1} \operatorname{vec} \Sigma_U. \tag{2.1.39}$$

For instance, for the example VAR(2) process (2.1.15) we get

$$\mathbf{A} = \begin{bmatrix} .5 & .1 & 0 & 0 \\ .4 & .5 & .25 & 0 \\ 1 & 0 & 0 & 0 \\ 0 & 1 & 0 & 0 \end{bmatrix} \tag{2.1.40}$$

and, assuming

$$\Sigma_u = \begin{bmatrix} .09 & 0 \\ 0 & .04 \end{bmatrix}, \tag{2.1.41}$$

we have

$$\Sigma_U = \begin{bmatrix} \Sigma_u & 0 \\ 0 & 0 \end{bmatrix} = \begin{bmatrix} .09 & 0 & 0 & 0 \\ 0 & .04 & 0 & 0 \\ 0 & 0 & 0 & 0 \\ 0 & 0 & 0 & 0 \end{bmatrix}.$$

Hence, using (2.1.39) and

$$\Gamma_Y(0) = \begin{bmatrix} \Gamma_y(0) & \Gamma_y(1) \\ \Gamma_y(1)' & \Gamma_y(0) \end{bmatrix}$$

gives

$$\Gamma_y(0) = \begin{bmatrix} .131 & .066 \\ .066 & .181 \end{bmatrix}, \qquad \Gamma_y(1) = \begin{bmatrix} .072 & .051 \\ .104 & .143 \end{bmatrix},$$

$$\Gamma_y(2) = A_1 \Gamma_y(1) + A_2 \Gamma_y(0) = \begin{bmatrix} .046 & .040 \\ .113 & .108 \end{bmatrix}, \tag{2.1.42}$$

$$\Gamma_y(3) = A_1 \Gamma_y(2) + A_2 \Gamma_y(1) = \begin{bmatrix} .035 & .031 \\ .093 & .083 \end{bmatrix},$$

and so on. A method for computing $\Gamma_y(0)$ without explicitly inverting $(I - \mathbf{A} \otimes \mathbf{A})$ is given by Barone (1987).

The *autocovariance function* of a stationary VAR(p) process *is positive semi-definite*, that is,

$$\sum_{j=0}^{n}\sum_{i=0}^{n} a_j'\Gamma_y(i-j)a_i$$

$$= (a_0', \dots, a_n')\begin{bmatrix} \Gamma_y(0) & \Gamma_y(1) & \dots & \Gamma_y(n) \\ \Gamma_y(-1) & \Gamma_y(0) & \dots & \Gamma_y(n-1) \\ \vdots & \vdots & \ddots & \vdots \\ \Gamma_y(-n) & \Gamma_y(-n+1) & \dots & \Gamma_y(0) \end{bmatrix}\begin{bmatrix} a_0 \\ a_1 \\ \vdots \\ a_n \end{bmatrix} \geq 0 \quad (2.1.43)$$

for any $n \geq 0$. Here the a_i are arbitrary $(K \times 1)$ vectors. This result follows because (2.1.43) is just the variance of

$$(a_0', \dots, a_n')\begin{bmatrix} y_t \\ y_{t-1} \\ \vdots \\ y_{t-n} \end{bmatrix}$$

which is always nonnegative.

2.1.4c Autocorrelations of a Stable VAR(p) Process

Since the autocovariances depend on the unit of measurement used for the variables of the system they are sometimes difficult to interpret. Therefore the autocorrelations

$$R_y(h) = D^{-1}\Gamma_y(h)D^{-1} \quad (2.1.44)$$

are usually more convenient to work with as they are scale invariant measures for the linear dependencies among the variables of the system. Here D is a diagonal matrix with the standard deviations of the components of y_t on the diagonal. That is, the diagonal elements of D are the square roots of the diagonal elements of $\Gamma_y(0)$. Denoting the covariance between $y_{i,t}$ and $y_{j,t-h}$ by $\gamma_{ij}(h)$ (i.e., $\gamma_{ij}(h)$ is the ij-th element of $\Gamma_y(h)$) the diagonal elements $\gamma_{11}(0), \dots, \gamma_{KK}(0)$ of $\Gamma_y(0)$ are the variances of $y_{1t}, \dots, y_{Kt}$. Thus,

$$D^{-1} = \begin{bmatrix} 1/\sqrt{\gamma_{11}(0)} & & 0 \\ & \ddots & \\ 0 & & 1/\sqrt{\gamma_{KK}(0)} \end{bmatrix}$$

and the correlation between $y_{i,t}$ and $y_{j,t-h}$ is

$$\rho_{ij}(h) = \frac{\gamma_{ij}(h)}{\sqrt{\gamma_{ii}(0)}\sqrt{\gamma_{jj}(0)}} \quad (2.1.45)$$

which is just the ij-th element of $R_y(h)$.

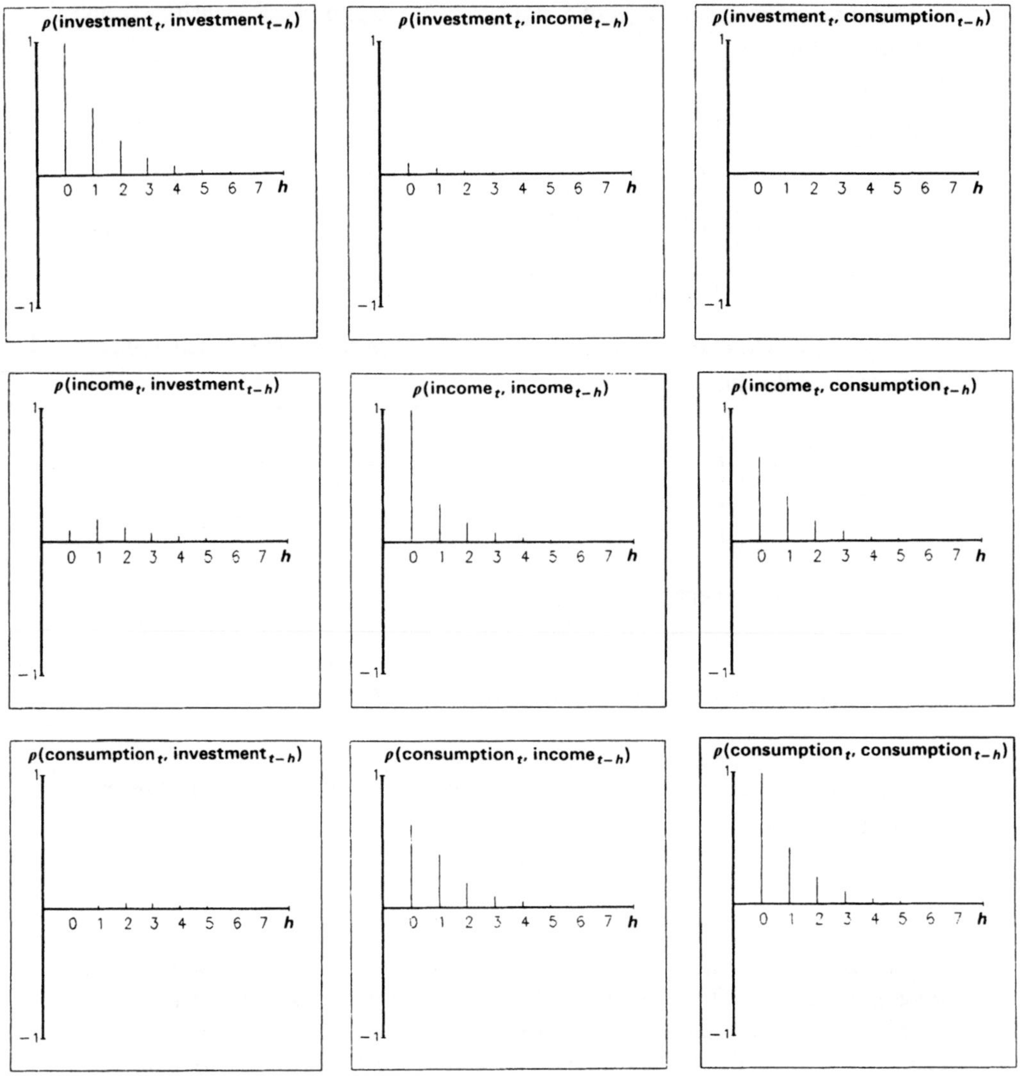

Fig. 2.4. Autocorrelations of the investment/income/consumption system.

For the VAR(1) example process (2.1.14) we get from (2.1.34),

$$D = \begin{bmatrix} \sqrt{3.000} & 0 & 0 \\ 0 & \sqrt{1.172} & 0 \\ 0 & 0 & \sqrt{.954} \end{bmatrix} = \begin{bmatrix} 1.732 & 0 & 0 \\ 0 & 1.083 & 0 \\ 0 & 0 & .977 \end{bmatrix}$$

and

$$R_y(0) = D^{-1}\Gamma_y(0)D^{-1} = \begin{bmatrix} 1 & .086 & .011 \\ .086 & 1 & .637 \\ .011 & .637 & 1 \end{bmatrix},$$

$$R_y(1) = D^{-1}\Gamma_y(1)D^{-1} = \begin{bmatrix} .500 & .043 & .005 \\ .172 & .286 & .336 \\ .022 & .413 & .441 \end{bmatrix}, \qquad (2.1.46)$$

$$R_y(2) = D^{-1}\Gamma_y(2)D^{-1} = \begin{bmatrix} .250 & .021 & .003 \\ .103 & .148 & .154 \\ .045 & .187 & .206 \end{bmatrix}.$$

A plot of some autocorrelations is shown in Figure 2.4. Assuming that the three variables of the system represent rates of change of investment, income, and consumption, respectively, it can, for instance, be seen that the contemporaneous and intertemporal correlations between consumption and investment are quite small, while the patterns of the autocorrelations of the individual series are very similar.

2.2 Forecasting

We have argued in the introduction that forecasting is one of the main objectives of multiple time series analysis. Therefore we will now discuss predictors based on VAR processes. Point forecasts and interval forecasts will be considered in turn. Before discussing particular predictors or forecasts (the two terms will be used interchangeably) we comment on the prediction problem in general.

2.2.1 The Loss Function

The forecaster usually finds himself in a situation where in a particular period t he has to make statements about the future values of variables $y_1, \dots, y_K$. For this purpose he has available a model for the data generation process and an information set, say Ω_t, containing the available information in period t. The data generation process may, for instance, be a VAR(p) process and Ω_t may contain the past and present variables of the system under consideration, that is, $\Omega_t = \{y_s | s \le t\}$, where $y_s = (y_{1s}, \dots, y_{Ks})'$. The period t, where the forecast is made, is the *forecast origin* and the number of periods into the future for which a forecast is desired is the *forecast horizon*. A predictor h periods ahead is an *h-step predictor*.

If forecasts are desired for a particular purpose a specific cost function may be associated with the forecast errors. A forecast will be optimal if it minimizes the cost. To find a forecast that is optimal in this sense is usually too ambitious a goal to be attainable in practice. Therefore minimizing the *expected* cost or loss is often used as an objective. In general it will depend on the particular loss function which forecast is optimal. On the other hand, forecasts of economic variables are often published for general use. In that case the specific cost or loss function of all potential users cannot be taken into account in computing a forecast. In this situation the statistical properties of the forecasts and perhaps interval forecasts are of interest to enable the user to draw proper conclusions for his or her particular needs. It may also be desirable to choose the forecast such that it minimizes a wide range of plausible loss functions.

In the context of VAR models, predictors that minimize the forecast mean squared errors (MSEs) are the most widely used ones. Arguments in favor of using the MSE as loss function are given by Granger (1969a) and Granger & Newbold (1986). They show that minimum MSE forecasts also minimize a range of loss functions other than the MSE. Moreover, for many loss functions the optimal predictors are simple functions of minimum MSE predictors. Furthermore, for an unbiased predictor the MSE is the forecast error variance which is useful in setting up interval forecasts. Therefore minimum MSE predictors will be of major interest in the following. If not otherwise stated the information set Ω_t is assumed to contain the variables of the system under consideration up to and including period t.

2.2.2 Point Forecasts

2.2.2a *Conditional Expectation*

Suppose $y_t = (y_{1t},\dots,y_{Kt})'$ is a K-dimensional stable VAR(p) process as in (2.1.1). Then the minimum MSE predictor for forecast horizon h at forecast origin t is the conditional expected value

$$E_t(y_{t+h}) := E(y_{t+h}|\Omega_t) = E(y_{t+h}|\{y_s|s \le t\}). \tag{2.2.1}$$

This predictor minimizes the MSE of each component of y_t. In other words, if $\bar{y}_t(h)$ is any h-step predictor at origin t,

$$\begin{aligned} \text{MSE}[\bar{y}_t(h)] &= E(y_{t+h} - \bar{y}_t(h))(y_{t+h} - \bar{y}_t(h))' \\ &\ge \text{MSE}[E_t(y_{t+h})] = E(y_{t+h} - E_t(y_{t+h}))(y_{t+h} - E_t(y_{t+h}))', \end{aligned} \tag{2.2.2}$$

where the inequality sign $\ge$ between two matrices means that the difference between the left-hand and the right-hand matrix is positive semidefinite. Equivalently, for any $(K \times 1)$ vector c,

$$\text{MSE}[c'\bar{y}_t(h)] \ge \text{MSE}[c'E_t(y_{t+h})].$$

The optimality of the conditional expectation can be seen by noting that

$$\begin{aligned} \text{MSE}[\bar{y}_t(h)] &= E[y_{t+h} - E_t(y_{t+h}) + E_t(y_{t+h}) - \bar{y}_t(h)][y_{t+h} - E_t(y_{t+h}) \\ &\qquad + E_t(y_{t+h}) - \bar{y}_t(h)]' \\ &= \text{MSE}[E_t(y_{t+h})] + E[E_t(y_{t+h}) - \bar{y}_t(h)][E_t(y_{t+h}) - \bar{y}_t(h)]', \end{aligned}$$

where $E[y_{t+h} - E_t(y_{t+h})][E_t(y_{t+h}) - \bar{y}_t(h)]' = 0$ has been used. The latter result holds because $[y_{t+h} - E_t(y_{t+h})]$ is a function of innovations after period t which are uncorrelated with the terms contained in $[E_t(y_{t+h}) - \bar{y}_t(h)]$ which is a function of y_s, $s \leq t$.

The optimality of the conditional expectation implies that

$$E_t(y_{t+h}) = \nu + A_1 E_t(y_{t+h-1}) + \cdots + A_p E_t(y_{t+h-p}) \tag{2.2.3}$$

is the optimal h-step predictor of a VAR(p) process y_t *provided* u_t *is independent white noise* so that u_t and u_s are independent for $s \neq t$ and hence $E_t(u_{t+h}) = 0$ for $h > 0$.

The formula (2.2.3) can be used for recursively computing the h-step predictors starting with $h = 1$:

$$E_t(y_{t+1}) = \nu + A_1 y_t + \cdots + A_p y_{t-p+1},$$

$$E_t(y_{t+2}) = \nu + A_1 E_t(y_{t+1}) + A_2 y_t + \cdots + A_p y_{t-p+2},$$

$$\vdots$$

By these recursions we get for a VAR(1) process,

$$E_t(y_{t+h}) = (I_K + A_1 + \cdots + A_1^{h-1})\nu + A_1^h y_t.$$

Assuming $y_t = (-6, 3, 5)'$ and $\nu = (0, 2, 1)'$ the following forecasts are obtained for the VAR(1) example process (2.1.14):

$$E_t(y_{t+1}) = \begin{bmatrix} 0 \\ 2 \\ 1 \end{bmatrix} + \begin{bmatrix} .5 & 0 & 0 \\ .1 & .1 & .3 \\ 0 & .2 & .3 \end{bmatrix} \begin{bmatrix} -6 \\ 3 \\ 5 \end{bmatrix} = \begin{bmatrix} -3.0 \\ 3.2 \\ 3.1 \end{bmatrix}, \tag{2.2.4a}$$

$$E_t(y_{t+2}) = (I_3 + A_1)\nu + A_1^2 y_t = \begin{bmatrix} -1.50 \\ 2.95 \\ 2.57 \end{bmatrix}, \tag{2.2.4b}$$

etc.. Similarly we get for the VAR(2) process (2.1.15) with $\nu = (.02, .03)'$, $y_t = (.06, .03)'$ and $y_{t-1} = (.055, .03)'$,

$$\begin{aligned} E_t(y_{t+1}) &= \begin{bmatrix} .02 \\ .03 \end{bmatrix} + \begin{bmatrix} .5 & .1 \\ .4 & .5 \end{bmatrix} \begin{bmatrix} .06 \\ .03 \end{bmatrix} + \begin{bmatrix} 0 & 0 \\ .25 & 0 \end{bmatrix} \begin{bmatrix} .055 \\ .03 \end{bmatrix} = \begin{bmatrix} .053 \\ .08275 \end{bmatrix}, \\ E_t(y_{t+2}) &= \begin{bmatrix} .02 \\ .03 \end{bmatrix} + \begin{bmatrix} .5 & .1 \\ .4 & .5 \end{bmatrix} \begin{bmatrix} .053 \\ .08275 \end{bmatrix} + \begin{bmatrix} 0 & 0 \\ .25 & 0 \end{bmatrix} \begin{bmatrix} .06 \\ .03 \end{bmatrix} = \begin{bmatrix} .0548 \\ .1076 \end{bmatrix}. \end{aligned} \tag{2.2.5}$$

The conditional expectation has the following properties:

(1) It is an unbiased predictor, that is, $E[y_{t+h} - E_t(y_{t+h})] = 0$.
(2) If u_t is independent white noise, $\text{MSE}[E_t(y_{t+h})] = \text{MSE}[E_t(y_{t+h}) | y_t, y_{t-1}, \ldots]$, that is, the MSE of the predictor equals the conditional MSE given y_t, $y_{t-1}, \ldots$.

The latter property follows by similar arguments as the optimality of the predictor $E_t(y_{t+h})$.

It must be emphasized that the prediction formula (2.2.3) relies on u_t being independent white noise. If u_t and u_s are not independent but just uncorrelated, $E_t(u_{t+h})$ will be nonzero in general. As an example consider the univariate AR(1) process $y_t = v + \alpha y_{t-1} + u_t$ with

$$u_t = \begin{cases} e_t & \text{for } t = 0, \pm 2, \pm 4, \ldots \\ (e_{t-1}^2 + 1)/\sqrt{2} & \text{for } t = \pm 1, \pm 3, \ldots \end{cases}$$

where the e_t are independent standard normal ($N(0, 1)$) random variables. The process u_t is easily seen to be uncorrelated but not independent white noise. For even t,

$$\begin{aligned} E_t(u_{t+1}) &= E[(e_t^2 + 1)/\sqrt{2} \mid y_t, y_{t-1}, \ldots] \\ &= (e_t^2 + 1)/\sqrt{2} \end{aligned}$$

since $e_t = y_t - v - \alpha y_{t-1}$.

2.2.2b Linear Minimum MSE Predictor

If u_t is not independent white noise additional assumptions are usually required to find the optimal predictor (conditional expectation) of a VAR(p) process. Without such assumptions we can achieve the less ambitious goal of finding the minimum MSE predictors among those that are *linear* functions of $y_t, y_{t-1}, \ldots$. Let us consider a zero mean VAR(1) process

$$y_t = A_1 y_{t-1} + u_t \tag{2.2.6}$$

first. As in (2.1.3) it follows that

$$y_{t+h} = A_1^h y_t + \sum_{i=0}^{h-1} A_1^i u_{t+h-i}.$$

Thus, for a predictor

$$y_t(h) = B_0 y_t + B_1 y_{t-1} + \cdots,$$

where the B_i are $(K \times K)$ coefficient matrices, we get a forecast error

$$y_{t+h} - y_t(h) = \sum_{i=0}^{h-1} A_1^i u_{t+h-i} + (A_1^h - B_0) y_t - \sum_{i=1}^{\infty} B_i y_{t-i}.$$

Using that u_{t+j}, for $j > 0$, is uncorrelated with y_{t-i}, for $i \geq 0$,

$$\begin{aligned} \text{MSE}[y_t(h)] = {} & E\left(\sum_{i=0}^{h-1} A_1^i u_{t+h-i}\right)\left(\sum_{i=0}^{h-1} A_1^i u_{t+h-i}\right)' \\ & + E\left[(A_1^h - B_0) y_t - \sum_{i=1}^{\infty} B_i y_{t-i}\right] \\ & \times \left[(A_1^h - B_0) y_t - \sum_{i=1}^{\infty} B_i y_{t-i}\right]'. \end{aligned}$$

Obviously this MSE matrix is minimal for $B_0 = A_1^h$ and $B_i = 0$ for $i > 0$. Thus, the optimal (*linear* minimum MSE) predictor for this special case is

$$y_t(h) = A_1^h y_t = A_1 y_t(h-1).$$

The forecast error is

$$\sum_{i=0}^{h-1} A_1^i u_{t+h-i}$$

and the MSE or forecast error covariance matrix is

$$\begin{aligned}\Sigma_y(h) := \mathrm{MSE}[y_t(h)] &= E\left(\sum_{i=0}^{h-1} A_1^i u_{t+h-i}\right)\left(\sum_{i=0}^{h-1} A_1^i u_{t+h-i}\right)' \\ &= \sum_{i=0}^{h-1} A_1^i \Sigma_u (A_1^i)' = \mathrm{MSE}[y_t(h-1)] + A_1^{h-1}\Sigma_u(A_1^{h-1})'.\end{aligned}$$

A general VAR(p) process with zero mean,

$$y_t = A_1 y_{t-1} + \cdots + A_p y_{t-p} + u_t$$

has a VAR(1) counterpart

$$Y_t = \mathbf{A} Y_{t-1} + U_t,$$

where Y_t, $\mathbf{A}$, and U_t are as defined in (2.1.8). Using the same arguments as above the optimal predictor of Y_{t+h} is seen to be

$$Y_t(h) = \mathbf{A}^h Y_t = \mathbf{A} Y_t(h-1).$$

It is easily seen by induction with respect to h that

$$Y_t(h) = \begin{bmatrix} y_t(h) \\ y_t(h-1) \\ \vdots \\ y_t(h-p+1) \end{bmatrix},$$

where $y_t(j) := y_{t+j}$ for $j \leq 0$. Defining the $(K \times Kp)$ matrix $J := [I_K\ 0 \dots 0]$ as in (2.1.11) we get the optimal h-step predictor of the process y_t at origin t as

$$\begin{aligned}y_t(h) &= J\mathbf{A} Y_t(h-1) = [A_1 \dots A_p]\, Y_t(h-1) \\ &= A_1 y_t(h-1) + \cdots + A_p y_t(h-p). \qquad (2.2.7)\end{aligned}$$

This formula may be used for recursively computing the forecasts. Obviously $y_t(h)$ is the conditional expectation $E_t(y_{t+h})$ if u_t is independent white noise since the recursion in (2.2.3) is the same as the one obtained here for a zero mean process with $\nu = 0$.

If the process y_t has nonzero mean, that is,

$$y_t = \nu + A_1 y_{t-1} + \cdots + A_p y_{t-p} + u_t$$

we define $x_t := y_t - \mu$, where $\mu := E(y_t) = (I - A_1 - \cdots - A_p)^{-1}\nu$. The process

x_t has zero mean and the optimal h-step predictor is

$$x_t(h) = A_1 x_t(h-1) + \cdots + A_p x_t(h-p).$$

Adding μ to both sides of this equation gives the optimal linear predictor of y_t,

$$\begin{aligned} y_t(h) = x_t(h) + \mu &= \mu + A_1(y_t(h-1) - \mu) + \cdots + A_p(y_t(h-p) - \mu) \\ &= \nu + A_1 y_t(h-1) + \cdots + A_p y_t(h-p). \end{aligned} \tag{2.2.8}$$

Henceforth we will refer to $y_t(h)$ as the *optimal predictor* irrespective of the properties of the white noise process u_t, that is, even if u_t is not independent but just uncorrelated white noise.

Using

$$Y_{t+h} = \mathbf{A}^h Y_t + \sum_{i=0}^{h-1} \mathbf{A}^i U_{t+h-i}$$

for a zero mean process, we get the forecast error

$$\begin{aligned} y_{t+h} - y_t(h) = J[Y_{t+h} - Y_t(h)] &= J\left[\sum_{i=0}^{h-1} \mathbf{A}^i U_{t+h-i}\right] \\ &= \sum_{i=0}^{h-1} J\mathbf{A}^i J' J U_{t+h-i} = \sum_{i=0}^{h-1} \Phi_i u_{t+h-i}, \end{aligned} \tag{2.2.9}$$

where the Φ_i are the MA coefficient matrices from (2.1.17). The forecast error is unchanged if y_t has nonzero mean since the mean term cancels. The forecast error representation (2.2.9) shows that the predictor $y_t(h)$ can also be expressed in terms of the MA representation (2.1.17),

$$y_t(h) = \mu + \sum_{i=h}^{\infty} \Phi_i u_{t+h-i} = \mu + \sum_{i=0}^{\infty} \Phi_{h+i} u_{t-i}. \tag{2.2.10}$$

From (2.2.9) the forecast error covariance or MSE matrix is easy to obtain,

$$\Sigma_y(h) := \mathrm{MSE}[y_t(h)] = \sum_{i=0}^{h-1} \Phi_i \Sigma_u \Phi_i' = \Sigma_y(h-1) + \Phi_{h-1} \Sigma_u \Phi_{h-1}'. \tag{2.2.11}$$

Hence the MSEs are monotonically nondecreasing and for $h \to \infty$ the MSE matrices approach the covariance matrix of y_t,

$$\Gamma_y(0) = \Sigma_y = \sum_{i=0}^{\infty} \Phi_i \Sigma_u \Phi_i'$$

(see (2.1.18)). That is,

$$\Sigma_y(h) \xrightarrow[h\to\infty]{} \Sigma_y. \tag{2.2.12}$$

If the process mean μ is used as a forecast the MSE matrix of that predictor is just the covariance matrix Σ_y of y_t. Hence, the optimal long range forecast ($h \to \infty$) is the process mean. In other words, the past of the process contains no information on the development of the process in the distant future. Zero mean processes with this property are *purely nondeterministic*, that is, $y_t - \mu$ is purely nondeterministic if the forecast MSEs satisfy (2.2.12).

For the example VAR(1) process (2.1.14) with Σ_u as in (2.1.33), using the MA coefficient matrices form (2.1.24), the forecast MSE matrices

$$\Sigma_y(1) = \Sigma_u = \begin{bmatrix} 2.25 & 0 & 0 \\ 0 & 1.0 & .5 \\ 0 & .5 & .74 \end{bmatrix},$$

$$\Sigma_y(2) = \Sigma_u + \Phi_1 \Sigma_u \Phi_1' = \begin{bmatrix} 2.813 & .113 & 0 \\ .113 & 1.129 & .632 \\ 0 & .632 & .907 \end{bmatrix}, \tag{2.2.13}$$

$$\Sigma_y(3) = \Sigma_y(2) + \Phi_2 \Sigma_u \Phi_2' = \begin{bmatrix} 2.953 & .146 & .011 \\ .146 & 1.161 & .663 \\ .011 & .663 & .943 \end{bmatrix}$$

are obtained. Similarly, for the VAR(2) example process (2.1.15) with white noise covariance matrix (2.1.41) we get from (2.1.25),

$$\Sigma_y(1) = \Sigma_u = \begin{bmatrix} .09 & 0 \\ 0 & .04 \end{bmatrix},$$

$$\Sigma_y(2) = \Sigma_u + \Phi_1 \Sigma_u \Phi_1' = \begin{bmatrix} .1129 & .02 \\ .02 & .0644 \end{bmatrix}. \tag{2.2.14}$$

2.2.3 Interval Forecasts and Forecast Regions

In order to set up interval forecasts or forecast intervals we need to make an assumption about the distributions of the y_t or the u_t. It is most common to consider *Gaussian processes* where $y_t, y_{t+1}, \ldots, y_{t+h}$ have a multivariate normal distribution for any t and h. Equivalently it may be assumed that u_t is Gaussian, that is, the u_t are multivariate normal, $u_t \sim N(0, \Sigma_u)$, and u_t and u_s are independent for $s \neq t$.

Under these conditions the forecast errors are also normally distributed as linear transformations of normal vectors,

$$y_{t+h} - y_t(h) = \sum_{i=0}^{h-1} \Phi_i u_{t+h-i} \sim N(0, \Sigma_y(h)). \tag{2.2.15}$$

This result implies that the forecast errors of the individual components are normal so that

$$\frac{y_{k,t+h} - y_{k,t}(h)}{\sigma_k(h)} \sim N(0, 1), \tag{2.2.16}$$

where $y_{k,t}(h)$ is the k-th component of $y_t(h)$ and $\sigma_k(h)$ is the square root of the k-th diagonal element of $\Sigma_y(h)$. Denoting by $z_{(\alpha)}$ the upper $\alpha 100$ percentage point of the standard normal distribution we get

$$1 - \alpha = \Pr\left\{-z_{(\alpha/2)} \leq \frac{y_{k,t+h} - y_{k,t}(h)}{\sigma_k(h)} \leq z_{(\alpha/2)}\right\}$$

$$= \Pr\{y_{k,t}(h) - z_{(\alpha/2)}\sigma_k(h) \leq y_{k,t+h} \leq y_{k,t}(h) + z_{(\alpha/2)}\sigma_k(h)\}.$$

Hence, a $(1 - \alpha)100\%$ interval forecast h periods ahead for the k-th component of y_t is

$$y_{k,t}(h) \pm z_{(\alpha/2)}\sigma_k(h) \tag{2.2.17a}$$

or

$$[y_{k,t}(h) - z_{(\alpha/2)}\sigma_k(h),\ y_{k,t}(h) + z_{(\alpha/2)}\sigma_k(h)]. \tag{2.2.17b}$$

If forecast intervals of this type are computed repeatedly from a large number of time series (realizations of the considered process) then about $(1 - \alpha)100\%$ of the intervals will contain the actual value of the random variable $y_{k,t+h}$.

Using (2.2.4) and (2.2.13), 95% forecast intervals for the components of the example VAR(1) process (2.1.14) are

$$\begin{aligned}
&y_{1,t}(1) \pm 1.96\sqrt{2.25} \text{ or } -3.0 \pm 2.94,\\
&y_{2,t}(1) \pm 1.96\sqrt{1.0} \text{ or } 3.2 \pm 1.96,\\
&y_{3,t}(1) \pm 1.96\sqrt{.74} \text{ or } 3.1 \pm 1.69,\\
&y_{1,t}(2) \pm 1.96\sqrt{2.813} \text{ or } -1.50 \pm 3.29,\\
&y_{2,t}(2) \pm 1.96\sqrt{1.129} \text{ or } 2.95 \pm 2.08,\\
&y_{3,t}(2) \pm 1.96\sqrt{.907} \text{ or } 2.57 \pm 1.87.
\end{aligned} \tag{2.2.18}$$

The result in (2.2.15) can also be used to establish joint forecast regions for two or more variables. For instance, if a joint forecast region for the first N components is desired, we define the $(N \times K)$ matrix $F := [I_N\ 0]$ and note that

$$[y_{t+h} - y_t(h)]'F'(F\Sigma_y(h)F')^{-1}F[y_{t+h} - y_t(h)] \sim \chi^2(N) \tag{2.2.19}$$

by a well-known result for multivariate normal vectors (see Appendix B). Hence, the $\chi^2(N)$-distribution can be used to determine a $(1 - \alpha)100\%$ forecast ellipsoid for the first N components of the process.

In practice the construction of the ellipsoid is quite demanding if N is greater than two or three. Therefore, a more practical approach is to use *Bonferroni's method* for constructing joint confidence regions. It is based on the fact that for events $E_1, \dots, E_N$ the following probability inequality holds:

$$\Pr(E_1 \cup \cdots \cup E_N) \leq \Pr(E_1) + \cdots + \Pr(E_N).$$

Hence,

$$\Pr\left[\bigcap_{i=1}^{N} E_i\right] \geq 1 - \sum_{i=1}^{N} \Pr(\bar{E}_i),$$

where $\bar{E}_i$ denotes the complement of E_i. Consequently, if E_i is the event that

$y_{i,t+h}$ falls within an interval H_i,

$$\Pr(Fy_{t+h} \in H_1 \times \cdots \times H_N) \geq 1 - \sum_{i=1}^{N} \Pr(\bar{E}_i). \tag{2.2.20}$$

In other words, if we choose an $\left[1 - \frac{\alpha}{N}\right]100\%$ forecast interval for each of the N components the resulting joint forecast region has probability at least $(1 - \alpha)100\%$ of containing all N variables jointly. For instance, for the VAR(1) example process considered previously,

$$\{(y_1, y_2) | -3.0 - 2.94 \leq y_1 \leq -3.0 + 2.94, 3.2 - 1.96 \leq y_2 \leq 3.2 + 1.96\}$$

is a joint forecast region of $(y_{1,t+1}, y_{2,t+1})$ with probability content at least 90%.

By the same method joint forecast regions for different horizons h can be obtained. For instance, a joint forecast region with probability content of at least $(1 - \alpha)100\%$ for $y_{k,t+1}, \ldots, y_{k,t+h}$ is

$$\{(y_{k,1}, \ldots, y_{k,h}) | y_{k,t}(i) - z_{(\alpha/2h)}\sigma_k(i) \leq y_{k,i} \leq y_{k,t}(i) + z_{(\alpha/2h)}\sigma_k(i), i = 1, \ldots, h\}. \tag{2.2.21}$$

Thus, for the example, a joint forecast region for $y_{2,t+1}, y_{2,t+2}$ with probability content of at least 90% is given by

$$\{(y_{2,1}, y_{2,2}) | 1.24 \leq y_{2,1} \leq 5.16, .87 \leq y_{2,2} \leq 5.03\}.$$

2.3 Structural Analysis with VAR Models

Since VAR models represent the correlations among a set of variables they are often used to analyze certain aspects of the relationships between the variables of interest. In the following, three ways to interpret a VAR model will be discussed. They are all closely related and they are all beset with problems that will be pointed out subsequently.

2.3.1 Granger-Causality and Instantaneous Causality

2.3.1a Definitions of Causality

Granger (1969b) has defined a concept of causality which, under suitable conditions, is fairly easy to deal with in the context of VAR models. Therefore it has become quite popular in recent years. The idea is that a cause cannot come after the effect. Thus, if a variable x affects a variable z, the former should help improving the predictions of the latter variable.

To formalize this idea, suppose that Ω_t is the information set containing all the relevant information in the universe available up to and including period t. Let $z_t(h|\Omega_t)$ be the optimal (minimum MSE) h-step predictor of the process z_t at origin t, based on the information in Ω_t. The corresponding forecast MSE will be denoted by $\Sigma_z(h|\Omega_t)$. The process x_t is said to *cause* z_t *in Granger's sense* if

$$\Sigma_z(h|\Omega_t) < \Sigma_z(h|\Omega_t \backslash \{x_s|s \leq t\}) \qquad \text{for at least one } h = 1, 2, \ldots \tag{2.3.1}$$

Alternatively we will say that x_t *Granger-causes* (or briefly causes) z_t or x_t is *Granger-causal* for z_t if (2.3.1) holds. In (2.3.1) $\Omega_t \backslash \{x_s|s \leq t\}$ is the set containing all the relevant information in the universe except for the information in the past and present of the x_t process. In other words, if z_t can be predicted more efficiently if the information in the x_t process is taken into account in addition to all other information in the universe, then x_t is *Granger-causal* for z_t.

The definition extends immediately to the case where z_t and x_t are M- and N-dimensional processes, respectively. In that case x_t is said to Granger-cause z_t if

$$\Sigma_z(h|\Omega_t) \neq \Sigma_z(h|\Omega_t \backslash \{x_s|s \leq t\}) \tag{2.3.2}$$

for some t and h. Alternatively this could be expressed by requiring the two MSEs to be different and

$$\Sigma_z(h|\Omega_t) \leq \Sigma_z(h|\Omega_t \backslash \{x_s|s \leq t\})$$

(i.e., the difference between the right-hand and the left-hand matrix is positive semidefinite). Since the null matrix is also positive semidefinite it is necessary to require in addition that the two matrices are not identical. If x_t causes z_t and z_t also causes x_t the process $(z_t', x_t')'$ is called a *feedback system*.

Sometimes the term "instantaneous causality" is used in economic analyses. We say that there is *instantaneous causality* between z_t and x_t if

$$\Sigma_z(1|\Omega_t \cup \{x_{t+1}\}) \neq \Sigma_z(1|\Omega_t). \tag{2.3.3}$$

In other words, in period t, adding x_{t+1} to the information set, helps to improve the forecast of z_{t+1}. We will see shortly that this concept of causality is really symmetric, that is, if there is instantaneous causality between z_t and x_t then there is also instantaneous causality between x_t and z_t (see Proposition 2.3). Therefore we do not use the notion "instantaneous causality *from* x_t *to* z_t" in the foregoing definition.

A possible criticism of the foregoing definitions could relate to the choice of the MSE as a measure of the forecast precision. Of course, the choice of another measure could lead to a different definition of causality. However, in the situations of interest in the following, equality of the MSEs will imply equality of the corresponding predictors. In that case a process z_t is not Granger-caused by x_t if the optimal predictor of z_t does not use information from the x_t process. This result is intuitively appealing.

A more serious practical problem is the choice of the information set Ω_t. Usually all the relevant information in the universe is not available to a forecaster and thus the optimal predictor given Ω_t cannot be determined. Therefore a less demanding definition of causality is often used in practice. Instead of all the information in the universe, only the information in the past and present of the process under study is considered relevant and Ω_t is replaced by $\{z_s, x_s|s \leq t\}$.

Furthermore, instead of *optimal* predictors, *optimal linear* predictors are compared. In other words, $z_t(h|\Omega_t)$ is replaced by the linear minimum MSE h-step predictor based on the information in $\{z_s, x_s|s \leq t\}$ and $z_t(h|\Omega_t \backslash \{x_s|s \leq t\})$ is replaced by the linear minimum MSE h-step predictor based on $\{z_s|s \leq t\}$. In the following, when the terms "Granger-causality" and "instantaneous causality" are used, these restrictive assumptions are implicitly used if not otherwise noted.

2.3.1b Characterization of Granger-Causality

In order to determine the Granger-causal relationships between the variables of the K-dimensional VAR process y_t, suppose it has the canonical MA representation

$$y_t = \mu + \sum_{i=0}^{\infty} \Phi_i u_{t-i} = \mu + \Phi(L)u_t, \qquad \Phi_0 = I_K, \tag{2.3.4}$$

where u_t is a white noise process with nonsingular covariance matrix Σ_u. Suppose that y_t consists of the M-dimensional process z_t and the $(K - M)$-dimensional process x_t and the MA representation is partitioned accordingly,

$$y_t = \begin{bmatrix} z_t \\ x_t \end{bmatrix} = \begin{bmatrix} \mu_1 \\ \mu_2 \end{bmatrix} + \begin{bmatrix} \Phi_{11}(L) & \Phi_{12}(L) \\ \Phi_{21}(L) & \Phi_{22}(L) \end{bmatrix} \begin{bmatrix} u_{1t} \\ u_{2t} \end{bmatrix}. \tag{2.3.5}$$

Using the prediction formula (2.2.10), the optimal 1-step forecast of z_t based on y_t is

$$\begin{aligned} z_t(1|\{y_s|s \leq t\}) &= [I_M \quad 0]y_t(1) \\ &= \mu_1 + \sum_{i=1}^{\infty} \Phi_{11,i} u_{1,t+1-i} + \sum_{i=1}^{\infty} \Phi_{12,i} u_{2,t+1-i}. \end{aligned} \tag{2.3.6}$$

Hence the forecast error is

$$z_{t+1} - z_t(1|\{y_s|s \leq t\}) = u_{1,t+1}. \tag{2.3.7}$$

We have mentioned in Section 2.1.3 that a subprocess of a stationary process also has a prediction error MA representation. Thus,

$$\begin{aligned} z_t &= \mu_1 + \sum_{i=0}^{\infty} \Phi_{11,i} u_{1,t-i} + \sum_{i=1}^{\infty} \Phi_{12,i} u_{2,t-i} \\ &= \mu_1 + \sum_{i=0}^{\infty} F_i v_{t-i}, \end{aligned} \tag{2.3.8}$$

where $F_0 = I_M$ and the last expression is a prediction error MA representation. Thus the optimal 1-step predictor based on z_t only is

$$z_t(1|\{z_s|s \leq t\}) = \mu_1 + \sum_{i=1}^{\infty} F_i v_{t+1-i} \tag{2.3.9}$$

and the corresponding forecast error is

$$z_{t+1} - z_t(1|\{z_s|s \le t\}) = v_{t+1}. \tag{2.3.10}$$

Consequently the predictors (2.3.6) and (2.3.9) are identical if and only if $v_t = u_{1,t}$ for all t. In other words, equality of the predictors is equivalent to z_t having the MA representation

$$\begin{aligned} z_t &= \mu_1 + \sum_{i=0}^{\infty} F_i u_{1,t-i} = \mu_1 + \sum_{i=0}^{\infty} [F_i \quad 0] u_{t-i} \\ &= \mu_1 + \sum_{i=0}^{\infty} [\Phi_{11,i} \quad \Phi_{12,i}] u_{t-i} = \mu_1 + \sum_{i=0}^{\infty} \Phi_{11,i} u_{1,t-i} + \sum_{i=1}^{\infty} \Phi_{12,i} u_{2,t-i}. \end{aligned}$$

Uniqueness of the canonical MA representation implies that $F_i = \Phi_{11,i}$ and $\Phi_{12,i} = 0$ for $i = 1, 2, \ldots$. Hence, we get the following proposition.

PROPOSITION 2.2 (*Characterization of Granger-Noncausality*)

Let y_t be a VAR process as in (2.3.4)/(2.3.5) with canonical MA operator $\Phi(z)$. Then

$$z_t(1|\{y_s|s \le t\}) = z_t(1|\{z_s|s \le t\}) \quad \Leftrightarrow \quad \Phi_{12,i} = 0 \quad \text{for } i = 1, 2, \ldots \tag{2.3.11}$$

■

Since we have just used the MA representation (2.3.4) and not its finite order VAR form the proposition is not only valid for VAR processes but more generally for processes having a canonical MA representation such as (2.3.4). From (2.2.10) it is obvious that equality of the 1-step predictors implies equality of the h-step predictors for $h = 2, 3, \ldots$. Hence, the proposition provides a necessary and sufficient condition for x_t being not Granger-causal for z_t, that is, z_t is not Granger-caused by x_t if and only if $\Phi_{12,i} = 0$ for $i = 1, 2, \ldots$. Thus, Granger-noncausality can be checked easily by looking at the MA representation of y_t. Since we are mostly concerned with VAR processes it is worth noting that for a stationary, stable VAR(p) process

$$y_t = \begin{bmatrix} z_t \\ x_t \end{bmatrix} = \begin{bmatrix} \nu_1 \\ \nu_2 \end{bmatrix} + \begin{bmatrix} A_{11,1} & A_{12,1} \\ A_{21,1} & A_{22,1} \end{bmatrix} \begin{bmatrix} z_{t-1} \\ x_{t-1} \end{bmatrix} + \cdots + \begin{bmatrix} A_{11,p} & A_{12,p} \\ A_{21,p} & A_{22,p} \end{bmatrix} \begin{bmatrix} z_{t-p} \\ x_{t-p} \end{bmatrix} + \begin{bmatrix} u_{1t} \\ u_{2t} \end{bmatrix} \tag{2.3.12}$$

the condition (2.3.11) is satisfied if and only if

$$A_{12,i} = 0 \qquad \text{for } i = 1, \ldots, p.$$

This result follows from the recursions in (2.1.22) or, alternatively, because the inverse of

$$\begin{bmatrix} \Phi_{11}(L) & 0 \\ \Phi_{21}(L) & \Phi_{22}(L) \end{bmatrix}$$

is

$$\begin{bmatrix} \Phi_{11}(L)^{-1} & 0 \\ -\Phi_{22}(L)^{-1}\Phi_{21}(L)\Phi_{11}(L)^{-1} & \Phi_{22}(L)^{-1} \end{bmatrix}.$$

Thus, we have the following result.

COROLLARY 2.2.1

If y_t is a stable VAR(p) process as in (2.3.12) with nonsingular white noise covariance matrix Σ_u, then

$$z_t(h|\{y_s|s \le t\}) = z_t(h|\{z_s|s \le t\}), \qquad h = 1, 2, \ldots$$

$$\Leftrightarrow \quad A_{12,i} = 0 \qquad \text{for } i = 1, \ldots, p. \tag{2.3.13}$$

Alternatively,

$$x_t(h|\{y_s|s \le t\}) = x_t(h|\{x_s|s \le t\}), \quad h = 1, 2, \ldots$$

$$\Leftrightarrow \quad A_{21,i} = 0 \quad \text{for } i = 1, \ldots, p. \tag{2.3.14}$$

■

This corollary implies that noncausalities can be determined by just looking at the VAR representation of the system. For instance, for the example process (2.1.14),

$$\begin{bmatrix} y_{1,t} \\ y_{2,t} \\ y_{3,t} \end{bmatrix} = v + \begin{bmatrix} .5 & 0 & 0 \\ .1 & .1 & .3 \\ 0 & .2 & .3 \end{bmatrix} \begin{bmatrix} y_{1,t-1} \\ y_{2,t-1} \\ y_{3,t-1} \end{bmatrix} + u_t,$$

$x_t = (y_{2t}, y_{3t})'$ does not Granger-cause $z_t = y_{1t}$ since $A_{12,1} = 0$ if the coefficient matrix is partitioned according to (2.3.12). On the other hand, z_t Granger-causes x_t. To give this discussion economic content let us assume that the variables in the system are rates of change of investment (y_1), income (y_2), and consumption (y_3). With these definitions the previous discussion shows that investment Granger-causes the consumption/income system whereas the converse is not true. It is also easy to check that consumption causes the income/investment system and vice versa. Note that we have defined Granger-causality only in terms of two groups of variables. Therefore, at this stage, we cannot talk about the Granger-causal relationship between consumption and income in the three-dimensional investment/income/consumption system.

Let us assume that the variables in the example VAR(2) process (2.1.15),

$$\begin{bmatrix} y_{1,t} \\ y_{2,t} \end{bmatrix} = v + \begin{bmatrix} .5 & .1 \\ .4 & .5 \end{bmatrix} \begin{bmatrix} y_{1,t-1} \\ y_{2,t-1} \end{bmatrix} + \begin{bmatrix} 0 & 0 \\ .25 & 0 \end{bmatrix} \begin{bmatrix} y_{1,t-2} \\ y_{2,t-2} \end{bmatrix} + u_t,$$

represent the inflation rate (y_1) and some interest rate (y_2). Using Corollary 2.2.1 it is immediately obvious that inflation causes the interest rate and vice versa. Hence the system is a feedback system. In the following we will refer to (2.1.15) as the inflation/interest rate system.

2.3.1c Characterization of Instantaneous Causality

In order to study the concept of instantaneous causality in the framework of the MA process (2.3.5) it is useful to rewrite that representation. Note that the positive definite symmetric matrix Σ_u can be written as the product $\Sigma_u = PP'$, where P is a lower triangular nonsingular matrix with positive diagonal elements (see Appendix A.9.3). Thus, (2.3.5) can be represented as

$$\begin{aligned} y_t &= \mu + \sum_{i=0}^{\infty} \Phi_i P P^{-1} u_{t-i} \\ &= \mu + \sum_{i=0}^{\infty} \Theta_i w_{t-i}, \end{aligned} \tag{2.3.15}$$

where $\Theta_i := \Phi_i P$ and $w_t := P^{-1}u_t$ is white noise with covariance matrix

$$\Sigma_w = P^{-1}\Sigma_u(P^{-1})' = I_K. \tag{2.3.16}$$

Since the white noise errors w_t have uncorrelated components they are often called *orthogonal* residuals or innovations.

Partitioning the representation (2.3.15) according to the partitioning of $y_t = (z_t', x_t')'$ gives

$$\begin{bmatrix} z_t \\ x_t \end{bmatrix} = \begin{bmatrix} \mu_1 \\ \mu_2 \end{bmatrix} + \begin{bmatrix} \Theta_{11,0} & 0 \\ \Theta_{21,0} & \Theta_{22,0} \end{bmatrix} \begin{bmatrix} w_{1,t} \\ w_{2,t} \end{bmatrix} + \begin{bmatrix} \Theta_{11,1} & \Theta_{12,1} \\ \Theta_{21,1} & \Theta_{22,1} \end{bmatrix} \begin{bmatrix} w_{1,t-1} \\ w_{2,t-1} \end{bmatrix} + \cdots$$

Hence,

$$z_{t+1} = \mu_1 + \Theta_{11,0} w_{1,t+1} + \Theta_{11,1} w_{1,t} + \Theta_{12,1} w_{2,t} + \cdots$$

and

$$x_{t+1} = \mu_2 + \Theta_{21,0} w_{1,t+1} + \Theta_{22,0} w_{2,t+1} + \Theta_{21,1} w_{1,t} + \Theta_{22,1} w_{2,t} + \cdots.$$

The optimal 1-step predictor of x_t based on $\{y_s | s \le t\}$ and, in addition, on z_{t+1} is equal to the 1-step predictor of x_t based on $\{w_s | s \le t\} \cup \{w_{1,t+1}\}$, that is,

$$\begin{aligned} x_t(1 | \{y_s | s \le t\} \cup \{z_{t+1}\}) &= x_t(1 | \{w_s = (w_{1,s}', w_{2,s}')' | s \le t\} \cup \{w_{1,t+1}\}) \\ &= \Theta_{21,0} w_{1,t+1} + x_t(1 | \{y_s | s \le t\}). \end{aligned} \tag{2.3.17}$$

Consequently,

$$x_t(1 | \{y_s | s \le t\} \cup \{z_{t+1}\}) = x_t(1 | \{y_s | s \le t\})$$

if and only if $\Theta_{21,0} = 0$. This condition, in turn, is easily seen to hold if and only if the covariance matrix Σ_u is block diagonal with a $((K - M) \times M)$ block of zeroes in the lower left-hand corner and an $(M \times (K - M))$ block of zeroes in the upper right-hand corner. Of course, this means that u_{1t} and u_{2t} in (2.3.5) have to be uncorrelated, i.e., $E(u_{1t}u_{2t}') = 0$.

PROPOSITION 2.3 (*Characterization of Instantaneous Causality*)

Let y_t be as in (2.3.5)/(2.3.15) with nonsingular innovation covariance matrix Σ_u. Then there is no instantaneous causality between z_t and x_t if and only if

$$E(u_{1t}u_{2t}') = 0. \tag{2.3.18}$$

■

This proposition provides a condition for instantaneous causality which is easy to check if the process is given in MA or VAR form. For instance, for the investment/income/consumption system with white noise covariance matrix (2.1.33),

$$\Sigma_u = \begin{bmatrix} 2.25 & 0 & 0 \\ 0 & 1.0 & .5 \\ 0 & .5 & .74 \end{bmatrix},$$

there is no instantaneous causality between (income, consumption) and investment.

From Propositions 2.2 and 2.3 it follows that $y_t = (z_t', x_t')'$ has a representation with orthogonal innovations as in (2.3.15) of the form

$$\begin{aligned}\begin{bmatrix} z_t \\ x_t \end{bmatrix} &= \begin{bmatrix} \mu_1 \\ \mu_2 \end{bmatrix} + \begin{bmatrix} \Theta_{11,0} & 0 \\ 0 & \Theta_{22,0} \end{bmatrix}\begin{bmatrix} w_{1,t} \\ w_{2,t} \end{bmatrix} + \begin{bmatrix} \Theta_{11,1} & 0 \\ \Theta_{21,1} & \Theta_{22,1} \end{bmatrix}\begin{bmatrix} w_{1,t-1} \\ w_{2,t-1} \end{bmatrix} + \cdots \\ &= \begin{bmatrix} \mu_1 \\ \mu_2 \end{bmatrix} + \begin{bmatrix} \Theta_{11}(L) & 0 \\ \Theta_{21}(L) & \Theta_{22}(L) \end{bmatrix}\begin{bmatrix} w_{1,t} \\ w_{2,t} \end{bmatrix}\end{aligned} \tag{2.3.19}$$

if x_t does not Granger-cause z_t and furthermore there is no instantaneous causation between x_t and z_t. In the absence of instantaneous causality, a similar representation with $\Theta_{21}(L) \equiv 0$ is obtained if z_t is not Granger-causal for x_t.

2.3.1d Interpretation and Critique of Instantaneous and Granger-Causality

At this point some words of caution seem appropriate. The term "causality" suggests a cause and effect relationship between two sets of variables. Proposition 2.3 shows that such an interpretation is problematic with respect to instantaneous causality because this term only describes a nonzero correlation between two sets of variables. It does not say anything about the cause and effect relation. The direction of instantaneous causation cannot be derived from the MA or VAR representation of the process but must be obtained from further knowledge on the relationship between the variables. Such knowledge may exist in the form of an economic theory.

Although a direction of causation has been defined in relation with Granger-causality it is problematic to interpret the absence of causality from x_t to z_t in the sense that variations in x_t will have no effect on z_t. To see this consider, for instance, the stable bivariate VAR(1) system

$$\begin{bmatrix} z_t \\ x_t \end{bmatrix} = \begin{bmatrix} \alpha_{11} & 0 \\ \alpha_{21} & \alpha_{22} \end{bmatrix} \begin{bmatrix} z_{t-1} \\ x_{t-1} \end{bmatrix} + \begin{bmatrix} u_{1t} \\ u_{2t} \end{bmatrix}. \tag{2.3.20}$$

In this system x_t does not Granger-cause z_t by Corollary 2.2.1. However, the system may be multiplied by some nonsingular matrix

$$B = \begin{bmatrix} 1 & \beta \\ 0 & 1 \end{bmatrix}$$

so that

$$\begin{bmatrix} z_t \\ x_t \end{bmatrix} = \begin{bmatrix} 0 & -\beta \\ 0 & 0 \end{bmatrix} \begin{bmatrix} z_t \\ x_t \end{bmatrix} + \begin{bmatrix} \gamma_{11} & \gamma_{12} \\ \gamma_{21} & \gamma_{22} \end{bmatrix} \begin{bmatrix} z_{t-1} \\ x_{t-1} \end{bmatrix} + \begin{bmatrix} v_{1t} \\ v_{2t} \end{bmatrix}, \tag{2.3.21}$$

where $\gamma_{11} = \alpha_{11} + \alpha_{21}\beta$, $\gamma_{12} = \alpha_{22}\beta$, $\gamma_{21} = \alpha_{21}$, $\gamma_{22} = \alpha_{22}$ and $(v_{1t}, v_{2t})' = B(u_{1t}, u_{2t})'$. Note that this is just another representation of the process $(z_t, x_t)'$ and not another process. (The reader may check that the process (2.3.21) has the same means and autocovariances as the one in (2.3.20).)

In other words, the stochastic interrelationships between the random variables of the system can either be characterized by (2.3.20) or by (2.3.21) although the two representations have quite different physical interpretations. If (2.3.21) happens to represent the actual ongoings in the system, changes in x_t may affect z_t through the term with coefficient $-\beta$ in the first equation. Thus, the lack of a Granger-causal relationship from one group of variables to the remaining variables cannot necessarily be interpreted as lack of a cause and effect relationship. It must be remembered that a VAR or MA representation characterizes the joint distribution of sets of random variables. In order to derive cause and effect relationships from it usually requires further assumptions regarding the relationship between the variables involved. We will return to this problem in the following subsections.

Further problems related to the interpretation of Granger-causality result from restricting the information set to contain only past and present variables of the system rather than all information in the universe. Only if all other information in the universe is irrelevant for the problem at hand the reduction of the information set is of no consequence. Let us discuss some resulting problems in terms of an inflation/interest rate system.

It may make a difference whether the information set contains monthly, quarterly or annual data. If a quarterly system is considered and no causality is found from the interest rate to inflation it does not follow that a corresponding monthly interest rate has no impact on the monthly inflation rate. In other words, the interest rate may Granger-cause inflation in a monthly system even if it does not in a quarterly system.

Similarly, if other relevant variables are omitted that may affect the system, including them into the information set may change the Granger-causal structure of the system. For instance, if a money stock variable such as $M1$ is added to the inflation/interest rate system a change in the causal structure is possible.

Furthermore, putting seasonally adjusted variables in the information set is not the same as using unadjusted variables. Consequently, if Granger-causality is found for the seasonally adjusted variables, it is still possible that in the actual seasonal system the interest rate is not Granger-causal for inflation. Similar comments apply in the presence of measurement errors. Finally, causality analyses are usually based on estimated rather than known systems. Additional problems result in that case. We will return to them in the next chapter.

The previous critical remarks are meant to caution the reader and multiple time series analyst against overinterpreting the evidence from a VAR model. Still, causality analyses are useful tools in practice if these critical points are kept in mind. At the very least a Granger-causality analysis tells the analyst whether a set of variables contains useful information for improving the predictions of another set of variables. Further discussions of causality issues and many further references may be found in Geweke (1984) and Granger (1982).

2.3.2 Impulse Response Analysis

In the previous subsection we have partitioned a set of variables in subsets z_t and x_t and we have defined Granger-causality from x_t to z_t and vice versa. The definition involves all the variables of the system. In applied work it is often of interest to know the response of *one* variable to an impulse in another variable in a system that involves a number of other variables as well. Thus, one would like to investigate the relationship between two variables in a higher dimensional system. Of course, if there is a reaction of one variable to an impulse in another variable we may call the latter causal for the former. In this subsection we will study this type of causality by tracing out the effect of an exogenous shock or innovation in one of the variables on some or all of the other variables. This kind of impulse response analysis is often called *multiplier analysis*. For instance, in a system consisting of an inflation rate and an interest rate the effect of an increase in the inflation rate may be of interest. In the real world such an increase may be induced exogenously from outside the system by events like the increase of the oil price in 1973/74 when the OPEC agreed on a joint action to raise prices. Alternatively, an increase or reduction in the interest rate may be administered by the central bank for reasons outside the simple two variable system under study.

2.3.2a Responses to Forecast Errors

Suppose the effect of an innovation in investment in a system containing investment (y_1), income (y_2), and consumption (y_3) is of interest. To isolate such an effect suppose that all three variables assume their mean value prior to time $t = 0$, $y_t = \mu$ for $t < 0$, and investment increases by one unit in period $t = 0$, that is, $u_{1,0} = 1$. Now we can trace out what happens to the system during periods $t = 1, 2, \ldots$ if no further shocks occur, that is, $u_{2,0} = u_{3,0} = 0$, $u_1 = 0$, $u_2 = 0, \ldots$. Since we are not interested in the mean of the system in such an exercise but just in the variations of the variables around their means we assume that all three variables have mean zero and set $\nu = 0$ in (2.1.14). Hence, $y_t = A_1 y_{t-1} + u_t$ or, more precisely,

$$\begin{bmatrix} y_{1,t} \\ y_{2,t} \\ y_{3,t} \end{bmatrix} = \begin{bmatrix} .5 & 0 & 0 \\ .1 & .1 & .3 \\ 0 & .2 & .3 \end{bmatrix} \begin{bmatrix} y_{1,t-1} \\ y_{2,t-1} \\ y_{3,t-1} \end{bmatrix} + \begin{bmatrix} u_{1,t} \\ u_{2,t} \\ u_{3,t} \end{bmatrix}. \tag{2.3.22}$$

Tracing a unit shock in the first variable in period $t = 0$ in this system we get

$$y_0 = \begin{bmatrix} y_{1,0} \\ y_{2,0} \\ y_{3,0} \end{bmatrix} = \begin{bmatrix} u_{1,0} \\ u_{2,0} \\ u_{3,0} \end{bmatrix} = \begin{bmatrix} 1 \\ 0 \\ 0 \end{bmatrix},$$

$$y_1 = \begin{bmatrix} y_{1,1} \\ y_{2,1} \\ y_{3,1} \end{bmatrix} = A_1 y_0 = \begin{bmatrix} .5 \\ .1 \\ 0 \end{bmatrix},$$

$$y_2 = \begin{bmatrix} y_{1,2} \\ y_{2,2} \\ y_{3,2} \end{bmatrix} = A_1 y_1 = A_1^2 y_0 = \begin{bmatrix} .25 \\ .06 \\ .02 \end{bmatrix}.$$

Continuing the procedure it turns out that $y_i = (y_{1,i}, y_{2,i}, y_{3,i})'$ is just the first column of A_1^i. An analogous line of arguments shows that a unit shock in y_{2t} (y_{3t}) at $t = 0$, after i periods, results in a vector y_i which is just the second (third) column of A_1^i. Thus, the elements of A_1^i represent the effects of unit shocks in the variables of the system after i periods. Therefore they are called impulse responses or dynamic multipliers.

Recall that $A_1^i = \Phi_i$ is just the i-th coefficient matrix of the MA representation of a VAR(1) process. Consequently the MA coefficient matrices contain the impulse responses of the system. This result holds more generally for higher order VAR(p) processes as well. To see this suppose y_t is a stationary VAR(p) process as in (2.1.1) with $\nu = 0$. This process has a corresponding VAR(1) process $Y_t = \mathbf{A} Y_{t-1} + U_t$ as in (2.1.8) with $\boldsymbol{\nu} = 0$. Under the assumptions of the previous example, $y_t = 0$ for $t < 0$, $u_t = 0$ for $t > 0$ and $y_0 = u_0$ is a K-dimensional unit vector e_k, say, with a one as the k-th coordinate and zeroes elsewhere. It follows that $Y_0 = (e_k' \quad 0 \dots 0)'$ and $Y_i = \mathbf{A}^i Y_0$. Hence, the impulse responses are the elements of the upper left-hand $(K \times K)$ block of $\mathbf{A}^i$. This matrix, however, was shown to be the i-th coefficient matrix Φ_i of the MA representation (2.1.17) of y_t, i.e. $\Phi_i = J\mathbf{A}^i J'$ with $J := [I_K \quad 0 \dots 0]$ a $(K \times Kp)$ matrix. In other words, $\phi_{jk,i}$, the jk-th element of Φ_i represents the reaction of the j-th variable of the system to a unit shock of variable k, i periods ago, provided, of course, the effect is not contaminated by other shocks to the system. Since the u_t are just the 1-step ahead forecast errors of the VAR process the shocks considered here may be regarded as forecast errors.

The response of variable j to a unit shock (forecast error) in variable k is sometimes depicted graphically to get a visual impression of the dynamic interrelationships within the system. Impulse responses of the investment/income/consumption system are plotted in Figure 2.5 and the dynamic responses of the inflation/interest rate system are depicted in Figure 2.6. For instance, in the latter figure an inflation innovation is seen to induce the interest rate to grow for two periods and then it tapers off to zero. In both systems the effect of a unit shock in any of the variables dies away quite rapidly due to the stability of the systems.

If the variables have different scales it is sometimes useful to consider innovations of one standard deviation rather than unit shocks. For instance, instead of tracing an unexpected unit increase in investment in the investment/income/consumption system with white noise covariance matrix (2.1.33) one may follow up on a shock of $\sqrt{2.25} = 1.5$ units because the standard deviation of u_{1t} is 1.5. Of course, this is just a matter of rescaling the impulse responses. In Figures 2.5 and 2.6 it suffices to choose the units at the vertical axes equal to the standard deviations of the residuals corresponding to the variables whose effects are considered. Such a rescaling may sometimes give a better picture of the dynamic relationships because the average size of the innovations occurring in a system depends on their standard deviation.

It follows from Proposition 2.2 that the impulse responses are zero if one of the variables does not Granger-cause the other variables taken as a group. More precisely, an innovation in variable k has no effect on the other variables if the former variable does not Granger-cause the set of the remaining variables. As we have mentioned previously, in applied work it is often of foremost interest whether one variable has an impact on one other variable. That is, one would like to know whether, for some $k \neq j$, $\phi_{jk,i} = 0$ for $i = 1, 2, \ldots$. If the $\phi_{jk,i}$ represent the actual reactions of variable j to a unit shock in variable k, we may call the latter *noncausal* for the j-th variable if $\phi_{jk,i} = 0$ for $i = 1, 2, \ldots$. In order to check the latter condition it is not necessary to compute infinitely many Φ_i matrices. The following proposition shows that it suffices to check the first $p(K-1)$ Φ_i matrices.

PROPOSITION 2.4 (*Zero Impulse Responses*)

If y_t is a K-dimensional stable VAR(p) process then, for $j \neq k$,

$$\phi_{jk,i} = 0 \qquad \text{for } i = 1, 2, \ldots$$

is equivalent to

$$\phi_{jk,i} = 0 \qquad \text{for } i = 1, \ldots, p(K-1).$$ ■

In other words, the proposition asserts that for a K-dimensional stationary, stable VAR(p), if the first $pK - p$ responses of variable j to an impulse in variable k are zero, all the following responses must also be zero. For instance, in the investment/income/consumption VAR(1) system, since the responses of investment for the next two periods after a consumption impulse are zero, we know that investment will not react at all to such an impulse. Note, that in a VAR(1) system of dimension greater than 2, it does not suffice to check, say, the upper

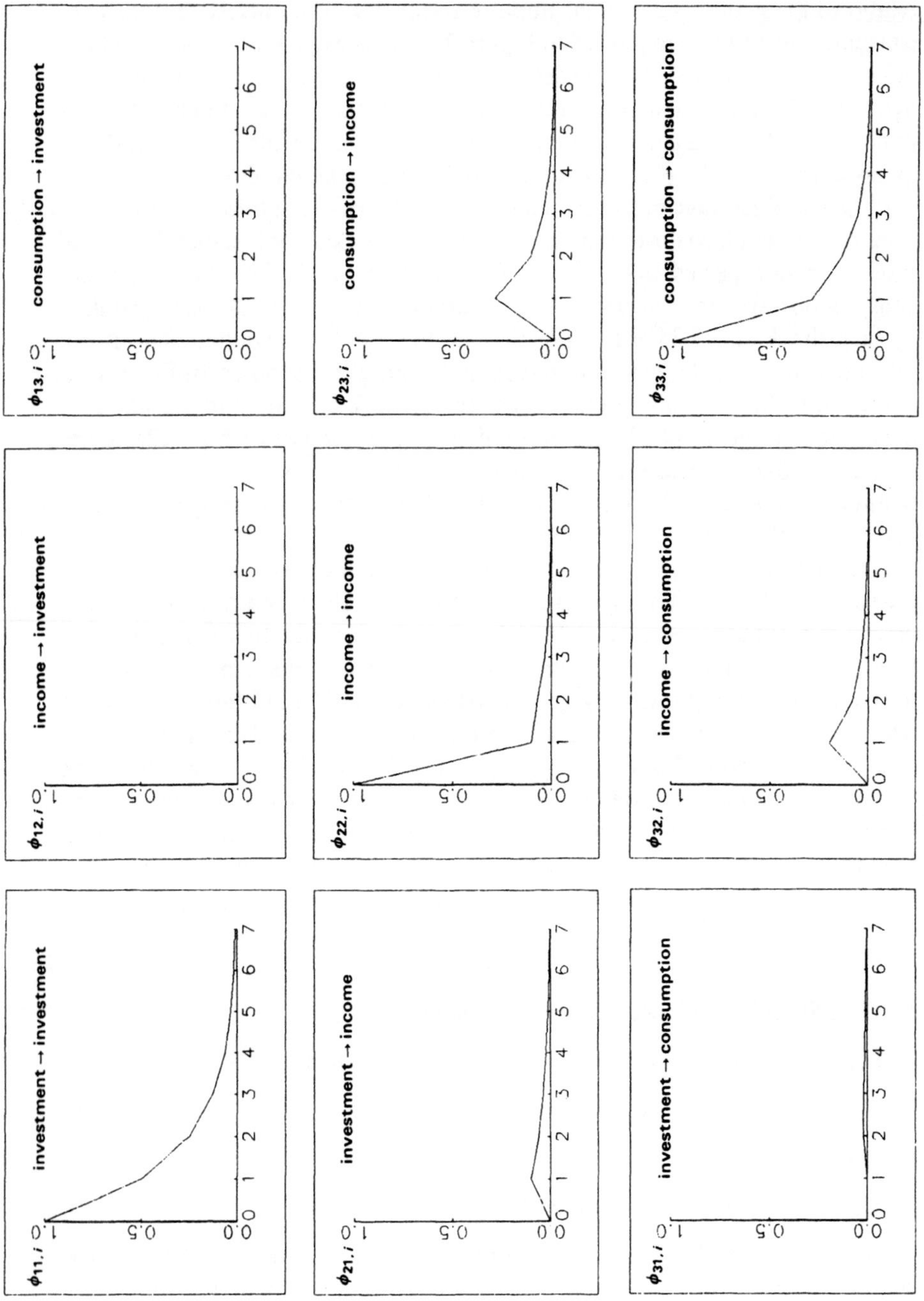

Fig. 2.5. Impulse responses of the investment/income/consumption system (impulse → response).

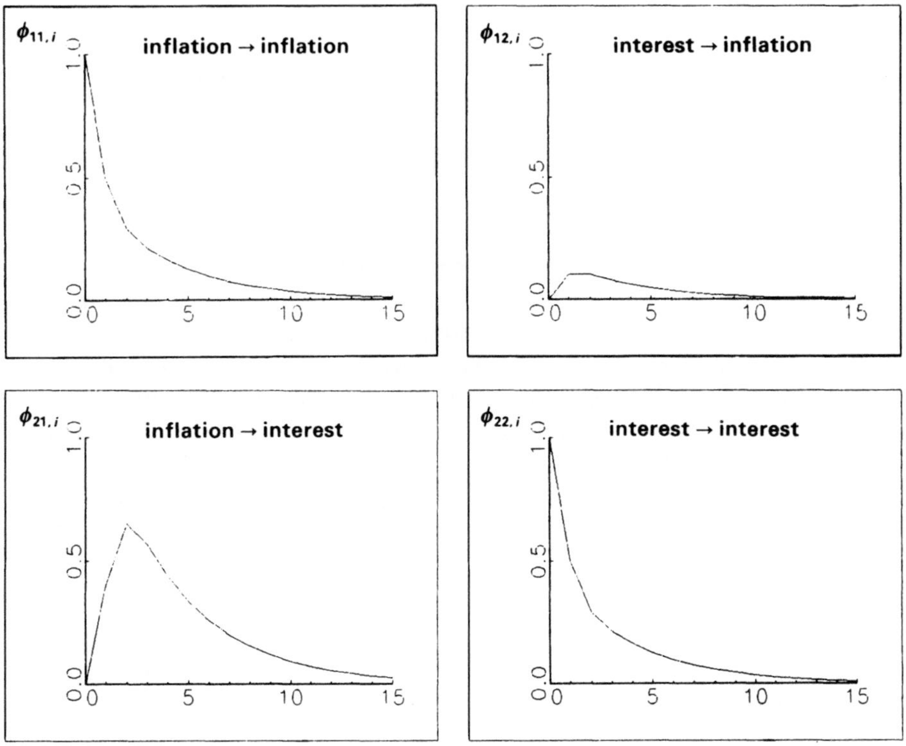

Fig. 2.6. Impulse responses of the inflation/interest rate system (impulse → response).

right-hand corner element of the coefficient matrix in order to determine whether the last variable is noncausal for the first variable. Proposition 2.4 will be helpful when testing of causal relations is discussed in the next chapter. We will now prove the proposition.

Proof of Proposition 2.4:

Returning to the lag operator notation of Section 2.1.2 we have

$$\Phi(L) = (\phi_{jk}(L))_{j,k} = A(L)^{-1} = A(L)^*/\det(A(L)),$$

where $A(L)^* = (A_{jk}(L))_{j,k}$ is the adjoint of $A(L) = I_K - A_1L - \cdots - A_pL^p$ (see Appendix A.4.1). Obviously, $\phi_{jk}(L) \equiv 0$ is equivalent to $A_{jk}(L) \equiv 0$. From the definition of a cofactor of a matrix in Appendix A.4.1 it is easy to see that $A_{jk}(L)$ has degree not greater than $pK - p$. Defining $\gamma(L) = [\det(A(L))]^{-1}$ we get for $k \neq j$,

$$\begin{aligned}\phi_{jk}(L) &= \phi_{jk,1}L + \phi_{jk,2}L^2 + \cdots \\ &= A_{jk}(L)\gamma(L) = (A_{jk,1}L + \cdots + A_{jk,pK-p}L^{pK-p})(1 + \gamma_1 L + \cdots).\end{aligned}$$

Hence,

$$\phi_{jk,1} = A_{jk,1} \qquad \text{and} \qquad \phi_{jk,i} = A_{jk,i} + \sum_{n=1}^{i-1} A_{jk,n}\gamma_{i-n} \quad \text{for } i > 1$$

with $A_{jk,n} = 0$ for $n > pK - p$. Consequently, $A_{jk,i} = 0$ for $i = 1, \ldots, pK - p$ is equivalent to $\phi_{jk,i} = 0$ for $i = 1, 2, \ldots, pK - p$, which proves the proposition. ■

Sometimes interest centers on the accumulated effect over several or more periods of a shock in one variable. This effect may be determined by summing up the MA coefficient matrices. For instance, the k-th column of $\Psi_n := \sum_{i=0}^{n} \Phi_i$ contains the *accumulated responses* over n periods to a unit shock in the k-th variable of the system. These quantities are sometimes called n-th *interim multipliers*. The total accumulated effects for all future periods are obtained by summing up all the MA coefficient matrices. $\Psi_\infty := \sum_{i=0}^{\infty} \Phi_i$ is sometimes called the matrix of *long-run effects* or *total multipliers*. Since the MA operator $\Phi(z)$ is the inverse of the VAR operator $A(z) = I_K - A_1 z - \cdots - A_p z^p$ the long-run effects are easily obtained as

$$\Psi_\infty = \Phi(1) = (I_K - A_1 - \cdots - A_p)^{-1}. \tag{2.3.23}$$

As an example accumulated responses for the investment/income/consumption system are depicted in Figure 2.7. Similarly interim and total multipliers of the inflation/interest rate system are shown in Figure 2.8.

2.3.2b *Responses to Orthogonal Impulses*

A problematic assumption in this type of impulse response analysis is that a shock occurs only in one variable at a time. Such an assumption may be reasonable if the shocks in different variables are independent. If they are not independent one may argue that the error terms consist of all the influences and variables that are not directly included in the set of y variables. Thus, in addition to forces that affect all the variables, there may be forces that affect variable 1, say, only. If a shock in the first variable is due to such forces it may again be reasonable to interpret the Φ_i coefficients as dynamic responses. On the other hand, correlation of the error terms may indicate that a shock in one variable is likely to be accompanied by a shock in another variable. In that case setting all other residuals to zero may provide a misleading picture of the actual dynamic relationships between the variables. For example, in the investment/income/consumption system the white noise or innovation covariance matrix is given in (2.1.33),

$$\Sigma_u = \begin{bmatrix} 2.25 & 0 & 0 \\ 0 & 1.0 & .5 \\ 0 & .5 & .74 \end{bmatrix}.$$

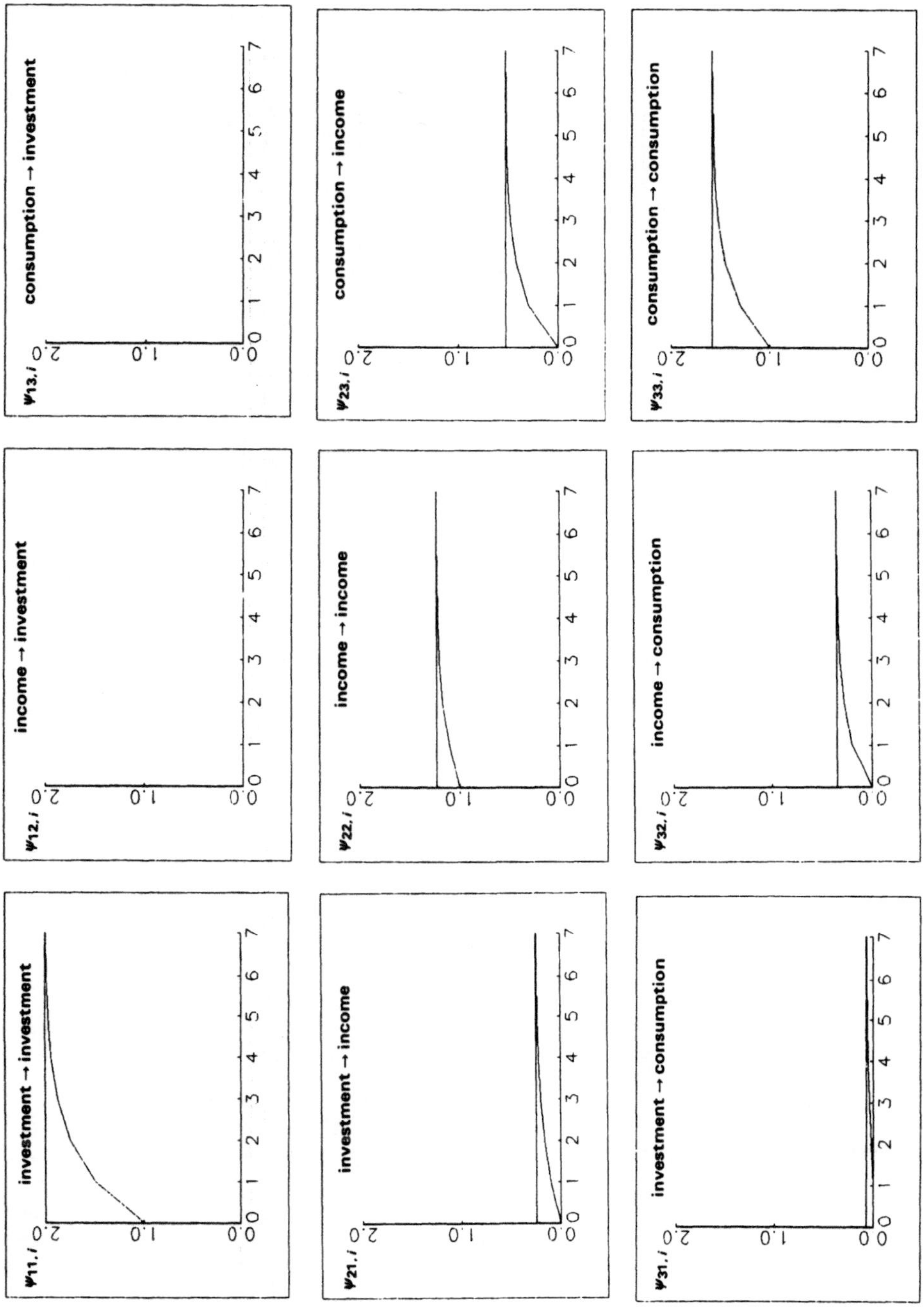

Fig. 2.7. Accumulated and long-run responses of the investment/income/consumption system (impulse → response).

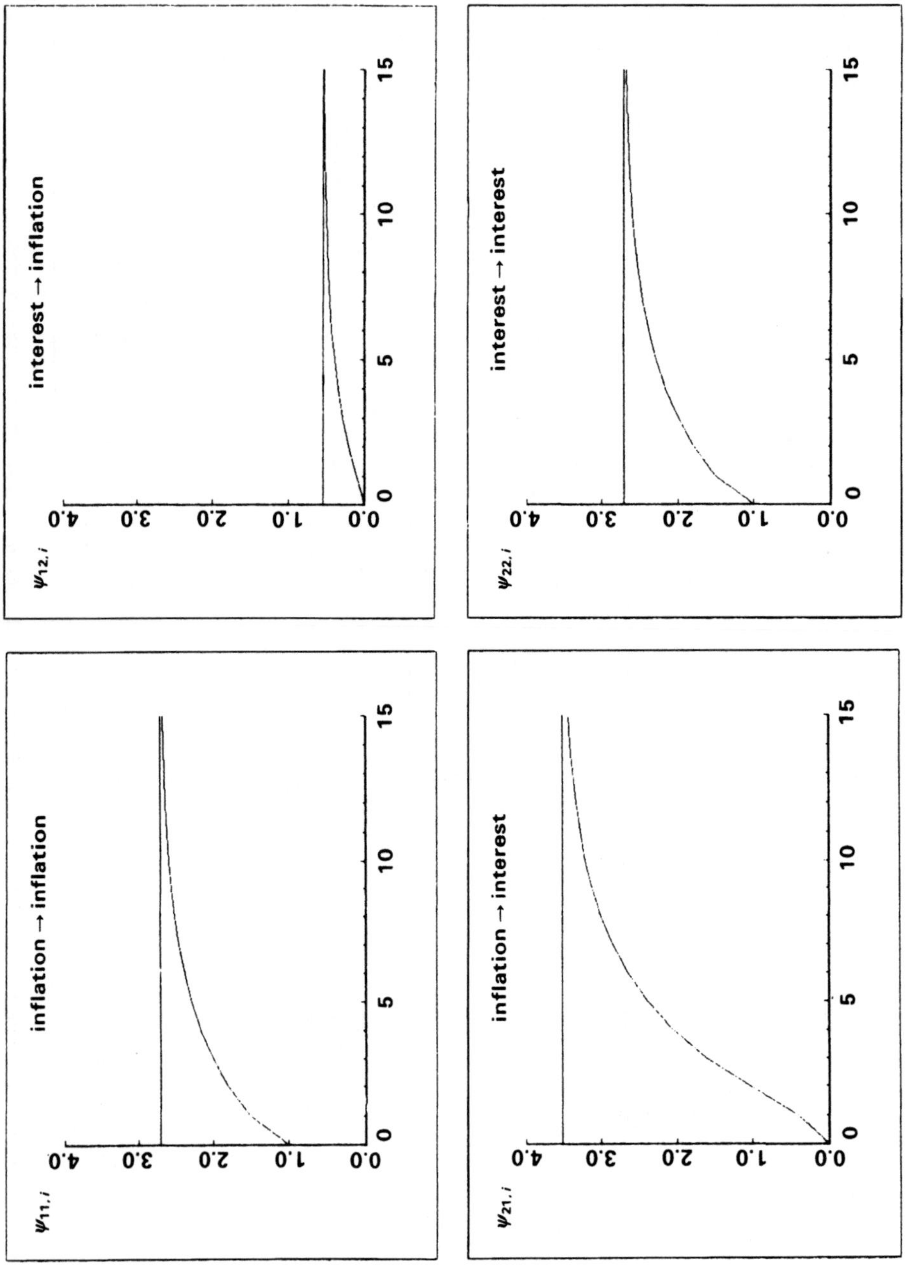

Fig. 2.8. Accumulated and total responses of the inflation/interest rate system (impulse → response).

Obviously, there is a quite strong positive correlation between $u_{2,t}$ and $u_{3,t}$, the residuals of the income and consumption equations, respectively. Consequently a shock in income may be accompanied by a shock in consumption in the same period. Therefore, forcing the consumption innovation to zero when the effect of an income shock is traced, as in the previous analysis, may in fact obscure the actual reaction of the system.

This is the reason why impulse response analysis is often performed in terms of the MA representation (2.3.15),

$$y_t = \sum_{i=0}^{\infty} \Theta_i w_{t-i}, \tag{2.3.24}$$

where the components of $w_t = (w_{1t}, \ldots, w_{Kt})'$ are uncorrelated and have unit variance, $\Sigma_w = I_K$. The mean term is dropped again because it is of no interest in the present analysis. Recall that the representation (2.3.24) is obtained by decomposing Σ_u as $\Sigma_u = PP'$, where P is a lower triangular matrix, and defining $\Theta_i = \Phi_i P$ and $w_t = P^{-1}u_t$. In (2.3.24) it is reasonable to assume that a change in one component of w_t has no effect on the other components because the components are orthogonal (uncorrelated). Since the variances of the components are one, a unit innovation is just an innovation of size one standard deviation. The elements of the Θ_i are interpreted as responses of the system to such innovations. More precisely, the jk-th element of Θ_i is assumed to represent the effect on variable j of a unit innovation in the k-th variable that has occurred i periods ago.

To relate these impulse responses to a VAR model we consider the zero mean VAR(p) process

$$y_t = A_1 y_{t-1} + \cdots + A_p y_{t-p} + u_t. \tag{2.3.25}$$

This process can be rewritten in such a way that the residuals of different equations are uncorrelated. For this purpose we choose a decomposition of the white noise covariance matrix $\Sigma_u = W\Lambda W'$, where Λ is a diagonal matrix with positive diagonal elements and W is a lower triangular matrix with unit diagonal. This decomposition is obtained from the Choleski decomposition $\Sigma_u = PP'$ by defining a diagonal matrix D which has the same diagonal as P and by specifying $W = PD^{-1}$ and $\Lambda = DD'$.

Premultiplying (2.3.25) by W^{-1} gives

$$W^{-1}y_t = B_1 y_{t-1} + \cdots + B_p y_{t-p} + v_t, \tag{2.3.26}$$

where $B_i := W^{-1}A_i$, $i = 1, \ldots, p$, and $v_t = (v_{1t}, \ldots, v_{Kt})' := W^{-1}u_t$ has diagonal covariance matrix,

$$\Sigma_v = E(v_t v_t') = W^{-1}E(u_t u_t')(W^{-1})' = \Lambda.$$

Adding $(I_K - W^{-1})y_t$ to both sides of (2.3.26) gives

$$y_t = B_0 y_t + B_1 y_{t-1} + \cdots + B_p y_{t-p} + v_t, \tag{2.3.27}$$

where $B_0 := I_K - W^{-1}$. Since W is lower triangular with unit diagonal the same is true for W^{-1}. Hence

$$B_0 = I_K - W^{-1} = \begin{bmatrix} 0 & 0 & \dots & 0 & 0 \\ \beta_{21} & 0 & \dots & 0 & 0 \\ \vdots & & & \ddots & \vdots \\ \beta_{K1} & \beta_{K2} & \dots & \beta_{K,K-1} & 0 \end{bmatrix}$$

is a lower triangular matrix with zero diagonal. For the representation (2.3.27) of our VAR(p) process this implies that the first equation contains no instantaneous y's on the right-hand side. The second equation may contain y_{1t} and otherwise lagged y's on the right-hand side. More generally, the k-th equation may contain $y_{1t}, \dots, y_{k-1,t}$ and not $y_{kt}, \dots, y_{Kt}$ on the right-hand side. Thus, if (2.3.27) reflects the actual ongoings in the system, y_{st} cannot have an instantaneous impact on y_{kt} for $k < s$. In the econometrics literature such a system is called a *recursive model* (see Theil (1971, Section 9.6)). Herman Wold has advocated these models where the researcher has to specify the instantaneous "causal" ordering of the variables. This type of causality is therefore sometimes referred to as *Wold-causality*. If we trace v_{it} innovations of size one standard error through the system (2.3.27) we just get the Θ impulse responses. This can be seen by solving the system (2.3.27) for y_t.

$$y_t = (I_K - B_0)^{-1} B_1 y_{t-1} + \cdots + (I_K - B_0)^{-1} B_p y_{t-p} + (I_K - B_0)^{-1} v_t.$$

Noting that $(I_K - B_0)^{-1} = W = PD^{-1}$ shows that the instantaneous effects of one-standard deviation shocks (v_{it} of size one standard deviation) to the system are represented by the elements of $WD = P = \Theta_0$ because the diagonal elements of D are just standard deviations of the components of v_t. The Θ_i may then be obtained by tracing these effects through the system.

The Θ_i may provide response functions that are quite different from the Φ_i responses. For the example VAR(1) system (2.3.22) with Σ_u as in (2.1.33) we get

$$\Theta_0 = P = \begin{bmatrix} 1.5 & 0 & 0 \\ 0 & 1 & 0 \\ 0 & .5 & .7 \end{bmatrix},$$

$$\Theta_1 = \Phi_1 P = \begin{bmatrix} .75 & 0 & 0 \\ .15 & .25 & .21 \\ 0 & .35 & .21 \end{bmatrix}, \qquad (2.3.28)$$

$$\Theta_2 = \Phi_2 P = \begin{bmatrix} .375 & 0 & 0 \\ .090 & .130 & .084 \\ .030 & .055 & .105 \end{bmatrix},$$

and so on. Some more innovation responses are depicted in Figure 2.9. Although they are similar to those given in Figure 2.5, there is an obvious difference in the response of consumption to an income innovation. While consumption responds with a time lag of one period in Figure 2.5, there is an instantaneous effect in Figure 2.9.

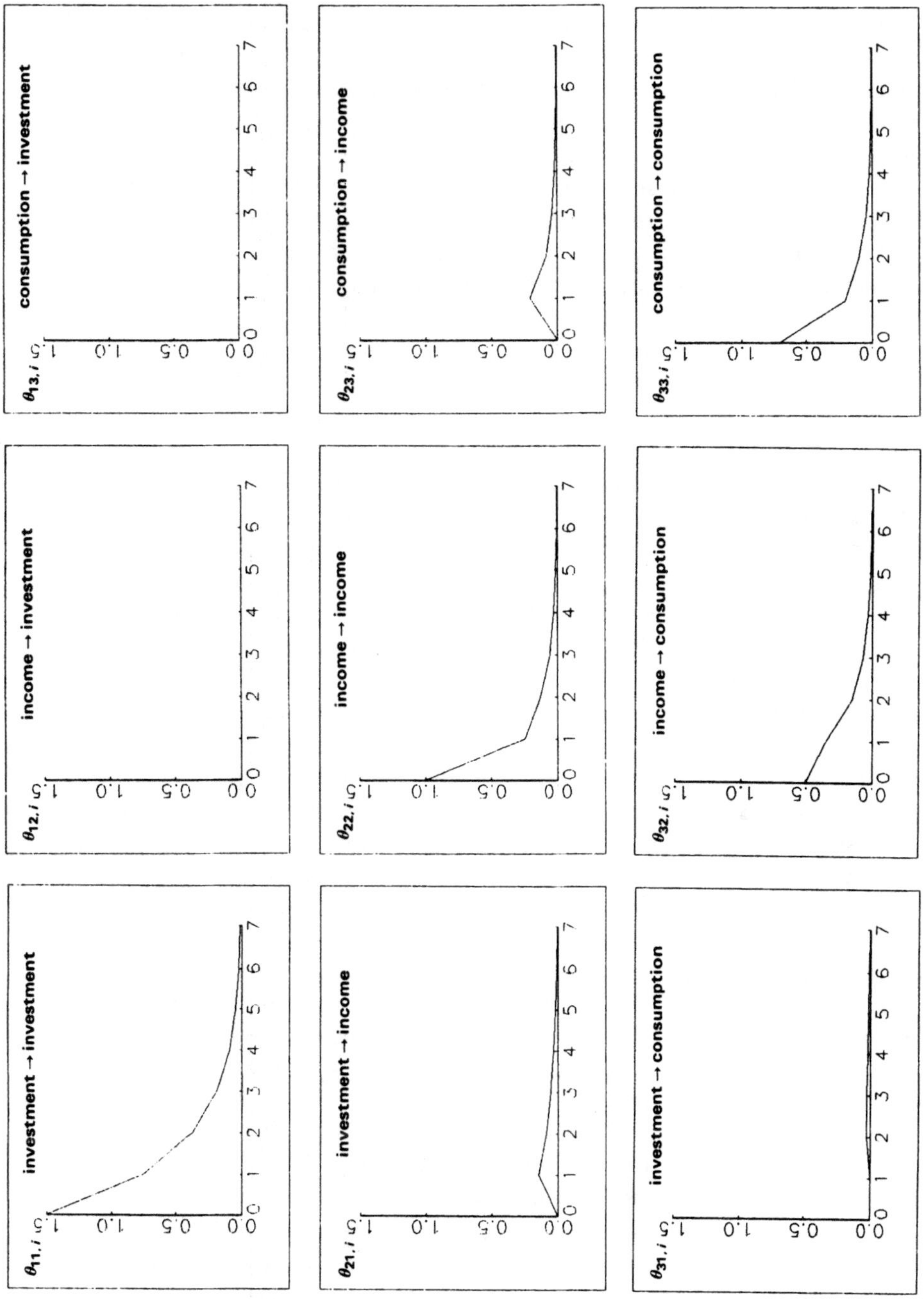

Fig. 2.9. Orthogonalized impulse responses of the investment/income/consumption system (impulse → response).

Note that $\Theta_0 = P$ is lower triangular and some elements below the diagonal will be nonzero if Σ_u has nonzero off-diagonal elements. For instance, for the investment/income/consumption example Θ_0 indicates that an income (y_2) innovation has an immediate impact on consumption (y_3). If the white noise covariance matrix Σ_u contains zeroes some components of $u_t = (u_{1t}, \dots, u_{Kt})'$ are contemporaneously uncorrelated. Suppose, for instance, that u_{1t} is uncorrelated with u_{it} for $i = 2, \dots, K$. In this case W^{-1} and, thus, B_0 has a block of zeroes so that y_1 has no instantaneous effect on $y_i, i = 2, \dots, K$. In the example, investment has no instantaneous impact on income and consumption because

$$\Sigma_u = \begin{bmatrix} 2.25 & 0 & 0 \\ 0 & 1.0 & .5 \\ 0 & .5 & .74 \end{bmatrix}$$

and hence u_{1t} is uncorrelated with u_{2t} and u_{3t}. This, of course, is reflected in the matrix of instantaneous effects Θ_0 given in (2.3.28). Since the elements of $P = \Theta_0$ represent the immediate responses of the system to unit innovations they are sometimes called *impact multipliers*.

The fact that Θ_0 is lower triangular shows that the ordering of the variables is of importance, that is, it is important which of the variables is called y_1 and which one is called y_2 and so on. One problem with this type of impulse response analysis is that the ordering of the variables cannot be determined with statistical methods but has to be specified by the analyst. The ordering has to be such that the first variable is the only one with a potential immediate impact on all other variables. The second variable may have an immediate impact on the last $K - 2$ components of y_t but not on y_{1t} and so on. To establish such an ordering may be a quite difficult exercise in practice. The choice of the ordering, the Wold causal ordering, may, to a large extent, determine the impulse responses and is therefore critical for the interpretation of the system. Currently we are dealing with known systems only. In this situation supposing that the ordering is known may not be a great restriction. For the investment/income/consumption example it may be reasonable to assume that an increase in income has an immediate effect on consumption while increased consumption stimulates the economy and hence income with some time lag.

Instead of choosing a different ordering of the variables of a given system we may equivalently consider a different P matrix. Note that a multitude of P matrices with $PP' = \Sigma_u$ exists. For instance, P could be chosen upper triangular. Such a choice may obviously be interpreted as considering a different ordering of the variables. If the appropriate choice of the impact multipliers or P matrix is unknown, only conclusions that do not depend on the P matrix may be drawn from an impulse response analysis. In terms of econometric simultaneous equations models the difference between the representation in terms of the Φ_i and the Θ_i corresponds to the difference between the *reduced form* and the *structural form* of the system. The latter must always be specified on the basis of a priori knowledge on the structure of the relationships between the variables of interest.

In order to determine whether there is no response at all of one variable to an impulse in one of the other variables it suffices to consider the first $pK - p$ response coefficients and the immediate effect. This result is stated formally in the next proposition where $\theta_{jk,i}$ denotes the jk-th element of Θ_i.

PROPOSITION 2.5 (*Zero Orthogonalized Impulse Responses*)

If y_t is a stable K-dimensional VAR(p) process then, for $j \neq k$,

$$\theta_{jk,i} = 0 \qquad \text{for } i = 0, 1, 2, \dots$$

is equivalent to

$$\theta_{jk,i} = 0 \qquad \text{for } i = 0, 1, \dots, p(K - 1). \qquad ■$$

The proof of this result is analogous to that of Proposition 2.4 and is left as an exercise (see Problem 2.2).

2.3.2c *Critique of Impulse Response Analysis*

Besides specifying the ordering of the variables there are a number of further problems that render the interpretation of impulse responses difficult. We have mentioned some of them in the context of Granger-causality. A major limitation of our systems is their potential incompleteness. Although in real economic systems almost everything depends on everything else we will usually work with low-dimensional VAR systems. All effects of omitted variables are assumed to be in the innovations. If important variables are omitted from the system this may lead to major distortions in the impulse responses and makes them worthless for structural interpretations. The system may still be useful for prediction, though.

To see the related problems more clearly, consider a system y_t which is partitioned in vectors z_t and x_t as in (2.3.5). If the z_t variables are considered only and the x_t variables are omitted from the analysis, we get a system

$$\begin{aligned} z_t &= \mu_1 + \sum_{i=0}^{\infty} \Phi_{11,i} u_{1,t-i} + \sum_{i=1}^{\infty} \Phi_{12,i} u_{2,t-i} \\ &= \mu_1 + \sum_{i=0}^{\infty} F_i v_{t-i} \end{aligned} \qquad (2.3.29)$$

as in (2.3.8). The actual reactions of the z_t components to innovations u_{1t} may be given by the $\Phi_{11,i}$ matrices. On the other hand, the F_i or corresponding orthogonalized "impulse responses" are likely to be interpreted as impulse responses if the analyst does not realize that important variables have been omitted. As we have seen in Section 2.3.1b, the F_i will be equal to the $\Phi_{11,i}$ if and only if x_t does not Granger-cause z_t.

Further problems related to the interpretation of the MA coefficients as dynamic multipliers or impulse responses result from measurement errors and the use of seasonally adjusted or temporally and/or contemporaneously aggregated variables. A detailed account of the aggregation problem is given by Lütkepohl (1987). We will discuss these problems in more detail in Chapter 6 in the context of more general models. These problems severely limit the interpretability of the MA coefficients of a VAR system as impulse responses. In the next subsection a further possibility to interpret VAR models will be considered.

2.3.3 Forecast Error Variance Decomposition

The MA representation (2.3.15) with orthogonal white noise innovations offers a further possibility to interpret a VAR(p) model. In terms of the representation

$$y_t = \mu + \sum_{i=0}^{\infty} \Theta_i w_{t-i} \tag{2.3.30}$$

with $\Sigma_w = I_K$, the error of the optimal h-step forecast is

$$\begin{aligned} y_{t+h} - y_t(h) &= \sum_{i=0}^{h-1} \Phi_i u_{t+h-i} = \sum_{i=0}^{h-1} \Phi_i P P^{-1} u_{t+h-i} \\ &= \sum_{i=0}^{h-1} \Theta_i w_{t+h-i}. \end{aligned} \tag{2.3.31}$$

Denoting the mn-th element of Θ_i by $\theta_{mn,i}$ as before, the h-step forecast error of the j-th component of y_t is

$$\begin{aligned} y_{j,t+h} - y_{j,t}(h) &= \sum_{i=0}^{h-1} (\theta_{j1,i} w_{1,t+h-i} + \cdots + \theta_{jK,i} w_{K,t+h-i}) \\ &= \sum_{k=1}^{K} (\theta_{jk,0} w_{k,t+h} + \cdots + \theta_{jk,h-1} w_{k,t+1}). \end{aligned} \tag{2.3.32}$$

Thus, the forecast error of the j-th component potentially consists of innovations of all other components of y_t as well. Of course, some of the $\theta_{mn,i}$ may be zero so that the innovations of some components may not appear in (2.3.32). Since the $w_{k,t}$ are uncorrelated and have variance one, the MSE of $y_{j,t}(h)$ is

$$E(y_{j,t+h} - y_{j,t}(h))^2 = \sum_{k=1}^{K} (\theta_{jk,0}^2 + \cdots + \theta_{jk,h-1}^2).$$

Therefore

$$\theta_{jk,0}^2 + \theta_{jk,1}^2 + \cdots + \theta_{jk,h-1}^2 = \sum_{i=0}^{h-1} (e_j' \Theta_i e_k)^2 \tag{2.3.33}$$

is sometimes interpreted as contribution of innovations in variable k to the forecast error variance or MSE of the h-step forecast of variable j. Here e_k is the k-th column of I_K. Dividing (2.3.33) by

$$\text{MSE}[y_{j,t}(h)] = \sum_{i=0}^{h-1} \sum_{k=1}^{K} \theta_{jk,i}^2$$

gives

$$\omega_{jk,h} = \sum_{i=0}^{h-1} (e_j' \Theta_i e_k)^2 / \text{MSE}[y_{j,t}(h)] \tag{2.3.34}$$

which is the proportion of the h-step forecast error variance of variable j accounted for by innovations in variable k. This way the forecast error variance is decomposed into components accounted for by innovations in the different variables of the system. From (2.3.31) the h-step forecast MSE matrix is seen to be

$$\Sigma_y(h) = \text{MSE}[y_t(h)] = \sum_{i=0}^{h-1} \Theta_i \Theta_i' = \sum_{i=0}^{h-1} \Phi_i \Sigma_u \Phi_i'.$$

The diagonal elements of this matrix are the MSEs of the y_{jt} variables which may be used in (2.3.34).

For the investment/income/consumption example forecast error variance decompositions of all three variables are given in Table 2.1. For instance, about 66% of the 1-step forecast error variance of consumption is accounted for by own innovations and about 34% is accounted for by income innovations. For long term forecasts, 57.5% and 42.3% of the error variance is accounted for by

Table 2.1. Forecast error variance decomposition of the investment/income/consumption system

forecast error in	forecast horizon h	proportions of forecast error variance h periods ahead accounted for by innovations in		
		investment	income	consumption
investment	1	1	0	0
	2	1	0	0
	3	1	0	0
	4	1	0	0
	5	1	0	0
	10	1	0	0
	∞	1	0	0
income	1	0	1.000	0
	2	.020	.941	.039
	3	.026	.930	.044
	4	.029	.926	.045
	5	.030	.925	.045
	10	.030	.925	.045
	∞	.030	.925	.045
consumption	1	0	.338	.662
	2	0	.411	.589
	3	.001	.421	.578
	4	.002	.423	.576
	5	.002	.423	.575
	10	.002	.423	.575
	∞	.002	.423	.575

consumption and income innovations, respectively. For any forecast horizon investment innovations contribute less than 1% to the forecast error variance of consumption. Moreover, only small fractions (less than 10%) of the forecast error variances of income are accounted for by innovations in the other variables of the system. This kind of analysis is sometimes called *innovation accounting.*

From Proposition 2.5 it is obvious that for a stationary, stable, K-dimensional VAR(p) process y_t all forecast error variance proportions of variable j accounted for by innovations in variable k will be zero if $\omega_{jk,h} = 0$ for $h = pK - p$. In this context it is perhaps worth pointing out the relationship between Granger-causality and forecast error variance components. For that purpose we consider a bivariate system $y_t = (z_t, x_t)'$ first. In such a system, if z_t does not Granger-cause x_t, the proportions of forecast error variances of x_t accounted for by innovations in z_t may still be nonzero. This follows directly from the definition of the Θ_i in (2.3.15). Granger-noncausality, by Proposition 2.2, implies zero contraints on the Φ_i which may disappear in the Θ_i if the error covariance matrix Σ_u is not diagonal. On the other hand, if Σ_u is diagonal, so that there is no instantaneous causation between z_t and x_t and if, in addition, z_t is not Granger-causal for x_t the lower left-hand elements of the Θ_i will be zero (see (2.3.19)). Therefore the proportion of forecast error variance of x_t accounted for by z_t innovations will also be zero.

In a higher-dimensional system suppose a set of variables z_t does not Granger-cause the remaining variables x_t and there is also no instantaneous causality between the two sets of variables. In that case the forecast MSE proportions of all x_t variables accounted for by z_t variables will be zero.

It is important to understand, however, that Granger-causality and forecast error variance decompositions are quite different concepts since Granger-causality and instantaneous causality are different concepts. While Granger-causality is a uniquely defined property of two subsets of variables of a given process, the forecast error variance decomposition is not unique as it depends on the Θ_i matrices and thus on the choice of the transformation matrix P. Therefore, the interpretation of a forecast error variance decomposition is subject to similar criticisms as the interpretation of impulse responses. In addition, all the critical points raised in the context of Granger-causality apply. That is, the forecast error variance components are conditional on the system under consideration. They may change if the system is expanded by adding further variables or if variables are deleted from the system. Also measurement errors, seasonal adjustment and the use of aggregates may contaminate the forecast error variance decompositions.

2.3.4 Remarks on the Interpretation of VAR Models

Innovation accounting and impulse response analysis in the framework of VAR models have been pioneered by Sims (1980, 1981) and others as an alternative to classical macroeconometric analyses. Sims' main criticism of the latter type of analysis is that macroeconometric models are often not based on sound economic theories or the available theories are not capable of providing a completely specified model. If economic theories are not available to specify the model, statistical tools must be applied. In this approach a fairly loose model is set up which does not impose rigid a priori restrictions on the data generation process. Statistical tools are then used to determine possible constraints. VAR models represent a class of loose models that may be used in such an approach. Of course, in order to interpret these models some restrictive assumptions need to be made. In particular, the ordering of the variables is essential for interpretations of the types discussed in the previous subsections. Sims (1981) suggests to try different orderings and investigate the sensitivity of the conclusions to the ordering of the variables.

So far we have assumed that a VAR model is given to us. Under this assumption we have discussed forecasting and interpretation of the system. In this situation it is of course unnecessary to use statistical tools in order to determine constraints of the system because all constraints are known. In practice we will virtually never be in such a fortunate situation but we have to determine the model from a given set of time series data. This problem will be treated in subsequent chapters. The purpose of this chapter is to identify some problems that are not related to estimation and model specification but are inherent to the types of models considered.

2.4 Exercises

Problem 2.1

Show that

$$\det(I_{Kp} - \mathbf{A}z) = \det(I_K - A_1 z - \cdots - A_p z^p)$$

where A_i, $i = 1, \ldots, p$, and $\mathbf{A}$ are as in (2.1.1) and (2.1.8), respectively.

Problem 2.2

Prove Proposition 2.5! (Hint: $\Theta(L) = \Phi(L)P = A(L)^* P/\det(A(L))$).

In the United States of Wonderland the growth rates of income (GNP) and money demand (M2) and an interest rate (IR) are related as in the following VAR(2) model:

$$\begin{bmatrix} \text{GNP}_t \\ \text{M2}_t \\ \text{IR}_t \end{bmatrix} = \begin{bmatrix} 2 \\ 1 \\ 0 \end{bmatrix} + \begin{bmatrix} .7 & .1 & 0 \\ 0 & .4 & .1 \\ .9 & 0 & .8 \end{bmatrix} \begin{bmatrix} \text{GNP}_{t-1} \\ \text{M2}_{t-1} \\ \text{IR}_{t-1} \end{bmatrix}$$
$$+ \begin{bmatrix} -.2 & 0 & 0 \\ 0 & .1 & .1 \\ 0 & 0 & 0 \end{bmatrix} \begin{bmatrix} \text{GNP}_{t-2} \\ \text{M2}_{t-2} \\ \text{IR}_{t-2} \end{bmatrix} + \begin{bmatrix} u_{1t} \\ u_{2t} \\ u_{3t} \end{bmatrix}$$

$$\Sigma_u = \begin{bmatrix} .26 & .03 & 0 \\ .03 & .09 & 0 \\ 0 & 0 & .81 \end{bmatrix} = PP', \qquad P = \begin{bmatrix} .5 & .1 & 0 \\ 0 & .3 & 0 \\ 0 & 0 & .9 \end{bmatrix}. \tag{2.4.1}$$

Problem 2.3

a) Show that the process $y_t = (\text{GNP}_t, \text{M2}_t, \text{IR}_t)'$ is stable.
b) Determine the mean vector of y_t.
c) Write the process y_t in VAR(1) form.
d) Compute the coefficient matrices $\Phi_1, \ldots, \Phi_5$ of the MA representation (2.1.17) of y_t.

Problem 2.4

Determine the autocovariances $\Gamma_y(0)$, $\Gamma_y(1)$, $\Gamma_y(2)$, $\Gamma_y(3)$ of the process defined in (2.4.1). Compute and plot the autocorrelations $R_y(0)$, $R_y(1)$, $R_y(2)$, $R_y(3)$.

Problem 2.5

a) Suppose that

$$y_{2000} = \begin{bmatrix} .7 \\ 1.0 \\ 1.5 \end{bmatrix} \quad \text{and} \quad y_{1999} = \begin{bmatrix} 1.0 \\ 1.5 \\ 3.0 \end{bmatrix}$$

and forecast y_{2001}, y_{2002}, and y_{2003}.
b) Determine the MSE matrices for forecast horizons $h = 1, 2, 3$.
c) Assume that y_t is a Gaussian process and construct 90% and 95% forecast intervals for $t = 2001, 2002, 2003$.
d) Using the Bonferroni method, determine a joint forecast region for GNP_{2001}, GNP_{2002}, GNP_{2003} with probability content at least 97%.

Problem 2.6

a) Is M2 Granger-causal for (GNP, IR) in the process (2.4.1)?
b) Is IR Granger-causal for (GNP, M2)?
c) Is there instantaneous causality between M2 and (GNP, IR)?
d) Is there instantaneous causality between IR and (GNP, M2)?

Problem 2.7

Plot the effect of a unit innovation in the interest rate (IR) on the three variables of the system (2.4.1) in terms of the MA representation (2.1.17). Consider only 5 periods following the innovation. Plot also the accumulated responses and interpret the plots.

Problem 2.8

For the system (2.4.1), derive the coefficient matrices $\Theta_0, \ldots, \Theta_5$ of the MA representation (2.3.15) using the *upper* triangular P matrix given in (2.4.1). Plot the effects of a unit innovation in IR in terms of that representation. Compare to the plots obtained in Problem 2.7 and interpret. Repeat the analysis with a lower triangular P matrix and comment on the results.

Problem 2.9

Decompose the MSE of $\text{GNP}_t(5)$ into the proportions accounted for by its own innovations and innovations in M2 and IR.

Chapter 3. Estimation of Vector Autoregressive Processes

3.1 Introduction

In this chapter it is assumed that a K-dimensional multiple time series $y_1, \ldots, y_T$ with $y_t = (y_{1t}, \ldots, y_{Kt})'$ is available that is known to be generated by a stationary, stable VAR(p) process

$$y_t = \nu + A_1 y_{t-1} + \cdots + A_p y_{t-p} + u_t. \qquad (3.1.1)$$

All symbols have their usual meanings, that is, $\nu = (\nu_1, \ldots, \nu_K)'$ is a $(K \times 1)$ vector of intercept terms, the A_i are $(K \times K)$ coefficient matrices and u_t is white noise with nonsingular covariance matrix Σ_u. In contrast to the assumptions of the previous chapter, the coefficients $\nu, A_1, \ldots, A_p$, and Σ_u are assumed unknown in the following. The time series data will be used to estimate the coefficients. Note that notationwise we do not distinguish between the stochastic process and a time series as a realization of a stochastic process. The particular meaning of a symbol should be obvious from the context.

In the next three sections different possibilities for estimating a VAR(p) process are discussed. In Section 3.5 the consequences of forecasting with estimated processes will be considered and in Section 3.6 tests for Granger-causality are described. The distribution of impulse responses obtained from estimated processes is considered in Section 3.7.

3.2 Multivariate Least Squares Estimation

In this section multivariate least squares (LS) estimation is discussed. The estimator obtained for the standard form (3.1.1) of a VAR(p) process is considered in Section 3.2.1. Some properties of the estimator are derived in Sections 3.2.2 and 3.2.4 and an example is given in Section 3.2.3.

3.2.1 The Estimator

It is assumed that a time series $y_1, \ldots, y_T$ of the y variables is available, that is, we have a sample of size T for each of the K variables. In addition, p presample values for each variable, $y_{-p+1}, \ldots, y_0$, are assumed to be available. Partitioning a multiple time series in sample and presample values is convenient in order to simplify the notation. We define

$$
\begin{aligned}
Y &:= (y_1, \dots, y_T) && (K \times T), \\
B &:= (\nu, A_1, \dots, A_p) && (K \times (Kp + 1)), \\
Z_t &:= \begin{bmatrix} 1 \\ y_t \\ \vdots \\ y_{t-p+1} \end{bmatrix} && ((Kp + 1) \times 1), \\
Z &:= (Z_0, \dots, Z_{T-1}) && ((Kp + 1) \times T), \\
U &:= (u_1, \dots, u_T) && (K \times T), \\
\mathbf{y} &:= \operatorname{vec}(Y) && (KT \times 1), \\
\boldsymbol{\beta} &:= \operatorname{vec}(B) && ((K^2 p + K) \times 1), \\
\mathbf{b} &:= \operatorname{vec}(B') && ((K^2 p + K) \times 1), \\
\mathbf{u} &:= \operatorname{vec}(U) && (KT \times 1).
\end{aligned}
\tag{3.2.1}
$$

Here vec is the column stacking operator as defined in Appendix A.12.

Using this notation, for $t = 1, \dots, T$, the VAR(p) model (3.1.1) can be written compactly as

$$Y = BZ + U \tag{3.2.2}$$

or

$$
\begin{aligned}
\operatorname{vec}(Y) &= \operatorname{vec}(BZ) + \operatorname{vec}(U) \\
&= (Z' \otimes I_K) \operatorname{vec}(B) + \operatorname{vec}(U)
\end{aligned}
$$

or

$$\mathbf{y} = (Z' \otimes I_K)\boldsymbol{\beta} + \mathbf{u}. \tag{3.2.3}$$

Note that the covariance matrix of $\mathbf{u}$ is

$$\Sigma_{\mathbf{u}} = I_T \otimes \Sigma_u. \tag{3.2.4}$$

Thus, multivariate LS estimation of $\boldsymbol{\beta}$ means to choose the estimator that minimizes

$$
\begin{aligned}
S(\boldsymbol{\beta}) &= \mathbf{u}'(I_T \otimes \Sigma_u)^{-1}\mathbf{u} = \mathbf{u}'(I_T \otimes \Sigma_u^{-1})\mathbf{u} \\
&= (\mathbf{y} - (Z' \otimes I_K)\boldsymbol{\beta})'(I_T \otimes \Sigma_u^{-1})(\mathbf{y} - (Z' \otimes I_K)\boldsymbol{\beta}) \\
&= \operatorname{vec}(Y - BZ)'(I_T \otimes \Sigma_u^{-1}) \operatorname{vec}(Y - BZ) \\
&= \operatorname{tr}[(Y - BZ)'\Sigma_u^{-1}(Y - BZ)].
\end{aligned}
\tag{3.2.5}
$$

In order to find the minimum of this function we note that

$$\begin{aligned} S(\boldsymbol{\beta}) &= \mathbf{y}'(I_T \otimes \Sigma_u^{-1})\mathbf{y} + \boldsymbol{\beta}'(Z \otimes I_K)(I_T \otimes \Sigma_u^{-1})(Z' \otimes I_K)\boldsymbol{\beta} \\ &\quad - 2\boldsymbol{\beta}'(Z \otimes I_K)(I_T \otimes \Sigma_u^{-1})\mathbf{y} \\ &= \mathbf{y}'(I_T \otimes \Sigma_u^{-1})\mathbf{y} + \boldsymbol{\beta}'(ZZ' \otimes \Sigma_u^{-1})\boldsymbol{\beta} - 2\boldsymbol{\beta}'(Z \otimes \Sigma_u^{-1})\mathbf{y}. \end{aligned}$$

Hence,

$$\frac{\partial S(\boldsymbol{\beta})}{\partial \boldsymbol{\beta}} = 2(ZZ' \otimes \Sigma_u^{-1})\boldsymbol{\beta} - 2(Z \otimes \Sigma_u^{-1})\mathbf{y}.$$

Equating to zero gives the *normal equations*

$$(ZZ' \otimes \Sigma_u^{-1})\hat{\boldsymbol{\beta}} = (Z \otimes \Sigma_u^{-1})\mathbf{y} \tag{3.2.6}$$

and consequently, the LS estimator is

$$\begin{aligned} \hat{\boldsymbol{\beta}} &= ((ZZ')^{-1} \otimes \Sigma_u)(Z \otimes \Sigma_u^{-1})\mathbf{y} \\ &= ((ZZ')^{-1}Z \otimes I_K)\mathbf{y}. \end{aligned} \tag{3.2.7}$$

The Hessian of $S(\boldsymbol{\beta})$,

$$\frac{\partial^2 S}{\partial \boldsymbol{\beta} \partial \boldsymbol{\beta}'} = 2(ZZ' \otimes \Sigma_u^{-1}),$$

is positive definite which confirms that $\hat{\boldsymbol{\beta}}$ is indeed a minimizing vector. Strictly speaking for these results to hold it has to be assumed that ZZ' is nonsingular. This will hold with probability 1 if y_t has a continuous distribution which will always be assumed in the following.

It may be worth noting that the multivariate LS estimator $\hat{\boldsymbol{\beta}}$ is identical to the ordinary LS (OLS) estimator obtained by minimizing

$$\bar{S}(\boldsymbol{\beta}) = \mathbf{u}'\mathbf{u} = (\mathbf{y} - (Z' \otimes I_K)\boldsymbol{\beta})'(\mathbf{y} - (Z' \otimes I_K)\boldsymbol{\beta}) \tag{3.2.8}$$

(see Problem 3.1).

The LS estimator can be written in different ways that will be useful later on:

$$\begin{aligned} \hat{\boldsymbol{\beta}} &= ((ZZ')^{-1}Z \otimes I_K)[(Z' \otimes I_K)\boldsymbol{\beta} + \mathbf{u}] \\ &= \boldsymbol{\beta} + ((ZZ')^{-1}Z \otimes I_K)\mathbf{u} \end{aligned} \tag{3.2.9}$$

or

$$\begin{aligned} \operatorname{vec}(\hat{B}) = \hat{\boldsymbol{\beta}} &= ((ZZ')^{-1}Z \otimes I_K) \operatorname{vec}(Y) \\ &= \operatorname{vec}(YZ'(ZZ')^{-1}). \end{aligned}$$

Thus,

$$\begin{aligned} \hat{B} &= YZ'(ZZ')^{-1} \\ &= (BZ + U)Z'(ZZ')^{-1} \\ &= B + UZ'(ZZ')^{-1}. \end{aligned} \tag{3.2.10}$$

Another possibility for deriving this estimator results from postmultiplying

$$y_t = BZ_{t-1} + u_t$$

by Z'_{t-1} and taking expectations:

$$E(y_t Z'_{t-1}) = BE(Z_{t-1}Z'_{t-1}). \tag{3.2.11}$$

Estimating $E(y_t Z'_{t-1})$ by

$$\frac{1}{T}\sum_{t=1}^{T} y_t Z'_{t-1} = \frac{1}{T} YZ'$$

and $E(Z_{t-1}Z'_{t-1})$ by

$$\frac{1}{T}\sum_{t=1}^{T} Z_{t-1}Z'_{t-1} = \frac{1}{T} ZZ'$$

we obtain the normal equations

$$\frac{1}{T} YZ' = \hat{B}\frac{1}{T} ZZ'$$

and hence $\hat{B} = YZ'(ZZ')^{-1}$. Note that (3.2.11) is similar but not identical to the system of Yule-Walker equations in (2.1.37). While central moments about the expectation $\mu = E(y_t)$ are considered in (2.1.37), moments about zero are used in (3.2.11).

Yet another possibility to write the LS estimator is

$$\hat{\mathbf{b}} = \operatorname{vec}(\hat{B}') = (I_K \otimes (ZZ')^{-1}Z) \operatorname{vec}(Y'). \tag{3.2.12}$$

In this form it is particularly easy to see that the multivariate LS estimation is equivalent to OLS estimation of each of the K equations in (3.1.1) separately. Let b'_k be the k-th row of B, that is, b_k contains all the parameters of the k-th equation. Obviously $\mathbf{b}' = (b'_1, \ldots, b'_K)$. Furthermore, let $y_{(k)} = (y_{k1}, \ldots, y_{kT})'$ be the time series available for the k-th variable, so that

$$\operatorname{vec}(Y') = \begin{bmatrix} y_{(1)} \\ \vdots \\ y_{(K)} \end{bmatrix}.$$

With this notation $\hat{b}_k = (ZZ')^{-1}Zy_{(k)}$ is the OLS estimator of the model $y_{(k)} = Z'b_k + u_{(k)}$, where $u_{(k)} = (u_{k1}, \ldots, u_{kT})'$ and $\hat{\mathbf{b}}' = (\hat{b}'_1, \ldots, \hat{b}'_K)$.

3.2.2 Asymptotic Properties of the Least Squares Estimator

Since small sample properties of the LS estimator are difficult to derive analytically we focus on asymptotic properties. Consistency and asymptotic normality of the LS estimator are easily established if the following results hold:

$$\Gamma := \operatorname{plim} ZZ'/T \text{ exists and is nonsingular} \tag{3.2.13}$$

and

$$\frac{1}{\sqrt{T}}\operatorname{vec}(UZ') = \frac{1}{\sqrt{T}}(Z \otimes I_K)\mathbf{u} \xrightarrow[T\to\infty]{d} N(0, \Gamma \otimes \Sigma_u), \tag{3.2.14}$$

where, as usual, $\xrightarrow{d}$ denotes convergence in distribution. It follows from a theorem due to Mann & Wald (1943) that these results are true under suitable conditions for u_t, if y_t is a stationary, stable VAR(p). For instance, the conditions stated in the following definition are sufficient.

DEFINITION 3.1 (*Standard White Noise*)

A white noise process $u_t = (u_{1t}, \ldots, u_{Kt})'$ is called *standard white noise* if the u_t are continuous random vectors satisfying $E(u_t) = 0$, $\Sigma_u = E(u_t u_t')$ is nonsingular, u_t and u_s are independent for $s \neq t$, and, for some finite constant c,

$E|u_{it}u_{jt}u_{kt}u_{mt}| \leq c$ for $i, j, k, m = 1, \ldots, K$ and all t. ■

The last condition means that all fourth moments exist and are bounded. Obviously, if the u_t are normally distributed (Gaussian) they satisfy the moment requirements. With this definition it is easy to state conditions for consistency and asymptotic normality of the LS estimator. The following lemma will be essential in proving these large sample results.

LEMMA 3.1

If y_t is a stable, K-dimensional VAR(p) process as in (3.1.1) with standard white noise innovations u_t, then (3.2.13) and (3.2.14) hold. ■

Proof: Similar to Theorem 8.2.3 of Fuller (1976, p. 340). ■

In the next proposition the asymptotic properties of the LS estimator are stated formally.

PROPOSITION 3.1 (*Asymptotic Properties of the LS Estimator*)

Let y_t be a stable, K-dimensional VAR(p) process as in (3.1.1) with standard white noise innovations, $\hat{B} = YZ'(ZZ')^{-1}$ is the LS estimator of the VAR coefficients B and all symbols are as defined in (3.2.1). Then

$$\operatorname{plim} \hat{B} = B$$

and

$$\sqrt{T}(\hat{\boldsymbol{\beta}} - \boldsymbol{\beta}) = \sqrt{T}\operatorname{vec}(\hat{B} - B) \xrightarrow{d} N(0, \Gamma^{-1} \otimes \Sigma_u) \tag{3.2.15}$$

or, equivalently,

$$\sqrt{T}(\hat{\mathbf{b}} - \mathbf{b}) = \sqrt{T}\operatorname{vec}(\hat{B}' - B') \xrightarrow{d} N(0, \Sigma_u \otimes \Gamma^{-1}), \tag{3.2.16}$$

where $\Gamma = \operatorname{plim} ZZ'/T$. ■

Proof: Using (3.2.10),

$$\operatorname{plim}(\hat{B} - B) = \operatorname{plim}\left[\frac{UZ'}{T}\right]\operatorname{plim}\left[\frac{ZZ'}{T}\right]^{-1} = 0$$

by Lemma 3.1, since (3.2.14) implies plim $UZ'/T = 0$. Thus, the consistency of $\hat{B}$ is established.

Using (3.2.9),

$$\begin{aligned}\sqrt{T}(\hat{\boldsymbol{\beta}} - \boldsymbol{\beta}) &= \sqrt{T}((ZZ')^{-1}Z \otimes I_K)\mathbf{u} \\ &= \left(\left(\frac{1}{T}ZZ'\right)^{-1} \otimes I_K\right)\frac{1}{\sqrt{T}}(Z \otimes I_K)\mathbf{u}.\end{aligned}$$

Thus, by Proposition C.2(4) of Appendix C, $\sqrt{T}(\hat{\boldsymbol{\beta}} - \boldsymbol{\beta})$ has the same asymptotic distribution as

$$\left[\operatorname{plim}\left(\frac{1}{T}ZZ'\right)^{-1} \otimes I_K\right]\frac{1}{\sqrt{T}}(Z \otimes I_K)\mathbf{u} = (\Gamma^{-1} \otimes I_K)\frac{1}{\sqrt{T}}(Z \otimes I_K)\mathbf{u}.$$

Hence the covariance matrix of the asymptotic distribution of $\sqrt{T}(\hat{\boldsymbol{\beta}} - \boldsymbol{\beta})$ is

$$(\Gamma^{-1} \otimes I_K)(\Gamma \otimes \Sigma_u)(\Gamma^{-1} \otimes I_K) = \Gamma^{-1} \otimes \Sigma_u.$$

The result (3.2.16) can be established with similar arguments (see Problem 3.2). ■

As mentioned previously, if u_t is *Gaussian* (normally distributed) white noise, it satisfies the conditions of Proposition 3.1 so that consistency and asymptotic normality of the LS estimator are ensured for stable Gaussian (normally distributed) VAR(p) processes y_t. Note that normality of u_t implies normality of the y_t for stable processes.

In order to assess the asymptotic dispersion of the LS estimator we need to know the matrices Γ and Σ_u. From (3.2.13) an obvious consistent estimator of Γ is

$$\hat{\Gamma} = ZZ'/T. \tag{3.2.17}$$

Since $\Sigma_u = E(u_t u_t')$, a plausible estimator for this matrix is

$$\begin{aligned}\tilde{\Sigma}_u &= \frac{1}{T}\sum_{t=1}^{T} \hat{u}_t\hat{u}_t' = \frac{1}{T}\hat{U}\hat{U}' = \frac{1}{T}(Y - \hat{B}Z)(Y - \hat{B}Z)' \\ &= \frac{1}{T}(Y - YZ'(ZZ')^{-1}Z)(Y - YZ'(ZZ')^{-1}Z)' \\ &= \frac{1}{T}Y(I_T - Z'(ZZ')^{-1}Z)(I_T - Z'(ZZ')^{-1}Z)'Y' \\ &= \frac{1}{T}Y(I_T - Z'(ZZ')^{-1}Z)Y'. \end{aligned}\tag{3.2.18}$$

Often an adjustment for degrees of freedom is desired because in a regression with fixed, nonstochastic regressors this leads to an unbiased estimator of the covariance matrix. Thus, an estimator

$$\hat{\Sigma}_u = \frac{T}{T - Kp - 1}\tilde{\Sigma}_u \tag{3.2.19}$$

may be considered. Note that there are $Kp + 1$ parameters in each of the K equations of (3.1.1) and hence there are $Kp + 1$ parameters in each equation of the system (3.2.2). Of course, $\hat{\Sigma}_u$ and $\tilde{\Sigma}_u$ are asymptotically equivalent. They are consistent estimators of Σ_u if the conditions of Proposition 3.1 hold. In fact, a bit more can be shown.

PROPOSITION 3.2 (*Asymptotic Properties of the White Noise Covariance Matrix Estimators*)

Let y_t be a stable, K-dimensional VAR(p) process as in (3.1.1) with standard white noise innovations and let $\bar{B}$ be an estimator of the VAR coefficients B so that $\sqrt{T}\,\text{vec}(\bar{B} - B)$ converges in distribution. Furthermore, using the symbols from (3.2.1), suppose that

$$\bar{\Sigma}_u = (Y - \bar{B}Z)(Y - \bar{B}Z)'/(T - c),$$

where c is a fixed constant. Then

$$\text{plim}\,\sqrt{T}(\bar{\Sigma}_u - UU'/T) = 0. \tag{3.2.20}$$

■

Proof:

$$\frac{1}{T}(Y - \bar{B}Z)(Y - \bar{B}Z)' = (B - \bar{B})\left[\frac{ZZ'}{T}\right](B - \bar{B})' + (B - \bar{B})\frac{ZU'}{T} + \frac{UZ'}{T}(B - \bar{B})' + \frac{UU'}{T}.$$

Under the conditions of the proposition $\text{plim}(B - \bar{B}) = 0$. Hence by Lemma 3.1,

$$\text{plim}(B - \bar{B})ZU'/\sqrt{T} = 0$$

and

$$\text{plim}\left[(B - \bar{B})\frac{ZZ'}{T}\sqrt{T}(B - \bar{B})'\right] = 0$$

(see Appendix C.1). Thus,

$$\text{plim}\,\sqrt{T}[(Y - \bar{B}Z)(Y - \bar{B}Z)'/T - UU'/T] = 0.$$

Therefore, the proposition follows by noting that $T/(T - c) \to 1$ as $T \to \infty$. ■

The proposition covers both estimators $\hat{\Sigma}_u$ and $\tilde{\Sigma}_u$. It implies that the feasible estimators $\tilde{\Sigma}_u$ and $\hat{\Sigma}_u$ have the same asymptotic properties as the estimator

$$\frac{UU'}{T} = \frac{1}{T}\sum_{t=1}^{T} u_t u_t'$$

which is based on the unknown true residuals and is therefore not feasible in practice. In particular, if $\sqrt{T}\operatorname{vec}(UU'/T - \Sigma_u)$ converges in distribution, $\sqrt{T}\operatorname{vec}(\hat{\Sigma}_u - \Sigma_u)$ and $\sqrt{T}\operatorname{vec}(\tilde{\Sigma}_u - \Sigma_u)$ will have the same limiting distribution (see Proposition C.2 of Appendix C.1). Moreover, it can be shown that the asymptotic distributions are independent of the limiting distribution of the LS estimator $\hat{B}$. Another immediate implication of Proposition 3.2 is that $\tilde{\Sigma}_u$ and $\hat{\Sigma}_u$ are consistent estimators of Σ_u. This result is established next.

COROLLARY 3.2.1

Under the conditions of Proposition 3.2,

$$\operatorname{plim}\tilde{\Sigma}_u = \operatorname{plim}\hat{\Sigma}_u = \operatorname{plim} UU'/T = \Sigma_u. \qquad \blacksquare$$

Proof: By Proposition 3.2 it suffices to show that plim $UU'/T = \Sigma_u$ which follows because

$$E\left(\frac{1}{T}UU'\right) = \frac{1}{T}\sum_{t=1}^{T} E(u_t u_t') = \Sigma_u$$

and

$$\operatorname{Var}\left(\frac{1}{T}\operatorname{vec}(UU')\right) = \frac{1}{T^2}\sum_{t=1}^{T}\operatorname{Var}[\operatorname{vec}(u_t u_t')] \le \frac{T}{T^2}g \xrightarrow[T\to\infty]{} 0$$

where g is a constant upper bound for $\operatorname{Var}[\operatorname{vec}(u_t u_t')]$ which exists since the fourth moments of u_t are bounded by Definition 3.1. ■

If y_t is stable with standard white noise Proposition 3.1 and Corollary 3.2.1 imply that $(\hat{\beta}_i - \beta_i)/\hat{s}_i$ has an asymptotic standard normal distribution. Here $\beta_i(\hat{\beta}_i)$ is the i-th component of $\boldsymbol{\beta}(\hat{\boldsymbol{\beta}})$ and $\hat{s}_i$ is the square root of the i-th diagonal element of

$$(ZZ')^{-1} \otimes \hat{\Sigma}_u. \tag{3.2.21}$$

This result means that we can use the "t-ratios" provided by common regression programs in setting up confidence intervals and tests for individual coefficients. The critical values and percentiles may be based on the asymptotic standard normal distribution. Since it has been found in simulation studies that the small sample distributions of the "t-ratios" have fatter tails than the standard normal distribution, one may want to approximate the small sample distribution by some t-distribution. The question is then what number of degrees of freedom (d.f.) should be used. The overall model (3.2.3) may suggest a choice of d.f. = $KT - K^2p - K$ because in a standard regression model with nonstochastic regressors the d.f. of the "t-ratios" are equal to the sample size minus the number of estimated parameters. In the present case it seems also reasonable to use d.f. $= T - Kp - 1$ because the multivariate LS estimator is identical to the LS estimator obtained for each of the K equations in (3.2.2) separately. In a separate regression for each individual equation we would have T observations and $Kp + 1$ parameters. If the sample size T is large, and thus the number of degrees of freedom is large, the corresponding t-distribution will be very close to the standard normal so that the choice between the two becomes irrelevant for large samples. Before we look a little further into the problem of choosing appropriate critical values let us illustrate the foregoing results by an example.

3.2.3 An Example

As an example we consider a three-dimensional system consisting of first differences of the logarithms of quarterly, seasonally adjusted West German fixed investment (y_1), disposable income (y_2), and consumption expenditures (y_3). We use data from 1960 to 1978. They are listed in Appendix E, Table E.1, and the original data and first differences of logarithms are plotted in Figures 3.1 and 3.2. The original data have a trend and are thus considered nonstationary. The trend is removed by taking first differences of logarithms. We will discuss this issue in some more detail in Chapter 11. Note that the value for 1960.1 is lost in the differenced series.

Let us assume that the data have been generated by a VAR(2) process. The

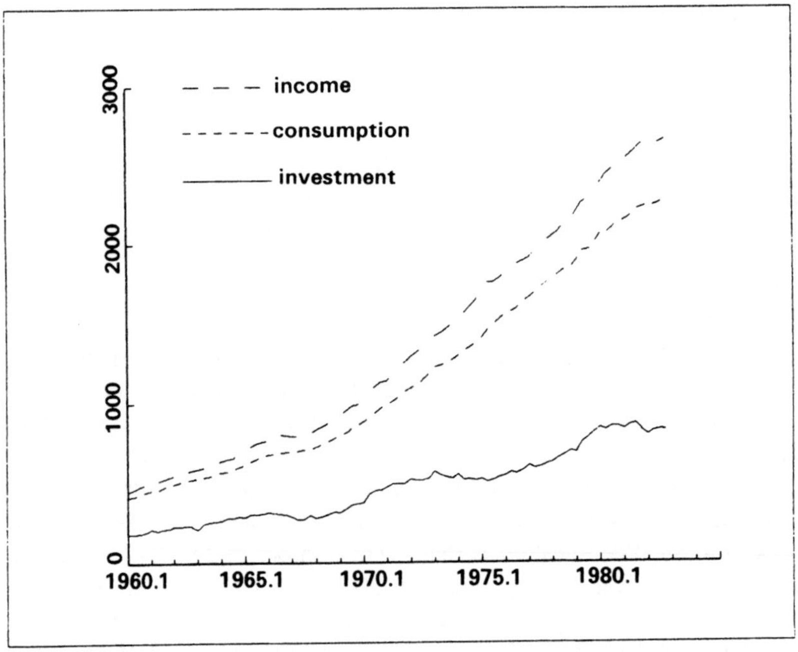

Fig. 3.1. West German investment, income, and consumption data.

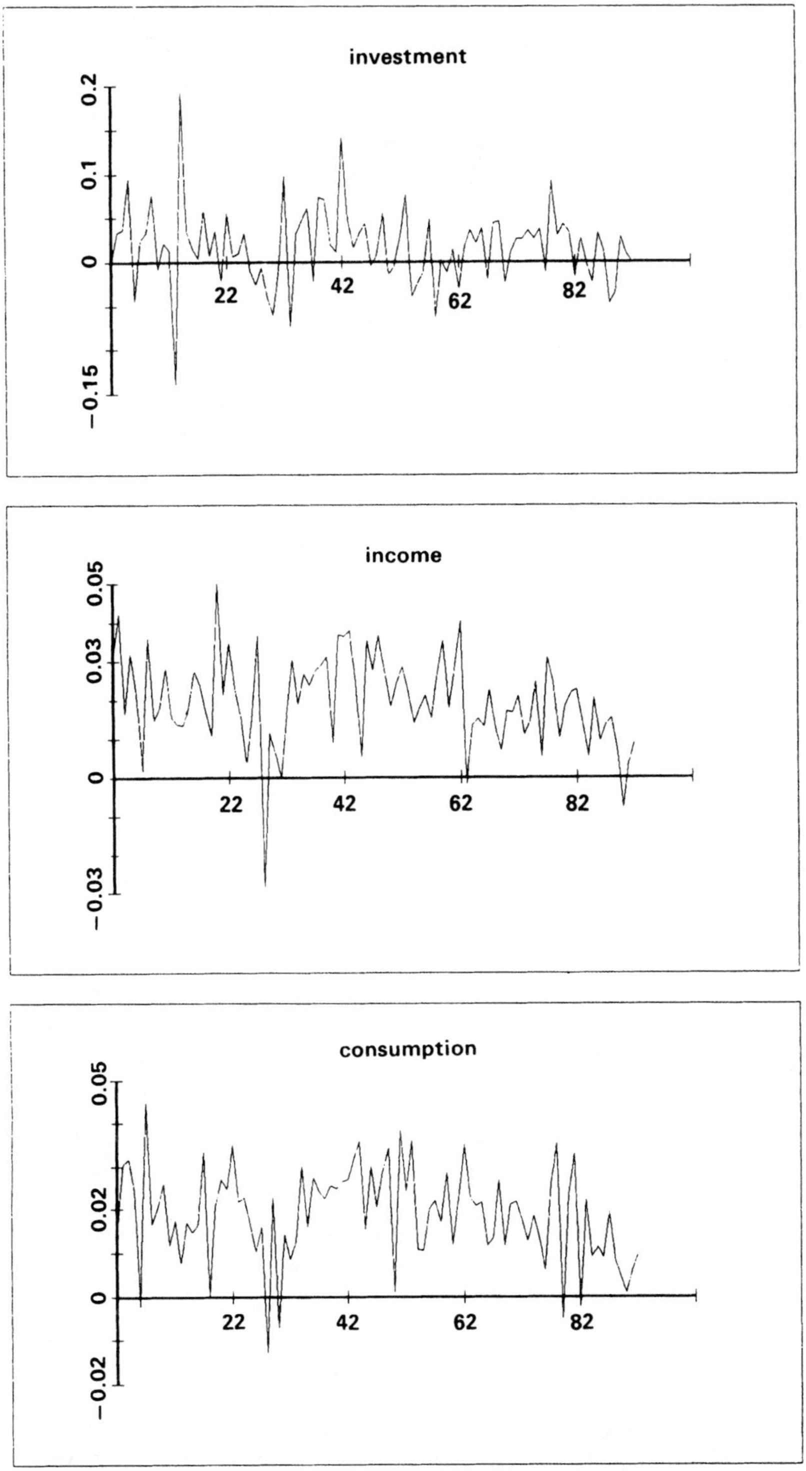

Fig. 3.2. First differences of logarithms of West German investment, income, and consumption.

choice of the VAR order $p = 2$ is arbitrary at this point. In the next chapter criteria for choosing the VAR order will be considered. Since the VAR order is two we keep the first two observations of the differenced series as presample values and use a sample size of $T = 73$. Thus we have a (3×73) matrix Y, $B = (\nu, A_1, A_2)$ is (3×7), Z is (7×73) and $\boldsymbol{\beta}$ and $\mathbf{b}$ are both (21×1) vectors.

The LS estimates are

$$\hat{B} = (\hat{\nu}, \hat{A}_1, \hat{A}_2) = YZ'(ZZ')^{-1}$$

$$= \begin{bmatrix} -.017 & -.320 & .146 & .961 & -.161 & .115 & .934 \\ .016 & .044 & -.153 & .289 & .050 & .019 & -.010 \\ .013 & -.002 & .225 & -.264 & .034 & .355 & -.022 \end{bmatrix}. \tag{3.2.22}$$

To check the stability of the estimated process we determine the roots of the polynomial $\det(I_3 - \hat{A}_1 z - \hat{A}_2 z^2)$ which is easily seen to have degree 6. Its roots are

$$z_1 = 1.753, \quad z_2 = -2.694, \quad z_{3/4} = -0.320 \pm 2.008i, \quad z_{5/6} = -1.285 \pm 1.280i.$$

Note that these roots have been computed using higher precision than the three digits in (3.2.22). They all have modulus greater than 1 and hence the stability condition is satisfied.

We get

$$\hat{\Sigma}_u = \frac{1}{T - Kp - 1}(YY' - YZ'(ZZ')^{-1}ZY')$$

$$= \begin{bmatrix} 21.30 & .72 & 1.23 \\ .72 & 1.37 & .61 \\ 1.23 & .61 & .89 \end{bmatrix} \times 10^{-4} \tag{3.2.23}$$

as estimate of the residual covariance matrix Σ_u. Furthermore,

$$\hat{\Gamma}^{-1} = (ZZ'/T)^{-1} = T \begin{bmatrix} .14 & .17 & -.69 & -2.51 & .10 & -.67 & -2.57 \\ . & 7.39 & 1.24 & -10.56 & 1.80 & 1.08 & -8.70 \\ . & . & 139.81 & -87.40 & -4.58 & 30.21 & -50.88 \\ . & . & . & 207.22 & .84 & -55.35 & 73.82 \\ . & . & . & . & 7.33 & -.03 & -9.31 \\ . & . & . & . & . & 134.19 & -82.64 \\ . & . & . & . & . & . & 207.71 \end{bmatrix}.$$

Dividing the elements of $\hat{B}$ by the square roots of the corresponding diagonal elements of $(ZZ')^{-1} \otimes \hat{\Sigma}_u$ we get the matrix of t-ratios:

$$\begin{bmatrix} -0.97 & -2.55 & 0.27 & 1.45 & -1.29 & 0.21 & 1.41 \\ 3.60 & 1.38 & -1.10 & 1.71 & 1.58 & 0.14 & -0.06 \\ 3.67 & -0.09 & 2.01 & -1.94 & 1.33 & 3.24 & -0.16 \end{bmatrix}. \tag{3.2.24}$$

We may compare these quantities with critical values from a t-distribution with d.f. $= KT - K^2p - K = 198$ or d.f. $= T - Kp - 1 = 66$. In both cases we get critical values of approximately ± 2 for a two-tailed test with significance level 5%. Thus, the critical values are approximately the same as those from a standard normal distribution.

Apparently quite a few coefficients are not significant under this criterion. This observation suggests that the model contains unnecessarily many free parameters. In subsequent chapters we will discuss the problem of choosing the VAR order and possible restrictions for the coefficients. Also, before an estimated model is used for forecasting and analysis purposes the assumptions underlying the analysis should be checked carefully. Checking the model adequacy will be treated in greater detail in Chapter 4.

3.2.4 Small Sample Properties of the LS Estimator

As mentioned earlier it is difficult to derive small sample properties of the LS estimator analytically. In such a case it is sometimes helpful to use *Monte Carlo methods* in order to get some idea about the small sample properties. In a Monte Carlo analysis specific processes are used to artificially generate a large number of time series. Then a set of estimates is computed for each multiple time series generated and the properties of the resulting empirical distributions of these estimates are studied (see Appendix D). Such an approach usually permits rather limited conclusions only because the findings may depend on the particular processes used for generating the time series. Nevertheless such exercises give some insight into the small sample properties of estimators.

In the following we use the bivariate VAR(2) example process (2.1.15),

$$y_t = \begin{bmatrix} .02 \\ .03 \end{bmatrix} + \begin{bmatrix} .5 & .1 \\ .4 & .5 \end{bmatrix} y_{t-1} + \begin{bmatrix} 0 & 0 \\ .25 & 0 \end{bmatrix} y_{t-2} + u_t \tag{3.2.25}$$

with error covariance matrix

$$\Sigma_u = \begin{bmatrix} 9 & 0 \\ 0 & 4 \end{bmatrix} \times 10^{-4} \tag{3.2.26}$$

to investigate the small sample properties of the multivariate LS estimator. With this process we have generated 1000 bivariate time series of length $T = 30$ plus 2 presample values using independent standard normal errors, that is, $u_t \sim N(0, \Sigma_u)$. Thus the 1000 bivariate time series are generated by a stable Gaussian process so that Propositions 3.1 and 3.2 provide the asymptotic properties of the LS estimators.

In Table 3.1 some empirical results are given. In particular the empirical mean, variance, and mean squared error (MSE) of each parameter estimator are given. Obviously the empirical means differ from the actual values of the coefficients. However, measuring the estimation precision by the empirical variance (average squared deviation from the mean in 1000 samples) or MSE (average squared deviation from the true parameter value), the coefficients are seen to be estimated quite precisely even with a sample size as small as $T = 30$. This is partly a consequence of the small Σ_u matrix.

Table 3.1. Empirical percentiles of t-ratios of parameter estimates and actual percentiles of t-distributions for sample size $T = 30$

parameter	empirical mean	variance	MSE	empirical percentiles of t-ratios: 1.	5.	10.	50.	90.	95.	99.
$\nu_1 = .02$	.041	.0011	.0015	−1.91	−1.04	−0.64	0.62	1.92	2.29	3.12
$\nu_2 = .03$	.038	.0005	.0006	−2.30	−1.40	−1.02	0.25	1.65	2.11	2.83
$\alpha_{11,1} = .5$	.41	.041	.049	−2.78	−2.18	−1.74	−0.43	0.92	1.28	2.01
$\alpha_{21,1} = .4$	.40	.018	.018	−2.61	−1.74	−1.28	0.04	1.28	1.71	2.65
$\alpha_{12,1} = .1$	.10	.078	.078	−2.27	−1.67	−1.35	−0.03	1.29	1.67	2.38
$\alpha_{22,1} = .5$	.44	.030	.034	−2.69	−1.97	−1.59	−0.35	0.89	1.30	2.06
$\alpha_{11,2} = 0$	−.05	.056	.058	−2.75	−1.93	−1.50	−0.24	1.02	1.38	2.09
$\alpha_{21,2} = .25$	.29	.023	.024	−1.99	−1.32	−0.99	0.20	1.45	1.81	2.48
$\alpha_{12,2} = 0$	−.07	.053	.058	−2.48	−1.91	−1.61	−0.28	0.97	1.39	2.03
$\alpha_{22,2} = 0$	−.01	.023	.024	−2.71	−1.72	−1.36	−0.03	1.18	1.53	2.18
	degrees of freedom (d.f.)			percentiles of t-distributions: 1.	5.	10.	50.	90.	95.	99.
	$T - Kp - 1 = 25$			−2.49	−1.71	−1.32	0	1.32	1.71	2.49
	$K(T - Kp - 1) = 50$			−2.41	−1.68	−1.30	0	1.30	1.68	2.41
	∞ (normal distribution)			−2.33	−1.65	−1.28	0	1.28	1.65	2.33

In Table 3.1 empirical percentiles of the t-ratios are also given together with the corresponding percentiles from the t- and standard normal distributions (d.f. $= \infty$). Even with the presently considered relatively small sample size the percentiles of the three distributions that might be used for inference do not differ much. Consequently it does not matter much which of the theoretical percentiles are used, in particular, since the empirical percentiles, in many cases, differ quite a bit from the corresponding theoretical quantities. This shows that the asymptotic results have to be used cautiously in setting up small sample tests and confidence intervals. On the other hand, this example also demonstrates that the asymptotic theory does provide some guidance for inference. For example, the empirical 95th percentiles of all coefficients lie between the 90th and the 99th percentile of the standard normal distribution given in the last row of the table. Of course, this is just one example and not a general finding.

In an extensive study Nankervis & Savin (1988) investigated the small sample distribution of the "t-statistic" for the parameter of a *univariate* AR(1) process. They find that it differs quite substantially from the corresponding t-distribution especially if the sample size is small ($T < 100$) and the parameter lies close to the instability region. Analytic results on the bias in estimating VAR models have been derived by Nicholls & Pope (1988) and Tjøstheim & Paulsen (1983). What should be learned from our Monte Carlo investigation and these remarks is that asymptotic distributions in the present context can only be used as rough guidelines for small sample inference. That, however, is much better than having no guidance at all.

3.3 Least Squares Estimation with Mean-Adjusted Data and Yule-Walker Estimation

3.3.1 Estimation when the Process Mean Is Known

Occasionally a VAR(p) model is given in *mean-adjusted form*,

$$(y_t - \mu) = A_1(y_{t-1} - \mu) + \cdots + A_p(y_{t-p} - \mu) + u_t. \tag{3.3.1}$$

Multivariate LS estimation of this model form is straightforward if the mean vector μ is known. Defining

$$\begin{aligned}
Y^0 &:= (y_1 - \mu, \dots, y_T - \mu) && (K \times T),\\
A &:= (A_1, \dots, A_p) && (K \times Kp),\\
Y_t^0 &:= \begin{bmatrix} y_t - \mu \\ \vdots \\ y_{t-p+1} - \mu \end{bmatrix} && (Kp \times 1),\\
X &:= (Y_0^0, \dots, Y_{T-1}^0) && (Kp \times T),\\
\mathbf{y}^0 &:= \operatorname{vec}(Y^0) && (KT \times 1),\\
\boldsymbol{\alpha} &:= \operatorname{vec}(A) && (K^2p \times 1),
\end{aligned} \tag{3.3.2}$$

we can write (3.3.1) for $t = 1, \ldots, T$ compactly as

$$Y^0 = AX + U \tag{3.3.3}$$

or

$$\mathbf{y}^0 = (X' \otimes I_K)\boldsymbol{\alpha} + \mathbf{u}, \tag{3.3.4}$$

where U and $\mathbf{u}$ are defined as in (3.2.1). The LS estimator is easily seen to be

$$\hat{\boldsymbol{\alpha}} = ((XX')^{-1}X \otimes I_K)\mathbf{y}^0 \tag{3.3.5}$$

or

$$\hat{A} = Y^0X'(XX')^{-1}. \tag{3.3.6}$$

If y_t is stable and u_t is standard white noise it can be shown that

$$\sqrt{T}(\hat{\boldsymbol{\alpha}} - \boldsymbol{\alpha}) \xrightarrow{d} N(0, \Sigma_{\hat{\alpha}}), \tag{3.3.7}$$

where

$$\Sigma_{\hat{\alpha}} = \Gamma_Y(0)^{-1} \otimes \Sigma_u \tag{3.3.8}$$

and $\Gamma_Y(0) := E(Y_t^0 Y_t^{0\prime})$.

3.3.2 Estimation of the Process Mean

Usually μ will not be known in advance. In that case it may be estimated by the vector of sample means,

$$\bar{y} = \frac{1}{T}\sum_{t=1}^{T} y_t. \tag{3.3.9}$$

Using (3.3.1) $\bar{y}$ can be written as

$$\bar{y} = \mu + A_1\left[\bar{y} + \frac{1}{T}(y_0 - y_T) - \mu\right] + \cdots +$$
$$+ A_p\left[\bar{y} + \frac{1}{T}(y_{-p+1} + \cdots + y_0 - y_{T-p+1} - \cdots - y_T) - \mu\right]$$
$$+ \frac{1}{T}\sum_{t=1}^{T} u_t.$$

Hence,

$$(I_K - A_1 - \cdots - A_p)(\bar{y} - \mu) = \frac{1}{T}z_T + \frac{1}{T}\sum_t u_t, \tag{3.3.10}$$

where

$$z_T = \sum_{i=1}^{p} A_i\left[\sum_{j=0}^{i-1}(y_{0-j} - y_{T-j})\right].$$

Evidently,

$$E(z_T/\sqrt{T}) = \frac{1}{\sqrt{T}} E(z_T) = 0$$

and

$$\operatorname{Var}(z_T/\sqrt{T}) = \frac{1}{T} \operatorname{Var}(z_T) \xrightarrow[T\to\infty]{} 0$$

since y_t is stable. In other words, $z_T/\sqrt{T}$ converges to zero in mean square. It follows that $\sqrt{T}(I_K - A_1 - \cdots - A_p)(\bar{y} - \mu)$ has the same asymptotic distribution as $\sum u_t/\sqrt{T}$ (see Appendix C, Proposition C.2). Hence, noting that, by a central limit theorem (e.g., Fuller (1976)),

$$\frac{1}{\sqrt{T}} \sum_{t=1}^{T} u_t \xrightarrow{d} N(0, \Sigma_u) \tag{3.3.11}$$

if u_t is standard white noise, we get the following result:

PROPOSITION 3.3 (*Asymptotic Properties of the Sample Mean*)

If the VAR(p) process y_t given in (3.3.1) is stable and u_t is standard white noise, then

$$\sqrt{T}(\bar{y} - \mu) \xrightarrow{d} N(0, \Sigma_{\bar{y}}) \tag{3.3.12}$$

where

$$\Sigma_{\bar{y}} = (I_K - A_1 - \cdots - A_p)^{-1} \Sigma_u (I_K - A_1 - \cdots - A_p)'^{-1}.$$

In particular, plim $\bar{y} = \mu$. ■

The proposition follows from (3.3.10), (3.3.11), and Proposition C.4 of Appendix C. The limiting distribution in (3.3.11) holds even in small samples for Gaussian white noise u_t.

Since $\mu = (I_K - A_1 - \cdots - A_p)^{-1}\nu$ (see Chapter 2, Section 2.1) an alternative estimator for the process mean is obtained from the LS estimator of the previous section:

$$\hat{\mu} = (I_K - \hat{A}_1 - \cdots - \hat{A}_p)^{-1}\hat{\nu}. \tag{3.3.13}$$

Using again Proposition C.4 of Appendix C this estimator is also consistent and has an asymptotic normal distribution,

$$\sqrt{T}(\hat{\mu} - \mu) \xrightarrow{d} N\left[0, \frac{\partial \mu}{\partial \boldsymbol{\beta}'}(\Gamma^{-1} \otimes \Sigma_u)\frac{\partial \mu'}{\partial \boldsymbol{\beta}}\right], \tag{3.3.14}$$

provided the conditions of Proposition 3.1 are satisfied. It can be shown that

$$\frac{\partial \mu}{\partial \boldsymbol{\beta}'}(\Gamma^{-1} \otimes \Sigma_u)\frac{\partial \mu'}{\partial \boldsymbol{\beta}} = \Sigma_{\bar{y}} \tag{3.3.15}$$

and hence the estimators $\hat{\mu}$ and $\bar{y}$ for μ are asymptotically equivalent (see Section 3.4). This result suggests that it does not matter asymptotically whether the mean is estimated separately or jointly with the other VAR coefficients. While this holds asymptotically it will usually matter in small samples which estimator is used. We will give an example shortly.

3.3.3 Estimation with Unknown Process Mean

If the mean vector μ is unknown it may be replaced by $\bar{y}$ in the vectors and matrices in (3.3.2) giving $\hat{X}$, $\hat{Y}^0$ and so on. The resulting LS estimator,

$$\hat{\hat{\boldsymbol{\alpha}}} = ((\hat{X}\hat{X}')^{-1}\hat{X} \otimes I_K)\hat{\mathbf{y}}^0,$$

is asymptotically equivalent to $\hat{\boldsymbol{\alpha}}$. More precisely, it can be shown that, under the conditions of Proposition 3.3,

$$\sqrt{T}(\hat{\hat{\boldsymbol{\alpha}}} - \boldsymbol{\alpha}) \xrightarrow{d} N(0, \Gamma_Y(0)^{-1} \otimes \Sigma_u), \tag{3.3.16}$$

where $\Gamma_Y(0) := E(Y_t^0 Y_t^{0\prime})$. This result will be discussed further in the next section on maximum likelihood estimation for Gaussian processes.

3.3.4 The Yule-Walker Estimator

The LS estimator can also be derived from the Yule-Walker equations given in Chapter 2, (2.1.37). They imply

$$\Gamma_y(h) = [A_1, \dots, A_p] \begin{bmatrix} \Gamma_y(h-1) \\ \vdots \\ \Gamma_y(h-p) \end{bmatrix}, \qquad h > 0,$$

or

$$[\Gamma_y(1), \dots, \Gamma_y(p)] = [A_1, \dots, A_p] \begin{bmatrix} \Gamma_y(0) & \dots & \Gamma_y(p-1) \\ \vdots & \ddots & \vdots \\ \Gamma_y(-p+1) & \dots & \Gamma_y(0) \end{bmatrix} = A\Gamma_Y(0) \tag{3.3.17}$$

and hence

$$A = [\Gamma_y(1), \dots, \Gamma_y(p)]\Gamma_Y(0)^{-1}.$$

Estimating $\Gamma_Y(0)$ by $\hat{X}\hat{X}'/T$ and $[\Gamma_y(1) \dots \Gamma_y(p)]$ by $\hat{Y}^0\hat{X}'/T$ the resulting estimator is just the LS estimator

$$\hat{\hat{A}} = \hat{Y}^0\hat{X}'(\hat{X}\hat{X}')^{-1}. \tag{3.3.18}$$

Alternatively the moment matrices $\Gamma_y(k)$ may be estimated using as many data as available including the presample values. Thus, if a sample $y_1, \dots, y_T$ and p presample observations $y_{-p+1}, \dots, y_0$ are available, μ may be estimated as

$$\bar{y}^* = \frac{1}{T+p} \sum_{t=-p+1}^{T} y_t$$

and $\Gamma_y(k)$ may be estimated as

$$\hat{\Gamma}_y(k) = \frac{1}{T+p-k} \sum_{t=-p+k+1}^{T} (y_t - \bar{y}^*)(y_{t-k} - \bar{y}^*)'. \tag{3.3.19}$$

Using these estimators in (3.3.18) the so-called *Yule-Walker estimator* for A is obtained. This estimator has the same asymptotic properties as the LS estimator. However, it may have less attractive small sample properties (e.g., Tjøstheim & Paulsen (1983)). Therefore the LS estimator is usually used in the following.

3.3.5 An Example

To illustrate the results of this section we use again the West German investment, income, and consumption data given in Table E.1 of Appendix E. The variables y_1, y_2, and y_3 are defined as in Section 3.2.3, the sample period ranges from 1960.4 to 1978.4, that is, $T = 73$ and the data for 1960.2 and 1960.3 are used as presample values. Using only the sample values we get

$$\bar{y} = \begin{bmatrix} .018 \\ .020 \\ .020 \end{bmatrix} \tag{3.3.20}$$

which is different, though not substantially so, from

$$\hat{\mu} = (I_3 - \hat{A}_1 - \hat{A}_2)^{-1}\hat{\nu} = \begin{bmatrix} .017 \\ .020 \\ .020 \end{bmatrix} \tag{3.3.21}$$

as obtained from the LS estimates in (3.2.22).

Subtracting the sample means from the data we get, based on (3.3.18),

$$\hat{\hat{A}} = (\hat{\hat{A}}_1, \hat{\hat{A}}_2) = \begin{bmatrix} -.319 & .143 & .960 & -.160 & .112 & .933 \\ .044 & -.153 & .288 & .050 & .019 & -.010 \\ -.002 & .224 & -.264 & .034 & .354 & -.023 \end{bmatrix}. \tag{3.3.22}$$

This estimate is clearly distinct from the corresponding part of (3.2.22) although the two estimates do not differ dramatically.

If the two presample values are used in estimating the process means and moment matrices we get

$$\hat{A}_{YW} = \begin{bmatrix} -.319 & .147 & .959 & -.160 & .115 & .932 \\ .044 & -.152 & .286 & .050 & .020 & -.012 \\ -.002 & .225 & -.264 & .034 & .355 & -.022 \end{bmatrix} \tag{3.3.23}$$

which is the Yule-Walker estimate. Although the sample size is moderate there is a slight difference between the estimates in (3.3.22) and (3.3.23).

3.4 Maximum Likelihood Estimation

3.4.1 The Likelihood Function

Assuming that the distribution of the process is known, maximum likelihood (ML) estimation is an alternative to LS estimation. We will consider ML estimation under the assumption that the VAR(p) process y_t is Gaussian. More precisely,

$$\mathbf{u} = \operatorname{vec}(U) = \begin{bmatrix} u_1 \\ \vdots \\ u_T \end{bmatrix} \sim N(0, I_T \otimes \Sigma_u). \tag{3.4.1}$$

In other words, the probability density of $\mathbf{u}$ is

$$f_{\mathbf{u}}(\mathbf{u}) = \frac{1}{(2\pi)^{KT/2}} |I_T \otimes \Sigma_u|^{-1/2} \exp\left[-\frac{1}{2} \mathbf{u}'(I_T \otimes \Sigma_u^{-1})\mathbf{u} \right]. \tag{3.4.2}$$

Moreover,

$$\mathbf{u} = \begin{bmatrix} I_K & 0 & \dots & 0 & \dots & 0 \\ -A_1 & I_K & & 0 & \dots & 0 \\ \vdots & \vdots & \ddots & \vdots & & \vdots \\ -A_p & -A_{p-1} & \dots & I_K & & 0 \\ 0 & -A_p & & & \ddots & \vdots \\ \vdots & & \ddots & & & \vdots \\ 0 & 0 & \dots & -A_p & \dots & I_K \end{bmatrix} (\mathbf{y} - \boldsymbol{\mu}^*)$$

$$+ \begin{bmatrix} -A_p & -A_{p-1} & \dots & -A_1 \\ 0 & -A_p & \dots & -A_2 \\ \vdots & & \ddots & \vdots \\ 0 & 0 & \dots & -A_p \\ \vdots & \vdots & & \vdots \\ 0 & 0 & \dots & 0 \end{bmatrix} (Y_0 - \boldsymbol{\mu}), \tag{3.4.3}$$

where $\mathbf{y} := \operatorname{vec}(Y)$ and $\boldsymbol{\mu}^* := (\mu', \dots, \mu')'$ are $(TK \times 1)$ vectors and $Y_0 := (y_0', \dots, y_{-p+1}')'$ and $\boldsymbol{\mu} := (\mu', \dots, \mu')'$ are $(Kp \times 1)$ vectors. Consequently $\partial\mathbf{u}/\partial\mathbf{y}$ is a lower triangular matrix with unit diagonal which has unit determinant. Hence, using that $\mathbf{u} = \mathbf{y} - \boldsymbol{\mu}^* - (X' \otimes I_K)\boldsymbol{\alpha}$,

$$\begin{aligned} f_{\mathbf{y}}(\mathbf{y}) &= \left| \frac{\partial \mathbf{u}}{\partial \mathbf{y}} \right| f_{\mathbf{u}}(\mathbf{u}) \\ &= \frac{1}{(2\pi)^{KT/2}} |I_T \otimes \Sigma_u|^{-1/2} \exp\left[-\frac{1}{2} (\mathbf{y} - \boldsymbol{\mu}^* - (X' \otimes I_K)\boldsymbol{\alpha})'(I_T \otimes \Sigma_u^{-1}) \right. \\ &\qquad \left. \times (\mathbf{y} - \boldsymbol{\mu}^* - (X' \otimes I_K)\boldsymbol{\alpha}) \right], \end{aligned} \tag{3.4.4}$$

where X and $\boldsymbol{\alpha}$ are as defined in (3.3.2). For simplicity the initial values Y_0 are assumed to be given fixed numbers. Hence we get a log-likelihood function

$$\begin{aligned}
\ln l(\mu, \boldsymbol{\alpha}, \Sigma_u) &= -\frac{KT}{2}\ln 2\pi - \frac{T}{2}\ln|\Sigma_u| \\
&\quad - \frac{1}{2}[\mathbf{y} - \boldsymbol{\mu}^* - (X' \otimes I_K)\boldsymbol{\alpha}]'(I_T \otimes \Sigma_u^{-1})[\mathbf{y} - \boldsymbol{\mu}^* - (X' \otimes I_K)\boldsymbol{\alpha}] \\
&= -\frac{KT}{2}\ln 2\pi - \frac{T}{2}\ln|\Sigma_u| - \frac{1}{2}\sum_{t=1}^{T}\left[(y_t - \mu) - \sum_{i=1}^{p} A_i(y_{t-i} - \mu)\right]' \\
&\quad \times \Sigma_u^{-1}\left[(y_t - \mu) - \sum_{i=1}^{p} A_i(y_{t-i} - \mu)\right] \\
&= -\frac{KT}{2}\ln 2\pi - \frac{T}{2}\ln|\Sigma_u| \\
&\quad - \frac{1}{2}\sum_t \left(y_t - \sum_i A_i y_{t-i}\right)' \Sigma_u^{-1}\left(y_t - \sum_i A_i y_{t-i}\right) \\
&\quad + \mu'\left(I_K - \sum_i A_i\right)' \Sigma_u^{-1} \sum_t \left(y_t - \sum_i A_i y_{t-i}\right) \\
&\quad - \frac{T}{2}\mu'\left(I_K - \sum_i A_i\right)' \Sigma_u^{-1}\left(I_K - \sum_i A_i\right)\mu \\
&= -\frac{KT}{2}\ln 2\pi - \frac{T}{2}\ln|\Sigma_u| - \frac{1}{2}\operatorname{tr}(Y^0 - AX)'\Sigma_u^{-1}(Y^0 - AX),
\end{aligned} \tag{3.4.5}$$

where $Y^0 := (y_1 - \mu, \dots, y_T - \mu)$ and $A := (A_1, \dots, A_p)$ are as defined in (3.3.2). These different expressions of the log-likelihood function will be useful in the following.

3.4.2 The ML Estimators

In order to determine the ML estimators of μ, $\boldsymbol{\alpha}$, and Σ_u the system of first partial derivatives is needed:

$$\begin{aligned}
\frac{\partial \ln l}{\partial \mu} &= \left(I_K - \sum_i A_i\right)' \Sigma_u^{-1} \sum_t \left(y_t - \sum_i A_i y_{t-i}\right) \\
&\quad - T\left(I_K - \sum_i A_i\right)' \Sigma_u^{-1}\left(I_K - \sum_i A_i\right)\mu \\
&= [I_K - A(j \otimes I_K)]'\Sigma_u^{-1}\left[\sum_t (y_t - \mu - AY_{t-1}^0)\right],
\end{aligned} \tag{3.4.6}$$

where Y_t^0 is as defined in (3.3.2) and $j := (1, \dots, 1)'$ is a $(p \times 1)$ vector of ones,

$$\frac{\partial \ln l}{\partial \boldsymbol{\alpha}} = (X \otimes I_K)(I_T \otimes \Sigma_u^{-1})[\mathbf{y} - \boldsymbol{\mu}^* - (X' \otimes I_K)\boldsymbol{\alpha}]$$

$$= (X \otimes \Sigma_u^{-1})(\mathbf{y} - \boldsymbol{\mu}^*) - (XX' \otimes \Sigma_u^{-1})\boldsymbol{\alpha}, \tag{3.4.7}$$

$$\frac{\partial \ln l}{\partial \Sigma_u} = -\frac{T}{2}\Sigma_u^{-1} + \frac{1}{2}\Sigma_u^{-1}(Y^0 - AX)(Y^0 - AX)'\Sigma_u^{-1}. \tag{3.4.8}$$

Equating to zero gives the system of normal equations which imply the estimators:

$$\tilde{\mu} = \frac{1}{T}\left[I_K - \sum_i \tilde{A}_i\right]^{-1} \sum_t \left[y_t - \sum_i \tilde{A}_i y_{t-i}\right], \tag{3.4.9}$$

$$\tilde{\boldsymbol{\alpha}} = ((\tilde{X}\tilde{X}')^{-1}\tilde{X} \otimes I_K)(\mathbf{y} - \tilde{\boldsymbol{\mu}}^*), \tag{3.4.10}$$

$$\tilde{\Sigma}_u = \frac{1}{T}(\tilde{Y}^0 - \tilde{A}\tilde{X})(\tilde{Y}^0 - \tilde{A}\tilde{X})', \tag{3.4.11}$$

where $\tilde{X}$ and $\tilde{Y}^0$ are obtained from X and Y^0, respectively, by replacing μ with $\tilde{\mu}$.

3.4.3 Properties of the ML Estimators

Comparing these results with the LS estimators obtained in Section 3.3 it turns out that the ML estimators of μ and $\boldsymbol{\alpha}$ are identical to the LS estimators. Thus, $\tilde{\mu}$ and $\tilde{\boldsymbol{\alpha}}$ are consistent estimators if y_t is a stationary, stable Gaussian VAR(p) process and $\sqrt{T}(\tilde{\mu} - \mu)$ and $\sqrt{T}(\tilde{\boldsymbol{\alpha}} - \boldsymbol{\alpha})$ are asymptotically normally distributed. This result also follows from a more general maximum likelihood theory (see Appendix C.4). In fact that theory implies that the covariance matrix of the asymptotic distribution of the ML estimators is the limit of T times the inverse information matrix. The information matrix is

$$I(\boldsymbol{\delta}) = -E\left[\frac{\partial^2 \ln l}{\partial \boldsymbol{\delta} \partial \boldsymbol{\delta}'}\right] \tag{3.4.12}$$

where $\boldsymbol{\delta}' := (\mu', \boldsymbol{\alpha}', \boldsymbol{\sigma}')$ with $\boldsymbol{\sigma} := \operatorname{vech}(\Sigma_u)$. Note that vech is a column stacking operator that stacks only the elements on and below the diagonal of Σ_u. It is related to the vec operator by the $(K(K+1)/2 \times K^2)$ elimination matrix $\mathbf{L}_K$, that is, $\operatorname{vech}(\Sigma_u) = \mathbf{L}_K \operatorname{vec}(\Sigma_u)$ or, defining $\boldsymbol{\omega} := \operatorname{vec}(\Sigma_u)$, $\boldsymbol{\sigma} = \mathbf{L}_K\boldsymbol{\omega}$ (see Appendix A.12). For instance, for $K = 3$,

$$\boldsymbol{\omega} = \operatorname{vec}(\Sigma_u) = \operatorname{vec}\begin{bmatrix} \sigma_{11} & \sigma_{12} & \sigma_{13} \\ \sigma_{12} & \sigma_{22} & \sigma_{23} \\ \sigma_{13} & \sigma_{23} & \sigma_{33} \end{bmatrix}$$

$$= (\sigma_{11}, \sigma_{12}, \sigma_{13}, \sigma_{12}, \sigma_{22}, \sigma_{23}, \sigma_{13}, \sigma_{23}, \sigma_{33})'$$

and

$$\boldsymbol{\sigma} = \text{vech}(\Sigma_u) = \mathbf{L}_3\boldsymbol{\omega} = \begin{bmatrix} \sigma_{11} \\ \sigma_{12} \\ \sigma_{13} \\ \sigma_{22} \\ \sigma_{23} \\ \sigma_{33} \end{bmatrix}. \tag{3.4.13}$$

Note that in $\boldsymbol{\delta}$ we want only the potentially different elements of Σ_u.

The asymptotic covariance matrix of the ML estimator $\tilde{\boldsymbol{\delta}}$ is known to be

$$\lim_{T\to\infty} [I(\boldsymbol{\delta})/T]^{-1}. \tag{3.4.14}$$

In order to determine this matrix we need the second order partial derivatives of the log-likelihood. From (3.4.6) to (3.4.8) we get

$$\frac{\partial^2 \ln l}{\partial \mu\, \partial \mu'} = -T\left(I_K - \sum_i A_i\right)' \Sigma_u^{-1}\left(I_K - \sum_i A_i\right), \tag{3.4.15}$$

$$\frac{\partial^2 \ln l}{\partial \boldsymbol{\alpha}\, \partial \boldsymbol{\alpha}'} = -(XX' \otimes \Sigma_u^{-1}), \tag{3.4.16}$$

$$\frac{\partial^2 \ln l}{\partial \boldsymbol{\omega}\, \partial \boldsymbol{\omega}'} = \frac{T}{2}(\Sigma_u^{-1} \otimes \Sigma_u^{-1}) - \frac{1}{2}(\Sigma_u^{-1} \otimes \Sigma_u^{-1} UU'\Sigma_u^{-1})$$
$$- \frac{1}{2}(\Sigma_u^{-1} UU'\Sigma_u^{-1} \otimes \Sigma_u^{-1}), \tag{3.4.17}$$

where $\boldsymbol{\omega} = \text{vec}(\Sigma_u)$ (see Problem 3.3),

$$\frac{\partial^2 \ln l}{\partial \mu\, \partial \boldsymbol{\alpha}'} = -[I_K - (j' \otimes I_K)A']\Sigma_u^{-1} \sum_t Y_{t-1}^{0\prime} \otimes I_K$$
$$- \left[\sum_t u_t'\Sigma_u^{-1} \otimes I_K\right](I_K \otimes j' \otimes I_K)\frac{\partial\, \text{vec}(A')}{\partial \boldsymbol{\alpha}'} \tag{3.4.18}$$

(see Problem 3.4),

$$\frac{\partial^2 \ln l}{\partial \boldsymbol{\omega}\, \partial \mu'} = \frac{1}{2}(\Sigma_u^{-1} \otimes \Sigma_u^{-1})\left[(I_K \otimes U)\frac{\partial\, \text{vec}(U')}{\partial \mu'} + (U \otimes I_K)\frac{\partial\, \text{vec}(U)}{\partial \mu'}\right] \tag{3.4.19}$$

(see Problem 3.5), and

$$\frac{\partial^2 \ln l}{\partial \boldsymbol{\omega}\, \partial \boldsymbol{\alpha}'} = -\frac{1}{2}(\Sigma_u^{-1} \otimes \Sigma_u^{-1})\left[(I_K \otimes UX')\frac{\partial\, \text{vec}(A')}{\partial \boldsymbol{\alpha}'} + (UX' \otimes I_K)\right] \tag{3.4.20}$$

(see Problem 3.6).

It is obvious from (3.4.18), (3.4.19), and (3.4.20) that

$$\lim T^{-1} E\left(\frac{\partial^2 \ln l}{\partial \mu\, \partial \boldsymbol{\alpha}'}\right) = 0 \tag{3.4.21}$$

because $E(\sum_t Y_{t-1}^0/T) \to 0$, and

$$E\left(\frac{\partial^2 \ln l}{\partial \boldsymbol{\omega}\, \partial \mu'}\right) = 0 \tag{3.4.22}$$

because $E(U) = 0$ and $\partial \operatorname{vec}(U')/\partial \mu'$ is a matrix of constants. Moreover,

$$\lim T^{-1} E\left(\frac{\partial^2 \ln l}{\partial \boldsymbol{\omega}\, \partial \boldsymbol{\alpha}'}\right) = 0 \tag{3.4.23}$$

because $E(UX'/T) \to 0$. Thus, $\lim(I(\boldsymbol{\delta})/T)$ is block diagonal and we get the asymptotic distributions of μ, $\boldsymbol{\alpha}$, and $\boldsymbol{\sigma}$ as follows.

Multiplying minus the inverse of (3.4.15) by T gives the asymptotic covariance matrix of the ML estimator for the mean vector μ, that is,

$$\sqrt{T}(\tilde{\mu} - \mu) \xrightarrow{d} N\left(0, \left(I_K - \sum_{i=1}^{p} A_i\right)^{-1} \Sigma_u \left(I_K - \sum_{i=1}^{p} A_i\right)'^{-1}\right). \tag{3.4.24}$$

Hence, $\tilde{\mu}$ has the same asymptotic distribution as $\bar{y}$ (see Proposition 3.3). In other words, the two estimators for μ are asymptotically equivalent and, under the present conditions, this implies that $\bar{y}$ is asymptotically efficient because the ML estimator is asymptotically efficient. The asymptotic equivalence of $\tilde{\mu}$ and $\bar{y}$ can also be seen from (3.4.9) (see the argument prior to Proposition 3.3 and Problem 3.7).

Taking the limit of T^{-1} times the expectation of minus (3.4.16) gives $\Gamma_Y(0) \otimes \Sigma_u^{-1}$. Note that $E(XX'/T)$ is not strictly equal to $\Gamma_Y(0)$ because we have assumed fixed initial values $y_{-p+1}, \dots, y_0$. However, asymptotically, as T goes to infinity, the impact of the initial values vanishes. Thus we get

$$\sqrt{T}(\tilde{\boldsymbol{\alpha}} - \boldsymbol{\alpha}) \xrightarrow{d} N[0, \Gamma_Y(0)^{-1} \otimes \Sigma_u]. \tag{3.4.25}$$

Of course, this result also follows from the equivalence of the ML and LS estimators.

Noting that $E(UU') = T\Sigma_u$ it follows from (3.4.17) that

$$E\left(\frac{\partial^2 \ln l}{\partial \boldsymbol{\omega}\, \partial \boldsymbol{\omega}'}\right) = -\frac{T}{2}(\Sigma_u^{-1} \otimes \Sigma_u^{-1}). \tag{3.4.26}$$

Denoting by $\mathbf{D}_K$ the $(K^2 \times K(K+1)/2)$ duplication matrix (see Appendix A.12) so that $\boldsymbol{\omega} = \mathbf{D}_K \boldsymbol{\sigma}$ we get

$$\frac{\partial^2 \ln l}{\partial \boldsymbol{\sigma}\, \partial \boldsymbol{\sigma}'} = \frac{\partial \boldsymbol{\omega}'}{\partial \boldsymbol{\sigma}} \frac{\partial^2 \ln l}{\partial \boldsymbol{\omega}\, \partial \boldsymbol{\omega}'} \frac{\partial \boldsymbol{\omega}}{\partial \boldsymbol{\sigma}'} = \mathbf{D}_K' \frac{\partial^2 \ln l}{\partial \boldsymbol{\omega}\, \partial \boldsymbol{\omega}'} \mathbf{D}_K$$

and, hence,

$$\sqrt{T}(\tilde{\boldsymbol{\sigma}} - \boldsymbol{\sigma}) \xrightarrow{d} N(0, \Sigma_{\tilde{\sigma}}) \tag{3.4.27}$$

with

$$\begin{aligned} \Sigma_{\tilde{\sigma}} &= -TE\left(\frac{\partial^2 \ln l}{\partial \boldsymbol{\sigma}\, \partial \boldsymbol{\sigma}'}\right)^{-1} = 2[\mathbf{D}_K'(\Sigma_u^{-1} \otimes \Sigma_u^{-1})\mathbf{D}_K]^{-1} \\ &= 2\mathbf{D}_K^{+}(\Sigma_u \otimes \Sigma_u)\mathbf{D}_K^{+\prime}, \end{aligned} \tag{3.4.28}$$

where $\mathbf{D}_K^+ = (\mathbf{D}_K'\mathbf{D}_K)^{-1}\mathbf{D}_K'$ is the Moore-Penrose generalized inverse of the duplication matrix $\mathbf{D}_K$ and Rule 17 from Appendix A.12 has been used. In summary we get the following proposition.

PROPOSITION 3.4 (*Asymptotic Properties of* ML *Estimators*)

Let y_t be a stationary, stable Gaussian VAR(p) process as in (3.3.1). Then the ML estimators $\tilde{\mu}$, $\tilde{\boldsymbol{\alpha}}$, and $\tilde{\boldsymbol{\sigma}} = \text{vech}(\tilde{\Sigma}_u)$ given in (3.4.9)–(3.4.11) are consistent and

$$\sqrt{T}\begin{bmatrix} \tilde{\mu} - \mu \\ \tilde{\boldsymbol{\alpha}} - \boldsymbol{\alpha} \\ \tilde{\boldsymbol{\sigma}} - \boldsymbol{\sigma} \end{bmatrix} \xrightarrow{d} N\left[0, \begin{bmatrix} \left(I_K - \sum_i A_i\right)^{-1} \Sigma_u \left(I_K - \sum_i A_i\right)'^{-1} & 0 & 0 \\ 0 & \Gamma_Y(0)^{-1} \otimes \Sigma_u & 0 \\ 0 & 0 & 2\mathbf{D}_K^+(\Sigma_u \otimes \Sigma_u)\mathbf{D}_K^{+\prime} \end{bmatrix}\right] \tag{3.4.29}$$

so that $\tilde{\mu}$ is asymptotically independent of $\tilde{\boldsymbol{\alpha}}$ and $\tilde{\Sigma}_u$, and $\tilde{\boldsymbol{\alpha}}$ is asymptotically independent of $\tilde{\mu}$ and $\tilde{\Sigma}_u$. The covariance matrix may be estimated consistently by replacing the unknown quantities by their ML estimators and estimating $\Gamma_Y(0)$ by $\tilde{X}\tilde{X}'/T$. ■

In this section we have chosen to consider the mean adjusted form of a VAR(p) process. Of course, it is possible to perform a similar derivation for the standard form given in (3.1.1). In that case the ML estimators of ν and $\boldsymbol{\alpha}$ are not asymptotically independent though. Their joint asymptotic distribution is identical to that of $\hat{\boldsymbol{\beta}}$ given in Proposition 3.1. From Proposition 3.2 we know that the asymptotic distribution of $\tilde{\boldsymbol{\sigma}}$ remains unaltered. In the next section we will investigate the consequences of forecasting with estimated rather than known processes.

3.5 Forecasting with Estimated Processes

3.5.1 General Assumptions and Results

In Chapter 2, Section 2.2, we have seen that the optimal h-step forecast of the process (3.1.1) is

$$y_t(h) = \nu + A_1 y_t(h-1) + \cdots + A_p y_t(h-p), \tag{3.5.1}$$

where $y_t(j) = y_{t+j}$ for $j \leq 0$. If the true coefficients $B = (\nu, A_1, \ldots A_p)$ are replaced by estimators $\hat{B} = (\hat{\nu}, \hat{A}_1, \ldots, \hat{A}_p)$ we get a forecast

$$\hat{y}_t(h) = \hat{\nu} + \hat{A}_1 \hat{y}_t(h-1) + \cdots + \hat{A}_p \hat{y}_t(h-p), \tag{3.5.2}$$

where $\hat{y}_t(j) = y_{t+j}$ for $j \leq 0$. Thus the forecast error is

$$\begin{aligned} y_{t+h} - \hat{y}_t(h) &= [y_{t+h} - y_t(h)] + [y_t(h) - \hat{y}_t(h)] \\ &= \sum_{i=0}^{h-1} \Phi_i u_{t+h-i} + [y_t(h) - \hat{y}_t(h)], \end{aligned} \tag{3.5.3}$$

where the Φ_i are the coefficient matrices of the canonical MA representation of y_t (see (2.2.9)). Under quite general conditions for the process y_t the forecast errors can be shown to have zero mean, $E[y_{t+h} - \hat{y}_t(h)] = 0$, so that the forecasts are unbiased even if the coefficients are estimated. Since we do not need this result in the following we refer to Dufour (1985) for the details and a proof. All the u_s in the first term on the right-hand side of the last equality sign in (3.5.3) are attached to periods $s > t$ whereas all the y_s in the second term correspond to periods $s \le t$. Therefore, the two terms are uncorrelated. Hence the MSE matrix of the forecast $\hat{y}_t(h)$ is of the form

$$\begin{aligned} \Sigma_{\hat{y}}(h) &= \text{MSE}[\hat{y}_t(h)] = E[y_{t+h} - \hat{y}_t(h)][y_{t+h} - \hat{y}_t(h)]' \\ &= \Sigma_y(h) + \text{MSE}[y_t(h) - \hat{y}_t(h)], \end{aligned} \tag{3.5.4}$$

where

$$\Sigma_y(h) = \sum_{i=0}^{h-1} \Phi_i \Sigma_u \Phi_i'$$

(see (2.2.11)). In order to evaluate the last term in (3.5.4) the distribution of the estimator $\hat{B}$ is needed. Since we have not been able to derive the small sample distributions of the estimators considered in the previous sections but have derived the asymptotic distributions instead, we cannot hope for more than an asymptotic approximation to the MSE of $y_t(h) - \hat{y}_t(h)$. Such an approximation will be derived in the following.

There are two alternative assumptions that can be made in order to facilitate the derivation of the desired result:

(1) Only data up to the forecast origin are used for estimation.
(2) Estimation is done using a realization (time series) of a process that is independent of the process used for prediction and has the same stochastic structure (for instance, it is Gaussian and has the same first and second moments as the process used for prediction).

The first assumption is the more realistic one from a practical point of view because estimation and forecasting are usually based on the same data set. In that case, since the sample size is assumed to go to infinity in deriving asymptotic results, either the forecast origin has to go to infinity too or it has to be assumed that more and more data at the beginning of the sample become available. Since the forecast uses only p vectors y_s prior to the forecast period, these variables will be asymptotically independent of the estimator $\hat{B}$ (they are asymptotically negligible in comparison with all the other observations going into the estimate). Thus asymptotically the first assumption amounts to the same thing as the second one. In the following, for simplicity, the second assumption will therefore be supposed to hold. Furthermore it will be assumed that for $\boldsymbol{\beta} = \text{vec}(B)$ and

$\hat{\boldsymbol{\beta}} = \text{vec}(\hat{B})$ we have

$$\sqrt{T}(\hat{\boldsymbol{\beta}} - \boldsymbol{\beta}) \xrightarrow{d} N(0, \Sigma_{\hat{\boldsymbol{\beta}}}). \tag{3.5.5}$$

Samaranayake & Hasza (1988) and Basu & Sen Roy (1986) give a formal proof of the result that the MSE approximation obtained in the following remains valid under assumption (1) above.

With the foregoing assumptions it follows that, conditional on a particular realization $Y_t = (y_t', \ldots, y_{t-p+1}')'$ of the process used for prediction,

$$\sqrt{T}[\hat{y}_t(h) - y_t(h) | Y_t] \xrightarrow{d} N\left[0, \frac{\partial y_t(h)}{\partial \boldsymbol{\beta}'} \Sigma_{\hat{\boldsymbol{\beta}}} \frac{\partial y_t(h)'}{\partial \boldsymbol{\beta}}\right] \tag{3.5.6}$$

because $y_t(h)$ is a differentiable function of $\boldsymbol{\beta}$ (see Appendix C, Proposition C.4(3)). Here T is the sample size (time series length) used for estimation. This result suggests the approximation of $\text{MSE}[\hat{y}_t(h) - y_t(h)]$ by $\Omega(h)/T$, where

$$\Omega(h) = E\left[\frac{\partial y_t(h)}{\partial \boldsymbol{\beta}'} \Sigma_{\hat{\boldsymbol{\beta}}} \frac{\partial y_t(h)'}{\partial \boldsymbol{\beta}}\right]. \tag{3.5.7}$$

In fact, for a Gaussian process y_t,

$$\sqrt{T}[\hat{y}_t(h) - y_t(h)] \xrightarrow{d} N(0, \Omega(h)). \tag{3.5.8}$$

Hence we get an approximation

$$\Sigma_{\hat{y}}(h) = \Sigma_y(h) + \frac{1}{T}\Omega(h) \tag{3.5.9}$$

for the MSE matrix of $\hat{y}_t(h)$.

From (3.5.7) it is obvious that $\Omega(h)$ and thus the approximate MSE $\Sigma_{\hat{y}}(h)$ can be reduced by using an estimator that is asymptotically more efficient than $\hat{\boldsymbol{\beta}}$ if such an estimator exists. In other words, efficient estimation is of importance in order to reduce the forecast uncertainty.

3.5.2 The Approximate MSE Matrix

To derive an explicit expression for $\Omega(h)$, the derivatives $\partial y_t(h)/\partial \boldsymbol{\beta}'$ are needed. They can be obtained easily by noting that

$$y_t(h) = J_1 \mathbf{B}^h Z_t, \tag{3.5.10}$$

where $Z_t := (1, y_t', \ldots, y_{t-p+1}')'$,

$$\mathbf{B} := \begin{bmatrix} 1 & 0 & 0 & \ldots & 0 & 0 \\ \nu & A_1 & A_2 & \ldots & A_{p-1} & A_p \\ 0 & I_K & 0 & \ldots & 0 & 0 \\ 0 & 0 & I_K & & 0 & 0 \\ \vdots & \vdots & & \ddots & & \vdots \\ 0 & 0 & 0 & \ldots & I_K & 0 \end{bmatrix} = \begin{bmatrix} 1 & 0 & \ldots & 0 \\ & B & & \\ 0 & I_{K(p-1)} & & 0 \end{bmatrix} \quad [(Kp+1) \times (Kp+1)]$$

and

$$J_1 := [\underbrace{0}_{(K\times 1)} \quad I_K \quad \underbrace{0 \quad \dots \quad 0}_{(K\times K(p-1))}] \qquad [K \times (Kp+1)].$$

The relation (3.5.10) follows by induction (see Problem 3.8). Using (3.5.10) we get

$$\begin{aligned}
\frac{\partial y_t(h)}{\partial \boldsymbol{\beta}'} &= \frac{\partial\, \mathrm{vec}(J_1 \mathbf{B}^h Z_t)}{\partial \boldsymbol{\beta}'} = (Z_t' \otimes J_1)\frac{\partial\, \mathrm{vec}(\mathbf{B}^h)}{\partial \boldsymbol{\beta}'} \\
&= (Z_t' \otimes J_1)\left[\sum_{i=0}^{h-1} (\mathbf{B}')^{h-1-i} \otimes \mathbf{B}^i\right]\frac{\partial\, \mathrm{vec}(\mathbf{B})}{\partial \boldsymbol{\beta}'} \qquad \text{(Appendix A.13, Rule 8)} \\
&= (Z_t' \otimes J_1)\left[\sum_{i=0}^{h-1} (\mathbf{B}')^{h-1-i} \otimes \mathbf{B}^i\right](I_{Kp+1} \otimes J_1') \qquad \text{(see the definition of } \mathbf{B}) \\
&= \sum_{i=0}^{h-1} Z_t'(\mathbf{B}')^{h-1-i} \otimes J_1 \mathbf{B}^i J_1' \\
&= \sum_{i=0}^{h-1} Z_t'(\mathbf{B}')^{h-1-i} \otimes \Phi_i,
\end{aligned} \tag{3.5.11}$$

where $\Phi_i = J_1 \mathbf{B}^i J_1'$ follows as in (2.1.17). Using the LS estimator $\hat{\boldsymbol{\beta}}$ with asymptotic covariance matrix $\Sigma_{\hat{\boldsymbol{\beta}}} = \Gamma^{-1} \otimes \Sigma_u$ (see Proposition 3.1), the matrix $\Omega(h)$ becomes

$$\begin{aligned}
\Omega(h) &= E\left[\frac{\partial y_t(h)}{\partial \boldsymbol{\beta}'}(\Gamma^{-1} \otimes \Sigma_u)\frac{\partial y_t(h)'}{\partial \boldsymbol{\beta}}\right] \\
&= \sum_{i=0}^{h-1}\sum_{j=0}^{h-1} E(Z_t'(\mathbf{B}')^{h-1-i}\Gamma^{-1}\mathbf{B}^{h-1-j}Z_t) \otimes \Phi_i \Sigma_u \Phi_j' \\
&= \sum_i \sum_j E[\mathrm{tr}(Z_t'(\mathbf{B}')^{h-1-i}\Gamma^{-1}\mathbf{B}^{h-1-j}Z_t)]\Phi_i \Sigma_u \Phi_j' \\
&= \sum_i \sum_j \mathrm{tr}[(\mathbf{B}')^{h-1-i}\Gamma^{-1}\mathbf{B}^{h-1-j}E(Z_t Z_t')]\Phi_i \Sigma_u \Phi_j' \\
&= \sum_{i=0}^{h-1}\sum_{j=0}^{h-1} \mathrm{tr}[(\mathbf{B}')^{h-1-i}\Gamma^{-1}\mathbf{B}^{h-1-j}\Gamma]\Phi_i \Sigma_u \Phi_j',
\end{aligned} \tag{3.5.12}$$

provided y_t is stable so that

$$\Gamma := \mathrm{plim}(ZZ'/T) = E(Z_t Z_t').$$

Here $Z := (Z_0, \dots, Z_{T-1})$ is the $((Kp+1) \times T)$ matrix defined in (3.2.1).

For example, for $h = 1$,

$$\Omega(1) = (Kp+1)\Sigma_u.$$

Hence, the approximation

$$\Sigma_{\hat{y}}(1) = \Sigma_u + \frac{Kp+1}{T}\Sigma_u = \frac{T+Kp+1}{T}\Sigma_u \tag{3.5.13}$$

of the MSE matrix of the 1-step forecast with estimated coefficients is obtained. This expression shows that the contribution of the estimation variability to the forecast MSE matrix $\Sigma_{\hat{y}}(1)$ depends on the dimension K of the process, the VAR order p, and the sample size T used for estimation. It can be quite substantial if the sample size is small or moderate. For instance, considering a three-dimensional process of order 8 which is estimated from 15 years of quarterly data (i.e. $T = 52$ plus 8 presample values needed for LS estimation), the 1-step forecast MSE matrix Σ_u for known processes is inflated by a factor $(T + Kp + 1)/T = 1.48$. Of course, this approximation is derived from asymptotic theory so that its small sample validity is not guaranteed. We will take a closer look into this problem shortly. Obviously, the inflation factor $(T + Kp + 1)/T \to 1$ for $T \to \infty$. Thus the MSE contribution due to sampling variability vanishes if the sample size gets large. This result is a consequence of estimating the VAR coefficients consistently. An expression for $\Omega(h)$ can also be derived on the basis of the mean-adjusted form of the VAR process (see Problem 3.9).

In practice, for $h > 1$, it will not be possible to evaluate $\Omega(h)$ without knowing the AR coefficients summarized in the matrix B. A consistent estimator $\hat{\Omega}(h)$ may be obtained by replacing all unknown parameters by their LS estimators, that is, $\mathbf{B}$ is replaced by $\hat{\mathbf{B}}$ which is obtained by using $\hat{B}$ for B, Σ_u is replaced by $\hat{\Sigma}_u$, Φ_i is estimated by $\hat{\Phi}_i = J_1 \hat{\mathbf{B}}^i J_1'$, and Γ is estimated by $\hat{\Gamma} = ZZ'/T$. The resulting estimator of $\Sigma_{\hat{y}}(h)$ will be denoted by $\hat{\Sigma}_{\hat{y}}(h)$ in the following.

The foregoing discussion is of importance in setting up interval forecasts. Assuming y_t is Gaussian, an approximate $(1 - \alpha)100\%$ interval forecast h periods ahead for the k-th component $y_{k,t}$ of y_t is

$$\hat{y}_{k,t}(h) \pm z_{(\alpha/2)}\hat{\hat{\sigma}}_k(h) \tag{3.5.14}$$

or

$$[\hat{y}_{k,t}(h) - z_{(\alpha/2)}\hat{\hat{\sigma}}_k(h);\ \hat{y}_{k,t}(h) + z_{(\alpha/2)}\hat{\hat{\sigma}}_k(h)], \tag{3.5.15}$$

where $z_{(\alpha)}$ is the upper $\alpha 100$-th percentile of the standard normal distribution and $\hat{\hat{\sigma}}_k(h)$ is the square root of the k-th diagonal element of $\hat{\Sigma}_{\hat{y}}(h)$. Using Bonferroni's inequality, approximate joint confidence regions for a set of forecasts can be obtained just as described in Section 2.2.3 of Chapter 2.

3.5.3 An Example

To illustrate the previous results we consider again the investment/income/consumption example of Section 3.2.3. Using the VAR(2) model with the coefficient estimates given in (3.2.22) and

$$y_{T-1} = y_{72} = \begin{bmatrix} .02551 \\ .02434 \\ .01319 \end{bmatrix} \quad \text{and} \quad y_T = y_{73} = \begin{bmatrix} .03637 \\ .00517 \\ .00599 \end{bmatrix}$$

results in forecasts

$$\hat{y}_T(1) = \hat{\nu} + \hat{A}_1 y_T + \hat{A}_2 y_{T-1} = \begin{bmatrix} -.011 \\ .020 \\ .022 \end{bmatrix},$$

$$\hat{y}_T(2) = \hat{\nu} + \hat{A}_1 \hat{y}_T(1) + \hat{A}_2 y_T = \begin{bmatrix} .011 \\ .020 \\ .015 \end{bmatrix}, \quad (3.5.16)$$

and so on.

The estimated forecast MSE matrix for $h = 1$ is

$$\hat{\Sigma}_{\hat{y}}(1) = \frac{T + Kp + 1}{T} \hat{\Sigma}_u = \frac{73 + 6 + 1}{73} \hat{\Sigma}_u = \begin{bmatrix} 23.34 & .785 & 1.351 \\ .785 & 1.505 & .674 \\ 1.351 & .674 & .978 \end{bmatrix} \times 10^{-4}, \quad (3.5.17)$$

where $\hat{\Sigma}_u$ from (3.2.23) has been used. We need $\hat{\Phi}_1$ for evaluating

$$\hat{\Sigma}_{\hat{y}}(2) = \hat{\Sigma}_y(2) + \frac{1}{T}\hat{\Omega}(2),$$

where

$$\hat{\Sigma}_y(2) = \hat{\Sigma}_u + \hat{\Phi}_1 \hat{\Sigma}_u \hat{\Phi}_1'$$

and

$$\hat{\Omega}(2) = \sum_{i=0}^{1} \sum_{j=0}^{1} \operatorname{tr}[(\hat{\mathbf{B}}')^{1-i}(ZZ'/T)^{-1}\hat{\mathbf{B}}^{1-j}(ZZ'/T)]\hat{\Phi}_i \hat{\Sigma}_u \hat{\Phi}_j'$$

$$= \operatorname{tr}[\hat{\mathbf{B}}'(ZZ')^{-1}\hat{\mathbf{B}}ZZ']\hat{\Sigma}_u + \operatorname{tr}(\hat{\mathbf{B}}')\hat{\Sigma}_u\hat{\Phi}_1' + \operatorname{tr}(\hat{\mathbf{B}})\hat{\Phi}_1\hat{\Sigma}_u + \operatorname{tr}(I_{Kp+1})\hat{\Phi}_1\hat{\Sigma}_u\hat{\Phi}_1'.$$

From (2.1.22) we know that $\Phi_1 = A_1$. Hence we use $\hat{\Phi}_1 = \hat{A}_1$ from (3.2.22). Now we get

$$\hat{\Sigma}_y(2) = \begin{bmatrix} 23.67 & .547 & 1.226 \\ .547 & 1.488 & .554 \\ 1.226 & .554 & .952 \end{bmatrix} \times 10^{-4}$$

and

$$\hat{\Omega}(2) = \begin{bmatrix} 10.59 & .238 & .538 \\ .238 & .675 & .233 \\ .538 & .233 & .422 \end{bmatrix} \times 10^{-3}.$$

Consequently,

$$\hat{\Sigma}_{\hat{y}}(2) = \begin{bmatrix} 25.12 & .580 & 1.300 \\ .580 & 1.581 & .586 \\ 1.300 & .586 & 1.009 \end{bmatrix} \times 10^{-4}. \quad (3.5.18)$$

Assuming that the data are generated by a Gaussian process we get the following approximate 95% interval forecasts:

$$\begin{aligned}
&\hat{y}_{1,T}(1) \pm 1.96\hat{\hat{\sigma}}_1(1) \quad \text{or} \quad -.011 \pm .095,\\
&\hat{y}_{2,T}(1) \pm 1.96\hat{\hat{\sigma}}_2(1) \quad \text{or} \quad .020 \pm .024,\\
&\hat{y}_{3,T}(1) \pm 1.96\hat{\hat{\sigma}}_3(1) \quad \text{or} \quad .022 \pm .019,\\
&\hat{y}_{1,T}(2) \pm 1.96\hat{\hat{\sigma}}_1(2) \quad \text{or} \quad .011 \pm .098,\\
&\hat{y}_{2,T}(2) \pm 1.96\hat{\hat{\sigma}}_2(2) \quad \text{or} \quad .020 \pm .025,\\
&\hat{y}_{3,T}(2) \pm 1.96\hat{\hat{\sigma}}_3(2) \quad \text{or} \quad .015 \pm .020.
\end{aligned} \tag{3.5.19}$$

In Figure 3.3 some more forecasts of the three variables with two-standard error bounds to each side are depicted. The intervals indicated by the dashed bounds may be interpreted as approximate 95% forecast intervals for the individual forecasts. If the region enclosed by the dashed lines is viewed as a joint confidence region for all 4 forecasts a lower bound for the (approximate) probability content is $(100 - 4 \times 5)\% = 80\%$. In the figure it can be seen that for investment and income the actually observed values for 1979 ($t = 77 - 80$) are well inside the forecast regions whereas two of the four consumption values are outside that region.

3.5.4 A Small Sample Investigation

It is not obvious that the MSE and interval forecast approximations derived in the foregoing are reasonable in small samples because the MSE modification has been based on asymptotic theory. To investigate the small sample behavior of the predictor with estimated coefficients we have used again 1000 realizations of the bivariate VAR(2) process (3.2.25)/(3.2.26) of Section 3.2.4 and we have computed forecast intervals for the period following the last sample period. In Table 3.2 the numbers of actual values falling in these intervals are reported for sample sizes of $T = 30$ and 100.

Obviously, for $T = 30$, the theoretical and actual percentages are in best agreement if the approximate MSEs $\Sigma_{\hat{y}}(h)$ are used in setting up the forecast intervals. On the other hand, only forecast intervals based on $\hat{\Sigma}_y(h) = \sum_{i=0}^{h-1} \hat{\Phi}_i \hat{\Sigma}_u \hat{\Phi}_i'$ and $\hat{\Sigma}_{\hat{y}}(h)$ are feasible in practice when the actual process coefficients are unknown and have to be estimated. Comparing only the results based on these two MSE matrices shows that it pays to use the asymptotic approximation $\hat{\Sigma}_{\hat{y}}(h)$.

In Table 3.2 we also give the corresponding results for $T = 100$. Since the estimation uncertainty decreases with increasing sample size one would expect that now the theoretical and actual percentages are in good agreement for all MSEs. This is precisely what can be observed in the table. Nevertheless, even now the use of the MSE adjustment in $\hat{\Sigma}_{\hat{y}}(1)$ gives slightly more accurate interval forecasts.

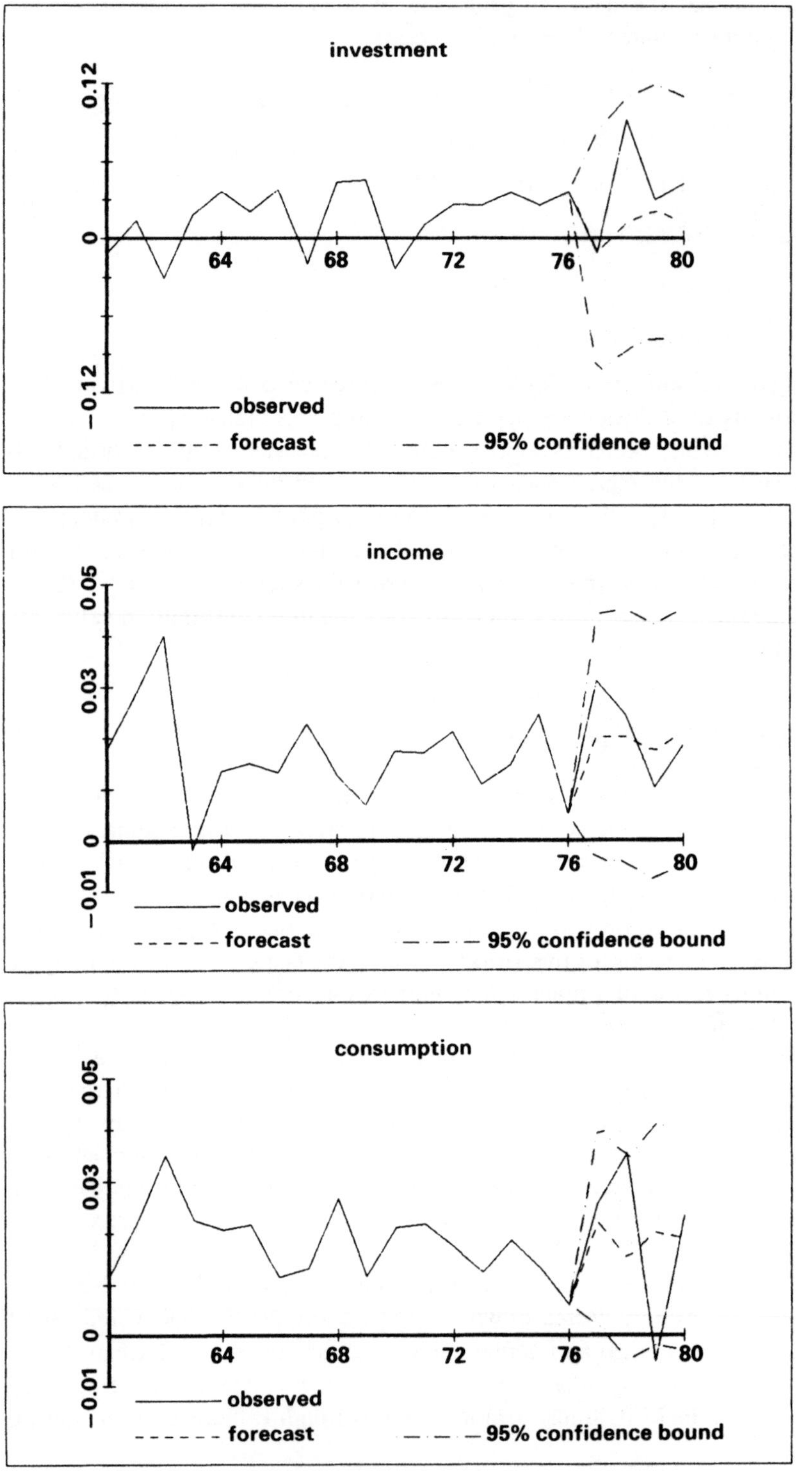

Fig. 3.3. Forecasts of the investment/income/consumption system.

Table 3.2. Accuracy of forecast intervals in small samples based on 1000 bivariate time series

MSE used in interval construction	% forecast interval	percent of actual values falling in the forecast interval			
		$T = 30$		$T = 100$	
		y_1	y_2	y_1	y_2
	90	86.5	85.7	89.7	89.4
$\Sigma_y(1)$	95	92.6	91.8	94.5	94.0
	99	98.1	98.0	99.0	98.5
	90	89.3	88.2	90.4	90.0
$\Sigma_{\hat{y}}(1)$	95	94.4	94.1	95.3	94.6
	99	99.0	98.4	99.3	98.8
	90	85.2	84.2	89.6	88.5
$\hat{\Sigma}_y(1)$	95	90.5	90.4	94.7	93.9
	99	98.4	96.5	98.9	98.3
	90	88.1	86.9	90.3	89.1
$\hat{\Sigma}_{\hat{y}}(1)$	95	93.4	92.7	95.2	94.0
	99	99.4	97.8	99.1	98.5

3.6 Testing for Granger-Causality and Instantaneous Causality

3.6.1 A Wald Test for Granger-Causality

In Chapter 2, Section 2.3.1, we have partitioned the VAR(p) process y_t in subprocesses z_t and x_t, that is, $y_t' = (z_t', x_t')$ and we have defined Granger-causality from x_t to z_t and vice versa. We have seen that this type of causality can be characterized by specific zero constraints on the VAR coefficients (see Corollary 2.2.1). Thus, in an estimated VAR(p) system, if we want to test for Granger-causality we need to test zero constraints for the coefficients. Given the results of Sections 3.2, 3.3, and 3.4 it is straightforward to derive *asymptotic* tests of such constraints.

More generally we consider testing

$$H_0\colon C\boldsymbol{\beta} = c \quad \text{against} \quad H_1\colon C\boldsymbol{\beta} \neq c, \tag{3.6.1}$$

where C is an $(N \times (K^2p + K))$ matrix of rank N and c is an $(N \times 1)$ vector. Assuming that

$$\sqrt{T}(\hat{\boldsymbol{\beta}} - \boldsymbol{\beta}) \xrightarrow{d} N(0, \Gamma^{-1} \otimes \Sigma_u) \tag{3.6.2}$$

as in LS/ML estimation, we get

$$\sqrt{T}(C\hat{\boldsymbol{\beta}} - C\boldsymbol{\beta}) \xrightarrow{d} N[0, C(\Gamma^{-1} \otimes \Sigma_u)C'] \tag{3.6.3}$$

(see Appendix C, Proposition C.4) and hence

$$T(C\hat{\boldsymbol{\beta}} - c)'[C(\Gamma^{-1} \otimes \Sigma_u)C']^{-1}(C\hat{\boldsymbol{\beta}} - c) \xrightarrow{d} \chi^2(N). \tag{3.6.4}$$

This statistic is the *Wald statistic* (see Appendix C.5).

Replacing Γ and Σ_u by their usual estimators $\hat{\Gamma} = ZZ'/T$ and $\hat{\Sigma}_u$ as given in (3.2.19) the resulting statistic

$$\lambda_w = (C\hat{\boldsymbol{\beta}} - c)'[C((ZZ')^{-1} \otimes \hat{\Sigma}_u)C']^{-1}(C\hat{\boldsymbol{\beta}} - c) \tag{3.6.5}$$

still has an asymptotic χ^2-distribution with N degrees of freedom provided y_t satisfies the conditions of Proposition 3.2 because under these conditions $[C((ZZ')^{-1} \otimes \hat{\Sigma}_u)C']^{-1}/T$ is a consistent estimator of $[C(\Gamma^{-1} \otimes \Sigma_u)C']^{-1}$. Hence we have the following result.

PROPOSITION 3.5 (*Asymptotic Distribution of the Wald Statistic*)

Suppose (3.6.2) holds. Furthermore, $\text{plim}(ZZ'/T) = \Gamma$, $\text{plim}\ \hat{\Sigma}_u = \Sigma_u$ are both nonsingular and H_0: $C\boldsymbol{\beta} = c$ is true, with C being an $(N \times (K^2p + K))$ matrix of rank N. Then

$$\lambda_w = (C\hat{\boldsymbol{\beta}} - c)'[C((ZZ')^{-1} \otimes \hat{\Sigma}_u)C']^{-1}(C\hat{\boldsymbol{\beta}} - c) \xrightarrow{d} \chi^2(N). \qquad \blacksquare$$

In practice it may be useful to make some adjustment to the statistic or the critical values of the test in order to compensate for the fact that the matrix $\Gamma^{-1} \otimes \Sigma_u$ is unknown and has been replaced by an estimator. Working in that direction we note that

$$NF(N, T) \xrightarrow[T\to\infty]{d} \chi^2(N), \tag{3.6.6}$$

where $F(N, T)$ denotes an F random variable with N and T degrees of freedom (d.f.) (Appendix C, Proposition C.3). Since an $F(N, T)$-distribution has a fatter tail than the $\chi^2(N)$-distribution divided by N, it seems reasonable to consider the test statistic

$$\lambda_F = \lambda_w/N \tag{3.6.7}$$

in conjunction with critical values from some F-distribution. The question is then what numbers of degrees of freedom should be used? From the foregoing discussion it is plausible to use N as the numerator degrees of freedom. On the other hand, any sequence that goes to infinity with the sample size qualifies as a candidate for the denominator d.f.. The usual F-statistic for a regression model with nonstochastic regressors has denominator d.f. equal to the sample size minus the number of estimated parameters. Therefore we may use this number here too. Note that in the model (3.2.3) we have a vector $\mathbf{y}$ with KT observations and $\boldsymbol{\beta}$ contains $K(Kp + 1)$ parameters. Alternatively, we will argue shortly that $T - Kp - 1$ is also a reasonable number for the denominator d.f.. Hence, we have the approximate distributions

$$\lambda_F \approx F(N, KT - K^2p - K) \approx F(N, T - Kp - 1). \tag{3.6.8}$$

3.6.2 An Example

To see how this result can be used in a test for Granger-causality let us consider again our example system from Section 3.2.3. The null hypothesis of no Granger-causality from income/consumption (y_2, y_3) to investment (y_1) may be expressed in terms of the coefficients of the VAR(2) process as

$$H_0: \alpha_{12,1} = \alpha_{13,1} = \alpha_{12,2} = \alpha_{13,2} = 0. \tag{3.6.9}$$

This null hypothesis may be written as in (3.6.1) by defining the (4×1) vector $c = 0$ and the (4×21) matrix

$$C = \begin{bmatrix} 0\,0\,0 & 0\,0\,0 & 1\,0\,0 & 0\,0\,0 & 0\,0\,0 & 0\,0\,0 & 0\,0\,0 \\ 0\,0\,0 & 0\,0\,0 & 0\,0\,0 & 1\,0\,0 & 0\,0\,0 & 0\,0\,0 & 0\,0\,0 \\ 0\,0\,0 & 0\,0\,0 & 0\,0\,0 & 0\,0\,0 & 0\,0\,0 & 1\,0\,0 & 0\,0\,0 \\ 0\,0\,0 & 0\,0\,0 & 0\,0\,0 & 0\,0\,0 & 0\,0\,0 & 0\,0\,0 & 1\,0\,0 \end{bmatrix}.$$

With this notation, using the estimation results from Section 3.2.3,

$$\lambda_F = \hat{\boldsymbol{\beta}}'C'[C((ZZ')^{-1} \otimes \hat{\Sigma}_u)C']^{-1}C\hat{\boldsymbol{\beta}}/4 = 1.59. \tag{3.6.10}$$

In contrast, the 95th percentile of the $F(4, 3 \cdot 73 - 9 \cdot 2 - 3) = F(4, 198) \approx F(4, 73 - 3 \cdot 2 - 1) = F(4, 66)$-distribution is about 2.5. Thus, in a 5% level test we cannot reject Granger-noncausality from income/consumption to investment.

In this example the denominator d.f. are so large (namely 198 or 66) that we could just as well use λ_w in conjunction with a critical value from a $\chi^2(4)$-distribution. The 95th percentile of that distribution is 9.49 and thus it is about four times that of the F-test while $\lambda_w = 4\lambda_F$.

In an example of this type it is quite reasonable to use $T - Kp - 1$ denominator d.f. for the F-test because all the restrictions are imposed on coefficients from one equation. Therefore λ_F actually reduces to an F-statistic related to one equation with $Kp + 1$ parameters which are estimated from T observations. The use of $T - Kp - 1$ d.f. may also be justified by arguments that do not rely on the restrictions being imposed on the parameters of one equation only, namely by appeal to the similarity between the λ_F statistic and Hotelling's T^2 (e.g., Anderson (1984)).

Many other tests for Granger-causality have been proposed and investigated (see e.g. Geweke, Meese & Dent (1983)). In the next chapter we will return to the testing of hypotheses and then an alternative test will be considered.

3.6.3 Testing for Instantaneous Causality

Tests for instantaneous causality can be developed in the same way as tests for Granger-causality because instantaneous causality can be expressed in terms of zero restrictions for $\boldsymbol{\sigma} = \text{vech}(\Sigma_u)$ (see Proposition 2.3). If y_t is a stable Gaussian VAR(p) process and we wish to test

$$H_0: C\boldsymbol{\sigma} = 0 \quad \text{against} \quad H_1: C\boldsymbol{\sigma} \neq 0 \tag{3.6.11}$$

we may use the asymptotic distribution of the ML estimator given in Proposition 3.4 to set up the Wald statistic

$$\lambda_w = T\tilde{\boldsymbol{\sigma}}'C'[2C\mathbf{D}_K^+(\tilde{\Sigma}_u \otimes \tilde{\Sigma}_u)\mathbf{D}_K^{+\prime}C']^{-1}C\tilde{\boldsymbol{\sigma}}, \tag{3.6.12}$$

where $\mathbf{D}_K^+$ is the Moore-Penrose inverse of the duplication matrix $\mathbf{D}_K$ and C is an $(N \times K(K+1)/2)$ matrix of rank N. Under H_0, λ_w has an asymptotic χ^2-distribution with N degrees of freedom.

Alternatively a Wald test of (3.6.11) could be based on the lower triangular matrix P which is obtained from a Choleski decomposition of Σ_u. Noting that instantaneous noncausality implies zero elements of Σ_u that correspond to zero elements of P we can write H_0 from (3.6.11) equivalently as

$$H_0: C \operatorname{vech}(P) = 0. \tag{3.6.13}$$

Since $\operatorname{vech}(P)$ is a continuously differentiable function of $\boldsymbol{\sigma}$ the asymptotic distribution of the estimator P obtained from decomposing $\tilde{\Sigma}_u$ follows from Proposition C.4(3) of Appendix C:

$$\sqrt{T}\operatorname{vech}(\tilde{P} - P) \xrightarrow{d} N(0, \bar{H}\Sigma_{\tilde{\sigma}}\bar{H}'), \tag{3.6.14}$$

where

$$\bar{H} = \frac{\partial \operatorname{vech}(P)}{\partial \boldsymbol{\sigma}'} = [\mathbf{L}_K(I_{K^2} + \mathbf{K}_{KK})(P \otimes I_K)\mathbf{L}_K']^{-1}$$

(see Appendix A.13, Rule 10). Here $\mathbf{K}_{mn}$ is the commutation matrix defined such that $\operatorname{vec}(G) = \mathbf{K}_{mn}\operatorname{vec}(G')$ for any $(m \times n)$ matrix G and $\mathbf{L}_K$ is the $(K(K+1)/2 \times K^2)$ elimination matrix defined such that $\operatorname{vech}(F) = \mathbf{L}_K\operatorname{vec}(F)$ for any $(K \times K)$ matrix F (see Appendix A.12.2). A Wald test of (3.6.13) may therefore be based on

$$\lambda_w = T\operatorname{vech}(\tilde{P})'C'[C\hat{\bar{H}}\hat{\Sigma}_{\tilde{\sigma}}\hat{\bar{H}}'C']^{-1}C\operatorname{vech}(\tilde{P}) \xrightarrow{d} \chi^2(N), \tag{3.6.15}$$

where carets denote the usual estimators. Although the two tests based on $\tilde{\boldsymbol{\sigma}}$ and $\tilde{P}$ are asymptotically equivalent they may differ in small samples. Of course, in the previous discussion we may replace $\tilde{\Sigma}_u$ by the asymptotically equivalent estimator $\hat{\Sigma}_u$.

In our investment/income/consumption example, suppose we wish to test for instantaneous causality from (income, consumption) to investment (or vice versa). Following Proposition 2.3, the null hypothesis of no causality is

$$H_0: \sigma_{21} = \sigma_{31} = 0 \qquad \text{or} \qquad C\boldsymbol{\sigma} = 0,$$

where σ_{ij} is a typical element of Σ_u and

$$C = \begin{bmatrix} 0 & 1 & 0 & 0 & 0 & 0 \\ 0 & 0 & 1 & 0 & 0 & 0 \end{bmatrix}.$$

For this hypothesis the test statistic in (3.6.12) assumes the value $\lambda_w = 5.46$. Equivalently we may test

$$H_0: p_{21} = p_{31} = 0 \qquad \text{or} \qquad C\operatorname{vech}(P) = 0,$$

where p_{ij} is a typical element of P. The corresponding value of the test statistic from (3.6.15) is $\lambda_w = 5.70$. Both tests are based on asymptotic $\chi^2(2)$-distributions and therefore do not reject the null hypotheses of no instantaneous causality at a 5% level. Note that the critical value for a 5% level test is 5.99.

3.7 The Asymptotic Distributions of Impulse Responses and Forecast Error Variance Decompositions

3.7.1 The Main Results

In Chapter 2, Section 2.3.2, we have seen that the coefficients of the MA representations

$$y_t = \mu + \sum_{i=0}^{\infty} \Phi_i u_{t-i}, \qquad \Phi_0 = I_K, \tag{3.7.1}$$

and

$$y_t = \mu + \sum_{i=0}^{\infty} \Theta_i w_{t-i} \tag{3.7.2}$$

are sometimes interpreted as impulse responses or dynamic multipliers of the system y_t. Here $\mu = E(y_t)$, the $\Theta_i = \Phi_i P$, $w_t = P^{-1}u_t$, and P is the lower triangular Choleski decomposition of Σ_u such that $\Sigma_u = PP'$. Hence $\Sigma_w = E(w_t w_t') = I_K$. In this section we will assume that the Φ_i and Θ_i are unknown and are estimated from the estimated VAR coefficients and error covariance matrix. We will derive the asymptotic distributions of the resulting estimated Φ_i and Θ_i. In these derivations we will not need the existence of MA representations (3.7.1) and (3.7.2). We will just assume that the Φ_i are obtained from given coefficient matrices $A_1, \ldots, A_p$ by recursions

$$\Phi_i = \sum_{j=1}^{i} \Phi_{i-j} A_j, \qquad i = 1, 2, \ldots$$

starting with $\Phi_0 = I_K$ and setting $A_j = 0$ for $j > p$. Furthermore the Θ_i are obtained from $A_1, \ldots, A_p$, and Σ_u as $\Theta_i = \Phi_i P$ where P is as specified above. In addition, the asymptotic distributions of the corresponding accumulated responses

$$\Psi_n = \sum_{i=0}^{n} \Phi_i, \qquad \Psi_\infty = \sum_{i=0}^{\infty} \Phi_i = (I_K - A_1 - \cdots - A_p)^{-1} \qquad \text{(if it exists)},$$

$$\Xi_n = \sum_{i=0}^{n} \Theta_i, \qquad \Xi_\infty = \sum_{i=0}^{\infty} \Theta_i = (I_K - A_1 - \cdots - A_p)^{-1} P \qquad \text{(if it exists)},$$

and the forecast error variance components

$$\omega_{jk,h} = \sum_{i=0}^{h-1} (e_j' \Theta_i e_k)^2 / \text{MSE}_j(h) \tag{3.7.3}$$

will be given. Here e_k is the k-th column of I_K and

$$\text{MSE}_j(h) = \sum_{i=0}^{h-1} e_j' \Phi_i \Sigma_u \Phi_i' e_j$$

is the j-th diagonal element of the MSE matrix $\Sigma_y(h)$ of an h-step forecast (see Chapter 2, Section 2.2.2).

The derivation of the asymptotic distributions is based on the following result from Appendix C, Proposition C.4(3). Suppose $\boldsymbol{\beta}$ is an $(n \times 1)$ vector of parameters and $\hat{\boldsymbol{\beta}}$ is an estimator such that

$$\sqrt{T}(\hat{\boldsymbol{\beta}} - \boldsymbol{\beta}) \xrightarrow{d} N(0, \Sigma_{\boldsymbol{\beta}}),$$

where T, as usual, denotes the sample size (time series length) used for estimation. Let $g(\boldsymbol{\beta})$ be a continuously differentiable function with values in m-dimensional Euclidean space and $\partial g_i/\partial \boldsymbol{\beta}' = (\partial g_i/\partial \beta_j)$ is nonzero at the true vector $\boldsymbol{\beta}$ for $i = 1, \dots, m$. Then

$$\sqrt{T}[g(\hat{\boldsymbol{\beta}}) - g(\boldsymbol{\beta})] \xrightarrow{d} N\left(0, \frac{\partial g}{\partial \boldsymbol{\beta}'} \Sigma_{\boldsymbol{\beta}} \frac{\partial g'}{\partial \boldsymbol{\beta}}\right).$$

In writing down the asymptotic distributions formally we use the notation

$$\boldsymbol{\alpha} := \text{vec}(A_1, \dots, A_p) \qquad (K^2 p \times 1),$$

$$\mathbf{A} := \begin{bmatrix} A_1 & A_2 & \dots & A_{p-1} & A_p \\ I_K & 0 & \dots & 0 & 0 \\ 0 & I_K & & 0 & 0 \\ \vdots & & \ddots & \vdots & \vdots \\ 0 & 0 & \dots & I_K & 0 \end{bmatrix} \qquad (Kp \times Kp),$$

$$\boldsymbol{\sigma} := \text{vech}(\Sigma_u) \qquad (K(K+1)/2 \times 1)$$

and the corresponding estimators are furnished with a caret. As before vec denotes the column stacking operator and vech is the corresponding operator that stacks the elements on and below the diagonal only. We also use the commutation matrix $\mathbf{K}_{mn}$ defined such that, for any $(m \times n)$ matrix G, $\mathbf{K}_{mn}\text{vec}(G) = \text{vec}(G')$, the $(m^2 \times m(m+1)/2)$ duplication matrix $\mathbf{D}_m$ defined such that $\mathbf{D}_m\text{vech}(F) = \text{vec}(F)$ for any symmetric $(m \times m)$ matrix F, and the $(m(m+1)/2 \times m^2)$ elimination matrix $\mathbf{L}_m$ defined such that, for any $(m \times m)$ matrix F, $\text{vech}(F) = \mathbf{L}_m\text{vec}(F)$ (see Appendix A.12.2). Furthermore $J := [I_K \;\; 0 \dots 0]$ is a $(K \times Kp)$ matrix. With this notation the following proposition from Lütkepohl (1990a) can be stated.

PROPOSITION 3.6 (*Asymptotic Distributions of Impulse Responses*)

Suppose

$$\sqrt{T}\begin{bmatrix} \hat{\boldsymbol{\alpha}} - \boldsymbol{\alpha} \\ \hat{\boldsymbol{\sigma}} - \boldsymbol{\sigma} \end{bmatrix} \xrightarrow{d} N\left[0, \begin{bmatrix} \Sigma_{\hat{\boldsymbol{\alpha}}} & 0 \\ 0 & \Sigma_{\hat{\boldsymbol{\sigma}}} \end{bmatrix}\right]. \tag{3.7.4}$$

Then

$$\sqrt{T}\,\mathrm{vec}(\hat{\Phi}_i - \Phi_i) \xrightarrow{d} N(0, G_i \Sigma_{\hat{\alpha}} G_i'), \qquad i = 1, 2, \ldots, \tag{3.7.5}$$

where

$$G_i := \frac{\partial\,\mathrm{vec}(\Phi_i)}{\partial \boldsymbol{\alpha}'} = \sum_{m=0}^{i-1} J(\mathbf{A}')^{i-1-m} \otimes \Phi_m.$$

$$\sqrt{T}\,\mathrm{vec}(\hat{\Psi}_n - \Psi_n) \xrightarrow{d} N(0, F_n \Sigma_{\hat{\alpha}} F_n'), \qquad n = 1, 2, \ldots, \tag{3.7.6}$$

where $F_n := G_1 + \cdots + G_n$.

If $(I_K - A_1 - \cdots - A_p)$ is nonsingular,

$$\sqrt{T}\,\mathrm{vec}(\hat{\Psi}_\infty - \Psi_\infty) \xrightarrow{d} N(0, F_\infty \Sigma_{\hat{\alpha}} F_\infty'), \tag{3.7.7}$$

where $F_\infty := \underbrace{(\Psi_\infty', \ldots, \Psi_\infty')}_{p\ \text{times}} \otimes \Psi_\infty$.

$$\sqrt{T}\,\mathrm{vec}(\hat{\Theta}_i - \Theta_i) \xrightarrow{d} N(0, C_i \Sigma_{\hat{\alpha}} C_i' + \bar{C}_i \Sigma_{\hat{\sigma}} \bar{C}_i'), \qquad i = 0, 1, 2, \ldots, \tag{3.7.8}$$

where

$$C_0 := 0, \quad C_i := (P' \otimes I_K) G_i, \quad i = 1, 2, \ldots, \quad \bar{C}_i := (I_K \otimes \Phi_i) H, \quad i = 0, 1, \ldots,$$

and

$$H := \frac{\partial\,\mathrm{vec}(P)}{\partial \boldsymbol{\sigma}'} = \mathbf{L}_K'\{\mathbf{L}_K[(I_K \otimes P)\mathbf{K}_{KK} + (P \otimes I_K)]\mathbf{L}_K'\}^{-1}$$

$$= \mathbf{L}_K'\{\mathbf{L}_K(I_{K^2} + \mathbf{K}_{KK})(P \otimes I_K)\mathbf{L}_K'\}^{-1}.$$

$$\sqrt{T}\,\mathrm{vec}(\hat{\Xi}_n - \Xi_n) \xrightarrow{d} N(0, B_n \Sigma_{\hat{\alpha}} B_n' + \bar{B}_n \Sigma_{\hat{\sigma}} \bar{B}_n'), \tag{3.7.9}$$

where $B_n := (P' \otimes I_K) F_n$ and $\bar{B}_n := (I_K \otimes \Psi_n) H$.

If $(I_K - A_1 - \cdots - A_p)$ is nonsingular,

$$\sqrt{T}\,\mathrm{vec}(\hat{\Xi}_\infty - \Xi_\infty) \xrightarrow{d} N(0, B_\infty \Sigma_{\hat{\alpha}} B_\infty' + \bar{B}_\infty \Sigma_{\hat{\sigma}} \bar{B}_\infty'), \tag{3.7.10}$$

where $B_\infty := (P' \otimes I_K) F_\infty$ and $\bar{B}_\infty := (I_K \otimes \Psi_\infty) H$.

Finally,

$$\sqrt{T}(\hat{\omega}_{jk,h} - \omega_{jk,h}) \xrightarrow{d} N(0, d_{jk,h} \Sigma_{\hat{\alpha}} d_{jk,h}' + \bar{d}_{jk,h} \Sigma_{\hat{\sigma}} \bar{d}_{jk,h}')$$
$$j, k = 1, \ldots, K;\ h = 1, 2, \ldots, \tag{3.7.11}$$

where

$$d_{jk,h} := \frac{2}{\mathrm{MSE}_j(h)^2} \sum_{i=0}^{h-1} [\mathrm{MSE}_j(h)(e_j' \Phi_i P e_k)(e_k' P' \otimes e_j') G_i$$
$$- (e_j' \Phi_i P e_k)^2 \sum_{m=0}^{h-1} (e_j' \Phi_m \Sigma_u \otimes e_j') G_m]$$

with $G_0 := 0$ and

$$\bar{d}_{jk,h} := \sum_{i=0}^{h-1} [2\,\mathrm{MSE}_j(h)(e_j'\Phi_i P e_k)(e_k' \otimes e_j'\Phi_i)H$$

$$- (e_j'\Phi_i P e_k)^2 \sum_{m=0}^{h-1} (e_j'\Phi_m \otimes e_j'\Phi_m)\mathbf{D}_K]/\mathrm{MSE}_j(h)^2. \qquad ■$$

In the next subsection the proof of the proposition is indicated. Some remarks are worthwhile now.

REMARK 1: In the proposition some matrices of partial derivatives may be zero. For instance, if a VAR(1) model is fitted although the true order is zero, that is, y_t is white noise, then

$$G_2 = J\mathbf{A}' \otimes I_K + JI_K \otimes \Phi_1 = 0$$

since $\mathbf{A} = A_1 = 0$ and $\Phi_1 = A_1 = 0$. Hence, a degenerate asymptotic distribution with zero covariance matrix is obtained for $\sqrt{T}(\hat{\Phi}_2 - \Phi_2)$. As explained in Appendix B we call such a distribution also multivariate normal. Otherwise it would be necessary to distinguish between cases with zero and nonzero partial derivatives or we have to assume that all partial derivatives are such that the covariance matrices have no zeroes on the diagonal. Note that estimators of the covariance matrices obtained by replacing unknown quantities by their usual estimators may be problematic when the asymptotic distribution is degenerate. In that case the usual t-ratios and confidence intervals may not be appropriate. ■

REMARK 2: In the proposition it is not explicitly assumed that y_t is stable. While the stability condition is partly introduced in (3.7.7) and (3.7.10) by requiring that $(I_K - A_1 - \cdots - A_p)$ be nonsingular so that

$$\det(I_K - A_1 z - \cdots - A_p z^p) \neq 0 \qquad \text{for } z = 1,$$

it is not needed for the other results to hold. The crucial condition is the asymptotic distribution of the process parameters in (3.7.4). Although we have used the stationarity and stability assumptions in Sections 3.2–3.4 in order to derive the asymptotic distribution of the process parameters, we will see in later chapters that asymptotic normality is also obtained for certain nonstationary, unstable processes. Therefore, at least parts of Proposition 3.6 will be useful in a nonstationary environment. ■

REMARK 3: The block-diagonal structure of the covariance matrix of the asymptotic distribution in (3.7.4) is in no way essential for the asymptotic normality of the impulse responses. In fact, the asymptotic distributions in (3.7.5)–(3.7.7) remain unchanged if the asymptotic covariance matrix of the parameter estimators is not block-diagonal. On the other hand, without the block-diagonal structure the simple additive structure of the asymptotic covariance matrices in (3.7.8)–(3.7.11) is lost. Although these asymptotic distributions are easily generalizable to the case of a general asymptotic covariance matrix of the VAR coefficients in (3.7.4) we have not stated the more general result here because it is not needed in subsequent chapters of this text. The slightly simpler version of the proposition is therefore preferred here. ■

REMARK 4: Under the conditions of Proposition 3.4 the covariance matrix of the asymptotic distribution of the parameters has precisely the block-diagonal structure required in (3.7.4) with

$$\Sigma_{\hat{\alpha}} = \Gamma_Y(0)^{-1} \otimes \Sigma_u$$

and

$$\Sigma_{\hat{\sigma}} = 2\mathbf{D}_K^+(\Sigma_u \otimes \Sigma_u)\mathbf{D}_K^{+\prime},$$

where $\mathbf{D}_K^+ = (\mathbf{D}_K'\mathbf{D}_K)^{-1}\mathbf{D}_K'$ is the Moore-Penrose inverse of the duplication matrix $\mathbf{D}_K$. Using these expressions in the proposition some simplifications of the covariance matrices can be obtained. For instance, the covariance matrix in (3.7.5) becomes

$$\begin{aligned} G_i \Sigma_{\hat{\alpha}} G_i' &= \left[\sum_{m=0}^{i-1} J(\mathbf{A}')^{i-1-m} \otimes \Phi_m\right](\Gamma_Y(0)^{-1} \otimes \Sigma_u)\left[\sum_{n=0}^{i-1} J(\mathbf{A}')^{i-1-n} \otimes \Phi_n\right]' \\ &= \sum_{m=0}^{i-1}\sum_{n=0}^{i-1} [J(\mathbf{A}')^{i-1-m}\Gamma_Y(0)^{-1}\mathbf{A}^{i-1-n}J'] \otimes [\Phi_m \Sigma_u \Phi_n'] \end{aligned}$$

which is computationally convenient since all matrices involved are of a relatively small size. The advantage of the general formulation is that it can be used with other $\Sigma_{\hat{\alpha}}$ matrices as well. We will see examples in subsequent chapters. ■

REMARK 5: In practice the unknown quantities in the covariance matrices in Proposition 3.6 may be replaced by their usual estimates given in Sections 3.2–3.4 for the case of a stationary, stable process y_t (see, however, Remark 1). ■

REMARK 6: Summing the forecast error variance components over k,

$$\sum_{k=1}^{K} \omega_{jk,h} = \sum_{k=1}^{K} \hat{\omega}_{jk,h} = 1$$

for each j and h. These restrictions are not taken into account in the derivation of the asymptotic distributions in (3.7.11). It is easily checked, however, that for dimension $K = 1$ the standard errors obtained from Proposition 3.6 are zero as they should be, because all forecast error variance components are 1 in that case. A problem in this context is that the asymptotic distribution of $\hat{\omega}_{jk,h}$ cannot be used in the usual way for tests of significance and setting up confidence intervals if $\omega_{jk,h} = 0$. In that case, from the definition of $d_{jk,h}$ and $\bar{d}_{jk,h}$, the variance of the asymptotic distribution is easily seen to be zero and, hence, estimating this quantity by replacing unknown parameters by their usual estimators may lead to t-ratios with nonnormal asymptotic distributions which cannot be used in the usual way. This state of affairs is unfortunate from a practical point of view because testing the significance of forecast error variance components is of particular interest in practice. Note, however, that

$$\omega_{jk,h} = 0 \quad \Leftrightarrow \quad \theta_{jk,i} = 0 \qquad \text{for } i = 0, \dots, h.$$

A Wald test of the latter hypothesis may be possible. ■

REMARK 7: Joint confidence regions and test statistics for testing hypotheses that involve several of the response coefficients can be obtained from Proposition 3.6 in the usual way. However, it has to be taken into account that, for instance, the elements of $\hat{\Phi}_i$ and $\hat{\Phi}_j$ will not be independent asymptotically. If elements from two or more MA matrices are involved the joint distribution of all the matrices must be determined. This distribution can be derived easily from the results given in the proposition. For instance, the covariance matrix of the joint asymptotic distribution of $\text{vec}(\hat{\Phi}_i, \hat{\Phi}_j)$ is

$$\frac{\partial \operatorname{vec}(\Phi_i, \Phi_j)}{\partial \boldsymbol{\alpha}'} \Sigma_{\hat{\alpha}} \frac{\partial \operatorname{vec}(\Phi_i, \Phi_j)'}{\partial \boldsymbol{\alpha}},$$

where

$$\frac{\partial \operatorname{vec}(\Phi_i, \Phi_j)}{\partial \boldsymbol{\alpha}'} = \begin{bmatrix} \dfrac{\partial \operatorname{vec}(\Phi_i)}{\partial \boldsymbol{\alpha}'} \\ \dfrac{\partial \operatorname{vec}(\Phi_j)}{\partial \boldsymbol{\alpha}'} \end{bmatrix}$$

etc.. We have chosen to state the proposition for individual MA coefficient matrices because that way all required matrices have relatively small dimensions and hence are easy to compute. ■

REMARK 8: Denoting the jk-th elements of Φ_i and Θ_i by $\phi_{jk,i}$ and $\theta_{jk,i}$, respectively, hypotheses of obvious interest, for $j \neq k$, are

$$H_0: \phi_{jk,i} = 0 \qquad \text{for } i = 1, 2, \dots \tag{3.7.12}$$

and

$$H_0: \theta_{jk,i} = 0 \qquad \text{for } i = 0, 1, 2, \dots \tag{3.7.13}$$

because they can be interpreted as hypotheses on noncausality from variable k to variable j, that is, an impulse in variable k does not induce any response of variable j. From Chapter 2, Propositions 2.4 and 2.5, we know that (3.7.12) is equivalent to

$$H_0: \phi_{jk,i} = 0 \qquad \text{for } i = 1, 2, \dots, p(K-1) \tag{3.7.14}$$

and (3.7.13) is equivalent to

$$H_0: \theta_{jk,i} = 0 \qquad \text{for } i = 0, 1, \dots, p(K-1). \tag{3.7.15}$$

Using Bonferroni's inequality (see Chapter 2, Section 2.2.3) a test of (3.7.14) with significance level at most $100\gamma\%$ is obtained by rejecting H_0 if

$$|\sqrt{T}\hat{\phi}_{jk,i}/\hat{\sigma}_{\phi_{jk}}(i)| > z_{(\gamma/2p(K-1))} \tag{3.7.16}$$

for at least one $i \in \{1, 2, \ldots, p(K-1)\}$. Here $z_{(\gamma)}$ is the upper 100γ percentage point of the standard normal distribution and $\hat{\sigma}_{\phi_{jk}}(i)$ is an estimate of the asymptotic standard deviation $\sigma_{\phi_{jk}}(i)$ of $\sqrt{T}\hat{\phi}_{jk,i}$ obtained via Proposition 3.6. In order to obtain a nondegenerate asymptotic standard normal distribution of the t-ratio $\sqrt{T}\hat{\phi}_{jk,i}/\hat{\sigma}_{\phi_{jk}}(i)$ the variance $\sigma^2_{\phi_{jk}}(i)$ must be nonzero, however.

A test of (3.7.15) with significance level at most γ is obtained by rejecting H_0 if

$$|\sqrt{T}\hat{\theta}_{jk,i}/\hat{\sigma}_{\theta_{jk}}(i)| \begin{cases} > z_{(\gamma/2(pK-p+1))} & \text{for at least one } i \in \{0, 1, 2, \ldots, p(K-1)\} \\ & \text{if } j > k \\ > z_{(\gamma/2(pK-p))} & \text{for at least one } i \in \{1, 2, \ldots, p(K-1)\} \\ & \text{if } j < k. \end{cases} \tag{3.7.17}$$

Here $\hat{\sigma}_{\theta_{jk}}(i)$ is a consistent estimator of the standard deviation of the asymptotic distribution of $\sqrt{T}\hat{\theta}_{jk,i}$ obtained from Proposition 3.6 and that standard deviation is assumed nonzero.

A test based on Bonferroni's principle may have quite low power because the actual significance level may be much smaller than the given upper bound. Therefore a test based on some χ^2- or F-statistic would be preferable. Unfortunately, such tests are not easily available for the present situation. For a more detailed discussion of this point see Lütkepohl (1990b). ■

3.7.2 Proof of Proposition 3.6

The proof of Proposition 3.6 is a straightforward application of the matrix differentiation rules given in Appendix A.13. It is sketched here for completeness and because it is distributed over a number of publications in the literature. Readers mainly interested in applying the proposition may skip this section without loss of continuity.

To prove (3.7.5) note that $\Phi_i = J\mathbf{A}^i J'$ (see Chapter 2, Section 2.1.2) and apply Rule 8 of Appendix A.13. The expression for F_n in (3.7.6) follows because

$$\frac{\partial\, \text{vec}(\Psi_n)}{\partial \boldsymbol{\alpha}'} = \sum_{i=1}^{n} \frac{\partial\, \text{vec}(\Phi_i)}{\partial \boldsymbol{\alpha}'}$$

and

$$F_\infty = \frac{\partial\, \text{vec}(\Psi_\infty)}{\partial \boldsymbol{\alpha}'} = \frac{\partial\, \text{vec}(\Psi_\infty)}{\partial\, \text{vec}(\Psi_\infty^{-1})'} \frac{\partial\, \text{vec}(\Psi_\infty^{-1})}{\partial \boldsymbol{\alpha}'}$$

$$= -(\Psi_\infty' \otimes \Psi_\infty) \frac{\partial\, \text{vec}(I_K - A_1 - \cdots - A_p)}{\partial \boldsymbol{\alpha}'}.$$

Furthermore,

$$C_i = \frac{\partial\, \text{vec}(\Theta_i)}{\partial \boldsymbol{\alpha}'} = \frac{\partial\, \text{vec}(\Phi_i P)}{\partial \boldsymbol{\alpha}'} = (P' \otimes I_K) \frac{\partial\, \text{vec}(\Phi_i)}{\partial \boldsymbol{\alpha}'}$$

and

$$\bar{C}_i = \frac{\partial\, \mathrm{vec}(\Theta_i)}{\partial \boldsymbol{\sigma}'} = (I_K \otimes \Phi_i)\frac{\partial\, \mathrm{vec}(P)}{\partial \boldsymbol{\sigma}'},$$

where

$$\frac{\partial\, \mathrm{vec}(P)}{\partial \boldsymbol{\sigma}'} = \mathbf{L}_K' \frac{\partial\, \mathrm{vech}(P)}{\partial \boldsymbol{\sigma}'} = H,$$

follows from Appendix A.13, Rule 10. The matrices B_n, $\bar{B}_n$, B_∞, and $\bar{B}_\infty$ are obtained in a like manner using the relations $\Xi_n = \Psi_n P$ and $\Xi_\infty = \Psi_\infty P$.

Finally, in (3.7.11),

$$\begin{aligned} d_{jk,h} &= \frac{\partial \omega_{jk,h}}{\partial \boldsymbol{\alpha}'} \\ &= \left[2 \sum_{i=0}^{h-1} (e_j'\Phi_i P e_k)(e_k' P' \otimes e_j') \frac{\partial\, \mathrm{vec}(\Phi_i)}{\partial \boldsymbol{\alpha}'} \mathrm{MSE}_j(h) \right. \\ &\qquad \left. - \sum_{i=0}^{h-1} (e_j'\Phi_i P e_k)^2 \frac{\partial\, \mathrm{MSE}_j(h)}{\partial \boldsymbol{\alpha}'} \right] \Big/ [\mathrm{MSE}_j(h)]^2, \end{aligned}$$

$$\begin{aligned} \frac{\partial\, \mathrm{MSE}_j(h)}{\partial \boldsymbol{\alpha}'} &= \sum_{m=0}^{h-1} \left[(e_j'\Phi_m \Sigma_u \otimes e_j') \frac{\partial\, \mathrm{vec}(\Phi_m)}{\partial \boldsymbol{\alpha}'} + (e_j' \otimes e_j'\Phi_m \Sigma_u) \frac{\partial\, \mathrm{vec}(\Phi_m')}{\partial \boldsymbol{\alpha}'} \right] \\ &= \sum_{m=0}^{h-1} [(e_j'\Phi_m \Sigma_u \otimes e_j') + (e_j' \otimes e_j'\Phi_m \Sigma_u)\mathbf{K}_{KK}] \frac{\partial\, \mathrm{vec}(\Phi_m)}{\partial \boldsymbol{\alpha}'} \\ &= \sum_{m=0}^{h-1} [(e_j'\Phi_m \Sigma_u \otimes e_j') + \mathbf{K}_{11}(e_j'\Phi_m \Sigma_u \otimes e_j')] G_m \\ &= 2 \sum_{m=0}^{h-1} (e_j'\Phi_m \Sigma_u \otimes e_j') G_m, \qquad \text{(see Appendix A.12.2, Rule 23)} \end{aligned}$$

$$\begin{aligned} \bar{d}_{jk,h} &= \frac{\partial \omega_{jk,h}}{\partial \boldsymbol{\sigma}'} \\ &= \sum_{i=0}^{h-1} \left[2(e_j'\Phi_i P e_k)(e_k' \otimes e_j'\Phi_i) \frac{\partial\, \mathrm{vec}(P)}{\partial \boldsymbol{\sigma}'} \mathrm{MSE}_j(h) \right. \\ &\qquad \left. - (e_j'\Phi_i P e_k)^2 \frac{\partial\, \mathrm{MSE}_j(h)}{\partial \boldsymbol{\sigma}'} \right] \Big/ [\mathrm{MSE}_j(h)]^2, \end{aligned}$$

and

$$\begin{aligned} \frac{\partial\, \mathrm{MSE}_j(h)}{\partial \boldsymbol{\sigma}'} &= \sum_{m=0}^{h-1} (e_j'\Phi_m \otimes e_j'\Phi_m) \frac{\partial\, \mathrm{vec}(\Sigma_u)}{\partial \boldsymbol{\sigma}'} \\ &= \sum_{m=0}^{h-1} (e_j'\Phi_m \otimes e_j'\Phi_m) \mathbf{D}_K \frac{\partial\, \mathrm{vech}(\Sigma_u)}{\partial \boldsymbol{\sigma}'}. \end{aligned}$$

Thereby Proposition 3.6 is proven. In the next section an example is discussed.

3.7.3 An Example

To illustrate the results of Section 3.7.1 we use again the investment/income/consumption example from Section 3.2.3. Since

$$\hat{\Phi}_1 = \hat{A}_1 = \begin{bmatrix} -.320 & .146 & .961 \\ .044 & -.153 & .289 \\ -.002 & .225 & -.264 \end{bmatrix},$$

the elements of $\hat{\Phi}_1$ must have the same standard errors as the elements of $\hat{A}_1$. Checking the covariance matrix in (3.7.5) it is seen that the asymptotic covariance matrix of $\hat{\Phi}_1$ is indeed the upper left hand $(K^2 \times K^2)$ block of $\Sigma_{\hat{\alpha}}$ since

$$G_1 = J \otimes I_K = [I_{K^2} \quad 0 \dots 0].$$

Thus, the square roots of the diagonal elements of

$$G_1 \hat{\Sigma}_{\hat{\alpha}} G_1'/T = \frac{1}{T}[I_9 \quad 0 \dots 0](\hat{\Gamma}_Y(0)^{-1} \otimes \hat{\Sigma}_u)\begin{bmatrix} I_9 \\ 0 \\ \vdots \\ 0 \end{bmatrix}$$

are estimates of the asymptotic standard errors of $\hat{\Phi}_1$. Note that here and in the following we use the LS estimators from the standard form of the VAR process (see Section 3.2) and not the mean-adjusted form. Accordingly the estimate $\hat{\Gamma}_Y(0)^{-1}$ is obtained from $(ZZ'/T)^{-1}$ by deleting the first row and column.

From (2.1.22) we get

$$\hat{\Phi}_2 = \hat{\Phi}_1 \hat{A}_1 + \hat{A}_2 = \begin{bmatrix} -.054 & .262 & .416 \\ .029 & .114 & -.088 \\ .045 & .261 & .110 \end{bmatrix}.$$

To estimate the corresponding standard errors we note that

$$G_2 = J\mathbf{A}' \otimes I_K + J \otimes \Phi_1.$$

Replacing the unknown quantities by the usual estimates gives

$$\frac{1}{T}\hat{G}_2 \hat{\Sigma}_{\hat{\alpha}} \hat{G}_2' = \frac{1}{T}[J\hat{\mathbf{A}}'\hat{\Gamma}_Y(0)^{-1}\hat{\mathbf{A}}J' \otimes \hat{\Sigma}_u + J\hat{\mathbf{A}}'\hat{\Gamma}_Y(0)^{-1}J' \otimes \hat{\Sigma}_u \hat{\Phi}_1'$$
$$+ J\hat{\Gamma}_Y(0)^{-1}\hat{\mathbf{A}}J' \otimes \hat{\Phi}_1 \hat{\Sigma}_u + J\hat{\Gamma}_Y(0)^{-1}J' \otimes \hat{\Phi}_1 \hat{\Sigma}_u \hat{\Phi}_1'].$$

The square roots of the diagonal elements of this matrix are estimates of the standard deviations of the elements of $\hat{\Phi}_2$ and so on. Some $\hat{\Phi}_i$ matrices together with estimated standard errors are given in Table 3.3. In Figures 3.4 and 3.5 some impulse responses are depicted graphically along with two-standard error bounds.

In Figure 3.4 consumption is seen to increase in response to a unit shock in income. However, under a two-standard error criterion (approximate 95% con-

Table 3.3. Estimates of impulse responses for the investment/income/consumption system with estimated standard errors in parentheses

i	$\hat{\Phi}_i$			$\hat{\Psi}_i$		
1	−0.320	0.146	0.961	0.680	0.146	0.961
	(0.125)	(0.562)	(0.657)	(0.125)	(0.562)	(0.657)
	0.044	−0.153	0.289	0.044	0.847	0.289
	(0.032)	(0.143)	(0.167)	(0.032)	(0.143)	(0.167)
	−0.002	0.225	−0.264	−0.002	0.225	0.736
	(0.025)	(0.115)	(0.134)	(0.025)	(0.115)	(0.134)
2	−0.054	0.262	0.416	0.626	0.408	1.377
	(0.129)	(0.546)	(0.663)	(0.148)	(0.651)	(0.755)
	0.029	0.114	−0.088	0.073	0.961	0.200
	(0.032)	(0.135)	(0.162)	(0.043)	(0.192)	(0.222)
	0.045	0.261	0.110	0.043	0.486	0.846
	(0.026)	(0.108)	(0.131)	(0.033)	(0.144)	(0.167)
3	0.119	0.353	−0.408	0.745	0.761	0.969
	(0.084)	(0.384)	(0.476)	(0.099)	(0.483)	(0.550)
	−0.009	0.071	0.120	0.064	1.033	0.320
	(0.016)	(0.078)	(0.094)	(0.037)	(0.176)	(0.203)
	−0.001	−0.098	0.091	0.042	0.388	0.937
	(0.017)	(0.078)	(0.102)	(0.033)	(0.156)	(0.183)
∞				0.756	0.836	1.295
				(0.133)	(0.661)	(0.798)
		0		0.076	1.076	0.344
				(0.048)	(0.236)	(0.285)
				0.053	0.505	0.964
				(0.043)	(0.213)	(0.257)

fidence bounds) only the second response coefficient is significantly different from zero. Of course, the large standard errors of the impulse response coefficients reflect the substantial estimation uncertainty in the VAR coefficient matrices A_1 and A_2.

In order to check the overall significance of the response coefficients of consumption to an income impulse we may use the procedure described in Remark 8 of Section 3.7.1. That is, we have to check the significance of the first $p(K-1) = 4$ response coefficients. Since one of them is individually significant at an asymptotic 5% level we may reject the null hypothesis of no response of consumption to income impulses at a significance level not greater than $4 \times 5\% = 20\%$. Of course, this is not a significance level we are used to in applied work. However, it becomes clear from Table 3.3 that the second response coefficient $\hat{\phi}_{32,2}$ is still significant if the individual significance levels are reduced to 2.5%. Note that the

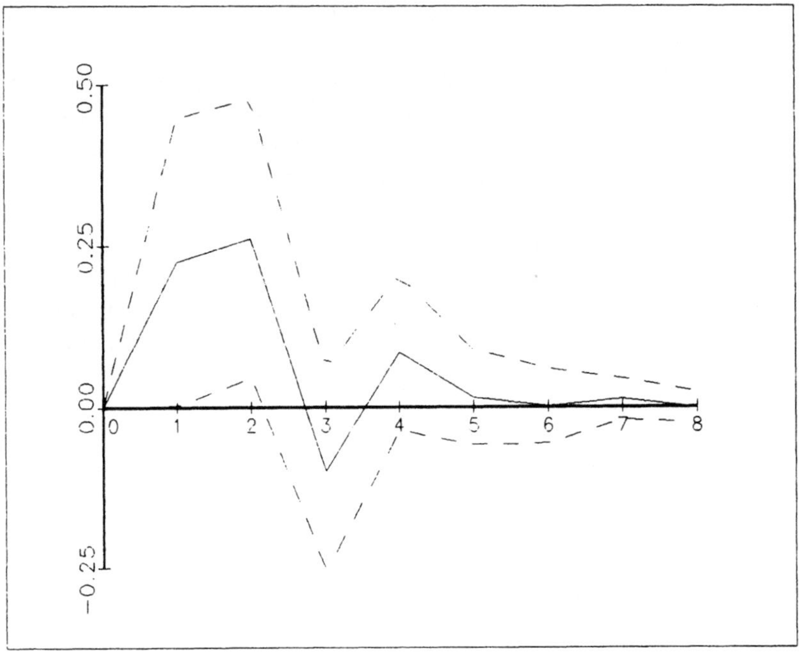

Fig. 3.4. Estimated responses of consumption to a forecast error impulse in income with two-standard error bounds.

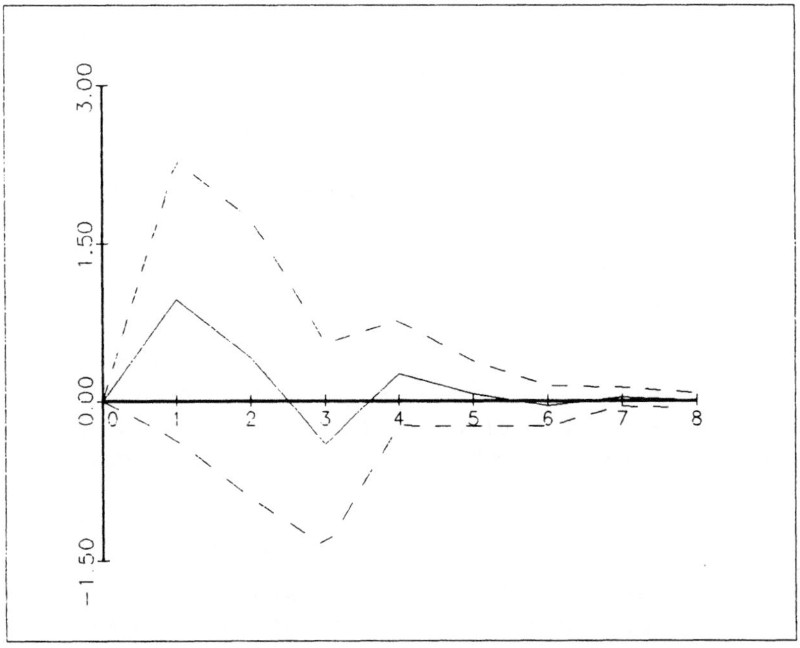

Fig. 3.5. Estimated responses of investment to a forecast error impulse in consumption with two-standard error bounds.

upper 1.25 percentage point of the standard normal distribution is $c_{0.0125} = 2.24$. Thus we may reject the no-response hypothesis at an overall $4 \times 2.5\% = 10\%$ level which is clearly a more common size for a test in applied work. Still, in this exercise the data do not reveal strong evidence for the intuitively appealing hypothesis that consumption responds to income impulses. In later chapters we will see how the coefficients can be estimated with more precision.

In Figure 3.5 the responses of investment to consumption impulses are depicted. None of them is significant under a two-standard error criterion. This result is in line with the Granger-causality analysis in Section 3.6. In that section we did not find evidence for Granger-causality from income/consumption to investment. Assuming that the test result describes the actual situation, the $\phi_{13,i}$ must be zero for $i = 1, 2, \ldots$ (see also Chapter 2, Section 2.3.1).

The covariance matrix of

$$\hat{\Psi}_1 = I_3 + \hat{\Phi}_1 = \begin{bmatrix} .680 & .146 & .961 \\ .044 & .847 & .289 \\ -.002 & .225 & .736 \end{bmatrix}$$

is, of course, the same as that of $\hat{\Phi}_1$ and an estimate of the covariance matrix of the elements of

$$\hat{\Psi}_2 = I_3 + \hat{\Phi}_1 + \hat{\Phi}_2 = \begin{bmatrix} .626 & .408 & 1.377 \\ .073 & .961 & .200 \\ .043 & .486 & .846 \end{bmatrix}$$

is obtained as $(G_1 + \hat{G}_2)\hat{\Sigma}_{\hat{\alpha}}(G_1 + \hat{G}_2)'/T$. Some accumulated impulse responses together with estimated standard errors are also given in Table 3.3 and accumulated responses of consumption to income impulses and of investment to consumption are shown in Figures 3.6 and 3.7, respectively. They reinforce the findings for the individual impulse responses in Figures 3.4 and 3.5.

An estimate of the asymptotic covariance matrix of the estimated long-run responses $\hat{\Psi}_\infty = (I_3 - \hat{A}_1 - \hat{A}_2)^{-1}$ is

$$\frac{1}{T}([\hat{\Psi}'_\infty, \hat{\Psi}'_\infty] \otimes \hat{\Psi}_\infty)\hat{\Sigma}_{\hat{\alpha}}\left[\begin{bmatrix} \hat{\Psi}_\infty \\ \hat{\Psi}_\infty \end{bmatrix} \otimes \hat{\Psi}'_\infty\right].$$

The matrix $\hat{\Psi}_\infty$ together with the resulting standard errors is also given in Table 3.3. For instance, the total long-run effect $\hat{\psi}_{13,\infty}$ of a consumption impulse on investment is 1.295 and its estimated asymptotic standard error is .798. Not surprisingly $\hat{\psi}_{13,\infty}$ is not significantly different from zero for any common level of significance (e.g. 10%). On the other hand $\hat{\psi}_{32,\infty}$, the long-run effect on consumption due to an impulse in income, is significant at an asymptotic 5% level.

For the interpretation of the $\hat{\Phi}_i$ the critical remarks at the end of Chapter 2 must be kept in mind. As explained there the $\hat{\Phi}_i$ and $\hat{\Psi}_n$ coefficients may not reflect the actual responses of the variables in the system. As an alternative one may want to determine the responses to orthogonal residuals. In order to obtain

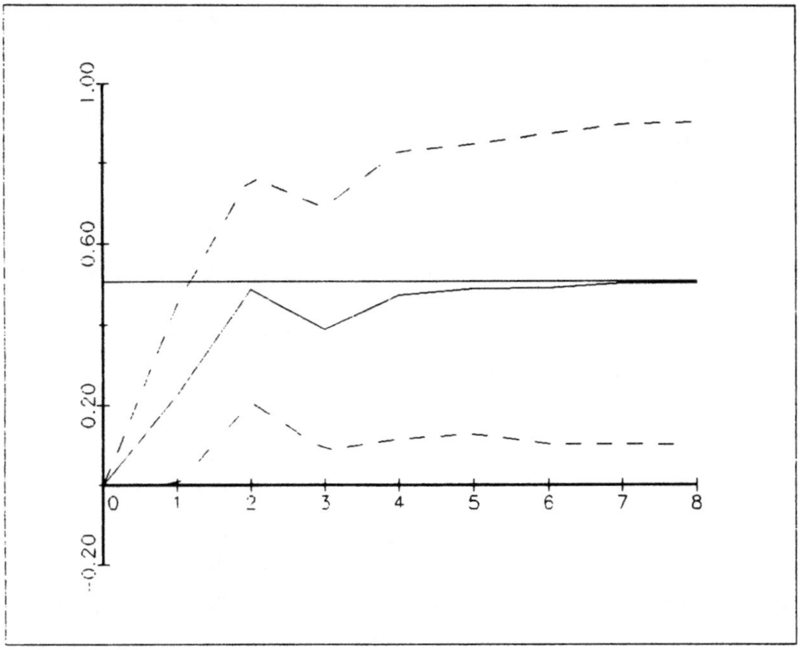

Fig. 3.6. Accumulated and long-run responses of consumption to a forecast error impulse in income with two-standard error bounds.

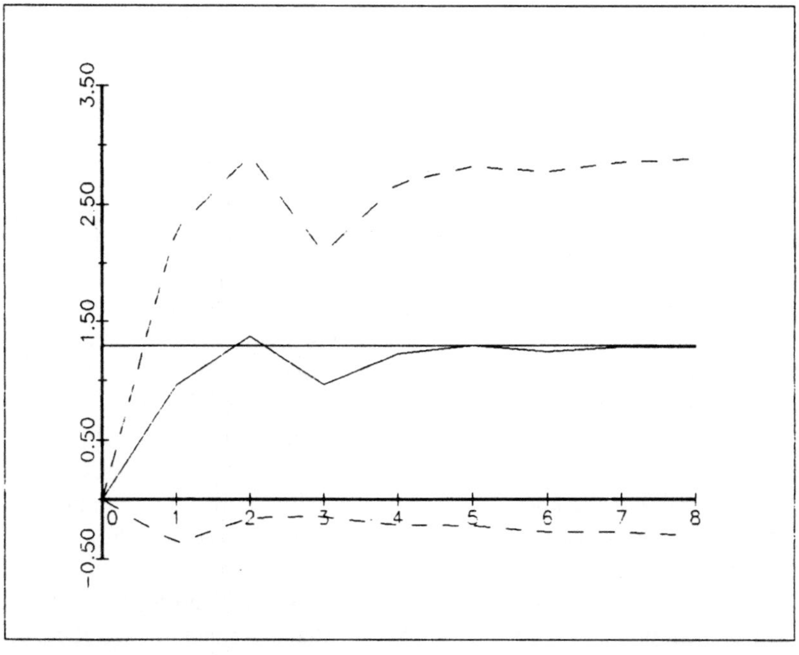

Fig. 3.7. Accumulated and long-run responses of investment to a forecast error impulse in consumption with two-standard error bounds.

the asymptotic covariance matrices of the $\hat{\Theta}_i$ and $\hat{\Xi}_n$, a decomposition of $\hat{\Sigma}_u$ is needed. For our example

$$\hat{P} = \begin{bmatrix} 4.61 & 0 & 0 \\ .16 & 1.16 & 0 \\ .27 & .49 & .76 \end{bmatrix} \times 10^{-2}$$

is the lower triangular matrix with positive diagonal elements satisfying $\hat{P}\hat{P}' = \hat{\Sigma}_u$ (Choleski decomposition). The asymptotic covariance matrix of $\text{vec}(\hat{P}) = \text{vec}(\hat{\Theta}_0)$ is a (9×9) matrix which is estimated as

$$\frac{1}{T}\hat{\bar{C}}_0 \hat{\Sigma}_{\hat{\sigma}} \hat{\bar{C}}_0' = \frac{2}{T}\hat{H}\mathbf{D}_K^+(\hat{\Sigma}_u \otimes \hat{\Sigma}_u)\mathbf{D}_K^{+\prime}\hat{H}',$$

where $\mathbf{D}_K^+ = (\mathbf{D}_K'\mathbf{D}_K)^{-1}\mathbf{D}_K'$ and

$$\hat{H} = \mathbf{L}_3'\{\mathbf{L}_3[(I_3 \otimes \hat{P})\mathbf{K}_{33} + (\hat{P} \otimes I_3)]\mathbf{L}_3'\}^{-1}.$$

The resulting estimated standard errors of the elements of $\hat{P}$ are given in Table 3.4. Note that the variances corresponding to elements above the diagonal of $\hat{P}$ are all zero because these elements are set at zero and are not estimated.

The asymptotic covariance matrix of the elements of

$$\hat{\Theta}_1 = \begin{bmatrix} -1.196 & .644 & .730 \\ .256 & -.035 & .219 \\ -.047 & .131 & -.201 \end{bmatrix} \times 10^{-2}$$

is obtained as the sum of the two matrices

$$\hat{C}_1 \hat{\Sigma}_{\hat{\alpha}} \hat{C}_1'/T = [(\hat{P}' \otimes I_3)G_1 \hat{\Sigma}_{\hat{\alpha}} G_1'(\hat{P} \otimes I_3)]/T$$

and

$$\hat{\bar{C}}_1 \hat{\Sigma}_{\hat{\sigma}} \hat{\bar{C}}_1'/T = (I_3 \otimes \hat{\Phi}_1)\hat{H}\hat{\Sigma}_{\hat{\sigma}}\hat{H}'(I_3 \otimes \hat{\Phi}_1')/T.$$

The resulting standard errors for the elements of $\hat{\Theta}_1$ are given in Table 3.4 along with some more $\hat{\Theta}_i$ and $\hat{\Xi}_n$ matrices.

Some responses and accumulated responses of consumption to income innovations with two-standard error bounds are depicted in Figures 3.8 and 3.9. The responses in Figures 3.4 and 3.8 are obviously a bit different. Note the (significant) immediate reaction of consumption in Figure 3.8. However, from period 1 onwards the response of consumption in both figures is qualitatively similar. The difference of scales is due to the different sizes of the shocks traced through the system. For instance, Figure 3.4 is based on a unit shock in income while Figure 3.8 is based on an innovation of size one standard deviation due to the transformation of the white noise residuals.

Again a test of overall significance of the impulse responses in Figure 3.8 could be performed using Bonferroni's principle. Now we have to check the significance of the $\hat{\theta}_{32,i}$ for $i = 0, 1, \ldots, 4 = p(K - 1)$. We reject the null hypothesis of no response if at least one of the coefficients is significantly different from zero. In

Table 3.4. Estimates of responses to orthogonal innovations for the investment/income/consumption system with estimated standard errors in parentheses

i	$\hat{\Theta}_i$	$\hat{\Xi}_i$
0	$\begin{bmatrix} 4.61 & 0 & 0 \\ (.38) & & \\ .16 & 1.16 & 0 \\ (.14) & (.10) & \\ .27 & .49 & .76 \\ (.11) & (.10) & (.06) \end{bmatrix} \times 10^{-2}$	$\begin{bmatrix} 4.61 & 0 & 0 \\ (.38) & & \\ .16 & 1.16 & 0 \\ (.14) & (.10) & \\ .27 & .49 & .76 \\ (.11) & (.10) & (.06) \end{bmatrix} \times 10^{-2}$
1	$\begin{bmatrix} -1.20 & .64 & .73 \\ (.57) & (.56) & (.50) \\ .26 & -.04 & .22 \\ (.14) & (.14) & (.13) \\ -.05 & .13 & -.20 \\ (.12) & (.12) & (.10) \end{bmatrix} \times 10^{-2}$	$\begin{bmatrix} 3.46 & .64 & .73 \\ (.63) & (.56) & (.50) \\ .41 & 1.13 & .22 \\ (.20) & (.17) & (.13) \\ .22 & .62 & .56 \\ (.15) & (.14) & (.11) \end{bmatrix} \times 10^{-2}$
2	$\begin{bmatrix} -.10 & .51 & .32 \\ (.58) & (.57) & (.50) \\ .13 & .09 & -.07 \\ (.14) & (.14) & (.12) \\ .28 & .36 & .08 \\ (.12) & (.12) & (.10) \end{bmatrix} \times 10^{-2}$	$\begin{bmatrix} 3.32 & 1.15 & 1.05 \\ (.74) & (.69) & (.58) \\ .54 & 1.22 & .15 \\ (.24) & (.22) & (.17) \\ .50 & .98 & .64 \\ (.20) & (.18) & (.14) \end{bmatrix} \times 10^{-2}$
∞	0	$\begin{bmatrix} 3.97 & 1.61 & .98 \\ (.82) & (.92) & (.61) \\ .61 & 1.42 & .26 \\ (.31) & (.34) & (.22) \\ .58 & 1.06 & .73 \\ (.28) & (.32) & (.20) \end{bmatrix} \times 10^{-2}$

this case we can reject at an asymptotic 5% level of significance because $\hat{\theta}_{32,0}$ is significant at the 1% level (see Table 3.4). Thus we may choose individual significance levels of 1% for each of the 5 coefficients and obtain 5% as an upper bound for the overall level. Of course, all these interpretations are based on the assumption that the actual asymptotic standard errors of the impulse responses are nonzero (see Section 3.7.1, Remark 1).

We have also performed forecast error variance decompositions and have computed the standard errors on the basis of the results given in Proposition 3.6. For some forecast horizons the decompositions are given in Table 3.5. The standard errors may be regarded as rough indications of the sampling uncertainty. It must be kept in mind, however, that they may be quite misleading if the true forecast error variance components are zero as explained in Remark 6 of Section 3.7.1. Obviously, this qualification limits their value in the present

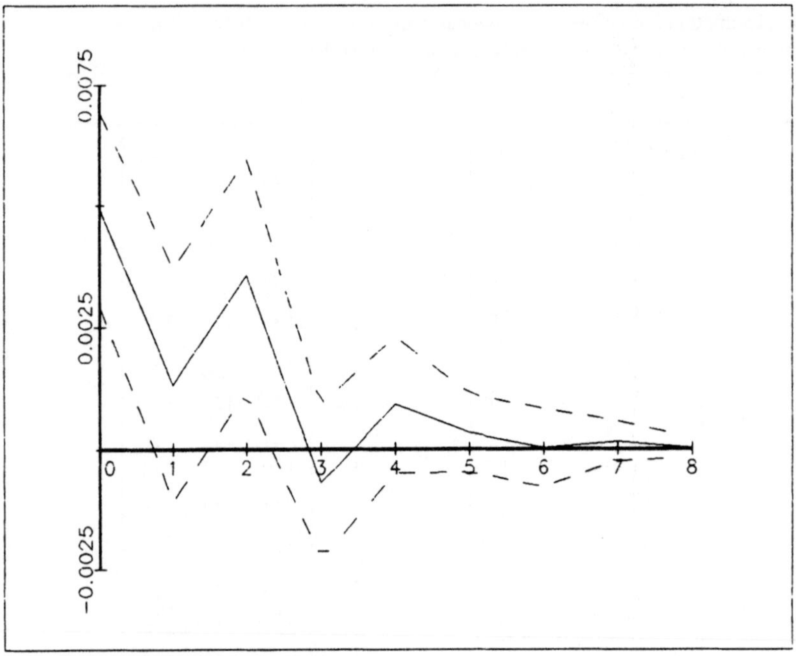

Fig. 3.8. Estimated responses of consumption to an orthogonalized impulse in income with two-standard error bounds.

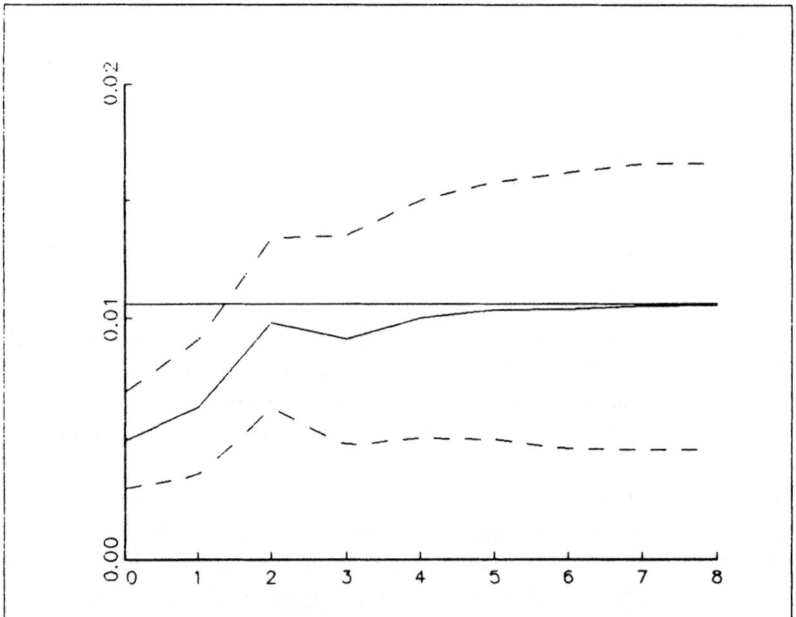

Fig. 3.9. Estimated accumulated and long-run responses of consumption to an orthogonalized impulse in income with two-standard error bounds.

Table 3.5. Forecast error variance decomposition of the investment/income/consumption system with estimated standard errors in parentheses

forecast error in	forecast horizon h	proportions of forecast error variance, h periods ahead, accounted for by innovations in investment $\hat{\omega}_{j1,h}$	income $\hat{\omega}_{j2,h}$	consumption $\hat{\omega}_{j3,h}$
investment	1	1.00(.00)	.00(.00)	.00(.00)
($j = 1$)	2	.96(.04)	.02(.03)	.02(.03)
	3	.95(.04)	.03(.03)	.03(.03)
	4	.94(.05)	.03(.03)	.03(.03)
	8	.94(.05)	.03(.03)	.03(.04)
income	1	.02(.03)	.98(.03)	.00(.00)
($j = 2$)	2	.06(.05)	.91(.06)	.03(.04)
	3	.07(.06)	.90(.07)	.03(.04)
	4	.07(.06)	.89(.07)	.04(.04)
	8	.07(.06)	.89(.07)	.04(.04)
consumption	1	.08(.06)	.27(.09)	.65(.09)
($j = 3$)	2	.08(.06)	.27(.08)	.65(.09)
	3	.13(.08)	.33(.09)	.54(.09)
	4	.13(.08)	.34(.09)	.54(.09)
	8	.13(.08)	.34(.09)	.53(.09)

example. Students are invited to reproduce the numbers in Table 3.5 and the previous tables of this section with the data given in Table E.1.

3.7.4 Investigating the Distributions of the Impulse Responses by Simulation Techniques

In the previous subsections it was indicated repeatedly that in some cases the small sample validity of the asymptotic results is problematic. In that situation one possibility is to use Monte Carlo or bootstrapping methods instead for investigating the sampling properties of the quantities of interest. Although these methods are quite expensive in terms of computer time they have been used in the past for evaluating the properties of impulse response functions. The general methodology is described in Appendix D.

In the present situation there are different approaches to simulation. One possibility is to assume a specific distribution of the white noise process, e.g., $u_t \sim N(0, \hat{\Sigma}_u)$, and generate a large number of time series realizations based on the estimated VAR coefficients. From these time series new sets of coefficients are then estimated and the corresponding impulse responses and/or forecast error variance components are computed. The empirical distributions obtained in this way may be used to investigate the actual distributions of the quantities of interest.

Alternatively, if an assumption regarding the white noise distribution cannot be made, new sets of residuals may be drawn from the estimation residuals and a large number of y_t time series is generated on the basis of these sets of disturbances (see Appendix D.3). The multiple time series obtained in this way are then used to compute estimates of the quantities of interest and study their properties.

3.8 Exercises

3.8.1 Algebraic Problems

The notation of Sections 3.2–3.5 is used in the following problems.

Problem 3.1

Show that $\hat{\boldsymbol{\beta}} = ((ZZ')^{-1}Z \otimes I_K)\mathbf{y}$ minimizes

$$\bar{S}(\boldsymbol{\beta}) = \mathbf{u}'\mathbf{u} = (\mathbf{y} - (Z' \otimes I_K)\boldsymbol{\beta})'(\mathbf{y} - (Z' \otimes I_K)\boldsymbol{\beta}).$$

Problem 3.2

Prove that

$$\sqrt{T}(\hat{\mathbf{b}} - \mathbf{b}) \xrightarrow{d} N(0, \Sigma_u \otimes \Gamma^{-1})$$

if y_t is stable and

$$\frac{1}{\sqrt{T}}\operatorname{vec}(ZU') = \frac{1}{\sqrt{T}}(I_K \otimes Z)\operatorname{vec}(U') \xrightarrow{d} N(0, \Sigma_u \otimes \Gamma).$$

Problem 3.3

Show (3.4.17). (Hint: Use the product rule for matrix differentiation and $\partial \operatorname{vec}(\Sigma_u^{-1})/\partial \operatorname{vec}(\Sigma_u)' = -\Sigma_u^{-1} \otimes \Sigma_u^{-1}$.)

Problem 3.4

Derive (3.4.18). (Hint: Use the last expression given in (3.4.6).)

Problem 3.5

Show (3.4.19).

Problem 3.6

Derive (3.4.20).

Problem 3.7

Prove that plim $\tilde{z}_T/\sqrt{T} = 0$, where

$$\tilde{z}_T = \sum_{i=1}^{p} \tilde{A}_i \sum_{j=0}^{i-1} (y_{-j} - y_{T-j}).$$

(Hint: Show that $E(\tilde{z}_T/\sqrt{T}) \to 0$ and $\operatorname{Var}(\tilde{z}_T/\sqrt{T}) \to 0$.)

Problem 3.8

Show that Equation (3.5.10) holds.
(Hint: Define

$$Z_t(h) := \begin{bmatrix} 1 \\ y_t(h) \\ \vdots \\ y_t(h-p+1) \end{bmatrix}$$

and show $Z_t(h) = \mathbf{B}Z_t(h-1)$ by induction.)

Problem 3.9

In the context of Section 3.5, suppose y_t is a stable Gaussian VAR(p) process which is estimated by ML in mean-adjusted form. Show that the forecast MSE correction term has the form

$$\Omega(h) = E\left(\frac{\partial y_t(h)}{\partial \mu'} \tilde{\Sigma}_{\tilde{\mu}} \frac{\partial y_t(h)'}{\partial \mu}\right) + E\left(\frac{\partial y_t(h)}{\partial \boldsymbol{\alpha}'} \tilde{\Sigma}_{\tilde{\boldsymbol{\alpha}}} \frac{\partial y_t(h)'}{\partial \boldsymbol{\alpha}}\right),$$

with

$$\frac{\partial y_t(h)}{\partial \mu'} = I_K - J\mathbf{A}^h \underset{(Kp \times K)}{\begin{bmatrix} I_K \\ \vdots \\ I_K \end{bmatrix}}$$

and

$$\frac{\partial y_t(h)}{\partial \boldsymbol{\alpha}'} = \sum_{i=0}^{h-1} (Y_t - \boldsymbol{\mu})'(\mathbf{A}')^{h-1-i} \otimes \Phi_i.$$

Here $\boldsymbol{\mu} := (\mu', \ldots, \mu')'$ is a $(Kp \times 1)$ vector, Y_t and $\mathbf{A}$ are defined as in (2.1.8), $J := [I_K \quad 0 \ldots 0]$ is a $(K \times Kp)$ matrix, and Φ_i is the i-th coefficient matrix of the prediction error MA representation (2.1.17).

3.8.2 Numerical Problems

The following problems require the use of a computer. In Appendix E, Table E.2, two quarterly, seasonally adjusted U.S. investment series are given. Use the variables

y_1 – first differences of fixed investment,
y_2 – first differences of change in business inventories,

in the following problems. Use the data from 1947 to 1968 only.

Problem 3.10

Plot the two time series y_{1t} and y_{2t} and comment on the stationarity and stability of the series.

Problem 3.11

Estimate the parameters of a VAR(1) model for $(y_{1t}, y_{2t})'$ using multivariate LS, that is, compute $\hat{B}$ and $\hat{\Sigma}_u$. Comment on the stability of the estimated process.

Problem 3.12

Use the mean-adjusted form of a VAR(1) model and estimate the coefficients. Assume that the data generation process is Gaussian and estimate the covariance matrix of the asymptotic distribution of the ML estimators.

Problem 3.13

Determine the Yule-Walker estimate of the VAR(1) coefficient matrix and compare to the LS estimate.

Problem 3.14

Use the LS estimate and compute point forecasts $\hat{y}_{86}(1)$, $\hat{y}_{86}(2)$ (that is, the forecast origin is the last quarter of 1968) and the corresponding MSE matrices $\hat{\Sigma}_y(1)$, $\hat{\Sigma}_y(2)$, $\hat{\Sigma}_{\hat{y}}(1)$, and $\hat{\Sigma}_{\hat{y}}(2)$. Use these estimates to set up approximate 95% interval forecasts assuming that the process y_t is Gaussian.

Problem 3.15

Test the hypothesis that y_2 does not Granger-cause y_1.

Problem 3.16

Estimate the coefficient matrices Φ_1 and Φ_2 from the LS estimates of the VAR(1) model for y_t and determine approximate standard errors of the estimates.

Problem 3.17

Determine the *upper* triangular matrix $\hat{P}$ with positive diagonal for which $\hat{P}\hat{P}' = \hat{\Sigma}_u$. Estimate the covariance matrix of the asymptotic distribution of $\hat{P}$ under the assumption that y_t is Gaussian. Test the hypothesis that the upper right-hand corner element of the underlying matrix P is zero.

Problem 3.18

Use the results of the previous problems to compute $\hat{\Theta}_0$, $\hat{\Theta}_1$, and $\hat{\Theta}_2$. Determine also estimates of the asymptotic standard errors of the elements of these three matrices.

Chapter 4. VAR Order Selection and Checking the Model Adequacy

4.1 Introduction

In the previous chapter we have assumed that we have given a K-dimensional multiple time series $y_1, \dots, y_T$, with $y_t = (y_{1t}, \dots, y_{Kt})'$, that is known to be generated by a VAR(p) process

$$y_t = \nu + A_1 y_{t-1} + \cdots + A_p y_{t-p} + u_t \tag{4.1.1}$$

and we have discussed estimation of the parameters ν, $A_1, \dots, A_p$, and $\Sigma_u = E(u_t u_t')$. In deriving the properties of the estimators a number of assumptions were made. In practice it will rarely be known with certainty whether the conditions hold that are required to derive the consistency and asymptotic normality of the estimators. Therefore statistical tools should be used in order to check the validity of the assumptions made. In this chapter some such tools will be provided.

In the next two sections it will be discussed what to do if the VAR order p is unknown. In practice the order will usually be unknown. In Chapter 3 we have assumed that a VAR(p) process as in (4.1.1) represents the data generation process. We have not assumed that all the A_i are nonzero. In particular A_p may be zero. In other words, p is just assumed to be an upper bound for the VAR order. On the other hand, from (3.5.13) we know that the approximate MSE matrix of the 1-step predictor will increase with the order p. Thus, choosing p unnecessarily large will reduce the forecast precision of the corresponding estimated VAR(p) model. Also, the estimation precision of the impulse responses depends on the precision of the parameter estimates. Therefore it is useful to have procedures or criteria for choosing an adequate VAR order.

In Sections 4.4 to 4.6 possibilities are discussed for checking some of the assumptions of the previous chapters. The asymptotic distribution of the residual autocorrelations and so-called portmanteau tests are considered in Section 4.4. The latter tests are popular tools for checking the whiteness of the residuals. More precisely, they are used to test for nonzero residual autocorrelations. In Section 4.5 tests for nonnormality are considered. The normality assumption was used in Chapter 3 in setting up forecast intervals.

One assumption underlying much of the previous analysis is the stationarity of the systems considered. Nonstationarities may have various forms. Not only trends indicate deviations from stationarity but also changes in the variability or variance of the system. Moreover exogenous shocks may affect various characteristics of the system. Tests for structural change are presented in Section 4.6.

4.2 A Sequence of Tests for Determining the VAR Order

Obviously there is not just one correct VAR order for the process (4.1.1). In fact, if (4.1.1) is a correct summary of the characteristics of the process y_t, then the same is true for

$$y_t = \nu + A_1 y_{t-1} + \cdots + A_p y_{t-p} + A_{p+1} y_{t-p-1} + u_t$$

with $A_{p+1} = 0$. In other words, if y_t is a VAR(p) process, in this sense it is also a VAR($p + 1$) process. In the assumptions of the previous chapter the possibility of zero coefficient matrices is not excluded. In this chapter it is practical to have a unique number that is called the order of the process. Therefore, in the following we will call y_t a VAR(p) process if $A_p \neq 0$ and $A_i = 0$ for $i > p$ so that p is the smallest possible order. This unique number will be called the VAR order.

4.2.1 The Impact of the Fitted VAR Order on the Forecast MSE

If y_t is a VAR(p) process it is useful to fit a VAR(p) model to the available multiple time series and not, for instance, a VAR($p + i$) because, under a mean squared error measure, forecasts from the latter process will be inferior to those based on an estimated VAR(p) model. This result follows from the approximate forecast MSE matrix $\Sigma_{\hat{y}}(h)$ derived in Section 3.5.2 of Chapter 3. For instance, for $h = 1$,

$$\Sigma_{\hat{y}}(1) = \frac{T + Kp + 1}{T} \Sigma_u$$

if a VAR(p) model is fitted to data generated by a K-dimensional VAR process with order not greater than p. Obviously, $\Sigma_{\hat{y}}(1)$ is an increasing function of the order of the model fitted to the data.

Since the approximate MSE matrix is derived from asymptotic theory it is of interest to know whether the result remains true in small samples. To get some feeling for the answer to this question we have generated 1000 Gaussian bivariate time series with the process (3.2.25),

$$y_t = \begin{bmatrix} .02 \\ .03 \end{bmatrix} + \begin{bmatrix} .5 & .1 \\ .4 & .5 \end{bmatrix} y_{t-1} + \begin{bmatrix} 0 & 0 \\ .25 & 0 \end{bmatrix} y_{t-2} + u_t,$$
$$\Sigma_u = \begin{bmatrix} .09 & 0 \\ 0 & .04 \end{bmatrix}. \tag{4.2.1}$$

We have fitted VAR(2), VAR(4), and VAR(6) models to the generated series and we have computed forecasts with the estimated models. Then we have compared these forecasts to generated post-sample values. The resulting average squared

Table 4.1. Average squared forecast errors for the estimated bivariate VAR(2) process based on 1000 realizations

sample size T	forecast horizon h	average squared forecast errors					
		VAR(2) model		VAR(4) model		VAR(6) model	
		y_1	y_2	y_1	y_2	y_1	y_2
30	1	.111	.052	.132	.062	.165	.075
	2	.155	.084	.182	.098	.223	.119
	3	.146	.141	.183	.166	.225	.202
50	1	.108	.043	.119	.048	.129	.054
	2	.132	.075	.144	.083	.161	.093
	3	.142	.120	.150	.130	.168	.145
100	1	.091	.044	.095	.046	.098	.049
	2	.120	.064	.125	.067	.130	.069
	3	.130	.108	.135	.113	.140	.113

forecasting errors for different forecast horizons h and sample sizes T are shown in Table 4.1. Obviously the forecasts based on estimated VAR(2) models are clearly superior to the VAR(4) and VAR(6) forecasts for sample sizes $T = 30$, 50, and 100. While the comparative advantage of the VAR(2) models is quite dramatic for $T = 30$ it diminishes with increasing sample size. This, of course, was to be expected given that the approximate forecast MSE matrix of an estimated process approaches that of the known process as the sample size increases (see Section 3.5).

Of course, the process considered in this example is a very special one. To see whether a similar result is obtained for other processes as well we have also generated 1000 three-dimensional time series with the VAR(1) process (2.1.14),

$$y_t = \begin{bmatrix} .01 \\ .02 \\ 0 \end{bmatrix} + \begin{bmatrix} .5 & 0 & 0 \\ .1 & .1 & .3 \\ 0 & .2 & .3 \end{bmatrix} y_{t-1} + u_t \quad \text{with} \quad \Sigma_u = \begin{bmatrix} 2.25 & 0 & 0 \\ 0 & 1.0 & .5 \\ 0 & .5 & .74 \end{bmatrix}. \tag{4.2.2}$$

We have fitted VAR(1), VAR(3), and VAR(6) models to these data and we have computed forecasts and forecast errors. Some average squared forecast errors are presented in Table 4.2. Again forecasts from lower order models are clearly superior to higher order models. In fact, in a large scale simulation study involving many more processes similar results were found (see Lütkepohl (1985)). Thus, it is useful to avoid fitting VAR models with unnecessarily great orders.

The question is then what to do if the true order is unknown and an upper bound, say M, for the order is known only. One possibility to check whether certain coefficient matrices may be zero is to set up a significance test. For our particular problem of determining the correct VAR order, we may set up a sequence of tests. First H_0: $A_M = 0$ is tested. If this null hypothesis cannot be

Table 4.2. Average squared forecast errors for the estimated three-dimensional VAR(1) process based on 1000 realizations

sample size T	forecast horizon h	average squared forecast errors								
		VAR(1) model			VAR(3) model			VAR(6) model		
		y_1	y_2	y_3	y_1	y_2	y_3	y_1	y_2	y_3
	1	.87	1.14	2.68	1.14	1.52	3.62	2.25	2.78	6.82
30	2	1.09	1.21	3.21	1.44	1.67	4.12	2.54	2.98	7.85
	3	1.06	1.31	3.32	1.35	1.58	4.23	2.59	2.79	8.63
	1	.81	1.03	2.68	.96	1.22	2.97	1.18	1.53	3.88
50	2	1.01	1.23	2.92	1.20	1.40	3.47	1.48	1.68	4.38
	3	1.01	1.29	3.11	1.12	1.44	3.48	1.42	1.77	4.66
	1	.73	.93	2.35	.77	1.00	2.62	.86	1.12	2.91
100	2	.94	1.15	2.86	1.00	1.24	3.12	1.12	1.38	3.53
	3	.90	1.15	3.02	.93	1.20	3.23	1.03	1.35	3.51

rejected we test H_0: $A_{M-1} = 0$ and so on until we can reject a null hypothesis. Before we discuss this procedure in more detail we will now introduce a possible test statistic.

4.2.2 The Likelihood Ratio Test Statistic

Since we just need to test zero restrictions on the coefficients of a VAR model we may use the Wald statistic discussed in Section 3.6 in the context of causality tests. To shed some more light on this type of statistic it may be instructive to consider the likelihood ratio testing principle. It is based on comparing the maxima of the log-likelihood function over the unrestricted and restricted parameter space. Specifically, the likelihood ratio statistic is

$$\lambda_{LR} = 2[\ln l(\tilde{\boldsymbol{\delta}}) - \ln l(\tilde{\boldsymbol{\delta}}_r)], \tag{4.2.3}$$

where $\tilde{\boldsymbol{\delta}}$ is the unrestricted ML estimator for a parameter vector $\boldsymbol{\delta}$ obtained by maximizing the likelihood function over the full feasible parameter space and $\tilde{\boldsymbol{\delta}}_r$ is the restricted ML estimator which is obtained by maximizing the likelihood function over that part of the parameter space where the restrictions of interest are satisfied (see Appendix C.5). For the case of interest here, where we have linear constraints on the coefficients of a VAR process, λ_{LR} can be shown to have an asymptotic χ^2-distribution with as many degrees of freedom as there are distinct linear restrictions.

To obtain this result let us assume for the moment that y_t is a stable Gaussian (normally distributed) VAR(p) process as in (4.1.1). Using the notation of Section 3.2.1 (as opposed to the mean-adjusted form considered in Section 3.4) the log-likelihood function is

$$\ln l(\boldsymbol{\beta}, \Sigma_u) = -\frac{KT}{2} \ln 2\pi - \frac{T}{2} \ln|\Sigma_u|$$
$$-\frac{1}{2}[\mathbf{y} - (Z' \otimes I_K)\boldsymbol{\beta}]'(I_T \otimes \Sigma_u^{-1})[\mathbf{y} - (Z' \otimes I_K)\boldsymbol{\beta}] \tag{4.2.4}$$

(see (3.4.5)). The first partial derivatives with respect to $\boldsymbol{\beta}$ are

$$\frac{\partial \ln l}{\partial \boldsymbol{\beta}} = (Z \otimes \Sigma_u^{-1})\mathbf{y} - (ZZ' \otimes \Sigma_u^{-1})\boldsymbol{\beta}. \tag{4.2.5}$$

Equating to zero and solving for $\boldsymbol{\beta}$ gives the unrestricted ML/LS estimator

$$\tilde{\boldsymbol{\beta}} = ((ZZ')^{-1}Z \otimes I_K)\mathbf{y}. \tag{4.2.6}$$

Suppose the restrictions for $\boldsymbol{\beta}$ are given in the form

$$C\boldsymbol{\beta} = c, \tag{4.2.7}$$

where C is a known $(N \times (K^2p + K))$ matrix of rank N and c is a known $(N \times 1)$ vector. Then the restricted ML estimator may be found by a Lagrangian approach (see Appendix A.14). The Lagrange function is

$$L(\boldsymbol{\beta}, \gamma) = \ln l(\boldsymbol{\beta}) + \gamma'(C\boldsymbol{\beta} - c), \tag{4.2.8}$$

where γ is an $(N \times 1)$ vector of Lagrange multipliers. Of course, L also depends on Σ_u. Since these parameters are not involved in the restrictions (4.2.7) we have skipped them here. The restricted maximum of the log-likelihood function with respect to $\boldsymbol{\beta}$ is known to be attained at a point where the first partial derivatives of L are zero.

$$\frac{\partial L}{\partial \boldsymbol{\beta}} = (Z \otimes \Sigma_u^{-1})\mathbf{y} - (ZZ' \otimes \Sigma_u^{-1})\boldsymbol{\beta} + C'\gamma, \tag{4.2.9a}$$

$$\frac{\partial L}{\partial \gamma} = C\boldsymbol{\beta} - c. \tag{4.2.9b}$$

Equating to zero and solving gives

$$\tilde{\boldsymbol{\beta}}_r = \tilde{\boldsymbol{\beta}} + [(ZZ')^{-1} \otimes \Sigma_u]C'[C((ZZ')^{-1} \otimes \Sigma_u)C']^{-1}(c - C\tilde{\boldsymbol{\beta}}) \tag{4.2.10}$$

(see Problem 4.1).

Since for any given coefficient matrix B^0 the maximum of $\ln l$ with respect to Σ_u is obtained for

$$\Sigma_u^0 = \frac{1}{T}(Y - B^0Z)(Y - B^0Z)'$$

(see Section 3.4.2, (3.4.8) and (3.4.11)), the maximum for the unrestricted case is attained for

$$\tilde{\Sigma}_u = \frac{1}{T}(Y - \tilde{B}Z)(Y - \tilde{B}Z)' \tag{4.2.11}$$

and for the restricted case we get

$$\tilde{\Sigma}_u^r = \frac{1}{T}(Y - \tilde{B}_r Z)(Y - \tilde{B}_r Z)'. \tag{4.2.12}$$

Here $\tilde{B}$ and $\tilde{B}_r$ are the coefficient matrices corresponding to $\tilde{\boldsymbol{\beta}}$ and $\tilde{\boldsymbol{\beta}}_r$ respectively, that is, $\tilde{\boldsymbol{\beta}} = \text{vec}(\tilde{B})$ and $\tilde{\boldsymbol{\beta}}_r = \text{vec}(\tilde{B}_r)$. Thus, for this particular situation the likelihood ratio statistic becomes

$$\lambda_{LR} = 2[\ln l(\tilde{\boldsymbol{\beta}}, \tilde{\Sigma}_u) - \ln l(\tilde{\boldsymbol{\beta}}_r, \tilde{\Sigma}_u^r)].$$

This statistic can be shown to have an asymptotic $\chi^2(N)$-distribution. In fact, this result also holds if y_t is not Gaussian but has a distribution from a larger family. If y_t is not Gaussian the estimators obtained by maximizing the Gaussian likelihood function in (4.2.4) are called *pseudo ML estimators*. We will now state the previous results formally and then present a proof.

PROPOSITION 4.1 (*Asymptotic Distribution of the LR Statistic*)

Let y_t be a stationary, stable VAR(p) process as in (4.1.1) with standard white noise u_t (see Definition 3.1). Suppose the true parameter vector $\boldsymbol{\beta}$ satisfies linear constraints $C\boldsymbol{\beta} = c$, where C is an $(N \times (K^2p + K))$ matrix of rank N and c is an $(N \times 1)$ vector. Moreover, let $\ln l$ denote the Gaussian log-likelihood function and let $\tilde{\boldsymbol{\beta}}$ and $\tilde{\boldsymbol{\beta}}_r$ be the (pseudo) ML and restricted (pseudo) ML estimators, respectively, with corresponding estimators $\tilde{\Sigma}_u$ and $\tilde{\Sigma}_u^r$ of the white noise covariance matrix Σ_u given in (4.2.11) and (4.2.12). Then

$$\begin{aligned}\lambda_{LR} &= 2[\ln l(\tilde{\boldsymbol{\beta}}, \tilde{\Sigma}_u) - \ln l(\tilde{\boldsymbol{\beta}}_r, \tilde{\Sigma}_u^r)] \\ &= T(\ln|\tilde{\Sigma}_u^r| - \ln|\tilde{\Sigma}_u|) & (4.2.13a) \\ &\simeq (C\tilde{\boldsymbol{\beta}} - c)'[C((ZZ')^{-1} \otimes \tilde{\Sigma}_u^r)C']^{-1}(C\tilde{\boldsymbol{\beta}} - c) \xrightarrow{d} \chi^2(N). & (4.2.13b)\end{aligned}$$

Here T is the sample size (time series length) and $Z := (Z_0, \ldots, Z_{T-1})$ with $Z_t' := (1, y_t', \ldots, y_{t-p+1}')$. ■

Note that in the proposition it is not assumed that y_t is Gaussian (normally distributed). It is just assumed that u_t is independent white noise with bounded fourth moments. Thus, $\ln l$ may not really be the log-likelihood function of $\mathbf{y} := \text{vec}[y_1, \ldots, y_T]$. It will only be the actual log-likelihood if $\mathbf{y}$ happens to be multivariate normal. In that case $\tilde{\boldsymbol{\beta}}$ and $\tilde{\boldsymbol{\beta}}_r$ are actual ML and restricted ML estimators. Otherwise they are pseudo ML estimators.

The second form of the LR statistic in (4.2.13a) is sometimes convenient for computing the actual test value. It is also useful for comparing the likelihood ratio tests to other procedures for VAR order selection as we will see in Section 4.3. Comparing the expression in (4.2.13b) to (3.6.5) shows that, for the special case considered here, the LR statistic is similar to the Wald statistic if in the latter $\hat{\Sigma}_u$ is replaced by $\tilde{\Sigma}_u^r$. In fact, the important difference between the Wald and LR statistics is that the former involves only estimators of the unrestricted model while both unrestricted and restricted estimators enter into λ_{LR} (see also (4.2.13a)).

As in the case of the Wald test one may consider using the statistic λ_{LR}/N in conjunction with the $F(N, T - Kp - 1)$-distribution in small samples. Another adjustment is suggested by Hannan (1970, p. 341).

Proof of Proposition 4.1:

We first show the equivalence of the three forms of the LR statistic given in (4.2.13). The equality in (4.2.13a) follows by noting that

$$\begin{aligned}[\mathbf{y} - (Z' \otimes I_K)\boldsymbol{\beta}]'(I_T \otimes \mathit{\Sigma}_u^{-1})[\mathbf{y} - (Z' \otimes I_K)\boldsymbol{\beta}] &= \operatorname{tr}(Y - BZ)'\mathit{\Sigma}_u^{-1}(Y - BZ) \\ &= \operatorname{tr} \mathit{\Sigma}_u^{-1}(Y - BZ)(Y - BZ)'.\end{aligned}$$

Replacing the matrices B and $\mathit{\Sigma}_u$ by $\tilde{B}$ and $\tilde{\mathit{\Sigma}}_u$, respectively, gives

$$\ln l(\tilde{\boldsymbol{\beta}}, \tilde{\mathit{\Sigma}}_u) = \text{constant} - \frac{T}{2}\ln|\tilde{\mathit{\Sigma}}_u|.$$

Similarly,

$$\ln l(\tilde{\boldsymbol{\beta}}_r, \tilde{\mathit{\Sigma}}_u^r) = \text{constant} - \frac{T}{2}\ln|\tilde{\mathit{\Sigma}}_u^r|$$

which gives the desired result.

In order to prove (4.2.13b) we observe that $\ln l$ is a quadratic function in $\boldsymbol{\beta}$. Thus, by Taylor's Theorem (Appendix A.13, Proposition A.3), for an arbitrary fixed vector $\boldsymbol{\beta}^0$,

$$\begin{aligned}\ln l(\boldsymbol{\beta}) = \ln l(\boldsymbol{\beta}^0) &+ \frac{\partial \ln l(\boldsymbol{\beta}^0)}{\partial \boldsymbol{\beta}'}(\boldsymbol{\beta} - \boldsymbol{\beta}^0) \\ &+ \frac{1}{2}(\boldsymbol{\beta} - \boldsymbol{\beta}^0)'\frac{\partial^2 \ln l(\boldsymbol{\beta}^0)}{\partial \boldsymbol{\beta}\, \partial \boldsymbol{\beta}'}(\boldsymbol{\beta} - \boldsymbol{\beta}^0).\end{aligned}$$

Taking $\tilde{\boldsymbol{\beta}}$ for $\boldsymbol{\beta}^0$ and $\tilde{\boldsymbol{\beta}}_r$ for $\boldsymbol{\beta}$, $\partial \ln l(\tilde{\boldsymbol{\beta}})/\partial \boldsymbol{\beta}' = 0$ so that

$$\lambda_{LR} = 2[\ln l(\tilde{\boldsymbol{\beta}}) - \ln l(\tilde{\boldsymbol{\beta}}_r)] = -(\tilde{\boldsymbol{\beta}}_r - \tilde{\boldsymbol{\beta}})'\frac{\partial^2 \ln l(\tilde{\boldsymbol{\beta}})}{\partial \boldsymbol{\beta}\, \partial \boldsymbol{\beta}'}(\tilde{\boldsymbol{\beta}}_r - \tilde{\boldsymbol{\beta}}). \tag{4.2.14}$$

As in Section 3.4 we can derive

$$\frac{\partial^2 \ln l}{\partial \boldsymbol{\beta}\, \partial \boldsymbol{\beta}'} = -(ZZ' \otimes \mathit{\Sigma}_u^{-1}).$$

Substituting $\tilde{\mathit{\Sigma}}_u^r$ for $\mathit{\Sigma}_u$ and using (4.2.10), we get from (4.2.14)

$$\begin{aligned}\lambda_{LR} &\simeq (C\tilde{\boldsymbol{\beta}} - c)'[C((ZZ')^{-1} \otimes \tilde{\mathit{\Sigma}}_u^r)C']^{-1}C((ZZ')^{-1} \otimes \tilde{\mathit{\Sigma}}_u^r)(ZZ' \otimes \tilde{\mathit{\Sigma}}_u^{r-1}) \\ &\quad \times ((ZZ')^{-1} \otimes \tilde{\mathit{\Sigma}}_u^r)C'[C((ZZ')^{-1} \otimes \tilde{\mathit{\Sigma}}_u^r)C']^{-1}(C\tilde{\boldsymbol{\beta}} - c) \\ &= (C\tilde{\boldsymbol{\beta}} - c)'[C((ZZ')^{-1} \otimes \tilde{\mathit{\Sigma}}_u^r)C']^{-1}(C\tilde{\boldsymbol{\beta}} - c)\end{aligned}$$

which is the desired result.

The asymptotic $\chi^2(N)$-distribution now follows from Proposition C.4(5) of Appendix C because $[C((ZZ'/T)^{-1} \otimes \tilde{\Sigma}_u^r)C']^{-1}$ is a consistent estimator of $[C(\Gamma^{-1} \otimes \Sigma_u)C']^{-1}$. ■

In the next subsection a sequential testing scheme based on LR tests is discussed.

4.2.3 A Testing Scheme for VAR Order Determination

Assuming that M is known to be an upper bound for the VAR order, the following sequence of null and alternative hypotheses may be tested using LR tests:

$$
\begin{aligned}
&H_0^1: A_M = 0 \quad \text{against} \quad H_1^1: A_M \neq 0 \\
&H_0^2: A_{M-1} = 0 \quad \text{against} \quad H_1^2: A_{M-1} \neq 0 \quad |A_M = 0 \\
&\quad \vdots \\
&H_0^i: A_{M-i+1} = 0 \quad \text{against} \quad H_1^i: A_{M-i+1} \neq 0 \quad |A_M = \cdots = A_{M-i+2} = 0 \\
&\quad \vdots \\
&H_0^M: A_1 = 0 \quad \text{against} \quad H_1^M: A_1 \neq 0 \quad |A_M = \cdots = A_2 = 0.
\end{aligned} \tag{4.2.15}
$$

In this scheme each null hypothesis is tested conditional on the previous ones being true. The procedure terminates and the VAR order is chosen accordingly if one of the null hypotheses is rejected. That is, if H_0^i is rejected, $\hat{p} = M - i + 1$ will be chosen as the estimate of the autoregressive order.

The likelihood ratio statistic for testing the i-th null hypothesis is

$$\lambda_{LR}(i) = T(\ln|\tilde{\Sigma}_u(M-i)| - \ln|\tilde{\Sigma}_u(M-i+1)|) \tag{4.2.16}$$

where $\tilde{\Sigma}_u(m)$ denotes the ML estimator of Σ_u when a VAR(m) model is fitted to a time series of length T. By Proposition 4.1 this statistic has an asymptotic $\chi^2(K^2)$-distribution if H_0^i and all previous null hypotheses are true. Note that K^2 parameters are set to zero in H_0^i. Hence we have K^2 restrictions and we use $\lambda_{LR}(i)$ in conjunction with critical values from a $\chi^2(K^2)$-distribution. Alternatively one may use $\lambda_{LR}(i)/K^2$ in conjunction with the $F(K^2, T - K(M-i+1) - 1)$-distribution.

Of course, the order chosen for a particular process will depend on the significance levels used in the tests. In this procedure it is important to realize that the significance levels of the individual tests must be distinguished from the Type I error of the whole procedure because rejection of H_0^i means that H_0^{i+1}, …, H_0^M are automatically rejected too. Thus, denoting by D_j the event that H_0^j is rejected in the j-th test when it is actually true, the probability of a Type I error for the i-th test in the sequence becomes

$$\varepsilon_i = \Pr(D_1 \cup D_2 \cup \cdots \cup D_i).$$

Since D_j is the event that $\lambda_{LR}(j)$ falls in the rejection region although H_0^j is true, $\gamma_j = \Pr(D_j)$ is just the significance level of the j-th individual test. Now, it can be shown that for $m \neq j$ and $m, j \leq i$, $\lambda_{LR}(m)$ and $\lambda_{LR}(j)$ are asymptotically independent statistics if $H_0^1, \ldots, H_0^i$ are true (see Paulsen & Tjøstheim (1985, pp. 223–224)). Hence, D_m and D_j are independent events in large samples so that

$$\begin{aligned}\varepsilon_i &= \Pr(D_1 \cup \cdots \cup D_{i-1}) + \Pr(D_i) - \Pr\{(D_1 \cup \cdots \cup D_{i-1}) \cap D_i\} \\ &= \varepsilon_{i-1} + \gamma_i - \varepsilon_{i-1}\gamma_i = \varepsilon_{i-1} + \gamma_i(1 - \varepsilon_{i-1}), \qquad i = 2, 3, \ldots, M. \end{aligned} \tag{4.2.17}$$

Of course, $\varepsilon_1 = \gamma_1$. Thus, it is easily seen by induction that

$$\varepsilon_i = 1 - (1 - \gamma_1) \cdots (1 - \gamma_i), \qquad i = 1, 2, \ldots, M. \tag{4.2.18}$$

If, for example, a 5% significance level is chosen for each individual test ($\gamma_i = .05$), then

$$\varepsilon_1 = .05, \qquad \varepsilon_2 = 1 - .95 \cdot .95 = .0975, \qquad \varepsilon_3 = .142625.$$

Hence, the actual rejection probability will become quite substantial if the sequence of null hypotheses to be tested is long.

It is difficult to decide on appropriate significance levels in the testing scheme (4.2.15). Whatever significance levels the researcher decides to use, she or he should keep in mind the distinction between the overall and the individual significance levels. Also, it must be kept in mind that we know the asymptotic distributions of the LR statistics only. Thus, the significance levels chosen will be approximate probabilities of Type I errors only.

Finally, in the literature another testing scheme was also suggested and used. In that scheme the first set of hypotheses ($i = 1$) is as in (4.2.15) and for $i > 1$ the following hypotheses are tested:

$$H_0^i\colon A_M = \cdots = A_{M-i+1} = 0 \quad \text{against} \quad H_1^i\colon A_M \neq 0 \text{ or } \ldots \text{ or } A_{M-i+1} \neq 0.$$

Here H_0^i is not tested conditional on the previous null hypotheses being true but is tested against the full VAR(M) model. Unfortunately the LR statistics to be used in such a sequence will not be independent so that the overall significance level (probability of Type I error) is difficult to determine.

4.2.4 An Example

To illustrate the sequential testing procedure described in the foregoing we use the investment/income/consumption example from Section 3.2.3. The variables y_1, y_2, and y_3 represent first differences of the logarithms of the investment, income, and consumption data in Table E.1 of Appendix E. We assume an upper bound of $M = 4$ for the VAR order and therefore we set aside the first 4 values as presample values. The data up to 1978.4 are used for estimation so that the sample size is $T = 71$ in each test. The estimated error covariance matrices and their determinants are given in Table 4.3. The corresponding χ^2- and F-test values are summarized in Table 4.4. Since the denominator degrees of freedom for the

Table 4.3. ML estimates of the error covariance matrix of the investment/income/consumption system

VAR order m	$\tilde{\Sigma}_u(m) \times 10^4$	$\lvert\tilde{\Sigma}_u(m)\rvert \times 10^{11}$
0	$\begin{bmatrix} 21.83 & .410 & 1.228 \\ . & 1.420 & .571 \\ . & . & 1.084 \end{bmatrix}$	2.473
1	$\begin{bmatrix} 20.14 & .493 & 1.173 \\ . & 1.318 & .625 \\ . & . & 1.018 \end{bmatrix}$	1.782
2	$\begin{bmatrix} 19.18 & .617 & 1.126 \\ . & 1.270 & .574 \\ . & . & .821 \end{bmatrix}$	1.255
3	$\begin{bmatrix} 19.08 & .599 & 1.126 \\ . & 1.235 & .543 \\ . & . & .784 \end{bmatrix}$	1.174
4	$\begin{bmatrix} 16.96 & .573 & 1.252 \\ . & 1.234 & .544 \\ . & . & .765 \end{bmatrix}$	.958

Table 4.4. LR Statistics for the investment/income/consumption system

i	H_0^i	VAR order under H_0^i	λ_{LR} [a]	$\lambda_{LR}/9$ [b]
1	$A_4 = 0$	3	14.44	1.60
2	$A_3 = 0$	2	4.76	.53
3	$A_2 = 0$	1	24.90	2.77
4	$A_1 = 0$	0	23.25	2.58

[a] Critical value for individual 5% level test: $\chi^2(9)_{.95} = 16.92$.
[b] Critical value for individual 5% level test: $F(9, 71 - 3(5 - i) - 1)_{.95} \approx 2$

F-statistics are quite high (ranging from 62 to 70) the F-tests are qualitatively similar to the χ^2-tests. Using individual significance levels of .05 in each test, H_0^3: $A_2 = 0$ is the first null hypothesis that is rejected. Thus, the estimated order from both tests is $\hat{p} = 2$. This supports the order chosen in the example in Chapter 3. Alternative procedures for choosing VAR orders are considered in the next section.

4.3 Criteria for VAR Order Selection

Although performing statistical tests is a common strategy for detecting nonzero parameters the approach described in the previous section is not completely satisfactory if a model is desired for a specific purpose. For instance, a VAR model is often constructed for prediction of the variables involved. In such a case we are not so much interested in finding the correct order of the underlying data generation process but we are interested in obtaining a good model for prediction. Hence it seems useful to take the objective of the analysis into account when choosing the VAR order. In the next subsection we will discuss criteria based on the forecasting objective.

If we really want to know the exact order of the data generation process (e.g., for analysis purposes) it is still questionable whether a testing procedure is the optimal strategy because that strategy has a positive probability of choosing an incorrect order even if the sample size (time series length) is large (see Section 4.3.3). In Section 4.3.2 we will present estimation procedures that choose the correct order with probability 1 at least in large samples.

4.3.1 Minimizing the Forecast MSE

If forecasting is the objective it makes sense to choose the order such that a measure of forecast precision is minimized. The forecast MSE (mean squared error) is such a measure. Therefore Akaike (1969, 1971) suggested to base the VAR order choice on the approximate 1-step forecast MSE given in Chapter 3, (3.5.13),

$$\Sigma_{\hat{y}}(1) = \frac{T + Km + 1}{T} \Sigma_u,$$

where m denotes the order of the VAR process fitted to the data, T is the sample size, and K the dimension of the time series. To make this criterion operational the white noise covariance matrix Σ_u has to be replaced by an estimate. Also, to obtain a unique solution we would like to have a scalar criterion rather than a matrix. Akaike suggests using the LS estimator with degrees of freedom adjustment,

$$\hat{\Sigma}_u(m) = \frac{T}{T - Km - 1} \tilde{\Sigma}_u(m),$$

for Σ_u and taking the determinant of the resulting expression. Here $\tilde{\Sigma}_u(m)$ is the ML estimator of Σ_u obtained by fitting a VAR(m) model, as in the previous section. The resulting criterion is called the final prediction error (FPE) criterion, that is,

$$\begin{aligned} \text{FPE}(m) &= \det\left[\frac{T + Km + 1}{T} \frac{T}{T - Km - 1} \tilde{\Sigma}_u(m)\right] \\ &= \left[\frac{T + Km + 1}{T - Km - 1}\right]^K \det(\tilde{\Sigma}_u(m)). \end{aligned} \tag{4.3.1}$$

We have written the criterion in terms of the ML estimator of the covariance matrix because in this form the FPE criterion has some intuitive appeal. If the order m is increased, $\det(\tilde{\Sigma}_u(m))$ declines while the multiplicative term $(T + Km + 1)/(T - Km - 1)$ increases. The VAR order estimate is obtained as that value for which the two forces are balanced optimally. Note that the determinant of the LS estimate $\hat{\Sigma}_u(m)$ may increase with increasing m. On the other hand, it is quite obvious that $|\tilde{\Sigma}_u(m)|$ cannot become larger when m increases because the maximum of the log-likelihood function is proportional to $-\ln|\tilde{\Sigma}_u(m)|$ apart from an additive constant and, for $m < n$, a VAR(m) model may be interpreted as a restricted VAR(n) model. Thus, $-\ln|\tilde{\Sigma}_u(m)| \leq -\ln|\tilde{\Sigma}_u(n)|$ or $|\tilde{\Sigma}_u(m)| \geq |\tilde{\Sigma}_u(n)|$.

Based on the FPE criterion the estimate $\hat{p}$(FPE) of p is chosen such that

$$\text{FPE}[\hat{p}(\text{FPE})] = \min\{\text{FPE}(m) | m = 0, 1, \ldots, M\}.$$

That is, VAR models of orders $m = 0, 1, \ldots, M$ are estimated and the corresponding FPE(m) values are computed. The order minimizing the FPE values is then chosen as estimate for p.

Akaike (1973, 1974), based on a quite different reasoning, has derived a very similar criterion usually abbreviated by AIC (*Akaike's Information Criterion*). For a VAR(m) process the criterion is defined as

$$\begin{aligned} \text{AIC}(m) &= \ln|\tilde{\Sigma}_u(m)| + \frac{2}{T}(\text{number of freely estimated parameters}) \\ &= \ln|\tilde{\Sigma}_u(m)| + \frac{2mK^2}{T}. \end{aligned} \tag{4.3.2}$$

The estimate $\hat{p}$(AIC) for p is chosen so that this criterion is minimized.

The similarity of the criteria AIC and FPE can be seen by noting that, for a constant N,

$$\frac{T + N}{T - N} = 1 + \frac{2N}{T} + O(T^{-2}).$$

The quantity $O(T^{-2})$ denotes a sequence of order T^{-2}, that is, a sequence indexed by T that remains bounded if multiplied by T^2. Thus the sequence goes to zero rapidly when $T \to \infty$. Hence,

$$\begin{aligned} \ln \text{FPE}(m) &= \ln|\tilde{\Sigma}_u(m)| + K \ln((T + Km + 1)/(T - Km - 1)) \\ &= \ln|\tilde{\Sigma}_u(m)| + K \ln(1 + 2(Km + 1)/T + O(T^{-2})) \\ &= \ln|\tilde{\Sigma}_u(m)| + K\frac{2(Km + 1)}{T} + O(T^{-2}) \\ &= \text{AIC}(m) + 2K/T + O(T^{-2}). \end{aligned} \tag{4.3.3}$$

The third equality sign follows from a Taylor series expansion of $\ln(1 + x)$. The term $2K/T$ does not depend on the order m and hence $\text{AIC}(m) + 2K/T$ and

Table 4.5. Estimation of the VAR order of the investment/income/consumption system

VAR order m	FPE$(m) \times 10^{11}$	AIC(m)	HQ(m)	SC(m)
0	2.691	−24.42	−24.42*	−24.42*
1	2.500	−24.50	−24.38	−24.21
2	2.272*	−24.59*	−24.37	−24.02
3	2.748	−24.41	−24.07	−23.55
4	2.910	−24.36	−23.90	−23.21

* Minimum.

AIC(m) assume their minimum for the same value of m. Consequently AIC and ln FPE differ essentially by a term of order $O(T^{-2})$ and thus the two criteria will be about equivalent for moderate and large T.

To illustrate these procedures for VAR order selection we use again the investment/income/consumption example. The determinants of the residual covariance matrices are given in Table 4.3. Using these determinants the FPE and AIC values presented in Table 4.5 are obtained. Both criteria reach their minimum for $\hat{p} = 2$, that is, $\hat{p}(\text{FPE}) = \hat{p}(\text{AIC}) = 2$. The other quantities given in the table will be discussed shortly.

4.3.2 Consistent Order Selection

If interest centers on the correct VAR order, it makes sense to choose an estimator that has desirable sampling properties. One problem of interest in this context is to determine the statistical properties of order estimators such as $\hat{p}(\text{FPE})$ and $\hat{p}(\text{AIC})$. Consistency is a desirable *asymptotic* property of an estimator. As usual, an estimator $\hat{p}$ of the VAR order p is called *consistent* if

$$\operatorname*{plim}_{T\to\infty} \hat{p} = p \qquad \text{or, equivalently,} \qquad \lim_{T\to\infty} \Pr\{\hat{p} = p\} = 1. \tag{4.3.4}$$

The latter definition of the plim may seem to differ slightly from the one given in Appendix C. However, it is easily checked that the two definitions are equivalent for integer valued random variables. Of course, a reasonable estimator for p should be integer valued. The estimator $\hat{p}$ is called *strongly consistent* if

$$\Pr\{\lim \hat{p} = p\} = 1. \tag{4.3.5}$$

Accordingly, a VAR order selection criterion will be called consistent or strongly consistent if the resulting estimator has these properties. The following proposition due to Hannan & Quinn (1979), Quinn (1980), and Paulsen (1984) is useful for investigating the consistency of criteria for order determination.

PROPOSITION 4.2 (*Consistency of VAR Order Estimators*)

Let y_t be a K-dimensional stationary, stable VAR(p) process with standard white noise (that is, u_t is independent white noise with bounded fourth moments). Suppose the maximum order $M \geq p$ and $\hat{p}$ is chosen so as to minimize a criterion

$$\text{Cr}(m) = \ln|\tilde{\Sigma}_u(m)| + mc_T/T \tag{4.3.6}$$

over $m = 0, 1, \ldots, M$. Here $\tilde{\Sigma}_u(m)$ denotes the (pseudo) ML estimator of Σ_u obtained for a VAR(m) model and c_T is a nondecreasing sequence of real numbers that depends on the sample size T. Then $\hat{p}$ is consistent if and only if

$$c_T \to \infty \quad \text{and} \quad c_T/T \to 0 \quad \text{as} \quad T \to \infty. \tag{4.3.7a}$$

$\hat{p}$ is a strongly consistent estimator if and only if (4.3.7a) holds and

$$c_T/2 \ln \ln T > 1 \tag{4.3.7b}$$

eventually as $T \to \infty$. ■

We will not prove this proposition here but refer the reader to Quinn (1980) and Paulsen (1984) for a proof. The basic idea of the proof is to show that, for $p > m$, the quantity $\ln|\tilde{\Sigma}_u(m)|/\ln|\tilde{\Sigma}_u(p)|$ will be greater than one in large samples since $\ln|\tilde{\Sigma}_u(m)|$ is essentially the minimum of minus the Gaussian log-likelihood function for a VAR(m) model. Consequently, since the penalty terms mc_T/T and pc_T/T go to zero as $T \to \infty$, $\text{Cr}(m) > \text{Cr}(p)$ for large T. Thus, the probability of choosing too small an order goes to zero as $T \to \infty$. Similarly, if $m > p$, $\ln|\tilde{\Sigma}_u(m)|/\ln|\tilde{\Sigma}_u(p)|$ approaches 1 in probability if $T \to \infty$ and the penalty term of the lower order model is smaller than that of a larger order process. Thus the lower order p will be chosen if the sample size is large. The following corollary is an easy consequence of the proposition.

COROLLARY 4.2.1

Under the conditions of Proposition 4.2, if $M > p$, $\hat{p}$(FPE) and $\hat{p}$(AIC) are not consistent. ■

Proof: Since FPE and AIC are asymptotically equivalent (see (4.3.3)), it suffices to prove the corollary for $\hat{p}$(AIC). Equating AIC(m) and Cr(m) from (4.3.6) gives

$$2mK^2/T = mc_T/T$$

or $c_T = 2K^2$. Obviously, this sequence does not satisfy (4.3.7a). ■

We will see shortly that the limiting probability for underestimating the VAR order is zero for both $\hat{p}$(AIC) and $\hat{p}$(FPE) so that asymptotically they overestimate the true order with positive probability. However, Paulsen and Tjøstheim (1985, p. 224) argue that the limiting probability for overestimating the order declines with increasing dimension K and is negligible for $K \geq 5$. In other words, asymptotically AIC and FPE choose the correct order almost with probability one if the underlying multiple time series has high dimension K.

Before we continue the investigation of AIC and FPE we shall introduce two *consistent* criteria that have been quite popular in recent applied work. The first one is due to Hannan & Quinn (1979) and Quinn (1980). It is often denoted by HQ (*Hannan-Quinn criterion*):

$$\begin{aligned}\mathrm{HQ}(m) &= \ln|\tilde{\Sigma}_u(m)| + \frac{2\ln\ln T}{T}\text{(number of freely estimated parameters)} \\ &= \ln|\tilde{\Sigma}_u(m)| + \frac{2\ln\ln T}{T}mK^2. \end{aligned} \tag{4.3.8}$$

The estimate $\hat{p}(\mathrm{HQ})$ is the order that minimizes HQ(m) for $m = 0, 1, \ldots, M$. Comparing to (4.3.6) shows that $c_T = 2K^2 \ln\ln T$ and thus, by (4.3.7), HQ is consistent for univariate processes and strongly consistent for $K > 1$ if the conditions of Proposition 4.2 are satisfied for y_t.

Using Bayesian arguments Schwarz (1978) derives the following criterion:

$$\begin{aligned}\mathrm{SC}(m) &= \ln|\tilde{\Sigma}_u(m)| + \frac{\ln T}{T}\text{(number of freely estimated parameters)} \\ &= \ln|\tilde{\Sigma}_u(m)| + \frac{\ln T}{T}mK^2. \end{aligned} \tag{4.3.9}$$

Again the order estimate $\hat{p}(\mathrm{SC})$ is chosen so as to minimize the value of the criterion. A comparison with (4.3.6) shows that for this criterion $c_T = K^2 \ln T$. Since

$$K^2 \ln T/2 \ln\ln T$$

approaches infinity for $T \to \infty$, (4.3.7b) is satisfied and SC is seen to be strongly consistent for any dimension K.

COROLLARY 4.2.2

Under the conditions of Proposition 4.2, SC is strongly consistent and HQ is consistent. If the dimension K of the process is greater than one, both criteria are strongly consistent. ■

In Table 4.5 the values of HQ and SC for the investment/income/consumption example are given. Both criteria assume the minimum for $m = 0$, that is, $\hat{p}(\mathrm{HQ}) = \hat{p}(\mathrm{SC}) = 0$.

4.3.3 Comparison of Order Selection Criteria

It is worth emphasizing that the foregoing results do not necessarily mean that AIC and FPE are inferior to HQ and SC. Only if consistency is the yardstick for evaluating the criteria the latter two are superior under the conditions of the previous section. So far we have not considered the small sample properties of the estimators. In small samples AIC and FPE may have better properties (choose the correct order more often) than HQ and SC. Also, the former two criteria are designed for minimizing the forecast error variance. Thus, in small as well as large samples, models based on AIC and FPE may produce superior forecasts although they may not estimate the orders correctly. In fact, Shibata (1980) derives asymptotic optimality properties of AIC and FPE for univariate processes. He shows that asymptotically they indeed minimize the 1-step forecast MSE.

Although it is difficult in general to derive *small sample properties* of the criteria some such properties can be obtained. The following proposition states small sample relations between the criteria.

PROPOSITION 4.3 (*Small Sample Comparison of AIC, HQ, and SC*)

Let $y_{-M+1}, \dots, y_0, y_1, \dots, y_T$ be any K-dimensional multiple time series and suppose that VAR(m) models, $m = 0, 1, \dots, M$, are fitted to $y_1, \dots, y_T$. Then the following relations hold:

$$\hat{p}(\text{SC}) \le \hat{p}(\text{AIC}) \quad \text{if } T \ge 8; \tag{4.3.10}$$

$$\hat{p}(\text{SC}) \le \hat{p}(\text{HQ}) \quad \text{for all } T; \tag{4.3.11}$$

$$\hat{p}(\text{HQ}) \le \hat{p}(\text{AIC}) \quad \text{if } T \ge 16. \tag{4.3.12}$$

■

Note that we do not require stationarity of y_t. In fact, we don't even require that the multiple time series is generated by a VAR process. Moreover, the proposition is valid in small samples and not just asymptotically. The proof is an easy consequence of the following lemma.

LEMMA 4.1

Let $a_0 < a_1 < \cdots < a_M$, $b_0 < b_1 < \cdots < b_M$ and $c_0 \ge c_1 \ge \cdots \ge c_M$ be real numbers. If

$$b_{m+1} - b_m < a_{m+1} - a_m, \qquad m = 0, 1, \dots, M-1, \tag{4.3.13a}$$

holds and if nonnegative integers n and k are chosen such that

$$c_n + a_n = \min\{c_m + a_m | m = 0, 1, \dots, M\} \tag{4.3.13b}$$

and

$$c_k + b_k = \min\{c_m + b_m | m = 0, 1, \dots, M\}, \tag{4.3.13c}$$

then $k \ge n$. ■

The proof of this lemma is left as an exercise (see Problem 4.2). It is now easy to prove Proposition 4.3.

Proof of Proposition 4.3:

Let $c_m = \ln|\tilde{\Sigma}_u(m)|$, $b_m = 2mK^2/T$ and $a_m = mK^2 \ln T/T$. Then $\text{AIC}(m) = c_m + b_m$ and $\text{SC}(m) = c_m + a_m$. The sequences a_m, b_m, and c_m satisfy the conditions of the lemma if

$$2K^2/T = 2(m+1)K^2/T - 2mK^2/T = b_{m+1} - b_m$$
$$< a_{m+1} - a_m = (m+1)K^2 \ln T/T - mK^2 \ln T/T = K^2 \ln T/T$$

or, equivalently, $\ln T > 2$ or $T > e^2 = 7.39$. Hence, choosing $k = \hat{p}(\text{AIC})$ and $n = \hat{p}(\text{SC})$ gives $\hat{p}(\text{SC}) \leq \hat{p}(\text{AIC})$ if $T \geq 8$. The relations (4.3.11) and (4.3.12) can be shown analogously. ■

An immediate consequence of Corollary 4.2.1 and Proposition 4.3 is that AIC and FPE asymptotically overestimate the true order with positive probability and underestimate the true order with probability zero.

COROLLARY 4.3.1

Under the conditions of Proposition 4.2, if $M > p$,

$$\lim_{T\to\infty} \Pr\{\hat{p}(\text{AIC}) < p\} = 0 \quad \text{and} \quad \lim \Pr\{\hat{p}(\text{AIC}) > p\} > 0 \tag{4.3.14}$$

and the same holds for $\hat{p}(\text{FPE})$. ■

Proof: By (4.3.10) and Corollary 4.2.2,

$$\Pr\{\hat{p}(\text{AIC}) < p\} \leq \Pr\{\hat{p}(\text{SC}) < p\} \to 0.$$

Since AIC is not consistent by Corollary 4.2.1, $\lim \Pr\{\hat{p}(\text{AIC}) = p\} < 1$. Hence (4.3.14) follows. The same holds for FPE because this criterion is asymptotically equivalent to AIC (see (4.3.3)). ■

The limitations of the asymptotic theory for the order selection criteria can be seen by considering the criterion obtained by setting c_T equal to $2 \ln \ln T$ in (4.3.6). This results in a criterion

$$C(m) = \ln|\tilde{\Sigma}_u(m)| + 2m \ln \ln T/T. \tag{4.3.15}$$

Under the conditions of Proposition 4.2 it is consistent. Yet, using Lemma 4.1 and the same line of reasoning as in the proof of Proposition 4.3, $\hat{p}(\text{AIC}) \leq \hat{p}(C)$ if $2 \ln \ln T \leq 2K^2$ or, equivalently, if $T \leq \exp(\exp K^2)$. For instance for a bivariate process ($K = 2$), $\exp(\exp K^2) \approx 5.14 \times 10^{23}$. Consequently, if $T < 5.14 \times 10^{23}$, the consistent criterion (4.3.15) chooses an order greater than or equal to $\hat{p}(\text{AIC})$ which in turn has a positive limiting probability for exceeding the true order. This example shows that large sample results sometimes are good approximations only if extreme sample sizes are available. The foregoing result was used by Quinn (1980) as an argument for making c_T a function of the dimension K of the process in the HQ criterion.

It is also of interest to compare the order selection criteria to the sequential testing procedure discussed in the previous section. We have mentioned in Section 4.2 that the order chosen in a sequence of tests will depend on the significance levels used. As a consequence a testing sequence may give the same order as a selection criterion if the significance levels are chosen accordingly. For instance, AIC chooses an order smaller than the maximum order M if $\mathrm{AIC}(M-1) < \mathrm{AIC}(M)$ or, equivalently, if

$$\lambda_{\mathrm{LR}}(1) = T(\ln|\tilde{\Sigma}_u(M-1)| - \ln|\tilde{\Sigma}_u(M)|) < 2MK^2 - 2(M-1)K^2 = 2K^2.$$

For $K = 2, 2K^2 = 8 \approx \chi^2(4)_{.90}$. Thus, for a bivariate process, in order to guarantee that AIC chooses an order less than M whenever the LR testing procedure does, we may use approximately a 10% significance level in the first test of the sequence provided the distribution of $\lambda_{\mathrm{LR}}(1)$ is well approximated by a $\chi^2(4)$-distribution.

The sequential testing procedure will not lead to a consistent order estimator if the sequence of significance levels is held constant. To see this, note that for $M > p$ and a fixed significance level γ, the null hypothesis H_0: $A_M = 0$ is rejected with probability γ. In other words, in the testing scheme M is incorrectly chosen as VAR order with probability γ. Thus, there is a positive probability of choosing too high an order. This problem can be circumvented by letting the significance level go to zero as $T \to \infty$.

4.3.4 Some Small Sample Simulation Results

As mentioned previously many of the small sample properties of interest in the context of VAR order selection are difficult to derive analytically. Therefore we have performed a small Monte Carlo experiment to get some feeling for the small sample behaviour of the estimators. Some results will now be reported.

We have simulated 1000 realizations of the VAR(2) process (4.2.1) and we have recorded the orders chosen by FPE, AIC, HQ, and SC for time series lengths of $T = 30$ and 100 and a maximum VAR order of $M = 6$. In addition we have determined the order by the sequence of LR tests described in Section 4.2 using a significance level of 5% in each individual test and corresponding critical values from χ^2-distributions. That is, we have used χ^2- rather than F-tests. The frequency distributions obtained with the five different procedures are displayed in Table 4.6. Obviously, for the sample sizes reported, none of the criteria is very successful in estimating the order $p = 2$ correctly. This may be due to the fact that A_2 contains only a single, small nonzero element. The similarity of AIC and FPE derived in (4.3.3) becomes evident for $T = 100$. The orders chosen by the LR testing procedures show that the actual significance levels are quite different from their asymptotic approximations especially for sample size $T = 30$. If λ_{LR} really had a $\chi^2(4)$-distribution the order $\hat{p} = M = 6$ should be chosen in about 5% of the cases while in the simulation experiment $\hat{p} = 6$ is chosen for about 25% of the realizations. Hence, the $\chi^2(4)$-distribution is hardly a good small sample approximation to the actual distribution of λ_{LR}.

Table 4.6. Simulation results based on 1000 realizations of a bivariate VAR(2) process

	FPE	AIC	HQ	SC	LR
VAR order	$T = 30$ frequency distributions of estimated VAR orders in %				
0	.1	.1	.7	2.6	.2
1	46.9	43.6	61.7	82.6	30.1
2	33.2	32.0	27.1	13.2	16.6
3	7.2	7.0	4.2	1.2	6.3
4	5.6	5.8	3.1	.3	9.5
5	3.5	4.1	1.2	.0	12.4
6	3.5	7.4	2.0	.1	24.9
forecast horizon	sums of average squared forecast errors of y_1 and y_2				
1	.175	.179	.163	.158	.199
2	.251	.254	.239	.235	.288
3	.309	.314	.296	.281	.360
VAR order	$T = 100$ frequency distributions of estimated VAR orders in %				
0	.0	.0	.0	.0	.0
1	17.8	17.6	43.8	74.4	21.4
2	69.8	69.6	54.0	25.4	52.6
3	8.6	8.9	2.0	.2	5.6
4	2.4	2.4	.2	.0	5.3
5	.7	.8	.0	.0	6.2
6	.7	.7	.0	.0	8.9
forecast horizon	sums of average squared forecast errors of y_1 and y_2				
1	.137	.137	.137	.137	.139
2	.185	.184	.186	.186	.187
3	.238	.238	.239	.240	.241

In Table 4.6 we also present the sum of the mean squared forecast errors of y_1 and y_2 obtained from post-sample forecasts with the estimated processes. Although SC often underestimates the true VAR order $p = 2$, the forecasts obtained with the SC models are superior to the other forecasts for $T = 30$. The reason is that not restricting the single nonzero coefficient in A_2 to zero does not sufficiently improve the forecasts to offset the additional sampling variability introduced by estimating all four elements of the A_2 coefficient matrix. For $T = 100$, corresponding forecast MSEs obtained with the different criteria and procedures are very similar although SC chooses the correct order much less often than the other criteria. This result indicates that choosing the correct VAR order and selecting a good forecasting model are objectives that may be reached by different VAR order selection procedures. Specifically, in this example, slight underestimation of the VAR order is not harmful to the forecast precision. In fact, for $T = 30$, the most parsimonious criterion which underestimates the true VAR order for about 85% of the realizations of our VAR(2) process provides the best forecasts on average.

It must be emphasized, however, that these results are very special and hold for the single bivariate VAR(2) process used in the simulations. Different results may be obtained for other processes. To substantiate this statement we have also simulated 1000 time series based on the VAR(1) process (4.2.2). Some results are given in Table 4.7. While for sample size $T = 30$ again none of the criteria and

Table 4.7. Simulation results based on 1000 realizations of a three-dimensional VAR(1) process

	FPE	AIC	HQ	SC	LR
VAR order	$T = 30$ frequency distributions of estimated VAR order in %				
0	27.1	19.9	45.3	81.1	1.2
1	49.4	33.9	40.0	18.6	4.2
2	8.3	6.2	3.3	.2	.9
3	3.0	2.1	1.5	.0	2.4
4	2.3	2.6	.7	.0	4.1
5	2.8	4.1	1.4	.0	13.8
6	7.1	31.2	7.8	.1	73.4
forecast horizon	sums of average squared forecast errors of y_1, y_2 and y_3				
1	5.78	7.84	5.76	5.31	11.12
2	6.69	9.12	6.62	5.64	12.55
3	6.69	9.36	6.57	5.74	13.38
VAR order	$T = 100$ frequency distributions of estimated VAR orders in %				
0	.0	.0	.5	7.7	.0
1	93.7	93.7	99.3	92.3	58.8
2	5.1	5.1	.2	.0	4.8
3	.8	.8	.0	.0	6.2
4	.3	.3	.0	.0	8.0
5	.0	.0	.0	.0	10.4
6	.1	.1	.0	.0	11.8
forecast horizon	sums of average squared forecast errors of y_1, y_2 and y_3				
1	4.03	4.03	4.01	4.13	4.39
2	4.98	4.98	4.95	4.97	5.39
3	5.07	5.07	5.07	5.07	5.38

procedures is very successful in detecting the correct VAR order $p = 1$, the four criteria FPE, AIC, HQ, and SC select the correct order in more than 90% of the replications for $T = 100$. The poor approximation of the small sample distribution of the LR statistic by a $\chi^2(9)$-distribution is evident. Note that we have used the critical values for 5% level individual tests from the χ^2-distribution. As in the VAR(2) example the prediction performance of the SC models is superior to the other models for $T = 30$. For both sample sizes the worst forecasts are obtained with the sequential testing procedure.

One word of caution with respect to the sums of the average squared forecast errors reported in the tables may be in order at this point. Taking the sum of two or three squared forecast errors as we did in Tables 4.6 and 4.7 may give a misleading impression of the actual forecast performance of the different models because the theoretical forecast error variances (MSEs), obtained when the VAR processes are assumed known, are different for different variables. In such a case a small percentage reduction in a large variance may be greater in nominal terms than a large percentage reduction in a small variance. However, in our examples, as far as the overall ranking is concerned, the results for the individual variables are the same as the aggregated results which are therefore given to save space.

After these two simulation experiments we still don't have a clear answer to the question which criterion to use in small sample situations. One conclusion that emerges from the two examples is that in very small samples slight underestimation of the true order is not necessarily harmful to the forecast precision. Moreover, both examples clearly demonstrate that the χ^2-approximation to the small sample distribution of the LR statistics is a poor one. In a simulation study based on many other processes Lütkepohl (1985) obtained similar results. In that study, for low order VAR processes, the most parsimonious SC criterion was found to do quite well in terms of choosing the right VAR order and providing good forecasting models. Unfortunately, in practice we often don't even know whether the underlying data generation law is of finite order VAR type. Sometimes we may just approximate an infinite order VAR scheme by a finite order model. In that case, for moderate sample sizes, some less parsimonious criterion like AIC may give superior results in terms of forecast precision. Therefore it may be a good strategy to compare the order estimates obtained with different criteria and possibly perform an analysis with different VAR orders.

4.4 Checking the Whiteness of the Residuals

In the previous sections we have considered procedures for choosing the order of a VAR model for the generation process of a given multiple time series. These procedures may be interpreted as methods for determining a filter that transforms the given data into a white noise series. In this context the criteria for model choice may be regarded as criteria for deciding whether the residuals are close enough to white noise to satisfy the investigator. Of course, if for example forecasting is the objective, it may not be of prime importance whether the residuals are really white noise as long as the model forecasts well. There are, however, situations where checking the white noise (whiteness) assumption for the residuals of a particular model is of interest. For instance, if the model order is chosen by nonstatistical methods (for example, on the basis of some economic theory) it may be useful to have statistical tools available for investigating the properties of the residuals. Moreover, because different criteria emphasize different aspects of the data generation process and may therefore all provide useful information for the analyst it is common not to rely on just one procedure or criterion for model choice but use a number of different statistical tools. Therefore, in this section, we shall discuss statistical tools for checking the autocorrelation properties of the residuals of a given VAR model.

In Sections 4.4.1 and 4.4.2 the asymptotic distributions of the residual autocovariances and autocorrelations are given under the assumption that the model residuals are indeed white noise. In Section 4.4.3 a popular statistic for checking the overall significance of the residual autocorrelations is discussed. The results of this section are adapted from Chitturi (1974), Hosking (1980, 1981a), Li & McLeod (1981), and Ahn (1988).

4.4.1 The Asymptotic Distributions of the Autocovariances and Autocorrelations of a White Noise Process

It is assumed that u_t is a K-dimensional white noise process with nonsingular covariance matrix Σ_u. For instance, u_t may represent the residuals of a VAR(p) process. Let $U := (u_1, \dots, u_T)$. The autocovariance matrices of u_t are estimated as

$$C_i := \hat{\Gamma}_u(i) := \frac{1}{T} \sum_{t=i+1}^{T} u_t u'_{t-i} = \frac{1}{T} U F_i U', \qquad i = 0, 1, \dots, h < T. \tag{4.4.1}$$

The $(T \times T)$ matrix F_i is defined in the obvious way. For instance, for $i = 2$,

$$F_i := \begin{bmatrix} 0 & 0 & \cdots & 0 & 0 & 0 \\ 0 & 0 & \cdots & 0 & 0 & 0 \\ 1 & 0 & \cdots & 0 & 0 & 0 \\ 0 & 1 & & 0 & 0 & 0 \\ \vdots & & \ddots & \vdots & \vdots & \vdots \\ 0 & 0 & \cdots & 1 & 0 & 0 \end{bmatrix}$$

$$= \begin{bmatrix} 0 & 0 & \cdots & 0 \\ 0 & 0 & \cdots & 0 \\ 1 & 0 & \cdots & 0 \\ 0 & 1 & & 0 \\ \vdots & & \ddots & \vdots \\ 0 & 0 & \cdots & 1 \end{bmatrix} \begin{bmatrix} 1 & 0 & \cdots & 0 \\ 0 & 1 & & 0 \\ \vdots & & \ddots & \vdots \\ 0 & 0 & \cdots & 1 \\ 0 & 0 & \cdots & 0 \\ 0 & 0 & \cdots & 0 \end{bmatrix}.$$

Of course, for $i = 0$, $F_0 = I_T$. In the following the precise form of F_i is not important. It is useful, though, to remember that F_i is defined such that

$$UF_iU' = \sum_{t=i+1}^{T} u_t u'_{t-i}.$$

Let

$$\mathbf{C}_h := (C_1, \ldots, C_h) = UF(I_h \otimes U'), \tag{4.4.2}$$

where $F := (F_1, \ldots, F_h)$ is a $(T \times hT)$ matrix that is understood to depend on h and T without this being indicated explicitly. Furthermore, let

$$\mathbf{c}_h := \operatorname{vec}(\mathbf{C}_h). \tag{4.4.3}$$

The estimated autocorrelation matrices of the u_t are denoted by R_i, that is,

$$R_i := D^{-1}C_iD^{-1}, \qquad i = 0, 1, \ldots, h, \tag{4.4.4}$$

where D is a $(K \times K)$ diagonal matrix, the diagonal elements being the square roots of the diagonal elements of C_0. In other words, a typical element of R_i is

$$r_{mn,i} = \frac{c_{mn,i}}{\sqrt{c_{mm,0}}\sqrt{c_{nn,0}}}$$

where $c_{mn,i}$ is the mn-th element of C_i. The matrix R_i in (4.4.4) is an estimator of the true autocorrelation matrix $R_u(i) = 0$ for $i \neq 0$. We use the notation

$$\mathbf{R}_h := (R_1, \ldots, R_h) \qquad \text{and} \qquad \mathbf{r}_h := \operatorname{vec}(\mathbf{R}_h) \tag{4.4.5}$$

and we denote by R_u the *true* correlation matrix corresponding to Σ_u. Now we can give the asymptotic distributions of $\mathbf{r}_h$ and $\mathbf{c}_h$.

PROPOSITION 4.4 (*Asymptotic Distributions of White Noise Autocovariances and Autocorrelations*)

Let u_t be a K-dimensional identically distributed standard white noise process, that is, u_t and u_s have the same multivariate distribution with nonsingular covariance matrix Σ_u and corresponding correlation matrix R_u. Then, for $h \geq 1$,

$$\sqrt{T}\mathbf{c}_h \xrightarrow{d} N(0, I_h \otimes \Sigma_u \otimes \Sigma_u) \tag{4.4.6}$$

and

$$\sqrt{T}\mathbf{r}_h \xrightarrow{d} N(0, I_h \otimes R_u \otimes R_u). \tag{4.4.7}$$

■

The result in (4.4.6) means that $\sqrt{T}\operatorname{vec}(C_i)$ has the same asymptotic distribution as $\sqrt{T}\operatorname{vec}(C_j)$, namely,

$$\sqrt{T}\operatorname{vec}(C_i), \sqrt{T}\operatorname{vec}(C_j) \xrightarrow{d} N(0, \Sigma_u \otimes \Sigma_u).$$

Moreover, for $i \neq j$ the two are asymptotically independent. By (4.4.7) the same holds for $\sqrt{T}\,\mathrm{vec}(R_i)$ and $\sqrt{T}\,\mathrm{vec}(R_j)$. In fact, (4.4.7) is a quite easy consequence of (4.4.6). From Proposition 3.2 we know that C_0 is a consistent estimator of Σ_u. Hence,

$$\sqrt{T}\,\mathrm{vec}(R_i) = \sqrt{T}(D^{-1} \otimes D^{-1})\,\mathrm{vec}(C_i) \xrightarrow{d} N(0, R_u \otimes R_u)$$

by Proposition C.4(1) of Appendix C and (4.4.6), because

$$\begin{aligned}\mathrm{plim}(D^{-1} \otimes D^{-1})(\Sigma_u \otimes \Sigma_u)(D^{-1} \otimes D^{-1}) &= \mathrm{plim}(D^{-1}\Sigma_u D^{-1} \otimes D^{-1}\Sigma_u D^{-1}) \\ &= R_u \otimes R_u.\end{aligned}$$

The result (4.4.6) follows from an appropriate central limit theorem. Proofs can be found in Fuller (1976, Chapter 6) and Hannan (1970, Chapter IV, Section 4) among others.

In practice the u_t and hence U will usually be unknown and the reader may wonder about the relevance of Proposition 4.4. The result is not only useful in proving other propositions but can also be used to check whether a given time series is white noise. Before we explain that procedure we mention that Proposition 4.4 remains valid if the considered white noise process is allowed to have nonzero mean and the mean vector is estimated by the sample mean vector. That is, we consider covariance matrices

$$C_i = \frac{1}{T}\sum_{t=i+1}^{T} (u_t - \bar{u})(u_{t-i} - \bar{u})',$$

where

$$\bar{u} = \frac{1}{T}\sum_{t=1}^{T} u_t.$$

Next we observe that the diagonal elements of $R_u \otimes R_u$ are all ones. Consequently, the variances of the asymptotic distributions of the elements of $\sqrt{T}\mathbf{r}_h$ are all unity. Hence, in large samples the $\sqrt{T}r_{mn,i}$ for $i > 0$ have approximate standard normal distributions. Denoting by $\rho_{mn}(i)$ the true correlation coefficients corresponding to the $r_{mn,i}$, a test, with level approximately 5%, of the null hypothesis

$$H_0\colon \rho_{mn}(i) = 0 \qquad \text{against} \qquad H_1\colon \rho_{mn}(i) \neq 0$$

rejects H_0 if $|\sqrt{T}r_{mn,i}| > 2$ or, equivalently, $|r_{mn,i}| > 2/\sqrt{T}$.

Now we have a test for checking the null hypothesis that a given multiple time series is generated by a white noise process. We simply compute the correlations of the original data (possibly after some stationarity transformation) and compare their absolute values with $2/\sqrt{T}$. In Section 4.3.2 we found that the SC and HQ estimate of the order for the generation process of the investment/income/consumption example data is $\hat{p} = 0$. Therefore one may want to check the white noise hypothesis for this example. The first two correlation matrices for the data from 1960.4 to 1978.4 are

$$R_1 = \begin{bmatrix} -.197 & .103 & .128 \\ .190 & .020 & .228 \\ -.047 & .150 & -.089 \end{bmatrix} \quad \text{and} \quad R_2 = \begin{bmatrix} -.045 & .067 & .097 \\ .119 & .079 & .009 \\ .255 & .355 & .279 \end{bmatrix}. \tag{4.4.8}$$

Comparing these quantities with $2/\sqrt{T} = 2/\sqrt{73} = .234$ we find that some are significantly different from zero and hence we reject the white noise hypothesis on the basis of this test.

In practice the estimated autocorrelations are often plotted and $\pm 2/\sqrt{T}$-bounds around zero are indicated. The white noise hypothesis is then rejected if any of the estimated correlation coefficients reach out of the area between the $\pm 2/\sqrt{T}$-bounds. In Figure 4.1 plots of some autocorrelations are provided for the example data. Some autocorrelations at lags 2, 4, 8, and 11 are seen to be significant under the aforementioned criterion.

There are several points that must be kept in mind in such a procedure. First, in an exact 5% level test, on average the test will reject one out of twenty times it is performed independently, even if the null hypothesis is correct. Thus, one would expect that one out of twenty autocorrelation estimates exceeds $2/\sqrt{T}$ in absolute value even if the underlying process is indeed white noise. Note, however, that although R_i and R_j are asymptotically independent for $i \neq j$ the same is not necessarily true for the elements of R_i. Thus, considering the individual correlation coefficients may provide a misleading picture of their significance as a group. A test for overall significance of groups of autocorrelations is discussed in Section 4.4.3.

Second, the tests we have considered here are just asymptotic tests. In other words, the actual sizes of the tests may differ from their nominal sizes. In fact, it has been shown by Dufour & Roy (1985) and others that in small samples the variances of the correlation coefficients may differ considerably from $1/T$. They will often be smaller so that the tests are conservative in that they reject the null hypothesis less often than is indicated by the significance level chosen.

Despite this criticism, this check for whiteness of a time series enjoys much popularity as it is very easy to carry out. It is a good idea, however, not to rely on this test exclusively.

4.4.2 The Asymptotic Distributions of the Residual Autocovariances and Autocorrelations of an Estimated VAR Process

4.4.2a Theoretical Results

If a VAR(p) model has been fitted to the data a procedure similar to that described in the previous subsection is often used to check the whiteness of the residuals. Instead of the actual u_t's the estimation residuals are used, however. We will now consider the consequences of that approach. For that purpose we assume that the model has been estimated by LS and, using the notation of Section 3.2, the coefficient estimator is $\hat{B}$ and the corresponding residuals are

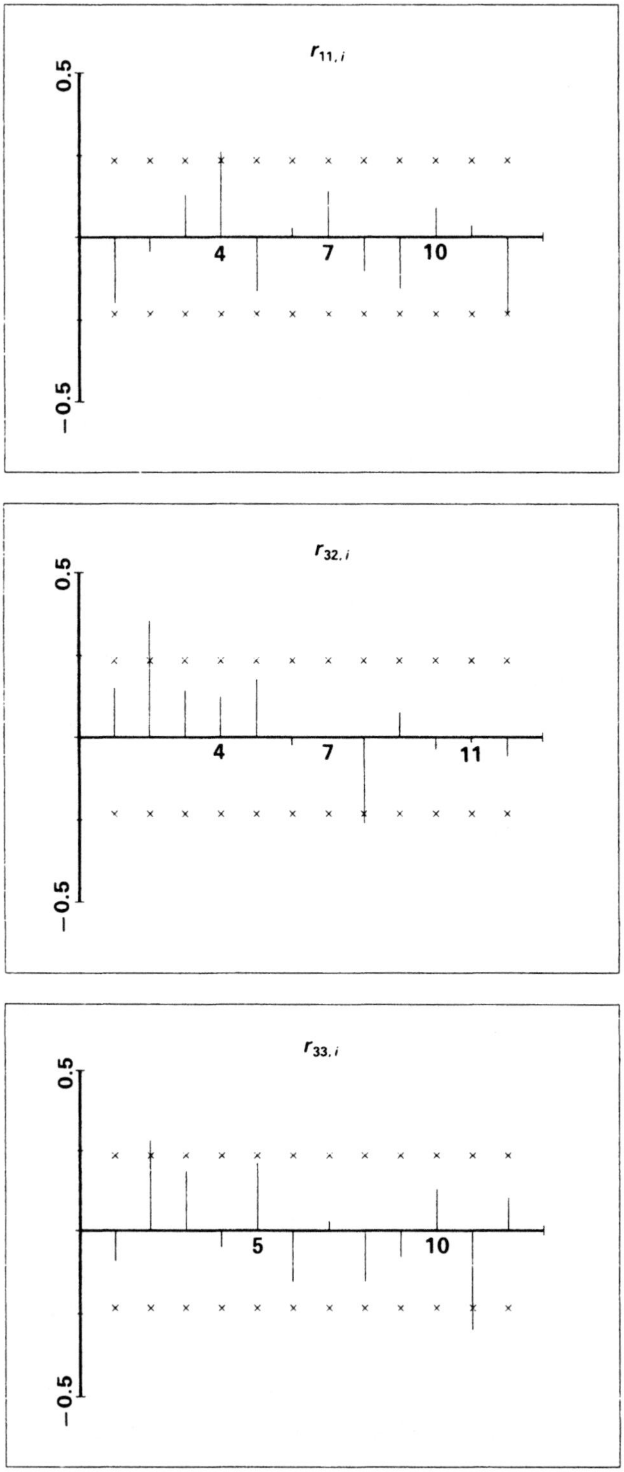

Fig. 4.1. Some estimated autocorrelations of the investment/income/consumption system.

$\hat{U} := (\hat{u}_1, \ldots, \hat{u}_T) = Y - \hat{B}Z$. Furthermore

$$\hat{C}_i := \frac{1}{T}\hat{U}F_i\hat{U}', \qquad i = 0, 1, \ldots h,$$

$$\hat{\mathbf{C}}_h := (\hat{C}_1, \ldots, \hat{C}_h) = \frac{1}{T}\hat{U}F(I_h \otimes \hat{U}'), \tag{4.4.9}$$

$$\hat{\mathbf{c}}_h := \operatorname{vec}(\hat{\mathbf{C}}_h),$$

and, correspondingly,

$$\hat{R}_i := \hat{D}^{-1}\hat{C}_i\hat{D}^{-1}, \qquad \hat{\mathbf{R}}_h := (\hat{R}_1, \ldots, \hat{R}_h), \qquad \hat{\mathbf{r}}_h := \operatorname{vec}(\hat{\mathbf{R}}_h), \tag{4.4.10}$$

where $\hat{D}$ is a diagonal matrix with the square roots of the diagonal elements of $\hat{C}_0$ on the diagonal. We will consider the asymptotic distribution of $\sqrt{T}\hat{\mathbf{c}}_h$ first. For that purpose the following lemma is helpful.

LEMMA 4.2

Let y_t be a stationary, stable VAR(p) process as in (4.1.1) with identically distributed standard white noise u_t and let $\hat{B}$ be a consistent estimator of $B = [\nu, A_1, \ldots, A_p]$ such that $\sqrt{T}\operatorname{vec}(\hat{B} - B)$ has an asymptotic normal distribution. Then $\sqrt{T}\hat{\mathbf{c}}_h$ has the same asymptotic distribution as

$$\sqrt{T}\mathbf{c}_h - \sqrt{T}G\operatorname{vec}(\hat{B} - B), \tag{4.4.11}$$

where $G := \tilde{G}' \otimes I_K$ with

$$\tilde{G} := \begin{bmatrix} 0 & 0 & \cdots & 0 \\ \Sigma_u & \Phi_1\Sigma_u & \cdots & \Phi_{h-1}\Sigma_u \\ 0 & \Sigma_u & \cdots & \Phi_{h-2}\Sigma_u \\ \vdots & & & \vdots \\ 0 & 0 & \cdots & \Phi_{h-p}\Sigma_u \end{bmatrix} ((Kp + 1) \times Kh). \tag{4.4.12}$$

■

Proof: Using the notation $Y = BZ + U$,

$$\hat{U} = Y - \hat{B}Z = BZ + U - \hat{B}Z = U - (\hat{B} - B)Z.$$

Hence,

$$\begin{aligned}\hat{U}F(I_h \otimes \hat{U}') = {} & UF(I_h \otimes U') - UF[I_h \otimes Z'(\hat{B} - B)'] \\ & - (\hat{B} - B)ZF(I_h \otimes U') + (\hat{B} - B)ZF[I_h \otimes Z'(\hat{B} - B)'].\end{aligned} \tag{4.4.13}$$

Dividing by T and applying the vec operator this expression becomes $\hat{\mathbf{c}}_h$. In order to obtain the expression in (4.4.11) we consider the terms on the right-hand side of (4.4.13) in turn. The first term becomes $\sqrt{T}\mathbf{c}_h$ upon division by $\sqrt{T}$ and application of the vec operator.

Dividing the second and last terms by $\sqrt{T}$ they can be shown to converge to zero in probability, that is,

$$\operatorname{plim} \sqrt{T} UF[I_h \otimes Z'(\hat{B} - B)']/T = 0 \tag{4.4.14}$$

and

$$\operatorname{plim} \sqrt{T}(\hat{B} - B)ZF[I_h \otimes Z'(\hat{B} - B)']/T = 0 \tag{4.4.15}$$

(see Problem 4.3). Thus it remains to show that dividing the third term in (4.4.13) by $\sqrt{T}$ and applying the vec operator yields an expression which is asymptotically equivalent to the last term in (4.4.11). To see this consider

$$ZF(I_h \otimes U') = (ZF_1 U', \dots, ZF_h U')$$

and

$$ZF_i U' = \sum_{t=i+1}^{T} Z_{t-1} u'_{t-i} = \sum_{t=i+1}^{T} \begin{bmatrix} 1 \\ y_{t-1} \\ \vdots \\ y_{t-p} \end{bmatrix} u'_{t-i},$$

$$= \sum_t \begin{bmatrix} 1 \\ \sum_{j=0}^{\infty} \Phi_j u_{t-1-j} \\ \vdots \\ \sum_{j=0}^{\infty} \Phi_j u_{t-p-j} \end{bmatrix} u'_{t-i},$$

where the Φ_i are the coefficient matrices of the canonical MA representation of y_t (see (2.1.17)). Upon division by T and application of the plim we get

$$\operatorname{plim} \frac{1}{T} ZF_i U' = \begin{bmatrix} 0 \\ \Phi_{i-1} \Sigma_u \\ \vdots \\ \Phi_{i-p} \Sigma_u \end{bmatrix},$$

where $\Phi_j = 0$ for $j < 0$. Hence,

$$\operatorname{plim} \frac{1}{T} ZF(I_h \otimes U') = \begin{bmatrix} 0 & 0 & \cdots & 0 \\ \Sigma_u & \Phi_1 \Sigma_u & \cdots & \Phi_{h-1} \Sigma_u \\ 0 & \Sigma_u & \cdots & \Phi_{h-2} \Sigma_u \\ \vdots & & & \vdots \\ 0 & 0 & \cdots & \Phi_{h-p} \Sigma_u \end{bmatrix} = \underset{((Kp+1) \times Kh)}{\tilde{G}}.$$

The lemma follows by noting that

$$\operatorname{vec}[(\hat{B} - B)ZF(I_h \otimes U')] = ([ZF(I_h \otimes U')]' \otimes I_K) \operatorname{vec}(\hat{B} - B). \qquad \blacksquare$$

The next lemma is also helpful later.

LEMMA 4.3

If y_t is a stable VAR(p) process as in (4.1.1) with identically distributed standard white noise, then

$$\begin{bmatrix} \frac{1}{\sqrt{T}} \operatorname{vec}(UZ') \\ \sqrt{T}\mathbf{c}_h \end{bmatrix} \xrightarrow{d} N\left[0, \begin{bmatrix} \Gamma & \tilde{G} \\ \tilde{G}' & I_h \otimes \Sigma_u \end{bmatrix} \otimes \Sigma_u\right], \tag{4.4.16}$$

where $\Gamma := \operatorname{plim} ZZ'/T$ and $\tilde{G}$ is as defined in (4.4.12). ■

We will not prove this result but refer the reader to Ahn (1988) for a proof. Now the asymptotic distribution of the residual autocovariances is easily obtained.

PROPOSITION 4.5 (*Asymptotic Distributions of Residual Autocovariances*)

Let y_t be a stationary, stable, K-dimensional VAR(p) process as in (4.1.1) with identically distributed standard white noise process u_t and let the coefficients be estimated by multivariate LS or an asymptotically equivalent procedure. Then

$$\sqrt{T}\hat{\mathbf{c}}_h \xrightarrow{d} N(0, \Sigma_c(h)),$$

where

$$\begin{aligned} \Sigma_c(h) &= (I_h \otimes \Sigma_u - \tilde{G}'\Gamma^{-1}\tilde{G}) \otimes \Sigma_u \\ &= (I_h \otimes \Sigma_u \otimes \Sigma_u) - \bar{G}(\Gamma_Y(0)^{-1} \otimes \Sigma_u)\bar{G}'. \end{aligned} \tag{4.4.17}$$

Here $\tilde{G}$ and Γ are the same matrices as in Lemma 4.3, $\Gamma_Y(0)$ is the covariance matrix of $Y_t = (y_t', \ldots, y_{t-p+1}')'$ and $\bar{G} := G^{\#\prime} \otimes I_K$, where $G^{\#}$ is a $(Kp \times Kh)$ matrix which has the same form as $\tilde{G}$ except that the first row of zeroes is eliminated. ■

Proof: Using Lemma 4.2, $\sqrt{T}\hat{\mathbf{c}}_h$ is known to have the same asymptotic distribution as

$$\begin{aligned} &\sqrt{T}\mathbf{c}_h - \sqrt{T}G \operatorname{vec}(\hat{B} - B) \\ &= [-\tilde{G}' \otimes I_K, I] \begin{bmatrix} \sqrt{T} \operatorname{vec}(\hat{B} - B) \\ \sqrt{T}\mathbf{c}_h \end{bmatrix} \\ &= [-\tilde{G}' \otimes I_K, I] \begin{bmatrix} \left[\frac{ZZ'}{T}\right]^{-1} \otimes I_K & 0 \\ 0 & I \end{bmatrix} \begin{bmatrix} \frac{1}{\sqrt{T}} \operatorname{vec}(UZ') \\ \sqrt{T}\mathbf{c}_h \end{bmatrix}. \end{aligned}$$

Noting that $\operatorname{plim}(ZZ'/T)^{-1} = \Gamma^{-1}$ the desired result follows from Lemma 4.3 and Proposition C.4(1) of Appendix C because

$$\begin{aligned} &[-\tilde{G}'\Gamma^{-1} \otimes I_K, I] \left[\begin{bmatrix} \Gamma & \tilde{G} \\ \tilde{G}' & I_h \otimes \Sigma_u \end{bmatrix} \otimes \Sigma_u\right] \begin{bmatrix} -\Gamma^{-1}\tilde{G} \otimes I_K \\ I \end{bmatrix} \\ &= (I_h \otimes \Sigma_u - \tilde{G}'\Gamma^{-1}\tilde{G}) \otimes \Sigma_u \end{aligned}$$

$$= I_h \otimes \Sigma_u \otimes \Sigma_u - (\tilde{G}' \otimes I_K)(\Gamma^{-1} \otimes \Sigma_u)(\tilde{G} \otimes I_K)$$
$$= I_h \otimes \Sigma_u \otimes \Sigma_u - \bar{G}(\Gamma_Y(0)^{-1} \otimes \Sigma_u)\bar{G}'. \qquad ■$$

The form (4.4.17) shows that the variances are lower than (not greater than) the diagonal elements of $I_h \otimes \Sigma_u \otimes \Sigma_u$. In other words, the variances of the asymptotic distribution of the white noise autocovariances are greater than or equal to the corresponding quantities of the estimation residuals. A similar result can also be shown for the autocorrelations of the estimation residuals.

PROPOSITION 4.6 (*Asymptotic Distributions of Residual Autocorrelations*)

Let D be the $(K \times K)$ diagonal matrix with the square roots of Σ_u on the diagonal and define $G_0 := \tilde{G}(I_h \otimes D^{-1})$. Then, under the conditions of Proposition 4.5,

$$\sqrt{T}\hat{\mathbf{r}}_h \xrightarrow{d} N(0, \Sigma_r(h)),$$

where

$$\Sigma_r(h) = ((I_h \otimes R_u) - G_0'\Gamma^{-1}G_0) \otimes R_u. \tag{4.4.18}$$

Specifically,

$$\sqrt{T}\,\mathrm{vec}(\hat{R}_j) \xrightarrow{d} N(0, \Sigma_R(j)), \qquad j = 1, 2, \ldots,$$

where

$$\Sigma_R(j) = \left[R_u - D^{-1}\Sigma_u[0 \quad \Phi_{j-1}' \ldots \Phi_{j-p}']\Gamma^{-1} \begin{bmatrix} 0 \\ \Phi_{j-1} \\ \vdots \\ \Phi_{j-p} \end{bmatrix} \Sigma_u D^{-1} \right] \otimes R_u \tag{4.4.19}$$

with $\Phi_i = 0$ for $i < 0$. ■

Proof: Noting that

$$\hat{\mathbf{r}}_h = \mathrm{vec}(\hat{\mathbf{R}}_h) = \mathrm{vec}[\hat{D}^{-1}\hat{\mathbf{C}}_h(I_h \otimes \hat{D}^{-1})]$$
$$= (I_h \otimes \hat{D}^{-1} \otimes \hat{D}^{-1})\hat{\mathbf{c}}_h$$

and $\hat{D}^{-1}$ is a consistent estimator of D^{-1}, we get from Proposition 4.5 that $\sqrt{T}\hat{\mathbf{r}}_h$ has an asymptotic normal distribution with mean zero and covariance matrix

$$(I_h \otimes D^{-1} \otimes D^{-1})\{(I_h \otimes \Sigma_u - \tilde{G}'\Gamma^{-1}\tilde{G}) \otimes \Sigma_u\}(I_h \otimes D^{-1} \otimes D^{-1})$$
$$= [(I_h \otimes R_u) - G_0'\Gamma^{-1}G_0] \otimes R_u$$

where $D^{-1}\Sigma_u D^{-1} = R_u$ has been used. ■

From (4.4.19) it is obvious that the diagonal elements of the asymptotic covariance matrices are not greater than 1 because a positive semidefinite matrix is subtracted from R_u. Hence, if estimation residual autocorrelations are used in a white noise test in a similar fashion as the autocorrelations of the original data, we will get a conservative test that rejects the null hypothesis less often than is indicated by the significance level, provided the asymptotic distribution is a correct indicator of the small sample behavior of the test. In particular for autocorrelations at small lags the variances will be less than 1, while the asymptotic variances approach one for elements of $\sqrt{T}\hat{R}_j$ with great j. This conclusion follows because $\Phi_{j-1}, \ldots, \Phi_{j-p}$ approach zero as $j \to \infty$. As a consequence, the matrix substracted from R_u goes to zero as $j \to \infty$.

In practice all unknown quantities are replaced by estimates in order to obtain standard errors of the residual autocorrelations and tests of specific hypotheses regarding the autocorrelations. It is perhaps worth noting, though, that if Γ is estimated by ZZ'/T, we have to use the ML estimator $\tilde{\Sigma}_u$ for Σ_u in order to ensure positive variances.

4.4.2b An Illustrative Example

As an example we consider the VAR(2) model for the investment/income/consumption system estimated in Section 3.2.3. For $j = 1$ we get

$$\hat{R}_1 = \begin{bmatrix} \underset{(.026)}{.015} & \underset{(.033)}{-.011} & \underset{(.049)}{-.010} \\ \underset{(.026)}{-.007} & \underset{(.033)}{-.002} & \underset{(.049)}{-.068} \\ \underset{(.026)}{-.024} & \underset{(.033)}{-.045} & \underset{(.049)}{-.096} \end{bmatrix},$$

where the estimated standard errors are given in parentheses. Obviously the standard errors of the elements of $\hat{R}_1$ are much smaller than $1/\sqrt{T} = .117$ which would be obtained if the variances of the elements of $\sqrt{T}\hat{R}_1$ were 1. In contrast, for $j = 6$ we get

$$\hat{R}_6 = \begin{bmatrix} \underset{(.117)}{.053} & \underset{(.116)}{-.008} & \underset{(.117)}{-.062} \\ \underset{(.117)}{.165} & \underset{(.116)}{.030} & \underset{(.117)}{-.051} \\ \underset{(.117)}{.068} & \underset{(.116)}{.026} & \underset{(.117)}{.020} \end{bmatrix},$$

where the standard errors are very close to .117.

In Figure 4.2 we have plotted the residual autocorrelations and *twice* their asymptotic standard errors (approximate 95% confidence bounds) around zero. Here it can be seen how the confidence bounds grow with increasing lag length. For a rough check of 5% level significance of autocorrelations at higher lags we may use the $\pm 2/\sqrt{T}$-bounds in practice which is convenient from a computational viewpoint.

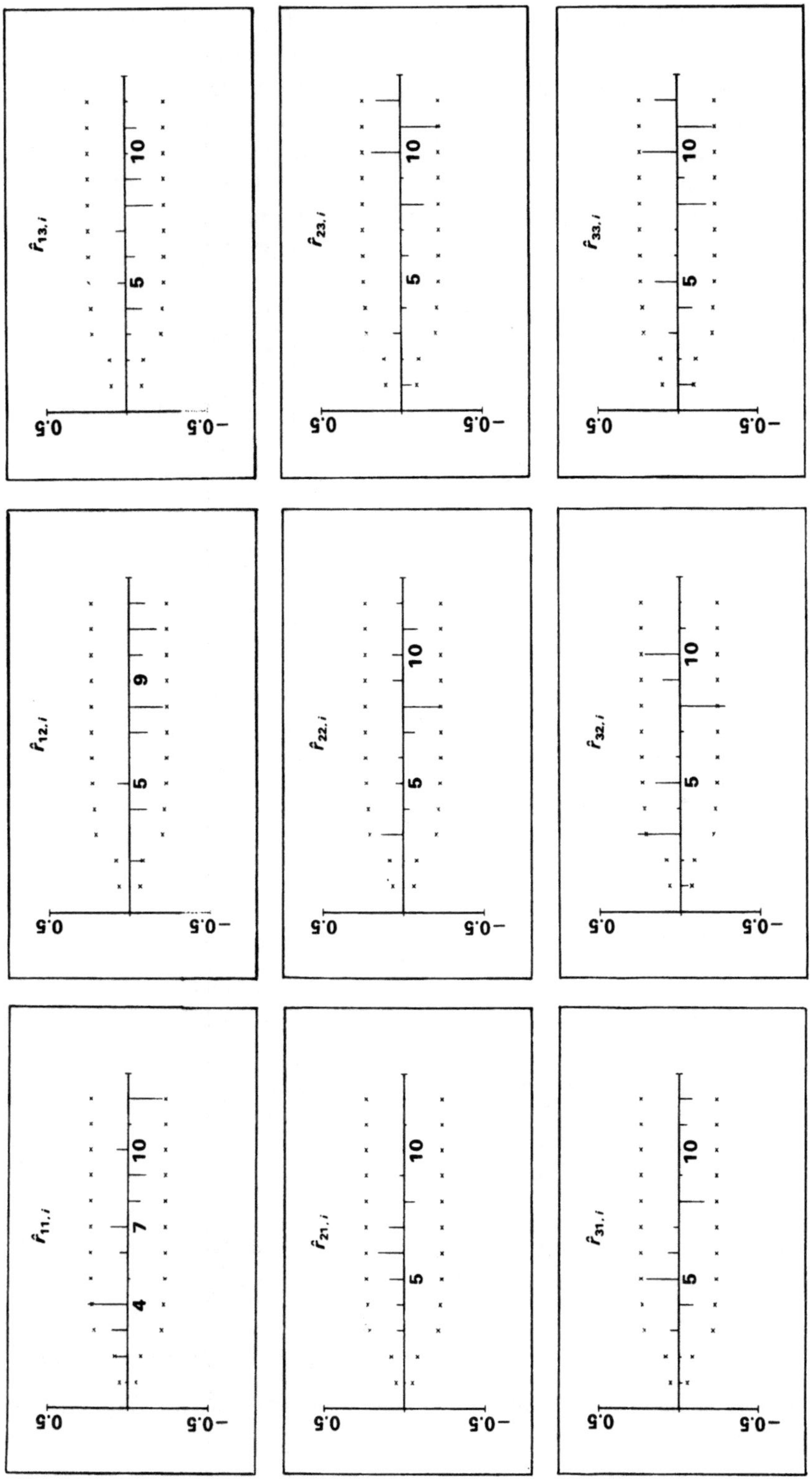

Fig. 4.2. Estimated residual autocorrelations of the investment/income/consumption VAR(2) model.

There are significant residual autocorrelations at lags 3, 4, 8, and 11. While the significant values at lags 3 and 4 may be a reason for concern one may not worry too much about the higher order lags because one may not be willing to fit a high order model if forecasting is the objective. As we have seen in Section 4.3.4, slight underfitting may even improve the forecast performance. In order to remove the significant residual autocorrelations at low lags it may help to fit a VAR(3) or VAR(4) model. Of course, this conflicts with choosing the model order on the basis of the model selection criteria. Thus, it has to be decided which criterion is given priority.

It may be worth noting that a plot like that in Figure 4.2 may give a misleading picture of the overall significance of the residual autocorrelations because they are not asymptotically independent. In particular at low lags there will not only be nonzero correlation between the elements of a specific $\hat{R}_j$ but also between $\hat{R}_j$ and $\hat{R}_i$ for $i \neq j$. Therefore it is desirable to have a test for overall significance of the residual autocorrelations of a VAR(p) model. Such a test is discussed in the next subsection.

4.4.3 Portmanteau Tests

The foregoing results may also be used to construct a popular test for the overall significance of the residual autocorrelations up to lag h. This test is commonly called *portmanteau test*. It is designed for testing

$$H_0: \mathbf{R}_h = (R_1, \dots, R_h) = 0 \qquad \text{against} \qquad H_1: \mathbf{R}_h \neq 0. \tag{4.4.20}$$

The test statistic is

$$\begin{aligned} P_h &:= T \sum_{i=1}^{h} \operatorname{tr}(\hat{R}_i' \hat{R}_u^{-1} \hat{R}_i \hat{R}_u^{-1}) \\ &= T \sum_{i=1}^{h} \operatorname{tr}(\hat{R}_i' \hat{R}_u^{-1} \hat{R}_i \hat{R}_u^{-1} \hat{D}^{-1} \hat{D}) \\ &= T \sum_{i=1}^{h} \operatorname{tr}(\hat{D} \hat{R}_i' \hat{D} \hat{D}^{-1} \hat{R}_u^{-1} \hat{D}^{-1} \hat{D} \hat{R}_i \hat{D} \hat{D}^{-1} \hat{R}_u^{-1} \hat{D}^{-1}) \\ &= T \sum_{i=1}^{h} \operatorname{tr}(\hat{C}_i' \hat{C}_0^{-1} \hat{C}_i \hat{C}_0^{-1}). \end{aligned} \tag{4.4.21}$$

Obviously, this statistic is very easy to compute from the estimation residuals. By Proposition 4.5 it has an approximate asymptotic χ^2-distribution.

PROPOSITION 4.7 (*Approximate Distribution of the Portmanteau Statistic*)

Under the conditions of Proposition 4.5 we have, approximately, for large T and h,

$$P_h = T \sum_{i=1}^{h} \operatorname{tr}(\hat{C}_i'\hat{C}_0^{-1}\hat{C}_i\hat{C}_0^{-1})$$
$$= T \operatorname{vec}(\hat{\mathbf{C}}_h)'(I_h \otimes \hat{C}_0^{-1} \otimes \hat{C}_0^{-1}) \operatorname{vec}(\hat{\mathbf{C}}_h) \approx \chi^2(K^2(h-p)). \tag{4.4.22}$$

■

Sketch of the proof: By Proposition C.4(5) of Appendix C, P_h has the same asymptotic distribution as

$$T\hat{\mathbf{c}}_h'(I_h \otimes \Sigma_u^{-1} \otimes \Sigma_u^{-1})\hat{\mathbf{c}}_h.$$

Defining the $(K \times K)$ matrix P such that $PP' = \Sigma_u$ and

$$\tilde{\mathbf{c}}_h := (I_h \otimes P \otimes P)^{-1}\hat{\mathbf{c}}_h$$

it is easily seen that P_h has the same asymptotic distribution as $T\tilde{\mathbf{c}}_h'\tilde{\mathbf{c}}_h$. Hence, by Proposition C.4(6) it suffices to show that $\sqrt{T}\tilde{\mathbf{c}}_h \xrightarrow{d} N(0, Q)$, where Q is an idempotent matrix of rank $K^2h - K^2p$. Since an approximate limiting χ^2-distribution of P_h is claimed only, we just show that Q is approximately equal to an idempotent matrix with rank $K^2(h - p)$.

Using Proposition 4.5 we get

$$Q = (I_h \otimes P^{-1} \otimes P^{-1})\Sigma_c(h)(I_h \otimes P'^{-1} \otimes P'^{-1})$$
$$= I_{hK^2} - \mathbf{P}\bar{G}(\Gamma_Y(0)^{-1} \otimes \Sigma_u)\bar{G}'\mathbf{P}',$$

where $\mathbf{P} = I_h \otimes P^{-1} \otimes P^{-1}$ and $\bar{G}$ is defined in Proposition 4.5. Noting that the ij-th block of $\Gamma_Y(0)$ is

$$\operatorname{Cov}(y_{t-i}, y_{t-j}) = \Gamma_y(j-i) = \sum_{n=0}^{\infty} \Phi_{n-i}\Sigma_u\Phi_{n-j}'$$

with $\Phi_k = 0$ for $k < 0$, we get approximately

$$\Gamma_y(0) \otimes \Sigma_u^{-1} \approx \left[\sum_{n=1}^{h} \Phi_{n-i}\Sigma_u\Phi_{n-j}'\right]_{i,j=1,\ldots,p} \otimes \Sigma_u^{-1}$$
$$= \left[\sum_{n=1}^{h} \Phi_{n-i}\Sigma_u P'^{-1}P^{-1}\Sigma_u\Phi_{n-j}'\right]_{i,j} \otimes \Sigma_u^{-1}$$
$$= \bar{G}'\mathbf{P}'\mathbf{P}\bar{G}.$$

Hence, if h is such that $\Phi_i \approx 0$ for $i > h - p$,

$$Q \approx I_{hK^2} - \mathbf{P}\bar{G}(\bar{G}'\mathbf{P}'\mathbf{P}\bar{G})^{-1}\bar{G}'\mathbf{P}'.$$

Thus, Q is approximately equal to an idempotent matrix with rank

$$\operatorname{tr}(I_{hK^2} - \mathbf{P}\bar{G}(\bar{G}'\mathbf{P}'\mathbf{P}\bar{G})^{-1}\bar{G}'\mathbf{P}') = hK^2 - pK^2,$$

as was to be shown. ■

Of course, this does not fully prove Proposition 4.7 because we have not shown that an approximately idempotent matrix Q leads to an approximate χ^2-distribution. To actually obtain the limiting χ^2-distribution we have to assume that h goes to infinity with the sample size. Since the sketch of the proof should suffice to show in what sense the result is approximate we do not persue this issue further and refer the reader to Ahn (1988) for details. For practical purposes it is important to remember that the χ^2-approximation to the distribution of the test statistic may be misleading for small values of h.

Like in previous sections we have discussed *asymptotic* distributions in this section. Not knowing the small sample distribution is clearly a shortcoming because in practice infinite samples are not available. Using Monte Carlo techniques it was found by some researchers that in small samples the nominal size of the portmanteau test tends to be lower than the significance level chosen (Davies, Triggs & Newbold (1977), Ljung & Box (1978), Hosking (1980)). As a consequence the test has low power against many alternatives. Therefore it has been suggested to use the modified test statistic

$$\bar{P}_h := T^2 \sum_{i=1}^{h} (T-i)^{-1} \operatorname{tr}(\hat{C}_i' \hat{C}_0^{-1} \hat{C}_i \hat{C}_0^{-1}). \tag{4.4.23}$$

The modification may be regarded as an adjustment for the number of terms in the sum in

$$\hat{C}_i = \frac{1}{T} \sum_{t=i+1}^{T} \hat{u}_t \hat{u}_{t-i}'.$$

For $T \to \infty$, $T/[T^2/(T-i)] \to 1$ and thus $\bar{P}_h$ has the same asymptotic distribution as P_h, that is, approximately in large samples and for large h,

$$\bar{P}_h \approx \chi^2(K^2(h-p)). \tag{4.4.24}$$

For our example model we obtained $\bar{P}_{12} = 81.9$. Comparing this value with $\chi^2(K^2(h-p))_{.95} = \chi^2(90)_{.95} \approx 113$ shows that we cannot reject the white noise hypothesis for the residuals.

As mentioned in the introduction to this section, these tests can also be used in a model selection/order estimation procedure. A sequence of hypotheses as in (4.2.15) is tested in such a procedure by checking whether the residuals are white noise. Hosking (1981b) has shown that the portmanteau test may be viewed as a Lagrange multiplier test for zero constraints on VAR coefficients.

4.5 Testing for Nonnormality

Normality of the underlying data generating process is needed, for instance, in setting up forecast intervals. Therefore testing this distributional assumption is desirable. We will present tests for multivariate normality of a white noise process first. In Subsection 4.5.2 it is then demonstrated that the tests remain valid if the true residuals are replaced by the residuals of an estimated VAR(p) process.

4.5.1 Tests for Nonnormality of a Vector White Noise Process

The tests developed in the following are based on the third and fourth central moments (skewness and kurtosis) of the normal distribution. If x is a univariate random variable with standard normal distribution, i.e., $x \sim N(0, 1)$, its third and fourth moments are known to be $E(x^3) = 0$ and $E(x^4) = 3$. Let u_t be a K-dimensional Gaussian white noise process with $u_t \sim N(\mu_u, \Sigma_u)$ and let P be a lower triangular matrix with positive diagonal satisfying $PP' = \Sigma_u$. Then

$$w_t = (w_{1t}, \dots, w_{Kt})' := P^{-1}(u_t - \mu_u) \sim N(0, I_K).$$

In other words, the components of w_t are independent standard normal random variables. Hence,

$$E\begin{bmatrix} w_{1t}^3 \\ \vdots \\ w_{Kt}^3 \end{bmatrix} = 0 \quad \text{and} \quad E\begin{bmatrix} w_{1t}^4 \\ \vdots \\ w_{Kt}^4 \end{bmatrix} = \begin{bmatrix} 3 \\ \vdots \\ 3 \end{bmatrix} =: 3_K. \tag{4.5.1}$$

This result will be utilized in checking the normality of the white noise process u_t. The idea is to compare the third and fourth moments of the transformed process with the theoretical values in (4.5.1) obtained for a Gaussian process.

For that purpose we assume to have observations $u_1, \dots, u_T$ and define

$$\bar{u} := \frac{1}{T}\sum_{t=1}^{T} u_t, \qquad S_u := \frac{1}{T-1}\sum_t (u_t - \bar{u})(u_t - \bar{u})',$$

and P_s is the lower triangular matrix with positive diagonal for which $P_sP_s' = S_u$. Moreover,

$$v_t := (v_{1t}, \dots, v_{Kt})' = P_s^{-1}(u_t - \bar{u}), \qquad t = 1, \dots, T,$$

$$b_1 := (b_{11}, \dots, b_{K1})' \qquad \text{with } b_{k1} = \frac{1}{T}\sum_t v_{kt}^3, \tag{4.5.2}$$

and

$$b_2 := (b_{12}, \dots, b_{K2})' \qquad \text{with } b_{k2} = \frac{1}{T}\sum_t v_{kt}^4. \tag{4.5.3}$$

Thus, b_1 and b_2 are estimators of the vectors in (4.5.1). In the next proposition the asymptotic distribution of b_1 and b_2 is given.

PROPOSITION 4.8 (*Asymptotic Distribution of Skewness and Kurtosis*)

If u_t is Gaussian white noise with nonsingular covariance matrix Σ_u and expectation μ_u, $u_t \sim N(\mu_u, \Sigma_u)$, then

$$\sqrt{T}\begin{bmatrix} b_1 \\ b_2 - 3_K \end{bmatrix} \xrightarrow{d} N\left[0, \begin{bmatrix} 6I_K & 0 \\ 0 & 24I_K \end{bmatrix}\right]. \qquad \blacksquare$$

In other words, b_1 and b_2 are asymptotically independent and normally distributed. The proposition implies that

$$\lambda_1 := Tb_1'b_1/6 \xrightarrow{d} \chi^2(K) \tag{4.5.4}$$

and

$$\lambda_2 := T(b_2 - 3_K)'(b_2 - 3_K)/24 \xrightarrow{d} \chi^2(K). \tag{4.5.5}$$

The first statistic can be used to test

$$H_0 : E\begin{bmatrix} w_{1t}^3 \\ \vdots \\ w_{Kt}^3 \end{bmatrix} = 0 \quad \text{against} \quad H_1 : E\begin{bmatrix} w_{1t}^3 \\ \vdots \\ w_{Kt}^3 \end{bmatrix} \neq 0 \tag{4.5.6}$$

and λ_2 may be used to test

$$H_0 : E\begin{bmatrix} w_{1t}^4 \\ \vdots \\ w_{Kt}^4 \end{bmatrix} = 3_K \quad \text{against} \quad H_1 : E\begin{bmatrix} w_{1t}^4 \\ \vdots \\ w_{Kt}^4 \end{bmatrix} \neq 3_K. \tag{4.5.7}$$

Furthermore

$$\lambda_3 := \lambda_1 + \lambda_2 \xrightarrow{d} \chi^2(2K), \tag{4.5.8}$$

which may be used for a joint test of the null hypotheses in (4.5.6) and (4.5.7).

Proof of Proposition 4.8

We state a helpful lemma first.

LEMMA 4.4

Let $z_t = (z_{1t}, \ldots, z_{Kt})'$ be a Gaussian white noise process with mean μ_z and covariance matrix I_K, i.e., $z_t \sim N(\mu_z, I_K)$. Furthermore, let

$$\bar{z} = (\bar{z}_1, \ldots, \bar{z}_K)' := \frac{1}{T}\sum_{t=1}^{T} z_t,$$

$b_{1,z}$ a $(K \times 1)$ vector with k-th component $b_{k1,z} := \frac{1}{T}\sum_t (z_{kt} - \bar{z}_k)^3$,

and

$b_{2,z}$ a $(K \times 1)$ vector with k-th component $b_{k2,z} := \frac{1}{T}\sum_t (z_{kt} - \bar{z}_k)^4$.

Then

$$\sqrt{T}\begin{bmatrix} b_{1,z} \\ b_{2,z} - 3_K \end{bmatrix} \xrightarrow{d} N\left[0, \begin{bmatrix} 6I_K & 0 \\ 0 & 24I_K \end{bmatrix}\right]. \tag{4.5.9}$$

■

The proof of this lemma is easily obtained, for instance, from results of Gasser (1975). Proposition 4.8 follows by noting that P_s is a consistent estimator of P (defined such that $PP' = \Sigma_u$) and by defining $z_t = P^{-1}u_t$. Hence,

$$\sqrt{T}(P_s^{-1} \otimes P_s^{-1} \otimes P_s^{-1})\frac{1}{T}\sum_t (u_t - \bar{u}) \otimes (u_t - \bar{u}) \otimes (u_t - \bar{u})$$

$$- \sqrt{T}\left[\frac{1}{T}\sum_t (z_t - \bar{z}) \otimes (z_t - \bar{z}) \otimes (z_t - \bar{z})\right]$$

$$= (P_s^{-1} \otimes P_s^{-1} \otimes P_s^{-1} - P^{-1} \otimes P^{-1} \otimes P^{-1})$$

$$\times \frac{1}{\sqrt{T}}\sum_t (u_t - \bar{u}) \otimes (u_t - \bar{u}) \otimes (u_t - \bar{u}) \xrightarrow{p} 0.$$

An analogous result is obtained for the fourth moments. Consequently,

$$\sqrt{T}\begin{bmatrix} b_1 - b_{1,z} \\ b_2 - b_{2,z} \end{bmatrix} \xrightarrow{p} 0$$

and the proposition follows from Proposition C.2 of Appendix C. ■

REMARK 1: In Proposition 4.8 the white noise process is not required to have zero mean. Thus, tests based on λ_1, λ_2, or λ_3 may be applied if the original observations are generated by a VAR(0) process. ■

REMARK 2: It is known that in the univariate case tests based on the skewness and kurtosis (third and fourth moments) have small sample distributions that differ substantially from their asymptotic counterparts (see, e.g., White & MacDonald (1980), Jarque & Bera (1987) and the references given there). Therefore tests based on λ_1, λ_2, and λ_3 in conjunction with the asymptotic χ^2-distributions in (4.5.4), (4.5.5), and (4.5.8) must be interpreted cautiously. They should be regarded as rough checks of normality only. ■

REMARK 3: Tests based on λ_1, λ_2, and λ_3 cannot be expected to possess power against distributions having the same first four moments as the normal distribution. Thus, if higher order moment characteristics are of interest, these tests cannot be recommended. Other tests for multivariate normality are described by Mardia (1980), Baringhaus & Henze (1988), and others. ■

4.5.2 Tests for Nonnormality of a VAR Process

A stationary, stable VAR(p) process, say

$$y_t - \mu = A_1(y_{t-1} - \mu) + \cdots + A_p(y_{t-p} - \mu) + u_t, \qquad (4.5.10)$$

is Gaussian (normally distributed) if and only if the white noise process u_t is Gaussian. Therefore the normality of the y_t may be checked via the u_t. In practice the u_t are replaced by estimation residuals. In the following we will demonstrate that this is of no consequence for the *asymptotic* distributions of the λ_i statistics considered in the previous subsection.

The reader may wonder why normality tests are based on estimation residuals rather than the original observations y_t. The reason is that tests based on the latter may be less powerful than those based on the estimation residuals. For the univariate case this point was demonstrated by Lütkepohl & Schneider (1989). It is also worth recalling that the forecast errors used in the construction of forecast intervals are weighted sums of the u_t. Therefore checking the normality of these quantities makes sense if the aim is to establish interval forecasts. The next proposition states that Proposition 4.8 remains valid if the true white noise innovations u_t are replaced by estimation residuals.

PROPOSITION 4.9 (*Asymptotic Distribution of Residual Skewness and Kurtosis*)

Let y_t be a K-dimensional stationary, stable Gaussian VAR(p) process as in (4.5.10), where u_t is zero mean white noise with nonsingular covariance matrix Σ_u and let $\hat{A}_1, \ldots, \hat{A}_p$ be consistent and asymptotically normally distributed estimators of the coefficients based on a sample $y_1, \ldots, y_T$ and possibly some presample values. Define

$$\hat{u}_t := (y_t - \bar{y}) - \hat{A}_1(y_{t-1} - \bar{y}) - \cdots - \hat{A}_p(y_{t-p} - \bar{y}), \qquad t = 1, \ldots, T,$$

$$\hat{\Sigma}_u := \frac{1}{T - Kp - 1} \sum_{t=1}^{T} \hat{u}_t \hat{u}_t',$$

and let $\hat{P}$ be the lower triangular matrix with positive diagonal satisfying $\hat{P}\hat{P}' = \hat{\Sigma}_u$. Furthermore, define

$$\hat{w}_t = (\hat{w}_{1t}, \ldots, \hat{w}_{Kt})' := \hat{P}^{-1}\hat{u}_t,$$

$$\hat{b}_1 = (\hat{b}_{11}, \ldots, \hat{b}_{K1})' \qquad \text{with } \hat{b}_{k1} := \frac{1}{T} \sum_{t=1}^{T} \hat{w}_{kt}^3,$$

and

$$\hat{b}_2 = (\hat{b}_{12}, \ldots, \hat{b}_{K2})' \qquad \text{with } \hat{b}_{k2} := \frac{1}{T} \sum_{t=1}^{T} \hat{w}_{kt}^4.$$

Then

$$\sqrt{T}\begin{bmatrix} \hat{b}_1 \\ \hat{b}_2 - 3_K \end{bmatrix} \xrightarrow{d} N\left[0, \begin{bmatrix} 6I_K & 0 \\ 0 & 24I_K \end{bmatrix}\right]. \qquad \blacksquare$$

Although the proposition is formulated in terms of the mean-adjusted form (4.5.10) of the process it also holds if estimation residuals from the standard intercept form are used instead. The parameter estimators may be unconstrained ML or LS estimators. However, the proposition does not require this. In other words, the proposition remains valid if, for instance, restricted LS or generalized LS estimators are used as discussed in the next chapter. The following lemma will be helpful in proving Proposition 4.9.

LEMMA 4.5

Under the conditions of Proposition 4.9,

$$\operatorname{plim}\left[\frac{1}{\sqrt{T}}\sum_{t=1}^{T}\hat{u}_t\otimes\hat{u}_t\otimes\hat{u}_t-\frac{1}{\sqrt{T}}\sum_t(u_t-\bar{u})\otimes(u_t-\bar{u})\otimes(u_t-\bar{u})\right]=0 \tag{4.5.11}$$

and

$$\operatorname{plim}\left[\frac{1}{\sqrt{T}}\sum_{t=1}^{T}\hat{u}_t\otimes\hat{u}_t\otimes\hat{u}_t\otimes\hat{u}_t - \frac{1}{\sqrt{T}}\sum_t(u_t-\bar{u})\otimes(u_t-\bar{u})\otimes(u_t-\bar{u})\otimes(u_t-\bar{u})\right]=0 \tag{4.5.12}$$

■

Proof: A proof for the special case of a VAR(1) process y_t is given and the generalization is left to the reader. Also, we just show the first result. The second one follows with analogous arguments. For the special VAR(1) case

$$\begin{aligned}\hat{u}_t &= (y_t-\bar{y})-\hat{A}_1(y_{t-1}-\bar{y})\\ &=(u_t-\bar{u})+(A_1-\hat{A}_1)(y_{t-1}-\bar{y})+a_T,\end{aligned}$$

where $a_T = A_1(y_T - y_0)/T$. Hence,

$$\frac{1}{\sqrt{T}}\sum_t[\hat{u}_t\otimes\hat{u}_t\otimes\hat{u}_t]=\frac{1}{\sqrt{T}}\sum_t[(u_t-\bar{u})\otimes(u_t-\bar{u})\otimes(u_t-\bar{u})]+d_T,$$

where d_T is a sum of expressions of the type

$$\begin{aligned}&\frac{1}{\sqrt{T}}\sum_t[(A_1-\hat{A}_1)(y_{t-1}-\bar{y})+a_T]\otimes(u_t-\bar{u})\otimes(u_t-\bar{u})\\ &\quad=\sqrt{T}[(A_1-\hat{A}_1)\otimes I_{2K}]\frac{1}{T}\sum_t[(y_{t-1}-\bar{y})\otimes(u_t-\bar{u})\otimes(u_t-\bar{u})]\\ &\qquad+\sqrt{T}a_T\otimes\frac{1}{T}\sum_t[(u_t-\bar{u})\otimes(u_t-\bar{u})],\end{aligned} \tag{4.5.13}$$

that is, d_T consists of sums of Kronecker products involving $(A_1 - \hat{A}_1)(y_{t-1} - \bar{y})$, $(u_t - \bar{u})$, and a_T. Therefore, $\operatorname{plim} d_T = 0$. For instance, (4.5.13) goes to zero in probability because

$$\operatorname{plim}\frac{1}{T}\sum_t(u_t-\bar{u})\otimes(u_t-\bar{u})\quad\text{exists and}\quad\operatorname{plim}\sqrt{T}a_T=0$$

so that the last term in (4.5.13) vanishes. Moreover, the elements of $\sqrt{T}(A_1 - \hat{A}_1)$ converge in distribution and

$$\text{plim}\,\frac{1}{T}\sum_t (y_{t-1} - \bar{y}) \otimes (u_t - \bar{u}) \otimes (u_t - \bar{u}) = 0 \tag{4.5.14}$$

(see Problem 4.4). Hence the first term in (4.5.13) vanishes. ■

Proof of Proposition 4.9

By Proposition C.2 of Appendix C and Proposition 4.8 it suffices to show that

$$(\hat{P}^{-1} \otimes \hat{P}^{-1} \otimes \hat{P}^{-1})\frac{1}{\sqrt{T}}\sum_t \hat{u}_t \otimes \hat{u}_t \otimes \hat{u}_t$$
$$- (P_s^{-1} \otimes P_s^{-1} \otimes P_s^{-1})\frac{1}{\sqrt{T}}\sum_t (u_t - \bar{u}) \otimes (u_t - \bar{u}) \otimes (u_t - \bar{u}) \xrightarrow{p} 0 \tag{4.5.15}$$

and the fourth moments possess a similar property. The result (4.5.15) follows from Lemma 4.5 by noting that $\hat{P}$ and P_s are both consistent estimators of P and, for stochastic vectors h_T, g_T and stochastic matrices H_T, G_T with

$$\text{plim}(h_T - g_T) = 0, \qquad h_T \xrightarrow{d} h,$$

and

$$\text{plim}\, H_T = \text{plim}\, G_T = H,$$

we get

$$H_T h_T - G_T g_T = (H_T - H)h_T + H(h_T - g_T) + (H - G_T)g_T \xrightarrow{p} 0.$$ ■

Proposition 4.9 implies that

$$\hat{\lambda}_1 := T\hat{b}_1'\hat{b}_1/6 \xrightarrow{d} \chi^2(K), \tag{4.5.16}$$

$$\hat{\lambda}_2 := T(\hat{b}_2 - 3_K)'(\hat{b}_2 - 3_K)/24 \xrightarrow{d} \chi^2(K), \tag{4.5.17}$$

and

$$\hat{\lambda}_3 := \hat{\lambda}_1 + \hat{\lambda}_2 \xrightarrow{d} \chi^2(2K). \tag{4.5.18}$$

Thus, all three statistics may be used in testing for nonnormality.

For illustrative purposes we use our standard investment/income/consumption example from Section 3.2.3. Using the least squares residuals from the VAR(2) model with intercepts yields

$$\hat{\lambda}_1 = 3.15 \qquad \text{and} \qquad \hat{\lambda}_2 = 4.69$$

which are both smaller than $\chi^2(3)_{.90} = 6.25$, the critical value of an asymptotic 10% level test. Also

$$\hat{\lambda}_3 = 7.84 < \chi^2(6)_{.90} = 10.64.$$

Thus, based on these asymptotic tests we cannot reject the null hypothesis of a Gaussian data generation process.

4.6 Tests for Structural Change

Time invariance or stationarity of the data generation process is an important condition that was used in deriving the properties of estimators and in computing forecasts and forecast intervals. Recall that stationarity is a property that ensures constant means, variances, and autocovariances of the process through time. As we have seen in the investment/income/consumption example, economic time series often have characteristics that do not conform with the assumption of stationarity of the underlying data generation process. For instance, economic time series often have trends or pronounced seasonal components and time varying variances. While these components can sometimes be eliminated by simple transformations there remains another important source of nonstationarity, namely events that cause turbulences in economic systems in particular time periods. For instance, wars usually change the economic conditions in some areas or countries markedly. Also new tax legislation may have a major impact on some economic variables. Furthermore, the oil price shocks in 1973/74 and 1979/80 are events that have caused drastic changes in some variables (notably the price for gasoline). Such events may be sources of structural change in economic systems.

Since stability and hence stationarity is an important assumption in our analysis it is desirable to have tools for checking this assumed property of the data generation process. In this section we consider possible tests that are based on comparing forecasts with actually observed values. More precisely, forecasts are made prior to a period of possible structural change and are compared to the values actually observed during that period. The stability or stationarity hypothesis is rejected if the forecasts differ too much from the actually observed values. The test statistics are given in Section 4.6.1 and 4.6.2 and an example is considered in Section 4.6.3. Other tests will be considered in later chapters.

4.6.1 A Test Statistic Based on one Forecast Period

Suppose y_t is a K-dimensional stationary, stable *Gaussian* VAR(p) process as in (4.1.1). The optimal h-step forecast at time T is denoted by $y_T(h)$ and the corresponding forecast error is

$$e_T(h) := y_{T+h} - y_T(h) = \sum_{i=0}^{h-1} \Phi_i u_{T+h-i} = [\Phi_{h-1} \dots \Phi_1 \quad I_K]\mathbf{u}_{T,h} \tag{4.6.1}$$

where $\mathbf{u}_{T,h} := (u'_{T+1}, \dots, u'_{T+h})'$, the Φ_i are the coefficient matrices of the canonical MA representation (see Section 2.2.2). Since $\mathbf{u}_{T,h} \sim N(0, I_h \otimes \Sigma_u)$, the forecast error is a linear transformation of a multivariate normal distribution and consequently (see Appendix B),

$$e_T(h) \sim N(0, \Sigma_y(h)), \tag{4.6.2}$$

where

$$\Sigma_y(h) = \sum_{i=0}^{h-1} \Phi_i \Sigma_u \Phi_i'$$

is the forecast MSE matrix (see (2.2.11)). Hence,

$$\tau_h := e_T(h)'\Sigma_y(h)^{-1}e_T(h) \sim \chi^2(K) \tag{4.6.3}$$

by Proposition B.3 of Appendix B.

This derivation assumes that y_{T+h} is generated by the same VAR(p) process that has generated the y_t for $t \leq T$. If this process does not prevail in period $T + h$ the statistic τ_h will in general not have a central χ^2-distribution. Hence, τ_h may be used to test the null hypothesis

H_0: (4.6.2) is true, that is, y_{T+h} is generated by the same Gaussian VAR(p) process that has generated $y_1, \dots, y_T$.

The alternative hypothesis is that y_{T+h} is *not* generated by the same process as $y_1, \dots, y_T$. The null hypothesis is rejected if the forecast errors are large so that τ_h exceeds a prespecified critical value from the $\chi^2(K)$-distribution. Such a test may be performed for $h = 1, 2, \dots$.

It may be worth noting that in these tests we also check the normality assumption for y_t. Even if the same process has generated y_{T+h} and $y_1, \dots, y_T$, (4.6.2) will not hold if that process is not Gaussian. Thus, the normality assumption for y_t is part of H_0. Other possible deviations from the null hypothesis include changes in the mean and changes in the variance of the process.

In practice the tests are not feasible in their present form because τ_h involves unknown quantities. The forecast errors $e_T(h)$ and $\Sigma_y(h)$ are both unknown and must be replaced by estimators. For the forecast errors we use

$$\hat{e}_T(h) := y_{T+h} - \hat{y}_t(h) = \sum_{i=0}^{h-1} \hat{\Phi}_i \hat{u}_{T+h-i}, \tag{4.6.4}$$

where the $\hat{\Phi}_i$ are obtained from the coefficient estimators $\hat{A}_i$ in the usual way (see Section 3.5.2) and

$$\hat{u}_t := y_t - \hat{\nu} - \hat{A}_1 y_{t-1} - \cdots - \hat{A}_p y_{t-p}.$$

The MSE matrix may be estimated by

$$\hat{\Sigma}_y(h) := \sum_{i=0}^{h-1} \hat{\Phi}_i \hat{\Sigma}_u \hat{\Phi}_i', \tag{4.6.5}$$

where $\hat{\Sigma}_u$ is the LS estimator of Σ_u. As usual, we use only data up to period T for estimation and not the data from the forecast period. If the conditions for consistency of the estimators are satisfied, that is,

$$\text{plim } \hat{\nu} = \nu, \qquad \text{plim } \hat{A}_i = A_i \qquad \text{and} \qquad \text{plim } \hat{\Sigma}_u = \Sigma_u,$$

then plim $\hat{\Phi}_i = \Phi_i$, plim $\hat{\Sigma}_y(h) = \Sigma_y(h)$ and

$$\begin{aligned}\text{plim}(\hat{u}_t - u_t) &= \text{plim}(\nu - \hat{\nu}) + \text{plim}(A_1 - \hat{A}_1)y_{t-1} + \cdots + \text{plim}(A_p - \hat{A}_p)y_{t-p} \\ &= 0.\end{aligned}$$

Hence, defining

$$\hat{\tau}_h := \hat{e}_T(h)'\hat{\mathit{\Sigma}}_y(h)\hat{e}_T(h),$$

we get $\text{plim}(\hat{\tau}_h - \tau_h) = 0$ and thus, by Proposition C.2(2) of Appendix C,

$$\hat{\tau}_h \xrightarrow{d} \chi^2(K). \tag{4.6.6}$$

In other words, if the unknown coefficients are replaced by consistent estimators, the resulting test statistics $\hat{\tau}_h$ have the same *asymptotic* distributions as the τ_h.

Of course, it is desirable to know whether the $\chi^2(K)$-distribution is a good approximation to the distribution of $\hat{\tau}_h$ in small samples. This, however, is not likely because in Section 3.5.1,

$$\mathit{\Sigma}_{\hat{y}}(h) = \mathit{\Sigma}_y(h) + \frac{1}{T}\Omega(h) \tag{4.6.7}$$

was found to be a better approximation to the MSE matrix than $\mathit{\Sigma}_y(h)$ if the forecasts are based on an estimated process. While asymptotically, as $T \to \infty$, the term $\Omega(h)/T$ vanishes it seems plausible to include this term in small samples. For univariate processes it was confirmed in a simulation study by Lütkepohl (1988a) that inclusion of the term results in a better agreement between the small sample and asymptotic distributions. For multivariate vector processes the simulation results of Section 3.5.4 point in the same direction. Thus, in small samples a statistic of the type

$$\hat{e}_T(h)'\hat{\mathit{\Sigma}}_{\hat{y}}(h)^{-1}\hat{e}_T(h)$$

is more plausible than $\hat{\tau}_h$. Here $\hat{\mathit{\Sigma}}_{\hat{y}}(h)$ is the estimator given in Section 3.5.2. In addition to this adjustment it is useful to adjust the statistic for using an estimated rather than known forecast error covariance matrix. Such an adjustment is often done by dividing by the degrees of freedom and using the statistic in conjunction with critical values from an F-distribution. That is, we use

$$\bar{\tau}_h := \hat{e}_T(h)'\hat{\mathit{\Sigma}}_{\hat{y}}(h)^{-1}\hat{e}_T(h)/K \approx F(K, T - Kp - 1). \tag{4.6.8}$$

The approximate F-distribution follows from Proposition C.3(2) of Appendix C and the denominator degrees of freedom are chosen by analogy with a result due to Hotelling (e.g., Anderson (1984)). Other choices are possible. Proposition C.3(2) requires however that the denominator degrees of freedom go to infinity with the sample size T.

4.6.2 A Test Based on Several Forecast Periods

Another set of stationarity tests is obtained by observing that the errors of forecasts 1- to h-steps ahead are also jointly normally distributed under the null hypothesis of structural stability,

$$\mathbf{e}_T(h) := \begin{bmatrix} e_T(1) \\ \vdots \\ e_T(h) \end{bmatrix} = \mathbf{\Phi}_h \mathbf{u}_{T,h} \sim N(0, \underline{\mathit{\Sigma}}_y(h)), \tag{4.6.9}$$

where

$$\mathbf{\Phi}_h := \begin{bmatrix} I_K & 0 & \dots & 0 \\ \Phi_1 & I_K & & 0 \\ \vdots & \vdots & \ddots & \vdots \\ \Phi_{h-1} & \Phi_{h-2} & \dots & I_K \end{bmatrix} \tag{4.6.10}$$

so that

$$\underline{\Sigma}_y(h) := \mathbf{\Phi}_h(I_K \otimes \Sigma_u)\mathbf{\Phi}_h'. \tag{4.6.11}$$

Using again Proposition B.3 of Appendix B,

$$\begin{aligned} \lambda_h &:= \mathbf{e}_T(h)'\underline{\Sigma}_y(h)^{-1}\mathbf{e}_T(h) = \mathbf{u}_{T,h}'(I_h \otimes \Sigma_u^{-1})\mathbf{u}_{T,h} \\ &= \sum_{i=1}^{h} u_{T+i}'\Sigma_u^{-1}u_{T+i} = \lambda_{h-1} + u_{T+h}'\Sigma_u^{-1}u_{T+h} \sim \chi^2(hK). \end{aligned} \tag{4.6.12}$$

Thus, λ_h may be used to check whether a structural change has occurred during the periods $T+1, \dots, T+h$.

In order to make this test feasible it is necessary to replace unknown quantities by estimators just as in the case of the τ-tests. Denoting the test statistics based on estimated VAR processes by $\hat{\lambda}_h$,

$$\hat{\lambda}_h \xrightarrow{d} \chi^2(hK) \tag{4.6.13}$$

follows with the same arguments used for $\hat{\tau}_h$, provided consistent estimators are used.

Again it seems plausible to make small sample adjustments to the statistics to take into account the fact that estimated quantities are used. The last expression in (4.6.12) suggests that a closer look at the terms

$$u_{T+i}'\Sigma_u^{-1}u_{T+i} \tag{4.6.14}$$

is useful in order to find a small sample adjustment. This expression involves the 1-step forecast errors $u_{T+i} = y_{T+i} - y_{T+i-1}(1)$. If estimated coefficients are used in the 1-step forecast the MSE or forecast error covariance matrix is approximately inflated by a factor $(T + Kp + 1)/T$ (see (3.5.13)). Since λ_h is the sum of terms of the form (4.6.14) it may be useful to replace Σ_u by $(T + Kp + 1)\hat{\Sigma}_u/T$ when estimated quantities are used. Note, however, that such an adjustment ignores possible dependencies between the estimated $\hat{u}_{T+i}$ and $\hat{u}_{T+j}$. Nevertheless, it leads to a computationally extremely simple form and was therefore suggested in the literature (Lütkepohl (1989a)). Furthermore it is suggested to divide by the degrees of freedom of the asymptotic χ^2-distribution and, by appeal to Proposition C.3(2) of Appendix C, use the resulting statistic, $\bar{\lambda}_h$ say, in conjunction with critical values from F-distributions to adjust for the fact that Σ_u is replaced by an estimator. In other words,

$$\bar{\lambda}_h := T\sum_{i=1}^{h} \hat{u}_{T+i}'\hat{\Sigma}_u^{-1}\hat{u}_{T+i}/[(T + Kp + 1)Kh] \approx F(Kh, T - Kp - 1). \tag{4.6.15}$$

The denominator degrees of freedom are chosen by the same arguments used in (4.6.8). Obviously, $\bar{\lambda}_1 = \bar{\tau}_1$.

Now we have two sets of stationarity tests and the question arises which ones to use in practice. To answer this question it would be useful to know the power characteristics of the tests because it is desirable to use the most powerful test available. For some alternatives the τ- and λ-statistics have noncentral χ^2-distributions (Lütkepohl (1988a, 1989a)). In these cases it is possible to investigate and compare their power. It turns out that for some alternatives the τ-tests are more powerful than the λ-tests and for other alternatives the opposite is true. Since we usually don't know the exact form of the alternative (the exact form of the structural change) it may be a good idea to apply both tests in practice.

4.6.3 An Example

To illustrate the use of the two tests for stationarity we use the first differences of logarithms of the West German investment, income, and consumption data and test for a possible structural change caused by the oil price shocks in 1973/74 and 1979/80. Since the first drastic price increase occurred in late 1973 we estimated a VAR(2) model using the sample period 1960.4–1973.2 and presample values from 1960.2 and 1960.3. Thus $T = 51$. It is important to note that the data from the forecast period are not used for estimation. We have used the estimated process to compute the $\bar{\tau}_h$ and $\bar{\lambda}_h$ for $h = 1, \ldots, 8$. The results are given in Table 4.8 together with the p-values of the tests. The p-value is the probability that the test statistic assumes a value greater than the observed test value if the null hypothesis is true. Thus, p-values smaller than .10 or .05 would be of concern. Obviously, in this case none of the test values is statistically significant at the 10% level. Thus, the tests do not give rise to concern about the stationarity of the underlying data generation process during the period in question. Although we have given the $\bar{\tau}_h$ and $\bar{\lambda}_h$ values for various forecast horizons h in Table 4.8

Table 4.8. Stability tests for 1973–1975

quarter	forecast horizon h	$\bar{\tau}_h$	p-value	$\bar{\lambda}_h$	p-value
1973.3	1	.872	.46	.872	.46
4	2	.271	.85	.717	.64
1974.1	3	.206	.89	.517	.85
2	4	.836	.48	.627	.81
3	5	.581	.63	.785	.69
4	6	.172	.91	.832	.65
1975.1	7	.126	.94	.863	.63
2	8	1.450	.24	1.041	.44

Table 4.9. Stability tests for 1979–1980

quarter	forecast horizon h	$\bar{\tau}_h$	p-value	$\bar{\lambda}_h$	p-value
1979.1	1	.277	.84	.277	.84
2	2	2.003	.12	1.077	.38
3	3	2.045	.12	1.464	.18
4	4	.203	.89	1.245	.27
1980.1	5	.630	.60	1.339	.20
2	6	1.898	.86	1.374	.17
3	7	.188	.90	1.204	.28
4	8	.535	.66	1.124	.34

we emphasize that the tests are not independent for different h. Thus, the evidence from the set of tests should not lead to overrating the confidence we may have in this result.

To check the possibility of a structural instability due to the 1979/80 oil price increases we used the VAR(2) model of Section 3.2.3 which is based on data up to the fourth quarter of 1978. The resulting values of the test statistics for $h = 1, \ldots, 8$ are presented in Table 4.9. Again none of the values is significant at the 10% level. However, in Section 3.5.2 we found that the observed consumption values in 1979 fall outside a 95% forecast interval. Hence, looking at the three series individually a possible nonstationarity would be detected by a prediction test. This possible instability in 1979 was a reason for using only data up to 1978 in the examples of previous chapters and sections. The example indicates what can also be demonstrated theoretically, namely that the power of a test based on joint forecasts of various variables may be lower than the power of a test based on forecasts for individual variables (see Lütkepohl (1989a)).

4.7 Exercises

4.7.1 Algebraic Problems

Problem 4.1

Show that the restricted ML estimator $\tilde{\boldsymbol{\beta}}_r$ can be written in the form (4.2.10).

Problem 4.2

Prove Lemma 4.1.

Problem 4.3

Show (4.4.14) and (4.4.15).

[Hint: $\text{vec}(\sqrt{T}UF[I_h \otimes Z'(\hat{B}-B)']/T)$

$$= \sqrt{T}\,\text{vec}\left[\frac{1}{T}UF(I_h \otimes Z')(I_h \otimes (\hat{B}-B)')\right]$$

$$= \left[I_K \otimes \frac{1}{T}UF(I_h \otimes Z')\right]\sqrt{T}\,\text{vec}(I_h \otimes (\hat{B}-B)')$$

and

$$\sqrt{T}\,\text{vec}\left(\frac{1}{T}(\hat{B}-B)ZF[I_h \otimes Z'(\hat{B}-B)']\right)$$

$$= \left\{[I_h \otimes (\hat{B}-B)]\frac{(I_h \otimes Z)F'Z'}{T} \otimes I_K\right\}\sqrt{T}\,\text{vec}(\hat{B}-B).\Big]$$

Problem 4.4

Show (4.5.14).

[Hint: Note that

$$(y_{t-1}-\bar{y}) \otimes (u_t-\bar{u}) \otimes (u_t-\bar{u}) = (y_{t-1}-\mu) \otimes u_t \otimes u_t - (y_{t-1}-\mu) \otimes u_t \otimes \bar{u} + \cdots,$$

define new variables of the type

$$z_t = (y_{t-1}-\mu) \otimes u_t \otimes u_t$$

and use that

$$\text{plim}\,\frac{1}{T}\sum_t z_t = E(z_t) = 0.\Big]$$

4.7.2 Numerical Problems

The following problems require the use of a computer. They refer to the bivariate series $y_t = (y_{1t}, y_{2t})'$ of first differences of the U.S. investment data given in Table E.2 of Appendix E.

Problem 4.5

Set up a sequence of tests for the correct VAR order of the data generating process using a maximum order of $M = 4$. Compute the required χ^2 and F likelihood ratio statistics. Which order would you choose?

Problem 4.6

Determine VAR order estimates on the basis of the four criteria FPE, AIC, HQ, and SC. Use a maximum VAR order of $M = 4$ in a first estimation round and $M = 8$ in a second estimation round. Compare the results.

Problem 4.7

Compute the residual autocorrelations $\hat{R}_1, \ldots, \hat{R}_{12}$ and estimate their standard errors using the VAR(1) model obtained in Problem 3.11. Interpret your results.

Problem 4.8

Compute portmanteau test values P_h and $\bar{P}_h$ for $h = 10$ and 12 on the basis of the residual autocorrelations from Problem 4.7. Test the whiteness of the residuals.

Problem 4.9

On the basis of a VAR(1) model, perform a test for nonnormality of the example data.

Problem 4.10

Investigate whether there was a structural change in U.S. investment after 1965 (possibly due to the increasing U.S. engagement in Vietnam).

Chapter 5. VAR Processes with Parameter Constraints

5.1 Introduction

In Chapter 3 we have discussed estimation of the parameters of a K-dimensional stationary, stable VAR(p) process of the form

$$y_t = \nu + A_1 y_{t-1} + \cdots + A_p y_{t-p} + u_t, \tag{5.1.1}$$

where all the symbols have their usual meanings. In the investment/income/consumption example considered throughout Chapter 3 we found that many of the coefficient estimates were not significantly different from zero. This observation may be interpreted in two ways. First, some of the coefficients may actually be zero and this fact may be reflected in the estimation results. For instance, if some variables are not Granger-causal for the remaining variables zero coefficients are encountered. Second, insignificant coefficient estimates are found if the information in the data is not rich enough to provide sufficiently precise estimates with confidence intervals that do not contain zero.

In the latter case one may want to think about better ways to extract the information from the data because, as we have seen in Chapter 3, a large estimation uncertainty for the VAR coefficients leads to poor forecasts (large forecast intervals) and imprecise estimates of the impulse responses and forecast error variance components. Getting imprecise parameter estimates in a VAR analysis is a common practical problem because the number of parameters is often quite substantial relative to the available sample size or time series length. Various cures for this problem have been proposed in the literature. They all amount to putting constraints on the coefficients.

For instance, in the previous chapter choosing the VAR order p has been discussed. Selecting an order that is less than the maximum order amounts to placing zero constraints on VAR coefficient matrices. This way complete coefficient matrices are eliminated. In the present chapter we will discuss putting zero constraints on individual coefficients. Such constraints are but one form of linear restrictions which will be treated in Section 5.2. Nonlinear constraints are considered in Section 5.3 and Bayesian estimation is the subject of Section 5.4.

5.2 Linear Constraints

In this section the consequences of estimating the VAR coefficients subject to linear constraints will be considered. Different estimation procedures are treated in Subsections 5.2.2–5.2.5; forecasting and impulse response analysis are discussed in Subsections 5.2.6 and 5.2.7; strategies for model selection or the choice of constraints are dealt with in Subsection 5.2.8; model checking follows in Subsection 5.2.9; and, finally, an example is discussed in Subsection 5.2.10.

5.2.1 The Model and the Constraints

We consider the model (5.1.1) for $t = 1, \dots, T$, written in compact form

$$Y = BZ + U, \tag{5.2.1}$$

where

$$Y := [y_1, \dots, y_T], \qquad Z := [Z_0, \dots, Z_{T-1}] \quad \text{with} \quad Z_t := \begin{bmatrix} 1 \\ y_t \\ \vdots \\ y_{t-p+1} \end{bmatrix},$$

$$B := [\nu, A_1, \dots, A_p], \qquad U := [u_1, \dots, u_T].$$

Suppose that linear constraints for B are given in the form

$$\boldsymbol{\beta} := \operatorname{vec}(B) = R\gamma + r, \tag{5.2.2}$$

where $\boldsymbol{\beta} = \operatorname{vec}(B)$ is a $(K(Kp + 1) \times 1)$ vector, R is a known $(K(Kp + 1) \times M)$ matrix of rank M, γ is an unrestricted $(M \times 1)$ vector of unknown parameters, and r is a $K(Kp + 1)$-dimensional vector of known constants. All the linear restrictions of interest can be expressed in this form. For instance, the restriction $A_p = 0$ can be written as in (5.2.2) by choosing $M = K^2(p - 1) + K$,

$$R = \begin{bmatrix} I_M \\ 0 \end{bmatrix}, \qquad \gamma = \operatorname{vec}(\nu, A_1, \dots, A_{p-1}),$$

and $r = 0$.

Although (5.2.2) is not the most conventional form of representing linear constraints, it is used here because it is particularly useful for our purposes. Often the constraints are expressed as

$$C\boldsymbol{\beta} = c, \tag{5.2.3}$$

where C is a known $(N \times (K^2p + K))$ matrix of rank N and c is a known $(N \times 1)$ vector (see Chapter 4, Section 4.2.2). Since $\operatorname{rk}(C) = N$ the matrix C has N linearly independent columns. For simplicity we assume that the first N columns are linearly independent and partition C as $C = [C_1 \quad C_2]$, where C_1 is $(N \times N)$ nonsingular and C_2 is $(N \times (K^2p + K - N))$. Partitioning $\boldsymbol{\beta}$ conformably gives

$$[C_1 \quad C_2]\begin{bmatrix}\boldsymbol{\beta}_1\\ \boldsymbol{\beta}_2\end{bmatrix} = C_1\boldsymbol{\beta}_1 + C_2\boldsymbol{\beta}_2 = c$$

or

$$\boldsymbol{\beta}_1 = -C_1^{-1}C_2\boldsymbol{\beta}_2 + C_1^{-1}c.$$

Therefore, choosing

$$R = \begin{bmatrix} -C_1^{-1}C_2 \\ I_{pK^2+K-N}\end{bmatrix}, \qquad \boldsymbol{\gamma} = \boldsymbol{\beta}_2, \qquad \text{and} \qquad r = \begin{bmatrix} C_1^{-1}c \\ 0 \end{bmatrix}$$

the constraints (5.2.3) can be written in the form (5.2.2). Also, it is not difficult to see that restrictions written as in (5.2.2) can be expressed in the form $C\boldsymbol{\beta} = c$ for suitable C and c. Thus, the two forms are equivalent.

The representation (5.2.2) permits to impose the constraints by a simple reparameterization of the original model. Vectorizing (5.2.1) and replacing $\boldsymbol{\beta}$ by $R\boldsymbol{\gamma} + r$ gives

$$\begin{aligned}\mathbf{y} := \operatorname{vec}(Y) &= (Z' \otimes I_K)\operatorname{vec}(B) + \operatorname{vec}(U)\\ &= (Z' \otimes I_K)(R\boldsymbol{\gamma} + r) + \mathbf{u}\end{aligned}$$

or

$$\mathbf{z} = (Z' \otimes I_K)R\boldsymbol{\gamma} + \mathbf{u}, \tag{5.2.4}$$

where $\mathbf{z} := \mathbf{y} - (Z' \otimes I_K)r$ and $\mathbf{u} := \operatorname{vec}(U)$. This form of the model allows us to derive the estimators and their properties just like in the original unconstrained model. Estimation of $\boldsymbol{\gamma}$ and $\boldsymbol{\beta}$ will be discussed in the following subsections.

5.2.2 LS, GLS, and EGLS Estimation

5.2.2a Asymptotic Properties

Denoting by Σ_u the covariance matrix of u_t, the vector $\hat{\boldsymbol{\gamma}}$ minimizing

$$\begin{aligned} S(\boldsymbol{\gamma}) &= \mathbf{u}'(I_T \otimes \Sigma_u^{-1})\mathbf{u} \\ &= (\mathbf{z} - (Z' \otimes I_K)R\boldsymbol{\gamma})'(I_T \otimes \Sigma_u^{-1})(\mathbf{z} - (Z' \otimes I_K)R\boldsymbol{\gamma}) \end{aligned} \tag{5.2.5}$$

with respect to $\boldsymbol{\gamma}$ is easily seen to be

$$\begin{aligned} \hat{\boldsymbol{\gamma}} &= [R'(ZZ' \otimes \Sigma_u^{-1})R]^{-1}R'(Z \otimes \Sigma_u^{-1})\mathbf{z} \\ &= [R'(ZZ' \otimes \Sigma_u^{-1})R]^{-1}R'(Z \otimes \Sigma_u^{-1})[(Z' \otimes I_K)R\boldsymbol{\gamma} + \mathbf{u}] \\ &= \boldsymbol{\gamma} + [R'(ZZ' \otimes \Sigma_u^{-1})R]^{-1}R'(I_{Kp+1} \otimes \Sigma_u^{-1})\operatorname{vec}(UZ') \end{aligned} \tag{5.2.6}$$

(see Chapter 3, Section 3.2.1). This estimator is commonly called a *generalized LS (GLS)* estimator because it minimizes the generalized sum of squared errors $S(\boldsymbol{\gamma})$ rather than the sum of squared errors $\mathbf{u}'\mathbf{u}$. We will see shortly that in contrast to the unrestricted case considered in Chapter 3, it may make a difference here whether $S(\gamma)$ or $\mathbf{u}'\mathbf{u}$ is used as objective function. The GLS estimator is in general asymptotically more efficient than the multivariate LS estimator and is therefore preferred here. We will see in Section 5.2.3 that, under Gaussian assumptions, the GLS estimator is equivalent to the ML estimator. From (5.2.6),

$$\sqrt{T}(\hat{\gamma} - \gamma) = \left[R'\left(\frac{ZZ'}{T} \otimes \Sigma_u^{-1}\right) R \right]^{-1} R'(I_{Kp+1} \otimes \Sigma_u^{-1}) \frac{1}{\sqrt{T}} \operatorname{vec}(UZ') \qquad (5.2.7)$$

and the asymptotic properties of $\hat{\gamma}$ are obtained as in Proposition 3.1.

PROPOSITION 5.1 (*Asymptotic Properties of the GLS Estimator*)

Suppose the conditions of Proposition 3.1 are satisfied, that is, y_t is a K-dimensional stable, stationary VAR(p) process and u_t is independent white noise with bounded fourth moments. If $\boldsymbol{\beta} = R\gamma + r$ as in (5.2.2) with $\operatorname{rk}(R) = M$, then $\hat{\gamma}$ given in (5.2.6) is a consistent estimator of γ and

$$\sqrt{T}(\hat{\gamma} - \gamma) \xrightarrow{d} N(0, [R'(\Gamma \otimes \Sigma_u^{-1})R]^{-1}), \qquad (5.2.8)$$

where $\Gamma := E(Z_t Z_t') = \operatorname{plim} ZZ'/T$ as in Chapter 3. ■

Proof: Under the conditions of the proposition $\operatorname{plim}(ZZ'/T) = \Gamma$ and

$$\frac{1}{\sqrt{T}} \operatorname{vec}(UZ') \xrightarrow{d} N(0, \Gamma \otimes \Sigma_u)$$

(see Lemma 3.1). Hence, by results stated in Appendix C, Proposition C.4(1), using (5.2.7), $\sqrt{T}(\hat{\gamma} - \gamma)$ has an asymptotic normal distribution with covariance matrix

$$\begin{aligned}&[R'(\Gamma \otimes \Sigma_u^{-1})R]^{-1} R'(I \otimes \Sigma_u^{-1})(\Gamma \otimes \Sigma_u)(I \otimes \Sigma_u^{-1})R[R'(\Gamma \otimes \Sigma_u^{-1})R]^{-1} \\ &\quad = [R'(\Gamma \otimes \Sigma_u^{-1})R]^{-1}.\end{aligned}$$ ■

Unfortunately the estimator $\hat{\gamma}$ is of limited value in practice because its computation requires knowledge of Σ_u. Since this matrix is usually unknown it has to be replaced by an estimator. Using any consistent estimator $\bar{\Sigma}_u$ instead of Σ_u in (5.2.6) we get an *EGLS* (*estimated GLS*) estimator

$$\hat{\hat{\gamma}} = [R'(ZZ' \otimes \bar{\Sigma}_u^{-1})R]^{-1} R'(Z \otimes \bar{\Sigma}_u^{-1})\mathbf{z} \qquad (5.2.9)$$

which has the same asymptotic properties as the GLS estimator $\hat{\gamma}$. This result is an easy consequence of the representation (5.2.7) and Proposition C.4(1) of Appendix C.

PROPOSITION 5.2 (*Asymptotic Properties of the EGLS Estimator*)

Under the conditions of Proposition 5.1, if $\operatorname{plim} \bar{\Sigma}_u = \Sigma_u$, the EGLS estimator $\hat{\hat{\gamma}}$ in (5.2.9) is asymptotically equivalent to the GLS estimator $\hat{\gamma}$ in (5.2.6), that is, plim $\hat{\hat{\gamma}} = \gamma$ and

$$\sqrt{T}(\hat{\hat{\gamma}} - \gamma) \xrightarrow{d} N(0, [R'(\Gamma \otimes \Sigma_u^{-1})R]^{-1}). \tag{5.2.10}$$

■

Once an estimator for γ is available an estimator for $\boldsymbol{\beta}$ is obtained by substituting in (5.2.2), that is,

$$\hat{\hat{\boldsymbol{\beta}}} = R\hat{\hat{\gamma}} + r. \tag{5.2.11}$$

The asymptotic properties of this estimator follow immediately from Appendix C, Proposition C.4(2). We get

PROPOSITION 5.3 (*Asymptotic Properties of the Implied Restricted EGLS Estimator*)

Under the conditions of Proposition 5.2 the estimator $\hat{\hat{\boldsymbol{\beta}}} = R\hat{\hat{\gamma}} + r$ is consistent and asymptotically normally distributed,

$$\sqrt{T}(\hat{\hat{\boldsymbol{\beta}}} - \boldsymbol{\beta}) \xrightarrow{d} N(0, R[R'(\Gamma \otimes \Sigma_u^{-1})R]^{-1}R'). \tag{5.2.12}$$

■

To make these EGLS estimators operational we need a consistent estimator of Σ_u. From Chapter 3, Corollary 3.2.1, we know that, under the conditions of Proposition 5.1,

$$\begin{aligned}\hat{\Sigma}_u &= \frac{1}{T - Kp - 1}(Y - \hat{B}Z)(Y - \hat{B}Z)' \\ &= \frac{1}{T - Kp - 1}Y(I_T - Z'(ZZ')^{-1}Z)Y'\end{aligned} \tag{5.2.13}$$

is a consistent estimator of Σ_u which may thus be used in place of $\bar{\Sigma}_u$. Here $\hat{B} = YZ'(ZZ')^{-1}$ is the unconstrained multivariate LS estimator of the coefficient matrix B.

Alternatively, the restricted LS estimator minimizing $\mathbf{u}'\mathbf{u}$ with respect to γ may be determined in a first step. The minimizing γ-vector is easily seen to be

$$\gamma^{\#} = [R'(ZZ' \otimes I_K)R]^{-1}R'(Z \otimes I_K)\mathbf{z} \tag{5.2.14}$$

(see Problem 5.1). As this LS estimator does not involve the white noise covariance matrix Σ_u it is generally distinct from the GLS estimator. We denote the corresponding $\boldsymbol{\beta}$-vector by $\boldsymbol{\beta}^{\#}$, that is, $\boldsymbol{\beta}^{\#} = R\gamma^{\#} + r$. Furthermore $B^{\#}$ is the corresponding coefficient matrix, that is, $\mathrm{vec}(B^{\#}) = \boldsymbol{\beta}^{\#}$. Then we may choose

$$\Sigma_u^{\#} = \frac{1}{T}(Y - B^{\#}Z)(Y - B^{\#}Z)' \tag{5.2.15}$$

as an estimator for Σ_u. The consistency of this estimator is a consequence of Proposition 3.2 and the fact that $B^{\#}$ is a consistent estimator of B with asymptotic normal distribution. This result follows from the asymptotic normality of $\gamma^{\#}$

which in turn follows by replacing $\tilde{\Sigma}_u$ with I_K in (5.2.6) and (5.2.7). Thus, $\boldsymbol{\beta}^{\#} = R\boldsymbol{\gamma}^{\#} + r$ is asymptotically normal. Consequently we have from Proposition 3.2 and Corollary 3.2.1:

PROPOSITION 5.4 (*Asymptotic Properties of the White Noise Covariance Estimator*)

Under the conditions of Proposition 5.1, $\tilde{\Sigma}_u^{\#}$ is consistent and

$$\text{plim}\sqrt{T}(\tilde{\Sigma}_u^{\#} - UU'/T) = 0. \qquad \blacksquare$$

In (5.2.15) T may be replaced by $T - Kp - 1$ without affecting the consistency of the covariance matrix estimator. However, there is little justification for subtracting $Kp + 1$ from T in the present situation because, due to zero restrictions, some or all of the K equations of the system may contain fewer than $Kp + 1$ parameters.

Of course, in practice one would like to know which one of the possible covariance estimators leads to an EGLS estimator $\hat{\hat{\boldsymbol{\gamma}}}$ with best small sample properties. Although we cannot give a general answer to this question it seems plausible to use an estimator that takes into account the nonsample information concerning the VAR coefficients, provided the restrictions are correct. Thus, if one is confident about the validity of the restrictions, the covariance matrix estimator $\tilde{\Sigma}_u^{\#}$ may be used.

As an alternative to the EGLS estimator described in the foregoing, an iterated EGLS estimator may be used. It is obtained by computing a new covariance matrix estimator from the EGLS residuals. This estimator is then used in place of $\bar{\Sigma}_u$ in (5.2.9) and again a new covariance matrix estimator is computed from the corresponding residuals and so on. The procedure is continued until convergence. We will not persue it here. From Propositions 5.2 and 3.2 it follows that the asymptotic properties of the resulting iterated EGLS estimator are the same as those of the EGLS estimator wherever the iteration is terminated.

5.2.2b Comparison of LS and Restricted EGLS Estimators

A question of interest in this context is how the covariance matrix in (5.2.12) compares with the asymptotic covariance matrix $\Gamma^{-1} \otimes \Sigma_u$ of the unrestricted multivariate LS estimator $\hat{\boldsymbol{\beta}}$. To see that the restricted estimator has smaller or at least not greater asymptotic variances than the unrestricted estimator it is helpful to write the restrictions in the form (5.2.3). In that case the restricted EGLS estimator of $\boldsymbol{\beta}$ turns out to be

$$\hat{\hat{\boldsymbol{\beta}}} = \hat{\boldsymbol{\beta}} + [(ZZ')^{-1} \otimes \bar{\Sigma}_u]C'[C((ZZ')^{-1} \otimes \bar{\Sigma}_u)C']^{-1}(c - C\hat{\boldsymbol{\beta}}) \tag{5.2.16}$$

(see Chapter 4, Section 4.2.2, and Problem 5.2). Noting that $C\boldsymbol{\beta} - c = 0$, subtracting $\boldsymbol{\beta}$ from both sides, and multiplying by $\sqrt{T}$ gives

$$\sqrt{T}(\hat{\hat{\boldsymbol{\beta}}} - \boldsymbol{\beta}) = \sqrt{T}(\hat{\boldsymbol{\beta}} - \boldsymbol{\beta}) - F_T\sqrt{T}(\hat{\boldsymbol{\beta}} - \boldsymbol{\beta}) = (I_{K^2p+K} - F_T)\sqrt{T}(\hat{\boldsymbol{\beta}} - \boldsymbol{\beta}),$$

where

$$F_T := \left[\left(\frac{ZZ'}{T}\right)^{-1} \otimes \bar{\Sigma}_u\right] C' \left[C\left(\left(\frac{ZZ'}{T}\right)^{-1} \otimes \bar{\Sigma}_u\right) C'\right]^{-1} C$$

so that

$$F := \text{plim } F_T = (\Gamma^{-1} \otimes \Sigma_u)C'[C(\Gamma^{-1} \otimes \Sigma_u)C']^{-1}C.$$

Thus, the covariance matrix of the asymptotic distribution of $\sqrt{T}(\hat{\hat{\boldsymbol{\beta}}} - \boldsymbol{\beta})$ is

$$\begin{aligned}(I - F)(\Gamma^{-1} \otimes \Sigma_u)(I - F)' \\ &= \Gamma^{-1} \otimes \Sigma_u - (\Gamma^{-1} \otimes \Sigma_u)F' - F(\Gamma^{-1} \otimes \Sigma_u) + F(\Gamma^{-1} \otimes \Sigma_u)F' \\ &= \Gamma^{-1} \otimes \Sigma_u - (\Gamma^{-1} \otimes \Sigma_u)C'[C(\Gamma^{-1} \otimes \Sigma_u)C']^{-1}C(\Gamma^{-1} \otimes \Sigma_u).\end{aligned}$$

In other words, a positive semidefinite matrix is subtracted from the covariance matrix $\Gamma^{-1} \otimes \Sigma_u$ to obtain the asymptotic covariance matrix of the restricted estimator. Hence, the asymptotic variances of the latter will be smaller than or at most equal to those of the unrestricted multivariate LS estimator. Since the two ways of writing the restrictions in (5.2.3) and (5.2.2) are equivalent, the EGLS estimator of $\boldsymbol{\beta}$ subject to restrictions $\boldsymbol{\beta} = R\boldsymbol{\gamma} + r$ must also be asymptotically superior to the unconstrained estimator. In other words,

$$\Gamma^{-1} \otimes \Sigma_u - R[R'(\Gamma \otimes \Sigma_u^{-1})R]^{-1}R'$$

is positive semidefinite. This result shows that imposing restrictions is advantageous in terms of asymptotic efficiency. It must be kept in mind, however, that the restrictions are assumed to be valid in the foregoing derivations. In practice there is usually some uncertainty with respect to the validity of the constraints.

5.2.3 Maximum Likelihood Estimation

So far in this chapter no specific distribution of the process y_t is assumed. If the precise distribution of the process is known, ML estimation of the VAR coefficients is possible. In the following we assume that y_t is Gaussian (normally distributed). The ML estimators of $\boldsymbol{\gamma}$ and Σ_u are found by equating to zero the first partial derivatives of the log-likelihood function and solving for $\boldsymbol{\gamma}$ and Σ_u. The partial derivatives are found as in Section 3.4 of Chapter 3. Note that

$$\frac{\partial \ln l}{\partial \boldsymbol{\gamma}} = \frac{\partial \boldsymbol{\beta}'}{\partial \boldsymbol{\gamma}} \frac{\partial \ln l}{\partial \boldsymbol{\beta}} = R' \frac{\partial \ln l}{\partial \boldsymbol{\beta}}$$

by the chain rule for vector differentiation (Appendix A.13). Proceeding as in Section 3.4 the ML estimator of $\boldsymbol{\gamma}$ is easily seen to be

$$\tilde{\boldsymbol{\gamma}} = [R'(ZZ' \otimes \tilde{\Sigma}_u^{-1})R]^{-1}R'(Z \otimes \tilde{\Sigma}_u^{-1})\mathbf{z}, \tag{5.2.17}$$

where $\tilde{\Sigma}_u$ is the ML estimator of Σ_u (see Problem 5.3). The resulting ML estimator of $\boldsymbol{\beta}$ is

$$\tilde{\boldsymbol{\beta}} = R\tilde{\boldsymbol{\gamma}} + r. \tag{5.2.18}$$

Furthermore the ML estimator of Σ_u is seen to be

$$\tilde{\Sigma}_u = \frac{1}{T}(Y - \tilde{B}Z)(Y - \tilde{B}Z)', \tag{5.2.19}$$

where $\tilde{B}$ is the $(K \times (Kp + 1))$ matrix satisfying $\text{vec}(\tilde{B}) = \tilde{\boldsymbol{\beta}}$.

An immediate consequence of the consistency of the ML estimator $\tilde{\Sigma}_u$ and of Proposition 5.2 is that the EGLS estimator $\hat{\hat{\boldsymbol{\gamma}}}$ and the ML estimator $\tilde{\boldsymbol{\gamma}}$ are asymptotically equivalent. In addition, it follows as in Section 3.2.2, Chapter 3, that $\tilde{\Sigma}_u$ has the same asymptotic properties as in the unrestricted case (see Proposition 3.2) and $\tilde{\boldsymbol{\beta}}$ and $\tilde{\Sigma}_u$ are asymptotically independent. In summary we get:

PROPOSITION 5.5 (*Asymptotic Properties of the Restricted ML Estimators*)

Let y_t be a Gaussian stable K-dimensional VAR(p) process as in (5.1.1) and $\boldsymbol{\beta} = \text{vec}(B) = R\boldsymbol{\gamma} + r$ as in (5.2.2). Then the ML estimators $\tilde{\boldsymbol{\beta}}$ and $\tilde{\boldsymbol{\sigma}} = \text{vech}(\tilde{\Sigma}_u)$ are consistent and asymptotically normally distributed,

$$\sqrt{T}\begin{bmatrix} \tilde{\boldsymbol{\beta}} - \boldsymbol{\beta} \\ \tilde{\boldsymbol{\sigma}} - \boldsymbol{\sigma} \end{bmatrix} \xrightarrow{d} N\left[0, \begin{bmatrix} R[R'(\Gamma \otimes \Sigma_u^{-1})R]^{-1}R' & 0 \\ 0 & 2\mathbf{D}_K^+(\Sigma_u \otimes \Sigma_u)\mathbf{D}_K^{+\prime} \end{bmatrix}\right]$$

where $\mathbf{D}_K^+ = (\mathbf{D}_K'\mathbf{D}_K)^{-1}\mathbf{D}_K'$ is the Moore-Penrose inverse of the $(K^2 \times K(K+1)/2)$ duplication matrix $\mathbf{D}_K$. ∎

Of course, we could have stated the proposition in terms of the joint distribution of $\tilde{\boldsymbol{\gamma}}$ and $\tilde{\boldsymbol{\sigma}}$ instead. In the following the distribution given in the proposition will turn out to be more useful, though.

Both EGLS and ML estimation could be discussed in terms of the mean-adjusted model considered in Section 3.3. However, the present discussion includes restrictions for the intercept terms in a convenient way. If the restrictions are equivalent in the different versions of the model, the asymptotic properties of the estimators of $\boldsymbol{\alpha} := \text{vec}(A_1, \ldots, A_p)$ will not be affected. For instance, the asymptotic covariance matrix of $\sqrt{T}(\tilde{\boldsymbol{\alpha}} - \boldsymbol{\alpha})$, where $\tilde{\boldsymbol{\alpha}}$ is the ML estimator, is just the lower right-hand $(K^2p \times K^2p)$ block of $R[R'(\Gamma \otimes \Sigma_u^{-1})R]^{-1}R'$ from Proposition 5.5. If the sample means are subtracted from all variables and the constraints are given in the form $\boldsymbol{\alpha} = R\boldsymbol{\gamma} + r$ for a suitable matrix R and vectors $\boldsymbol{\gamma}$ and r, the covariance matrix of the asymptotic distribution of $\sqrt{T}(\tilde{\boldsymbol{\alpha}} - \boldsymbol{\alpha})$ can be written as

$$R[R'(\Gamma_Y(0) \otimes \Sigma_u^{-1})R]^{-1}R', \tag{5.2.20}$$

where $\Gamma_Y(0) := \Sigma_Y = \text{Cov}(Y_t)$ with $Y_t := (y_t', \ldots, y_{t-p+1}')'$.

5.2.4 Constraints for Individual Equations

In practice parameter restrictions are often formulated for the K equations of the system (5.1.1) separately. In that case it may be easier to write the restrictions in terms of the vector $\mathbf{b} := \text{vec}(B')$ which contains the parameters of the first equation in the first $Kp + 1$ positions and those of the second equation in the second $Kp + 1$ positions etc.. If the constraints are expressed as

$$\mathbf{b} = \bar{R}\mathbf{c} + \bar{r}, \tag{5.2.21}$$

where $\bar{R}$ is a known $((K^2p + K) \times M)$ matrix of rank M, $\mathbf{c}$ is an unknown $(M \times 1)$ parameter vector, and $\bar{r}$ is a known $(K^2p + K)$-dimensional vector, the restricted EGLS and ML estimators of $\mathbf{b}$ and their properties are easily derived. We get the following proposition:

PROPOSITION 5.6 (*EGLS Estimator of Parameters Arranged Equationwise*)

Under the conditions of Proposition 5.2, if $\mathbf{b} = \text{vec}(B')$ satisfies (5.2.21), the EGLS estimator of $\mathbf{c}$ is

$$\hat{\hat{\mathbf{c}}} = [\bar{R}'(\bar{\Sigma}_u^{-1} \otimes ZZ')\bar{R}]^{-1}\bar{R}'(\bar{\Sigma}_u^{-1} \otimes Z)[\text{vec}(Y') - (Z \otimes I_K)\bar{r}], \tag{5.2.22}$$

where $\bar{\Sigma}_u$ is a consistent estimator of Σ_u. The corresponding estimator of $\mathbf{b}$ is

$$\hat{\hat{\mathbf{b}}} = \bar{R}\hat{\hat{\mathbf{c}}} + \bar{r} \tag{5.2.23}$$

which is consistent and asymptotically normally distributed,

$$\sqrt{T}(\hat{\hat{\mathbf{b}}} - \mathbf{b}) \xrightarrow{d} N(0, \bar{R}[\bar{R}'(\bar{\Sigma}_u^{-1} \otimes \Gamma)\bar{R}]^{-1}\bar{R}'). \tag{5.2.24}$$

■

The proof is left as an exercise (see Problem 5.4). An estimator of $\boldsymbol{\beta}$ is obtained from $\hat{\hat{\mathbf{b}}}$ by premultiplying with the commutation matrix $\mathbf{K}_{Kp+1,K}$. If the restrictions in (5.2.21) are equivalent to those in (5.2.2), the estimator for $\boldsymbol{\beta}$ obtained in this way is identical to $\hat{\hat{\boldsymbol{\beta}}}$ given in (5.2.11).

5.2.5 Restrictions on the White Noise Covariance Matrix

Occasionally restrictions for the white noise covariance matrix Σ_u are available. For instance, in Chapter 2, Section 2.3.1, we have seen that instantaneous noncausality is equivalent to Σ_u being block-diagonal. Thus, in that case there are zero off-diagonal elements. Zero constraints are, in fact, the most common constraints for the off-diagonal elements of Σ_u. Therefore we will focus on such restrictions in the following.

Estimation under zero restrictions for Σ_u is often most easily performed in the context of the recursive model introduced in Chapter 2, Section 2.3.2. In order to obtain the recursive form corresponding to the standard VAR system

$$y_t = \nu + A_1 y_{t-1} + \cdots + A_p y_{t-p} + u_t,$$

Σ_u is decomposed as $\Sigma_u = W\Lambda W'$, where W is lower triangular with unit main diagonal and Λ is a diagonal matrix. Then, premultiplying with W^{-1} gives the recursive system

$$y_t = \eta + B_0 y_t + B_1 y_{t-1} + \cdots + B_p y_{t-p} + v_t,$$

where $\eta := W^{-1}\nu$, $B_0 := I_K - W^{-1}$ is a lower triangular matrix with zero diagonal, $B_i := W^{-1}A_i$, $i = 1, \ldots, p$, and $v_t = (v_{1t}, \ldots, v_{Kt})' := W^{-1}u_t$ has diagonal covariance matrix, $\Sigma_v := E(v_t v_t') = \Lambda$. The characteristic feature of the recursive representation of our process is that the k-th equation may involve $y_{1,t}, \ldots, y_{k-1,t}$ (current values of $y_1, \ldots, y_{k-1}$) on the right-hand side and the components of the white noise process v_t are uncorrelated.

Many zero restrictions for the off-diagonal elements of Σ_u are equivalent to simple zero restrictions on B_0 which are easy to impose in equationwise LS estimation. For instance, if Σ_u is block-diagonal, say

$$\Sigma_u = \begin{bmatrix} \Sigma_{11} & 0 \\ 0 & \Sigma_{22} \end{bmatrix},$$

then Σ_{11} and Σ_{22} can be decomposed in the form

$$\Sigma_{ii} = W_i \Lambda_i W_i', \qquad i = 1, 2,$$

where W_i is lower triangular with unit diagonal and Λ_i is a diagonal matrix. Hence,

$$B_0 = I_K - \begin{bmatrix} W_1^{-1} & 0 \\ 0 & W_2^{-1} \end{bmatrix} =: \begin{bmatrix} B_{01} & 0 \\ 0 & B_{02} \end{bmatrix},$$

where

$$B_{0i} = \begin{bmatrix} 0 & \cdots & \cdots & 0 \\ * & \ddots & & \vdots \\ \vdots & \ddots & \ddots & \\ * & \cdots & * & 0 \end{bmatrix}, \qquad i = 1, 2,$$

are lower triangular with zero diagonal. In summary, if Σ_u is block-diagonal with an $(m \times n)$ block of zeroes in its lower left-hand corner the same holds for B_0.

Since the error terms of the K equations of the recursive system are uncorrelated it can be shown that estimating each equation separately does not result in a loss of asymptotic efficiency (see Problem 5.6). Using the notation

$$y_{(k)} := \begin{bmatrix} y_{k1} \\ \vdots \\ y_{kT} \end{bmatrix}, \qquad v_{(k)} := \begin{bmatrix} v_{k1} \\ \vdots \\ v_{kT} \end{bmatrix}$$

and denoting by $b_{(k)}$ the vector of all nonzero coefficients and by $Z_{(k)}$ the corresponding matrix of regressors in the k-th equation of the recursive form of the system, we may write the k-th equation as

$$y_{(k)} = Z_{(k)} b_{(k)} + v_{(k)}.$$

The LS estimator of $b_{(k)}$ is

$$\hat{b}_{(k)} = (Z_{(k)}' Z_{(k)})^{-1} Z_{(k)}' y_{(k)}.$$

Under Gaussian assumptions it is equivalent to the ML estimator and is thus asymptotically efficient. Obviously this framework makes it easy to take into account zero restrictions by just eliminating regressors.

Generally restrictions on Σ_u imply restrictions for B_0 and vice versa. Unfortunately, zero restrictions on Σ_u do not always imply zero restrictions for B_0. Consider, for instance, the covariance matrix

$$\Sigma_u = \begin{bmatrix} 1 & 1 & -1 \\ 1 & 2 & 0 \\ -1 & 0 & 3 \end{bmatrix} = \begin{bmatrix} 1 & 0 & 0 \\ 1 & 1 & 0 \\ -1 & 1 & 1 \end{bmatrix} I_3 \begin{bmatrix} 1 & 1 & -1 \\ 0 & 1 & 1 \\ 0 & 0 & 1 \end{bmatrix} (= W\Lambda W').$$

Hence,

$$B_0 = I_3 - W^{-1} = I_3 - \begin{bmatrix} 1 & 0 & 0 \\ -1 & 1 & 0 \\ 2 & -1 & 1 \end{bmatrix} = \begin{bmatrix} 0 & 0 & 0 \\ 1 & 0 & 0 \\ -2 & 1 & 0 \end{bmatrix}.$$

Thus, although Σ_u has a zero off-diagonal element, all elements of B_0 below the main diagonal are nonzero.

In practice, subject matter theory is often more likely to provide restrictions for the B_0 matrix than for Σ_u because, as we have seen in Section 2.3.2b, the elements of B_0 can sometimes be interpreted as impact multipliers that represent the instantaneous effects of impulses in the variables. For this reason the recursive form of the system has considerable appeal.

Note, however, that if restrictions are available on the coefficients A_i of the standard VAR form the implied constraints for the B_i matrices should be taken into account in the estimation. Such restrictions may be cross-equation restrictions that involve coefficients from different equations. Taking them into account may require simultaneous estimation of all or some equations of the system rather than single equation LS estimation. In the following sections we return to the standard form of the VAR model.

5.2.6 Forecasting

Forecasting with estimated processes was discussed in Section 3.5 of Chapter 3. The general results of that section remain valid even if parameter restrictions are imposed in the estimation procedure. Some differences in details will be pointed out in the following.

We focus on the standard form (5.1.1) of the VAR system and denote the parameter estimators by $\hat{\nu}$, $\hat{A}_1, \ldots, \hat{A}_p$, and $\hat{\boldsymbol{\beta}}$. These estimators may be EGLS or ML estimators. The resulting h-step forecast at origin t is

$$\hat{y}_t(h) = \hat{\nu} + \hat{A}_1\hat{y}_t(h-1) + \cdots + \hat{A}_p\hat{y}_t(h-p) \tag{5.2.25}$$

with $\hat{y}_t(j) := y_{t+j}$ for $j \leq 0$ as in Section 3.5. In line with that section we assume that forecasting and parameter estimation are based on independent processes with identical stochastic structure. Then we get the approximate MSE matrix

$$\Sigma_{\hat{y}}(h) = \Sigma_y(h) + \frac{1}{T}\Omega(h), \tag{5.2.26}$$

where

$$\Sigma_y(h) := E[y_{t+h} - y_t(h)][y_{t+h} - y_t(h)]' = \sum_{i=0}^{h-1} \Phi_i \Sigma_u \Phi_i',$$

Φ_i being, as usual, the i-th coefficient matrix of the canonical MA representation of y_t, and

$$\Omega(h) := E\left[\frac{\partial y_t(h)}{\partial \boldsymbol{\beta}'} \Sigma_{\hat{\boldsymbol{\beta}}} \frac{\partial y_t(h)'}{\partial \boldsymbol{\beta}}\right],$$

where $\Sigma_{\hat{\boldsymbol{\beta}}}$ is the covariance matrix of the asymptotic distribution of $\sqrt{T}(\hat{\boldsymbol{\beta}} - \boldsymbol{\beta})$.

In Chapter 3 the matrix $\Omega(h)$ has a particularly simple form because in that chapter $\Sigma_{\hat{\boldsymbol{\beta}}} = \Gamma^{-1} \otimes \Sigma_u$. In the present situation where parameter restrictions are imposed,

$$\Sigma_{\hat{\boldsymbol{\beta}}} = R[R'(\Gamma \otimes \Sigma_u^{-1})R]^{-1}R'$$

and the form of $\Omega(h)$ is not quite so simple. Since the covariance matrix $\Sigma_{\hat{\boldsymbol{\beta}}}$ is now smaller than in Chapter 3, $\Omega(h)$ will also become smaller (not greater). Using

$$\frac{\partial y_t(h)}{\partial \boldsymbol{\beta}'} = \sum_{i=0}^{h-1} Z_t'(\mathbf{B}')^{h-1-i} \otimes \Phi_i$$

from Chapter 3, (3.5.11), we may now estimate $\Omega(h)$ by

$$\hat{\Omega}(h) = \frac{1}{T}\sum_{t=1}^{T}\left[\sum_{i=0}^{h-1} Z_t'(\mathbf{B}')^{h-1-i} \otimes \Phi_i\right] \Sigma_{\hat{\boldsymbol{\beta}}} \left[\sum_{i=0}^{h-1} Z_t'(\mathbf{B}')^{h-1-i} \otimes \Phi_i\right]'. \tag{5.2.27}$$

Here

$$\mathbf{B} := \begin{bmatrix} 1 & 0 & \cdots & 0 \\ & B & & \\ 0 & I_{K(p-1)} & & 0 \end{bmatrix}, \qquad ((Kp+1) \times (Kp+1))$$

(see Section 3.5.2). In practice the unknown matrices $\mathbf{B}$, Φ_i, and $\Sigma_{\hat{\boldsymbol{\beta}}}$ are replaced by consistent estimators. Of course, if T is large we may simply ignore the term $\Omega(h)/T$ in (5.2.26) because it approaches zero as $T \to \infty$. An estimator of $\Sigma_{\hat{y}}(h)$ is then obtained by simply replacing the unknown quantities in $\Sigma_y(h)$ by estimators. Assuming that y_t is Gaussian, forecast intervals and regions can be determined exactly as in Section 3.5.

5.2.7 Impulse Response Analysis and Forecast Error Variance Decomposition

Impulse response analysis and forecast error variance decomposition with restricted VAR models can be done as described in Section 3.7. Proposition 3.6 is formulated sufficiently general to accommodate the case of restricted estimation. The impulse responses are then estimated from the restricted estimators of $A_1, \ldots, A_p$. As mentioned earlier, the covariance matrix of the restricted estimator of $\boldsymbol{\alpha} := \mathrm{vec}(A_1, \ldots, A_p)$ is obtained by considering the lower right-hand $(K^2p \times K^2p)$ block of

$$\Sigma_{\hat{\boldsymbol{\beta}}} = R[R'(\Gamma \otimes \Sigma_u^{-1})R]^{-1}R'.$$

As we have seen in Subsection 5.2.3, Proposition 5.5, the asymptotic covariance matrix $\Sigma_{\tilde{\boldsymbol{\sigma}}}$ of $\sqrt{T}(\tilde{\boldsymbol{\sigma}} - \boldsymbol{\sigma})$ is not affected by the restrictions for $\boldsymbol{\beta}$. However, the estimator of

$$\Sigma_{\tilde{\boldsymbol{\sigma}}} = 2\mathbf{D}_K^+(\Sigma_u \otimes \Sigma_u)\mathbf{D}_K^{+\prime}$$

may be affected. As discussed in Section 5.2.2, we have the choice of different consistent estimators for Σ_u that may or may not take into account the parameter constraints. In other words, we may estimate Σ_u from the residuals of an unrestricted estimation or we may use the residuals of the restricted LS or EGLS estimation. The lower triangular matrix P that is used in estimating the impulse responses for orthogonal innovations is estimated accordingly. In the examples considered below we will usually base the estimators of Σ_u and P on the residuals of the restricted EGLS estimation. In contrast Γ will usually be estimated by ZZ'/T as in the unrestricted case. Of course, instead of the intercept version of the process we may use the mean-adjusted form for estimation as mentioned in Section 5.2.3.

5.2.8 Specification of Subset VAR Models

A VAR model with zero constraints on the coefficients is called a *subset VAR model*. Formally zero restrictions can be written as in (5.2.2) or (5.2.21) with $r = \bar{r} = 0$. We have encountered such models in previous chapters. For instance, when Granger-causality restrictions are imposed we get subset VAR models. This example suggests possibilities how to obtain such restrictions, namely, from prior nonsample information and/or from tests of particular hypotheses. Subject matter theory often implies a set of restrictions on a VAR model that can be taken into account using the estimation procedures outlined in the foregoing. However, in many cases generally accepted a priori restrictions are not available. In that case statistical procedures may be used to detect or confirm possible zero constraints. In the following we will discuss such procedures.

If little or no a priori knowledge of possible zero constraints is available one may want to compare various different processes or models and choose the one which is optimal under a specific criterion. Using hypothesis tests in such a situation may create problems because the different possible models may not be nested. In that case statistical tests may not lead to a unique answer as to which model to use. Therefore, in subset VAR modeling it is not uncommon to base the model choice on model selection criteria. For instance, appropriately modified versions of AIC, SC, or HQ may be employed. Generally speaking, in such an approach the subset VAR model is chosen that optimizes some prespecified criterion.

Suppose it is just known that the order of the process is not greater than some number p and otherwise no prior knowledge of possible zero constraints is available. In that situation one would ideally fit all possible subset VAR models and select the one that optimizes the criterion chosen. The practicability of such a procedure is limited by its tremendous computational expense. Note that for a K-dimensional VAR(p) process, even if we do not take into account the intercept terms for the moment, there exist K^2p coefficients from which $\binom{K^2p}{j}$ subsets with j elements can be chosen. Thus, there is a total of

$$\sum_{j=0}^{K^2p-1} \binom{K^2p}{j} = 2^{K^2p} - 1$$

subset VAR models, not counting the full VAR(p) model which is also a possible candidate. For instance, for a bivariate VAR(4) process there are as many as $2^{16} - 1 = 65{,}535$ subset models plus the full VAR(4) model. Of course, in practice the dimension and order of the process will often be greater than in this example and there may be drastically more subset VAR models. Therefore specific strategies for subset VAR modeling have been proposed that avoid fitting all potential candidates. Some possibilities will be described briefly in the following.

5.2.8a Elimination of Complete Matrices

Penm & Terrell (1982) consider subset models where complete coefficient matrices A_i rather than individual coefficients are set at zero. Such a strategy reduces the models to be compared to $\sum_{j=0}^{p}\binom{p}{j} = 2^p$. For instance, for a VAR(4) process only 16 models need to be compared.

An obvious advantage of the procedure is its relatively small computational expense. Deleting complete coefficient matrices may be reasonable if seasonal data with strong seasonal components are considered for which only coefficients at seasonal lags are different from zero. On the other hand, there may still be potential for further parameter restrictions. Moreover, some of the deleted coefficient matrices may contain elements that would not have been deleted had they been checked individually. Therefore the following strategy may be more useful.

5.2.8b Top-down Strategy

The top-down strategy starts from the full VAR(p) model and coefficients are deleted in the K equations separately. The k-th equation may be written as

$$\begin{aligned} y_{kt} = \nu_k &+ \alpha_{k1,1}y_{1,t-1} + \cdots + \alpha_{kK,1}y_{K,t-1} + \\ &\vdots \\ &+ \alpha_{k1,p}y_{1,t-p} + \cdots + \alpha_{kK,p}y_{K,t-p} + u_{kt}. \end{aligned} \tag{5.2.28}$$

The goal is to find the zero restrictions for the coefficients of this equation that lead to the *minimum* value of a prespecified criterion. For this purpose the equation is estimated by LS and the corresponding value of the criterion is evaluated. Then the last coefficient $\alpha_{kK,p}$ is set at zero (i.e., $y_{K,t-p}$ is deleted from the equation) and the equation is estimated again with this restriction. If the value of the criterion for the restricted model is greater than for the unrestricted model, $y_{K,t-p}$ is kept in the equation. Otherwise it is eliminated. Then the same procedure is repeated for the second last coefficient $\alpha_{k,K-1,p}$ or variable $y_{K-1,t-p}$ and so on up to ν_k. In each step a lag of a variable is deleted if the criterion does not increase by that additional constraint compared to the smallest value obtained in the previous steps.

Criteria that may be used in this procedure are

$$\text{AIC} = \ln \tilde{\sigma}^2 + \frac{2}{T}(\text{number of parameters}), \tag{5.2.29}$$

$$\text{HQ} = \ln \tilde{\sigma}^2 + \frac{2 \ln \ln T}{T}(\text{number of parameters}), \tag{5.2.30}$$

or

$$\text{SC} = \ln \tilde{\sigma}^2 + \frac{\ln T}{T}(\text{number of parameters}). \tag{5.2.31}$$

Here $\tilde{\sigma}^2$ stands for the sum of squared estimation residuals divided by the sample size T. For instance, the AIC value for a model with or without zero restrictions is computed by estimating the k-th equation, computing the residual sum of squares and dividing by T to obtain $\tilde{\sigma}^2$. Then two times the number of parameters contained in the estimated equation is divided by T and added to the natural logarithm of $\tilde{\sigma}^2$. In the final equation only those variables and coefficients are retained that lead to the minimum AIC value.

In a more formal manner this procedure can be described as follows. The k-th equation of the system may be written as

$$y_{(k)} = \begin{bmatrix} y_{k1} \\ \vdots \\ y_{kT} \end{bmatrix} = Z'b_k + u_{(k)} = Z'\bar{R}_k c_k + u_{(k)},$$

where $b_k = \bar{R}_k c_k$ reflects the zero restrictions imposed on the parameters b_k of the k-th equation. $\bar{R}_k$ is the restriction matrix. The LS estimator of c_k is

$$\hat{c}_k = (\bar{R}_k' ZZ' \bar{R}_k)^{-1} \bar{R}_k' Z y_{(k)}$$

and the implied restricted LS estimator for b_k is

$$\hat{b}_k = \bar{R}_k \hat{c}_k.$$

Furthermore, a corresponding estimator of the residual variance is

$$\tilde{\sigma}^2(\bar{R}_k) = (y_{(k)} - Z'\hat{b}_k)'(y_{(k)} - Z'\hat{b}_k)/T.$$

Thus, the AIC value for a model with these restrictions is

$$\text{AIC}(\bar{R}_k) = \ln \tilde{\sigma}^2(\bar{R}_k) + \frac{2}{T}\text{rk}(\bar{R}_k).$$

The other criteria are determined in a similar way.

In the foregoing subset procedure based on AIC the unrestricted model with $\bar{R}_k = I_{Kp+1}$ is estimated first and the corresponding value $\text{AIC}(I_{Kp+1})$ is determined. Then the last column of I_{Kp+1} is eliminated. Let us denote the resulting restriction matrix by $\bar{R}_k^{(1)}$. If

$$\text{AIC}(\bar{R}_k^{(1)}) \leq \text{AIC}(I_{Kp+1})$$

the next restriction matrix $\bar{R}_k^{(2)}$, say, is obtained by deleting the last column of $\bar{R}_k^{(1)}$ and $\text{AIC}(\bar{R}_k^{(2)})$ is compared to $\text{AIC}(\bar{R}_k^{(1)})$. If

$$\text{AIC}(\bar{R}_k^{(1)}) > \text{AIC}(I_{Kp+1})$$

the restriction matrix $\bar{R}_k^{(2)}$ is obtained by deleting the second last column of I_{Kp+1} and the next restriction matrix is decided upon by comparing $\text{AIC}(\bar{R}_k^{(2)})$ to $\text{AIC}(I_{Kp+1})$. In each step a column is deleted if that leads to a reduction or at least not to an increase of the AIC criterion. Otherwise the column is retained.

The procedure is repeated for each of the K equations of the K-dimensional system, that is, a restriction matrix, $\bar{R}_k$ say, is determined for each equation separately. Once all zero restrictions have been determined by this strategy the K equations of the restricted model with overall restriction matrix

$$\bar{R} = \begin{bmatrix} \bar{R}_1 & & 0 \\ & \ddots & \\ 0 & & \bar{R}_K \end{bmatrix}$$

can be estimated simultaneously by EGLS as described in Sections 5.2.2 and 5.2.4. Note that SC tends to choose the most parsimonious models with the fewest coefficients whereas AIC has a tendency to select the most profligate models.

The advantage of this top-down procedure, starting from the top (largest model) and then working down gradually, is that it permits to check all individual coefficients. Also, the computational expense is very reasonable. The disadvantage of the method is that it requires estimation of each full equation in the initial selection step. This may exhaust the available degrees of freedom if a high order model is deemed necessary for some high-dimensional system. Therefore, a slightly more elaborate bottom-up strategy may be preferred occasionally.

5.2.8c Bottom-up Strategy

Again the restrictions are chosen for each equation separately. In the k-th equation only lags of the first variable are considered initially and an optimal lag length p_1, say, for that variable is selected. That is, we select the optimal model of the form

$$y_{kt} = \nu_k + \alpha_{k1,1} y_{1,t-1} + \cdots + \alpha_{k1,p_1} y_{1,t-p_1} + u_{kt}$$

by fitting models

$$y_{kt} = \nu_k + \alpha_{k1,1} y_{1,t-1} + \cdots + \alpha_{k1,n} y_{1,t-n} + u_{kt},$$

where n ranges from zero to some prespecified upper bound p for the order. p_1 is that order for which the selection criterion, e.g. AIC, HQ, or SC, is minimized.

In the next step p_1 is held fixed and lags of y_2 are added into the equation. Denoting the optimal lag length for y_2 by p_2 gives

$$y_{kt} = \nu_k + \alpha_{k1,1} y_{1,t-1} + \cdots + \alpha_{k1,p_1} y_{1,t-p_1} + \alpha_{k2,1} y_{2,t-1} + \cdots$$
$$+ \alpha_{k2,p_2} y_{2,t-p_2} + u_{kt}.$$

Note that p_2 may, of course, be zero in which case y_2 does not enter the equation.

In the third step p_1 and p_2 are both held fixed and the third variable y_3 is absorbed into the equation in the same way. This procedure is continued until an optimal lag length for each of the K variables is obtained conditional on the "optimal" lags of the previous variables.

Due to omitted variables effects some of the lag lengths may be overstated in the final equation. For instance, when none of the other variables enters the equation, lags of y_1 may be useful in explaining y_{kt} and in reducing the selection criterion. In contrast, lags of y_1 may not contribute to explaining y_k when lags of all the other variables are present too. Therefore, once $p_1, \ldots, p_K$ are chosen, a top-down run as described in the previous subsection may complete the search for zero restrictions on the k-th equation. After zero constraints have been obtained for each equation in this fashion, the K restricted equations may be estimated as one system using EGLS or ML procedures.

Obviously, it is possible in this bottom-up approach that the largest model where all K variables enter with p lags in each equation is never fitted. Thereby considerable savings of degrees of freedom may be possible especially if the maximum order p is substantial. Procedures similar to the one discussed here were, for instance, applied by Hsiao (1979, 1982) and Lütkepohl (1987). Other subset VAR strategies are proposed by Penm & Terrell (1984, 1986).

5.2.8d *Monte Carlo Comparison of Strategies for Subset VAR Modeling*

In order to illustrate the different strategies for subset modeling we have performed a Monte Carlo experiment based on the bivariate process (2.1.15). More precisely,

$$\begin{bmatrix} y_{1t} \\ y_{2t} \end{bmatrix} = \begin{bmatrix} .02 \\ .03 \end{bmatrix} + \begin{bmatrix} .5 & .1 \\ .4 & .5 \end{bmatrix} \begin{bmatrix} y_{1,t-1} \\ y_{2,t-1} \end{bmatrix} + \begin{bmatrix} 0 & 0 \\ .25 & 0 \end{bmatrix} \begin{bmatrix} y_{1,t-2} \\ y_{2,t-2} \end{bmatrix} + \begin{bmatrix} u_{1t} \\ u_{2t} \end{bmatrix}$$

Table 5.1. Proportions of zero coefficients obtained for 1000 realizations of length $T = 30$ of the VAR(2) process

subset strategy	model selection criterion	proportions of zero coefficients				normalized forecast MSE	
		ν	A_1	A_2	A_3	1-step	5-step
top-down	AIC	$\begin{bmatrix} .61 \\ .64 \end{bmatrix}$	$\begin{bmatrix} .20 & .65 \\ .06 & .11 \end{bmatrix}$	$\begin{bmatrix} .79 & .73 \\ .31 & .76 \end{bmatrix}$	$\begin{bmatrix} .80 & .77 \\ .76 & .80 \end{bmatrix}$	1.23	1.33
	HQ	$\begin{bmatrix} .66 \\ .68 \end{bmatrix}$	$\begin{bmatrix} .21 & .72 \\ .08 & .12 \end{bmatrix}$	$\begin{bmatrix} .83 & .79 \\ .35 & .82 \end{bmatrix}$	$\begin{bmatrix} .84 & .82 \\ .81 & .84 \end{bmatrix}$	1.22	1.30
bottom-up	AIC	$\begin{bmatrix} .61 \\ .63 \end{bmatrix}$	$\begin{bmatrix} .20 & .80 \\ .06 & .15 \end{bmatrix}$	$\begin{bmatrix} .86 & .87 \\ .30 & .81 \end{bmatrix}$	$\begin{bmatrix} .95 & .93 \\ .77 & .89 \end{bmatrix}$	1.19	1.27
	HQ	$\begin{bmatrix} .65 \\ .67 \end{bmatrix}$	$\begin{bmatrix} .22 & .85 \\ .08 & .18 \end{bmatrix}$	$\begin{bmatrix} .90 & .91 \\ .34 & .85 \end{bmatrix}$	$\begin{bmatrix} .96 & .96 \\ .80 & .92 \end{bmatrix}$	1.20	1.24

and

$$\Sigma_u = \begin{bmatrix} .09 & 0 \\ 0 & .04 \end{bmatrix}.$$

We have generated 1000 realizations (time series) with this data generation mechanism and we have applied different subset strategies and model selection criteria to a VAR(3) model. In other words, the last coefficient matrix contains zeroes only and A_2 contains three zeroes and one nonzero element. In Table 5.1 the proportions of zeroes determined by different subset strategies are presented, based on time series of length $T = 30$. Since the sample information is not rich it is not surprising that none of the criteria and strategies detects the correct zero elements with high probability. In particular, the small nonzero coefficient in the upper right-hand corner of A_1 is set to zero quite often by all subset strategies. Overall HQ tends to choose smaller models with fewer nonzero coefficients than AIC. Given the discussion in Section 4.3.3 this, of course, was to be expected.

With so little sample information it is also not surprising that the forecast MSEs from the more parsimonious HQ models are overall smaller than those from the AIC models. The normalized forecast MSEs in Table 5.1 are computed by dividing the squared forecast errors of each component by its theoretical forecast MSE so that they are comparable in size and should ideally be 1. More precisely, denoting by $\hat{y}_{i,T}(h)_n$ the h-step forecast obtained for the i-th component of the n-th generated time series and denoting by $\sigma_i^2(h)$ the i-th diagonal element of $\Sigma_y(h)$, the normalized forecast MSE in Table 5.1 is

$$\frac{1}{2}\sum_{i=1}^{2}\frac{1}{1000}\sum_{n=1}^{1000}\frac{(y_{i,T+h,n} - \hat{y}_{i,T}(h)_n)^2}{\sigma_i^2(h)}$$

Table 5.2. Proportions of zero coefficients obtained for 1000 realizations of length $T = 100$ of the VAR(2) process

subset strategy	model selection criterion	proportions of zero coefficients				normalized forecast MSE	
		ν	A_1	A_2	A_3	1-step	5-step
top-down	AIC	$\begin{bmatrix} .64 \\ .50 \end{bmatrix}$	$\begin{bmatrix} .00 & .58 \\ .00 & .00 \end{bmatrix}$	$\begin{bmatrix} .82 & .81 \\ .05 & .81 \end{bmatrix}$	$\begin{bmatrix} .81 & .84 \\ .81 & .85 \end{bmatrix}$	1.02	1.04
	HQ	$\begin{bmatrix} .73 \\ .63 \end{bmatrix}$	$\begin{bmatrix} .00 & .69 \\ .00 & .00 \end{bmatrix}$	$\begin{bmatrix} .90 & .90 \\ .08 & .90 \end{bmatrix}$	$\begin{bmatrix} .89 & .90 \\ .90 & .92 \end{bmatrix}$	1.02	1.03
bottom-up	AIC	$\begin{bmatrix} .63 \\ .50 \end{bmatrix}$	$\begin{bmatrix} .00 & .73 \\ .00 & .00 \end{bmatrix}$	$\begin{bmatrix} .84 & .91 \\ .05 & .84 \end{bmatrix}$	$\begin{bmatrix} .90 & .94 \\ .81 & .92 \end{bmatrix}$	1.01	1.03
	HQ	$\begin{bmatrix} .73 \\ .63 \end{bmatrix}$	$\begin{bmatrix} .00 & .81 \\ .00 & .00 \end{bmatrix}$	$\begin{bmatrix} .91 & .97 \\ .08 & .92 \end{bmatrix}$	$\begin{bmatrix} .96 & .98 \\ .90 & .98 \end{bmatrix}$	1.01	1.03

where $y_{i,T+h,n}$ is the generated value of the i-th component series of the n-th multiple time series.

For $T = 30$ the bottom-up strategy leads to slightly smaller forecast MSEs than the top-down strategy. To see whether the subset strategies have an advantage over unrestricted VAR modeling we have also computed the normalized forecast MSEs for the full VAR(3) model. The 1-step and 5-step MSEs turned out to be 1.27 and 1.40, respectively. Thus they clearly exceed those of the subset models given in Table 5.1. In other words, from a forecasting point of view, application of either subset strategy is advantageous relative to full VAR modeling.

We have repeated the experiment with $T = 100$ and give the results in Table 5.2. Now the zero and nonzero coefficients are identified correctly much more often by both modeling strategies and selection criteria. Moreover the forecast MSEs are all close together and also close to the ideal value of 1. In this case the normalized 1-step and 5-step forecast MSEs from the full VAR(3) model are 1.03 and 1.04, respectively. Hence, they are also close to the values obtained from the subset models. Consequently, as far as forecast precision is concerned, there is little to choose between the different subset strategies and criteria and unrestricted full VAR modeling. In summary, the simulation indicates that, from a forecasting point of view, subset modeling is most useful if the sample information is scarce.

As with previous Monte Carlo experiments the present results are not necessarily generalizable to other processes. They are just meant to illustrate the subset VAR methodology. In practice it is not uncommon to apply different modeling strategies and investigate the sensitivity of the main conclusions with respect to the modeling procedure or selection criterion.

5.2.9 Model Checking

After a subset VAR model has been fitted some checks of the model adequacy are in order. Of course, one check is incorporated in the model selection procedure if some criterion is optimized. By definition the best model is the one that leads to the optimum criterion value. In practice the choice of the criterion is often ad hoc or even arbitrary and, in fact, several competing criteria are often applied. It is then left to the applied researcher to decide on the final model to be used for forecasting or economic analysis. In some cases statistical tests of restrictions may aid in that decision. F-tests as described in Section 4.2.2 may be helpful for that purpose.

5.2.9a *Residual Autocovariances and Autocorrelations*

Another possible check is a test of the white noise assumption for the u_t process. For that purpose the residual autocovariances and autocorrelations may be considered. In analogy with Section 4.4 of Chapter 4 we use the following notation:

$$C_i := \frac{1}{T} \sum_{t=i+1}^{T} u_t u'_{t-i}, \qquad i = 0, 1, \dots, h,$$

$$\mathbf{C}_h := (C_1, \dots, C_h),$$

$$\mathbf{c}_h := \mathrm{vec}(\mathbf{C}_h),$$

$\hat{u}_t$ is the t-th estimation residual of a restricted estimation,

$$\hat{C}_i := \frac{1}{T} \sum_{t=i+1}^{T} \hat{u}_t \hat{u}'_{t-i}, \qquad i = 0, 1, \dots, h,$$

$$\hat{\mathbf{C}}_h := (\hat{C}_1, \dots, \hat{C}_h),$$

$$\hat{\mathbf{c}}_h := \mathrm{vec}(\hat{\mathbf{C}}_h),$$

$\hat{D}$ is the diagonal matrix with the square roots of the diagonal elements of $\hat{C}_0$ on the diagonal,

$$\hat{R}_i := \hat{D}^{-1} \hat{C}_i \hat{D}^{-1}, \qquad i = 0, 1, \dots, h,$$

$$\hat{\mathbf{R}}_h := (\hat{R}_1, \dots, \hat{R}_h),$$

$$\hat{\mathbf{r}}_h := \mathrm{vec}(\hat{\mathbf{R}}_h).$$

In the following proposition the asymptotic distributions of $\hat{\mathbf{c}}_h$ and $\hat{\mathbf{r}}_h$ are given under the assumption of a correctly specified model.

PROPOSITION 5.7 (*Asymptotic Distributions of Residual Autocovariances and Autocorrelations*)

Suppose y_t is a stable, stationary, K-dimensional VAR(p) process with identically distributed standard white noise u_t and the parameter vector $\boldsymbol{\beta}$ satisfies the restrictions $\boldsymbol{\beta} = R\gamma + r$ with R being a known $(K(Kp + 1) \times M)$ matrix of rank M. Furthermore suppose that $\boldsymbol{\beta}$ is estimated by EGLS such that $\hat{\hat{\boldsymbol{\beta}}} = R\hat{\hat{\gamma}} + r$. Then

$$\sqrt{T}\hat{\mathbf{c}}_h \xrightarrow{d} N(0, \Sigma_c^r(h)), \tag{5.2.32}$$

where

$$\Sigma_c^r(h) = I_h \otimes \Sigma_u \otimes \Sigma_u - GR[R'(\Gamma \otimes \Sigma_u^{-1})R]^{-1}R'G'$$

and $G := \tilde{G}' \otimes I_K$ is the matrix defined in Chapter 4, Lemma 4.2. Furthermore,

$$\sqrt{T}\hat{\mathbf{r}}_h \xrightarrow{d} N(0, \Sigma_r^r(h)), \tag{5.2.33}$$

where

$$\Sigma_r^r(h) = I_h \otimes R_u \otimes R_u - (G_0' \otimes D^{-1})R[R'(\Gamma \otimes \Sigma_u^{-1})R]^{-1}R'(G_0 \otimes D^{-1})$$

and $G_0 := \tilde{G}(I_h \otimes D^{-1})$ is defined in Proposition 4.6, D is the diagonal matrix with the square roots of the diagonal elements of Σ_u on the diagonal, and $R_u := D^{-1}\Sigma_u D^{-1}$ is the correlation matrix corresponding to Σ_u. ■

Proof: The proof is similar to that of Propositions 4.5 and 4.6. Defining $\tilde{G}$ as in Lemma 4.2, the lemma implies that $\sqrt{T}\hat{\mathbf{c}}_h$ is known to have the same asymptotic distribution as

$$\begin{aligned}
&\sqrt{T}\mathbf{c}_h - \sqrt{T}G\,\mathrm{vec}(\hat{\hat{B}} - B) \\
&= [-\tilde{G}' \otimes I_K, I]\begin{bmatrix} \sqrt{T}\,\mathrm{vec}(\hat{\hat{B}} - B) \\ \sqrt{T}\mathbf{c}_h \end{bmatrix} \\
&= [-(\tilde{G}' \otimes I_K)R, I]\begin{bmatrix} \sqrt{T}(\hat{\hat{\gamma}} - \gamma) \\ \sqrt{T}\mathbf{c}_h \end{bmatrix} \\
&= [-(\tilde{G}' \otimes I_K)R, I]\begin{bmatrix} \left[R'\left(\dfrac{ZZ'}{T} \otimes \bar{\Sigma}_u^{-1}\right)R\right]^{-1} R'(I_{Kp+1} \otimes \bar{\Sigma}_u^{-1}) & 0 \\ 0 & I \end{bmatrix}\begin{bmatrix} \dfrac{1}{\sqrt{T}}\mathrm{vec}(UZ') \\ \sqrt{T}\mathbf{c}_h \end{bmatrix}
\end{aligned}$$

(see (5.2.7)). The asymptotic distribution in (5.2.32) then follows from Lemma 4.3 and Proposition C.4(1) by noting that $\Gamma = \mathrm{plim}(ZZ'/T)$ and $\Sigma_u = \mathrm{plim}\,\bar{\Sigma}_u$. The limiting distribution of $\sqrt{T}\hat{\mathbf{r}}_h$ follows as in the proof of Proposition 4.6. ■

The results in Proposition 5.7 can be used to check the white noise assumption for the u_t's. As in Section 4.4, residual autocorrelations are often plotted and evaluated on the basis of two-standard error bounds about zero. Estimators of the standard errors are obtained by replacing all unknown quantities in $\Sigma_r^r(h)$ by consistent estimators. Specifically Σ_u may be estimated by $\hat{C}_0$. We will demonstrate the resulting white noise test in Section 5.2.10 by an example.

5.2.9b Portmanteau Tests

For the portmanteau statistic

$$\begin{aligned}P_h &:= T \sum_{i=1}^{h} \operatorname{tr}(\hat{C}_i'\hat{C}_0^{-1}\hat{C}_i\hat{C}_0^{-1}) \\ &= T\hat{\mathbf{c}}_h'(I_h \otimes \hat{C}_0^{-1} \otimes \hat{C}_0^{-1})\hat{\mathbf{c}}_h \end{aligned} \tag{5.2.34}$$

we get the following result:

PROPOSITION 5.8 (*Approximate Distribution of the Portmanteau Statistic*)

Suppose the conditions of Proposition 5.7 are satisfied and there are no restrictions linking the intercept terms to the $A_1, \dots, A_p$ coefficients, that is,

$$R = \begin{bmatrix} R_{(1)} & 0 \\ 0 & R_{(2)} \end{bmatrix}$$

is block-diagonal with $R_{(1)}$ and $R_{(2)}$ having row-dimensions K and K^2p, respectively. Then P_h has an approximate limiting χ^2-distribution with $K^2h - \operatorname{rk}(R_{(2)})$ degrees of freedom for large T and h. ■

Proof: Under the conditions of the proposition the covariance matrix of the asymptotic distribution in (5.2.32) is

$$\Sigma_c^r(h) = I_h \otimes \Sigma_u \otimes \Sigma_u - GR_{(2)}[R_{(2)}'(\Gamma_Y(0) \otimes \Sigma_u^{-1})R_{(2)}]^{-1}R_{(2)}'G',$$

where G is the matrix defined in Lemma 4.2. Using this fact, Proposition 5.8 can be proven just as Proposition 4.7 by replacing $\bar{G}$ in that proof by $GR_{(2)}$ (see Section 4.4.3). ■

The degrees of freedom in Proposition 5.8 are obtained by substracting the number of unconstrained A_i coefficients from K^2h. As in Section 4.4.3, the modified portmanteau statistic

$$\bar{P}_h := T^2 \sum_{i=1}^{h} (T-i)^{-1} \operatorname{tr}(\hat{C}_i'\hat{C}_0^{-1}\hat{C}_i\hat{C}_0^{-1}) \tag{5.2.35}$$

may be preferable for testing the white noise assumption in small samples. In other words, under the white noise hypothesis, the small sample distribution of $\bar{P}_h$ may be closer to the limiting χ^2-distribution than that of P_h.

5.2.9c Other Checks of Restricted Models

It must also be kept in mind that our discussion has been based on a number of further assumptions that should be checked. Prominent among them are stationarity, stability, and normality. The latter is used in setting up forecast intervals and regions and the former properties are basic conditions underlying much of our analysis (see, for instance, Propositions 5.1–5.6). The stability tests based on predictions and described in Section 4.6 of Chapter 4 may be applied in the same way as for full unrestricted VAR processes. Of course, now the forecasts and MSE matrix estimators should be based on the restricted coefficient estimators as discussed in Section 5.2.6. Also, it is easy to see from Section 4.5 that the tests for nonnormality remain valid when true restrictions are placed on the VAR coefficient matrices.

5.2.10 An Example

As an example we use again the same data as in Section 3.2.3 and some other previous sections. That is, y_1, y_2, and y_3 are first differences of logarithms of investment, income, and consumption, respectively. We keep four presample values and use sample values from the period 1961.2–1978.4. Hence, the time series length is $T = 71$. We have applied the top-down strategy with selection criteria AIC, HQ, and SC and a VAR order of $p = 4$. In other words, we use the same maximum order as in the order selection procedure for full VAR models in Chapter 4. Since HQ and SC choose the same model we get two different models which are shown in Table 5.3. As usual, the HQ-SC model is more parsimonious than the AIC model.

In Table 5.3 modified portmanteau statistics with corresponding p-values are also given for both models. Obviously, none of the test values gives rise to concern about the models. In Figure 5.1 residual autocorrelations of the HQ-SC model with estimated two-standard error bounds about zero are depicted. The rather unusual looking estimated two-standard error bounds for some low lags are a consequence of the zero elements in the estimated VAR coefficient matrices. Although some individual autocorrelations fall outside the two-standard error bounds about zero this is not necessarily a reason for modifying the model. As in Chapter 4, such a decision depends on which criterion is given priority.

We have also produced forecasts with the HQ-SC model and give them in Table 5.4 together with forecasts from a full VAR(4) model. In this example the forecasts from the two models are quite close and the estimated forecast intervals from the subset model are all smaller than those of a full VAR(4) model. Although theoretically the more parsimonious subset model produces more precise forecasts if the restrictions are correct, it must be kept in mind that in the present case the restrictions, the forecasts and forecast intervals are estimated on the basis of a single realization of an unknown data generation process. Under these circumstances a subset model may produce less precise forecasts than a heavily parameterized full VAR model. Note that in the present subset model the income forecasts are the same for all forecast horizons because the income is a white noise process in the HQ-SC model.

Table 5.3. Subset VAR models for the investment/income/consumption data

model selection criterion	EGLS estimates of coefficient matrices					modified portmanteau tests	
	$\hat{\hat{\nu}}$	$\hat{\hat{A}}_1$	$\hat{\hat{A}}_2$	$\hat{\hat{A}}_3$	$\hat{\hat{A}}_4$	values	*p*-values
AIC	$\begin{bmatrix} .015 \\ (.006) \\ .015 \\ (.003) \\ .013 \\ (.003) \end{bmatrix}$	$\begin{bmatrix} -.219 & 0 & 0 \\ (.104)^* & & \\ 0 & 0 & .235 \\ & & (.133) \\ 0 & .274 & -.391 \\ & (.082) & (.116) \end{bmatrix}$	$\begin{bmatrix} 0 & 0 & 0 \\ .010 & 0 & 0 \\ (.024) & & \\ 0 & .335 & 0 \\ & (.073) & \end{bmatrix}$	$\begin{bmatrix} 0 & 0 & 0 \\ 0 & 0 & 0 \\ 0 & .095 & 0 \\ & (.076) & \end{bmatrix}$	$\begin{bmatrix} .340 & 0 & 0 \\ (.103) & & \\ 0 & 0 & 0 \\ 0 & 0 & 0 \end{bmatrix}$	$\bar{P}_{12} = 79.3$ $\bar{P}_{20} = 144$	.937 .943
HQ-SC	$\begin{bmatrix} .015 \\ (.006) \\ .020 \\ (.001) \\ .016 \\ (.003) \end{bmatrix}$	$\begin{bmatrix} -.225 & 0 & 0 \\ (.104) & & \\ 0 & 0 & 0 \\ 0 & .261 & -.439 \\ & (.081) & (.095) \end{bmatrix}$	$\begin{bmatrix} 0 & 0 & 0 \\ 0 & 0 & 0 \\ 0 & .329 & 0 \\ & (.074) & \end{bmatrix}$	$\begin{bmatrix} 0 & 0 & 0 \\ 0 & 0 & 0 \\ 0 & 0 & 0 \end{bmatrix}$	$\begin{bmatrix} .331 & 0 & 0 \\ (.103) & & \\ 0 & 0 & 0 \\ 0 & 0 & 0 \end{bmatrix}$	$\bar{P}_{12} = 85.5$ $\bar{P}_{20} = 152$	.893 .898

* Estimated standard errors.

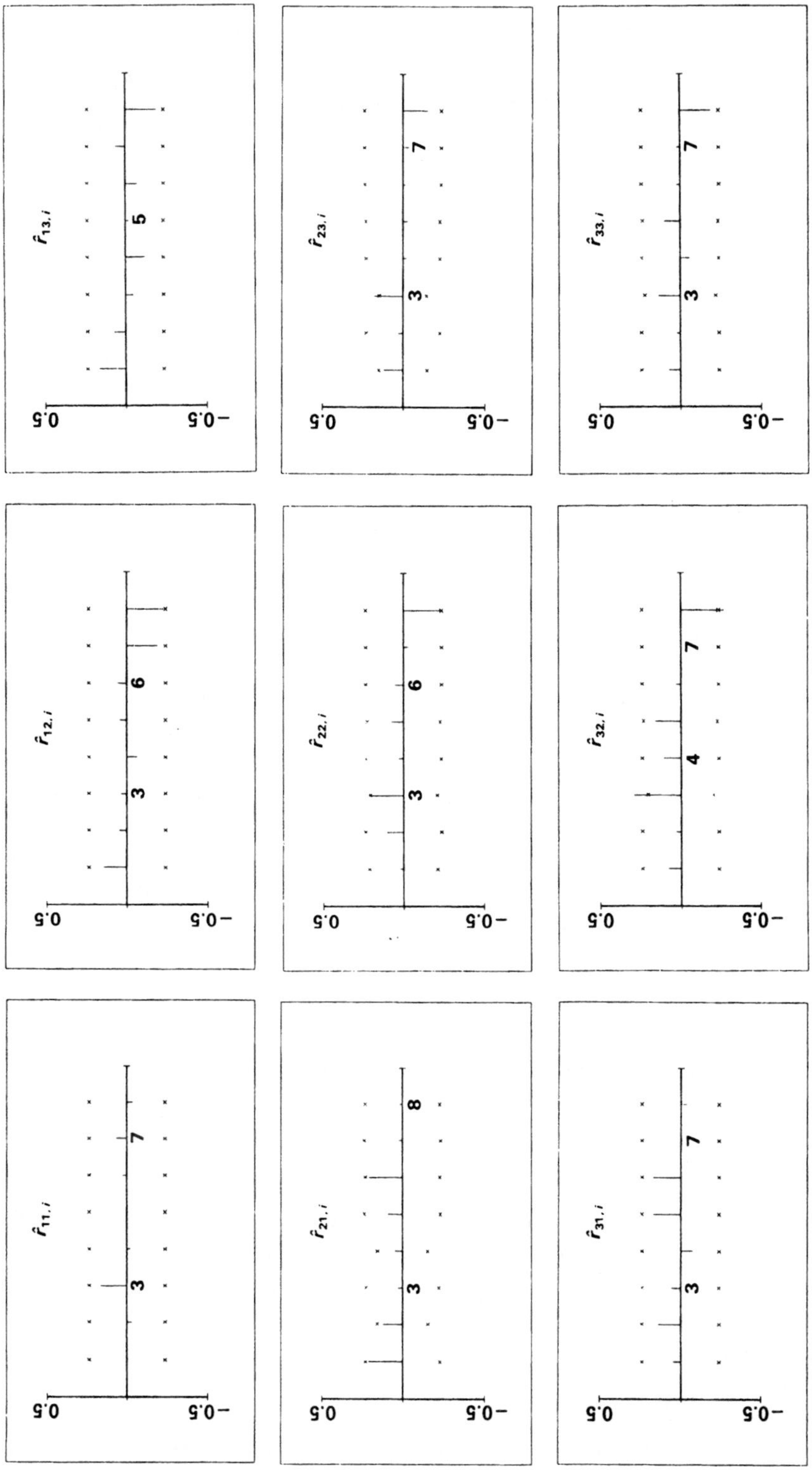

Fig. 5.1. Estimated residual autocorrelations of the investment/income/consumption HQ-SC subset VAR model.

Table 5.4. Point and interval forecasts from full and subset VAR(4) models

variable	forecast horizon	full VAR(4) model point forecast	full VAR(4) model 95% interval forecast	HQ-SC subset VAR(4) model point forecast	HQ-SC subset VAR(4) model 95% interval forecast
investment	1	.006	[−.091, .103]	.015	[−.074, .105]
	2	.025	[−.075, .125]	.023	[−.068, .115]
	3	.028	[−.071, .126]	.018	[−0.73, .110]
	4	.026	[−.074, .125]	.023	[−.069, .115]
income	1	.021	[−.005, .047]	.020	[−.004, .044]
	2	.022	[−.004, .049]	.020	[−.004, .044]
	3	.017	[−.009, .043]	.020	[−.004, .044]
	4	.022	[−.004, .049]	.020	[−.004, .044]
consumption	1	.022	[.001, .042]	.023	[.004, .042]
	2	.015	[−.006, .036]	.013	[−.007, .033]
	3	.020	[−.004, .043]	.022	[.001, .044]
	4	.019	[−.004, .042]	.018	[−.004, .040]

We have also computed impulse responses from the HQ-SC subset VAR model. The Θ_i responses of consumption to an impulse in income based on orthogonalized residuals are depicted in Figure 5.2. Comparing them with Figure 3.8 shows that they are qualitatively similar to the impulse responses from the full VAR(2) model. Considering the responses of investment to a consumption innovation reveals that they are all zero in the subset VAR model. A closer look at Table 5.3 shows that income/consumption are not Granger-causal for investment in both subset models. This result was also obtained in the full VAR model (see Section 3.6.2). However, now it is directly seen in the model without further causality testing. In other words, the causality testing is built into the model selection procedure.

5.3 VAR Processes with Nonlinear Parameter Restrictions

5.3.1 Some Types of Nonlinear Constraints

Some authors have suggested nonlinear constraints for the coefficients of a VAR model. For instance, multiplicative models with VAR operator

$$\begin{aligned} A(L) &= I_K - A_1 L - \cdots - A_p L^p \\ &= (I_K - B_1 L^s - \cdots - B_Q L^{sQ})(I_K - C_1 L - \cdots - C_q L^q) \\ &= B(L^s)C(L) \end{aligned}$$

have been considered. Here L is the lag operator defined in Chapter 2, Section

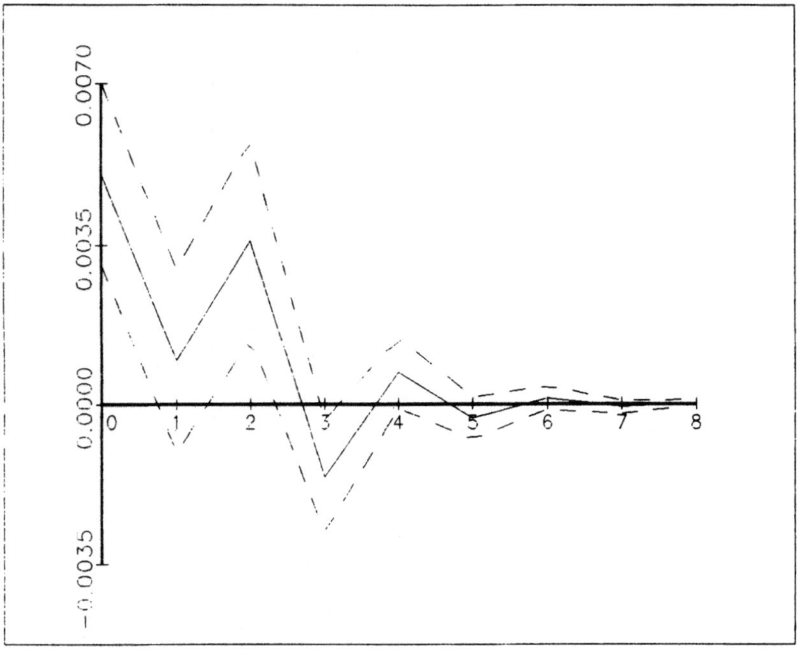

Fig. 5.2. Estimated responses of consumption to an orthogonalized impulse in income with two-standard error bounds based on a subset VAR model.

2.1.2, the B_i and C_j are $(K \times K)$ coefficient matrices and $B(L^s)$ contains "seasonal" powers of L only. Such models may be useful for seasonal data. For instance, for quarterly data, a multiplicative seasonal operator may have the form

$$(I_K - B_1 L^4)(I_K - C_1 L - C_2 L^2).$$

The corresponding VAR operator is

$$\begin{aligned} A(L) &= I_K - A_1 L - \cdots - A_6 L^6 \\ &= I_K - C_1 L - C_2 L^2 - B_1 L^4 + B_1 C_1 L^5 + B_1 C_2 L^6, \end{aligned}$$

so that $A_1 = C_1, A_2 = C_2, A_3 = 0, A_4 = B_1, A_5 = -B_1 C_1, A_6 = -B_1 C_2$. Hence, the coefficients $\boldsymbol{\alpha} := \text{vec}[A_1, \ldots, A_p]$ are determined by $\boldsymbol{\gamma} := \text{vec}[B_1, C_1, C_2]$, that is,

$$\boldsymbol{\alpha} = g(\boldsymbol{\gamma}). \tag{5.3.1}$$

There are also other types of nonlinear constraints that may be written in this way. For example, the VAR operator may have the form $A(L) = B(L)C(L)$, where

$$C(L) = \begin{bmatrix} c_1(L) & & 0 \\ & \ddots & \\ 0 & & c_K(L) \end{bmatrix}$$

is a diagonal operator with $c_k(L) = 1 + c_{k1}L + \cdots + c_{kq}L^q$, which represents the individual dynamics of the variables and $B(L) = I_K - B_1 L - \cdots - B_n L^n$ takes care of joint relations. Again the implied restrictions for $\boldsymbol{\alpha}$ can easily be cast in the form (5.3.1).

In principle, under general conditions, if restrictions are given in the form (5.3.1), the analysis can proceed analogously to the linear restriction case. That is, we need to find an estimator $\hat{\boldsymbol{\gamma}}$ of $\boldsymbol{\gamma}$, for instance, by minimizing

$$S(\boldsymbol{\gamma}) = [\mathbf{y} - (Z' \otimes I_K)g(\boldsymbol{\gamma})]'(I_T \otimes \Sigma_u^{-1})[\mathbf{y} - (Z' \otimes I_K)g(\boldsymbol{\gamma})],$$

where $\mathbf{y}$, Z, and Σ_u are as defined in Section 5.2. The minimization may require an iterative algorithm. Such algorithms are described in Section 7.3.2 in the context of estimating VARMA models. Once we have an estimator $\hat{\boldsymbol{\gamma}}$ we may estimate $\boldsymbol{\alpha}$ as $\hat{\boldsymbol{\alpha}} = g(\hat{\boldsymbol{\gamma}})$. Under similar conditions as for the linear case, the estimators will be consistent and asymptotically normally distributed, e.g.,

$$\sqrt{T}(\hat{\boldsymbol{\alpha}} - \boldsymbol{\alpha}) \xrightarrow{d} N(0, \Sigma_{\hat{\boldsymbol{\alpha}}}). \tag{5.3.2}$$

The corresponding estimators $\hat{A}_1, \ldots, \hat{A}_p$ may be used in computing forecasts and impulse responses etc. The asymptotic properties of these quantities then follow exactly as in the previous sections (see in particular Sections 5.2.6 and 5.2.7).

Another type of "multiplicative" VAR operator has the form

$$A(L) = I_K - B(L)C(L), \tag{5.3.3}$$

where

$$B(L) = B_0 + B_1 L + \cdots + B_q L^q$$

is of dimension $(K \times r)$, that is, the B_i have dimension $(K \times r)$, and

$$C(L) = C_1 L + \cdots + C_p L^p$$

is of dimension $(r \times K)$, with $r < K$. For $p = 1$, neglecting the intercept terms, the process becomes

$$y_t = B_0 C_1 y_{t-1} + \cdots + B_q C_1 y_{t-q-1} + u_t$$

which is sometimes called an *index model* because y_t is represented in terms of lagged values of the "index" $C_1 y_t$. In the extreme case where $r = 1$, $C_1 y_t$ is simply a weighted sum or index of the components of y_t which justifies the name of the model. Such models have been investigated by Reinsel (1983) in some detail.

Alternatively, if $q = 0$, the process is called a reduced rank (RR) VAR process. This type of model and its problems will be discussed in more detail in the next subsections. We will see that in general the implied nonlinear restrictions for the VAR coefficients are not easily cast in the form (5.3.1). They are discussed here in some detail because they have a close relationship to the nonstationary models considered in Chapter 11.

5.3.2 Reduced Rank VAR Models

To simplify matters we assume now that the mean has been subtracted from all variables and we denote the mean-adjusted variables by y_t. Thus, we get a VAR(p) process

$$\begin{aligned} y_t &= A_1 y_{t-1} + \cdots + A_p y_{t-p} + u_t \\ &= AX_{t-1} + u_t, \end{aligned} \tag{5.3.4}$$

where $A := [A_1, \ldots, A_p]$ and

$$X_t := \begin{bmatrix} y_t \\ \vdots \\ y_{t-p+1} \end{bmatrix}.$$

Later we will comment on the consequences of using an estimated rather than known mean term. Suppose that $q = 0$ in (5.3.3) and A can be written as

$$A = B[C_1, \ldots, C_p] = BC,$$

where $B = B_0$ is $(K \times r)$ and the C_i are $(r \times K)$ so that $C := [C_1, \ldots, C_p]$ is an $(r \times Kp)$ matrix.

The idea behind this reduced rank model is that, except for multiplicative constants contained in B, there may be linear combinations of the X_{t-1} vector that contain all the relevant information regarding y_t and thus it is unnecessary to write each of the K equations in (5.3.4) as a separate linear combination of X_{t-1}. Suppose, for instance, that one linear combination of X_{t-1}, say CX_{t-1}, suffices for all K equations so that (5.3.4) becomes

$$y_t = BCX_{t-1} + u_t \tag{5.3.5}$$

with B a $(K \times 1)$ vector and C a $(1 \times Kp)$ vector. This representation of the model is an enormous simplification over the full model (5.3.4). The latter contains K^2p coefficients whereas (5.3.5) contains only $K + Kp$ coefficients. For example, for a 5-dimensional process of order 4, $K^2p = 100$ and $K + Kp = 25$.

In (5.3.5) the matrix A is written as the product of two vectors, $A = BC$. In this case A has rank 1. More generally, if the $(K \times Kp)$ matrix A has rank $r < K$ it may be represented as the product $A = BC$, where B and C are $(K \times r)$ and $(r \times Kp)$ matrices, respectively, of rank r. The number of elements in B is Kr and that in C is rKp. The sum of these two numbers may be substantially lower than K^2p. In fact, we may reduce the parameter space even further by noting that B and C are not uniquely determined without further restrictions. For any nonsingular $(r \times r)$ matrix F,

$$A = BFF^{-1}C = \bar{B}\bar{C}, \tag{5.3.6}$$

so that $\bar{B} := BF$ and $\bar{C} := F^{-1}C$ could be used instead of B and C. Thus, we may impose further constraints on B and/or C to make the matrices unique. More precisely, we may impose restrictions so that the only possible matrix F in (5.3.6) is the identity matrix I_r. For instance, if B is such that its upper $(r \times r)$ submatrix is I_r, that is,

$$B = \begin{bmatrix} I_r \\ B_1 \end{bmatrix}, \tag{5.3.7}$$

then the representation $A = BC$ is unique. A similar constraint could be chosen for C.

Restrictions of this type can easily be written in the form (5.3.1) and it is quite convenient to work with them, as we have argued in Section 5.3.1. However, here they have the disadvantage of imposing constraints on the system that are somewhat stronger than just the restriction $\text{rk}(A) = r$. To see this, consider the bivariate VAR(1) model with coefficient matrix

$$A = \begin{bmatrix} 0 & 0 \\ \alpha_{21} & \alpha_{22} \end{bmatrix}.$$

If α_{21} or α_{22} is nonzero, $\text{rk}(A) = 1$, although A cannot be written in the form BC, where B and C' are (2×1) vectors with

$$B = \begin{bmatrix} 1 \\ \beta \end{bmatrix}.$$

Note that for this choice of B the zeroes in the first row of A require $C = 0$ which is incompatible with $\beta C = [\alpha_{21} \quad \alpha_{22}]$.

Even though the normalization in (5.3.7) is unsuitable in general if we just know the constraint $\text{rk}(A) = r$, it does show that we may impose r^2 constraints on B and C. Typical constraints that are imposed are normalizations and orthogonality restrictions for the columns of B and the rows of C or related matrices. Define

$$Y := [y_1, \ldots, y_T],$$

$$X := [X_0, \ldots, X_{T-1}],$$

$$\Sigma_{XX} := \text{plim}\, XX'/T = \Gamma_Y(0), \tag{5.3.8}$$

$$\Sigma_{YX} := \Sigma'_{XY} := \text{plim}\, YX'/T, \tag{5.3.9}$$

$$\Sigma_u := E(u_t u_t'),$$

and let V be the $(K \times r)$ matrix containing the orthonormal eigenvectors of

$$\Sigma_u^{-1/2} \Sigma_{YX} \Sigma_{XX}^{-1} \Sigma_{XY} \Sigma_u^{-1/2} \tag{5.3.10}$$

corresponding to the r largest eigenvalues in decreasing order. Then we may choose

$$B = \Sigma_u^{1/2} V \tag{5.3.11}$$

and

$$C = V' \Sigma_u^{-1/2} \Sigma_{YX} \Sigma_{XX}^{-1}. \tag{5.3.12}$$

In this case,

$$B'\Sigma_u^{-1}B = V'V = I_r \tag{5.3.13}$$

and

$$C\Sigma_{XX}C' = V'\Sigma_u^{-1/2}\Sigma_{YX}\Sigma_{XX}^{-1}\Sigma_{XY}\Sigma_u^{-1/2}V = \begin{bmatrix} \lambda_1 & & 0 \\ & \ddots & \\ 0 & & \lambda_r \end{bmatrix}, \tag{5.3.14}$$

where λ_i is the i-th largest eigenvalue of (5.3.10). The latter result follows from Appendix A.9.2. Since $B'\Sigma_u^{-1}B$ and $C\Sigma_{XX}C'$ are symmetric we have $r(r+1)/2$ restrictions in (5.3.13) and $r(r-1)/2$ restrictions in (5.3.14) which gives a total of r^2 constraints on B and C.

At this point the choice of B and C may seem quite arbitrary. However, it is obvious that $A = BC$ has rank r with this choice of B and C. Moreover, no constraints are imposed on A in addition to the rank restriction. In the following subsections it will become clear that this choice of B and C is useful in deriving asymptotic properties of estimators.

An observation that we can make at this point is that the total number of free parameters in an RRVAR model of dimension K, order p, and rank r is obtained by subtracting r^2 from the sum of the coefficients of B and C, that is, we have $Kpr + Kr - r^2$ free parameters. In the next subsection the multivariate LS estimator for the coefficients is derived.

5.3.3 Multivariate LS Estimation of Reduced Rank VAR Models

Given the model

$$Y = AX + U, \tag{5.3.15}$$

where

$$Y := [y_1, \dots, y_T], \qquad A := [A_1, \dots, A_p], \qquad X := [X_0, \dots, X_{T-1}],$$
$$U := [u_1, \dots, u_T],$$

the objective is to find the $(K \times Kp)$ matrix $\hat{A}$ that minimizes

$$\operatorname{tr}(Y - AX)'\Sigma_u^{-1}(Y - AX) \tag{5.3.16}$$

with respect to A, subject to the constraint $\operatorname{rk}(A) = r < K$. Equivalently, we will try to find matrices $\hat{B}$ and $\hat{C}$ of dimensions $(K \times r)$ and $(r \times Kp)$ that minimize

$$\operatorname{tr}(Y - BCX)'\Sigma_u^{-1}(Y - BCX). \tag{5.3.17}$$

A solution to the minimization problem is given in the following proposition.

PROPOSITION 5.9 (*LS Estimator of the Reduced Rank VAR Model*)

Let Y, X, Σ_u, B, and C be matrices of dimensions $(K \times T)$, $(Kp \times T)$, $(K \times K)$, $(K \times r)$, and $(r \times Kp)$, respectively, with Σ_u positive definite, $\operatorname{rk}(B) = \operatorname{rk}(C) = r$,

$\text{rk}(X) = Kp$, and $\text{rk}(Y) = K$. Then a minimum of (5.3.17) with respect to B and C is obtained for

$$B = \hat{B} := \Sigma_u^{1/2}\hat{V} \quad \text{and} \quad C = \hat{C} := \hat{V}'\Sigma_u^{-1/2}YX'(XX')^{-1}, \tag{5.3.18}$$

where $\hat{V} = [\hat{v}_1, \ldots, \hat{v}_r]$ is the $(K \times r)$ matrix of the orthonormal eigenvectors corresponding to the r largest eigenvalues of

$$\frac{1}{T}\Sigma_u^{-1/2}YX'(XX')^{-1}XY'\Sigma_u^{-1/2}$$

in nonincreasing order. ■

Proof: See Proposition A.5 of Appendix A.14. ■

Since in the proposition Σ_u is any positive definite $(K \times K)$ matrix, it may be replaced by

$$\tilde{\Sigma}_u = \frac{1}{T}Y(I_T - X'(XX')^{-1}X)Y'.$$

We always assume that this estimator is substituted for Σ_u if not otherwise noted.

It may be worth emphasizing again that the solution of the minimization problem given in Proposition 5.9 is by no means unique. Nevertheless we will refer to this particular solution as the LS estimator of B and C in the following. In the next subsection the asymptotic distributions of $\hat{B}$, $\hat{C}$, and the resulting estimator for A are given.

5.3.4 Asymptotic Properties of Reduced Rank LS Estimators

Under our usual conditions (for a restatement see Proposition 3.1) the estimators $\hat{B}$ and $\hat{C}$ in (5.3.18) have well-defined probability limits because an eigenvector v of a positive semidefinite matrix Ω is a continuous function of that matrix (see Magnus & Neudecker (1988, Chapter 8)). Loosely speaking, small changes in Ω result in small changes in the eigenvector v corresponding to a particular eigenvalue, say the largest. Hence, since

$$\text{plim}\frac{1}{T}\Sigma_u^{-1/2}YX'(XX')^{-1}XY'\Sigma_u^{-1/2} = \Sigma_u^{-1/2}\Sigma_{YX}\Sigma_{XX}^{-1}\Sigma_{XY}\Sigma_u^{-1/2}, \tag{5.3.19}$$

Slutsky's Theorem (Appendix C, Proposition C.1(5)) implies $\text{plim}\ \hat{V} = V$ and, consequently, $\text{plim}\ \hat{B} = B$ and $\text{plim}\ \hat{C} = C$, where V, B, and C are the matrices defined in (5.3.10)–(5.3.12). We will give the asymptotic distributions of $\sqrt{T}\,\text{vec}(\hat{B} - B)$ and $\sqrt{T}\,\text{vec}(\hat{C} - C)'$ in the next proposition which is due to Velu, Reinsel & Wichern (1986). From these results it will be easy to obtain the asymptotic properties of the estimator $\hat{A}_{(r)} := \hat{B}\hat{C}$ of A.

PROPOSITION 5.10 (*Asymptotic Properties of Reduced Rank LS Estimators*)

Let y_t be a zero mean K-dimensional stable VAR(p) process as in (5.3.4) and let u_t be standard white noise, that is, u_t is mean zero independent white noise with bounded fourth moments and positive definite covariance matrix Σ_u. Suppose that $\operatorname{rk}(A) = r < K$ so that $A = BC$, where

$$B = [\beta_1, \dots, \beta_r] \qquad \text{and} \qquad C = \begin{bmatrix} \gamma_1 \\ \vdots \\ \gamma_r \end{bmatrix}.$$

are the $(K \times r)$ and $(r \times Kp)$ matrices respectively given in (5.3.11) and (5.3.12), that is, the β_i are $(K \times 1)$ and the γ_i are $(1 \times Kp)$ vectors. Moreover, let $\hat{B}$ and $\hat{C}$ be the LS estimators given in Proposition 5.9 and suppose that the r largest eigenvalues $\lambda_1 > \lambda_2 > \cdots > \lambda_r$ of the matrix (5.3.19) are all distinct and the remaining eigenvalues are $\lambda_{r+1} \geq \cdots \geq \lambda_K$. Then, $\hat{B}$ and $\hat{C}$ are consistent estimators of B and C and

$$\sqrt{T}\begin{bmatrix} \operatorname{vec}(\hat{B} - B) \\ \operatorname{vec}(\hat{C}' - C') \end{bmatrix} \xrightarrow{d} N\left[0, \begin{bmatrix} \Sigma_B & \Sigma_{BC} \\ \Sigma_{CB} & \Sigma_C \end{bmatrix}\right], \tag{5.3.20}$$

where Σ_B is a $(Kr \times Kr)$ matrix with typical $(K \times K)$ block

$$\Sigma_{B,ij} = \begin{cases} \displaystyle\sum_{\substack{k=1 \\ k \neq i}}^{K} \frac{\lambda_i + \lambda_k}{(\lambda_i - \lambda_k)^2} \beta_k \beta_k' & \text{for } i = j \\[2ex] \displaystyle -\frac{\lambda_i + \lambda_j}{(\lambda_i - \lambda_j)^2} \beta_j \beta_i' & \text{for } i \neq j, \end{cases}$$

with $\beta_k = \Sigma_u^{1/2} v_k$, v_k being the eigenvector corresponding to the k-th largest eigenvalue of the matrix (5.3.10), $\Sigma_{BC} = \Sigma_{CB}'$ is a $(rK \times rKp)$ matrix with typical $(K \times Kp)$ block

$$\Sigma_{BC,ij} = \begin{cases} \displaystyle 2\lambda_i \sum_{\substack{k=1 \\ k \neq i}}^{r} \frac{1}{(\lambda_i - \lambda_k)^2} \beta_k \gamma_k & \text{for } i = j \\[2ex] \displaystyle -\frac{2\lambda_j}{(\lambda_i - \lambda_j)^2} \beta_j \gamma_i & \text{for } i \neq j, \end{cases}$$

and Σ_C is an $(rKp \times rKp)$ matrix with typical $(Kp \times Kp)$ block

$$\Sigma_{C,ij} = \begin{cases} \displaystyle\sum_{\substack{k=1 \\ k \neq i}}^{r} \frac{3\lambda_i - \lambda_k}{(\lambda_i - \lambda_k)^2} \gamma_k' \gamma_k + \Sigma_{XX}^{-1} & \text{for } i = j \\[2ex] \displaystyle -\frac{\lambda_i + \lambda_j}{(\lambda_i - \lambda_j)^2} \gamma_j' \gamma_i & \text{for } i \neq j. \end{cases}$$

■

Proof: See Velu, Reinsel & Wichern (1986). ■

The asymptotic properties of $\hat{A}_{(r)} := \hat{B}\hat{C}$ follow easily from this proposition.

COROLLARY 5.10.1

Under the conditions of Proposition 5.10, $\hat{A}_{(r)} := \hat{B}\hat{C}$ is a consistent estimator of $A = BC$, and

$$\sqrt{T}\,\mathrm{vec}(\hat{A}_{(r)} - A) \xrightarrow{d} N(0, \Sigma_{(r)}), \tag{5.3.21}$$

where

$$\begin{aligned}\Sigma_{(r)} = {} & (C' \otimes I_K)\Sigma_B(C \otimes I_K) + (I_{Kp} \otimes B)\Sigma_{CB}(C \otimes I_K) \\ & + (C' \otimes I_K)\Sigma_{BC}\mathbf{K}'_{Kp,r}(I_{Kp} \otimes B') + (I_{Kp} \otimes B)\mathbf{K}_{Kp,r}\Sigma_C\mathbf{K}'_{Kp,r}(I_{Kp} \otimes B')\end{aligned}$$

and $\mathbf{K}_{Kp,r}$ denotes a commutation matrix (Appendix A.12.2). ■

Proof:

$$\mathrm{plim}\,\hat{A}_{(r)} = \mathrm{plim}(\hat{B})\,\mathrm{plim}(\hat{C}) = BC$$

and

$$\hat{B}\hat{C} - BC = (\hat{B} - B)C + \hat{B}(\hat{C} - C).$$

Applying the vec operator and multiplying by $\sqrt{T}$ gives

$$\begin{aligned}\sqrt{T}\,\mathrm{vec}(\hat{A}_{(r)} - A) &= \sqrt{T}(C' \otimes I_K)\,\mathrm{vec}(\hat{B} - B) \\ &\quad + \sqrt{T}(I_{Kp} \otimes \hat{B})\mathbf{K}_{Kp,r}\,\mathrm{vec}(\hat{C}' - C') \\ &= \sqrt{T}[(C' \otimes I_K), (I_{Kp} \otimes \hat{B})\mathbf{K}_{Kp,r}]\begin{bmatrix}\mathrm{vec}(\hat{B} - B) \\ \mathrm{vec}(\hat{C}' - C')\end{bmatrix}\end{aligned}$$

and the corollary follows from Proposition C.4 of Appendix C. ■

REMARK 1: The matrix $[(C' \otimes I_K), (I_{Kp} \otimes B)\mathbf{K}_{Kp,r}]$ has dimensions $(K^2p \times (rK + rKp))$ and has rank less than K^2p if $r < K$. Therefore, as a consequence of the rank restriction for A, $\Sigma_{(r)}$ is a singular covariance matrix and the asymptotic distribution in (5.3.21) is singular multivariate normal. ■

REMARK 2: It can be shown that replacing Σ_u by $\tilde{\Sigma}_u$ has no consequence for the asymptotic properties of $\hat{B}$, $\hat{C}$, and $\hat{A}_{(r)}$, if the conditions of Proposition 5.10 are satisfied. ■

REMARK 3: If the original data are generated by a process with unknown and possibly nonzero mean, then the sample mean must be substracted from each variable. Again the asymptotic properties of $\hat{B}$, $\hat{C}$, and $\hat{A}_{(r)}$ are not affected. Furthermore, the sample mean vector $\bar{y}$ is asymptotically independent of $\hat{B}$, $\hat{C}$, and $\hat{A}_{(r)}$ and its asymptotic distribution is precisely the one stated in Proposition 3.3. ■

REMARK 4: If the process y_t is Gaussian it is an easy consequence of Proposition 3.2 and Corollary 5.10.1 that the estimator

$$\tilde{\Sigma}_u(r, p) := (Y - \hat{A}_{(r)}X)(Y - \hat{A}_{(r)}X)'/T$$

is asymptotically independent of $\hat{A}_{(r)}$ and normally distributed,

$$\sqrt{T}\,\mathrm{vech}(\tilde{\Sigma}_u(r, p) - \Sigma_u) \xrightarrow{d} N(0, 2\mathbf{D}_K^+(\Sigma_u \otimes \Sigma_u)\mathbf{D}_K^{+\prime}),$$

where $\mathbf{D}_K^+$ is, as usual, the Moore-Penrose inverse of the duplication matrix $\mathbf{D}_K$.

■

The estimator $\hat{A}_{(r)}$ and its asymptotic covariance matrix $\Sigma_{(r)}$ may be used for forecasting and structural analyses in the way described in Sections 5.2.6 and 5.2.7. In the next subsection model specification and checking the model adequacy are considered.

5.3.5 Specification and Checking of Reduced Rank VAR Models

Specifying a reduced rank VAR model requires selection of the VAR order p and the rank r of the coefficient matrix A. In a sequential procedure one may first choose the order and then determine r.

Since we have discussed order selection in Chapter 4 let us assume for the moment that the order p is known or has been specified in a previous step. Then, if some rank r has been selected one may want to test the null hypothesis

$$H_0\colon \mathrm{rk}(A) \le r \quad \text{against} \quad H_1\colon \mathrm{rk}(A) > r, \tag{5.3.22}$$

that is, one may want to check whether the rank has been chosen too small. Assuming that y_t is a stable Gaussian VAR(p) process the likelihood ratio statistic for this null hypothesis can be shown to be

$$\lambda_{\mathrm{LR}} = -T \sum_{i=r+1}^{K} \ln(1 - \hat{\lambda}_i), \tag{5.3.23}$$

where $\hat{\lambda}_1 > \cdots > \hat{\lambda}_K$ are the ordered eigenvalues of the matrix

$$(YY')^{-1/2}YX'(XX')^{-1}XY'(YY')^{-1/2}. \tag{5.3.24}$$

As usual the LR test statistic has an asymptotic χ^2-distribution with degrees of freedom equal to the number of restrictions imposed under H_0, provided H_0 is true. As we have seen earlier, if $\mathrm{rk}(A) = r$, the matrix A contains $rKp + rK - r^2$ free parameters. Thus, the number of restrictions imposed on the full rank matrix is

$$K^2p - rKp - rK + r^2 = (K - r)(Kp - r).$$

Hence,

$$\lambda_{\mathrm{LR}} \xrightarrow{d} \chi^2((K - r)(Kp - r)) \tag{5.3.25}$$

(see, e.g., Velu, Reinsel & Wichern (1986)).

If $\text{rk}(A)$ is unknown the hypotheses

$$H_0^1: r \leq K - 1 \quad \text{against} \quad H_1^1: r = K$$

$$H_0^2: r \leq K - 2 \quad \text{against} \quad H_1^2: r > K - 2$$

and so on, may be tested sequentially. The procedure terminates when the null hypothesis is rejected for the first time and the upper bound for the rank specified in the previous null hypothesis is chosen. For this sequential procedure similar comments apply as in Section 4.2.3. In particular, it must be kept in mind that the overall significance level of the procedure is different from the significance levels of the individual tests.

Alternatively the choice of the rank may be based on some model selection criterion. In that case the order and rank may be chosen simultaneously. Examples of possible model selection criteria are

$$\text{AIC}(p, r) = \ln|\tilde{\Sigma}_u(r, p)| + \frac{2}{T} r(Kp + K - r), \tag{5.3.26}$$

$$\text{HQ}(p, r) = \ln|\tilde{\Sigma}_u(r, p)| + \frac{2 \ln \ln T}{T} r(Kp + K - r), \tag{5.3.27}$$

and

$$\text{SC}(p, r) = \ln|\tilde{\Sigma}_u(r, p)| + \frac{\ln T}{T} r(Kp + K - r), \tag{5.3.28}$$

where

$$\tilde{\Sigma}_u(r, p) = (Y - \hat{A}_{(r)}X)(Y - \hat{A}_{(r)}X)'/T,$$

that is, $\tilde{\Sigma}_u(r, p)$ is the estimator of the white noise covariance matrix Σ_u obtained from the residuals of a RRVAR(p) model of rank r. In each of the three criteria $r(Kp + K - r)$ represents the number of free parameters in a model of order p, rank r, and dimension K. The order and rank are chosen so as to minimize the preferred criterion.

Once a model is specified the prediction tests for nonstationarity described in Section 4.6 may be applied in the context of RRVAR models with critical values from the same distributions as in Chapter 4 even if the forecasts are based on reduced rank coefficient estimators. Also the tests for nonnormality discussed in Section 4.5 may be applied to the residuals of RRVAR models. In the following subsection we discuss an illustrative example.

5.3.6 An Illustrative Example

We use again the investment/income/consumption data from Table E.1, Appendix E, up to 1978 to illustrate reduced rank VAR modeling. In Table 5.5 the values of the three criteria AIC, HQ, and SC are given for orders up to $p = 4$. The models were fitted to the mean-adjusted data. Both AIC and HQ choose order 2 and rank 1 while the optimal SC model is obtained for order and rank zero. Recall that in the unrestricted VAR case AIC has chosen order 2 whereas HQ and SC

Table 5.5. Selection of VAR order and rank of RRVAR model for the investment/income/consumption system

VAR order	rank	AIC	HQ	SC
0	0	−24.42	−24.42	−24.42*
1	1	−24.51	−24.45	−24.35
	2	−24.51	−24.41	−24.25
	3	−24.50	−24.38	−24.21
2	1	−24.64*	−24.54*	−24.38
	2	−24.64	−24.46	−24.19
	3	−24.59	−24.37	−24.02
3	1	−24.58	−24.44	−24.23
	2	−24.51	−24.26	−23.88
	3	−24.41	−24.06	−23.55
4	1	−24.59	−24.41	−24.14
	2	−24.52	−24.19	−23.69
	3	−24.36	−23.90	−23.21

* Minimum.

have selected order 0. This result can be recovered by comparing only the rank-3 models in Table 5.5. Allowing reduced rank models in the competition HQ modifies its choice of the VAR order. In the following analysis we will go with the AIC-HQ RRVAR model.

Reestimating the model with a sample of size $T = 73$ and two presample values only gives

$$\hat{B} = \begin{bmatrix} .0040 \\ (.0104) \\ -.0020 \\ (.0024) \\ .0064 \\ (.0013) \end{bmatrix}, \qquad \hat{C}_1 = \begin{bmatrix} -2.59 & 41.31 & -60.49 \\ (2.72) & (11.82) & (14.39) \end{bmatrix}, \qquad \hat{C}_2 = \begin{bmatrix} 1.35 & 46.07 & -5.30 \\ (2.71) & (11.58) & (14.41) \end{bmatrix},$$

where the estimated asymptotic standard errors obtaind from Proposition 5.10 are given in parentheses. These estimates lead to

$$\hat{A}_{(1)} = \hat{B}[\hat{C}_1 \quad \hat{C}_2] = \begin{bmatrix} -.010 & .165 & -.241 & .005 & .184 & -.021 \\ (.029) & (.433) & (.633) & (.018) & (.493) & (.080) \\ .005 & -.084 & .123 & -.003 & -.093 & .011 \\ (.008) & (.101) & (.147) & (.006) & (.112) & (.032) \\ -.017 & .263 & -.385 & .009 & .293 & -.034 \\ (.018) & (.093) & (.122) & (.017) & (.096) & (.092) \end{bmatrix} \tag{5.3.29}$$

Table 5.6. Point and interval forecasts from full rank and reduced rank VAR(2) models

		full rank model		reduced rank model	
variable	forecast horizon	point forecast	95% interval forecast	point forecast	95% interval forecast
investment	1	−.011	[−.105, .084]	.020	[−.074, .114]
	2	.011	[−.087, .109]	.015	[−.078, .108]
	3	.021	[−.076, .119]	.020	[−.073, .112]
	4	.012	[−.085, .110]	.018	[−.075, .111]
income	1	.020	[−.004, .044]	.019	[−.004, .043]
	2	.020	[−.004, .045]	.022	[−.002, .045]
	3	.017	[−.008, .042]	.020	[−.004, .043]
	4	.021	[−.004, .045]	.020	[−.003, .044]
consumption	1	.022	[.002, .041]	.022	[.004, .041]
	2	.015	[−.005, .034]	.015	[−.005, .034]
	3	.020	[−.002, .041]	.022	[.001, .042]
	4	.019	[−.003, .040]	.019	[−.001, .040]

with estimated asymptotic standard errors obtained from Corollary 5.10.1. Comparing these estimates with unrestricted estimates of the VAR coefficient matrices (e.g. in (3.2.22) and (3.2.24)) reveals that there are some differences. However, the significant coefficients with t-ratios greater than two have the same signs in (3.2.22) and (5.3.29).

In Table 5.6 we compare point and interval forecasts from an unrestricted VAR(2) model with those from a reduced rank VAR(2) model with rank 1. There is a considerable overlap of corresponding forecast intervals obtained from the two models. Also the size of corresponding intervals is very similar, the RRVAR intervals being overall slightly smaller. The point forecasts, especially for income and consumption are quite close. These results indirectly confirm the restrictions placed on the RRVAR model because they demonstrate that with respect to forecasting they don't imply drastic changes.

We have also computed impulse responses from the RRVAR model although in structural analyses subset VAR models will often be preferable because they include zero restrictions with possibly direct causal interpretations. The estimated Φ_i consumption responses to an income impulse are depicted in Figure 5.3. Qualitatively they look quite similar to the consumption responses obtained from the full VAR(2) model (see Figure 3.4), that is, after an initial increase the response of consumption tapers off rapidly. In Figure 5.4 the response of investment to an impulse in consumption is seen to be almost zero. This brings out a result more clearly which was also obtained in the full rank model (see Figure 3.5). Obviously the result is not quite so clear here as in the subset VAR analysis where the investment response to a consumption impulse was precisely zero.

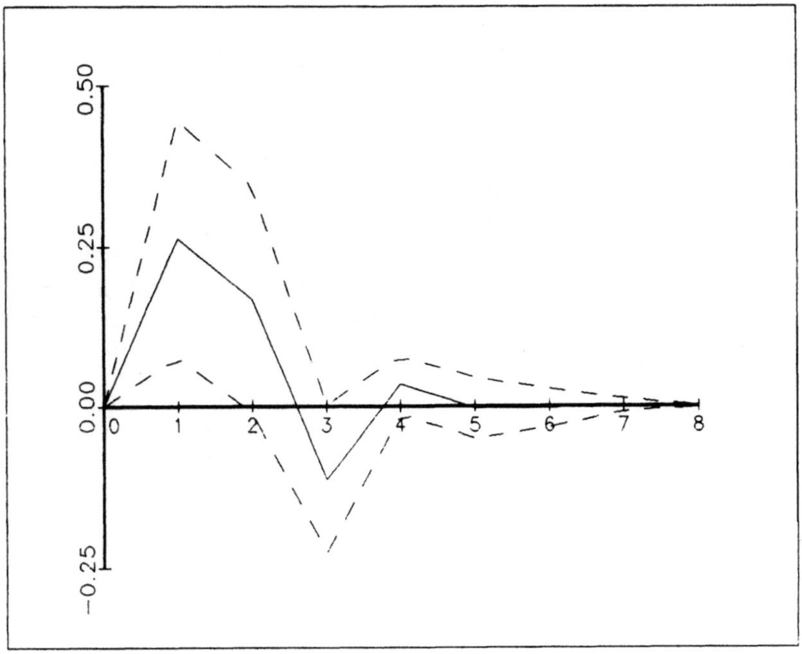

Fig. 5.3. Estimated responses of consumption to a forecast error impulse in income with two-standard error bounds based on a reduced rank VAR model.

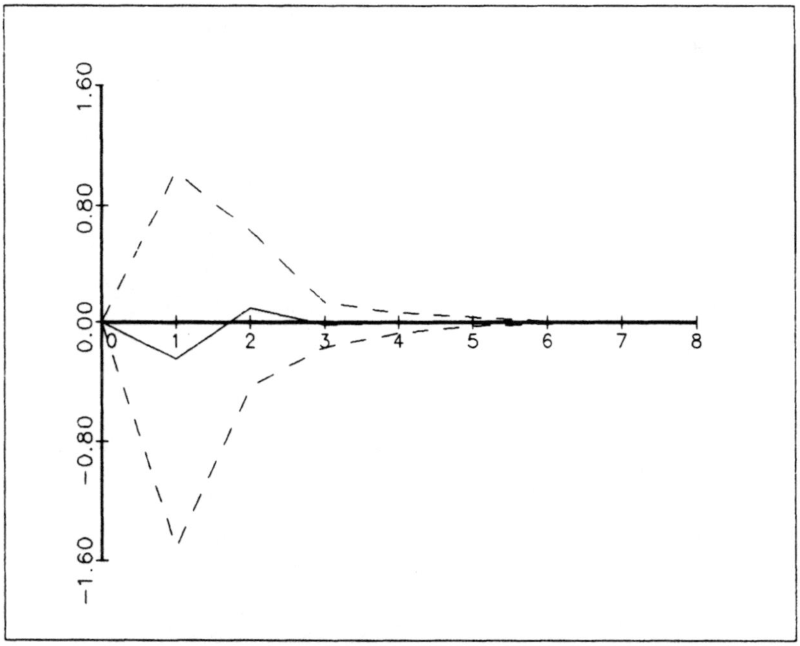

Fig. 5.4. Estimated responses of investment to a forecast error impulse in consumption with two-standard error bounds based on a reduced rank VAR model.

5.4 Bayesian Estimation

5.4.1 Basic Terms and Notations

Although the reader is assumed familiar with Bayesian estimation we summarize some basics here. In the Bayesian approach it is assumed that the nonsample or prior information is available in the form of a density. Denoting the parameters of interest by $\boldsymbol{\alpha}$ let us assume that the prior information is summarized in the *prior p.d.f.* $g(\boldsymbol{\alpha})$. The sample information is summarized in the sample p.d.f., say $f(\mathbf{y}|\boldsymbol{\alpha})$, which is algebraically identical to the likelihood function $l(\boldsymbol{\alpha}|\mathbf{y})$. The two types of information are combined via Bayes' Theorem which states that

$$g(\boldsymbol{\alpha}|\mathbf{y}) = \frac{f(\mathbf{y}|\boldsymbol{\alpha})g(\boldsymbol{\alpha})}{f(\mathbf{y})},$$

where $f(\mathbf{y})$ denotes the unconditional sample density which, for a given sample, is just a normalizing constant. In other words, the distribution of $\boldsymbol{\alpha}$, given the sample information contained in $\mathbf{y}$, can be summarized by $g(\boldsymbol{\alpha}|\mathbf{y})$ which is proportional to the likelihood function times the prior density $g(\boldsymbol{\alpha})$,

$$g(\boldsymbol{\alpha}|\mathbf{y}) \propto f(\mathbf{y}|\boldsymbol{\alpha})g(\boldsymbol{\alpha}) = l(\boldsymbol{\alpha}|\mathbf{y})g(\boldsymbol{\alpha}). \tag{5.4.1}$$

The conditional density $g(\boldsymbol{\alpha}|\mathbf{y})$ is the *posterior p.d.f.* It contains all the information available on the parameter vector $\boldsymbol{\alpha}$. Point estimators of $\boldsymbol{\alpha}$ may be derived from the posterior distribution. For instance, the mean of that distribution, called the *posterior mean*, is often used as an estimator for $\boldsymbol{\alpha}$. In the next subsection this general framework is specialized to VAR models.

5.4.2 Normal Priors for the Parameters of a Gaussian VAR Process

Suppose y_t is a zero mean, stable, stationary Gaussian VAR(p) process of the form (5.3.4) and the prior distribution for $\boldsymbol{\alpha} := \operatorname{vec}(A) = \operatorname{vec}(A_1, \dots, A_p)$ is a multivariate normal with known mean $\boldsymbol{\alpha}^*$ and covariance matrix V_α,

$$g(\boldsymbol{\alpha}) = \left(\frac{1}{2\pi}\right)^{K^2p/2} |V_\alpha|^{-1/2} \exp\left[-\frac{1}{2}(\boldsymbol{\alpha} - \boldsymbol{\alpha}^*)'V_\alpha^{-1}(\boldsymbol{\alpha} - \boldsymbol{\alpha}^*)\right]. \tag{5.4.2}$$

Combining this information with the sample information summarized in the Gaussian likelihood function,

$$l(\boldsymbol{\alpha}|\mathbf{y}) = \left(\frac{1}{2\pi}\right)^{KT/2} |I_T \otimes \Sigma_u|^{-1/2}$$

$$\times \exp\left[-\frac{1}{2}(\mathbf{y} - (X' \otimes I_K)\boldsymbol{\alpha})'(I_T \otimes \Sigma_u^{-1})(\mathbf{y} - (X' \otimes I_K)\boldsymbol{\alpha})\right]$$

(see Section 3.4, Chapter 3), gives the posterior density

$$g(\boldsymbol{\alpha}|\mathbf{y}) \propto g(\boldsymbol{\alpha})l(\boldsymbol{\alpha}|\mathbf{y})$$
$$\propto \exp\{-\tfrac{1}{2}[(V_{\alpha}^{-1/2}(\boldsymbol{\alpha}-\boldsymbol{\alpha}^*))'(V_{\alpha}^{-1/2}(\boldsymbol{\alpha}-\boldsymbol{\alpha}^*)) + ((I_T\otimes \Sigma_u^{-1/2})\mathbf{y}$$
$$-(X'\otimes \Sigma_u^{-1/2})\boldsymbol{\alpha})'((I_T\otimes \Sigma_u^{-1/2})\mathbf{y}-(X'\otimes \Sigma_u^{-1/2})\boldsymbol{\alpha})]\}. \tag{5.4.3}$$

Here the white noise covariance matrix Σ_u is assumed known for the moment. Defining

$$w := \begin{bmatrix} V_{\alpha}^{-1/2}\boldsymbol{\alpha}^* \\ (I_T\otimes \Sigma_u^{-1/2})\mathbf{y}\end{bmatrix} \quad \text{and} \quad W := \begin{bmatrix} V_{\alpha}^{-1/2} \\ X'\otimes \Sigma_u^{-1/2}\end{bmatrix}$$

the exponent in (5.4.3) can be rewritten as

$$-\tfrac{1}{2}(w-W\boldsymbol{\alpha})'(w-W\boldsymbol{\alpha}) = -\tfrac{1}{2}[(\boldsymbol{\alpha}-\bar{\boldsymbol{\alpha}})'W'W(\boldsymbol{\alpha}-\bar{\boldsymbol{\alpha}}) + (w-W\bar{\boldsymbol{\alpha}})'(w-W\bar{\boldsymbol{\alpha}})], \tag{5.4.4}$$

where

$$\bar{\boldsymbol{\alpha}} := (W'W)^{-1}W'w = [V_{\alpha}^{-1} + (XX'\otimes \Sigma_u^{-1})]^{-1}[V_{\alpha}^{-1}\boldsymbol{\alpha}^* + (X\otimes \Sigma_u^{-1})\mathbf{y}]. \tag{5.4.5}$$

Since the second term on the right-hand side of (5.4.4) does not contain $\boldsymbol{\alpha}$ it may be absorbed into the constant of proportionality. Hence,

$$g(\boldsymbol{\alpha}|\mathbf{y}) \propto \exp[-\tfrac{1}{2}(\boldsymbol{\alpha}-\bar{\boldsymbol{\alpha}})'\bar{\Sigma}_{\alpha}^{-1}(\boldsymbol{\alpha}-\bar{\boldsymbol{\alpha}})],$$

where $\bar{\boldsymbol{\alpha}}$ is given in (5.4.5) and

$$\bar{\Sigma}_{\alpha} := (W'W)^{-1} = [V_{\alpha}^{-1} + (XX'\otimes \Sigma_u^{-1})]^{-1}. \tag{5.4.6}$$

Thus, the posterior density is easily recognizable as the density of a multivariate normal with mean $\bar{\boldsymbol{\alpha}}$ and covariance matrix $\bar{\Sigma}_{\alpha}$, that is, the posterior distribution of $\boldsymbol{\alpha}$ is $N(\bar{\boldsymbol{\alpha}}, \bar{\Sigma}_{\alpha})$. This distribution may be used for inference regarding $\boldsymbol{\alpha}$.

Sometimes one would like to leave some of the coefficients without any restrictions because no prior information is available. In the above framework this case could be handled by setting the corresponding prior variance to infinity. Unfortunately such a choice is inconvenient here because algebraic operations have to be performed with the elements of V_{α} in order to compute $\bar{\boldsymbol{\alpha}}$ and $\bar{\Sigma}_{\alpha}$. Therefore, in such cases it is preferable to write the prior information in the form

$$C\boldsymbol{\alpha} = c + e \quad \text{with} \quad e \sim N(0, I). \tag{5.4.7}$$

Here C is a fixed matrix and c is a fixed vector. If C is a $(K^2p \times K^2p)$ nonsingular matrix,

$$\boldsymbol{\alpha} \sim N(C^{-1}c, C^{-1}C^{-1\prime}).$$

That is, the prior information is given in the form of a multivariate normal distribution with mean $C^{-1}c$ and covariance matrix $(C'C)^{-1}$. From (5.4.5), under Gaussian assumptions, the resulting posterior mean is

$$\bar{\boldsymbol{\alpha}} = [C'C + (XX'\otimes \Sigma_u^{-1})]^{-1}[C'c + (X\otimes \Sigma_u^{-1})\mathbf{y}]. \tag{5.4.8}$$

A practical advantage of this representation of the posterior mean is that it does not require the inversion of V_α. Moreover, this form can also be used if no prior information is available for some of the coefficients. For instance, if no prior information on the first coefficient is available we may simply eliminate one row from C and put zeroes in the first column. Although the prior information cannot be represented in the form of a proper multivariate normal distribution in this case, the estimator $\bar{\alpha}$ in (5.4.8) can still be used.

In order to make these concepts useful the prior mean α^* and covariance matrix V_α or C and c must be specified. In the next subsection possible choices are considered.

5.4.3 The Minnesota or Litterman Priors

In Litterman (1986) and Doan, Litterman & Sims (1984) a specific prior, often referred to as Minnesota prior or Litterman prior, for the parameters of a VAR model is described. A similar prior will be considered here as an example. The so-called Minnesota prior was suggested for certain nonstationary processes. We will translate it for the stationary case because we are still dealing with stationary, stable processes. The nonstationary version of the Minnesota prior will be treated in Chapter 11.

If the intertemporal dependence of the variables is believed to be weak one way to describe this is to set the prior mean of the VAR coefficients to zero with nonzero prior variances. In other words, $\alpha^* = 0$ and $V_\alpha \neq 0$. With this choice of α^* the posterior mean in (5.4.5) reduces to

$$\bar{\alpha} = [V_\alpha^{-1} + (XX' \otimes \Sigma_u^{-1})]^{-1}(X \otimes \Sigma_u^{-1})\mathbf{y}. \tag{5.4.9}$$

This estimator for α looks like the multivariate LS estimator except for the inverse covariance matrix V_α^{-1}.

In the spirit of Litterman (1986) the prior covariance matrix V_α may be specified as a diagonal matrix with diagonal elements

$$v_{ij,l} = \begin{cases} (\lambda/l)^2 & \text{if } i = j \\ (\lambda\theta\sigma_i/l\sigma_j)^2 & \text{if } i \neq j, \end{cases} \tag{5.4.10}$$

where $v_{ij,l}$ is the prior variance of $\alpha_{ij,l}$, λ is the prior standard deviation of the coefficients $\alpha_{kk,1}$, $k = 1, \dots, K$, $0 < \theta < 1$, and σ_i^2 is the i-th diagonal element of Σ_u. For each equation λ controls how tightly the coefficient of the first lag of the dependent variable is believed concentrated around zero. For instance, in the k-th equation of the system it is the prior standard deviation of $\alpha_{kk,1}$. In practice different values of λ are sometimes tried. Using different λ's in different equations may also be considered.

Since it is believed that coefficients of high order lags are likely to be close to zero the prior variance decreases with increasing lag length l. Furthermore, it is believed that most of the variation in each of the variables is accounted for by own lags. Therefore coefficients of variables other than the dependent variable are assigned a smaller variance in relative terms by choosing θ between 0 and 1, for instance, $\theta = .2$. The ratio σ_i^2/σ_j^2 is included to take care of the differences in the variability of the different variables. Here the residual variances are preferred over the y_k variances because it is assumed that the response of one variable to another is largely determined by the unexpected movements reflected in the residual variance. Finally the assumption of a diagonal V_α matrix means that independent prior distributions of the different coefficients are specified. This mainly reflects our inability to model dependencies between the coefficients.

As an example consider a bivariate VAR(2) system consisting of the two equations

$$\begin{aligned}
y_{1t} &= \underset{(\lambda)}{\alpha_{11,1}y_{1,t-1}} + \underset{(\lambda\theta\sigma_1/\sigma_2)}{\alpha_{12,1}y_{2,t-1}} + \underset{(\lambda/2)}{\alpha_{11,2}y_{1,t-2}} + \underset{(\lambda\theta\sigma_1/2\sigma_2)}{\alpha_{12,2}y_{2,t-2}} + u_{1t},\\
y_{2t} &= \underset{(\lambda\theta\sigma_2/\sigma_1)}{\alpha_{21,1}y_{1,t-1}} + \underset{(\lambda)}{\alpha_{22,1}y_{2,t-1}} + \underset{(\lambda\theta\sigma_2/2\sigma_1)}{\alpha_{21,2}y_{1,t-2}} + \underset{(\lambda/2)}{\alpha_{22,2}y_{2,t-2}} + u_{2t},
\end{aligned} \tag{5.4.11}$$

where the prior standard deviations are given in parentheses. The prior covariance matrix of the eight coefficients of this system is

$$V_\alpha = \begin{bmatrix} \lambda^2 & & & & & & & \\ & (\lambda\theta\sigma_2/\sigma_1)^2 & & & & & 0 & \\ & & (\lambda\theta\sigma_1/\sigma_2)^2 & & & & & \\ & & & \lambda^2 & & & & \\ & & & & (\lambda/2)^2 & & & \\ & & & & & (\lambda\theta\sigma_2/2\sigma_1)^2 & & \\ & & 0 & & & & (\lambda\theta\sigma_1/2\sigma_2)^2 & \\ & & & & & & & (\lambda/2)^2 \end{bmatrix}.$$

In terms of (5.4.7) this prior may be specified by choosing $c = 0$ and C an (8×8) diagonal matrix with the square roots of the reciprocals of the diagonal elements of V_α on the diagonal.

5.4.4 Practical Considerations

In specifying the Minnesota priors, even if λ and θ are chosen appropriately, there remain some practical problems. The first results from the fact that Σ_u is usually unknown. In a strict Bayesian approach a prior p.d.f. for the elements of Σ_u would be chosen. However, that would lead to a more difficult posterior distribution for α. Therefore, a more practical approach is to replace the σ_i by the square roots of the diagonal elements of the LS and ML estimator of Σ_u, e.g.,

$$\tilde{\Sigma}_u = Y(I_T - X'(XX')^{-1}X)Y'/T.$$

A second problem is the computational expense that may result from the inversion of the matrix $V_\alpha^{-1} + (XX' \otimes \Sigma_u^{-1})$ or $C'C + (XX' \otimes \Sigma_u^{-1})$ in the posterior mean $\bar{\boldsymbol{\alpha}}$ which is usually used as an estimator for $\boldsymbol{\alpha}$. This matrix has dimension $(K^2p \times K^2p)$. Since in a Bayesian analysis sometimes one may want to choose a high order p and put tight zero priors on the coefficients of high lags rather than make them zero with probability 1 like in an order selection approach, the dimension of the matrix to be inverted in computing $\bar{\boldsymbol{\alpha}}$ may be quite substantial. Therefore Bayesian estimation is sometimes applied to each of the K equations of the system individually. For instance, for the k-th equation,

$$\bar{a}_k := [V_k^{-1} + \sigma_k^{-2}XX']^{-1}[V_k^{-1}a_k^* + \sigma_k^{-2}Xy_{(k)}] \tag{5.4.12}$$

is used as an estimator of the parameters a_k (the transpose of the k-th row of $A = [A_1, \ldots, A_p]$). Here a_k^* is the prior mean and V_k is the prior covariance of a_k and $y'_{(k)}$ is the k-th row of Y as in Section 3.2. Using (5.4.12) instead of (5.4.5) reduces the computational expense a bit.

A further problem is related to the zero mean assumption made in the foregoing for the process y_t. In practice one may simply subtract the sample mean from each variable and then perform a Bayesian analysis for the mean-adjusted data. This amounts to assuming that no prior information exists for the mean terms. Alternatively, intercept terms may be included in the analysis. If the prior information is specified in terms of (5.4.7) it is easy to leave the intercept terms unrestricted if desired.

5.4.5 An Example

To illustrate the Bayesian approach we have computed estimates $\bar{a}_k$ as in (5.4.12) for the investment/income/consumption example data using different values of λ and θ. Again we use first differences of logarithms of the data from Table E.1 for the years 1960–1978. In Table 5.7 we give estimates for the investment equation of a VAR(2) model. In a Bayesian analysis one would usually choose a higher VAR order. For illustrative purposes the VAR(2) model is helpful, however.

In the investment equation the parameter λ controls the overall prior variance

Table 5.7. Bayesian estimates of the investment equation from the investment/income/consumption system

λ	θ	ν_1	$\alpha_{11,1}$	$\alpha_{12,1}$	$\alpha_{13,1}$	$\alpha_{11,2}$	$\alpha_{12,2}$	$\alpha_{13,2}$
∞	1	−.017	−.320	.146	.961	−.161	.115	.934
1	.99	−.015	−.309	.159	.921	−.147	.135	.854
.1	.99	.008	−.096	.150	.297	−.011	.062	.100
.01	.99	.018	−.001	.003	.005	−.000	.000	.001
1	.50	−.013	−.301	.194	.847	−.141	.165	.718
1	.10	.009	−.245	.190	.369	−.099	.074	.137
1	.01	.023	−.208	.004	.007	−.078	.001	.002

of all VAR coefficients while θ controls the tightness of the variances of the coefficients of lagged income and consumption. Roughly speaking θ specifies the fraction of the prior standard deviation λ attached to the coefficients of lagged income and consumption. Thus, a value of θ close to one means that all coefficients of lag 1 have about the same prior variance except for a scaling factor that takes care of the different variability of different variables. Note that the intercept terms are not restricted (prior variance $= \infty$).

We assume a prior mean of zero for all coefficients, $a_k^* = 0$, and thus shrink towards zero by tightening the prior standard deviation λ. This effect is clearly reflected in Table 5.7. For $\theta = .99$ and $\lambda = 1$ we get coefficient estimates which are quite similar to unrestricted LS estimates ($\lambda = \infty$, $\theta = 1$). Decreasing λ to zero tightens the prior variance and shrinks all VAR coefficients to zero. For $\lambda = .01$ they are quite close to zero already. On the other hand, moving the variance fraction θ towards zero shrinks the consumption and income coefficients ($\alpha_{12,i}, \alpha_{13,i}$) towards zero. In Table 5.7, for $\lambda = 1$ and $\theta = .01$ they are seen to be almost zero. This, of course, has some impact on the investment coefficients too.

5.4.6 Classical versus Bayesian Interpretation of $\bar{\boldsymbol{\alpha}}$ in Forecasting and Structural Analyses

If the coefficients of a VAR process are estimated by a Bayesian procedure the estimated process may be used for prediction and economic analyses as described in the previous sections. Again one question of interest concerns the statistical properties of the resulting forecasts and impulse responses. It is possible to interpret $\bar{\boldsymbol{\alpha}}$ in (5.4.5) or (5.4.8) as an estimator in the classical sense and to answer this question in terms of asymptotic theory as in the previous sections. In the classical context $\bar{\boldsymbol{\alpha}}$ may be interpreted as a shrinkage estimator or under the heading of estimation with stochastic restrictions (e.g., Judge et al. (1985, Chapter 3)). In regression models with nonstochastic regressors such estimators, under suitable conditions, have smaller mean squared errors then ML estimators in small samples. In the present framework the small sample properties are unknown in general.

In order to derive asymptotic properties let us consider the representation (5.4.8). It is easily seen that, under our standard conditions,

$$\begin{aligned}
\operatorname{plim} \bar{\boldsymbol{\alpha}} &= \operatorname{plim}\left(\frac{C'C}{T} + \frac{XX'}{T} \otimes \Sigma_u^{-1}\right)^{-1} \operatorname{plim}\left[\frac{C'c}{T} + \operatorname{vec}\left(\frac{\Sigma_u^{-1} YX'}{T}\right)\right] \\
&= \left[\operatorname{plim}\left(\frac{XX'}{T}\right)^{-1} \otimes \Sigma_u\right] \operatorname{plim} \operatorname{vec}\left(\frac{\Sigma_u^{-1} YX'}{T}\right) \\
&= \boldsymbol{\alpha}.
\end{aligned}$$

Here $\operatorname{plim} C'C/T = \lim C'C/T = 0$ and $\operatorname{plim} C'c/T = 0$ has been used. Moreover, viewing $\bar{\boldsymbol{\alpha}}$ as an estimator in the classical sense it has the same asymptotic distribution as the unconstrained multivariate LS estimator

$$\hat{\boldsymbol{\alpha}} = \operatorname{vec}(YX'(XX')^{-1}),$$

because

$$\sqrt{T}(\bar{\boldsymbol{\alpha}} - \hat{\boldsymbol{\alpha}}) = \left[\frac{C'C}{T} + \frac{XX'}{T} \otimes \bar{\Sigma}_u^{-1}\right]^{-1} \left[\frac{C'c}{\sqrt{T}} + \frac{1}{\sqrt{T}} \operatorname{vec}(\bar{\Sigma}_u^{-1} YX')\right]$$
$$- \left[\left(\frac{XX'}{T}\right)^{-1} \otimes \bar{\Sigma}_u\right] \frac{1}{\sqrt{T}} \operatorname{vec}(\bar{\Sigma}_u^{-1} YX') \xrightarrow{p} 0.$$

Thus, $\bar{\boldsymbol{\alpha}}$ and $\hat{\boldsymbol{\alpha}}$ have the same asymptotic distribution by Proposition C.2(2) of Appendix C. This result is intuitively appealing as it shows that the contribution of the prior information becomes negligible when the sample size approaches infinity and the sample information becomes exhaustive. Yet the result is not very helpful when a small sample is given in a practical situation.

Consequently it may be preferable to base the analysis on the posterior distribution of $\boldsymbol{\alpha}$. In general it will be difficult to derive the distribution of, say, the impulse responses from the posterior distribution of $\boldsymbol{\alpha}$. In that case one may obtain, for instance, confidence intervals of these quantities from a simulation. That is, a large number of samples is drawn from the posterior distribution of $\boldsymbol{\alpha}$ and the corresponding impulse response coefficients are computed. The required percentage points are then estimated from the empirical distributions of the estimated impulse responses (see Appendix D). Of course, such a procedure is computationally quite demanding.

5.5 Exercises

In the following exercises the notation of the previous sections of this chapter is used.

5.5.1 Algebraic Exercises

Problem 5.1

Show that $\boldsymbol{\gamma}^{\#}$ given in (5.2.14) minimizes

$$(\mathbf{z} - (Z' \otimes I_K)R\boldsymbol{\gamma})'(\mathbf{z} - (Z' \otimes I_K)R\boldsymbol{\gamma})$$

with respect to $\boldsymbol{\gamma}$.

Problem 5.2

Prove that $\hat{\hat{\boldsymbol{\beta}}}$ given in (5.2.16) minimizes

$$(\mathbf{y} - (Z' \otimes I_K)\boldsymbol{\beta})'(I_T \otimes \bar{\Sigma}_u^{-1})(\mathbf{y} - (Z' \otimes I_K)\boldsymbol{\beta})$$

subject to the restriction $C\boldsymbol{\beta} = c$, where C is $(N \times K(Kp + 1))$ of rank N and c is $(N \times 1)$. (*Hint:* Specify the appropriate Lagrange function and find its stationary point as described in Appendix A.14.)

Problem 5.3

Show that $\tilde{\gamma}$ given in (5.2.17) is the ML estimator of γ. (*Hint:* Use the partial derivatives from Section 3.4.)

Problem 5.4

Prove Proposition 5.6.

Problem 5.5

Derive the asymptotic distribution of the EGLS estimator of $\boldsymbol{\alpha} := \text{vec}(A_1, \dots, A_p)$ based on mean-adjusted variables subject to restrictions $\boldsymbol{\alpha} = R\gamma + r$, where R, γ, and r have suitable dimensions.

Problem 5.6

Consider the recursive system of Section 5.2.5,

$$y_t = \eta + B_0 y_t + \cdots + B_p y_{t-p} + v_t,$$

where v_t has diagonal covariance matrix Σ_v. Show that $\sum_t v_t' \Sigma_v^{-1} v_t$ and $\sum_t v_t' v_t$ assume their minima with respect to the unknown parameters for the same values of $\eta, B_0, \dots, B_p$.
(*Hint:* Note that

$$\sum_{t=1}^{T} v_t' \Sigma_v^{-1} v_t = \sum_{k=1}^{K} \sum_{t=1}^{T} v_{kt}^2 / \sigma_{v_k}^2$$

and consider the partial derivatives with respect to the coefficients of the k-th equation. Here v_{kt} is the k-th element of v_t and $\sigma_{v_k}^2$ is the k-th diagonal element of Σ_v.)

5.5.2 Numerical Problems

The following problems require the use of a computer. They are based on the bivariate time series $y_t = (y_{1t}, y_{2t})'$ of first differences of the U.S. investment data given in Table E.2 of Appendix E.

Problem 5.7

Fit a VAR(2) model to the first differences of the data from Table E.2 subject to the restrictions $\alpha_{12,i} = 0, i = 1, 2$. Determine the EGLS parameter estimates and estimates of their asymptotic standard errors. Perform an F-test to check the restrictions.

Problem 5.8

Based on the results of the previous problem, perform an impulse response analysis for y_1 and y_2.

Problem 5.9

Use a maximum order of 4 and the AIC criterion to determine an optimal subset VAR model for y_t with the top-down strategy described in Section 5.2.8b. Repeat the exercise with the HQ criterion. Compare the two models and interpret.

Problem 5.10

Fit a reduced rank VAR(2) model of rank 1 to y_t. Compute the resulting estimates $\hat{A}_1$ and $\hat{A}_2$ and give their asymptotic standard errors. Compare the results with those of Problem 5.7. Test the rank restriction with the likelihood ratio statistic (5.3.23).

Problem 5.11

Based on the results of Problem 5.10, perform an impulse response analysis for y_1 and y_2 and compare the results to those of Problem 5.8.

Problem 5.12

Assume a maximum VAR order of 4 and determine the optimal RRVAR model for y_t using AIC, HQ, and SC. Compare the three models and interpret.

Problem 5.13

Use the Minnesota prior with $\lambda = 1$ and $\theta = .2$ and compute the posterior mean of the coefficients of a VAR(4) model for the mean-adjusted y_t. Compare this estimator to the unconstrained multivariate LS estimator of a VAR(4) model for the mean-adjusted data. Repeat the exercise with a VAR(4) model that contains intercept terms.

Part II
Infinite Order Vector Autoregressive Processes

So far we have considered finite order VAR processes. A more flexible and perhaps more realistic class of processes is obtained by allowing for an infinite VAR order. Of course, having only a finite string of time series data the infinitely many VAR coefficients cannot be estimated without further assumptions. There are two competing approaches that have been used in practice in order to overcome this problem. In one approach it is assumed that the infinite number of VAR coefficients depend on finitely many parameters. In Chapter 6 vector autoregressive moving average (VARMA) processes are introduced that may be viewed as finite parameterizations of potentially infinite order VAR processes. Estimation and specification of these processes are discussed in Chapters 7 and 8, respectively. In Chapter 9 another approach is persued. In that approach the infinite order VAR operator is truncated at some finite lag and the resulting finite order VAR model is estimated. It is assumed, however, that the truncation point depends on the time series length available for estimation. A suitable asymptotic theory for the resulting estimators is discussed.

Infinite Order Vector Autoregressive Processes

Chapter 6. Vector Autoregressive Moving Average Processes

6.1 Introduction

In this chapter we extend our standard finite order VAR model

$$y_t = \nu + A_1 y_{t-1} + \cdots + A_p y_{t-p} + \varepsilon_t$$

by allowing the error terms, here ε_t, to be autocorrelated rather than white noise. The autocorrelation structure is assumed to be of a relatively simple type so that ε_t has a finite order moving average (MA) representation,

$$\varepsilon_t = u_t + M_1 u_{t-1} + \cdots + M_q u_{t-q},$$

where, as usual, u_t is zero mean white noise with nonsingular covariance matrix Σ_u. A finite order VAR process with finite order MA error term is called a VARMA (*vector autoregressive moving average*) process.

Before we study VARMA processes in general we will discuss some properties of finite order MA processes in Section 6.2. In Section 6.3 we consider the more general VARMA processes and we will learn that generally they have infinite order pure VAR and MA representations. Their autocovariance and autocorrelation properties are treated in Section 6.4 and forecasting VARMA processes is discussed in Section 6.5. In Section 6.6 transforming and aggregating these processes is considered. In that section we will see that a linearly transformed finite order VAR(p) process, in general, does not admit a finite order VAR representation but becomes a VARMA process. Since transformations of variables are quite common in practice this result is a powerful argument in favor of the more general VARMA class. Finally, Section 6.7 contains discussions of causality issues and impulse response analysis in the context of VARMA systems.

6.2 Finite Order Moving Average Processes

In Chapter 2 we have introduced MA processes of possibly infinite order. Specifically we have seen that stationary, stable finite order VAR processes can be represented as MA processes. Now we deal explicitly with *finite order* MA processes. Let us begin with the simplest case of a K-dimensional MA process of order 1 (MA(1) process), $y_t = \mu + u_t + M_1 u_{t-1}$, where $y_t = (y_{1t}, \ldots, y_{Kt})'$, u_t is zero mean white noise with nonsingular covariance matrix Σ_u and $\mu =$

$(\mu_1, \ldots, \mu_K)'$ is the mean vector of y_t, i.e., $E(y_t) = \mu$ for all t. For notational simplicity we will assume in the following that $\mu = 0$, that is, y_t is a zero mean process. Thus we consider

$$y_t = u_t + M_1 u_{t-1} \tag{6.2.1}$$

which may be rewritten as

$$u_t = y_t - M_1 u_{t-1}.$$

By successive substitution we get

$$\begin{aligned} u_t &= y_t - M_1(y_{t-1} - M_1 u_{t-2}) = y_t - M_1 y_{t-1} + M_1^2 u_{t-2} \\ &= \cdots = y_t - M_1 y_{t-1} + \cdots + (-M_1)^n y_{t-n} + (-M_1)^{n+1} u_{t-n-1} \\ &= y_t + \sum_{i=1}^{\infty} (-M_1)^i y_{t-i} \end{aligned}$$

if $M_1^i \to 0$ as $i \to \infty$. Hence,

$$y_t = -\sum_{i=1}^{\infty} (-M_1)^i y_{t-i} + u_t \tag{6.2.2}$$

which is the potentially infinite order VAR representation of the process. Since $(-M_1)^i$ may equal zero for i greater than some finite number p the process may in fact be a finite order VAR(p). For instance, we get $p = 1$ for a bivariate process with

$$M_1 = \begin{bmatrix} 0 & m \\ 0 & 0 \end{bmatrix}.$$

For the representation (6.2.2) to be meaningful M_1^i must approach zero as $i \to \infty$, which in turn requires that the eigenvalues of M_1 are all less than 1 in modulus or, equivalently,

$$\det(I_K + M_1 z) \neq 0 \qquad \text{for } z \in \mathbb{C}, |z| \leq 1.$$

This condition is analogous to the stability condition for a VAR(1) process. It guarantees that the infinite sum in (6.2.2) exists as a mean square limit. More generally it can be shown that a (zero mean) MA(q) process (moving average process of order q)

$$y_t = u_t + M_1 u_{t-1} + \cdots + M_q u_{t-q} \tag{6.2.3}$$

has a pure VAR representation

$$y_t = \sum_{i=1}^{\infty} \Pi_i y_{t-i} + u_t \tag{6.2.4}$$

if

$$\det(I_K + M_1 z + \cdots + M_q z^q) \neq 0 \qquad \text{for } z \in \mathbb{C}, |z| \leq 1. \tag{6.2.5}$$

An MA(q) process with this property is called *invertible* in the following because we can invert from the MA to a VAR representation. Writing the process in lag operator notation as

$$y_t = (I_K + M_1 L + \cdots + M_q L^q)u_t = M(L)u_t$$

the MA operator $M(L) := I_K + M_1 L + \cdots + M_q L^q$ is invertible if it satisfies (6.2.5) and we may formally write

$$M(L)^{-1} y_t = u_t.$$

The actual computation of the coefficient matrices Π_i in

$$M(L)^{-1} = \Pi(L) = I_K - \sum_{i=1}^{\infty} \Pi_i L^i$$

can be done recursively using $\Pi_1 = M_1$ and

$$\Pi_i = M_i - \sum_{j=1}^{i-1} \Pi_{i-j} M_j, \qquad i = 2, 3, \ldots, \tag{6.2.6}$$

where $M_j := 0$ for $j > q$. These recursions follow immediately from the corresponding recursions used to compute the MA coefficients of a pure VAR process (see Chapter 2, (2.1.22)).

The autocovariances of the MA(q) process (6.2.3) are particularly easy to obtain. They follow directly from those of an infinite order MA process given in Chapter 2, Section 2.1.2, (2.1.18):

$$\Gamma_y(h) = E(y_t y_{t-h}') = \begin{cases} \sum_{i=0}^{q-h} M_{i+h} \Sigma_u M_i', & h = 0, 1, \ldots, q \\ 0, & h = q+1, q+2, \ldots \end{cases} \tag{6.2.7}$$

with $M_0 := I_K$. As before, $\Gamma_y(-h) = \Gamma_y(h)'$. Thus, vectors y_t and y_{t-h} are uncorrelated if $h > q$. Obviously, the process (6.2.3) is stationary since the $\Gamma_y(h)$ do not depend on t and the mean $E(y_t) = 0$ for all t.

It can be shown that a noninvertible MA(q) process violating (6.2.5) also has a pure VAR representation if the determinantal polynomial in (6.2.5) has no roots on the complex unit circle, i.e., if

$$\det(I_K + M_1 z + \cdots + M_q z^q) \neq 0 \qquad \text{for } |z| = 1. \tag{6.2.8}$$

The VAR representation will, however, not be of the type (6.2.4) in that the white noise process will not be the one appearing in (6.2.3). The reason is that for any noninvertible MA(q) process satisfying (6.2.8) there is an equivalent invertible MA(q) satisfying (6.2.5) which has an identical autocovariance structure (see Hannan & Deistler (1988, Chapter 1, Section 3)). For instance, for the univariate MA(1) process

$$y_t = u_t + m u_{t-1} \tag{6.2.9}$$

the invertibility condition requires that $1 + mz$ has no roots for $|z| \leq 1$ or, equivalently, $|m| < 1$. For any m the process has autocovariances

$$E(y_t y_{t-h}) = \begin{cases} (1+m^2)\sigma_u^2 & \text{for } h = 0, \\ m\sigma_u^2 & \text{for } h = \pm 1, \\ 0 & \text{otherwise,} \end{cases}$$

where $\sigma_u^2 := \mathrm{Var}(u_t)$. It is easy to check that the process $v_t + \frac{1}{m}v_{t-1}$, where v_t is a white noise process with $\sigma_v^2 := \mathrm{Var}(v_t) = m^2\sigma_u^2$, has the very same autocovariance structure. Thus, if $|m| > 1$ we may choose the invertible MA(1) representation

$$y_t = v_t + \frac{1}{m}v_{t-1} \tag{6.2.10}$$

with

$$\begin{aligned} v_t &= \left[1 + \frac{1}{m}L\right]^{-1} y_t = \sum_{i=0}^{\infty}\left[\frac{-1}{m}\right]^i y_{t-i} \\ &= \left[1 + \frac{1}{m}L\right]^{-1}(1 + mL)u_t. \end{aligned}$$

The reader is invited to check that v_t is indeed a white noise process with $\sigma_v^2 = m^2\sigma_u^2$ (see Problem 6.10). Only if $|m| = 1$ and hence $1 + mz = 0$ for some z on the unit circle ($z = 1$ or -1), an invertible representation does not exist.

Although for higher order and higher-dimensional processes, where roots inside and outside the unit circle may exist, it is more complicated to find the invertible representation, it can be done whenever (6.2.8) is satisfied. In the remainder of this chapter we will therefore assume without notice that all MA processes are invertible unless stated otherwise. It should be understood that this assumption implies a slight loss of generality because MA processes with roots on the complex unit circle are excluded.

6.3 VARMA Processes

6.3.1 The Pure MA and Pure VAR Representations of a VARMA Process

As mentioned in the introduction to this chapter allowing finite order VAR processes to have finite order MA instead of white noise error terms results in the broad and flexible class of vector autoregressive moving average (VARMA) processes. The general form of a process from this class with VAR order p and MA order q is

$$y_t = \nu + A_1 y_{t-1} + \cdots + A_p y_{t-p} + u_t + M_1 u_{t-1} + \cdots + M_q u_{t-q}, \quad t = 0, \pm 1, \pm 2, \ldots \tag{6.3.1}$$

Such a process is briefly called a VARMA(p, q) process. As before, u_t is zero mean white noise with nonsingular covariance matrix Σ_u.

It may be worth elaborating a bit on this specification. What kind of process

y_t is defined by the VARMA(p, q) model (6.3.1)? To look into this question let us denote the MA part by ε_t, that is, $\varepsilon_t = u_t + M_1 u_{t-1} + \cdots + M_q u_{t-q}$ and

$$y_t = \nu + A_1 y_{t-1} + \cdots + A_p y_{t-p} + \varepsilon_t.$$

If this process is *stable*, that is, if

$$\det(I_K - A_1 z - \cdots - A_p z^p) \neq 0 \qquad \text{for } |z| \leq 1, \tag{6.3.2}$$

then by the same arguments used in Chapter 2, Section 2.1.2, and by Proposition C.7 of Appendix C.3,

$$\begin{aligned} y_t &= \mu + \sum_{i=0}^{\infty} D_i \varepsilon_{t-i} \\ &= \mu + \sum_{i=0}^{\infty} D_i(u_{t-i} + M_1 u_{t-i-1} + \cdots + M_q u_{t-i-q}) \\ &= \mu + \sum_{i=0}^{\infty} \Phi_i u_{t-i} \end{aligned} \tag{6.3.3}$$

is well-defined as a limit in mean square given a well-defined white noise process u_t. Here

$$\mu := (I_K - A_1 - \cdots - A_p)^{-1}\nu,$$

the D_i are $(K \times K)$ matrices satisfying

$$\sum_{i=0}^{\infty} D_i z^i = (I_K - A_1 z - \cdots - A_p z^p)^{-1},$$

and the Φ_i are $(K \times K)$ matrices satisfying

$$\sum_{i=0}^{\infty} \Phi_i z^i = \left[\sum_{i=0}^{\infty} D_i z^i\right](I_K + M_1 z + \cdots + M_q z^q).$$

In the following, when we call y_t a stable VARMA(p, q) process we mean the well-defined process given in (6.3.3). For instance, if u_t is Gaussian white noise it can be shown that y_t is a Gaussian process with all finite subcollections of vectors $y_t, \ldots, y_{t+h}$ having joint multivariate normal distributions. The representation (6.3.3) is a pure MA or simply MA representation of y_t.

To make the derivation of the MA representation more transparent let us write the process (6.3.1) in lag operator notation,

$$A(L)y_t = \nu + M(L)u_t, \tag{6.3.4}$$

where $A(L) := I_K - A_1 L - \cdots - A_p L^p$ and $M(L) := I_K + M_1 L + \cdots + M_q L^q$. A pure MA representation of y_t is obtained by premultiplying with $A(L)^{-1}$,

$$y_t = A(1)^{-1}\nu + A(L)^{-1}M(L)u_t = \mu + \sum_{i=0}^{\infty} \Phi_i u_{t-i}.$$

Hence, multiplying from the left by $A(L)$ gives

$$(I_K - A_1 L - \cdots - A_p L^p)\left[\sum_{i=0}^{\infty} \Phi_i L^i\right] = I_K + \sum_{i=1}^{\infty}\left[\Phi_i - \sum_{j=1}^{i} A_j \Phi_{i-j}\right] L^i$$
$$= I_K + M_1 L + \cdots + M_q L^q$$

and, thus, comparing coefficients results in

$$M_i = \Phi_i - \sum_{j=1}^{i} A_j \Phi_{i-j}, \qquad i = 1, 2, \ldots$$

with $\Phi_0 := I_K$, $A_j := 0$ for $j > p$ and $M_i := 0$ for $i > q$. Rearranging gives

$$\Phi_i = M_i - \sum_{j=1}^{i} A_j \Phi_{i-j}, \qquad i = 1, 2, \ldots. \tag{6.3.5}$$

If the MA operator $M(L)$ satisfies the invertibility condition (6.2.5) the *VARMA process* (6.3.4) is called *invertible*. In that case it has a pure VAR representation,

$$y_t - \sum_{i=1}^{\infty} \Pi_i y_{t-i} = M(L)^{-1} A(L) y_t = M(1)^{-1} v + u_t$$

and the Π_i coefficient matrices are obtained by comparing coefficients in

$$I_K - \sum_{i=1}^{\infty} \Pi_i L^i = M(L)^{-1} A(L).$$

Alternatively, multiplying this expression from the left by $M(L)$ gives

$$(I_K + M_1 L + \cdots + M_q L^q)\left[I_K - \sum_{i=1}^{\infty} \Pi_i L^i\right] = I_K + \sum_{i=1}^{\infty}\left[M_i - \sum_{j=1}^{i} M_{i-j} \Pi_j\right] L^i$$
$$= I_K - A_1 L - \cdots - A_p L^p,$$

where $M_0 := I_K$ and $M_i := 0$ for $i > q$. Setting $A_i := 0$ for $i > p$ and comparing coefficients gives

$$-A_i = M_i - \sum_{j=1}^{i-1} M_{i-j} \Pi_j - \Pi_i$$

or

$$\Pi_i = A_i + M_i - \sum_{j=1}^{i-1} M_{i-j} \Pi_j \qquad \text{for } i = 1, 2, \ldots \tag{6.3.6}$$

As usual the sum is defined to be zero when the lower bound for the summation index exceeds its upper bound.

For instance, for the zero mean VARMA(1, 1) process

$$y_t = A_1 y_{t-1} + u_t + M_1 u_{t-1} \tag{6.3.7}$$

we get

$$\Pi_1 = A_1 + M_1$$

$$\Pi_2 = A_2 + M_2 - M_1\Pi_1 = -M_1A_1 - M_1^2$$
$$\vdots$$
$$\Pi_i = (-1)^{i-1}(M_1^i + M_1^{i-1}A_1), \qquad i = 1, 2, \ldots$$

and the coefficients of the pure MA representation are

$$\Phi_0 = I_K$$
$$\Phi_1 = M_1 + A_1$$
$$\Phi_2 = M_2 + A_1\Phi_1 + A_2\Phi_0 = A_1(M_1 + A_1)$$
$$\vdots$$
$$\Phi_i = A_1^{i-1}M_1 + A_1^i, \qquad i = 1, 2, \ldots.$$

If y_t is a stable and invertible VARMA process the pure MA representation (6.3.3) is called the *canonical* or *prediction error MA representation* in accordance with the terminology used in the finite order VAR case. In addition to the pure MA and VAR representations considered in this section a VARMA process also has VAR(1) representations. One such representation is introduced next.

6.3.2 A VAR(1) Representation of a VARMA Process

Suppose y_t has the VARMA(p, q) representation (6.3.1). For simplicity we assume that its mean is zero and hence $\nu = 0$. Let

$$Y_t := \underset{(K(p+q)\times 1)}{\begin{bmatrix} y_t \\ \vdots \\ y_{t-p+1} \\ u_t \\ \vdots \\ u_{t-q+1} \end{bmatrix}}, \qquad U_t := \begin{bmatrix} u_t \\ 0 \\ \vdots \\ 0 \\ u_t \\ 0 \\ \vdots \\ 0 \end{bmatrix} \begin{matrix} \left.\begin{matrix} \\ \\ \\ \\ \end{matrix}\right\} (Kp \times 1) \\ \left.\begin{matrix} \\ \\ \\ \\ \end{matrix}\right\} (Kq \times 1) \end{matrix}$$

and

$$\mathbf{A} := \begin{bmatrix} \mathbf{A}_{11} & \mathbf{A}_{12} \\ \mathbf{A}_{21} & \mathbf{A}_{22} \end{bmatrix} \quad [K(p+q) \times K(p+q)],$$

where

$$\mathbf{A}_{11} := \underset{(Kp \times Kp)}{\begin{bmatrix} A_1 & \ldots & A_{p-1} & A_p \\ I_K & & 0 & 0 \\ & \ddots & & \vdots \\ 0 & \ldots & I_K & 0 \end{bmatrix}}, \qquad \mathbf{A}_{12} := \underset{(Kp \times Kq)}{\begin{bmatrix} M_1 & \ldots & M_{q-1} & M_q \\ 0 & \ldots & 0 & 0 \\ \vdots & & \vdots & \vdots \\ 0 & \ldots & 0 & 0 \end{bmatrix}},$$

$$\underset{(Kq \times Kp)}{\mathbf{A}_{21} := 0}, \qquad \mathbf{A}_{22} := \underset{(Kq \times Kq)}{\begin{bmatrix} 0 & \dots & 0 & 0 \\ I_K & & 0 & 0 \\ & \ddots & & \vdots \\ 0 & \dots & I_K & 0 \end{bmatrix}}.$$

With this notation we get the VAR(1) representation of Y_t,

$$Y_t = \mathbf{A} Y_{t-1} + U_t. \tag{6.3.8}$$

If the VAR order is zero ($p = 0$) we choose $p = 1$ and set $A_1 = 0$ in this representation.

The $K(p+q)$-dimensional VAR(1) process in (6.3.8) is stable if and only if y_t is stable. This result follows since

$$\begin{aligned} \det(I_{K(p+q)} - \mathbf{A}z) &= \det(I_{Kp} - \mathbf{A}_{11}z)\det(I_{Kq} - \mathbf{A}_{22}z) \\ &= \det(I_K - A_1 z - \cdots - A_p z^p). \end{aligned} \tag{6.3.9}$$

Here the rules for the determinant of a partitioned matrix from Appendix A.10 have been used and we have also used that $I_{Kq} - \mathbf{A}_{22}z$ is a lower triangular matrix with ones on the main diagonal which has determinant 1. Furthermore $\det(I_{Kp} - \mathbf{A}_{11}z) = \det(I_K - A_1 z - \cdots - A_p z^p)$ follows as in Section 2.1.1.

From Chapter 2 we know that if y_t and hence Y_t is stable, the latter process has an MA representation

$$Y_t = \sum_{i=0}^{\infty} \mathbf{A}^i U_{t-i}.$$

Left-multiplying by the $(K \times K(p+q))$ matrix $J := [I_K \quad 0 \dots 0]$ gives

$$y_t = \sum_{i=0}^{\infty} J\mathbf{A}^i U_{t-i} = \sum_{i=0}^{\infty} J\mathbf{A}^i HJU_{t-i} = \sum_{i=0}^{\infty} J\mathbf{A}^i H u_{t-i} = \sum_{i=0}^{\infty} \Phi_i u_{t-i},$$

where

$$H = \begin{bmatrix} I_K \\ 0 \\ \vdots \\ 0 \\ I_K \\ 0 \\ \vdots \\ 0 \end{bmatrix} \begin{matrix} \left.\vphantom{\begin{matrix} I_K \\ 0 \\ \vdots \\ 0 \end{matrix}}\right\} (Kp \times K) \\ \left.\vphantom{\begin{matrix} I_K \\ 0 \\ \vdots \\ 0 \end{matrix}}\right\} (Kq \times K) \end{matrix}.$$

Thus,

$$\Phi_i = J\mathbf{A}^i H. \tag{6.3.10}$$

As an example consider the zero mean VARMA(1, 1) process from (6.3.7),

$$y_t = A_1 y_{t-1} + u_t + M_1 u_{t-1}.$$

For this process

$$Y_t = \begin{bmatrix} y_t \\ u_t \end{bmatrix}, \qquad \mathbf{A} = \begin{bmatrix} A_1 & M_1 \\ 0 & 0 \end{bmatrix}, \qquad U_t = \begin{bmatrix} u_t \\ u_t \end{bmatrix},$$

$$J = (I_K \quad 0) \quad (K \times 2K),$$

and

$$H = \begin{bmatrix} I_K \\ I_K \end{bmatrix} \quad (2K \times K).$$

Hence,

$$\Phi_0 = JH = I_K,$$

$$\Phi_1 = J\mathbf{A}H = [A_1 \quad M_1]H = A_1 + M_1,$$

$$\Phi_2 = J\mathbf{A}^2 H = J\begin{bmatrix} A_1^2 & A_1 M_1 \\ 0 & 0 \end{bmatrix} H = A_1^2 + A_1 M_1,$$

$$\vdots$$

$$\Phi_i = J\mathbf{A}^i H = J\begin{bmatrix} A_1^i & A_1^{i-1} M_1 \\ 0 & 0 \end{bmatrix} H = A_1^i + A_1^{i-1} M_1, \qquad i = 1, 2, \ldots \qquad (6.3.11)$$

This, of course, is precisely the same formula obtained from the recursions in (6.3.5).

The foregoing method of computing the MA matrices is just another way of computing the coefficient matrices of the power series

$$I_K + \sum_{i=1}^{\infty} \Phi_i L^i = (I_K - A_1 L - \cdots - A_p L^p)^{-1}(I_K + M_1 L + \cdots + M_q L^q).$$

Therefore it can just as well be used to compute the Π_i coefficient matrices of the pure VAR representation of a VARMA process. Recall that

$$I_K - \sum_{i=1}^{\infty} \Pi_i L^i = (I_K + M_1 L + \cdots + M_q L^q)^{-1}(I_K - A_1 L - \cdots - A_p L^p).$$

Hence, if we define

$$\mathbf{M} := \begin{bmatrix} \mathbf{M}_{11} & \mathbf{M}_{12} \\ \mathbf{M}_{21} & \mathbf{M}_{22} \end{bmatrix}, \qquad (6.3.12)$$

where

$$\mathbf{M}_{11} := \begin{bmatrix} -M_1 & \ldots & -M_{q-1} & -M_q \\ I_K & & 0 & 0 \\ & \ddots & \vdots & \vdots \\ 0 & \ldots & I_K & 0 \end{bmatrix}, \quad (Kq \times Kq)$$

$$\mathbf{M}_{12} := \underset{(Kq \times Kp)}{\begin{bmatrix} -A_1 & \dots & -A_{p-1} & -A_p \\ 0 & \dots & 0 & 0 \\ \vdots & & \vdots & \vdots \\ 0 & \dots & 0 & 0 \end{bmatrix}},$$

$$\underset{(Kp \times Kq)}{\mathbf{M}_{21} := 0,} \qquad \mathbf{M}_{22} := \underset{(Kp \times Kp)}{\begin{bmatrix} 0 & \dots & 0 & 0 \\ I_K & & 0 & 0 \\ & \ddots & \vdots & \vdots \\ 0 & \dots & I_K & 0 \end{bmatrix}},$$

we get

$$-\Pi_i = J\mathbf{M}^i H \tag{6.3.13}$$

with

$$H := \begin{bmatrix} I_K \\ 0 \\ \vdots \\ 0 \\ I_K \\ 0 \\ \vdots \\ 0 \end{bmatrix} \begin{matrix} \left.\vphantom{\begin{matrix} I_K \\ 0 \\ \vdots \\ 0 \end{matrix}}\right\} (Kq \times K) \\ \left.\vphantom{\begin{matrix} I_K \\ 0 \\ \vdots \\ 0 \end{matrix}}\right\} (Kp \times K) \end{matrix}$$

6.4 The Autocovariances and Autocorrelations of a VARMA(p, q) Process

For the K-dimensional, zero mean, stable VARMA(p, q) process

$$y_t = A_1 y_{t-1} + \cdots + A_p y_{t-p} + u_t + M_1 u_{t-1} + \cdots + M_q u_{t-q} \tag{6.4.1}$$

the autocovariances can be obtained formally from its pure MA representation as in Section 2.1.2. For instance, if y_t has canonical MA representation

$$y_t = \sum_{i=0}^{\infty} \Phi_i u_{t-i},$$

the autocovariance matrices are

$$\Gamma_y(h) := E(y_t y'_{t-h}) = \sum_{i=0}^{\infty} \Phi_{h+i} \Sigma_u \Phi'_i.$$

For the actual computation of the autocovariance matrices the following approach is more convenient. Postmultiplying (6.4.1) by y'_{t-h} and taking expectations gives

$$E(y_t y_{t-h}') = A_1 E(y_{t-1} y_{t-h}') + \cdots + A_p E(y_{t-p} y_{t-h}') + E(u_t y_{t-h}') + \cdots$$
$$+ M_q E(u_{t-q} y_{t-h}').$$

From the pure MA representation of the process it can be seen that $E(u_t y_s') = 0$ for $s < t$. Hence, we get for $h > q$,

$$\Gamma_y(h) = A_1 \Gamma_y(h-1) + \cdots + A_p \Gamma_y(h-p). \tag{6.4.2}$$

If $p > q$ and $\Gamma_y(0), \dots, \Gamma_y(p-1)$ are available, this relation can be used to compute the autocovariances recursively for $h = p, p+1, \dots$.

The initial matrices can be obtained from the VAR(1) representation (6.3.8) just as in Chapter 2, Section 2.1.4. In that section we obtained the relation

$$\Gamma_Y(0) = \mathbf{A} \Gamma_Y(0) \mathbf{A}' + \Sigma_U \tag{6.4.3}$$

for the covariance matrix of the VAR(1) process Y_t. Here $\Sigma_U = E(U_t U_t')$ is the covariance matrix of the white noise process in (6.3.8). Applying the vec operator to (6.4.3) and rearranging gives

$$\operatorname{vec} \Gamma_Y(0) = (I_{K^2(p+q)^2} - \mathbf{A} \otimes \mathbf{A})^{-1} \operatorname{vec}(\Sigma_U), \tag{6.4.4}$$

where the existence of the inverse follows again from the stability of the process as in Section 2.1.4 by appealing to the determinantal relation (6.3.9).

Having computed $\Gamma_Y(0)$ as in (6.4.4) we may collect $\Gamma_y(0), \dots, \Gamma_y(p-1)$ from

$$\Gamma_Y(0) = \begin{bmatrix} \Gamma_y(0) & \Gamma_y(1) & \dots & \Gamma_y(p-1) & E(y_t u_t') & E(y_t u_{t-1}') & \dots & E(y_t u_{t-q+1}') \\ \Gamma_y(-1) & \Gamma_y(0) & \dots & \Gamma_y(p-2) & 0 & E(y_{t-1} u_{t-1}') & \dots & E(y_{t-1} u_{t-q+1}') \\ \vdots & \vdots & \ddots & \vdots & \vdots & & & \vdots \\ \Gamma_y(-p+1) & \Gamma_y(-p+2) & \dots & \Gamma_y(0) & 0 & 0 & & E(y_{t-p+1} u_{t-q+1}') \\ & & & & \Sigma_u & 0 & \dots & 0 \\ & & & & 0 & \Sigma_u & & 0 \\ & & & & \vdots & & \ddots & \vdots \\ & & & & 0 & 0 & \dots & \Sigma_u \end{bmatrix}.$$

As mentioned previously the recursions (6.4.2) are valid for $h > q$ only. Thus, this way of computing the autocovariances requires that $p > q$. If the VAR order is less than q it may be increased artificially by adding lags of y_t with zero coefficient matrices until the VAR order p exceeds the MA order q. Then the aforementioned procedure can be applied. A computationally more efficient method of computing the autocovariances of a VARMA process is described by Mittnik (1990).

The autocorrelations of a VARMA(p, q) process are obtained from its autocovariances as in Chapter 2, Section 2.1.4. That is,

$$R_y(h) = D^{-1} \Gamma_y(h) D^{-1}, \tag{6.4.5}$$

where D is a diagonal matrix with the square roots of the diagonal elements of $\Gamma_y(0)$ on the diagonal.

To illustrate the computation of the covariance matrices we consider the VARMA(1, 1) process (6.3.7). Since $p = q$ we add a second lag of y_t so that

$$y_t = A_1 y_{t-1} + A_2 y_{t-2} + u_t + M_1 u_{t-1}$$

with $A_2 := 0$. Thus, in this case

$$Y_t = \begin{bmatrix} y_t \\ y_{t-1} \\ u_t \end{bmatrix}, \quad \mathbf{A} = \begin{bmatrix} A_1 & 0 & M_1 \\ I_K & 0 & 0 \\ 0 & 0 & 0 \end{bmatrix}, \quad U_t = \begin{bmatrix} u_t \\ 0 \\ u_t \end{bmatrix}, \quad \Sigma_U = \begin{bmatrix} \Sigma_u & 0 & \Sigma_u \\ 0 & 0 & 0 \\ \Sigma_u & 0 & \Sigma_u \end{bmatrix}.$$

With this notation we get from (6.4.4),

$$\operatorname{vec} \begin{bmatrix} \Gamma_y(0) & \Gamma_y(1) & \Sigma_u \\ \Gamma_y(-1) & \Gamma_y(0) & 0 \\ \Sigma_u & 0 & \Sigma_u \end{bmatrix} = (I_{9K^2} - \mathbf{A} \otimes \mathbf{A})^{-1} \operatorname{vec}(\Sigma_U).$$

Now, since we have the starting-up matrices $\Gamma_y(0)$, $\Gamma_y(1)$, the recursions (6.4.2) may be applied giving

$$\Gamma_y(h) = A_1 \Gamma_y(h-1) \qquad \text{for } h = 2, 3, \ldots$$

In stating the assumptions for the VARMA(p, q) process at the beginning of this section invertibility has not been mentioned. This is no accident because this condition is actually not required for computing the autocovariances of a VARMA(p, q) process. The same formulas may be used for invertible and noninvertible processes.

6.5 Forecasting VARMA Processes

Suppose the K-dimensional zero mean VARMA(p, q) process

$$y_t = A_1 y_{t-1} + \cdots + A_p y_{t-p} + u_t + M_1 u_{t-1} + \cdots + M_q u_{t-q} \tag{6.5.1}$$

is stable and invertible. As we have seen in Section 6.3.1 it has a pure VAR representation

$$y_t = \sum_{i=1}^{\infty} \Pi_i y_{t-i} + u_t \tag{6.5.2}$$

and a pure MA representation

$$y_t = \sum_{i=0}^{\infty} \Phi_i u_{t-i}. \tag{6.5.3}$$

Formulas for optimal forecasts can be given in terms of each of these representations.

Assuming that u_t is *independent* white noise and applying the conditional expectation operator E_t, given information up to time t, to (6.5.1) gives an optimal h-step forecast

$$y_t(h) = \begin{cases} A_1 y_t(h-1) + \cdots + A_p y_t(h-p) + M_h u_t + \cdots + M_q u_{t+h-q} & \text{for } h \le q, \\ A_1 y_t(h-1) + \cdots + A_p y_t(h-p) & \text{for } h > q, \end{cases} \tag{6.5.4}$$

where, as usual, $y_t(j) := y_{t+j}$ for $j \le 0$. Analogously we get from (6.5.2),

$$y_t(h) = \sum_{i=1}^{\infty} \Pi_i y_t(h-i), \tag{6.5.5}$$

and in Chapter 2, Section 2.2.2, we have seen that the optimal predictor in terms of the infinite order MA representation is

$$y_t(h) = \sum_{i=h}^{\infty} \Phi_i u_{t+h-i} = \sum_{i=0}^{\infty} \Phi_{h+i} u_{t-i} \tag{6.5.6}$$

(see (2.2.10)). Although in Chapter 2 this result was derived in the slightly more special setting of finite order VAR processes it is not difficult to see that it carries over to the present situation. All three formulas (6.5.4)–(6.5.6) result, of course, in equivalent predictors or forecasts. They are different representations of the *linear* minimum MSE predictors if u_t is uncorrelated but not necessarily independent white noise.

A forecasting formula can also be obtained from the VAR(1) representation (6.3.8) of the VARMA(p, q) process. From Section 2.2.2 the optimal h-step forecast of a VAR(1) process at origin t is known to be

$$Y_t(h) = \mathbf{A}^h Y_t = \mathbf{A} Y_t(h-1). \tag{6.5.7}$$

Premultiplying with the $(K \times K(p+q))$ matrix $J = [I_K \quad 0 \dots 0]$ results precisely in the recursive relation (6.5.4) (see Problem 6.4).

The forecasts at origin t are based on the information set

$$\Omega_t = \{y_s | s \le t\}.$$

This information set has the drawback of being unavailable in practice. Usually a finite stretch of y_t data is given only and, hence, the u_t cannot be determined exactly. Thus, even if the parameters of the process are known the prediction formulas (6.5.4)–(6.5.6) cannot be used. However, the invertibility of the process implies that the Π_i coefficient matrices go to zero exponentially with increasing i and we have the approximation

$$\sum_{i=1}^{\infty} \Pi_i y_t(h-i) \approx \sum_{i=1}^{n} \Pi_i y_t(h-i)$$

for large n. Consequently, in practice, if the information set is

$$\{y_1, \dots, y_T\} \tag{6.5.8}$$

and T is large, the forecast

$$\bar{y}_T(h) = \sum_{i=1}^{T} \Pi_i \bar{y}_T(h-i) \tag{6.5.9}$$

where $\bar{y}_T(j) := y_{T+j}$ for $j \leq 0$, will be almost identical to the optimal forecast. For a low order process, as they are commonly used in practice, for which the roots of

$$\det(I_K + M_1 z + \cdots + M_q z^q)$$

are not close to the unit circle, $T > 50$ will usually result in forecasts that cannot be distinguished from the optimal forecasts. It is worth noting, however, that the optimal forecasts based on the finite information set (6.5.8) can be determined. The resulting forecast formulas are for instance given by Brockwell & Davis (1987, Chapter 11, § 11.4). A similar problem is not encountered in forecasting finite order VAR processes because there the optimal forecast depends on a finite string of past variables only.

In the presently considered theoretical setting the forecast MSE matrices are most easily obtained from the representation (6.5.6). The forecast error is

$$y_{t+h} - y_t(h) = \sum_{i=0}^{h-1} \Phi_i u_{t+h-i}$$

and hence the forecast MSE matrix turns out to be

$$\begin{aligned} \Sigma_y(h) &= E[(y_{t+h} - y_t(h))(y_{t+h} - y_t(h))'] \\ &= \sum_{i=0}^{h-1} \Phi_i \Sigma_u \Phi_i' \end{aligned} \tag{6.5.10}$$

as in the finite order VAR case. Note, however, that in the present case the M_i coefficient matrices enter in computing the Φ_i matrices. Since the forecasts are unbiased, that is, the forecast errors have mean zero, the MSE matrix is the forecast error covariance matrix. Consequently, if the process is Gaussian, i.e., for all t and h, $y_t, \ldots, y_{t+h}$ have a multivariate normal distribution and also the u_t are normally distributed, the forecast errors are normally distributed,

$$y_{t+h} - y_t(h) \sim N(0, \Sigma_y(h)). \tag{6.5.11}$$

This result may be used in the usual fashion in setting up forecast intervals.

If a process with nonzero mean vector μ is considered the mean vector may simply be added to the prediction formula for the mean-adjusted process. For example, if y_t has zero mean and $x_t = y_t + \mu$ then the optimal h-step forecast of x_t is

$$x_t(h) = y_t(h) + \mu.$$

The forecast MSE matrix is not affected, that is, $\Sigma_x(h) = \Sigma_y(h)$.

6.6 Transforming and Aggregating VARMA Processes

In practice the original variables of interest are often transformed before their generation process is modeled. For example, data are often seasonally adjusted prior to an analysis. Also, sometimes they are temporally aggregated. For instance, quarterly data may have been obtained by adding up the corresponding monthly values or by taking their averages. Moreover, contemporaneous aggregation over a number of households, regions or sectors of the economy is quite common. For example, the GNP (gross national product) value for some period is the sum of private consumption, investment expenditures, net exports, and government spending for that period. It is often of interest to see what these transformations do to the generation processes of the variables in order to assess the consequences of transformations for forecasting and structural analysis. In the following we assume that the original data are generated by a VARMA process and we study the consequences of linear transformations. These results are of importance because temporal as well as contemporaneous aggregation procedures can be represented as linear transformations.

6.6.1 Linear Transformations of VARMA Processes

We shall begin with the result that a linear transformation of a process possessing an MA(q) representation gives a process that also has a finite order MA representation with order not greater than q.

PROPOSITION 6.1 (*Linear Transformation of an* MA(q) *Process*)

Let u_t be a K-dimensional white noise process with nonsingular covariance matrix Σ_u and let

$$y_t = \mu + u_t + M_1 u_{t-1} + \cdots + M_q u_{t-q}$$

be a K-dimensional invertible MA(q) process. Furthermore let F be an $(M \times K)$ matrix of rank M. Then the M-dimensional process $z_t = Fy_t$ has an invertible MA($\bar{q}$) representation

$$z_t = F\mu + v_t + N_1 v_{t-1} + \cdots + N_{\bar{q}} v_{t-\bar{q}}$$

where v_t is M-dimensional white noise with nonsingular covariance matrix Σ_v, the N_i are $(M \times M)$ coefficient matrices and $\bar{q} \leq q$. ■

We will not give a proof of this result here but refer the reader to Lütkepohl (1984 or 1987, Chapter 4). The proposition is certainly not surprising because considering the autocovariance matrices of z_t it is seen that

$$\Gamma_z(h) = E(Fy_t - F\mu)(Fy_{t-h} - F\mu)' = F\Gamma_y(h)F'$$

$$= \begin{cases} \sum_{i=0}^{q-h} FM_{i+h}\Sigma_u M_i' F', & h = 0, 1, \ldots, q \\ 0, & h = q+1, q+2, \ldots \end{cases}$$

by (6.2.7). Thus, the autocovariances of z_t for lags greater than q are all zero. This result is a necessary requirement for the proposition to be true. It also helps to understand that the MA order of z_t may be lower than that of y_t because $\Gamma_z(h) = F\Gamma_y(h)F'$ may be zero even if $\Gamma_y(h)$ is nonzero.

The proposition has some interesting implications. As we will see in the following (Corollary 6.1.1) it implies that a linearly transformed VARMA(p, q) process has again a finite order VARMA representation. Thus, the VARMA class is closed with respect to linear transformations. The same is not true for the class of finite order VAR processes because, as we will see shortly, a linearly transformed VAR(p) process may not admit a finite order VAR representation. This, of course, is an argument in favor of considering the VARMA class rather than restricting the analysis to finite order VAR processes.

COROLLARY 6.1.1

Let y_t be a K-dimensional, stable, invertible VARMA(p, q) process and let F be an $(M \times K)$ matrix of rank M. Then the process $z_t = Fy_t$ has a VARMA$(\bar{p}, \bar{q})$ representation with

$$\bar{p} \leq Kp$$

and

$$\bar{q} \leq (K-1)p + q. \qquad \blacksquare$$

Proof: We write the process y_t in lag operator notation as

$$A(L)y_t = M(L)u_t, \tag{6.6.1}$$

where the mean is set to zero without loss of generality as y_t may represent deviations from the mean. Left-multiplying by the adjoint of $A(L)$ gives

$$|A(L)|\, y_t = A(L)^* M(L) u_t, \tag{6.6.2}$$

where $A(L)^*A(L) = |A(L)|$ has been used. It is easy to check that $|A(z)^*| \neq 0$ for $|z| \leq 1$. Thus, (6.6.2) is a stable and invertible VARMA representation of y_t. Pre-multiplying (6.6.2) with F results in

$$|A(L)|\, z_t = FA(L)^* M(L) u_t. \tag{6.6.3}$$

The operator $A(L)^*M(L)$ is easily seen to have degree at most $p(K-1) + q$ and, thus, the right-hand side of (6.6.3) is just a linearly transformed finite order MA process which, by Proposition 6.1, has an MA$(\bar{q})$ representation with

$$\bar{q} \leq p(K-1) + q.$$

The degree of the AR-operator $|A(L)|$ is at most Kp because the determinant is just a sum of products involving one operator from each row and each column of $A(L)$. This proves the corollary. $\blacksquare$

The corollary gives upper bounds for the VARMA orders of a linearly transformed VARMA process. For instance, if y_t is a VAR(p) = VARMA$(p, 0)$ process a linear transformation $z_t = Fy_t$ has a VARMA($\bar{p}, \bar{q}$) representation with $\bar{p} \leq Kp$ and $\bar{q} \leq (K-1)p$. For some linear transformations $\bar{q}$ will be zero. We will see in the following, however, that generally for some transformations the upper bounds for the orders are attained and a representation with lower orders does not exist. This result implies that a linear transformation of a finite order VAR(p) process may not admit a finite order VAR representation. Specifically, the subprocesses or marginal processes of a K-dimensional process y_t are obtained by using transformation matrices such as $F = [I_M \quad 0]$. Hence a subprocess of a VAR(p) process may not have a finite order VAR but just a mixed VARMA representation. An example will be given shortly.

Other bounds for the VARMA orders than those provided in Corollary 6.1.1 for a linearly transformed VARMA process and bounds for special linear transformations are given in various articles in the literature. For further results and references see Lütkepohl (1987, Chapter 4; 1986, Kapitel 2).

To illustrate Corollary 6.1.1 we consider the bivariate VAR(1) process

$$\begin{bmatrix} 1-0.5L & 0.66L \\ 0.5L & 1+0.3L \end{bmatrix} \begin{bmatrix} y_{1t} \\ y_{2t} \end{bmatrix} = \begin{bmatrix} u_{1t} \\ u_{2t} \end{bmatrix} \quad \text{with } \Sigma_u = I_2. \tag{6.6.4}$$

Here $K = 2$, $p = 1$, and $q = 0$. Thus, $z_t = [1, 0]y_t = y_{1t}$ as a univariate ($M = 1$) marginal process has an ARMA representation with orders not greater than (2, 1). The precise form of the process can be determined with the help of the representation (6.6.3). Using that representation gives

$$[(1+0.3L)(1-0.5L) - 0.66 \cdot 0.5L^2]z_t = [1, 0] \begin{bmatrix} 1+0.3L & -0.66L \\ -0.5L & 1-0.5L \end{bmatrix} \begin{bmatrix} u_{1t} \\ u_{2t} \end{bmatrix}$$

$$= (1+0.3L)u_{1t} - 0.66Lu_{2t}. \tag{6.6.5}$$

The right-hand side, say w_{1t}, is the sum of an MA(1) process and a white noise process. Thus, by Proposition 6.1 it is known to have an MA(1) representation, say $w_{1t} = v_{1t} + \gamma v_{1,t-1}$. To determine γ and $\sigma_1^2 = \mathrm{Var}(v_{1t})$ we use

$$\begin{aligned} E(w_{1t}^2) &= E(v_{1t} + \gamma v_{1,t-1})^2 = (1+\gamma^2)\sigma_1^2 \\ &= E[(1+0.3L)u_{1t} - 0.66Lu_{2t}]^2 = 1.5 \end{aligned}$$

and

$$\begin{aligned} E(w_t w_{t-1}) &= E[(v_{1t} + \gamma v_{1,t-1})(v_{1,t-1} + \gamma v_{1,t-2})] = \gamma\sigma_1^2 \\ &= E[((1+0.3L)u_{1t} - 0.66u_{2,t-1})((1+0.3L)u_{1,t-1} - 0.66u_{2,t-2})] \\ &= -0.5. \end{aligned}$$

Solving this nonlinear system of two equations for γ and σ_1^2 gives

$$\gamma = 0.205 \qquad \text{and} \qquad \sigma_1^2 = 1.46.$$

Note that we have picked the invertible solution with $|\gamma| < 1$. Thus, from (6.6.5) we get a marginal process

$$(1 - 0.2L - 0.48L^2)y_{1t} = (1 + 0.205L)v_{1t} \qquad \text{with } \sigma_1^2 = 1.46.$$

In other words, y_{1t} has indeed an ARMA(2, 1) representation and it is easy to check that cancellation of the AR and MA operators is not possible. Hence, the ARMA orders are minimal in this case.

As another example consider again the bivariate VAR(1) process (6.6.4) and suppose we are interested in the process $z_t = y_{1t} + y_{2t}$. Thus, $F = [1, 1]$ is again a (1×2) vector. Multiplying (6.6.4) by the adjoint of the VAR operator gives

$$(1 - 0.2L - 0.48L^2)\begin{bmatrix} y_{1t} \\ y_{2t} \end{bmatrix} = \begin{bmatrix} 1 + 0.3L & -0.66L \\ -0.5L & 1 - 0.5L \end{bmatrix}\begin{bmatrix} u_{1t} \\ u_{2t} \end{bmatrix}.$$

Hence, multiplying by F gives

$$(1 - 0.2L - 0.48L^2)(y_{1t} + y_{2t}) = (1 - 0.2L)u_{1t} + (1 - 1.16L)u_{2t}.$$

Using similar arguments as in (6.6.5) it can be shown that the right-hand side of this expression is a process with MA(1) representation $v_t - 0.504v_{t-1}$, where $\sigma_v^2 = \text{Var}(v_t) = 2.70$. Consequently, the process of interest has the ARMA(2, 1) representation

$$(1 - 0.2L - 0.48L^2)z_t = (1 - 0.504L)v_t \qquad \text{with } \sigma_v^2 = 2.70. \tag{6.6.6}$$

The following result is of interest if forecasting is the objective of the analysis.

PROPOSITION 6.2 (*Forecast Efficiency of Linearly Transformed VARMA Processes*)

Let y_t be a stable, invertible, K-dimensional VARMA(p, q) process, let F be an $(M \times K)$ matrix of rank M and let $z_t = Fy_t$. Furthermore, denote the MSE matrices of the optimal h-step predictors of y_t and z_t by $\Sigma_y(h)$ and $\Sigma_z(h)$, respectively. Then

$$\Sigma_z(h) - F\Sigma_y(h)F'$$

is positive semidefinite. ■

This result means that $Fy_t(h)$ is generally a better predictor of z_{t+h} with smaller (at least not greater) MSEs than $z_t(h)$. In other words, forecasting the original process y_t and transforming the forecasts is generally better than forecasting the transformed process directly. For a proof and references of related results see Lütkepohl (1987, Chapter 4). To see the point more clearly consider again the example process (6.6.4) and suppose we are interested in the sum of its components $z_t = y_{1t} + y_{2t}$. Forecasting the bivariate process one step ahead results in a forecast MSE matrix $\Sigma_y(1) = \Sigma_u = I_2$. Thus the corresponding 1-step forecast of z_t has MSE

$$[1, 1]\Sigma_y(1)\begin{bmatrix} 1 \\ 1 \end{bmatrix} = 2.$$

In contrast, if a univariate forecast is obtained on the basis of the ARMA(2, 1) representation (6.6.6) the 1-step forecast MSE becomes $\sigma_v^2 = 2.70$. Clearly, the latter forecast is inferior in terms of MSE.

Of course, these results hold for VARMA processes for which all the parameters are known. They do not necessarily carry over to estimated processes, a case which is also investigated and reviewed by Lütkepohl (1987).

6.6.2 Aggregation of VARMA Processes

There is little to be added to the foregoing results for the case of contemporaneous aggregation. Suppose $y_t = (y_{1t}, \dots, y_{Kt})'$ consists of K variables. If all or some of them are contemporaneously aggregated by taking their sum or average this just means that y_t is transformed linearly and the foregoing results apply directly. In particular, the aggregated process has a finite order VARMA representation if the original process does. Moreover, if forecasts for the aggregated variables are desired it is generally preferable to forecast the disaggregate process and aggregate the forecasts rather than forecast the aggregated process directly.

The foregoing results are also helpful in studying the consequences of temporal aggregation. Suppose we wish to aggregate the variables y_t generated by

$$y_t = A_1 y_{t-1} + A_2 y_{t-2} + u_t + M_1 u_{t-1}$$

over, say, $m = 3$ subsequent periods. To be able to use the previous framework we specify

$$\begin{bmatrix} I_K & 0 & 0 \\ -A_1 & I_K & 0 \\ -A_2 & -A_1 & I_K \end{bmatrix} \begin{bmatrix} y_{m(\tau-1)+1} \\ y_{m(\tau-1)+2} \\ y_{m\tau} \end{bmatrix}$$

$$= \begin{bmatrix} 0 & A_2 & A_1 \\ 0 & 0 & A_2 \\ 0 & 0 & 0 \end{bmatrix} \begin{bmatrix} y_{m(\tau-2)+1} \\ y_{m(\tau-2)+2} \\ y_{m(\tau-1)} \end{bmatrix} + \begin{bmatrix} I_K & 0 & 0 \\ M_1 & I_K & 0 \\ 0 & M_1 & I_K \end{bmatrix} \begin{bmatrix} u_{m(\tau-1)+1} \\ u_{m(\tau-1)+2} \\ u_{m\tau} \end{bmatrix}$$

$$+ \begin{bmatrix} 0 & 0 & M_1 \\ 0 & 0 & 0 \\ 0 & 0 & 0 \end{bmatrix} \begin{bmatrix} u_{m(\tau-2)+1} \\ u_{m(\tau-2)+2} \\ u_{m(\tau-1)} \end{bmatrix}.$$

Defining

$$\underset{\sim}{y}_\tau := \begin{bmatrix} y_{m(\tau-1)+1} \\ y_{m(\tau-1)+2} \\ y_{m\tau} \end{bmatrix} \quad \text{and} \quad \underset{\sim}{u}_\tau := \begin{bmatrix} u_{m(\tau-1)+1} \\ u_{m(\tau-1)+2} \\ u_{m\tau} \end{bmatrix}$$

we get

$$\underset{\sim}{A}_0 \underset{\sim}{y}_\tau = \underset{\sim}{A}_1 \underset{\sim}{y}_{\tau-1} + \underset{\sim}{M}_0 \underset{\sim}{u}_\tau + \underset{\sim}{M}_1 \underset{\sim}{u}_{\tau-1}, \tag{6.6.7}$$

where $\underset{\sim}{A}_0$, $\underset{\sim}{A}_1$, $\underset{\sim}{M}_0$, and $\underset{\sim}{M}_1$ have the obvious definitions. This is a VARMA(1, 1) representation of the $3K$-dimensional process $\underset{\sim}{y}_\tau$. Our standard form of a

VARMA(1, 1) process can be obtained from this form by left-multiplying with $\underset{\sim}{A}_0^{-1}$ and defining $\underset{\sim}{v}_\tau = \underset{\sim}{A}_0^{-1}\underset{\sim}{M}_0\underset{\sim}{u}_\tau$ which gives

$$\underset{\sim}{y}_\tau = \underset{\sim}{A}_0^{-1}\underset{\sim}{A}_1\underset{\sim}{y}_{\tau-1} + \underset{\sim}{v}_\tau + \underset{\sim}{A}_0^{-1}\underset{\sim}{M}_1\underset{\sim}{M}_0^{-1}\underset{\sim}{A}_0\underset{\sim}{v}_{\tau-1}.$$

Now temporal aggregation over $m = 3$ periods can be represented as a linear transformation of the process $\underset{\sim}{y}_\tau$. Clearly, it is not difficult to see that this method generalizes for higher order processes and temporal aggregation over more than three periods. Moreover different types of temporal aggregation can be handled. For instance, the aggregate may be the sum of subsequent values or it may be their average. Furthermore, temporal and contemporaneous aggregation can be dealt with simultaneously. In all of these cases the aggregate has a VARMA representation if the original variables are generated by a finite order VARMA process and its structure can be studied using the foregoing framework. Moreover, by Proposition 6.2, if forecasts of the aggregate are of interest, it is in general preferable to forecast the original disaggregate process and aggregate the forecasts rather than forecast the aggregate directly. A detailed discussion of these issues and also of forecasting with estimated processes can be found in Lütkepohl (1987).

6.7 Interpretation of VARMA Models

The same tools and concepts that we have used for interpreting VAR models may also be applied in the VARMA case. We will consider Granger-causality and impulse response analysis in turn.

6.7.1 Granger-Causality

In order to study Granger-causality in the context of VARMA processes we partition y_t in two groups of variables, z_t and x_t, and we partition the VAR and MA operators as well as the white noise process u_t accordingly. Hence, we get

$$\begin{bmatrix} A_{11}(L) & A_{12}(L) \\ A_{21}(L) & A_{22}(L) \end{bmatrix} \begin{bmatrix} z_t \\ x_t \end{bmatrix} = \begin{bmatrix} M_{11}(L) & M_{12}(L) \\ M_{21}(L) & M_{22}(L) \end{bmatrix} \begin{bmatrix} u_{1t} \\ u_{2t} \end{bmatrix}, \tag{6.7.1}$$

where again a zero mean is assumed for simplicity and without loss of generality. The results derived in the following are not affected by a nonzero mean term. The process (6.7.1) is assumed to be stable and invertible and its pure, canonical MA representation is

$$\begin{bmatrix} z_t \\ x_t \end{bmatrix} = \begin{bmatrix} \Phi_{11}(L) & \Phi_{12}(L) \\ \Phi_{21}(L) & \Phi_{22}(L) \end{bmatrix} \begin{bmatrix} u_{1t} \\ u_{2t} \end{bmatrix}.$$

From Proposition 2.2 we know that x_t is not Granger-causal for z_t if and only if $\Phi_{12}(L) \equiv 0$. Although the proposition is stated for VAR processes it is easy to see that it remains correct for the presently considered VARMA case. We also know that

$$\begin{bmatrix} \Phi_{11}(L) & \Phi_{12}(L) \\ \Phi_{21}(L) & \Phi_{22}(L) \end{bmatrix}$$

$$= \begin{bmatrix} A_{11}(L) & A_{12}(L) \\ A_{21}(L) & A_{22}(L) \end{bmatrix}^{-1} \begin{bmatrix} M_{11}(L) & M_{12}(L) \\ M_{21}(L) & M_{22}(L) \end{bmatrix}$$

$$= \begin{bmatrix} D(L) & -D(L)A_{12}(L)A_{22}(L)^{-1} \\ -A_{22}(L)^{-1}A_{21}(L)D(L) & A_{22}(L)^{-1} + A_{22}(L)^{-1}A_{21}(L)D(L)A_{12}(L)A_{22}(L)^{-1} \end{bmatrix}$$

$$\cdot \begin{bmatrix} M_{11}(L) & M_{12}(L) \\ M_{21}(L) & M_{22}(L) \end{bmatrix},$$

where

$$D(L) := [A_{11}(L) - A_{12}(L)A_{22}(L)^{-1}A_{21}(L)]^{-1}$$

and the rules for the partitioned inverse have been used (see Appendix A, A.10). Consequently, x_t is not Granger-causal for z_t if and only if

$$0 \equiv D(L)M_{12}(L) - D(L)A_{12}(L)A_{22}(L)^{-1}M_{22}(L)$$

or, equivalently,

$$M_{12}(L) - A_{12}(L)A_{22}(L)^{-1}M_{22}(L) \equiv 0.$$

Moreover, it follows as in Proposition 2.3 that there is no instantaneous causation between x_t and z_t if and only if $E(u_{1t}u'_{2t}) = 0$. We state these results as a proposition.

PROPOSITION 6.3 (*Characterization of Noncausality*)

Let

$$y_t = \begin{bmatrix} z_t \\ x_t \end{bmatrix}$$

be a stable and invertible VARMA(p, q) process as in (6.7.1) with possibly nonzero mean. Then x_t is not Granger-causal for z_t if and only if

$$M_{12}(L) \equiv A_{12}(L)A_{22}(L)^{-1}M_{22}(L). \tag{6.7.2}$$

There is no instantaneous causation between z_t and x_t if and only if $E(u_{1t}u'_{2t}) = 0$.

■

REMARK 1: Obviously, the restrictions characterizing Granger-noncausality are not quite so easy here as in the VAR(p) case. Consider, for instance, a bivariate VARMA(1, 1) process

$$\begin{bmatrix} z_t \\ x_t \end{bmatrix} = \begin{bmatrix} \alpha_{11,1} & \alpha_{12,1} \\ \alpha_{21,1} & \alpha_{22,1} \end{bmatrix} \begin{bmatrix} z_{t-1} \\ x_{t-1} \end{bmatrix} + \begin{bmatrix} u_{1t} \\ u_{2t} \end{bmatrix} + \begin{bmatrix} m_{11,1} & m_{12,1} \\ m_{21,1} & m_{22,1} \end{bmatrix} \begin{bmatrix} u_{1,t-1} \\ u_{2,t-1} \end{bmatrix}.$$

For this process the restrictions (6.7.2) reduce to

$$m_{12,1}L = (-\alpha_{12,1}L)(1 - \alpha_{22,1}L)^{-1}(1 + m_{22,1}L)$$

or

$$(1 - \alpha_{22,1}L)m_{12,1}L = -(1 + m_{22,1}L)\alpha_{12,1}L$$

or

$$m_{12,1} = -\alpha_{12,1} \quad \text{and} \quad \alpha_{22,1}m_{12,1} = \alpha_{12,1}m_{22,1}.$$

This, of course, is a set of nonlinear restrictions whereas only linear constraints were required to characterize Granger-noncausality in the pure VAR(p) case. However, a sufficient condition for (6.7.2) to hold is

$$M_{12}(L) \equiv A_{12}(L) \equiv 0, \tag{6.7.3}$$

which is again a set of linear constraints. Occasionally these sufficient conditions may be easier to test than (6.7.2). ■

REMARK 2: To turn the arguments put forward prior to Proposition 6.3 into a formal proof requires that we convince ourselves that all the operations performed with the matrices of lag polynomials are feasible and correct. Since we have not proven these results the arguments should be taken as just an indication of how a proof may proceed. ■

6.7.2 Impulse Response Analysis

The impulse responses and forecast error variance decompositions of a VARMA model are obtained from its pure MA representation as in the finite order VAR case. Thus, the discussion of Sections 2.3.2 and 2.3.3 carries over to the present case except that the Φ_i's are computed with different formulas. Also, Propositions 2.4 and 2.5 need modification. We will not give the details here but refer the reader to the exercises (see Problem 6.9).

It may be worth re-iterating some caveats of impulse response analysis which may be more apparent now after the discussion of transformations in Section 6.6. In particular we have seen there that dropping variables (considering subprocesses) or aggregating the components of a VARMA process temporally and/or contemporaneously results in possibly quite different VARMA structures. They will in general have quite different coefficients in their pure MA representations. In other words, the impulse responses may change drastically if important variables are excluded from a system or if the level of aggregation is altered, for instance, if quarterly instead of monthly data are considered. Again, this does not necessarily render impulse response analysis useless. It should caution the reader against overinterpreting the evidence from VARMA models, though. Some thought must be given to the choice of variables, the level of aggregation and other transformations of variables.

6.8 Exercises

Problem 6.1

Write the MA(1) process $y_t = u_t + M_1 u_{t-1}$ in VAR(1) form $Y_t = \mathbf{A} Y_{t-1} + U_t$ and determine $\mathbf{A}^i$ for $i = 1, 2$.

Problem 6.2

Suppose $y_t = A_1 y_{t-1} + u_t + M_1 u_{t-1} + M_2 u_{t-2}$ is a stable and invertible VARMA(1, 2) process. Determine the coefficient matrices Π_i, $i = 1, 2, 3, 4$, of its pure VAR representation and the coefficient matrices Φ_i, $i = 1, 2, 3, 4$, of its pure MA representation.

Problem 6.3

Evaluate the autocovariances $\Gamma_y(h)$, $h = 1, 2, 3$, of the bivariate VARMA(2, 1) process

$$y_t = \begin{bmatrix} .3 \\ .5 \end{bmatrix} + \begin{bmatrix} .5 & .1 \\ .4 & .5 \end{bmatrix} y_{t-1} + \begin{bmatrix} 0 & 0 \\ .25 & 0 \end{bmatrix} y_{t-2} + u_t + \begin{bmatrix} .6 & .2 \\ 0 & .3 \end{bmatrix} u_{t-1}. \tag{6.8.1}$$

(Hint: The use of a computer will greatly simplify this problem.)

Problem 6.4

Write the VARMA(1, 1) process $y_t = A_1 y_{t-1} + u_t + M_1 u_{t-1}$ in VAR(1) form $Y_t = \mathbf{A} Y_{t-1} + U_t$. Determine forecasts $Y_t(h) = \mathbf{A}^h Y_t$ for $h = 1, 2, 3$ and compare them to forecasts obtained from the recursive formula (6.5.4).

Problem 6.5

Derive a univariate ARMA representation of the second component, y_{2t}, of the process given in (6.6.4).

Problem 6.6

Provide upper bounds for the ARMA orders of the process $z_t = y_{1t} + y_{2t} + y_{3t}$, where $y_t = (y_{1t}, y_{2t}, y_{3t}, y_{4t})'$ is a 4-dimensional VARMA(3, 3) process.

Problem 6.7

Write the VARMA(1, 1) process y_t from Problem 6.4 in a form such as (6.6.7) that permits to analyze temporal aggregation over four periods in the framework of Section 6.6.2. Give upper bounds for the orders of a VARMA representation of the process obtained by temporally aggregating y_t over four periods.

Problem 6.8

Write down explicitly the restrictions characterizing Granger-noncausality for a bivariate VARMA(2, 1) process. Is y_{1t} Granger-causal for y_{2t} in the process (6.8.1)?

Problem 6.9

Generalize Propositions 2.4 and 2.5 to the VARMA(p, q) case. (Hint: Show that for a K-dimensional VARMA(p, q) process

$$\phi_{jk,i} = 0 \qquad \text{for } i = 1, 2, \dots$$

is equivalent to

$$\phi_{jk,i} = 0 \qquad \text{for } i = 1, 2, \dots, p(K-1) + q;$$

and

$$\theta_{jk,i} = 0 \qquad \text{for } i = 0, 1, 2, \dots$$

is equivalent to

$$\theta_{jk,i} = 0 \qquad \text{for } i = 0, 1, \dots, p(K-1) + q.)$$

Problem 6.10

Suppose m is a real number with $|m| > 1$ and u_t is a white noise process. Show that the process

$$v_t = \left[1 + \frac{1}{m}L\right]^{-1}(1 + mL)u_t$$

is also white noise with $\mathrm{Var}(v_t) = m^2\,\mathrm{Var}(u_t)$.

Chapter 7. Estimation of VARMA Models

In this chapter maximum likelihood estimation of the coefficients of a VARMA model is considered. Before we can proceed to the actual estimation a unique set of parameters must be specified. In this context the problem of nonuniqueness of a VARMA representation becomes of importance. This problem is often referred to as the identification problem, that is, the problem of identifying a unique structure among many equivalent ones. It is treated in Section 7.1. In Section 7.2 the Gaussian likelihood function of a VARMA model is considered. A numerical algorithm for maximizing it and thus for computing the actual estimates is discussed in Section 7.3. The asymptotic properties of the ML estimators are the subject of Section 7.4. Forecasting with estimated processes and impulse response analysis are dealt with in Sections 7.5 and 7.6, respectively.

7.1 The Identification Problem

7.1.1 Nonuniqueness of VARMA Representations

In the previous chapter we have considered K-dimensional, stationary processes y_t with VARMA(p, q) representation

$$y_t = A_1 y_{t-1} + \cdots + A_p y_{t-p} + u_t + M_1 u_{t-1} + \cdots + M_q u_{t-q} \tag{7.1.1}$$

Since the mean term is of no importance for the presently considered problem we have set it to zero. Therefore no intercept term appears in (7.1.1). This model can be written in lag operator notation as

$$A(L)y_t = M(L)u_t, \tag{7.1.2}$$

where

$$A(L) := I_K - A_1 L - \cdots - A_p L^p \quad \text{and} \quad M(L) := I_K + M_1 L + \cdots + M_q L^q.$$

Assuming that the VARMA representation is stable and invertible the well-defined process described by the model (7.1.1) or (7.1.2) is given by

$$y_t = \sum_{i=0}^{\infty} \Phi_i u_{t-i} = \Phi(L)u_t = A(L)^{-1}M(L)u_t.$$

In practice it is sometimes useful to consider a slightly more general type of

VARMA models by attaching nonidentity coefficient matrices to y_t and u_t, that is, one may want to consider representations of the type

$$A_0 y_t = A_1 y_{t-1} + \cdots + A_p y_{t-p} + M_0 \bar{u}_t + M_1 \bar{u}_{t-1} + \cdots + M_q \bar{u}_{t-q}. \tag{7.1.3}$$

Such a form may be suggested by subject matter theory which may imply instantaneous effects of some variables on other variables. It will also turn out to be useful in finding unique structures for VARMA models. By the specification (7.1.3) we mean the well-defined process

$$y_t = (A_0 - A_1 L - \cdots - A_p L^p)^{-1}(M_0 + M_1 L + \cdots + M_q L^q)\bar{u}_t.$$

Such a process has a standard VARMA(p, q) representation with identity coefficient matrices attached to the instantaneous y_t and u_t if A_0 and M_0 are nonsingular. To see this we left-multiply (7.1.3) by A_0^{-1} and define $u_t = A_0^{-1} M_0 \bar{u}_t$ which gives

$$\begin{aligned} y_t = {} & A_0^{-1} A_1 y_{t-1} + \cdots + A_0^{-1} A_p y_{t-p} + u_t + A_0^{-1} M_1 M_0^{-1} A_0 u_{t-1} + \cdots \\ & + A_0^{-1} M_q M_0^{-1} A_0 u_{t-q}. \end{aligned}$$

Redefining the matrices appropriately this, of course, is a representation of the type (7.1.1) with identity coefficient matrices at lag zero which describes the same process as (7.1.3). The assumption that both A_0 and M_0 are nonsingular does not entail any loss of generality as long as none of the components of y_t can be written as a linear combination of the other components. We call a stable and invertible representation as in (7.1.1) a *VARMA representation in standard form* or a *standard VARMA representation* to distinguish it from representations with nonidentity matrices at lag zero as in (7.1.3). This discussion shows that VARMA representations are not unique, that is, a given process y_t can be written in standard form or in nonstandard form by just left-multiplying by any nonsingular $(K \times K)$ matrix. We have encountered a similar problem in dealing with finite order VAR processes which can, for instance, be written in standard form and in recursive form. However, once we consider standard VAR models only we have unique representations. This is in sharp contrast to the presently considered VARMA case where in general a standard form is not a unique representation as we will see shortly.

It may be useful at this stage to emphasize what we mean by equivalent representations of a process. Generally, two representations of a process y_t are equivalent if they give rise to the same realizations (except on a set of measure zero) and thus to the same multivariate distributions of any finite subcollection of variables $y_t, y_{t+1}, \ldots, y_{t+h}$ for arbitrary integers t and h. Of course, this just says that equivalent representations really represent the same process. If y_t is a zero mean process with canonical MA representation

$$\begin{aligned} y_t &= \sum_{i=0}^{\infty} \Phi_i u_{t-i}, \qquad \Phi_0 = I_K, \\ &= \Phi(L) u_t, \end{aligned} \tag{7.1.4}$$

where $\Phi(L) := \sum_{i=0}^{\infty} \Phi_i L^i$, then any VARMA model $A(L)y_t = M(L)u_t$ for which

$$A(L)^{-1}M(L) = \Phi(L) \tag{7.1.5}$$

is an equivalent representation of the process y_t. In other words, all VARMA models are equivalent for which $A(L)^{-1}M(L)$ results in the same operator $\Phi(L)$. Thus, in order to ensure uniqueness of a VARMA representation we must impose restrictions on the VAR and MA operators such that there is precisely one feasible set of operators $A(L)$ and $M(L)$ satisfying (7.1.5) for a given $\Phi(L)$.

Obviously, given some stable, invertible VARMA representation $A(L)y_t = M(L)u_t$ an equivalent representation results if we left-multiply by any nonsingular matrix A_0. Therefore, to remove this source of nonuniqueness let us for the moment focus on VARMA representations in standard form. As mentioned earlier, even then uniqueness is not ensured. To see this more clearly let us consider a bivariate VARMA(1, 1) process in standard form,

$$y_t = A_1 y_{t-1} + u_t + M_1 u_{t-1}. \tag{7.1.6}$$

From Section 6.3.1 we know that this process has the canonical MA representation

$$y_t = \sum_{i=0}^{\infty} \Phi_i u_{t-i} = u_t + \sum_{i=1}^{\infty} (A_1^i + A_1^{i-1} M_1) u_{t-i}. \tag{7.1.7}$$

Thus, for example any VARMA(1, 1) representation with $M_1 = -A_1$ will result in the same canonical MA representation. In other words, if it turns out that y_t is such that $M_1 = -A_1$ for some set of coefficients, then any choice of A_1 matrix that gives rise to a stable VAR operator can be matched by an M_1 matrix that leads to an equivalent VARMA(1, 1) representation of y_t. Of course, in this case the MA coefficient matrices in (7.1.7) are in fact all zero and $y_t = u_t$ is really white noise, that is, y_t actually has a VARMA(0, 0) structure. This fact is also quite easy to see from the lag operator representation of (7.1.6),

$$(I_2 - A_1 L)y_t = (I_2 + M_1 L)u_t.$$

Of course, if $M_1 = -A_1$, the MA operator cancels against the VAR operator. This type of parameter indeterminacy is also known from univariate ARMA processes and is usually ruled out by the assumption that the AR and MA operators have no common factors. Let us make a similar assumption in the presently considered multivariate case by requiring that y_t is not white noise, i.e., $M_1 \neq -A_1$.

Unfortunately, in the multivariate case the nonuniqueness problem is not solved by this assumption. To see this suppose that

$$A_1 = \begin{bmatrix} 0 & \alpha \\ 0 & 0 \end{bmatrix} \quad \text{and} \quad M_1 = 0,$$

where $\alpha \neq 0$. In this case the canonical MA representation (7.1.4) has coefficient matrices

$$\Phi_1 = A_1, \quad \Phi_2 = \Phi_3 = \cdots = 0, \tag{7.1.8}$$

because $A_1^i = 0$ for $i > 1$. The same MA representation results if

$$A_1 = 0 \quad \text{and} \quad M_1 = \begin{bmatrix} 0 & \alpha \\ 0 & 0 \end{bmatrix}.$$

More generally, a canonical MA representation with coefficient matrices as in (7.1.8) is obtained if

$$A_1 = \begin{bmatrix} 0 & \alpha + m \\ 0 & 0 \end{bmatrix} \quad \text{and} \quad M_1 = \begin{bmatrix} 0 & -m \\ 0 & 0 \end{bmatrix},$$

whatever the value of m. Note also that the VARMA representation will be stable and invertible for any value of m.

To understand where the parameter indeterminacy comes from consider the VAR operator

$$I_2 - \begin{bmatrix} 0 & \alpha \\ 0 & 0 \end{bmatrix} L. \tag{7.1.9}$$

The inverse of this operator is

$$I_2 + \begin{bmatrix} 0 & \alpha \\ 0 & 0 \end{bmatrix} L \tag{7.1.10}$$

which is easily checked by multiplying the two operators together. Thus, the operator (7.1.9) has a finite order inverse. Operators of this type are precisely the ones that cause trouble in setting up a uniquely parameterized VARMA representation of a given process because multiplying by such an operator may cancel part of one operator (VAR or MA) while at the same time the finite order of the other operator is maintained.

To get a better sense for this problem let us look at the following VARMA(1, 1) process:

$$A(L)y_t = M(L)u_t,$$

where

$$A(L) = \begin{bmatrix} 1 - \alpha_{11}L & -\alpha_{12}L \\ 0 & 1 \end{bmatrix} \quad \text{and} \quad M(L) = \begin{bmatrix} 1 + m_{11}L & m_{12}L \\ 0 & 1 \end{bmatrix}.$$

The two operators do not cancel if $\alpha_{11} \neq -m_{11}$ and $\alpha_{12} \neq -m_{12}$. Still we can factor an operator

$$D(L) := \begin{bmatrix} 1 & 0 \\ 0 & 1 \end{bmatrix} + \begin{bmatrix} 0 & \gamma \\ 0 & 0 \end{bmatrix} L = \begin{bmatrix} 1 & \gamma L \\ 0 & 1 \end{bmatrix}$$

from both operators without changing their general structure:

$$A(L) = D(L) \begin{bmatrix} 1 - \alpha_{11}L & -(\gamma + \alpha_{12})L \\ 0 & 1 \end{bmatrix},$$

$$M(L) = D(L) \begin{bmatrix} 1 + m_{11}L & (m_{12} - \gamma)L \\ 0 & 1 \end{bmatrix}.$$

Cancelling $D(L)$ gives operators

$$\begin{bmatrix} 1-\alpha_{11}L & -(\gamma+\alpha_{12})L \\ 0 & 1 \end{bmatrix} = D(L)\begin{bmatrix} 1+\alpha_{11}L & -(2\gamma+\alpha_{12})L \\ 0 & 1 \end{bmatrix}$$

and

$$\begin{bmatrix} 1+m_{11}L & (m_{12}-\gamma)L \\ 0 & 1 \end{bmatrix} = D(L)\begin{bmatrix} 1+m_{11}L & (m_{12}-2\gamma)L \\ 0 & 1 \end{bmatrix}.$$

Thus we can again factor and cancel $D(L)$. In fact, we can cancel $D(L)$ as often as we like without changing the general structure of the process. Hence, even if the orders of both operators cannot be reduced simultaneously by cancellation it may still be possible to factor some operator from both $A(L)$ and $M(L)$ which does not change their general structure. Note that the troubling operator $D(L)$ is again one with finite order inverse,

$$D(L)^{-1} = \begin{bmatrix} 1 & -\gamma L \\ 0 & 1 \end{bmatrix}.$$

Finite order operators that have a finite order inverse are characterized by the property that their determinant is a nonzero constant, that is, it does not involve L or powers of L. Operators with this property are called *unimodular*. For instance, the operator (7.1.9) has determinant,

$$\left| I_2 - \begin{bmatrix} 0 & \alpha \\ 0 & 0 \end{bmatrix} L \right| = \begin{vmatrix} 1 & -\alpha L \\ 0 & 1 \end{vmatrix} = 1$$

and hence it is unimodular. The property of a unimodular operator to have a finite order inverse follows because the inverse of an operator $A(L)$ is its adjoint divided by its determinant,

$$A(L)^{-1} = A(L)^*/|A(L)| = |A(L)|^{-1}A(L)^*.$$

The determinant is a univariate operator. A finite order invertible univariate operator, however, has an infinite order inverse unless its degree is zero, that is, unless it is a constant.

In order to state uniqueness conditions for a VARMA representation we will first of all require that a representation is chosen for which further cancellation is not possible in the sense that there are no common factors in the VAR and MA parts, except for unimodular operators. Operators $A(L)$ and $M(L)$ with this property are *left-coprime*. This property may be defined by calling the matrix operator $[A(L), M(L)]$ left-coprime if the existence of operators $D(L)$, $\bar{A}(L)$, and $\bar{M}(L)$ satisfying

$$D(L)[\bar{A}(L), \bar{M}(L)] = [A(L), M(L)] \tag{7.1.11}$$

implies that $D(L)$ is unimodular, that is, $|D(L)|$ is a nonzero constant. From the foregoing examples it should be understood that in general factoring unimodular operators from $A(L)$ and $M(L)$ is unavoidable if no further constraints are imposed. Thus, to obtain uniqueness of left-coprime operators we have to impose restrictions ensuring that the only feasible unimodular operator $D(L)$ in (7.1.11) is $D(L) = I_K$. We will now give two sets of conditions that ensure uniqueness of a VARMA representation.

7.1.2 Final Equations Form and Echelon Form

Suppose y_t is a stationary zero mean process that has a stable, invertible VARMA representation

$$A(L)y_t = M(L)u_t, \tag{7.1.12}$$

where

$$A(L) := A_0 - A_1 L - \cdots - A_p L^p \quad \text{and} \quad M(L) := M_0 + M_1 L + \cdots + M_q L^q.$$

Further suppose that $A(L)$ and $M(L)$ are left-coprime and the white noise covariance matrix Σ_u is nonsingular.

DEFINITION 7.1 (*Final Equations Form*)

The VARMA representation (7.1.12) is said to be in *final equations form* if $M_0 = I_K$ and $A(L) = \alpha(L)I_K$, where $\alpha(L) := 1 - \alpha_1 L - \cdots - \alpha_p L^p$ is a scalar (one-dimensional) operator with $\alpha_p \neq 0$. ■

For instance, the bivariate VARMA(3, 1) model

$$(1 - \alpha_1 L - \alpha_2 L^2 - \alpha_3 L^3)\begin{bmatrix} y_{1t} \\ y_{2t} \end{bmatrix} = \begin{bmatrix} 1 + m_{11,1}L & m_{12,1}L \\ m_{21,1}L & 1 + m_{22,1}L \end{bmatrix}\begin{bmatrix} u_{1t} \\ u_{2t} \end{bmatrix} \tag{7.1.13}$$

with $\alpha_3 \neq 0$, is in final equations form. The label "final equations form" for this type of VARMA representation will be motivated in Chapter 10.

Uniqueness of the final equations form

$$\alpha(L)y_t = M(L)u_t$$

is seen by noting that $D(L) = I_K$ is the only operator that retains the scalar AR part upon multiplication. For the operator $D(L)\alpha(L)I_K$ to maintain order p, the operator $D(L)$ must have degree zero, that is, $D(L) = D$. However, the only possible matrix D that guarantees a zero order matrix I_K for the VAR operator is $D = I_K$.

DEFINITION 7.2 (*Echelon Form*)

The VARMA representation (7.1.12) is said to be in *echelon form* if the VAR and MA operators $A(L) = [\alpha_{ki}(L)]_{k,i=1,\ldots,K}$ and $M(L) = [m_{ki}(L)]_{k,i}$ satisfy the following conditions: All operators $\alpha_{ki}(L)$ and $m_{kj}(L)$ in the k-th row of $A(L)$ and $M(L)$ have the same degree p_k and have the form

$$\alpha_{kk}(L) = 1 - \sum_{j=1}^{p_k} \alpha_{kk,j} L^j, \qquad \text{for } k = 1, \ldots, K,$$

$$\alpha_{ki}(L) = - \sum_{j=p_k-p_{ki}+1}^{p_k} \alpha_{ki,j} L^j, \qquad \text{for } k \neq i,$$

and

$$m_{ki}(L) = \sum_{j=0}^{p_k} m_{ki,j} L^j \qquad \text{for } k, i = 1, \ldots, K, \qquad \text{with } M_0 = A_0.$$

In the VAR operators $\alpha_{ki}(L)$,

$$p_{ki} := \begin{cases} \min(p_k + 1, p_i) & \text{for } k \geq i \\ \min(p_k, p_i) & \text{for } k < i \end{cases} \qquad k, i = 1, \ldots, K. \tag{7.1.14}$$

That is, p_{ki} specifies the number of free coefficients in the operator $\alpha_{ki}(L)$ for $i \neq k$. The row orders $(p_1, \ldots, p_K)$ are the *Kronecker indices* and their sum $\sum_{k=1}^{K} p_i$ is the *McMillan degree*. Obviously, for the VARMA orders we have, in general, $p = q = \max(p_1, \ldots, p_K)$. ■

The following model is an example of a bivariate VARMA process in echelon form:

$$\begin{bmatrix} 1 - \alpha_{11,1}L - \alpha_{11,2}L^2 & -\alpha_{12,2}L^2 \\ -\alpha_{21,0} - \alpha_{21,1}L & 1 - \alpha_{22,1}L \end{bmatrix} \begin{bmatrix} y_{1t} \\ y_{2t} \end{bmatrix}$$

$$= \begin{bmatrix} 1 + m_{11,1}L + m_{11,2}L^2 & m_{12,1}L + m_{12,2}L^2 \\ -\alpha_{21,0} + m_{21,1}L & 1 + m_{22,1}L \end{bmatrix} \begin{bmatrix} u_{1t} \\ u_{2t} \end{bmatrix} \tag{7.1.15a}$$

or

$$\begin{bmatrix} 1 & 0 \\ -\alpha_{21,0} & 1 \end{bmatrix} \begin{bmatrix} y_{1,t} \\ y_{2,t} \end{bmatrix} = \begin{bmatrix} \alpha_{11,1} & 0 \\ \alpha_{21,1} & \alpha_{22,1} \end{bmatrix} \begin{bmatrix} y_{1,t-1} \\ y_{2,t-1} \end{bmatrix} + \begin{bmatrix} \alpha_{11,2} & \alpha_{12,2} \\ 0 & 0 \end{bmatrix} \begin{bmatrix} y_{1,t-2} \\ y_{2,t-2} \end{bmatrix}$$

$$+ \begin{bmatrix} 1 & 0 \\ -\alpha_{21,0} & 1 \end{bmatrix} \begin{bmatrix} u_{1,t} \\ u_{2,t} \end{bmatrix} + \begin{bmatrix} m_{11,1} & m_{12,1} \\ m_{21,1} & m_{22,1} \end{bmatrix} \begin{bmatrix} u_{1,t-1} \\ u_{2,t-1} \end{bmatrix}$$

$$+ \begin{bmatrix} m_{11,2} & m_{12,2} \\ 0 & 0 \end{bmatrix} \begin{bmatrix} u_{1,t-2} \\ u_{2,t-2} \end{bmatrix}. \tag{7.1.15b}$$

In this model the Kronecker indices (row degrees) are $p_1 = 2$ and $p_2 = 1$. Thus, the McMillan degree is 3. The p_{ki} numbers are

$$\begin{bmatrix} p_{11} & p_{12} \\ p_{21} & p_{22} \end{bmatrix} = \begin{bmatrix} 2 & 1 \\ 2 & 1 \end{bmatrix}$$

(see (7.1.14)). The off-diagonal elements p_{12} and p_{21} of this matrix indicate the numbers of parameters contained in the operators $\alpha_{12}(L)$ and $\alpha_{21}(L)$, respectively. Since $\alpha_{12}(L)$ belongs to the first row or first equation of the system it has degree $p_1 = 2$. Hence, since it has just one free coefficient ($p_{12} = 1$) it must have the form $\alpha_{12}(L) = -\alpha_{12,2}L^2$. Similarly, $\alpha_{21}(L)$ belongs to the second row of the system and thus it has degree $p_2 = 1$. Since it has $p_{21} = 2$ free coefficients it must be of the form $\alpha_{21}(L) = -\alpha_{21,0} - \alpha_{21,1}L$. Another characteristic feature of the echelon form is that A_0 is lower-triangular and has ones on the main diagonal. Moreover, the zero order MA coefficient matrix is identical to the zero order VAR matrix, $M_0 = A_0$.

Some free coefficients of the echelon form of a VARMA model may be zero and hence p or q may be less than $\max(p_1, \ldots, p_K)$. For instance, in the example process (7.1.15), $m_{11,2}$ and $m_{12,2}$ may be zero. In that case $q = 1 < \max(p_1, p_2) = 2$. In order for a representation to be an echelon form with Kronecker indices $(p_1, \ldots, p_K)$ at least one operator in the k-th row of $[A(L), M(L)]$ must have degree p_k, with nonzero coefficient at lag p_k.

An echelon is a certain positioning of an army in the form of steps. Similarly, the nonzero parameters in an echelon VARMA representation are positioned in a specific way. In particular the positioning of freely varying parameters in the k-th equation depends only on Kronecker indices $p_i \leq p_k$ and not on Kronecker indices $p_j > p_k$. More precisely, if $p_j > p_k$, the positioning of the free parameters in the k-th equation will be the same for any value p_j. For the example process (7.1.15) it is easy to check that the positions of the free parameters in the second equation will remain the same if the row degree of the first equation is increased to $p_1 = 3$. In other words, p_{21} does not change.

It can be shown that the echelon form just like the final equations form guarantees uniqueness of the VARMA representation. In other words, if a VARMA representation is in echelon form the representation is unique within the class of all echelon representations. A similar statement applies for the final equations form. Also, for any stable, invertible VARMA(p, q) representation there exists an equivalent echelon form and an equivalent final equations form.

The reader may wonder why we consider the complicated looking echelon representation although the final equations form serves the same purpose. The reason is that the echelon form is usually preferable in practice because it often involves fewer free parameters than the equivalent final equations form. We will see an example of this phenomenon shortly. Having as few free parameters as possible is important to ease the numerical problems in maximizing the likelihood function and to gain efficiency of the parameter estimators.

There are a number of other unique or *identified* parameterizations of VARMA models. We have chosen to present the final equations form and the echelon form because these two forms will play a role when we discuss the issue of specifying VARMA models in Chapter 8. For proofs of the uniqueness of the echelon form and for other identification conditions we refer to Hannan (1969, 1970, 1976, 1979), Deistler & Hannan (1981) and Hannan & Deistler (1988). We now proceed with illustrations of the final equations form and the echelon form.

7.1.3 Illustrations

Starting from some VARMA(p, q) representation $A(L)y_t = M(L)u_t$ one strategy for finding the corresponding final equations form results from premultiplying with the adjoint $A(L)^*$ of the VAR operator $A(L)$ which gives

$$|A(L)|y_t = A(L)^*M(L)u_t, \tag{7.1.16}$$

where $A(L)^*A(L) = |A(L)|$ has been used. Obviously, (7.1.16) has a scalar VAR operator and hence is in final equations form if all superfluous terms are cancelled.

In order to find the echelon form corresponding to a given VARMA model we have to cancel as much as possible so as to make the VAR and MA operators left-coprime. Then a unimodular matrix operator has to be determined which upon left-multiplication transforms the given model into echelon form. It usually helps to determine the Kronecker indices (row degrees) and the corresponding numbers p_{ki} first. We will now consider examples.

Let us begin with the simple bivariate process

$$\left[I_2 - \begin{bmatrix} 0 & \alpha \\ 0 & 0 \end{bmatrix} L\right] y_t = u_t \tag{7.1.17}$$

with $\alpha \neq 0$. Noting that

$$|A(L)| = \begin{vmatrix} 1 & -\alpha L \\ 0 & 1 \end{vmatrix} = 1 \qquad \text{and} \qquad A(L)^* = \begin{bmatrix} 1 & \alpha L \\ 0 & 1 \end{bmatrix},$$

the final equations form is seen to be

$$y_t = \left[I_2 + \begin{bmatrix} 0 & \alpha \\ 0 & 0 \end{bmatrix} L\right] u_t. \tag{7.1.18}$$

To find the echelon representation we first determine the Kronecker indices or row degrees and the implied p_{ki} from Definition 7.2. The first row of (7.1.17) has degree $p_1 = 1$ and the second row has degree $p_2 = 0$. Hence,

$$p_{11} = 1, \qquad p_{12} = 0, \qquad p_{21} = 1, \qquad p_{22} = 0,$$

so that

$$\alpha_{11}(L) = 1 - \alpha_{11,1}L, \quad \alpha_{12}(L) = 0, \quad \alpha_{21}(L) = -\alpha_{21,0}, \quad \text{and} \quad \alpha_{22}(L) = 1.$$

Thus, the echelon form is

$$\begin{bmatrix} 1 - \alpha_{11,1}L & 0 \\ -\alpha_{21,0} & 1 \end{bmatrix} y_t = \begin{bmatrix} 1 + m_{11,1}L & m_{12,1}L \\ -\alpha_{21,0} & 1 \end{bmatrix} u_t. \tag{7.1.19}$$

The unique parameter values in this representation corresponding to the specific process (7.1.17) are easily seen to be

$$\alpha_{11,1} = \alpha_{21,0} = m_{11,1} = 0 \qquad \text{and} \qquad m_{12,1} = \alpha.$$

Thus, in this particular case, the final equations form and the echelon form coincide.

As another example we consider a 3-dimensional process with VARMA(2, 1) representation

$$\begin{bmatrix} 1-\theta_1 L & -\theta_2 L & 0 \\ 0 & 1-\theta_3 L-\theta_4 L^2 & -\theta_5 L \\ 0 & 0 & 1 \end{bmatrix} y_t = \begin{bmatrix} 1-\eta_1 L & 0 & 0 \\ 0 & 1-\eta_2 L & 0 \\ 0 & 0 & 1-\eta_3 L \end{bmatrix} u_t. \tag{7.1.20}$$

Using (7.1.16), its final equations form is

$$\begin{aligned}(1-\theta_1 L)&(1-\theta_3 L-\theta_4 L^2)y_t \\ &= \begin{bmatrix} 1-\theta_3 L-\theta_4 L^2 & \theta_2 L & \theta_2\theta_5 L^2 \\ 0 & 1-\theta_1 L & \theta_5 L-\theta_1\theta_5 L^2 \\ 0 & 0 & (1-\theta_1 L)(1-\theta_3 L-\theta_4 L^2) \end{bmatrix} \\ &\quad\times \begin{bmatrix} 1-\eta_1 L & 0 & 0 \\ 0 & 1-\eta_2 L & 0 \\ 0 & 0 & 1-\eta_3 L \end{bmatrix} u_t\end{aligned}$$

which is easily recognizable as a VARMA(3, 4) structure with scalar VAR operator.

The Kronecker indices, that is, the row degrees of (7.1.20) are $(p_1, p_2, p_3) = (1, 2, 1)$ and the implied p_{ki}-numbers from (7.1.14) are collected in the following matrix

$$[p_{ki}]_{k,i=1,2,3} = \begin{bmatrix} 1 & 1 & 1 \\ 1 & 2 & 1 \\ 1 & 2 & 1 \end{bmatrix}.$$

Consequently, the VAR operator of the echelon form becomes

$$\begin{bmatrix} 1-\alpha_{11,1}L & -\alpha_{12,1}L & -\alpha_{13,1}L \\ -\alpha_{21,2}L^2 & 1-\alpha_{22,1}L-\alpha_{22,2}L^2 & -\alpha_{23,2}L^2 \\ -\alpha_{31,1}L & -\alpha_{32,0}-\alpha_{32,1}L & 1-\alpha_{33,1}L \end{bmatrix}. \tag{7.1.21}$$

Hence, in the echelon representation,

$$A_0 = \begin{bmatrix} 1 & 0 & 0 \\ 0 & 1 & 0 \\ 0 & -\alpha_{32,0} & 1 \end{bmatrix}$$

is different from I_3, if $\alpha_{32,0} \neq 0$, and thus $M_0 = A_0$ is also not the identity matrix. The MA operator is

$$\begin{bmatrix} 1+m_{11,1}L & m_{12,1}L & m_{13,1}L \\ m_{21,1}L+m_{21,2}L^2 & 1+m_{22,1}L+m_{22,2}L^2 & m_{23,1}L+m_{23,2}L^2 \\ m_{31,1}L & -\alpha_{32,0}+m_{32,1}L & 1+m_{33,1}L \end{bmatrix}. \tag{7.1.22}$$

The reader may be puzzled by the fact that the last element in the second row of

(7.1.21) does not involve a term with first power of L while such a term appears in (7.1.20). This simply means that there is a VARMA representation equivalent to (7.1.20) with the second but not the first power of L in the last operator in the second row of $A(L)$. The fact, that there always exists an equivalent echelon representation does not mean there is always an immediately obvious relation between the coefficients of any given VARMA representation and its equivalent echelon form. However, in the present case it is fairly easy to relate the representations (7.1.20) and (7.1.21)/(7.1.22). Just left-multiply (7.1.20) by the operator

$$\begin{bmatrix} 1 & 0 & 0 \\ 0 & 1 & \theta_5 L \\ 0 & 0 & 1 \end{bmatrix}. \tag{7.1.23}$$

The result is a VAR operator

$$\begin{bmatrix} 1-\theta_1 L & -\theta_2 L & 0 \\ 0 & 1-\theta_3 L-\theta_4 L^2 & 0 \\ 0 & 0 & 1 \end{bmatrix}$$

and the MA operator changes accordingly. Notice that the operator (7.1.23) has constant determinant and, of course, the resulting VARMA model is equivalent to (7.1.20). The relation between its coefficients and those of the echelon representation (7.1.21)/(7.1.22) is obvious:

$$\alpha_{11,1}=\theta_1, \qquad \alpha_{12,1}=\theta_2, \qquad \alpha_{13,1}=0,$$

$$\alpha_{21,2}=0, \qquad \alpha_{22,1}=\theta_3, \qquad \alpha_{22,2}=\theta_4, \qquad \alpha_{23,2}=0,$$

$$\alpha_{31,1}=\alpha_{32,0}=\alpha_{32,1}=\alpha_{33,1}=0,$$

and the relation between (7.1.22) and the coefficients of (7.1.20) is also apparent. Of course, if the zero coefficients are known this knowledge may be used to reduce the number of free coefficients in the echelon form.

In this example the unrestricted final equations form has 3 AR coefficients and 36 MA coefficients. Thus, the unrestricted form contains 39 parameters apart from white noise covariance coefficients. In contrast, the unrestricted echelon form (7.1.21)/(7.1.22) has only 23 free parameters and is therefore preferable in terms of parameter parsimony. Note that in practice the true coefficient values are unknown and we pick an identified structure, for example, a final equations form or an echelon form. At that stage further parameter restrictions may not be available. Hence, if (7.1.20) is the actual data generation process we may pick a VARMA(3, 4) model with scalar AR operator if we decide to go with a final equations representation and we may choose the model (7.1.21)/(7.1.22) if we decide on an echelon form representation. Obviously, the latter choice results in a more parsimonious parameterization. As mentioned earlier, for estimation purposes the more parsimonious representation is advantageous.

Although $A_0 \neq I$ in the previous example it should be understood that in many echelon representations $A_0 = M_0 = I_K$. In particular if the row degrees $p_1 = \ldots = p_K = p$, all $p_{ki} = p$, $i, k = 1, \ldots, K$, and the echelon form is easily seen to be a standard VARMA(p, p) model with $A_0 = M_0 = I_K$. We are now ready to turn to the actual estimation of the parameters of an identified VARMA model and we shall discuss its Gaussian likelihood function next.

7.2 The Gaussian Likelihood Function

For maximum likelihood (ML) estimation the likelihood function is needed. We will now derive useful approximations to the likelihood function of a Gaussian VARMA(p, q) process. Special case MA processes will be considered first.

7.2.1 The Likelihood Function of an MA(1) Process

Since a zero mean MA(1) process is the simplest member of the finite order MA family we use that as a starting point. Hence, we assume to have a sample $y_1, \ldots, y_T$ which is generated by the Gaussian, K-dimensional, invertible MA(1) process

$$y_t = u_t + M_1 u_{t-1}, \tag{7.2.1}$$

where u_t is a Gaussian white noise process with covariance matrix Σ_u. Thus,

$$\mathbf{y} := \begin{bmatrix} y_1 \\ \vdots \\ y_T \end{bmatrix} = \overline{\mathfrak{M}}_1 \begin{bmatrix} u_0 \\ u_1 \\ \vdots \\ u_T \end{bmatrix},$$

where

$$\overline{\mathfrak{M}}_1 := \begin{bmatrix} M_1 & I_K & 0 & \ldots & 0 & 0 \\ 0 & M_1 & I_K & \ddots & 0 & 0 \\ \vdots & & \ddots & \ddots & \ddots & \vdots \\ 0 & 0 & 0 & \ldots & M_1 & I_K \end{bmatrix} \tag{7.2.2}$$

is a $(KT \times K(T+1))$ matrix. Using that u_t is Gaussian white noise and thus

$$\begin{bmatrix} u_0 \\ u_1 \\ \vdots \\ u_T \end{bmatrix} \sim N(0, I_{T+1} \otimes \Sigma_u),$$

it follows that

$$\mathbf{y} \sim N(0, \overline{\mathfrak{M}}_1 (I_{T+1} \otimes \Sigma_u) \overline{\mathfrak{M}}_1')$$

and the likelihood function is seen to be

$$l(M_1, \Sigma_u | \mathbf{y}) \propto |\overline{\mathfrak{M}}_1 (I_{T+1} \otimes \Sigma_u) \overline{\mathfrak{M}}_1'|^{-1/2} \exp\{-\tfrac{1}{2}\mathbf{y}'[\overline{\mathfrak{M}}_1 (I_{T+1} \otimes \Sigma_u) \overline{\mathfrak{M}}_1']^{-1}\mathbf{y}\}, \tag{7.2.3}$$

where $\propto$ stands for "is proportional to", that is, we have dropped a multiplicative constant from the likelihood function which does not change the maximizing values of M_1 and Σ_u.

It is inconvenient that this function involves the determinant and the inverse of a $(KT \times KT)$ matrix. A simpler form is obtained if u_0 is set at zero, that is, the MA(1) process is assumed to be started up with a nonrandom fixed vector $u_0 = 0$. In that case

$$\mathbf{y} = \mathfrak{M}_1 \mathbf{u},$$

where

$$\underset{(KT \times KT)}{\mathfrak{M}_1 := \begin{bmatrix} I_K & 0 & \dots & 0 & 0 \\ M_1 & I_K & \ddots & 0 & 0 \\ & \ddots & \ddots & & \vdots \\ 0 & 0 & \dots & M_1 & I_K \end{bmatrix}} \quad \text{and} \quad \underset{(KT \times 1)}{\mathbf{u} := \begin{bmatrix} u_1 \\ \vdots \\ u_T \end{bmatrix}}. \tag{7.2.4}$$

The likelihood function is then proportional to

$$\begin{aligned} l_0(M_1, \Sigma_u | \mathbf{y}) &= |\mathfrak{M}_1 (I_T \otimes \Sigma_u) \mathfrak{M}_1'|^{-1/2} \exp\{-\tfrac{1}{2}\mathbf{y}'[\mathfrak{M}_1 (I_T \otimes \Sigma_u) \mathfrak{M}_1']^{-1}\mathbf{y}\} \\ &= |\Sigma_u|^{-T/2} \exp\{-\tfrac{1}{2}\mathbf{y}'\mathfrak{M}_1'^{-1}(I_T \otimes \Sigma_u^{-1})\mathfrak{M}_1^{-1}\mathbf{y}\} \\ &= |\Sigma_u|^{-T/2} \exp\left\{-\frac{1}{2}\sum_{t=1}^{T} u_t' \Sigma_u^{-1} u_t\right\}, \end{aligned} \tag{7.2.5}$$

where it has been used that $|\mathfrak{M}_1| = 1$ and

$$\begin{aligned} \mathfrak{M}_1^{-1} &= \begin{bmatrix} I_K & 0 & \dots & 0 & 0 \\ -M_1 & I_K & \ddots & 0 & 0 \\ (-M_1)^2 & -M_1 & \ddots & 0 & 0 \\ \vdots & & \ddots & \ddots & \vdots \\ (-M_1)^{T-1} & (-M_1)^{T-2} & \dots & -M_1 & I_K \end{bmatrix} \\ &= \begin{bmatrix} I_K & 0 & \dots & 0 \\ -\Pi_1 & I_K & & 0 \\ \vdots & & \ddots & \vdots \\ -\Pi_{T-1} & -\Pi_{T-2} & \dots & I_K \end{bmatrix}, \end{aligned}$$

where the $\Pi_i = -(-M_1)^i$ are the coefficients of the pure VAR representation of the process. Moreover, by successive substitution the MA(1) process in (7.2.1) can be rewritten as

$$y_t + \sum_{i=1}^{t-1} (-M_1)^i y_{t-i} + (-M_1)^t u_0 = u_t. \tag{7.2.6}$$

Thus if $u_0 = 0$,

$$u_t = y_t + \sum_{i=1}^{t-1} (-M_1)^i y_{t-i}$$

and the last expression in (7.2.5) is obtained.

The equation (7.2.6) also shows that for large t the assumption regarding u_0 becomes inconsequential because, for an invertible process, M_1^t approaches zero as $t \to \infty$. The impact of u_0 disappears more rapidly for processes for which M_1^t goes to zero more rapidly as t gets large. In other words, if all eigenvalues of M_1 are close to zero or, equivalently, all roots of $\det(I_K + M_1 z)$ are far outside the unit circle the impact of u_0 is lower than for processes with roots close to the unit circle. In summary, the likelihood approximation in (7.2.5) will improve as the sample size gets large and will become exact as $T \to \infty$. In small samples it is better for processes with roots of $\det(I_K + M_1 z)$ far away from the unit circle than for those with roots close to the noninvertibility region. Since we will be concerned predominantly with large sample properties in the following we will indulge in the luxury of working with likelihood approximations such as l_0 in (7.2.5).

7.2.2 The MA(q) Case

A similar reasoning as for MA(1) processes can also be employed for higher order MA processes. Suppose the generation process of y_t has a zero mean MA(q) representation

$$y_t = u_t + M_1 u_{t-1} + \cdots + M_q u_{t-q}. \tag{7.2.7}$$

Then

$$\mathbf{y} = \overline{\mathfrak{M}}_q \begin{bmatrix} u_{-q+1} \\ \vdots \\ u_0 \\ u_1 \\ \vdots \\ u_T \end{bmatrix},$$

where

$$\overline{\mathfrak{M}}_q := \begin{bmatrix} M_q & M_{q-1} & \dots & M_1 & I_K & 0 & \dots & & 0 \\ 0 & M_q & \dots & M_2 & M_1 & I_K & \ddots & & 0 \\ \vdots & & \ddots & & \ddots & \ddots & \ddots & & \\ 0 & 0 & \dots & & M_q & \dots & M_2 & M_1 & I_K \end{bmatrix} \tag{7.2.8}$$

is a $(KT \times K(T+q))$ matrix and the exact likelihood for a sample of size T becomes

$$l(M_1, \dots, M_q, \Sigma_u | \mathbf{y}) \propto |\overline{\mathfrak{M}}_q (I_{T+q} \otimes \Sigma_u) \overline{\mathfrak{M}}_q'|^{-1/2} \times \exp\{-\tfrac{1}{2}\mathbf{y}'[\overline{\mathfrak{M}}_q (I_{T+q} \otimes \Sigma_u) \overline{\mathfrak{M}}_q']^{-1}\mathbf{y}\}. \tag{7.2.9}$$

Again a convenient approximation to the likelihood function is obtained by setting $u_{-q+1} = \cdots = u_0 = 0$. In that case the likelihood is, apart from a multiplicative constant,

$$l_0(M_1, \ldots, M_q, \Sigma_u | \mathbf{y}) = |\Sigma_u|^{-T/2} \exp\{-\tfrac{1}{2}\mathbf{y}'[\mathfrak{M}_q'^{-1}(I_T \otimes \Sigma_u^{-1})\mathfrak{M}_q^{-1}]\mathbf{y}\}, \quad (7.2.10)$$

where

$$\mathfrak{M}_q := \begin{bmatrix} I_K & 0 & \ldots & 0 & 0 \\ M_1 & I_K & & 0 & 0 \\ M_2 & M_1 & \ddots & 0 & 0 \\ \vdots & \vdots & \ddots & \ddots & \vdots \\ M_q & M_{q-1} & & \ddots & \\ 0 & M_q & & \ddots & \\ \vdots & & \ddots & & \ddots \\ 0 & 0 & \ldots & M_q & \ldots & M_1 & I_K \end{bmatrix} \quad (7.2.11)$$

and, hence,

$$\mathfrak{M}_q^{-1} = \begin{bmatrix} I_K & 0 & \ldots & 0 \\ -\Pi_1 & I_K & & 0 \\ \vdots & \vdots & \ddots & \vdots \\ -\Pi_{T-1} & -\Pi_{T-2} & \ldots & I_K \end{bmatrix}.$$

Here the Π_i are the coefficient matrices of the pure VAR representation of the process y_t. Thus, the Π_i can be computed recursively as in Section 6.2 of Chapter 6.

An alternative expression for the approximate likelihood is easily seen to be

$$l_0(M_1, \ldots, M_q, \Sigma_u | \mathbf{y}) = |\Sigma_u|^{-T/2} \exp\left\{-\frac{1}{2}\sum_{t=1}^{T} u_t'\Sigma_u^{-1}u_t\right\}, \quad (7.2.12)$$

where

$$u_t = y_t - \sum_{i=1}^{t-1} \Pi_i y_{t-i}.$$

Again the likelihood approximation will be quite precise if T is reasonably large and the roots of $\det(I_K + M_1 z + \cdots + M_q z^q)$ are not close to the unit circle.

Although we will work with likelihood approximations in the following it is perhaps worth noting that an expression for the exact likelihood of an MA(q) process can be derived that is more manageable than the one in (7.2.9) (see, e.g., Hillmer & Tiao (1979), Kohn (1981)).

7.2.3 The VARMA(1, 1) Case

Before we tackle general mixed VARMA models we shall consider the simplest candidate, namely a Gaussian zero mean, stationary, stable, and invertible VARMA(1, 1) process

$$y_t = A_1 y_{t-1} + u_t + M_1 u_{t-1}. \tag{7.2.13}$$

Assuming we have a sample $y_1, \ldots, y_T$ generated by this process and defining

$$\mathfrak{A}_p := \begin{bmatrix} I_K & 0 & \cdots & & 0 & 0 \\ -A_1 & I_K & & & 0 & 0 \\ -A_2 & -A_1 & \ddots & & 0 & 0 \\ \vdots & \vdots & \ddots & \ddots & \vdots & \vdots \\ -A_p & -A_{p-1} & & \ddots & 0 & 0 \\ 0 & -A_p & & & 0 & 0 \\ \vdots & & \ddots & & \vdots & \vdots \\ 0 & 0 & & \ddots & I_K & 0 \\ 0 & 0 & \cdots & -A_p \quad \cdots & -A_1 & I_K \end{bmatrix} \tag{7.2.14}$$

we get

$$\mathfrak{A}_1 \begin{bmatrix} y_1 \\ \vdots \\ y_T \end{bmatrix} + \begin{bmatrix} -A_1 y_0 \\ 0 \\ \vdots \\ 0 \end{bmatrix} = \overline{\mathfrak{M}}_1 \begin{bmatrix} u_0 \\ u_1 \\ \vdots \\ u_T \end{bmatrix}.$$

Hence, for given, fixed presample values y_0,

$$\mathbf{y} = \begin{bmatrix} y_1 \\ \vdots \\ y_T \end{bmatrix} \sim N(\mathfrak{A}_1^{-1}\mathbf{y}_0, \mathfrak{A}_1^{-1}\overline{\mathfrak{M}}_1(I_{T+1} \otimes \Sigma_u)\overline{\mathfrak{M}}_1' \mathfrak{A}_1'^{-1}), \tag{7.2.15}$$

where

$$\mathbf{y}_0 := \begin{bmatrix} A_1 y_0 \\ 0 \\ \vdots \\ 0 \end{bmatrix}.$$

The corresponding likelihood function conditional on y_0 is

$$\begin{aligned} &l(A_1, M_1, \Sigma_u | \mathbf{y}, y_0) \\ &\quad \propto |\mathfrak{A}_1^{-1}\overline{\mathfrak{M}}_1(I_{T+1} \otimes \Sigma_u)\overline{\mathfrak{M}}_1' \mathfrak{A}_1'^{-1}|^{-1/2} \\ &\qquad \times \exp\{-\tfrac{1}{2}(\mathbf{y} - \mathfrak{A}_1^{-1}\mathbf{y}_0)'\mathfrak{A}_1'[\overline{\mathfrak{M}}_1(I_{T+1} \otimes \Sigma_u)\overline{\mathfrak{M}}_1']^{-1}\mathfrak{A}_1(\mathbf{y} - \mathfrak{A}_1^{-1}\mathbf{y}_0)\} \\ &\quad = |\overline{\mathfrak{M}}_1(I_{T+1} \otimes \Sigma_u)\overline{\mathfrak{M}}_1'|^{-1/2} \exp\{-\tfrac{1}{2}(\mathfrak{A}_1\mathbf{y} - \mathbf{y}_0)'[\overline{\mathfrak{M}}_1(I_{T+1} \otimes \Sigma_u)\overline{\mathfrak{M}}_1']^{-1} \\ &\qquad\qquad \times (\mathfrak{A}_1\mathbf{y} - \mathbf{y}_0)\}, \end{aligned} \tag{7.2.16}$$

where $|\mathfrak{A}_1| = 1$ has been used.

With the same arguments as in the pure MA case a simple approximation is obtained by setting $u_0 = y_0 = 0$. Then we get

$$l_0(A_1, M_1, \Sigma_u) = |\Sigma_u|^{-T/2} \exp\{-\tfrac{1}{2}(\mathfrak{M}_1^{-1}\mathfrak{A}_1\mathbf{y})'(I_T \otimes \Sigma_u^{-1})\mathfrak{M}_1^{-1}\mathfrak{A}_1\mathbf{y}\}$$
$$= |\Sigma_u|^{-T/2} \exp\left\{-\frac{1}{2}\sum_{t=1}^{T} u_t'\Sigma_u^{-1}u_t\right\}, \tag{7.2.17}$$

where

$$u_t = y_t - \sum_{i=1}^{t-1} \Pi_i y_{t-i} \tag{7.2.18}$$

and the Π_i are the coefficient matrices of the pure VAR representation, that is, for the present case $\Pi_i = (-1)^{i-1}(M_1^i + M_1^{i-1}A_1)$, $i = 1, 2, \ldots$ (see Section 6.3.1). Note that in writing the likelihood approximation l_0 we have dropped the conditions $\mathbf{y}$ and y_0 for notational simplicity.

The effect of starting up the process with $y_0 = u_0 = 0$ is quite easily seen in (7.2.18), namely, the infinite order pure VAR representation is truncated at lag $t - 1$ for y_t. Such a truncation has little effect if the sample size is large and the roots of the MA operator are not close to the unit circle.

7.2.4 The General VARMA(p, q) Case

Now suppose a sample $y_1, \ldots, y_T$ is generated by the Gaussian K-dimensional, stable, invertible VARMA(p, q) process

$$A_0(y_t - \mu) = A_1(y_{t-1} - \mu) + \cdots + A_p(y_{t-p} - \mu)$$
$$+ A_0 u_t + M_1 u_{t-1} + \cdots + M_q u_{t-q} \tag{7.2.19}$$

with mean vector μ and nonsingular white noise covariance matrix Σ_u. Notice that A_0 appears as the coefficient matrix of y_t and of u_t as in the echelon form. Thus, the echelon form is covered by our treatment of the general VARMA(p, q) case. We have chosen the mean-adjusted form of the process because this form has certain advantages in ML estimation as we will see later.

Usually some elements of the coefficient matrices will be zero or obey some other type of restrictions. Therefore, to be realistic, we define

$$\alpha_0 := \operatorname{vec}(A_0) \qquad \text{and} \qquad \boldsymbol{\beta} := \operatorname{vec}[A_1, \ldots, A_p, M_1, \ldots, M_q] \tag{7.2.20}$$

and assume that these coefficients are linearly related to an $(N \times 1)$ parameter vector γ, that is,

$$\begin{bmatrix} \alpha_0 \\ \boldsymbol{\beta} \end{bmatrix} = R\gamma + r \tag{7.2.21}$$

for a suitable, known $(K^2(p + q + 1) \times N)$ matrix R and a known $K^2(p + q + 1)$-vector r. For example, for a bivariate process with echelon representation

$$\begin{bmatrix} 1 - \alpha_{11,1}L & 0 \\ -\alpha_{21,0} & 1 \end{bmatrix}(y_t - \mu) = \begin{bmatrix} 1 + m_{11,1}L & m_{12,1}L \\ -\alpha_{21,0} & 1 \end{bmatrix}u_t$$

or

$$\begin{bmatrix} 1 & 0 \\ -\alpha_{21,0} & 1 \end{bmatrix}(y_t - \mu) = \begin{bmatrix} \alpha_{11,1} & 0 \\ 0 & 0 \end{bmatrix}(y_{t-1} - \mu) + \begin{bmatrix} 1 & 0 \\ -\alpha_{21,0} & 1 \end{bmatrix} u_t + \begin{bmatrix} m_{11,1} & m_{12,1} \\ 0 & 0 \end{bmatrix} u_{t-1},$$

that is, the Kronecker indices are $p_1 = 1$ and $p_2 = 0$, we have

$$\alpha_0 = \begin{bmatrix} 1 \\ -\alpha_{21,0} \\ 0 \\ 1 \end{bmatrix}, \quad \boldsymbol{\beta} = \begin{bmatrix} \alpha_{11,1} \\ 0 \\ 0 \\ 0 \\ m_{11,1} \\ 0 \\ m_{12,1} \\ 0 \end{bmatrix}, \quad R = \begin{bmatrix} 0 & 0 & 0 & 0 \\ -1 & 0 & 0 & 0 \\ 0 & 0 & 0 & 0 \\ 0 & 0 & 0 & 0 \\ 0 & 1 & 0 & 0 \\ 0 & 0 & 0 & 0 \\ 0 & 0 & 0 & 0 \\ 0 & 0 & 0 & 0 \\ 0 & 0 & 1 & 0 \\ 0 & 0 & 0 & 0 \\ 0 & 0 & 0 & 1 \\ 0 & 0 & 0 & 0 \end{bmatrix}, \quad \gamma = \begin{bmatrix} \alpha_{21,0} \\ \alpha_{11,1} \\ m_{11,1} \\ m_{12,1} \end{bmatrix}, \quad \text{and} \quad r = \begin{bmatrix} 1 \\ 0 \\ 0 \\ 1 \\ 0 \\ 0 \\ \vdots \\ 0 \end{bmatrix}.$$

Similarly, for the final equations form

$$(1 - \alpha_1 L)(y_t - \mu) = \begin{bmatrix} 1 + m_{11}L & m_{12}L \\ m_{21}L & 1 + m_{22}L \end{bmatrix} u_t$$

or

$$\begin{bmatrix} 1 & 0 \\ 0 & 1 \end{bmatrix}(y_t - \mu) = \begin{bmatrix} \alpha_1 & 0 \\ 0 & \alpha_1 \end{bmatrix}(y_{t-1} - \mu) + \begin{bmatrix} 1 & 0 \\ 0 & 1 \end{bmatrix} u_t + \begin{bmatrix} m_{11} & m_{12} \\ m_{21} & m_{22} \end{bmatrix} u_{t-1}$$

we get

$$\alpha_0 = \begin{bmatrix} 1 \\ 0 \\ 0 \\ 1 \end{bmatrix}, \quad \boldsymbol{\beta} = \begin{bmatrix} \alpha_1 \\ 0 \\ 0 \\ \alpha_1 \\ m_{11} \\ m_{21} \\ m_{12} \\ m_{22} \end{bmatrix}, \quad R = \begin{bmatrix} 0 & 0 & 0 & 0 & 0 \\ 0 & 0 & 0 & 0 & 0 \\ 0 & 0 & 0 & 0 & 0 \\ 0 & 0 & 0 & 0 & 0 \\ 1 & 0 & 0 & 0 & 0 \\ 0 & 0 & 0 & 0 & 0 \\ 0 & 0 & 0 & 0 & 0 \\ 1 & 0 & 0 & 0 & 0 \\ 0 & 1 & 0 & 0 & 0 \\ 0 & 0 & 1 & 0 & 0 \\ 0 & 0 & 0 & 1 & 0 \\ 0 & 0 & 0 & 0 & 1 \end{bmatrix},$$

$$\gamma = \begin{bmatrix} \alpha_1 \\ m_{11} \\ m_{21} \\ m_{12} \\ m_{22} \end{bmatrix}, \quad \text{and} \quad r = \begin{bmatrix} 1 \\ 0 \\ 0 \\ 1 \\ 0 \\ \vdots \\ 0 \end{bmatrix}.$$

The likelihood function is a function of μ, γ, and Σ_u. Its exact form, given fixed initial values $y_{-p+1}, \ldots, y_0$, can be derived analogously to the previously considered special cases (see Problem 7.4 and Hillmer & Tiao (1979)). Here we will just give the likelihood approximation obtained by assuming

$$y_{-p+1} - \mu = \ldots = y_0 - \mu = u_{-q+1} = \ldots = u_0 = 0.$$

Apart from a constant we get

$$l_0(\mu, \gamma, \Sigma_u) = |\Sigma_u|^{-T/2} \exp\left\{-\frac{1}{2}\sum_{t=1}^{T} u_t(\mu, \gamma)'\Sigma_u^{-1}u_t(\mu, \gamma)\right\}, \tag{7.2.22}$$

where

$$u_t(\mu, \gamma) = (y_t - \mu) - \sum_{i=1}^{t-1} \Pi_i(\gamma)(y_{t-i} - \mu) \tag{7.2.23}$$

with the $\Pi_i(\gamma)$ being again the coefficient matrices of the pure VAR representation of y_t. We have indicated that these matrices are determined by the parameter vector γ. Formally the likelihood approximation has the same appearance as in the special cases. Of course the u_t are now potentially more complicated functions of the parameters.

It is perhaps worth noting that the uniqueness or identification problem discussed in Section 7.1 is reflected in the likelihood function. If the model is parameterized in a unique way, for instance, in final equations form or echelon form, the likelihood function has a locally unique maximum. This property is of obvious importance to guarantee unique ML estimators. Note, however, that the likelihood function in general has more than one local maximum. A more detailed discussion of the properties of the likelihood function can be found in Deistler & Pötscher (1984).

The next section focuses on the maximization of the approximate likelihood function (7.2.22) or, equivalently the maximization of its logarithm

$$\ln l_0(\mu, \gamma, \Sigma_u) = -\frac{T}{2}\ln|\Sigma_u| - \frac{1}{2}\sum_{t=1}^{T} u_t(\mu, \gamma)'\Sigma_u^{-1}u_t(\mu, \gamma). \tag{7.2.24}$$

7.3 Computation of the ML Estimates

In the pure finite order VAR case in Chapters 3 and 5 we have obtained the ML estimates by solving the normal equations. In the presently considered VARMA(p, q) case we may use the same principle. In other words, we determine the first partial derivatives of the log-likelihood function or rather its approximation given in (7.2.24) and equate them to zero. We will obtain the normal equations in Section 7.3.1. It turns out that they are nonlinear in the parameters and we discuss algorithms for solving the ML optimization problem in Section 7.3.2. The optimization procedures are iterative algorithms that require starting-up values or preliminary estimates for the parameters. A possible choice of initial estimates is proposed in Section 7.3.4. One of the optimization algorithms involves the information matrix which is given in Section 7.3.3. An example is discussed in Section 7.3.5.

7.3.1 The Normal Equations

In order to set up the normal equations corresponding to the approximate log-likelihood given in (7.2.24) we derive the first partial derivatives with respect to all the parameters μ, γ, and Σ_u.

$$\frac{\partial \ln l_0}{\partial \mu'} = -\sum_{t=1}^{T} u_t' \Sigma_u^{-1} \frac{\partial u_t}{\partial \mu'}$$

$$= \sum_{t=1}^{T} u_t' \Sigma_u^{-1} \left[I_K - \sum_{i=1}^{t-1} \Pi_i(\gamma) \right], \tag{7.3.1}$$

$$\frac{\partial \ln l_0}{\partial \gamma'} = -\sum_{t=1}^{T} u_t' \Sigma_u^{-1} \frac{\partial u_t}{\partial \gamma'}. \tag{7.3.2}$$

A recursive formula for computing the $\partial u_t / \partial \gamma'$ is given in the following lemma.

LEMMA 7.1

Let

$$u_t = y_t - A_0^{-1}[A_1 y_{t-1} + \cdots + A_p y_{t-p} + M_1 u_{t-1} + \cdots + M_q u_{t-q}], \tag{7.3.3}$$

$$\alpha_0 := \text{vec}(A_0),$$

$$\beta := \text{vec}[A_1, \ldots, A_p, M_1, \ldots, M_q],$$

and suppose

$$\begin{bmatrix} \alpha_0 \\ \beta \end{bmatrix} = R\gamma + r, \tag{7.3.4}$$

where R is a known $(K^2(p + q + 1) \times N)$ matrix, r is a known $K^2(p + q + 1)$-dimensional vector, and γ is an $(N \times 1)$ vector of unknown parameters. Then, defining $\partial u_0 / \gamma' = \partial u_{-1} / \partial \gamma' = \cdots = \partial u_{-q+1} / \partial \gamma' = 0$ and $y_0 = \cdots = y_{-p+1} = u_0 = \cdots = u_{-q+1} = 0$,

$$\frac{\partial u_t}{\partial \gamma'} = \{(A_0^{-1}[A_1 y_{t-1} + \cdots + A_p y_{t-p} + M_1 u_{t-1} + \cdots + M_q u_{t-q}])' \otimes A_0^{-1}\}$$
$$\times [I_{K^2} \quad 0 \ldots 0]R - [(y'_{t-1}, \ldots, y'_{t-p}, u'_{t-1}, \ldots, u'_{t-q}) \otimes A_0^{-1}][0 \; I_{K^2(p+q)}]R$$
$$- A_0^{-1}\left[M_1 \frac{\partial u_{t-1}}{\partial \gamma'} + \cdots + M_q \frac{\partial u_{t-q}}{\partial \gamma'}\right] \tag{7.3.5}$$

for $t = 1, \ldots, T$. ■

Replacing y_t with $y_t - \mu$ in this lemma, the expression in (7.3.5) can be used for recursively computing the $\partial u_t/\partial \gamma'$ required in (7.3.2).

Proof:

$$\frac{\partial u_t}{\partial \gamma'} = -[(A_1 y_{t-1} + \cdots + A_p y_{t-p} + M_1 u_{t-1} + \cdots + M_q u_{t-q})' \otimes I_K] \frac{\partial \operatorname{vec}(A_0^{-1})}{\partial \gamma'}$$
$$- [(y'_{t-1}, \ldots, y'_{t-p}, u'_{t-1}, \ldots, u'_{t-q}) \otimes A_0^{-1}] \frac{\partial \operatorname{vec}[A_1, \ldots, A_p, M_1, \ldots, M_q]}{\partial \gamma'}$$
$$- A_0^{-1}[A_1, \ldots, A_p, M_1, \ldots, M_q] \left[\partial \begin{bmatrix} y_{t-1} \\ \vdots \\ y_{t-p} \\ u_{t-1} \\ \vdots \\ u_{t-q} \end{bmatrix} \Big/ \partial \gamma' \right]. \tag{7.3.6}$$

The lemma follows by noting that

$$\frac{\partial \operatorname{vec}(A_0^{-1})}{\partial \gamma'} = \frac{\partial \operatorname{vec}(A_0^{-1})}{\partial \alpha'_0} \frac{\partial \alpha_0}{\partial \gamma'} = -[(A_0^{-1})' \otimes A_0^{-1}][I_{K^2} \quad 0 \ldots 0]R \tag{7.3.7}$$

(see Rule (9) of Appendix A.13). ■

The partial derivatives of the approximate log-likelihood with respect to the elements of Σ_u are

$$\frac{\partial \ln l_0}{\partial \Sigma_u} = -\frac{T}{2}\Sigma_u^{-1} + \frac{1}{2}\Sigma_u^{-1}\left[\sum_{t=1}^{T} u_t u_t'\right]\Sigma_u^{-1} \tag{7.3.8}$$

(see Problem 7.5). Setting this expression to zero and solving for Σ_u gives

$$\tilde{\Sigma}_u(\mu, \gamma) = \frac{1}{T}\sum_{t=1}^{T} u_t(\mu, \gamma) u_t(\mu, \gamma)'. \tag{7.3.9}$$

Substituting for Σ_u in (7.3.1) and (7.3.2) and setting to zero results in a generally nonlinear set of normal equations which may be solved by numerical methods. Before we discuss a possible algorithm it may be worth pointing out that by substituting $\tilde{\Sigma}_u(\mu, \gamma)$ for Σ_u in $\ln l_0$ we get

$$\ln l_0(\mu, \gamma) = -\frac{T}{2}\ln|\tilde{\Sigma}_u(\mu, \gamma)| - \frac{1}{2}\operatorname{tr}\left[\tilde{\Sigma}_u(\mu, \gamma)^{-1}\sum_{t=1}^{T} u_t(\mu, \gamma)u_t(\mu, \gamma)'\right]$$

$$= -\frac{T}{2}\ln|\tilde{\Sigma}_u(\mu, \gamma)| - \frac{TK}{2}. \tag{7.3.10}$$

Thus, instead of maximizing $\ln l_0$ we may equivalently minimize

$$\ln|\tilde{\Sigma}_u(\mu, \gamma)| \quad \text{or} \quad |\tilde{\Sigma}_u(\mu, \gamma)|. \tag{7.3.11}$$

7.3.2 Optimization Algorithms

The problem of optimizing (minimizing or maximizing) a function arises not only in ML estimation but also in various other contexts. Therefore general algorithms have been developed. Following Judge et al. (1985, Section B.2) we will give a brief introduction to so-called gradient algorithms and then address the specific problem at hand. With the objective in mind that we want to find the coefficient values that minimize $-\ln l_0$ or $\ln|\tilde{\Sigma}_u(\mu, \gamma)|$ we assume that the problem is to minimize a twice continuously differentiable, scalar valued function $h(\gamma)$, where γ is some $(N \times 1)$ vector.

Given a vector γ_i in the parameter space, we are looking for a direction (vector) $\mathbf{d}$ in which the objective function declines. Then we can perform a step of length s, say, in that direction which will take us downhill. In other words, we seek an appropriate *step direction* $\mathbf{d}$ and a step length s such that

$$h(\gamma_i + s\mathbf{d}) < h(\gamma_i). \tag{7.3.12}$$

If $\mathbf{d}$ is a downhill direction, a small step in that direction will always decrease the objective function. Thus, we are seeking a $\mathbf{d}$ such that $h(\gamma_i + s\mathbf{d})$ is a decreasing function of s for s sufficiently close to zero. In other words, $\mathbf{d}$ must be such that

$$0 > \left.\frac{dh(\gamma_i + s\mathbf{d})}{ds}\right|_{s=0} = \left[\left.\frac{\partial h(\gamma)}{\partial \gamma'}\right|_{\gamma_i}\right]\left[\left.\frac{\partial(\gamma_i + s\mathbf{d})}{\partial s}\right|_{s=0}\right] = \left[\left.\frac{\partial h(\gamma)}{\partial \gamma'}\right|_{\gamma_i}\right]\mathbf{d}.$$

Using the abbreviation

$$\mathbf{h}_i := \left.\frac{\partial h(\gamma)}{\partial \gamma}\right|_{\gamma_i}$$

for the gradient of $h(\gamma)$ at γ_i a possible choice of $\mathbf{d}$ is

$$\mathbf{d} = -D_i\mathbf{h}_i,$$

where D_i is any positive definite matrix. With this choice of $\mathbf{d}$,

$$\mathbf{h}_i'\mathbf{d} = -\mathbf{h}_i'D_i\mathbf{h}_i < 0$$

if $\mathbf{h}_i \neq 0$. Since the gradient is zero at a local minimum of the function we hope to have reached the minimum once $\mathbf{h}_i = 0$ and hence, $\mathbf{d} = 0$. The general form of an iteration of a *gradient algorithm* is therefore

$$\gamma_{i+1} = \gamma_i - s_i D_i \mathbf{h}_i, \tag{7.3.13}$$

where s_i denotes the step length in the i-th iteration and D_i is a positive definite *direction matrix*. The name "gradient algorithm" stems from the fact that the gradient is involved in the choice of the step direction. Many such algorithms have been proposed in the literature (see, for example, Judge et al. (1985, Section B.2)). They differ in their choice of the direction matrix D_i and the step length s_i.

To motivate the choice of the D_i matrix that will be considered in the ML algorithm presented below we expand the objective function $h(\gamma)$ in a Taylor series about γ_i (see Appendix A.13, Proposition A.3),

$$h(\gamma) \approx h(\gamma_i) + \mathbf{h}_i'(\gamma - \gamma_i) + \tfrac{1}{2}(\gamma - \gamma_i)' H_i(\gamma - \gamma_i), \tag{7.3.14}$$

where

$$H_i := \left. \frac{\partial^2 h}{\partial \gamma \partial \gamma'} \right|_{\gamma_i}$$

is the Hessian matrix of second order partial derivatives of $h(\gamma)$ evaluated at γ_i. If $h(\gamma)$ were a quadratic function the right-hand side of (7.3.14) were exactly equal to $h(\gamma)$ and the first order conditions for a minimum would result by taking first partial derivatives of the right-hand side and setting to zero:

$$\mathbf{h}_i' + H_i(\gamma - \gamma_i)' = 0$$

or

$$\gamma = \gamma_i - H_i^{-1}\mathbf{h}_i.$$

Thus, if $h(\gamma)$ were a quadratic function, starting from any vector γ_i we would reach the minimum in one step of length $s_i = 1$ by choosing the inverse Hessian as direction matrix. In general, if $h(\gamma)$ is not a quadratic function, the choice $D_i = H_i^{-1}$ is still reasonable once we are close to the minimum. Recall that a positive definite Hessian is the second order condition for a local minimum. Therefore the inverse Hessian qualifies as direction matrix. A gradient algorithm with the inverse Hessian as direction matrix is called a *Newton* or *Newton-Raphson algorithm.*

From the previous subsection we know that the first partial derivatives of our objective function $-\ln l_0$ are quite complicated and, thus, finding the Hessian matrix of second order partial derivatives is even more complicated. Therefore we approximate the Hessian by an estimate of the information matrix,

$$I(\gamma) = E\left[\frac{\partial^2 - \ln l_0}{\partial \gamma \partial \gamma'}\right], \tag{7.3.15}$$

which is the expected value of the Hessian matrix. The estimate of $I(\gamma)$ will be denoted by $\hat{I}(\gamma)$. A computable expression will be given in the next subsection. Since the true parameter vector γ is unknown $\hat{I}(\gamma_i)$ is used as estimate of $I(\gamma)$ in the i-th iteration step. Hence, for given mean vector μ and white noise covariance matrix Σ_u, we get a minimization algorithm with i-th iteration step

$$\gamma_{i+1} = \gamma_i - s_i \hat{I}(\gamma_i)^{-1} \left[\frac{\partial - \ln l_0}{\partial \gamma} \bigg|_{\gamma_i} \right]. \tag{7.3.16}$$

This algorithm is called *scoring algorithm*.

As it stands we still need some more information before we can execute this algorithm. First, we need a starting-up vector γ_1 for the first iteration. This vector should be close to the minimizing vector to ensure that $\hat{I}(\gamma_1)$ is positive definite and we make good progress towards the minimum even in the first iteration. We will consider one possible choice in Section 7.3.4.

Second, we have to choose the step length s_i. There are various possible alternatives (see, e.g., Judge et al. (1985, Section B.2)). Since we are just interested in the main principles of the algorithm we will ignore the problem here and choose $s_i = 1$.

Third, the algorithm provides an ML estimate of γ conditional on some given Σ_u matrix and mean vector μ since both the information matrix and the gradient involve these quantities. They are usually also unknown. As in the pure finite order VAR case it can be shown that the sample mean

$$\bar{y} = \frac{1}{T} \sum_{t=1}^{T} y_t$$

is an estimator for μ which has the same asymptotic properties as the ML estimator. Therefore it is common practice to perform the ML estimation of γ and Σ_u conditional on $\mu = \bar{y}$. In other words, the sample mean is subtracted from the data before the VARMA coefficients are estimated.

There are different ways to handle the unknown Σ_u matrix. From (7.3.9) we know that

$$\tilde{\Sigma}_u(\mu, \gamma) = \frac{1}{T} \sum_{t=1}^{T} u_t(\mu, \gamma) u_t(\mu, \gamma)'.$$

Therefore, one possibility is to use $\Sigma_i := \tilde{\Sigma}_u(\bar{y}, \gamma_i)$ in the i-th iteration.

A number of computer program packages contain exact or approximate ML algorithms which may be used in practice. The foregoing algorithm is just meant to demonstrate some basic principles. Modifications in actual applications may result in improved convergence properties. Slow convergence or no convergence at all may be the consequence of using too high VARMA orders or Kronecker indices.

7.3.3 The Information Matrix

In the scoring algorithm described previously an estimate of the information matrix is needed. To see how that can be obtained we consider the second order partial derivatives of $-\ln l_0$,

$$\frac{\partial^2 - \ln l_0}{\partial\gamma\partial\gamma'} = \partial\left[\sum_{t=1}^{T}\frac{\partial u_t'}{\partial\gamma}\Sigma_u^{-1}u_t\right]\Big/\partial\gamma' \qquad \text{(see (7.3.2))}$$

$$= \sum_{t=1}^{T}\frac{\partial u_t'}{\partial\gamma}\Sigma_u^{-1}\frac{\partial u_t}{\partial\gamma'} + (u_t'\Sigma_u^{-1}\otimes I)\frac{\partial\,\mathrm{vec}[\partial u_t'/\partial\gamma]}{\partial\gamma'}.$$

Taking the expectation of this expression the last term vanishes because $E(u_t) = 0$ and $u_t'\Sigma_u^{-1}\otimes I$ is independent of

$$\frac{\partial\,\mathrm{vec}[\partial u_t'/\partial\gamma]}{\partial\gamma'}$$

since this term does not contain current y_t or u_t variables (see Lemma 7.1). Hence,

$$E\left[\frac{\partial^2 - \ln l_0}{\partial\gamma\partial\gamma'}\right] = \sum_{t=1}^{T} E\left[\frac{\partial u_t'}{\partial\gamma}\Sigma_u^{-1}\frac{\partial u_t}{\partial\gamma'}\right].$$

Estimating the expected value in the usual way by the sample average gives an estimator

$$\frac{1}{T}\sum_{t=1}^{T}\frac{\partial u_t'}{\partial\gamma}\Sigma_u^{-1}\frac{\partial u_t}{\partial\gamma'}$$

for

$$E\left[\frac{\partial u_t'}{\partial\gamma}\Sigma_u^{-1}\frac{\partial u_t}{\partial\gamma'}\right].$$

This suggests the estimator

$$\hat{I}(\gamma) = \sum_{t=1}^{T}\frac{\partial u_t(\bar{y},\gamma)'}{\partial\gamma}\Sigma_u^{-1}\frac{\partial u_t(\bar{y},\gamma)}{\partial\gamma'} \qquad (7.3.17)$$

for the information matrix $I(\gamma)$. In the i-th iteration of the scoring algorithm we evaluate this estimator for $\gamma = \gamma_i$. The quantities $\partial u_t/\partial\gamma'$ may be obtained recursively as in Lemma 7.1 to make this estimator operational.

If γ is the true parameter value, the asymptotic information matrix equals $\mathrm{plim}(\hat{I}(\gamma)/T)$. Thus, if we have a consistent estimator $\tilde{\gamma}$ of γ, $\hat{I}(\tilde{\gamma})/T$ is a consistent estimator of the asymptotic information matrix, that is,

$$I_a(\gamma) = \mathrm{plim}\,\hat{I}(\tilde{\gamma})/T. \qquad (7.3.18)$$

In Section 7.4 we will see that the inverse of this matrix, if it exists, is the asymptotic covariance matrix of the ML estimator for γ. If a nonidentified structure is used this is reflected in the asymptotic information matrix being singular. This is one reason why it is important at this stage to have an identified version of a VARMA model.

7.3.4 Preliminary Estimation

The coefficients of a VARMA(p, q) model in standard form

$$y_t = A_1 y_{t-1} + \cdots + A_p y_{t-p} + u_t + M_1 u_{t-1} + \cdots + M_q u_{t-q}$$

could be estimated by multivariate LS if the lagged u_t were given. We assume that the sample mean $\bar{y}$ has been substracted previously. It is therefore neglected here. In deriving preliminary estimators for the other parameters the idea is to fit a long pure autoregression first and then use estimated residuals in place of the true residuals. Hence, we fit a VAR(n) model

$$y_t = \sum_{i=1}^{n} \Pi_i(n) y_{t-i} + u_t(n),$$

where n is a bit greater than p and q. From that estimation we compute estimation residuals

$$\hat{u}_t(n) := y_t - \sum_{i=1}^{n} \hat{\Pi}_i(n) y_{t-i}, \tag{7.3.19}$$

where $\hat{\Pi}_i(n)$ are the multivariate LS estimators. Then we set up a multivariate regression model

$$Y = [A \quad M] X_n + U^0, \tag{7.3.20}$$

where $Y := [y_1, \ldots, y_T]$, $A := [A_1, \ldots, A_p]$, $M := [M_1, \ldots, M_q]$,

$$X_n := [Y_{0,n}, \ldots, Y_{T-1,n}] \qquad \text{with } Y_{t,n} := \begin{bmatrix} y_t \\ \vdots \\ y_{t-p+1} \\ \hat{u}_t(n) \\ \vdots \\ \hat{u}_{t-q+1}(n) \end{bmatrix} (K(p+q) \times 1)$$

and U^0 is a $(K \times T)$ matrix of residuals. Usually restrictions will be imposed on the parameters A and M of the model, for instance, if the model is given in final equations form. Additional restrictions may also be available. Suppose the restrictions are such that there exists a matrix R and a vector γ satisfying

$$\text{vec}[A \quad M] = R\gamma. \tag{7.3.21}$$

Applying the vec operator to (7.3.20) and substituting $R\gamma$ for vec$[A \quad M]$ gives

$$\text{vec}(Y) = (X_n' \otimes I_K) R\gamma + \text{vec}(U^0) \tag{7.3.22}$$

and the LS estimator of γ is known to be

$$\hat{\gamma}(n) = [R'(X_n X_n' \otimes I_K) R]^{-1} R'(X_n \otimes I_K) \, \text{vec}(Y) \tag{7.3.23}$$

(see Chapter 5, Section 5.2). This estimator may be used as initial vector γ_1 in the ML algorithm described in the previous subsection.

Using this estimator a new set of residuals is obtained as

$$\text{vec}(\hat{U}^0) = \text{vec}(Y) - (X_n' \otimes I_K) R\hat{\gamma}(n)$$

which may be used to obtain a white noise covariance estimator

$$\tilde{\Sigma}_u(n) = \hat{U}^0 \hat{U}^{0\prime}/T. \tag{7.3.24}$$

This estimator may be used in place of Σ_u in the initial round of the iterative optimization algorithm described earlier.

The echelon form of a VARMA(p, q) process may be of the more general type

$$A_0 y_t = A_1 y_{t-1} + \cdots + A_p y_{t-p} + A_0 u_t + M_1 u_{t-1} + \cdots + M_q u_{t-q}, \qquad (7.3.25)$$

where A_0 is a lower triangular matrix with unit diagonal. To handle this case we proceed in a similar manner as in the standard case and substitute the residuals $\hat{u}_t(n)$ for the lagged u_t and for current residuals from other equations. In other words, in the k-th equation we substitute estimation residuals for u_{it}, $i < k$. Since A_0 is the coefficient matrix for both y_t and u_t we define

$$X_n^c := [Y_{0,n}^c, \dots, Y_{T-1,n}^c], \qquad \text{where } Y_{t,n}^c := \begin{bmatrix} \hat{u}_{t+1}(n) - y_{t+1} \\ Y_{t,n} \end{bmatrix}$$

and we pick a restriction matrix R_c and a vector γ_c such that

$$R_c \gamma_c = \operatorname{vec}[A_0 - I_K, A, M].$$

Hence

$$\operatorname{vec}(Y) = (X_n^{c\prime} \otimes I_K) R_c \gamma_c + \operatorname{vec}(U^0)$$

and the LS estimator of γ_c becomes

$$\hat{\gamma}_c(n) = [R_c'(X_n^c X_n^{c\prime} \otimes I_K) R_c]^{-1} R_c'(X_n^c \otimes I_K)\operatorname{vec}(Y).$$

The starting-up estimator of Σ_u is then obtained from the residuals of this regression. It is possible that the VARMA process corresponding to these coefficients is unstable or noninvertible. Especially in the latter case modifications are desirable (see Hannan & Kavalieris (1984), Hannan & Deistler (1988)).

To see more clearly what is being done in this preliminary estimation procedure let us look at an example. Suppose the bivariate VARMA(1, 1) echelon form model from (7.1.19) with Kronecker indices $(p_1, p_2) = (1, 0)$ is to be estimated:

$$\begin{aligned} y_{1,t} &= \alpha_{11,1} y_{1,t-1} + u_{1,t} + m_{11,1} u_{1,t-1} + m_{12,1} u_{2,t-1}, \\ y_{2,t} &= \alpha_{21,0} y_{1,t} - \alpha_{21,0} u_{1,t} + u_{2,t} = \alpha_{21,0}(-u_{1,t} + y_{1,t}) + u_{2,t}. \end{aligned} \qquad (7.3.26)$$

We assume that the sample mean has been removed previously. The parameters in the first equation are estimated by applying the LS method to

$$\begin{bmatrix} y_{1,1} \\ \vdots \\ y_{1,T} \end{bmatrix} = \begin{bmatrix} y_{1,0} & \hat{u}_{1,0}(n) & \hat{u}_{2,0}(n) \\ \vdots & \vdots & \vdots \\ y_{1,T-1} & \hat{u}_{1,T-1}(n) & \hat{u}_{2,T-1}(n) \end{bmatrix} \begin{bmatrix} \alpha_{11,1} \\ m_{11,1} \\ m_{12,1} \end{bmatrix} + \begin{bmatrix} u_{1,1} \\ \vdots \\ u_{1,T} \end{bmatrix},$$

or, using obvious notation, to

$$y_{(1)} = X_{(1)} \gamma_1 + u_{(1)}.$$

Here the $\hat{u}_{i,t}(n)$ are the residuals from the estimated long VAR model of order n. The LS estimator of γ_1 is $\hat{\gamma}_1 = (X'_{(1)}X_{(1)})^{-1}X'_{(1)}y_{(1)}$.

Similarly $\alpha_{21,0}$ is estimated by applying the LS method to

$$\begin{bmatrix} y_{2,1} \\ \vdots \\ y_{2,T} \end{bmatrix} = \begin{bmatrix} -\hat{u}_{1,1}(n) + y_{1,1} \\ \vdots \\ -\hat{u}_{1,T}(n) + y_{1,T} \end{bmatrix} \alpha_{21,0} + \begin{bmatrix} u_{2,1} \\ \vdots \\ u_{2,T} \end{bmatrix}.$$

In this case it would be possible to use the residuals of the first regression instead of the $\hat{u}_{1,t}(n)$ which are the residuals from the long VAR. However, we have chosen to use the latter in the preliminary estimation procedure.

In the foregoing we have so far ignored the problem of choosing presample values for the estimation. Two alternative choices are reasonable. Either all presample values are replaced by zero or some y_t values at the beginning of the sample are set aside as presample values and the presample values for the residuals are replaced by zero.

The initial estimators obtained in the foregoing procedure can be shown to be consistent under general conditions if n goes to infinity with the sample size (see Hannan & Kavalieris (1984), Hannan & Deistler (1988), Poskitt (1989a)). We will discuss the situation where VAR processes of increasing order are fitted to a potentially infinite order process in Chapter 9 and therefore we do not give details here.

7.3.5 An Illustration

We illustrate the estimation procedure using the income (y_1) and consumption (y_2) data from Table E.1 of Appendix E. As in previous chapters we use first differences of logarithms of the data from 1960 to 1978. In this case we subtract the sample mean at an initial stage and denote the mean-adjusted income and consumption variables by y_{1t} and y_{2t}, respectively. We assume a VARMA(2, 2) model in echelon form with Kronecker indices $\mathbf{p} = (p_1, p_2)' = (0, 2)'$,

$$\begin{bmatrix} y_{1,t} \\ y_{2,t} \end{bmatrix} = \begin{bmatrix} 0 & 0 \\ 0 & \alpha_{22,1} \end{bmatrix} \begin{bmatrix} y_{1,t-1} \\ y_{2,t-1} \end{bmatrix} + \begin{bmatrix} 0 & 0 \\ 0 & \alpha_{22,2} \end{bmatrix} \begin{bmatrix} y_{1,t-2} \\ y_{2,t-2} \end{bmatrix} + \begin{bmatrix} u_{1,t} \\ u_{2,t} \end{bmatrix}$$

$$+ \begin{bmatrix} 0 & 0 \\ m_{21,1} & m_{22,1} \end{bmatrix} \begin{bmatrix} u_{1,t-1} \\ u_{2,t-1} \end{bmatrix} + \begin{bmatrix} 0 & 0 \\ m_{21,2} & m_{22,2} \end{bmatrix} \begin{bmatrix} u_{1,t-2} \\ u_{2,t-2} \end{bmatrix}. \qquad (7.3.27)$$

In the next chapter it will become apparent why this model is chosen. It implies that the first variable (income) is white noise ($y_{1t} = u_{1t}$). Given the subset VAR models of Chapter 5 (Table 5.3) this specification does not appear to be totally unreasonable. The second equation in (7.3.27) describes consumption as a function of lagged consumption, lagged income ($u_{1,t-i} = y_{1,t-i}$), and a moving average term involving lagged residuals $u_{2,t}$.

Eventually we use a sample from 1960.II ($t = 1$) to 1978.IV ($t = 75$), that is, $T = 75$. In the preliminary estimation of the model (7.3.27) we estimate a VAR(8) model first using 8 presample values. Then using two more presample values we run a regression of y_{2t} on its own lags and lagged $\hat{u}_{it}(8)$. More precisely the regression model is

$$\begin{bmatrix} y_{2,11} \\ \vdots \\ y_{2,T} \end{bmatrix} = \begin{bmatrix} y_{2,10} & y_{2,9} & \hat{u}_{1,10}(8) & \hat{u}_{2,10}(8) & \hat{u}_{1,9}(8) & \hat{u}_{2,9}(8) \\ \vdots & \vdots & \vdots & \vdots & \vdots & \vdots \\ y_{2,T-1} & y_{2,T-2} & \hat{u}_{1,T-1}(8) & \hat{u}_{2,T-1}(8) & \hat{u}_{1,T-2}(8) & \hat{u}_{2,T-2}(8) \end{bmatrix} \gamma + \begin{bmatrix} u_{2,11} \\ \vdots \\ u_{2,T} \end{bmatrix},$$

where $\gamma := (\alpha_{22,1}, \alpha_{22,2}, m_{21,1}, m_{22,1}, m_{21,2}, m_{22,2})'$. In this particular case we could have substituted y_{1t} for $\hat{u}_{1t}(8)$ because the model implies $y_{1t} = u_{1t}$. We have not done so, however, but we have used the residuals from the long autoregression. The resulting preliminary parameter estimates

$$\tilde{\gamma}_1 = (\tilde{\alpha}_{22,1}(1), \ldots, \tilde{m}_{22,2}(1))'$$

are given in Table 7.1.

We use these estimates to start the scoring algorithm. For our particular example the i-th iteration proceeds as follows:

(1) Compute residuals

$$\tilde{u}_t(i) = y_t - \tilde{A}_1(i)y_{t-1} - \tilde{A}_2(i)y_{t-2} - \tilde{M}_1(i)\tilde{u}_{t-1}(i) - \tilde{M}_2(i)\tilde{u}_{t-2}(i)$$

recursively, for $t = 1, 2, \ldots, T$, with $\tilde{u}_{-1}(i) = \tilde{u}_0(i) = y_{-1} = y_0 = 0$ and

$$\tilde{A}_1(i) = \begin{bmatrix} 0 & 0 \\ 0 & \tilde{\alpha}_{22,1}(i) \end{bmatrix}, \quad \tilde{A}_2(i) = \begin{bmatrix} 0 & 0 \\ 0 & \tilde{\alpha}_{22,2}(i) \end{bmatrix},$$

$$\tilde{M}_1(i) = \begin{bmatrix} 0 & 0 \\ \tilde{m}_{21,1}(i) & \tilde{m}_{22,1}(i) \end{bmatrix}, \quad \tilde{M}_2(i) = \begin{bmatrix} 0 & 0 \\ \tilde{m}_{21,2}(i) & \tilde{m}_{22,2}(i) \end{bmatrix}.$$

(2) Compute the partial derivatives $\tilde{\partial} u_t/\partial\gamma$ recursively as

Table 7.1. Iterative estimates of the income/consumption system

i	$\tilde{\gamma}_i$: $\alpha_{22,1}$	$\alpha_{22,2}$	$m_{21,1}$	$m_{22,1}$	$m_{21,2}$	$m_{22,2}$	$\lvert\tilde{\Sigma}_u(\tilde{\gamma}_i)\rvert \times 10^8$
1	0.020	0.395	0.296	−0.367	0.181	−0.224	0.872564
2	−0.178	0.492	0.331	−0.527	0.175	−0.015	0.942791
3	0.072	0.117	0.305	−0.589	0.191	0.065	0.779788
4	0.202	0.078	0.311	−0.731	0.146	0.147	0.776107
5	0.219	0.063	0.312	−0.744	0.142	0.158	0.775959
6	0.224	0.062	0.313	−0.748	0.140	0.159	0.775952
⋮							
10	0.225	0.061	0.313	−0.750	0.140	0.160	0.775951

$$\frac{\tilde{\partial} u_t}{\partial \gamma'}(i) = -\begin{bmatrix} 0 & 0 & 0 & 0 & 0 & 0 \\ y_{2,t-1} & y_{2,t-2} & \tilde{u}_{1,t-1}(i) & \tilde{u}_{2,t-1}(i) & \tilde{u}_{1,t-2}(i) & \tilde{u}_{2,t-2}(i) \end{bmatrix}$$

$$- \tilde{M}_1(i)\frac{\tilde{\partial} u_{t-1}}{\partial \gamma'}(i) - \tilde{M}_2(i)\frac{\tilde{\partial} u_{t-2}}{\partial \gamma'}(i)$$

for $t = 1, 2, \ldots, T$, with

$$\frac{\tilde{\partial} u_{-1}}{\partial \gamma'}(i) = \frac{\tilde{\partial} u_0}{\partial \gamma'}(i) = 0.$$

(3) Compute

$$\tilde{\Sigma}_u(\tilde{\gamma}_i) = \frac{1}{T}\sum_{t=1}^{T} \tilde{u}_t(i)\tilde{u}_t(i)',$$

$$\hat{I}(\tilde{\gamma}_i) = \sum_{t=1}^{T} \frac{\tilde{\partial} u_t'}{\partial \gamma}(i)\tilde{\Sigma}_u(\tilde{\gamma}_i)^{-1}\frac{\tilde{\partial} u_t}{\partial \gamma'}(i),$$

and

$$\left.\frac{\partial - \ln l_0}{\partial \gamma'}\right|_{\tilde{\gamma}_i} = \sum_{t=1}^{T} \tilde{u}_t(i)'\tilde{\Sigma}_u(\tilde{\gamma}_i)^{-1}\frac{\tilde{\partial} u_t}{\partial \gamma'}(i).$$

(4) Perform the iteration step

$$\tilde{\gamma}_{i+1} = \tilde{\gamma}_i - \hat{I}(\tilde{\gamma}_i)^{-1}\left[\left.\frac{\partial - \ln l_0}{\partial \gamma}\right|_{\tilde{\gamma}_i}\right].$$

Some estimates obtained in these iterations are also given in Table 7.1 together with $|\tilde{\Sigma}_u(\tilde{\gamma}_i)|$. After a few iterations the latter quantity approximately reaches its minimum and thus $-\ln l_0$ obtains its minimum. After the tenth iteration there is not much change in the $\tilde{\gamma}_i$ and $|\tilde{\Sigma}_u(\tilde{\gamma}_i)|$ in further steps. We work with $\tilde{\gamma}_{10}$ in the following.

The determinantal polynomial of the MA operator for $i = 10$ is

$$\begin{aligned} |I_2 + \tilde{M}_1(10)z + \tilde{M}_2(10)z^2| &= 1 + \tilde{m}_{22,1}(10)z + \tilde{m}_{22,2}(10)z^2 \\ &= 1 - .750z + .160z^2 \end{aligned}$$

which has roots that are clearly outside the unit circle. Thus, the estimated MA operator is invertible. Also, the determinant of the estimated VAR polynomial,

$$\begin{aligned} |I_2 - \tilde{A}_1(10)z - \tilde{A}_2(10)z^2| &= 1 - \tilde{\alpha}_{22,1}(10)z - \tilde{\alpha}_{22,2}(10)z^2 \\ &= 1 - .225z - .061z^2 \end{aligned}$$

is easily seen to have its roots outside the unit circle. Hence, the estimated VARMA process is stable and invertible.

7.4 Asymptotic Properties of the ML Estimators

7.4.1 Theoretical Results

In this section the asymptotic properties of the ML estimators are given. We will not prove the main result but refer the reader to Hannan (1979), Dunsmuir & Hannan (1976), Hannan & Deistler (1988), and Kohn (1979) for further discussions and proofs.

PROPOSITION 7.1 (*Asymptotic Properties of ML Estimators*)

Let y_t be a K-dimensional, stationary Gaussian process with stable and invertible VARMA(p, q) representation

$$A_0(y_t - \mu) = A_1(y_{t-1} - \mu) + \cdots + A_p(y_{t-p} - \mu) + A_0 u_t + M_1 u_{t-1} + \cdots + M_q u_{t-q}, \tag{7.4.1}$$

where u_t is Gaussian white noise with nonsingular covariance matrix Σ_u. Suppose the VAR and MA operators are left-coprime and either in final equations form or in echelon form with possibly linear restrictions on the coefficients so that the coefficient matrices $A_0, A_1, \ldots, A_p, M_1, \ldots, M_q$ depend on a set of unrestricted parameters γ as in (7.2.21). Let $\tilde{\mu}$, $\tilde{\gamma}$, and $\tilde{\Sigma}_u$ be the ML estimators of μ, γ, and Σ_u, respectively, and denote vech(Σ_u) and vech$(\tilde{\Sigma}_u)$ by σ and $\tilde{\sigma}$, respectively. Then all three ML estimators are consistent and asymptotically normally distributed,

$$\sqrt{T}\begin{bmatrix} \tilde{\mu} - \mu \\ \tilde{\gamma} - \gamma \\ \tilde{\sigma} - \sigma \end{bmatrix} \xrightarrow{d} N\left[0, \begin{bmatrix} \Sigma_{\tilde{\mu}} & 0 & 0 \\ 0 & \Sigma_{\tilde{\gamma}} & 0 \\ 0 & 0 & \Sigma_{\tilde{\sigma}} \end{bmatrix}\right], \tag{7.4.2}$$

where

$$\Sigma_{\tilde{\mu}} = A(1)^{-1} M(1) \Sigma_u M(1)' A(1)'^{-1},$$

$$\Sigma_{\tilde{\gamma}} = I_a(\gamma)^{-1} = \operatorname{plim}\left[\frac{1}{T}\sum_{t=1}^{T} \frac{\partial u_t'}{\partial \gamma} \Sigma_u^{-1} \frac{\partial u_t}{\partial \gamma'}\right]^{-1}$$

with $\partial u_t/\partial \gamma'$ as given in Lemma 7.1, and

$$\Sigma_{\tilde{\sigma}} = 2\mathbf{D}_K^+(\Sigma_u \otimes \Sigma_u)\mathbf{D}_K^{+\prime}$$

with $\mathbf{D}_K^+ = (\mathbf{D}_K'\mathbf{D}_K)^{-1}\mathbf{D}_K'$ and $\mathbf{D}_K$ being the $(K^2 \times K(K+1)/2)$ duplication matrix. The covariance matrix in (7.4.2) is consistently estimated by replacing the unknown quantities by their ML estimators. ■

Some remarks on this proposition may be worthwhile.

REMARK 1: The results of the proposition do not change if the ML estimator $\tilde{\mu}$ is replaced by $\bar{y}$ and $\tilde{\gamma}$ and $\tilde{\sigma}$ are ML estimators conditional on $\bar{y}$, that is, $\tilde{\gamma}$ and

$\tilde{\sigma}$ are obtained by replacing μ by $\bar{y}$ in the ML algorithm. One consequence of this result is that asymptotically the sample mean is a fully efficient estimator of μ. ■

REMARK 2: The proposition is formulated in terms of final equations form or echelon form VARMA models. Its statement remains true for other uniquely identified structures. ■

REMARK 3: Since the covariance matrix of the asymptotic distribution in (7.4.2) is block-diagonal the estimators of μ, γ, and $\tilde{\Sigma}_u$ are asymptotically independent. ■

REMARK 4: Much of the proposition remains valid even if the y_t are not normally distributed. In that case the estimators obtained by maximizing the Gaussian likelihood function are pseudo ML estimators. If u_t is independent standard white noise (see Chapter 3, Definition 3.1), $\tilde{\gamma}$ and $\bar{y}$ maintain their asymptotic properties. The covariance matrix of $\tilde{\sigma}$ may be different from the one given in Proposition 7.1. ■

REMARK 5: The results of the proposition remain valid under general conditions if instead of the ML estimator $\tilde{\gamma}$ an estimator is used which is obtained from one iteration of the scoring algorithm outlined in Section 7.3.2 starting from the preliminary estimator of Section 7.3.4. Thus, one possibility to proceed in estimating the parameters of a VARMA model is to compute the sample mean $\bar{y}$ first and use that as an estimator of μ. Then the preliminary estimator for γ may be computed as described in Section 7.3.4 and that estimator is used as initial vector in the optimization algorithm of Section 7.3.2. Then just one step of the form (7.3.16) is performed with $s_i = s_1 = 1$. The resulting estimators $\tilde{\gamma}_2$ and $\tilde{\Sigma}_u(\bar{y}, \tilde{\gamma}_2)$ may then be used instead of $\tilde{\gamma}$ and $\tilde{\Sigma}_u$ in Proposition 7.1. Under general conditions they have the same *asymptotic* distributions as the actual ML estimators. Of course, this is a computationally attractive way to estimate the coefficients of a VARMA model. In general the small sample properties of the resulting estimators are not the same as those of the ML estimators, however. ■

REMARK 6: Since often the final equations form involves more parameters than the echelon form, unrestricted estimation of the former may result in inefficient estimators. Intuitively, if we start from the echelon form and determine the corresponding final equations form, the coefficients of the latter are seen to satisfy restrictions that could be imposed to obtain more efficient estimators. ■

In the following sections we will occasionally be interested in the asymptotic distribution of the coefficients of the standard representation of the process,

$$(y_t - \mu) = A_1(y_{t-1} - \mu) + \cdots + A_p(y_{t-p} - \mu) + u_t + M_1 u_{t-1} + \cdots + M_q u_{t-q}. \tag{7.4.3}$$

The coefficients are functions of γ and their asymptotic distribution follows in the usual way. Let

$$\boldsymbol{\alpha} := \operatorname{vec}[A_1, \dots, A_p] \qquad \text{and} \qquad \mathbf{m} := \operatorname{vec}[M_1, \dots, M_q]$$

then

$$\begin{bmatrix} \boldsymbol{\alpha} \\ \mathbf{m} \end{bmatrix} = \begin{bmatrix} \boldsymbol{\alpha}(\gamma) \\ \mathbf{m}(\gamma) \end{bmatrix}.$$

The ML estimators are

$$\begin{bmatrix} \tilde{\boldsymbol{\alpha}} \\ \tilde{\mathbf{m}} \end{bmatrix} = \begin{bmatrix} \boldsymbol{\alpha}(\tilde{\gamma}) \\ \mathbf{m}(\tilde{\gamma}) \end{bmatrix}.$$

They are consistent and asymptotically normal,

$$\sqrt{T}\left[\begin{bmatrix} \tilde{\boldsymbol{\alpha}} \\ \tilde{\mathbf{m}} \end{bmatrix} - \begin{bmatrix} \boldsymbol{\alpha} \\ \mathbf{m} \end{bmatrix}\right] \xrightarrow{d} N\left[0, \Sigma_{\left[\begin{smallmatrix} \tilde{\boldsymbol{\alpha}} \\ \tilde{\mathbf{m}} \end{smallmatrix}\right]} = \begin{bmatrix} \dfrac{\partial \boldsymbol{\alpha}}{\partial \gamma'} \\ \dfrac{\partial \mathbf{m}}{\partial \gamma'} \end{bmatrix} \Sigma_{\tilde{\gamma}} \left[\frac{\partial \boldsymbol{\alpha}'}{\partial \gamma}, \frac{\partial \mathbf{m}'}{\partial \gamma}\right]\right]. \tag{7.4.4}$$

If $A_0 = I_K$, $\boldsymbol{\alpha}$ and $\mathbf{m}$ will often be linearly related to γ and we get the following corollary of Proposition 7.1.

COROLLARY 7.1.1

Under the conditions of Proposition 7.1, if

$$\begin{bmatrix} \boldsymbol{\alpha} \\ \mathbf{m} \end{bmatrix} = R\gamma + r,$$

$$\sqrt{T}\left[\begin{bmatrix} \tilde{\boldsymbol{\alpha}} \\ \tilde{\mathbf{m}} \end{bmatrix} - \begin{bmatrix} \boldsymbol{\alpha} \\ \mathbf{m} \end{bmatrix}\right] \xrightarrow{d} N(0, R\Sigma_{\tilde{\gamma}}R')$$

and $\tilde{\boldsymbol{\alpha}}$ and $\tilde{\mathbf{m}}$ are asymptotically independent of $\bar{y}$, $\tilde{\mu}$, and $\tilde{\boldsymbol{\sigma}}$. ∎

The remarks following the proposition also apply for the corollary. For illustrative purposes consider the bivariate VARMA(1, 1) model in echelon form with Kronecker indices (0, 1),

$$\begin{bmatrix} 1 & 0 \\ 0 & 1 - \alpha_{22,1}L \end{bmatrix} y_t = \begin{bmatrix} 1 & 0 \\ m_{21,1}L & 1 + m_{22,1}L \end{bmatrix} u_t \tag{7.4.5}$$

or

$$y_t = \begin{bmatrix} 0 & 0 \\ 0 & \alpha_{22,1} \end{bmatrix} y_{t-1} + u_t + \begin{bmatrix} 0 & 0 \\ m_{21,1} & m_{22,1} \end{bmatrix} u_{t-1}.$$

In this case

$$\boldsymbol{\alpha} = \begin{bmatrix} 0 \\ 0 \\ 0 \\ \alpha_{22,1} \end{bmatrix}, \quad \mathbf{m} = \begin{bmatrix} 0 \\ m_{21,1} \\ 0 \\ m_{22,1} \end{bmatrix}, \quad \boldsymbol{\gamma} = \begin{bmatrix} \alpha_{22,1} \\ m_{21,1} \\ m_{22,1} \end{bmatrix},$$

$$R = \begin{bmatrix} 0 & 0 & 0 \\ 0 & 0 & 0 \\ 0 & 0 & 0 \\ 1 & 0 & 0 \\ 0 & 0 & 0 \\ 0 & 1 & 0 \\ 0 & 0 & 0 \\ 0 & 0 & 1 \end{bmatrix}, \quad \text{and} \quad r = \begin{bmatrix} 0 \\ 0 \\ 0 \\ 0 \\ 0 \\ 0 \\ 0 \\ 0 \end{bmatrix}.$$

If the VARMA model is not in standard form originally we left-multiply by A_0^{-1} to get

$$\begin{aligned} y_t - \mu = {} & A_0^{-1}A_1(y_{t-1} - \mu) + \cdots + A_0^{-1}A_p(y_{t-p} - \mu) \\ & + u_t + A_0^{-1}M_1u_{t-1} + \cdots + A_0^{-1}M_qu_{t-q}. \end{aligned} \tag{7.4.6}$$

In this case it is more reasonable to assume that

$$\boldsymbol{\beta}_0 := \operatorname{vec}[A_0, A_1, \ldots, A_p, M_1, \ldots, M_q] \tag{7.4.7}$$

is linearly related to $\boldsymbol{\gamma}$, say,

$$\boldsymbol{\beta}_0 = R\boldsymbol{\gamma} + r. \tag{7.4.8}$$

Then it follows for

$$\boldsymbol{\alpha} := \operatorname{vec}[A_0^{-1}A_1, \ldots, A_0^{-1}A_p] = \operatorname{vec}(A_0^{-1}[A_1, \ldots, A_p]) \tag{7.4.9}$$

and

$$\mathbf{m} := \operatorname{vec}(A_0^{-1}[M_1, \ldots, M_q]), \tag{7.4.10}$$

that

$$\begin{bmatrix} \dfrac{\partial \boldsymbol{\alpha}}{\partial \boldsymbol{\gamma}'} \\ \dfrac{\partial \mathbf{m}}{\partial \boldsymbol{\gamma}'} \end{bmatrix} = \begin{bmatrix} \dfrac{\partial \boldsymbol{\alpha}}{\partial \boldsymbol{\beta}_0'} \\ \dfrac{\partial \mathbf{m}}{\partial \boldsymbol{\beta}_0'} \end{bmatrix} \frac{\partial \boldsymbol{\beta}_0}{\partial \boldsymbol{\gamma}'} = \begin{bmatrix} \dfrac{\partial \boldsymbol{\alpha}}{\partial \boldsymbol{\beta}_0'} \\ \dfrac{\partial \mathbf{m}}{\partial \boldsymbol{\beta}_0'} \end{bmatrix} R.$$

Hence, we need to evaluate $\partial\boldsymbol{\alpha}/\partial\boldsymbol{\beta}_0'$ and $\partial\mathbf{m}/\partial\boldsymbol{\beta}_0'$ in order to obtain the asymptotic covariance matrix of the standard form coefficients.

$$\frac{\partial \boldsymbol{\alpha}}{\partial \boldsymbol{\beta}_0'} = (I_{Kp} \otimes A_0^{-1}) \frac{\partial \operatorname{vec}[A_1, \ldots, A_p]}{\partial \boldsymbol{\beta}_0'} + \left[\begin{bmatrix} A_1' \\ \vdots \\ A_p' \end{bmatrix} \otimes I_K \right] \frac{\partial \operatorname{vec}(A_0^{-1})}{\partial \boldsymbol{\beta}_0'}$$

$$= (I_{Kp} \otimes A_0^{-1})[0 \quad I_{K^2p} \quad 0] - \left[\begin{bmatrix} A_1' \\ \vdots \\ A_p' \end{bmatrix} \otimes I_K \right]((A_0^{-1})' \otimes A_0^{-1}) \frac{\partial \operatorname{vec}(A_0)}{\partial \boldsymbol{\beta}_0'}$$

(see Rule 9 of Appendix A.13)

$$= (I_{Kp} \otimes A_0^{-1})[0 \quad I_{K^2p} \quad 0] - \left[\begin{bmatrix} (A_0^{-1}A_1)' \\ \vdots \\ (A_0^{-1}A_p)' \end{bmatrix} \otimes A_0^{-1} \right][I_{K^2} \quad 0]. \tag{7.4.11}$$

A similar expression is obtained for $\partial \mathbf{m}/\partial \boldsymbol{\beta}_0'$. This result is summarized in the next corollary.

COROLLARY 7.1.2

Under the conditions of Proposition 7.1, if $\boldsymbol{\beta}_0$ is as defined in (7.4.7) and satisfies the restrictions in (7.4.8) and $\boldsymbol{\alpha}$ and $\mathbf{m}$ are the coefficients of the standard form VARMA representation defined in (7.4.9) and (7.4.10), respectively, with ML estimators $\tilde{\boldsymbol{\alpha}}$ and $\tilde{\mathbf{m}}$, then

$$\sqrt{T}\left[\begin{bmatrix} \tilde{\boldsymbol{\alpha}} \\ \tilde{\mathbf{m}} \end{bmatrix} - \begin{bmatrix} \boldsymbol{\alpha} \\ \mathbf{m} \end{bmatrix}\right] \xrightarrow{d} N\left[0, \Sigma_{\left[\begin{smallmatrix} \tilde{\boldsymbol{\alpha}} \\ \tilde{\mathbf{m}} \end{smallmatrix}\right]} = \begin{bmatrix} H_{\boldsymbol{\alpha}} \\ H_{\mathbf{m}} \end{bmatrix} R \Sigma_{\tilde{\gamma}} R' [H_{\boldsymbol{\alpha}}' \; H_{\mathbf{m}}'] \right],$$

where

$$H_{\boldsymbol{\alpha}} := \frac{\partial \boldsymbol{\alpha}}{\partial \boldsymbol{\beta}_0'} = (I_{Kp} \otimes A_0^{-1})[\underbrace{\;0\;}_{(K^2p \times K^2)} \quad I_{K^2p} \quad \underbrace{\;0\;}_{(K^2p \times K^2q)}]$$

$$- \underbrace{\left[\begin{bmatrix} (A_0^{-1}A_1)' \\ \vdots \\ (A_0^{-1}A_p)' \end{bmatrix} \otimes A_0^{-1} \right][I_{K^2} \quad 0]}_{(K^2p \times K^2(p+q+1))}$$

and

$$H_{\mathbf{m}} := \frac{\partial \mathbf{m}}{\partial \boldsymbol{\beta}_0'} = (I_{Kq} \otimes A_0^{-1})[0 \quad I_{K^2q}] - \underbrace{\left[\begin{bmatrix} (A_0^{-1}M_1)' \\ \vdots \\ (A_0^{-1}M_q)' \end{bmatrix} \otimes A_0^{-1} \right][I_{K^2} \quad 0]}_{(K^2q \times K^2(p+q+1))}. \qquad \blacksquare$$

Again an example may be worthwhile. Consider the bivariate VARMA(2, 1) process in echelon form with Kronecker indices (2, 1) and some zero restrictions placed on the coefficients (see also Problem 7.3):

$$\begin{bmatrix} 1 - \alpha_{11,1}L - \alpha_{11,2}L^2 & 0 \\ -\alpha_{21,0} - \alpha_{21,1}L & 1 - \alpha_{22,1}L \end{bmatrix} y_t = \begin{bmatrix} 1 & 0 \\ -\alpha_{21,0} & 1 + m_{22,1}L \end{bmatrix} u_t \tag{7.4.12}$$

or

$$\begin{bmatrix} 1 & 0 \\ -\alpha_{21,0} & 1 \end{bmatrix} y_t = \begin{bmatrix} \alpha_{11,1} & 0 \\ \alpha_{21,1} & \alpha_{22,1} \end{bmatrix} y_{t-1} + \begin{bmatrix} \alpha_{11,2} & 0 \\ 0 & 0 \end{bmatrix} y_{t-2}$$

$$+ \begin{bmatrix} 1 & 0 \\ -\alpha_{21,0} & 1 \end{bmatrix} u_t + \begin{bmatrix} 0 & 0 \\ 0 & m_{22,1} \end{bmatrix} u_{t-1}.$$

Hence,

$$\boldsymbol{\beta}_0 = \begin{bmatrix} 1 \\ -\alpha_{21,0} \\ 0 \\ 1 \\ \alpha_{11,1} \\ \alpha_{21,1} \\ 0 \\ \alpha_{22,1} \\ \alpha_{11,2} \\ 0 \\ 0 \\ 0 \\ 0 \\ 0 \\ 0 \\ m_{22,1} \end{bmatrix}, \quad R = \begin{bmatrix} 0 & 0 & 0 & 0 & 0 & 0 \\ -1 & 0 & 0 & 0 & 0 & 0 \\ 0 & 0 & 0 & 0 & 0 & 0 \\ 0 & 0 & 0 & 0 & 0 & 0 \\ 0 & 1 & 0 & 0 & 0 & 0 \\ 0 & 0 & 1 & 0 & 0 & 0 \\ 0 & 0 & 0 & 0 & 0 & 0 \\ 0 & 0 & 0 & 1 & 0 & 0 \\ 0 & 0 & 0 & 0 & 1 & 0 \\ 0 & 0 & 0 & 0 & 0 & 0 \\ 0 & 0 & 0 & 0 & 0 & 0 \\ 0 & 0 & 0 & 0 & 0 & 0 \\ 0 & 0 & 0 & 0 & 0 & 0 \\ 0 & 0 & 0 & 0 & 0 & 0 \\ 0 & 0 & 0 & 0 & 0 & 0 \\ 0 & 0 & 0 & 0 & 0 & 1 \end{bmatrix}, \quad \boldsymbol{\gamma} = \begin{bmatrix} \alpha_{21,0} \\ \alpha_{11,1} \\ \alpha_{21,1} \\ \alpha_{22,1} \\ \alpha_{11,2} \\ m_{22,1} \end{bmatrix}, \quad r = \begin{bmatrix} 1 \\ 0 \\ 0 \\ 1 \\ 0 \\ 0 \\ 0 \\ 0 \\ 0 \\ 0 \\ 0 \\ 0 \\ 0 \\ 0 \\ 0 \\ 0 \end{bmatrix}.$$

Furthermore,

$$A_0^{-1} = \begin{bmatrix} 1 & 0 \\ -\alpha_{21,0} & 1 \end{bmatrix}^{-1} = \begin{bmatrix} 1 & 0 \\ \alpha_{21,0} & 1 \end{bmatrix}.$$

Thus,

$$\boldsymbol{\alpha} = \operatorname{vec}[A_0^{-1}A_1, A_0^{-1}A_2]$$

$$= \operatorname{vec}\left[\begin{array}{cc|cc} \alpha_{11,1} & 0 & \alpha_{11,2} & 0 \\ \alpha_{11,1}\alpha_{21,0} + \alpha_{21,1} & \alpha_{22,1} & \alpha_{11,2}\alpha_{21,0} & 0 \end{array}\right] = \begin{bmatrix} \alpha_{11,1} \\ \alpha_{11,1}\alpha_{21,0} + \alpha_{21,1} \\ 0 \\ \alpha_{22,1} \\ \alpha_{11,2} \\ \alpha_{11,2}\alpha_{21,0} \\ 0 \\ 0 \end{bmatrix}$$

and

$$\mathbf{m} = \text{vec}[A_0^{-1}M_1] = \text{vec}\begin{bmatrix} 0 & 0 \\ 0 & m_{22,1} \end{bmatrix} = \begin{bmatrix} 0 \\ 0 \\ 0 \\ m_{22,1} \end{bmatrix}.$$

Consequently,

$$\frac{\partial \boldsymbol{\alpha}}{\partial \boldsymbol{\gamma}'} = H_{\boldsymbol{\alpha}} R = \begin{bmatrix} 0 & 1 & 0 & 0 & 0 & 0 \\ \alpha_{11,1} & \alpha_{21,0} & 1 & 0 & 0 & 0 \\ 0 & 0 & 0 & 0 & 0 & 0 \\ 0 & 0 & 0 & 1 & 0 & 0 \\ 0 & 0 & 0 & 0 & 1 & 0 \\ \alpha_{11,2} & 0 & 0 & 0 & \alpha_{21,0} & 0 \\ 0 & 0 & 0 & 0 & 0 & 0 \\ 0 & 0 & 0 & 0 & 0 & 0 \end{bmatrix} \tag{7.4.13}$$

and

$$\frac{\partial \mathbf{m}}{\partial \boldsymbol{\gamma}'} = H_{\mathbf{m}} R = \begin{bmatrix} 0 & 0 & 0 & 0 & 0 & 0 \\ 0 & 0 & 0 & 0 & 0 & 0 \\ 0 & 0 & 0 & 0 & 0 & 0 \\ 0 & 0 & 0 & 0 & 0 & 1 \end{bmatrix} \tag{7.4.14}$$

(see also Problem 7.7).

7.4.2 A Real Data Example

In their general form the results may look more complicated than they usually are. Therefore, considering our income/consumption example from Section 7.3.5 again may be helpful. For the VARMA(2, 2) model with Kronecker indices (0, 2) given in (7.3.27) the parameters are

$$\boldsymbol{\gamma} = (\alpha_{22,1}, \alpha_{22,2}, m_{21,1}, m_{22,1}, m_{21,2}, m_{22,2})'.$$

The ML estimates are given in Table 7.1. Using $\tilde{\boldsymbol{\gamma}} = \tilde{\boldsymbol{\gamma}}_{10}$ from that table an estimate of $I(\boldsymbol{\gamma})$ is obtained from the iterations described in Section 7.3.5, that is, we use $\hat{I}(\tilde{\boldsymbol{\gamma}}_{10}) = \hat{I}(\tilde{\boldsymbol{\gamma}})$. The square roots of the diagonal elements of $\hat{I}(\tilde{\boldsymbol{\gamma}})^{-1}$ are estimates of the standard errors of the elements of $\tilde{\boldsymbol{\gamma}}$. Giving the estimated standard errors in parentheses we get

$$\tilde{\boldsymbol{\gamma}} = \begin{bmatrix} .225 & (.252) \\ .061 & (.166) \\ .313 & (.090) \\ -.750 & (.274) \\ .140 & (.141) \\ .160 & (.233) \end{bmatrix}. \tag{7.4.15}$$

As mentioned in Remark 5 of Section 7.4.1 an alternative, *asymptotically* equivalent estimator is obtained by iterating just once. In the present example that leads to estimates

$$\tilde{\gamma}_2 = \begin{bmatrix} -.178 & (.165) \\ .492 & (.133) \\ .331 & (.099) \\ -.527 & (.172) \\ .175 & (.127) \\ -.015 & (.152) \end{bmatrix}. \tag{7.4.16}$$

These estimates are somewhat different from those in (7.4.15). However, given the sampling variability reflected in the estimated standard errors the differences in most of the parameter estimates are not substantial.

Under a two-standard error criterion only two of the coefficients in (7.4.15) are significantly different from zero. As a consequence one may wish to restrict some of the coefficients to zero and thereby further reduce the parameter space. We will not do so at this stage but consider the implied estimates of $\boldsymbol{\alpha}$ and $\mathbf{m}$ (see, however, Problem 7.10). In this case $\tilde{\boldsymbol{\alpha}}$ and $\tilde{\mathbf{m}}$ are simply obtained by adding zeroes to $\tilde{\boldsymbol{\gamma}}$:

$$\tilde{\boldsymbol{\alpha}} = \operatorname{vec}(\tilde{A}_1, \tilde{A}_2) = \begin{bmatrix} 0 \\ 0 \\ 0 \\ .225\,(.252) \\ 0 \\ 0 \\ 0 \\ .061\,(.166) \end{bmatrix}, \quad \tilde{\mathbf{m}} = \operatorname{vec}(\tilde{M}_1, \tilde{M}_2) = \begin{bmatrix} 0 \\ .313\,(.090) \\ 0 \\ -.750\,(.274) \\ 0 \\ .140\,(.141) \\ 0 \\ .160\,(.233) \end{bmatrix}. \tag{7.4.17}$$

The standard errors are, of course, not affected by adding a few zero elements. A more elaborate but still simple computation becomes necessary to obtain the standard errors if $A_0 \neq I_K$ (see Corollary 7.1.2).

7.5 Forecasting Estimated VARMA Processes

With respect to forecasting with estimated processes, in principle, the same arguments apply for VARMA models that have been put forward for pure VAR models. Suppose that the generation process of a multiple time series of interest admits a VARMA(p, q) representation

$$y_t - \mu = A_1(y_{t-1} - \mu) + \cdots + A_p(y_{t-p} - \mu) + u_t + M_1 u_{t-1} + \cdots + M_q u_{t-q} \tag{7.5.1}$$

and denote by $\hat{y}_t(h)$ the h-step forecast at origin t given in Section 6.5, based on estimated rather than known coefficients. For instance, using the pure VAR representation of the process,

$$\hat{y}_t(h) = \hat{\mu} + \sum_{i=1}^{h-1} \hat{\Pi}_i(\hat{y}_t(h-i) - \hat{\mu}) + \sum_{i=h}^{\infty} \hat{\Pi}_i(y_{t+h-i} - \hat{\mu}). \tag{7.5.2}$$

For practical purposes one would, of course, truncate the infinite sum at $i = T$. For the moment we will, however, consider the infinite sum. For this predictor the forecast error is

$$y_{t+h} - \hat{y}_t(h) = [y_{t+h} - y_t(h)] + [y_t(h) - \hat{y}_t(h)],$$

where $y_t(h)$ is the optimal forecast based on known coefficients and the two terms on the right-hand side are uncorrelated as the first one can be written in terms of u_s with $s > t$ and the second one contains only y_s with $s \leq t$. Thus the forecast MSE becomes

$$\begin{aligned}\Sigma_{\hat{y}}(h) &= \mathrm{MSE}[y_t(h)] + \mathrm{MSE}[y_t(h) - \hat{y}_t(h)] \\ &= \Sigma_y(h) + E[y_t(h) - \hat{y}_t(h)][y_t(h) - \hat{y}_t(h)]'.\end{aligned} \tag{7.5.3}$$

Formally this is the same expression that was obtained for finite order VAR processes and, using the same arguments as in that case, we approximate $\mathrm{MSE}[y_t(h) - \hat{y}_t(h)]$ by $\Omega(h)/T$, where

$$\Omega(h) = E\left[\frac{\partial y_t(h)}{\partial \boldsymbol{\eta}'} \Sigma_{\tilde{\boldsymbol{\eta}}} \frac{\partial y_t(h)'}{\partial \boldsymbol{\eta}}\right], \tag{7.5.4}$$

$\boldsymbol{\eta}$ is the vector of estimated coefficients, and $\Sigma_{\tilde{\boldsymbol{\eta}}}$ is its asymptotic covariance matrix. If ML estimation is used and

$$\boldsymbol{\eta} = \begin{bmatrix} \mu \\ \boldsymbol{\alpha} \\ \mathbf{m} \end{bmatrix},$$

where $\boldsymbol{\alpha} = \mathrm{vec}[A_1, \ldots, A_p]$ and $\mathbf{m} = \mathrm{vec}[M_1, \ldots, M_q]$, we have from Proposition 7.1 and Corollaries 7.1.1 and 7.1.2,

$$\Sigma_{\tilde{\boldsymbol{\eta}}} = \begin{bmatrix} \Sigma_{\tilde{\mu}} & 0 \\ 0 & \Sigma_{\left[\begin{smallmatrix}\tilde{\boldsymbol{\alpha}} \\ \tilde{\mathbf{m}}\end{smallmatrix}\right]} \end{bmatrix}.$$

Thus,

$$\frac{\partial y_t(h)}{\partial \boldsymbol{\eta}'} \Sigma_{\tilde{\boldsymbol{\eta}}} \frac{\partial y_t(h)'}{\partial \boldsymbol{\eta}} = \frac{\partial y_t(h)}{\partial \mu'} \Sigma_{\tilde{\mu}} \frac{\partial y_t(h)'}{\partial \mu} + \frac{\partial y_t(h)}{\partial [\boldsymbol{\alpha}', \mathbf{m}']} \Sigma_{\left[\begin{smallmatrix}\tilde{\boldsymbol{\alpha}} \\ \tilde{\mathbf{m}}\end{smallmatrix}\right]} \frac{\partial y_t(h)'}{\partial \begin{bmatrix} \boldsymbol{\alpha} \\ \mathbf{m} \end{bmatrix}}.$$

Hence, in order to get an expression for $\Omega(h)$ we need the partial derivatives of $y_t(h)$ with respect to μ, $\boldsymbol{\alpha}$, and $\mathbf{m}$. They are given in the next lemma.

LEMMA 7.2

If y_t is a process with stable and invertible VARMA(p, q) representation (7.5.1) and pure VAR representation

$$y_t = \mu + \sum_{i=1}^{\infty} \Pi_i(y_{t-i} - \mu) + u_t,$$

we have

$$\frac{\partial y_t(h)}{\partial \mu'} = \begin{cases} \left(I_K - \sum_{i=1}^{\infty} \Pi_i\right) & \text{for } h = 1, \\ \left(I_K - \sum_{i=1}^{\infty} \Pi_i\right) + \sum_{i=1}^{h-1} \Pi_i \frac{\partial y_t(h-i)}{\partial \mu'}, & h = 2, 3, \ldots, \end{cases}$$

$$\frac{\partial y_t(h)}{\partial[\boldsymbol{\alpha}', \mathbf{m}']} = \sum_{i=1}^{h-1} ((y_t(h-i) - \mu)' \otimes I_K) \frac{\partial \operatorname{vec}(\Pi_i)}{\partial[\boldsymbol{\alpha}', \mathbf{m}']} + \sum_{i=1}^{h-1} \Pi_i \frac{\partial y_t(h-i)}{\partial[\boldsymbol{\alpha}', \mathbf{m}']}$$

$$+ \sum_{i=h}^{\infty} ((y_{t+h-i} - \mu)' \otimes I_K) \frac{\partial \operatorname{vec}(\Pi_i)}{\partial[\boldsymbol{\alpha}', \mathbf{m}']}, \qquad \text{for } h = 1, 2, \ldots,$$

with

$$\frac{\partial \operatorname{vec}(\Pi_i)}{\partial[\boldsymbol{\alpha}', \mathbf{m}']} = -\sum_{j=0}^{i-1} (H'(\mathbf{M}')^{i-1-j} \otimes \mathbf{J}\mathbf{M}^j) \begin{bmatrix} 0 & I_{Kq} \otimes J' \\ I_{Kp} \otimes J' & 0 \end{bmatrix},$$

where H, $\mathbf{M}$, and J are as defined in Chapter 6, Section 6.3.2, (6.3.12). In other words, H, $\mathbf{M}$, and J are defined so that $-\Pi_i = J\mathbf{M}^i H$. ■

The proof of this lemma is left as an exercise (see Problem 7.8). The formulas given in this lemma can be used for recursively computing the partial derivatives of $y_t(h)$ with respect to the VARMA coefficients for $h = 1, 2, \ldots$.

An estimator of $\Omega(h)$ is obtained by replacing all unknown quantities by their respective estimators and truncating the infinite sum at $i = T$. Denoting the resulting estimated partial derivatives by

$$\frac{\hat{\partial} y_t(h)}{\partial \mu'} \quad \text{and} \quad \frac{\hat{\partial} y_t(h)}{\partial[\boldsymbol{\alpha}', \mathbf{m}']},$$

an estimator for $\Omega(h)$ is

$$\hat{\Omega}(h) = \frac{1}{T}\sum_{t=1}^{T} \left[\frac{\hat{\partial} y_t(h)}{\partial \mu'} \hat{\Sigma}_{\tilde{\mu}} \frac{\hat{\partial} y_t(h)'}{\partial \mu} + \frac{\hat{\partial} y_t(h)}{\partial[\boldsymbol{\alpha}', \mathbf{m}']} \hat{\Sigma}_{\left[\begin{smallmatrix}\tilde{\boldsymbol{\alpha}}\\ \tilde{\mathbf{m}}\end{smallmatrix}\right]} \frac{\hat{\partial} y_t(h)'}{\partial \begin{bmatrix}\boldsymbol{\alpha}\\ \mathbf{m}\end{bmatrix}}\right], \tag{7.5.5}$$

where $\hat{\Sigma}_{\tilde{\mu}}$ and $\hat{\Sigma}_{\left[\begin{smallmatrix}\tilde{\boldsymbol{\alpha}}\\ \tilde{\mathbf{m}}\end{smallmatrix}\right]}$ are estimators of $\Sigma_{\tilde{\mu}}$ and $\Sigma_{\left[\begin{smallmatrix}\tilde{\boldsymbol{\alpha}}\\ \tilde{\mathbf{m}}\end{smallmatrix}\right]}$, respectively (see Corollaries 7.1.1 and 7.1.2 for the latter matrix). An estimator of the forecast MSE matrix (7.5.3) is then

$$\hat{\Sigma}_{\hat{y}}(h) = \hat{\Sigma}_y(h) + \frac{1}{T}\hat{\Omega}(h), \tag{7.5.6}$$

where the estimator $\hat{\Sigma}_y(h)$ is again obtained by replacing unknown quantities by their respective estimators.

With these results in hand, forecast intervals can be set up under Gaussian assumptions just as in the finite order VAR case discussed in Chapters 2 and 3.

7.6 Estimated Impulse Responses

As mentioned in Section 6.7.2 the impulse responses of a VARMA(p, q) process are the coefficients of pure MA representations. For instance, if the process is in standard form, the forecast error impulse responses are

$$\Phi_i = J\mathbf{A}^i H \tag{7.6.1}$$

with J, $\mathbf{A}$, and H as defined in Section 6.3.2 (see (6.3.10)). Other quantities of interest may be the elements of $\Theta_i = \Phi_i P$, where P is a lower triangular Choleski decomposition of Σ_u, the white noise covariance matrix. Also forecast error variance components and accumulated impulse responses may be of interest. All these quantities are estimated in the usual way from the estimated coefficients of the process. For example, $\hat{\Phi}_i = J\hat{\mathbf{A}}^i H$, where $\hat{\mathbf{A}}$ is obtained from $\mathbf{A}$ by replacing the A_i and M_j by estimators $\hat{A}_i$ and $\hat{M}_j$. The asymptotic distributions of the estimated quantities follow immediately from Propositions 3.6 which is formulated for the finite order VAR case. The only modifications that we have to make to accommodate the VARMA(p, q) case are to replace $\boldsymbol{\alpha}$ by

$$\boldsymbol{\beta} := \text{vec}[A_1, \ldots, A_p, M_1, \ldots, M_q] = \begin{bmatrix} \boldsymbol{\alpha} \\ \mathbf{m} \end{bmatrix},$$

replace $\Sigma_{\hat{\boldsymbol{\alpha}}}$ by $\Sigma_{\hat{\boldsymbol{\beta}}}$ and specify

$$\begin{aligned} G_i = \frac{\partial \,\text{vec}(\Phi_i)}{\partial \boldsymbol{\beta}'} &= (H' \otimes J)\frac{\partial \,\text{vec}(\mathbf{A}^i)}{\partial \boldsymbol{\beta}'} \\ &= (H' \otimes J)\left[\sum_{m=0}^{i-1} (\mathbf{A}')^{i-1-m} \otimes \mathbf{A}^m\right]\frac{\partial \,\text{vec}(\mathbf{A})}{\partial \boldsymbol{\beta}'} \\ &= \sum_{m=0}^{i-1} H'(\mathbf{A}')^{i-1-m} \otimes J\mathbf{A}^m J'. \end{aligned} \tag{7.6.2}$$

With these modifications of Proposition 3.6 the asymptotic distributions of all the quantities of interest are available. Of course, all the caveats of Proposition 3.6 apply here too.

7.7 Exercises

Problem 7.1

Are the operators

$$\begin{bmatrix} 1 - 0.5L & 0.3L \\ 0 & 1 \end{bmatrix} \quad \text{and} \quad \begin{bmatrix} 1 - 0.2L & 1.3L - 0.44L^2 \\ 0.5L & 1 + 0.2L \end{bmatrix}$$

left-coprime? (Hint: Show that the first operator is a common factor.)

Problem 7.2

Write the bivariate process

$$\begin{bmatrix} 1 - \beta_1 L & 0 \\ \beta_2 L^2 & 1 - \beta_3 L \end{bmatrix} y_t = \begin{bmatrix} 1 - \beta_1 L & 0 \\ \beta_4 L & 1 \end{bmatrix} u_t$$

in final equations form and in echelon form.

Problem 7.3

Show that (7.4.12) is an echelon representation.

Problem 7.4

Derive the likelihood function for a general Gaussian VARMA(p, q) model given fixed but not necessarily zero initial vectors $y_{-p+1}, \ldots, y_0$. Do not assume that $u_{-q+1} = \cdots = u_0 = 0$!

Problem 7.5

Identify the rules from Appendix A that are used in deriving the partial derivatives in (7.3.8).

Problem 7.6

Suppose $\ln|\tilde{\Sigma}_u(\tilde{\mu}, \gamma)|$ given in (7.3.11) is to be minimized with respect to γ. Show that the resulting normal equations are

$$\frac{\partial \ln|\tilde{\Sigma}_u(\mu, \gamma)|}{\partial \gamma'} = \frac{2}{T} \sum_{t=1}^{T} u_t' \tilde{\Sigma}_u^{-1} \frac{\partial u_t}{\partial \gamma'}.$$

Thus, the normal equations are equivalent to those obtained from the log-likelihood function.

Problem 7.7

Consider the bivariate VARMA(2, 1) process given in (7.4.12) and set up the matrices $H_{\boldsymbol{\alpha}}$ and $H_{\mathbf{m}}$ according to their general form given in Corollary 7.1.2. Show that $H_{\boldsymbol{\alpha}}R$ and $H_{\mathbf{m}}R$ are identical to the matrices specified in (7.4.13) and (7.4.14), respectively.

Problem 7.8

Prove Lemma 7.2. (Hint: Use Rule (8) of Appendix A.13.)

Problem 7.9

Derive the asymptotic covariance matrices of the impulse responses and forecast error variance components obtained from an estimated VARMA process. (Hint: Use the suggestion given in Section 7.6.)

Problem 7.10

Consider the income/consumption example of Section 7.3.5 and determine preliminary and full ML estimates for the parameters of the model

$$\begin{bmatrix} y_{1,t} \\ y_{2,2} \end{bmatrix} = \begin{bmatrix} 0 & 0 \\ 0 & \alpha_{22,2} \end{bmatrix} \begin{bmatrix} y_{1,t-2} \\ y_{2,t-2} \end{bmatrix} + \begin{bmatrix} u_{1,t} \\ u_{2,t} \end{bmatrix} + \begin{bmatrix} 0 & 0 \\ m_{21,1} & m_{22,1} \end{bmatrix} \begin{bmatrix} u_{1,t-1} \\ u_{2,t-1} \end{bmatrix}$$
$$+ \begin{bmatrix} 0 & 0 \\ m_{21,2} & 0 \end{bmatrix} \begin{bmatrix} u_{1,t-2} \\ u_{2,t-2} \end{bmatrix}.$$

Chapter 8. Specification and Checking the Adequacy of VARMA Models

8.1 Introduction

A great number of strategies has been suggested for specifying VARMA models. There is not a single one that has become a standard like the Box-Jenkins (1976) approach in the univariate case. None of the multivariate procedures is in widespread use for modeling moderate or high-dimensional economic time series. Some are mainly based on a subjective assessment of certain characteristics of a process such as the autocorrelations and partial autocorrelations. A decision on specific orders and constraints on the coefficient matrices is then based on these quantities. Other methods rely on a mixture of statistical testing, use of model selection criteria and personal judgement of the analyst. Again other procedures are based predominantly on statistical model selection criteria and, in principle, they could be performed automatically by a computer. Automatic procedures have the advantage that their statistical properties can possibly be derived rigorously. In actual applications some kind of mixture of different approaches is often applied. In other words, the expertise and prior knowledge of an analyst will usually not be abolished in favor of purely statistical procedures. Models suggested by different types of criteria and procedures will be judged and evaluated by an expert before one or more candidates are put to a specific use such as forecasting. The large amount of information in a moderate number of moderately long time series makes it usually necessary to condense the information considerably before essential features of a system become visible.

In the following we will outline procedures for specifying the final equations form and the echelon form of a VARMA process. We do not claim that these procedures are superior to other approaches. They are just meant to illustrate what is involved in the specification of VARMA models. The specification strategies for both forms could be turned into automatic algorithms. On the other hand, they also leave room for human intervention if desired. In Section 8.4 some references for other specification strategies are given and model checking is discussed briefly in Section 8.5. A critique of VARMA modeling is given in Section 8.6.

8.2 Specification of the Final Equations Form

8.2.1 A Specification Procedure

Historically procedures for specifying final equations VARMA representations have been among the earlier strategies for modeling systems of economic time series (see, for example, Zellner & Palm (1974), Wallis (1977)). The objective is to find the orders p and q of the representation

$$\alpha(L)y_t = M(L)u_t, \tag{8.2.1}$$

where

$$\alpha(L) = 1 - \alpha_1 L - \cdots - \alpha_p L^p$$

is a (1×1) scalar operator,

$$M(L) = I_K + M_1 L + \cdots + M_q L^q$$

is a $(K \times K)$ operator and it is assumed that the process mean has been removed in a previous step of the analysis.

If a K-dimensional system $y_t = (y_{1t}, \ldots, y_{Kt})'$ has a VARMA representation of the form (8.2.1) it follows that each component series has a univariate ARMA representation

$$\alpha(L)y_{kt} = \bar{m}_k(L)v_{kt}, \qquad k = 1, \ldots, K,$$

where $\bar{m}_k(L)$ is an operator of degree at most q because the k-th row of $M(L)u_t$ is

$$m_{k1}(L)u_{1t} + \cdots + m_{kK}(L)u_{Kt}.$$

In other words, it is a sum of MA(q) processes which is known to have an MA representation of degree at most q (see Proposition 6.1). Thus, each component series of y_t has the same AR operator and an MA operator of degree at most q. In general at least one of the component series will have MA degree q because a reduction of the MA order of all component series requires a very special set of parameters which is not regarded likely in practice. This fact is used in specifying the final form VARMA representation by first determining univariate component models and then putting them together in a joint model. Specifically the following specification strategy is used.

STAGE I: Specify univariate models

$$\alpha_k(L)y_{kt} = m_k(L)v_{kt}$$

for the components of y_t. Here

$$\alpha_k(L) = 1 - \alpha_{k1}L - \cdots - \alpha_{kp_k}L^{p_k}$$

is of order p_k,

$$m_k(L) = 1 + m_{k1}L + \cdots + m_{kq_k}L^{q_k}$$

is of order q_k, and v_{kt} is a univariate white noise process. ■

The Box-Jenkins (1976) strategy for specifying univariate ARMA models may be used at this stage. Alternatively, some automatic procedure or criterion such as the one proposed by Hannan & Rissanen (1982) or Poskitt (1987) may be applied.

STAGE II: Determine a common AR operator $\alpha(L)$ for all component processes, specify the corresponding MA orders and choose the degree q of the joint MA operator as the maximum of the individual MA degrees obtained in this way. ■

At this stage a common AR operator may, for example, be obtained as the product of the individual operators, that is,

$$\alpha(L) = \alpha_1(L)\cdots\alpha_K(L).$$

In this case the k-th component process is multiplied by

$$\prod_{\substack{i=1\\i\neq k}}^{K} \alpha_i(L)$$

and $\alpha(L)$ has degree $p = \sum_{i=1}^{K} p_i$ while the MA operator

$$\bar{m}_k(L) = m_k(L)\prod_{\substack{i=1\\i\neq k}}^{K} \alpha_i(L)$$

has degree

$$q_k + \sum_{\substack{i=1\\i\neq k}}^{K} p_i.$$

The joint MA operator of the VARMA representation (8.2.1) is then assumed to have degree

$$\max_k \left[q_k + \sum_{\substack{i=1\\i\neq k}}^{K} p_i \right]. \tag{8.2.2}$$

Of course, the $\alpha_k(L)$, $k = 1, \ldots, K$, may have common factors. In that case a joint AR operator $\alpha(L)$ with degree much lower than $\sum_{i=1}^{K} p_i$ may be possible. Correspondingly the joint MA operator may have degree lower than (8.2.2). Suppose, for instance, that $K = 3$ and

$$\alpha_1(L) = 1 - \alpha_{11}L, \qquad m_1(L) = 1 + m_{11}L,$$

$$\alpha_2(L) = 1 - \alpha_{21}L - \alpha_{22}L^2, \qquad m_2(L) = 1 + m_{21}L,$$

$$\alpha_3(L) = 1 - \alpha_{31}L, \qquad m_3(L) = 1.$$

Now a joint AR operator is

$$\alpha(L) = \alpha_1(L)\alpha_2(L)\alpha_3(L),$$

which has degree 4. However, if $\alpha_2(L)$ can be factored as

$$\alpha_2(L) = (1 - \alpha_{11}L)(1 - \alpha_{31}L) = \alpha_1(L)\alpha_3(L),$$

a common AR operator $\alpha(L) = \alpha_2(L)$ may be chosen and we get univariate models

$$\begin{aligned}
\alpha(L)y_{1t} &= \alpha_3(L)m_1(L)v_{1t} \qquad &(\text{ARMA}(2, 2)),\\
\alpha(L)y_{2t} &= m_2(L)v_{2t} \qquad &(\text{ARMA}(2, 1)),\\
\alpha(L)y_{3t} &= \alpha_1(L)m_3(L)v_{3t} \qquad &(\text{ARMA}(2, 1)).
\end{aligned} \tag{8.2.3}$$

The maximum of the individual MA degrees is chosen as the joint MA degree, that is, $q = 2$ and, of course,

$$p = \text{degree}(\alpha(L)) = 2.$$

A problem that should be noticed from this discussion and example is that the degrees p and q determined in this way may be quite large. It is conceivable that $p = \sum_{i=1}^{K} p_i$ is the smallest possible AR order for the final equations form representation and the corresponding MA degree may be quite substantial too. This, clearly, can be a disadvantage as unduely many parameters can cause trouble in a final estimation algorithm and may lead to imprecise forecasts and impulse responses.

Often it may be possible to impose restrictions on the AR and MA operators in (8.2.1). This may either be done in a third stage of the procedure or it may be incorporated in Stages I and/or II depending on the type of information available. Restrictions may be obtained with the help of statistical tools such as testing the significance of single coefficients or groups of parameters. Alternatively restrictions may be implied by subject matter theory. Zellner & Palm (1974) give a detailed example where both types of restrictions are used.

Perhaps because of the potentially great number of parameters, final form modeling seems to have lost popularity recently. It can only be recommended if it results in a reasonably parsimonious parameterization.

8.2.2 An Example

For illustrative purposes we consider a bivariate system consisting of first differences of logarithms of income (y_1) and consumption (y_2). We use again the data from Table E.1 up to the fourth quarter of 1978. If the 3-dimensional system involving investment in addition really were generated by a VAR(2) process as assumed in Chapters 3 and 4, it is quite possible that the subprocess consisting of income and consumption only has a mixed VARMA generation process with nontrivial MA part (see Section 6.6.1). Moreover one would expect that the marginal univariate processes for y_1 and y_2 are of a mixed ARMA type. However, we found that the subset AR(3) models

$$\underset{}{(1 - \underset{(.113)}{.245}L^3)}y_{1t} = \underset{(.003)}{.015} + v_{1t} \tag{8.2.4}$$

and

$$(1 - .309L^2 - .187L^3)y_{2t} = .010 + v_{2t} \tag{8.2.5}$$
$$\quad\;\; (.111) \qquad (.111) \qquad\qquad (.004)$$

fit the data quite well. For illustrative purposes we will therefore proceed from these models. The reader may try to find better models and repeat the analysis with them.

Generally a (1×1) scalar operator

$$\gamma(L) = 1 - \gamma_1 L - \cdots - \gamma_p L^p$$

of degree p can be factored in p components,

$$\gamma(L) = (1 - \lambda_1 L)\cdots(1 - \lambda_p L),$$

where $\lambda_1, \ldots, \lambda_p$ are the reciprocals of the roots of $\gamma(z)$. Thus, the two AR operators from (8.2.4) and (8.2.5) can be factored as

$$\begin{aligned}\alpha_1(L) &= 1 - .245L^3 \\ &= (1 - .626L)(1 + (.313 + .542i)L)(1 + (.313 - .542i)L)\end{aligned} \tag{8.2.6}$$

and

$$\begin{aligned}\alpha_2(L) &= 1 - .309L^2 - .187L^3 \\ &= (1 - .747L)(1 + (.374 + .332i)L)(1 + (.374 - .332i)L),\end{aligned} \tag{8.2.7}$$

where i denotes the imaginary part of the complex numbers. None of the factors in (8.2.6) is very close to any of the factors in (8.2.7). Thus, models with common AR operator may be of the form

$$\alpha_1(L)\alpha_2(L)y_{1t} = \alpha_2(L)v_{1t}$$

and

$$\alpha_1(L)\alpha_2(L)y_{2t} = \alpha_1(L)v_{2t}.$$

With the arguments of the previous subsection the resulting bivariate final equations model is a VARMA(6, 3) process,

$$\alpha_1(L)\alpha_2(L)\begin{bmatrix} y_{1t} \\ y_{2t} \end{bmatrix} = (I_3 + M_1 L + M_2 L^2 + M_3 L^3)\begin{bmatrix} u_{1t} \\ u_{2t} \end{bmatrix}. \tag{8.2.8}$$

Obviously, this model involves very many parameters and is therefore unattractive. In fact, such a heavily parameterized model may cause numerical problems when full maximum likelihood estimation is attempted. It is possible, if not likely, that some parameters turn out to be insignificant and could be set at zero. However, the significance of parameters is commonly judged on the basis of their standard errors or t-ratios. These quantities become available in the ML estimation round which, as we have argued, may be problematic in the present case.

Given the estimation uncertainty one may argue that the real factors in the operators $\alpha_1(L)$ and $\alpha_2(L)$ may be identical. Proceeding under that assumption results in a VARMA(5, 2) final equations form. Such a model is more parsimonious and has therefore more appeal than (8.2.8). Still it involves a considerable number of parameters. This illustrates why final equations modeling, although relatively simple, does not enjoy general popularity. For higher-dimensional models the problem of heavy parameterization becomes even more severe because the number of parameters is likely to increase rapidly with the dimension of the system. We will now present procedures for specifying echelon forms.

8.3 Specification of Echelon Forms

In specifying an echelon VARMA representation the objective is to find the Kronecker indices and possibly impose some further restrictions on the parameters. For a K-dimensional process there are K Kronecker indices. Different strategies have been proposed for their specification. We will discuss some of them in the following. Once the Kronecker indices are determined further restrictions may be imposed, for instance, on the basis of significance tests for individual coefficients or groups of parameters.

In the first subsection below we will discuss a procedure for specifying the Kronecker indices which is usually not feasible in practice. It is nevertheless useful to study that procedure because the feasible strategies considered in Subsections 8.3.2–8.3.4 may be regarded as approximations or short-cuts of that procedure with similar asymptotic properties. In Subsection 8.3.2 we present a procedure which is easy to carry out for systems with small Kronecker indices and low dimension. It is quite costly for higher-dimensional systems, though. For such systems a specification strategy inspired by Hannan & Kavalieris (1984) or a procedure due to Poskitt (1989a) may be more appropriate. These approaches are considered in Subsections 8.3.3 and 8.3.4, respectively. The material discussed in this section is covered in more depth and more rigorously in Hannan & Deistler (1988, Chapters 5, 6, 7) and Poskitt (1989a).

8.3.1 A Procedure for Small Systems

If it is known that the generation process of a given K-dimensional multiple time series admits an echelon VARMA representation with Kronecker indices $p_k \leq p_{max}$, $k = 1, \ldots, K$, where p_{max} is a prespecified number, then, in theory, it is possible to evaluate the maximum log-likelihood for all sets of Kronecker indices $\mathbf{p} = (p_1, \ldots, p_K)'$ with $p_k \leq p_{max}$ and choose the set $\hat{\mathbf{p}}$ that optimizes a specific criterion. This approach is completely analogous to the specification of the VAR order in the finite order VAR case considered in Section 4.3. In that section we have discussed the possibility to consistently estimate the VAR order with such an approach. It turns out that a similar result can be obtained for the present more general VARMA case.

Before we give further details it may be worth emphasizing, however, that in the VARMA case such a specification strategy is generally not feasible in practice because the maximization of the log-likelihood is usually quite costly and, for systems with moderate or high dimension an enormous number of likelihood maximizations would be required. For instance, for a five-dimensional system, evaluating the maximum of the log-likelihood for all vectors of Kronecker indices $\mathbf{p} = (p_1, \dots, p_5)'$ with $p_k \leq 8$ requires $9^5 = 59{,}049$ likelihood maximizations. Despite this practical problem we discuss the theoretical properties of this procedure to provide a basis for the following subsections.

Let us denote by $\tilde{\Sigma}_u(\mathbf{p})$ the ML estimator of the white noise covariance matrix Σ_u obtained for a set of Kronecker indices $\mathbf{p}$. Furthermore, let

$$Cr(\mathbf{p}) := \ln|\tilde{\Sigma}_u(\mathbf{p})| + c_T d(\mathbf{p})/T \tag{8.3.1}$$

be a criterion to be minimized over all sets of Kronecker indices $\mathbf{p} = (p_1, \dots, p_K)'$, $p_k \leq p_{\max}$. Here $d(\mathbf{p})$ is the number of freely varying parameters in the echelon form with Kronecker indices $\mathbf{p}$. For example, for a bivariate system with Kronecker indices $\mathbf{p} = (p_1, p_2)' = (1, 0)'$, the echelon form is

$$\begin{bmatrix} 1 & 0 \\ -\alpha_{21,0} & 1 \end{bmatrix} \begin{bmatrix} y_{1,t} \\ y_{2,t} \end{bmatrix} = \begin{bmatrix} \alpha_{11,1} & 0 \\ 0 & 0 \end{bmatrix} \begin{bmatrix} y_{1,t-1} \\ y_{2,t-1} \end{bmatrix} + \begin{bmatrix} 1 & 0 \\ -\alpha_{21,0} & 1 \end{bmatrix} \begin{bmatrix} u_{1,t} \\ u_{2,t} \end{bmatrix}$$
$$+ \begin{bmatrix} m_{11,1} & m_{12,1} \\ 0 & 0 \end{bmatrix} \begin{bmatrix} u_{1,t-1} \\ u_{2,t-1} \end{bmatrix}.$$

Thus, $d((1, 0)') = 4$. In (8.3.1) c_T is a sequence indexed by the sample size T.

In general, if models are included in the search procedure for which all Kronecker indices exceed the true ones, the estimation of unidentified models is required for which cancellation of the VAR and MA operators is possible. This is not a problem for evaluating the criterion in (8.3.1) since we only need the maximum log-likelihood or rather $\ln|\tilde{\Sigma}_u(\mathbf{p})|$ in that criterion. That quantity can be determined even if the corresponding VARMA coefficients are meaningless. The coefficients cannot and should not be interpreted, however.

Note that the criterion (8.3.1) is very similar to that considered in Proposition 4.2 of Chapter 4. In that proposition the consistency or inconsistency of a criterion is seen to depend on the choice of the sequence c_T. Hannan (1981) and Hannan & Deistler (1988, Chapter 5, Section 5) show that a criterion such as the one in (8.3.1) provides a consistent estimator of the true set of Kronecker indices if c_T is a nondecreasing function of T satisfying

$$c_T \to \infty \qquad \text{and} \qquad c_T/T \to 0 \qquad \text{as } T \to \infty, \tag{8.3.2}$$

and the true data generation process satisfies some weak conditions. If, in addition,

$$c_T/2 \ln \ln T > 1 \tag{8.3.3}$$

eventually as $T \to \infty$, the procedure provides a strongly consistent estimator of the true Kronecker indices. The conditions for the VARMA process are, for instance, satisfied if the white noise process u_t is identically distributed standard white noise (see Definition 3.1) and the true data generation process admits a stable and invertible echelon VARMA representation with Kronecker indices not greater than $p_{\max}$. This result extends Proposition 4.2 to the VARMA case.

Implications of this result are that the Schwarz criterion with $c_T = \ln T$,

$$\mathrm{SC}(\mathbf{p}) := \ln|\tilde{\Sigma}_u(\mathbf{p})| + d(\mathbf{p}) \ln T/T \tag{8.3.4}$$

is strongly consistent and that the Hannan-Quinn criterion using the broderline penalty term $c_T = 2 \ln \ln T$,

$$\mathrm{HQ}(\mathbf{p}) := \ln|\tilde{\Sigma}_u(\mathbf{p})| + 2d(\mathbf{p}) \ln \ln T/T \tag{8.3.5}$$

is consistent. Hannan & Deistler (1988) also show that

$$\mathrm{AIC}(\mathbf{p}) := \ln|\tilde{\Sigma}_u(\mathbf{p})| + 2d(\mathbf{p})/T \tag{8.3.6}$$

with $c_T = 2$ is not a consistent criterion. Again these results are similar to those for the finite order VAR case.

As in that case it is worth emphasizing that these results do not necessarily imply the inferiority of AIC or HQ. In small samples these criteria may be preferable. They may, in fact, provide superior forecasting models. Also, in practice the actual data generation mechanism will usually not really admit a VARMA representation. Recall that the best we can hope for is that our model is a good approximation to the true process. In that case the relevance of the consistency property is of course doubtful.

Again the specification strategy presented in the foregoing is not likely to have much practical appeal as it is computationally too burdensome. In the next subsections more practical modifications are discussed.

8.3.2 A Full Search Procedure Based on Linear Least Squares Computations

8.3.2a The Procedure

A major obstacle for using the procedure described in the previous subsection is the requirement to maximize the log-likelihood various times. This is costly because for mixed VARMA models the log-likelihood function is nonlinear in the parameters and iterative optimization algorithms have to be employed. Since we just need an estimator of Σ_u for the evaluation of model selection criteria such as (8.3.1) an obvious modification of the procedure would use an estimator that avoids the nonlinear optimization problem. Such an estimator may be obtained from the preliminary estimation procedure described in Chapter 7, Section 7.3.4. Therefore a specification of the Kronecker indices may proceed in the following stages.

STAGE I: Fit a long VAR process of order n, say, to the data and obtain the residual vectors $\hat{u}_t(n)$, $t = 1, \ldots, T$. ■

The choice of n could be based on an order selection criterion such as AIC. In any case, n has to be greater than the greatest Kronecker index $p_{\max}$ that is being considered in the next stage of the procedure.

STAGE II: Using the residuals $\hat{u}_t(n)$ from Stage I, compute the preliminary estimator of Section 7.3.4 for all sets of Kronecker indices $\mathbf{p}$ with $p_k \le p_{\max}$, where the latter number is a prespecified upper bound for the Kronecker indices. Determine all corresponding estimators $\tilde{\Sigma}_u(\mathbf{p})$ based on the residuals of the preliminary estimations (see (7.3.24)). (Here we suppress the order n from the first stage for notational convenience because the same n is used for all $\tilde{\Sigma}_u(\mathbf{p})$ at this stage.) Choose the estimator $\hat{\mathbf{p}}$ which minimizes a prespecified criterion such as (8.3.1). ■

The choice of the criterion $Cr(\mathbf{p})$ is left to the researcher. SC, HQ, and AIC from (8.3.4)–(8.3.6) are possible candidates. Stage II could be iterated by using the residuals from a previous run through Stage II instead of the residuals from Stage I. Once an estimate $\hat{\mathbf{p}}$ of the Kronecker indices is determined the ML estimates conditional an $\hat{\mathbf{p}}$ may be computed in a final stage.

STAGE III: Estimate the echelon form VARMA model with Kronecker indices $\hat{\mathbf{p}}$ by maximizing the Gaussian log-likelihood function or by just one step of the scoring algorithm (see Section 7.3.2). ■

Hannan & Deistler (1988, Chapter 6) discuss conditions under which this procedure provides a consistent estimator of the Kronecker indices and VARMA parameter estimators that have the same asymptotic properties as the estimators obtained for given, known, true Kronecker indices (see Proposition 7.1). In addition to our usual assumptions for the VARMA process such as stability and invertibility the required assumptions relate to the criteria for choosing the VAR order in Stage I and the Kronecker indices in Stage II. These conditions are asymptotic conditions that leave some room for the actual choice in small samples. Any criterion from (8.3.4)–(8.3.6) may be a reasonable choice in practice.

The procedure still involves extensive computations unless the dimension K of the underlying multiple time series and $p_{\max}$ are small. For example, for a five-dimensional system with $p_{\max} = 8$ we still have to perform $9^5 = 59{,}049$ estimations in order to compare all feasible models. Although these estimations involve linear least squares computations only the computational costs are prohibitive. Therefore we outline two less costly procedures in the following subsections. For small systems the present procedure is a reasonable choice. We give an example next.

8.3.2b An Example

We consider again the income/consumption example from Section 8.2.2. In the first stage of our procedure we fit a VAR($n = 8$) model and we use the residuals

$\hat{u}_t(8)$ in the next stage. The choice of $n = 8$ is to some extent arbitrary. We have chosen a fairly high order to gain flexibility for the Kronecker indices considered in Stage II. Recall that n must exceed all Kronecker indices to be considered subsequently. Using the procedure described in Stage II we have estimated models with Kronecker indices $p_k \le p_{\max} = 4$ and we have determined the corresponding values of the criteria AIC and HQ. They are given in Tables 8.1 and 8.2, respectively. Both criteria reach their minimum for $\mathbf{p} = (p_1, p_2)' = (0, 2)'$. This is precisely the model estimated in Chapter 7, Section 7.3.5. Replacing the parameters by their ML estimates with estimated standard errors in parentheses we have

$$\begin{bmatrix} y_{1,t} \\ y_{2,t} \end{bmatrix} = \begin{bmatrix} 0 & 0 \\ 0 & \underset{(.252)}{.225} \end{bmatrix} \begin{bmatrix} y_{1,t-1} \\ y_{2,t-1} \end{bmatrix} + \begin{bmatrix} 0 & 0 \\ 0 & \underset{(.166)}{.061} \end{bmatrix} \begin{bmatrix} y_{1,t-2} \\ y_{2,t-2} \end{bmatrix} + \begin{bmatrix} \hat{u}_{1,t} \\ \hat{u}_{2,t} \end{bmatrix}$$

$$+ \begin{bmatrix} 0 & 0 \\ \underset{(.090)}{.313} & \underset{(.274)}{-.750} \end{bmatrix} \begin{bmatrix} \hat{u}_{1,t-1} \\ \hat{u}_{2,t-1} \end{bmatrix} + \begin{bmatrix} 0 & 0 \\ \underset{(.141)}{.140} & \underset{(.233)}{.160} \end{bmatrix} \begin{bmatrix} \hat{u}_{1,t-2} \\ \hat{u}_{2,t-2} \end{bmatrix}.$$

Obviously, some of the parameters are quite small compared to their estimated standard errors. In such a situation one may want to impose further zero con-

Table 8.1. AIC values of echelon form VARMA models with Kronecker indices (p_1, p_2) for the income/consumption data

p_2 \ p_1	0	1	2	3	4
0	−16.83	−18.41	−18.30	−18.25	−18.15
1	−18.50	−18.42	−18.30	−18.23	−18.13
2	−18.64*	−18.55	−18.42	−18.29	−18.19
3	−18.57	−18.50	−18.37	−18.27	−18.19
4	−18.47	−18.38	−18.27	−18.20	−18.05

* Minimum

Table 8.2. HQ values of echelon form VARMA models with Kronecker indices (p_1, p_2) for the income/consumption data

p_2 \ p_1	0	1	2	3	4
0	−16.83	−18.35	−18.21	−18.12	−17.98
1	−18.46	−18.31	−18.14	−18.03	−17.89
2	−18.56*	−18.41	−18.21	−18.03	−17.88
3	−18.45	−18.32	−18.12	−17.95	−17.82
4	−18.31	−18.16	−17.98	−17.84	−17.63

* Minimum

straints on the parameters. Since $\hat{\alpha}_{22,1}$, $\hat{\alpha}_{22,2}$ and $\hat{m}_{22,2}$ have the smallest t-ratios in absolute terms, we restrict these estimates to zero and reestimate the model. The resulting system obtained by ML estimation is

$$\begin{bmatrix} y_{1,t} \\ y_{2,t} \end{bmatrix} = \begin{bmatrix} \hat{u}_{1,t} \\ \hat{u}_{2,t} \end{bmatrix} + \begin{bmatrix} 0 & 0 \\ .308 & -.475 \\ (.088) & (.104) \end{bmatrix} \begin{bmatrix} \hat{u}_{1,t-1} \\ \hat{u}_{2,t-1} \end{bmatrix}$$

$$+ \begin{bmatrix} 0 & 0 \\ .302 & 0 \\ (.076) & \end{bmatrix} \begin{bmatrix} \hat{u}_{1,t-2} \\ \hat{u}_{2,t-2} \end{bmatrix}.$$

Now all parameters are significant under a two-standard error criterion.

8.3.3 Hannan-Kavalieris Procedure

A full search procedure for the optimal Kronecker indices as in Stage II of the previous subsection involves a prohibitive amount of computation work if the dimension K of the time series considered is large or if the upper bound $p_{\max}$ for the Kronecker indices is high. For instance, if monthly data are considered and lags of at least one year are deemed necessary, $p_{\max} \geq 12$ is required. Even if the system involves just three variables ($K = 3$) the number of models to be compared is vast, namely $13^3 = 2197$. Therefore shortcuts for Stage II of the previous subsection have been proposed. The first one we present here is inspired by discussions of Hannan & Kavalieris (1984). Therefore we call it the Hannan-Kavalieris procedure although these authors propose a more sophisticated approach. In particular they discuss a number of computational simplifications (see also Hannan & Deistler (1988, Chapter 6)). The following modification of Stage II may be worth trying.

STAGE II (HK): Based on the covariance estimators obtained from the preliminary estimation procedure of Section 7.3.4 find the Kronecker indices, say $\mathbf{p}^{(1)} = p^{(1)}(1, \dots, 1)'$, that minimize a prespecified criterion of the type $Cr(\mathbf{p})$ in (8.3.1) over $\mathbf{p} = p(1, \dots, 1)'$, $p = 0, \dots, p_{\max}$, that is, all Kronecker indices are identical in this first step. Then the last index p_K is varied between 0 and $p^{(1)}$ while all other indices are fixed at $p^{(1)}$. We denote the optimal value of p_K by $\hat{p}_K$, that is, $\hat{p}_K$ minimizes the prespecified criterion. Then we proceed in the same way with p_{K-1} and so on. More generally, $\hat{p}_k$ is chosen such that

$$Cr((p^{(1)}, \dots, p^{(1)}, \hat{p}_k, \dots, \hat{p}_K)')$$
$$= \min\{Cr((p^{(1)}, \dots, p^{(1)}, p, \hat{p}_{k+1}, \dots, \hat{p}_K)') | p = 0, \dots, p^{(1)}\}. \quad \blacksquare$$

This modification reduces the computational burden considerably. Just to give an example, for $K = 5$ and $p_{\max} = 8$, at most $9 + 5 \cdot 9 = 54$ models have to be estimated. If $p^{(1)}$ is small the number may be substantially lower. For comparison purposes we repeat that the number of estimations in a full search procedure would be $9^5 = 59{,}049$.

To illustrate the procedure consider the following panel of criterion values for Kronecker indices (p_1, p_2):

p_2 \ p_1	0	1	2	3
0	3.48	3.28	3.26	3.27
1	3.25	3.23	3.14	3.20
2	3.23	3.21	3.15	3.19
3	3.24	3.20	3.21	3.18

The minimum value on the main diagonal is obtained for $p^{(1)} = 2$ with $Cr(p^{(1)}, p^{(1)}) = 3.15$. Going upward from $(p_1, p_2) = (2, 2)$ the minimum is seen to be $Cr(2, 1) = 3.14$. Turning left from $(p_1, p_2) = (2, 1)$ a further reduction of the criterion value is not obtained. Therefore the estimate for the Kronecker indices is $\hat{\mathbf{p}} = (2, 1)'$.

In this case we have actually found the overall minimum of all criterion values in the panel, that is,

$$Cr(2, 1) = \min\{Cr(p_1, p_2) | p_i = 0, 1, 2, 3\}.$$

In general the Hannan-Kavalieris procedure will not lead to the overall minimum. For instance, for our bivariate income/consumption example from Subsection 8.3.2b we can find the HK estimates $\hat{\mathbf{p}}$ from Tables 8.1 and 8.2. On the main diagonals of both panels the criteria assume their minima for $(p_1, p_2) = (1, 1)$ and $\hat{\mathbf{p}}(\text{AIC}) = (0, 1)'$ while $\hat{\mathbf{p}}(\text{HQ}) = (1, 0)'$. Both differ from the estimate $\hat{\mathbf{p}} = (0, 2)'$ that was obtained in the full search procedure.

Under suitable conditions for the model selection criteria the HK procedure is consistent. Hannan & Deistler (1988) also discuss the consequences of the true data generation process being not in the class of considered processes.

8.3.4 Poskitt's Procedure

Another short-cut version of Stage II of the model specification procedure was suggested by Poskitt (1989a). It capitalizes on the important property of echelon forms that the restrictions for the k-th equation implied by a set of Kronecker indices $\mathbf{p}$ are determined by the indices $p_i \leq p_k$. They do not depend on the specific values of the p_j which are greater than p_k. The proposed modification of Stage II is based on separate LS estimations of each of the K equations of the system. The estimation is similar to the preliminary estimation method outlined in Section 7.3.4, that is, it uses the residuals of a long autoregression from Stage I instead of the true u_t. A model selection criterion of the form

$$Cr_k(\mathbf{p}) := \ln \tilde{\sigma}_k^2(\mathbf{p}) + c_T d_k(\mathbf{p})/T \tag{8.3.7}$$

is then evaluated for each of the K equations separately. Here $T\tilde{\sigma}_k^2(\mathbf{p})$ is the residual sum of squares of the k-th equation in a system with Kronecker indices $\mathbf{p}$, $d_k(\mathbf{p})$ is the number of freely varying parameters in the k-th equation, and c_T is a number that depends on the sample size T. Of course, (8.3.7) is the single equation analogue of the systems criterion (8.3.1). Stage II of the procedure proceeds then as follows:

STAGE II (P): Determine the required values $Cr_k(\mathbf{p})$ and choose the estimates $\hat{p}_k$ of the Kronecker indices according to the following rule:

If $Cr_k([0, \dots, 0]) \geq Cr_k([1, \dots, 1])$ for all $k = 1, \dots, K$, compute $Cr_k([2, \dots, 2])$, $k = 1, \dots, K$, and compare to $Cr_k([1, \dots, 1])$. If the $Cr_k([2, \dots, 2])$ are all not greater than the corresponding $Cr_k([1, \dots, 1])$ proceed to $Cr_k([3, \dots, 3])$ and so on.

If at some stage

$$Cr_k([j-1, \dots, j-1]) \geq Cr_k([j, \dots, j])$$

does *not* hold *for all* k, choose $\hat{p}_k = j - 1$ for all k with

$$Cr_k([j-1, \dots, j-1]) < Cr_k([j, \dots, j]).$$

The $\hat{p}_k$ obtained in this way are fixed in all the following steps. We continue by increasing the remaining indices and comparing the criteria for those equations for which the Kronecker indices are not yet fixed. Here it is important that the restrictions for the k-th equation do not depend on the Kronecker indices $p_i > p_k$ which are chosen in subsequent steps. ■

To make the procedure a bit more transparent it may be helpful to consider an example. Suppose that interest centers on a three-dimensional system, that is, $K = 3$. First $Cr_k([0, 0, 0])$ and $Cr_k([1, 1, 1])$ are computed for $k = 1, 2, 3$. Suppose

$$Cr_k([0, 0, 0]) \geq Cr_k([1, 1, 1]) \qquad \text{for } k = 1, 2, 3.$$

Then we evaluate $Cr_k([2, 2, 2])$, $k = 1, 2, 3$. Suppose

$$Cr_1([1, 1, 1]) < Cr_1([2, 2, 2])$$

and

$$Cr_k([1, 1, 1]) \geq Cr_k([2, 2, 2]), \qquad \text{for } k = 2, 3.$$

Then $\hat{p}_1 = 1$ is fixed and $Cr_k([1, 2, 2])$ is compared to $Cr_k([1, 3, 3])$ for $k = 2, 3$. Suppose

$$Cr_2([1, 2, 2]) \geq Cr_2([1, 3, 3]) \qquad \text{and} \qquad Cr_3([1, 2, 2]) < Cr_3([1, 3, 3]).$$

Then we fix $\hat{p}_3 = 2$ and compare $Cr_2([1, 3, 2])$ to $Cr_2([1, 4, 2])$ and so on until p_2 can also be fixed because no further reduction of the criterion $Cr_2(\cdot)$ is obtained in one step. It is important to note that for each index only the first local minimum of the corresponding criterion is searched for. We are not seeking a global minimum over all $\mathbf{p}$ with p_k less than some prespecified upper bound. For moderate or large systems the present procedure has the advantage of involving a very reasonable amount of computations only.

Poskitt (1989a) derives the properties of the Kronecker indices and the VARMA coefficients estimated by this procedure. He gives conditions under which the Kronecker indices are estimated consistently and the final VARMA parameter estimators have the asymptotic distribution given in Proposition 7.1. Assuming that the true data generation process can indeed be described by a stable, invertible echelon VARMA representation with a finite set of Kronecker indices the conditions imposed by Poskitt relate to the distribution of the white noise process u_t, to the choice of n, and to the criteria $Cr_k(\mathbf{p})$.

With respect to the process or white noise distribution the assumptions are satisfied, for example, if u_t is Gaussian. In fact, for most results it suffices that u_t is standard white noise. An exception is the asymptotic distribution of the white noise covariance estimator $\tilde{\Sigma}_u$. It may change if u_t has a nonnormal distribution.

The VAR order n in Stage I is assumed to go to infinity with the sample size at a certain rate. In practice the order selection criteria AIC or HQ may be used in Stage I. It must be guaranteed, however, that n is greater than the Kronecker indices considered in Stage II(P).

Poskitt (1989a) also discusses a modification of his algorithm that appears to have some practical advantages. We do not go into that procedure here but recommend that the interested reader examines the relevant literature. The message from the present discussion should be that consistent and feasible strategies for estimating the Kronecker indices exist. Poskitt also discusses the case where the true data generation process is not in the class of VARMA processes considered in the specification procedure. He derives some asymptotic results for this case as well.

In summary, a full search procedure is feasible only for low-dimensional systems if the maximum for the Kronecker indices is small or moderate. For high-dimensional systems and/or large upper bound of the Kronecker indices the Hannan-Kavalieris procedure or Poskitt's specification strategy are preferable from a computational point of view. The relative performance in small samples is so far unknown in general. It is left to the individual researcher to decide on a specific specification procedure with his or her available resources and perhaps the objective of the analysis in mind. Of course, it is legitimate to try different strategies and criteria and compare the resulting models and the implications for the subsequent analysis.

8.4 Remarks on other Specification Strategies for VARMA Models

A number of other specification strategies for VARMA processes have been proposed and investigated in the literature based on representations other than the final equations and echelon forms. Examples are Quenouille (1957), Tiao &

Box (1981), Jenkins & Alavi (1981), Aoki (1987), Cooper & Wood (1982), Granger & Newbold (1986), Akaike (1976), Tiao & Tsay (1989), Tsay (1989a, b) to list just a few. Some of these strategies are based on subjective criteria. As mentioned earlier, none of these procedures seems to be in common use for analyzing economic time series and none of them has become *the* standard procedure. So far few VARMA analyses of higher-dimensional time series are reported in the literature. Given this state of affairs it is difficult to give well-founded recommendations as to which strategy to use in any particular situation. Those familiar with the Box-Jenkins approach for univariate time series modeling will be aware of the problems that can arise even in the univariate case if the investigator has to decide on a model on the basis of statistics such as the autocorrelations and partial autocorrelations. Therefore it is an obvious advantage to have an automatic or semiautomatic procedure if one feels uncertain about the interpretation of statistical quantities related to specific characteristics of a process and if little or no prior information is available. On the other hand, if firmly based prior information about the data generation process is available it may be advantageous to use that at an early stage and depart from automatic procedures.

8.5 Model Checking

Prominent candidates in the model checking tool-kid are tests of statistical hypotheses. All three testing principles, LR (likelihood ratio), LM (Lagrange multiplier), and Wald tests (see Appendix C.5) can be applied in principle in the VARMA context. Since estimation requires iterative procedures it is often desirable to estimate just one model. Hence, LR tests which require estimation under both the null and alternative hypotheses are often unattractive. In finite order VAR models the unrestricted version is usually relatively easy to estimate and therefore it makes sense to use Wald tests in the pure VAR case because these tests are based on the unconstrained estimator. In contrast, the restricted estimator is often easier to obtain in the VARMA context when models with nontrivial MA part are considered. In this situation LM tests have an obvious advantage because the LM statistic involves the restricted estimator only. Of course, the restricted estimator is especially easy to determine if the constrained model is a pure, finite order VAR process. We will briefly discuss LM tests in the following. For further discussions and proofs the reader is referred to Kohn (1979), Hosking (1981b), and Poskitt & Tremayne (1982).

8.5.1 LM Tests

Suppose we wish to test

$$H_0: \varphi(\boldsymbol{\beta}) = 0 \quad \text{against} \quad H_1: \varphi(\boldsymbol{\beta}) \neq 0, \tag{8.5.1}$$

where $\boldsymbol{\beta}$ is an M-dimensional parameter vector and $\varphi(\cdot)$ is a twice continuously differentiable function with values in the N-dimensional space. In other words, $\varphi(\boldsymbol{\beta})$ is an $(N \times 1)$ vector and we assume that the matrix $\partial\varphi/\partial\boldsymbol{\beta}'$ of first partial derivatives has rank N at the true parameter vector. In this set-up we consider the case where the restrictions relate to the VARMA coefficients only. Moreover, we assume that the conditions of Proposition 7.1 are satisfied.

For instance, in the bivariate zero mean VARMA(1, 1) model with Kronecker indices $(p_1, p_2) = (1, 1)$,

$$\begin{bmatrix} y_{1,t} \\ y_{2,t} \end{bmatrix} = \begin{bmatrix} \alpha_{11,1} & \alpha_{12,1} \\ \alpha_{21,1} & \alpha_{22,1} \end{bmatrix} \begin{bmatrix} y_{1,t-1} \\ y_{2,t-1} \end{bmatrix} + \begin{bmatrix} u_{1,t} \\ u_{2,t} \end{bmatrix} + \begin{bmatrix} m_{11,1} & m_{12,1} \\ m_{21,1} & m_{22,1} \end{bmatrix} \begin{bmatrix} u_{1,t-1} \\ u_{2,t-1} \end{bmatrix} \tag{8.5.2}$$

with $\boldsymbol{\beta}' = (\alpha_{11,1}, \alpha_{21,1}, \alpha_{12,1}, \alpha_{22,1}, m_{11,1}, m_{21,1}, m_{12,1}, m_{22,1})$ one may wish to test that the MA degree is zero, that is,

$$\varphi(\boldsymbol{\beta}) = \begin{bmatrix} m_{11,1} \\ m_{21,1} \\ m_{12,1} \\ m_{22,1} \end{bmatrix} = \begin{bmatrix} 0 \\ 0 \\ 0 \\ 0 \end{bmatrix}.$$

The corresponding matrix of partial derivatives is

$$\frac{\partial\varphi}{\partial\boldsymbol{\beta}'} = [0 \quad I_4]$$

which obviously has rank $N = 4$.

As another example, suppose we wish to test for Granger-causality from y_{2t} to y_{1t} in the model (8.5.2). In that case

$$\varphi(\boldsymbol{\beta}) = \begin{bmatrix} \alpha_{12,1} + m_{12,1} \\ \alpha_{22,1} m_{12,1} - \alpha_{12,1} m_{22,1} \end{bmatrix} = \begin{bmatrix} 0 \\ 0 \end{bmatrix} \tag{8.5.3}$$

(see Remark 1 of Section 6.7.1). The corresponding matrix of partial derivatives is

$$\frac{\partial\varphi}{\partial\boldsymbol{\beta}'} = \begin{bmatrix} 0 & 0 & 1 & 0 & 0 & 0 & 1 & 0 \\ 0 & 0 & -m_{22,1} & m_{12,1} & 0 & 0 & \alpha_{22,1} & -\alpha_{12,1} \end{bmatrix}.$$

This matrix may have rank 1 under special conditions. In particular, this occurs if $\alpha_{12,1} = m_{12,1} = 0$ and $\alpha_{22,1} = -m_{22,1}$.

The LM statistic for testing (8.5.1) is

$$\lambda_{LM} := s(\tilde{\boldsymbol{\beta}}_r)' \tilde{I}_a(\tilde{\boldsymbol{\beta}}_r, \tilde{\Sigma}_u^r)^{-1} s(\tilde{\boldsymbol{\beta}}_r)/T, \tag{8.5.4}$$

where

$$s(\tilde{\boldsymbol{\beta}}_r) = \left.\frac{\partial \ln l_0}{\partial\boldsymbol{\beta}}\right|_{\tilde{\boldsymbol{\beta}}_r} = \sum_{t=1}^{T} \left[\left.\frac{\partial u_t(\bar{y}, \boldsymbol{\beta})'}{\partial\boldsymbol{\beta}}\right|_{\tilde{\boldsymbol{\beta}}_r} \right] (\tilde{\Sigma}_u^r)^{-1} \tilde{u}_t(\bar{y}, \tilde{\boldsymbol{\beta}}_r) \tag{8.5.5}$$

is the score vector evaluated at the restricted estimator $\tilde{\boldsymbol{\beta}}_r$ and

$$\tilde{I}_a(\tilde{\boldsymbol{\beta}}_r, \tilde{\Sigma}_u^r) = \frac{1}{T}\sum_{t=1}^{T}\left[\frac{\partial u_t(\bar{y}, \boldsymbol{\beta})'}{\partial \boldsymbol{\beta}}\bigg|_{\tilde{\boldsymbol{\beta}}_r}\right](\tilde{\Sigma}_u^r)^{-1}\left[\frac{\partial u_t(\bar{y}, \boldsymbol{\beta})}{\partial \boldsymbol{\beta}'}\bigg|_{\tilde{\boldsymbol{\beta}}_r}\right] \tag{8.5.6}$$

is an estimator of the asymptotic information matrix based on the restricted estimator $\tilde{\boldsymbol{\beta}}_r$. Here

$$\tilde{\Sigma}_u^r = \frac{1}{T}\sum_{t=1}^{T} \tilde{u}_t(\bar{y}, \tilde{\boldsymbol{\beta}}_r)\tilde{u}_t(\bar{y}, \tilde{\boldsymbol{\beta}}_r)'.$$

Note that in contrast to Appendix C, Section C.5, an estimator of the *asymptotic* information matrix rather than the information matrix is used in (8.5.4). Therefore T appears in the denominator. If H_0 is true the statistic λ_{LM} has an asymptotic $\chi^2(N)$-distribution.

The LM test is especially suitable for model checking because testing larger VAR or MA orders against a maintained model is particularly easy. A new estimation is not required as long as the null hypothesis does not change. For instance, if we wish to test a given VARMA(p, q) specification against a VARMA($p + s, q$) or a VARMA($p, q + s$) model we just need an estimator of the coefficients of the VARMA(p, q) process. Note, however, that a VARMA(p, q) cannot be tested against a VARMA($p + s, q + s$), that is, we cannot increase both the VAR and MA orders simultaneously because the VARMA($p + s, q + s$) model will not be identified (cancellation is possible!) if the null hypothesis is true. In that case the LM statistic will not have its usual asymptotic χ^2-distribution.

8.5.2 Residual Autocorrelations and Portmanteau Tests

Alternative tools for model checking are the residual autocorrelations and portmanteau tests. The asymptotic distributions of the residual autocorrelations of estimated VARMA models are discussed by Hosking (1980), Li & McLeod (1981), and Poskitt & Tremayne (1982) among others. We do not give the details here but just mention that the resulting standard errors of autocorrelations at large lags obtained from asymptotic considerations are approximately $1/\sqrt{T}$ while they may be much smaller for low lags just as for pure finite order VAR processes.

The modified portmanteau statistic is

$$\bar{P}_h := T^2 \sum_{i=1}^{h} (T - i)^{-1} \operatorname{tr}(\hat{C}_i'\hat{C}_0^{-1}\hat{C}_i\hat{C}_0^{-1}), \tag{8.5.7}$$

where

$$\hat{C}_i := \frac{1}{T}\sum_{t=i+1}^{T} \tilde{u}_t(\bar{y}, \tilde{\boldsymbol{\beta}})\tilde{u}_{t-i}(\bar{y}, \tilde{\boldsymbol{\beta}})'$$

and the $\tilde{u}_t(\bar{y}, \tilde{\boldsymbol{\beta}})$ are the residuals of an estimated VARMA model as before. Under general conditions $\bar{P}_h$ has an *approximate* asymptotic χ^2-distribution. The degrees of freedom are obtained by subtracting the number of freely estimated VARMA coefficients from K^2h.

8.5.3 Prediction Tests for Structural Change

In the pure VAR case we have considered prediction tests for structural change as model checking devices. If the data generation process is Gaussian the two tests introduced in Chapter 4, Section 4.6, may be applied in the VARMA case as well with minor modifications.

The statistics based on h-step forecasts only are of the form

$$\bar{\tau}_h := \hat{e}_T(h)'\hat{\Sigma}_{\hat{y}}(h)^{-1}\hat{e}_T(h), \tag{8.5.8}$$

where $\hat{e}_T(h) = y_{T+h} - \hat{y}_T(h)$ is the error vector of an h-step forecast based on an estimated VARMA(p, q) process and $\hat{\Sigma}_{\hat{y}}(h)$ is an estimator of the corresponding MSE matrix (see Section 7.5). The statistic may be applied in conjunction with an $F(K, T - K(p + q) - 1)$-distribution. The denominator degrees of freedom may be used even if constraints are imposed on the VARMA coefficients because the F-distribution is just chosen as a small sample approximation to a $\chi^2(K)/K$. Its justification comes from the fact that $F(K, T - s)$ converges to $\chi^2(K)/K$ for any constant s, as T approaches infinity. Thus, any constant that is subtracted from T in the denominator degrees of freedom of the F-distribution is justified on the same asymptotic grounds. Presently it is not clear which choice is best from a small sample point of view.

The other statistic considered in Section 4.6.2 is based on 1- to h-step forecasts and, for the present case, it may be modified as

$$\bar{\lambda}_h := T \sum_{i=1}^{h} \hat{u}'_{T+i}\tilde{\Sigma}_u^{-1}\hat{u}_{T+i}/[(T + K(p + q) + 1)Kh] \tag{8.5.9}$$

and its approximate distribution for a structurally stable Gaussian process is $F(Kh, T - K(p + q) - 1)$. Here $\hat{u}_{T+i} = y_{T+i} - \hat{y}_{T+i-1}(1)$ and $\tilde{\Sigma}_u$ is the ML estimator of Σ_u. Note that the LS estimator of Σ_u was used in Section 4.6.2 instead. Again there is not much theoretical justification for the choice of the denominator in (8.5.9) and for the denominator degrees of freedom in the approximating F-distribution. More detailed investigations into the small sample distribution of $\bar{\lambda}_h$ are required before firmly based recommendations regarding modifications of the statistic are possible. Here we have just used the direct analogue of the finite order pure VAR case.

It is also possible to fit a finite order VAR process to data generated by a mixed VARMA process and base the prediction tests on forecasts from that model. In the next chapter it will be shown that such an approach is theoretically sound under general conditions.

8.6 Critique of VARMA Model Fitting

In this and the previous two chapters much of the analysis is based on the assumption that the true data generation mechanism is from the VARMA(p, q) class. In practice any such model is just an approximation to the actual data generation process. Therefore the model selection task is not really the problem of finding the true structure but of finding a good or useful approximation to the real life mechanism. Despite this fact it is sometimes helpful to assume a specific true process or process class to be able to derive, under ideal conditions, the statistical properties of the procedures used. One then hopes that the actual properties of a procedure in a particular practical situation are at least similar to those obtained under ideal conditions.

Against this background one may wonder whether it is sufficient or even preferable to approximate the generation process of a given multiple time series by a finite order VAR(p) process rather than go through the painstaking specification and estimation of a mixed VARMA model. Clearly the estimation of VARMA models is in general more complicated than that of finite order VAR models. Moreover, the specification of VAR models by statistical methods is much simpler than that of VARMA models. Are there still situations where it is reasonable to consider the more complicated VARMA models? The answer to this question is in the affirmative. For instance, if subject matter theory suggests a VARMA model with nontrivial MA part it is often necessary to work with such a specification in order to answer the questions of interest or derive the relevant results. Also, in some cases a VARMA approximation may be more parsimonious in terms of the parameters involved than an appropriate finite order VAR approximation. In such a case the VARMA approximation may, for instance, result in more efficient forecasts that justify the costly specification and estimation procedures. The future attractiveness of VARMA models will depend on the availability of efficient and robust estimation and specification procedures that reduce the costs to an acceptable level.

In the next chapter we will follow another road and explicitly assume that just an approximating and not a true VAR(p) model is fitted. Assumptions will be provided that allow the derivation of statistical properties in that case.

8.7 Exercises

Problem 8.1

At the first stage of a final form specification procedure the following two univariate models were obtained:

$$(1 + 0.3L - 0.4L^2)y_{1t} = (1 + 0.6L)v_{1t},$$

$$(1 - 0.5L)y_{2t} = (1 + 0.6L)v_{2t}.$$

Which orders do you choose for the bivariate final equations VARMA representation of $(y_{1t}, y_{2t})'$?

Problem 8.2

At Stage II of a specification procedure for an echelon form of a bivariate system the following values of the HQ criterion are obtained:

p_2 \ p_1	0	1	2	3	4
0	2.1	1.9	1.5	1.5	1.6
1	1.8	1.7	1.4	1.2	1.3
2	1.7	1.4	1.3	1.4	1.4
3	1.7	1.4	1.3	1.4	1.5
4	1.8	1.7	1.6	1.5	1.5

Choose an estimate $(\hat{p}_1, \hat{p}_2)'$ by the Hannan-Kavalieris procedure. Interpret the estimate in the light of a full search procedure.

Problem 8.3

At the second stage of the Poskitt procedure the specification criteria $Cr_1(p_1, p_2)$, $Cr_2(p_1, p_2)$ assume the following values:

$Cr_1(p_1, p_2)$

p_2 \ p_1	0	1	2	3
0	3.5	2.5	1.7	1.8
1	3.5	1.5	1.8	1.7
2	3.5	1.5	1.8	1.9
3	3.5	1.5	1.8	1.4

$Cr_2(p_1, p_2)$

p_2 \ p_1	0	1	2	3
0	4.2	3.2	3.2	3.2
1	3.5	1.8	1.9	1.9
2	3.1	1.9	1.7	1.6
3	3.4	2.1	1.8	1.9

Use the Poskitt strategy to find an estimate $(\hat{p}_1, \hat{p}_2)$ of the Kronecker indices.

The following problems require the use of a computer. They are based on the *first differences* of the U.S. investment data given in Table E.2 of Appendix E.

Problem 8.4

Determine a final form VARMA model for the U.S. investment data for the years 1947–1968 using the specification strategy described in Section 8.2.1.

Problem 8.5

Determine an echelon VARMA model for the U.S. investment data using the specification strategy described in Section 8.3.2 with $n = 6$ and based on the HQ criterion. Compare the model to the final form model from Problem 8.4.

Problem 8.6

Compute forecasts for the investment series for the years 1969 and 1970 based on (i) the final form VARMA model, (ii) the echelon VARMA model, and (iii) a bivariate VAR(1) model. Compare the forecasts to the true values and interpret.

Problem 8.7

Compare Φ_i and Θ_i impulse responses from the two models obtained in Problems 8.4 and 8.5, compare and interpret them.

Problem 8.8

Specify a univariate ARMA model for the sum $z_t = y_{1t} + y_{2t}$ of the two investment series for the years 1947–1968. Is the univariate ARMA model compatible with the bivariate echelon form model specified in Problem 8.5? (Hint: Use the results of Section 6.6.1.)

Problem 8.9

Evaluate forecasts for the z_t series of the previous problem for the years 1969–1970 and compare them to forecasts obtained by aggregating the bivariate forecasts from the echelon VARMA model of Problem 8.5.

Chapter 9. Fitting Finite Order VAR Models to Infinite Order Processes

9.1 Background

In the previous chapters we have derived properties of models, estimators, forecasts, and test statistics under the assumption of a true model. We have also argued that such an assumption is virtually never fulfilled in practice. In other words, in practice, all we can hope for is a model that provides a useful approximation to the actual data generation process of a given multiple time series. In this chapter we will, to some extent, take into account this state of affairs and assume that an approximating rather than a true model is fitted. Specifically we assume that the true data generation process is a stable, infinite order VAR process and, for a given sample size T, a finite order VAR(p) is fitted to the data.

In practice it is likely that a higher order VAR model is considered if the sample size or time series length is larger. In other words, the order p increases with the sample size T. If an order selection criterion is used in choosing the VAR order the maximum order to be considered is likely to depend on T. This again implies that the actual order chosen depends on the sample size because it will depend on the maximum order. In summary, the actual order selected may be regarded as a function of the sample size T. In order to derive statistical properties of estimators and forecasts we will make this assumption in the following. More precisely we will assume that p goes to infinity with the sample size. Under that assumption an asymptotic theory has been developed that will be discussed in this chapter.

In Section 9.2 the assumptions for the underlying true process and for the order of the process fitted to the data are specified in detail and asymptotic estimation results are provided. In Section 9.3 the consequences for forecasting are discussed and impulse response analysis is considered in Section 9.4. Our standard investment/income/consumption example is used to contrast the present approach to that considered in Chapter 3, where a true finite order process is assumed.

9.2 Multivariate Least Squares Estimation

Suppose the generation process of a multiple time series is a stationary, stable, K-dimensional, infinite order VAR process

$$y_t = \sum_{i=1}^{\infty} \Pi_i y_{t-i} + u_t \tag{9.2.1}$$

with absolutely summable Π_i, that is,

$$\sum_{i=1}^{\infty} \|\Pi_i\| < \infty \tag{9.2.2}$$

(see Appendix C.3) and canonical MA representation

$$y_t = \sum_{i=0}^{\infty} \Phi_i u_{t-i}, \qquad \Phi_0 = I_K, \tag{9.2.3}$$

satisfying

$$\det\left[\sum_{i=0}^{\infty} \Phi_i z^i\right] \neq 0 \qquad \text{for } |z| \leq 1 \quad \text{and} \quad \sum_{i=1}^{\infty} i^{1/2}\|\Phi_i\| < \infty. \tag{9.2.4}$$

The zero mean assumption implied by these conditions is not essential and is imposed for convenience only. Stable, invertible VARMA processes satisfy the foregoing conditions. The assumptions allow for more general processes, however. Of course, the generation process may also be a stable, finite order VAR(p) in which case $\Pi_i = 0$ for $i > p$.

We have argued in the previous section that in practice the true structure will usually be unknown and the investigator may consider fitting a finite order VAR process with the VAR order depending on the length T of the available time series. For this situation Lewis & Reinsel (1985) have shown consistency and asymptotic normality of the multivariate LS estimators. For univariate processes similar results were discussed earlier by Berk (1974) and Bhansali (1978).

To state these results formally we use the following notation:

$$\Pi(n) := [\Pi_1, \ldots, \Pi_n],$$

$$\pi(n) := \operatorname{vec} \Pi(n).$$

Fitting a VAR(n) process, the i-th estimated coefficient matrix is denoted by $\hat{\Pi}_i(n)$,

$$\hat{\Pi}(n) := [\hat{\Pi}_1(n), \ldots, \hat{\Pi}_n(n)],$$

and

$$\hat{\pi}(n) := \operatorname{vec} \hat{\Pi}(n).$$

Now we can state the result of Lewis & Reinsel (1985).

PROPOSITION 9.1 (*Properties of the LS Estimator of an Approximating VAR Model*)

Let the multiple time series $y_1, \ldots, y_T$ be generated by a potentially infinite order VAR process satisfying (9.2.1)–(9.2.4) with standard white noise u_t. Suppose finite order VAR(n_T) processes are fitted by multivariate LS and assume that the order n_T depends upon the sample size T such that

$$n_T \to \infty, \quad n_T^3/T \to 0, \quad \text{and} \quad \sqrt{T} \sum_{i=n_T+1}^{\infty} \|\Pi_i\| \to 0 \quad \text{as } T \to \infty. \tag{9.2.5}$$

Furthermore, let $\mathbf{f}(n)$ be a sequence of $(K^2 n \times 1)$ vectors and c_1, c_2 constants such that

$$0 < c_1 \leq \mathbf{f}(n)'\mathbf{f}(n) \leq c_2 < \infty \qquad \text{for } n = 1, 2, \ldots.$$

Then

$$\frac{\sqrt{T - n_T}\,\mathbf{f}(n_T)'[\hat{\boldsymbol{\pi}}(n_T) - \boldsymbol{\pi}(n_T)]}{[\mathbf{f}(n_T)'(\Gamma_{n_T}^{-1} \otimes \Sigma_u)\mathbf{f}(n_T)]^{1/2}} \xrightarrow[T\to\infty]{d} N(0, 1), \tag{9.2.6}$$

where

$$\Gamma_n := E\left[\begin{bmatrix} y_t \\ \vdots \\ y_{t-n+1} \end{bmatrix} [y_t', \ldots, y_{t-n+1}']\right]. \tag{9.2.7}$$

■

REMARK 1: The assumption (9.2.5) means that, although the VAR order has to go to infinity with the sample size, it has to do so at a much slower rate because $n_T^3/T \to 0$. The requirement

$$\sqrt{T} \sum_{i=n_T+1}^{\infty} \|\Pi_i\| \to 0 \tag{9.2.8}$$

is always satisfied if y_t is actually a finite order VAR process and $n_T \to \infty$. For infinite order VAR processes this condition implies a lower bound for the rate at which n_T goes to infinity. To see this consider the *univariate* MA(1) process

$$y_t = u_t - m u_{t-1},$$

where $0 < |m| < 1$ to ensure invertibility. Its AR representation is

$$y_t = -\sum_{i=1}^{\infty} m^i y_{t-i} + u_t$$

and condition (9.2.8) becomes

$$\begin{aligned}\sqrt{T} \sum_{i=n+1}^{\infty} |m^i| &= \sqrt{T}\,|m|^{n+1} \sum_{i=0}^{\infty} |m|^i \\ &= \sqrt{T}\,\frac{|m|^{n+1}}{1 - |m|} \xrightarrow[T\to\infty]{} 0.\end{aligned} \tag{9.2.9}$$

Here the subscript T has been dropped from n_T for notational simplicity. In this example $n_T = T^{1/\varepsilon}$ with $\varepsilon > 3$ is a possible choice for the sequence n_T that satisfies both (9.2.9) and $n_T^3/T \to 0$. On the other hand, $n_T = \ln \ln T$ is not a permissible choice because for this choice

$$\sqrt{T}\,|m|^{n_T+1}$$

does not approach zero as $T \to \infty$. This result is easily established by considering the logarithm of (9.2.9),

$\frac{1}{2}\ln T + (n_T + 1)\ln|m| - \ln(1 - |m|),$

which obviously goes to infinity for $n_T = \ln \ln T$.

In summary, (9.2.8) is a lower bound and $n_T^3/T \to 0$ establishes an upper bound for the rate at which n_T has to go to infinity with the sample size T. ■

REMARK 2: Proposition 9.1 implies that for fixed m,

$$\sqrt{T - n_T}\,\text{vec}([\hat{\Pi}_1(n_T), \ldots, \hat{\Pi}_m(n_T)] - [\Pi_1, \ldots, \Pi_m])$$

has an asymptotic multivariate normal distribution with mean zero and covariance matrix $V \otimes \Sigma_u$, where V is obtained as follows: Let V_n be the upper left-hand $(Km \times Km)$ block of the inverse of Γ_n for $n \geq m$. Then $V = \lim_{n\to\infty} V_n$. Loosely speaking V is the upper left-hand $(Km \times Km)$ block of the inverse of the infinite order matrix

$$E\left[\begin{bmatrix} y_t \\ y_{t-1} \\ \vdots \end{bmatrix} [y_t', y_{t-1}', \ldots]\right].$$ ■

REMARK 3: If the data generation process has nonzero mean originally the sample mean $\bar{y}$ may be subtracted initially from the data. It is asymptotically independent of the $\hat{\Pi}_i(n_T)$ and has an asymptotic normal distribution,

$$\sqrt{T}(\bar{y} - \mu) \xrightarrow{d} N(0, \Sigma_{\bar{y}}),$$

where

$$\Sigma_{\bar{y}} = \left[\sum_{i=0}^{\infty} \Phi_i\right] \Sigma_u \left[\sum_{i=0}^{\infty} \Phi_i\right]'.$$ ■

A corresponding result from Lütkepohl & Poskitt (1991) for the white noise covariance matrix is stated next.

PROPOSITION 9.2 (*Asymptotic Properties of the White Noise Covariance Matrix Estimator*)

Let

$$\hat{u}_t(n) := y_t - \sum_{i=1}^{n} \hat{\Pi}_i(n) y_{t-i}, \qquad t = 1, \ldots, T,$$

be the multivariate LS residuals from a VAR(n) model fitted to a multiple time series of length T, let

$$\tilde{\Sigma}_u(n) := \frac{1}{T} \sum_{t=1}^{T} \hat{u}_t(n)\hat{u}_t(n)'$$

be the corresponding estimator of the white noise covariance matrix and let $U := [u_1, \ldots, u_T]$ so that

$$UU'/T = \frac{1}{T}\sum_{t=1}^{T} u_t u_t'$$

is an estimator of Σ_u based on the true white noise process u_t. Then, under the conditions of Proposition 9.1,

$$\operatorname{plim} \sqrt{T}(\tilde{\Sigma}_u(n_T) - UU'/T) = 0. \quad \blacksquare$$

We know from Chapter 3, Propositions 3.2 and 3.4, that, for a Gaussian process, UU'/T has an asymptotic normal distribution,

$$\sqrt{T}\operatorname{vech}(UU'/T - \Sigma_u) \xrightarrow{d} N(0, 2\mathbf{D}_K^+(\Sigma_u \otimes \Sigma_u)\mathbf{D}_K^{+\prime}), \tag{9.2.10}$$

where, as usual, $\mathbf{D}_K^+ = (\mathbf{D}_K'\mathbf{D}_K)^{-1}\mathbf{D}_K'$ is the Moore-Penrose inverse of the $(K^2 \times K(K+1)/2)$ duplication matrix $\mathbf{D}_K$. Using Proposition C.2(2) of Appendix C.1, Proposition 9.2 implies that

$$\sqrt{T}\operatorname{vech}(\tilde{\Sigma}_u(n_T) - \Sigma_u)$$

has precisely the same asymptotic distribution. Obviously, this distribution does not depend on the VAR structure of y_t or the VAR coefficients. In addition, the estimator $\tilde{\Sigma}_u(n_T)$ is asymptotically independent of $\hat{\pi}(n_T)$. In the following the consequences of these results for prediction and impulse response analysis will be discussed.

9.3 Forecasting

9.3.1 Theoretical Results

Suppose the VAR(n_T) model estimated in the previous section is used for prediction. In that case the usual h-step forecast at origin T is

$$\tilde{y}_T(h) := \sum_{i=1}^{n_T} \hat{\Pi}_i(n_T)\tilde{y}_T(h-i), \tag{9.3.1}$$

where $\tilde{y}_T(j) := y_{T+j}$ for $j \le 0$ (see Section 3.5). We use the notation

$$\tilde{\mathbf{y}}_T(h) := \begin{bmatrix} \tilde{y}_T(1) \\ \vdots \\ \tilde{y}_T(h) \end{bmatrix}, \qquad \mathbf{y}_T(h) := \begin{bmatrix} y_T(1) \\ \vdots \\ y_T(h) \end{bmatrix}, \qquad \mathbf{y}_{T,h} := \begin{bmatrix} y_{T+1} \\ \vdots \\ y_{T+h} \end{bmatrix}$$

and

$$\underline{\Sigma}_y(h) = E[\mathbf{y}_{T,h} - \mathbf{y}_T(h)][\mathbf{y}_{T,h} - \mathbf{y}_T(h)]',$$

where $y_T(j)$, $j = 1, \ldots, h$, is the optimal j-step forecast at origin T based on the infinite past, that is,

$$y_T(j) = \sum_{i=1}^{\infty} \Pi_i y_T(j-i)$$

with $y_T(i) := y_{T+i}$ for $i \le 0$ (see Section 6.5). The following result is also essentially due to Lewis & Reinsel (1985) (see also Lütkepohl (1987, Section 3.3, Proposition 3.2)).

PROPOSITION 9.3 (*Asymptotic Distributions of Estimated Forecasts*)

Under the conditions of Proposition 9.1, if y_t is a Gaussian process and if independent processes with identical stochastic structures are used for estimation and forecasting, respectively, then

$$\sqrt{\frac{T}{n_T}}\,[\tilde{\mathbf{y}}_T(h) - \mathbf{y}_T(h)] \xrightarrow{d} N(0, K\underline{\Sigma}_{\mathbf{y}}(h))$$

for $h = 1, 2, \ldots$. ■

REMARK 1: The proposition implies that for large samples the forecast vector $\tilde{\mathbf{y}}_T(h)$ has approximate MSE matrix

$$\underline{\Sigma}_{\tilde{\mathbf{y}}}(h) = \left[1 + \frac{Kn_T}{T}\right]\underline{\Sigma}_{\mathbf{y}}(h). \tag{9.3.2}$$

This can be seen by noting that

$$\begin{aligned} &E[(\mathbf{y}_{T,h} - \tilde{\mathbf{y}}_T(h))(\mathbf{y}_{T,h} - \tilde{\mathbf{y}}_T(h))'] \\ &\quad = E[(\mathbf{y}_{T,h} - \mathbf{y}_T(h))(\mathbf{y}_{T,h} - \mathbf{y}_T(h))'] + E[(\mathbf{y}_T(h) - \tilde{\mathbf{y}}_T(h))(\mathbf{y}_T(h) - \tilde{\mathbf{y}}_T(h))'] \end{aligned}$$

and approximating the last term via the asymptotic result of Proposition 9.3. ■

REMARK 2: An approximation of the MSE matrix of an h-step forecast $\tilde{y}_T(h)$ follows directly from (9.3.2),

$$\Sigma_{\tilde{y}}(h) = \left[1 + \frac{Kn_T}{T}\right]\Sigma_y(h), \qquad h = 1, 2, \ldots \tag{9.3.3}$$

In Section 3.5.1 we have obtained an approximate MSE matrix

$$\Sigma_{\hat{y}}(h) = \Sigma_y(h) + \frac{1}{T}\Omega(h) \tag{9.3.4}$$

for an h-step forecast based on an estimated VAR process with known finite order. If in Chapter 3 the process mean is known to be zero and is not estimated, it can be shown that $\Omega(h)$ approaches zero as $h \to \infty$. In other words, the MSE part due to estimation variability goes to zero as the forecast horizon increases. The same does not hold in the present case. In fact, the $\Sigma_{\tilde{y}}(h)$ are monotonically nondecreasing for growing h, that is,

$$\Sigma_{\tilde{y}}(h) \ge \Sigma_{\tilde{y}}(i) \qquad \text{for } h \ge i.$$

The explanation for this result is that, under the present assumptions, increasingly many parameters are estimated with growing sample size. While for a zero mean VAR process with known finite order the optimal forecast approaches the process mean of zero when the forecast horizon gets large and thus the estimated VAR parameters do not contribute to the forecast uncertainty for long-run forecasts, the same is not true under the present conditions where the VAR order goes to infinity. ■

REMARK 3: We have also seen in Section 3.5.2 that $\Omega(1) = (Kp + 1)\Sigma_u$ for a K-dimensional VAR(p) process with estimated intercept term. It is easy to see that when the process mean is known to be zero and the mean term is not estimated, $\Omega(1) = Kp\Sigma_u$. Hence, in that case,

$$\Sigma_{\tilde{y}}(1) = \Sigma_y(1) + \frac{Kp}{T}\Sigma_u = \Sigma_u + \frac{Kp}{T}\Sigma_u = \Sigma_{\hat{y}}(1),$$

if $n_T = p$. In other words, for 1-step forecasts the two MSE approximations are identical if the same VAR orders are used in both approaches. It is easy to see that the same does not hold in general for predictions more than 1 step ahead (see Problem 9.2). ■

REMARK 4: Since forecasts can be obtained from finite order approximations to infinite order VAR processes we may also base the prediction tests for structural change considered in Sections 4.6 and 8.5.3 on such approximations. Of course, in that case the MSE approximation implied by Proposition 9.3 should be used in setting up the test statistics. For instance, a test statistic based on h-step forecasts would be

$$\bar{\tau}_h = (y_{T+h} - \tilde{y}_T(h))'\tilde{\Sigma}_{\tilde{y}}(h)^{-1}(y_{T+h} - \tilde{y}_T(h)),$$

where $\tilde{\Sigma}_{\tilde{y}}(h)$ is an estimator of $\Sigma_{\tilde{y}}(h)$. ■

REMARK 5: If y_t is a process with nonzero mean vector μ the sample mean may be subtracted from the original data and the previous analysis may be performed with the mean-adjusted data. If the sample mean is added to the forecasts an extra term should be added to the approximate MSE matrix. A term similar to that resulting from an estimated mean term in a finite order VAR setting with known order may be added (see Problem 3.9, Chapter 3). ■

9.3.2 An Example

To illustrate the effects of approximating a potentially infinite order VAR process by a finite order model we use again the West German investment/income/consumption data from Table E.1. The variables y_1, y_2, and y_3 are defined as in Chapter 3, Section 3.2.3, and we use the same sample period 1960–1978 and a VAR order $n_T = 2$. That is, we assume that the VAR order depends on the sample size in such a way that $n_T = 2$ for $T = 73$. Note that the condition (9.2.5) for the VAR order is an asymptotic condition that leaves open the actual choice in finite samples. Therefore we choose the VAR order that was suggested by the AIC criterion in Chapter 4 and thus, we use the same VAR order as in Chapter 3. As a consequence the point forecasts obtained under our present assumptions are the same one gets from a mean-adjusted model under the conditions of Chapter 3. The interval forecasts obtained under the different sets of assumptions are different for $h > 1$, however, because the approximate MSE matrices are different. We have estimated $\Sigma_{\tilde{y}}(h)$ by

$$\tilde{\Sigma}_{\tilde{y}}(h) = \left[1 + \frac{3n_T}{T}\right] \sum_{i=0}^{h-1} \hat{\Phi}_i \hat{\Sigma}_u \hat{\Phi}_i' + \frac{1}{T}\hat{G}_y(h), \tag{9.3.5}$$

where the $\hat{\Phi}_i$ and $\hat{\Sigma}_u$ are obtained from the VAR(2) estimates as in Section 3.5.3 and $\hat{G}_y(h)/T$ is a term that takes account of the fact that the mean term is estimated in addition to the VAR coefficients. It is the same term that is used if a VAR(2) process with true order $p = 2$ is assumed and the model is estimated in mean-adjusted form (see Problem 3.9).

We have used the approximate forecast MSEs from (9.3.5) to set up forecast intervals under Gaussian assumptions and give them in Table 9.1. For comparison purposes we also give forecast intervals obtained from a VAR(2) process in mean-adjusted form based on the asymptotic theory of Chapter 3 assuming that the true order is $p = 2$. As we know from Remark 3 above the 1-step forecast MSEs are the same under the two competing assumptions. For larger forecast horizons most of the intervals based on the infinite order assumption become slightly wider than those based on the known finite order assumption as expected

Table 9.1. Interval forecasts from a VAR(2) model based on different asymptotic theories

variable	forecast horizon	point forecast	95% interval forecasts: based on known order assumption	based on infinite order assumption
investment	1	−.010	[−.105, .085]	[−.105, .085]
	2	.012	[−.087, .110]	[−.088, .112]
	3	.022	[−.075, .119]	[−.078, .122]
	4	.013	[−.084, .111]	[−.088, .114]
income	1	.020	[−.004, .044]	[−.004, .044]
	2	.020	[−.004, .045]	[−.005, .045]
	3	.017	[−.007, .042]	[−.008, .042]
	4	.021	[−.004, .045]	[−.005, .047]
consumption	1	.022	[.002, .041]	[.002, .041]
	2	.015	[−.005, .035]	[−.005, .035]
	3	.020	[−.002, .042]	[−.002, .042]
	4	.019	[−.003, .041]	[−.003, .041]

on the basis of Remark 2. For our sample size the differences are quite small, though.

Which of the two sets of forecast intervals should we use in practice? This question is difficult to answer. Assuming a known finite VAR order is, of course, more restrictive and less realistic than the assumption of an unknown and possibly infinite order. The additional uncertainty introduced by the latter assumption is reflected in the wider forecast intervals. It may be worth noting, however, that such a result is not necessarily obtained in all practical situations. In other words, there may be time series and generation processes for which the infinite order assumption actually leads to smaller forecast intervals than the assumption of a known finite VAR order (see Problem 9.2). Under both sets of assumptions the MSE approximations are derived from asymptotic theory and little is known about the small sample quality of these approximations. Both approaches are based on a set of assumptions that may fail in practice. Notably the stationarity and normality assumptions may be doubtful in many practical situations. Given all these reservations there is still one argument in favor of the present approach assuming a potentially infinite VAR order. For $h > 1$, the MSE approximation in (9.3.3) is generally simpler to compute than the one obtained in Chapter 3.

9.4 Impulse Response Analysis and Forecast Error Variance Decompositions

9.4.1 Asymptotic Theory

For a researcher who does not know the true structure of the data generating process it is possible to base an impulse response analysis or forecast error variance decomposition on an approximating finite order VAR process. Given the results of Section 9.2 we can now study the consequences of such an approach. As in Sections 2.3.2 and 2.3.3 the quantities of interest are the forecast error impulse responses

$$\Phi_i = \sum_{j=1}^{i} \Phi_{i-j}\Pi_j, \qquad i = 1, 2, \ldots, \qquad \Phi_0 = I_K,$$

the accumulated forecast error impulse responses,

$$\Psi_m = \sum_{i=0}^{m} \Phi_i, \qquad m = 0, 1, \ldots,$$

the responses to orthogonalized impulses,

$$\Theta_i = \Phi_i P, \qquad i = 0, 1, \ldots,$$

where P is the lower triangular matrix obtained from a Choleski decomposition of Σ_u, the accumulated orthogonalized impulse responses,

$$\Xi_m = \sum_{i=0}^{m} \Theta_i, \qquad m = 0, 1, \ldots,$$

and the forecast error variance components

$$\omega_{jk,h} = \sum_{i=0}^{h-1} (e_j'\Theta_i e_k)^2/\mathrm{MSE}_j(h), \qquad h = 1, 2, \ldots,$$

where e_k is the k-th column of I_K and

$$\mathrm{MSE}_j(h) = \sum_{i=0}^{h-1} e_j'\Phi_i \Sigma_u \Phi_i' e_j$$

is the j-th diagonal element of the MSE matrix $\Sigma_y(h)$ of an h-step forecast.

Estimators of these quantities are obtained from the $\hat{\Pi}_i(n_T)$ and $\tilde{\Sigma}_u(n_T)$ in the obvious way. For instance, estimators for the Φ_i are obtained recursively as

$$\tilde{\Phi}_i(n_T) = \sum_{j=1}^{i} \tilde{\Phi}_{i-j}(n_T)\hat{\Pi}_j(n_T), \qquad i = 1, 2, \ldots, \qquad \tilde{\Phi}_0(n_T) = I_K,$$

and

$$\tilde{\Theta}_i(n_T) = \tilde{\Phi}_i(n_T)\tilde{P}(n_T), \qquad i = 0, 1, \ldots$$

are estimators of the Θ_i. Here $\tilde{P}(n_T)$ is the unique lower triangular matrix with positive diagonal for which

$$\tilde{P}(n_T)\tilde{P}(n_T)' = \tilde{\Sigma}_u(n_T).$$

The asymptotic distributions of all the estimators are given in the next proposition. Proofs, based on Propositions 9.1 and 9.2, are given by Lütkepohl (1988b) and Lütkepohl & Poskitt (1991).

PROPOSITION 9.4 (*Asymptotic Distributions of Impulse Responses*)

Under the conditions of Proposition 9.2 the impulse responses and forecast error variance components have the following asymptotic normal distributions:

$$\sqrt{T}\,\mathrm{vec}(\tilde{\Phi}_i(n_T) - \Phi_i) \xrightarrow{d} N\left(0, \Sigma_u^{-1} \otimes \sum_{j=0}^{i-1} \Phi_j \Sigma_u \Phi_j'\right), \qquad i = 1, 2, \ldots; \tag{9.4.1}$$

$$\sqrt{T}\,\mathrm{vec}(\tilde{\Psi}_m(n_T) - \Psi_m) \xrightarrow{d} N\left(0, \Sigma_u^{-1} \otimes \sum_{k=1}^{m}\sum_{l=1}^{m}\sum_{j=0}^{l-1} \Phi_j \Sigma_u \Phi_{k-l+j}'\right), \tag{9.4.2}$$

$m = 1, 2, \ldots,$ with $\Phi_j := 0$ for $j < 0$;

$$\sqrt{T}\,\mathrm{vec}(\tilde{\Theta}_i(n_T) - \Theta_i) \xrightarrow{d} N(0, \Omega_\theta(i)), \qquad i = 0, 1, \ldots, \tag{9.4.3}$$

where

$$\Omega_\theta(i) = \left(I_K \otimes \sum_{j=0}^{i-1} \Phi_j \Sigma_u \Phi_j'\right) + (I_K \otimes \Phi_i) H \Sigma_{\tilde{\sigma}} H'(I_K \otimes \Phi_i'),$$

$H = \mathbf{L}_K'[\mathbf{L}_K(I_{K^2} + \mathbf{K}_{KK})(P \otimes I_K)\mathbf{L}_K']^{-1}$,

$\mathbf{L}_K$ is the $(K(K+1)/2 \times K^2)$ elimination matrix,

$\mathbf{K}_{KK}$ is the $(K^2 \times K^2)$ commutation matrix,

and

$\Sigma_{\tilde{\sigma}}$ is the asymptotic covariance matrix of $\sqrt{T}\,\mathrm{vech}\left[\frac{1}{T}\sum_{t=1}^{T} u_t u_t' - \Sigma_u\right]$;

$$\sqrt{T}\,\mathrm{vec}(\tilde{\Xi}_m(n_T) - \Xi_m) \xrightarrow{d} N(0, \Omega_\xi(m)), \qquad m = 1, 2, \ldots, \tag{9.4.4}$$

where

$$\Omega_\xi(m) = \sum_{k=0}^{m}\sum_{l=0}^{m}\left[I_K \otimes \sum_{j=0}^{l-1} \Phi_j \Sigma_u \Phi_{k-l+j}' + (I_K \otimes \Phi_l) H \Sigma_{\tilde{\sigma}} H'(I_K \otimes \Phi_k')\right]$$

with $\Phi_j := 0$ for $j < 0$;

$$\sqrt{T}(\tilde{\omega}_{jk,h}(n_T) - \omega_{jk,h}) \xrightarrow{d} N(0, \sigma_{jk,h}^2), \quad h = 1, 2, \ldots, \quad j, k = 1, \ldots, K, \tag{9.4.5}$$

where

$$\sigma_{jk,h}^2 = \sum_{l=0}^{h-1}\sum_{m=0}^{h-1} g_{jk,h}(l)\left[I_K \otimes \sum_{i=0}^{m-1} \Phi_i \Sigma_u \Phi_{l-m+i}' \right.$$
$$\left. + (I_K \otimes \Phi_m) H \Sigma_{\tilde{\sigma}} H'(I_K \otimes \Phi_l')\right] g_{jk,h}(m)'$$

with

$$g_{jk,h}(m) = 2[(e_k' \otimes e_j')(e_j'\Theta_m e_k)\mathrm{MSE}_j(h) - (e_j'\Theta_m \otimes e_j')\sum_{i=0}^{h-1}(e_j'\Theta_i e_k)^2]/\mathrm{MSE}_j(h)^2.$$

■

REMARK 1: In the proposition it is ignored that $\sigma_{jk,h}^2$ may be zero in which case the asymptotic normal distribution is degenerate. In particular, $\sigma_{jk,h}^2 = 0$ if $\omega_{jk,h} = 0$. This is easily seen by noting that $\omega_{jk,h}$ is zero if and only if $\theta_{jk,0} = \cdots = \theta_{jk,h-1} = 0$, where $\theta_{jk,m}$ is the jk-th element of Θ_m. Thus, the asymptotic distribution in (9.4.5) is not immediately useful for testing

$$H_0: \omega_{jk,h} = 0 \quad \text{against} \quad H_1: \omega_{jk,h} \neq 0 \tag{9.4.6}$$

which is a set of hypotheses of particular interest in practice. The significance of $\omega_{jk,h}$ may be checked however by testing $\theta_{jk,0} = \cdots = \theta_{jk,h-1} = 0$. Using a minor generalization of (9.4.3) this hypothesis can be tested (see Lütkepohl & Poskitt (1991)). ■

REMARK 2: In sharp contrast to the case where the VAR order is assumed known and finite (see Proposition 3.6), the asymptotic variances of all impulse responses are nonzero in the present case. Another difference between the finite and infinite order VAR cases is that in the former the asymptotic standard errors of the Φ_i and Θ_i go to zero as i increases while the covariance matrix in (9.4.1) is a nondecreasing function of i and the covariance matrix in (9.4.3) is bounded away from zero, for $i > 0$. ■

REMARK 3: For $i = 1$, the asymptotic covariance matrix of $\tilde{\Phi}_1(n_T)$ in (9.4.1) is $\Sigma_u^{-1} \otimes \Sigma_u$. It can be shown that the same asymptotic covariance matrix is obtained for $\hat{\Phi}_1$ from Proposition 3.6 if a VAR(n) process is fitted with n greater than the true order p (see Lütkepohl (1988b)). A similar result is obtained for $\tilde{\Theta}_i(n_T)$ and $\hat{\Theta}_i$ for $i = 0, 1$ (see Problem 9.4). ■

9.4.2 An Example

To illustrate the consequences of the finite and infinite VAR order assumptions we use again the VAR(2) model for the investment/income/consumption data. Of course, the same estimated impulse responses are obtained as in Section 3.7. (The intercept form of the model is used now.) The standard errors are different, however. In Figures 9.1 and 9.2 consumption responses to income impulses are depicted and the two-standard error bounds obtained from both sets of assumptions are shown. In both figures the two-standard error bounds based on Prop-

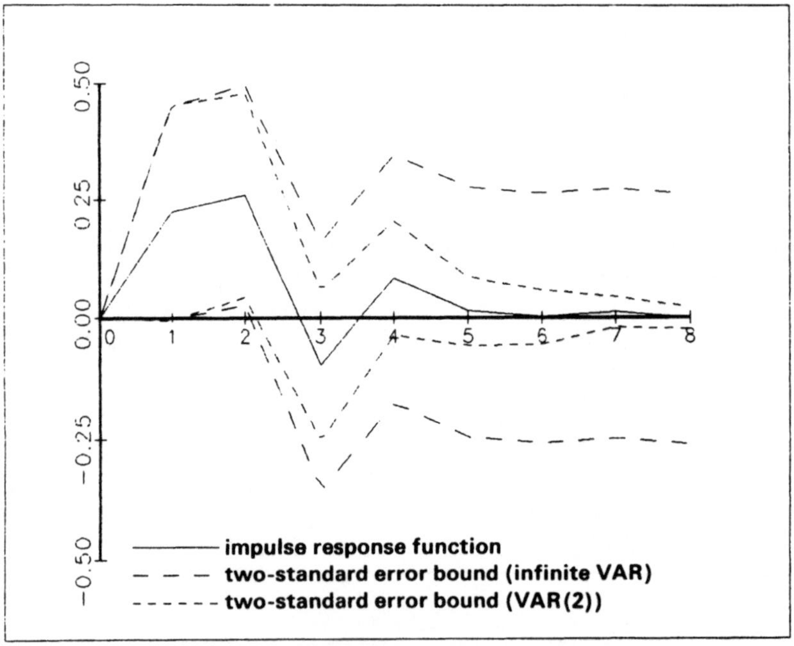

Fig. 9.1. Estimated responses of consumption to a forecast error impulse in income with two-standard error bounds based on finite and infinite VAR order assumptions.

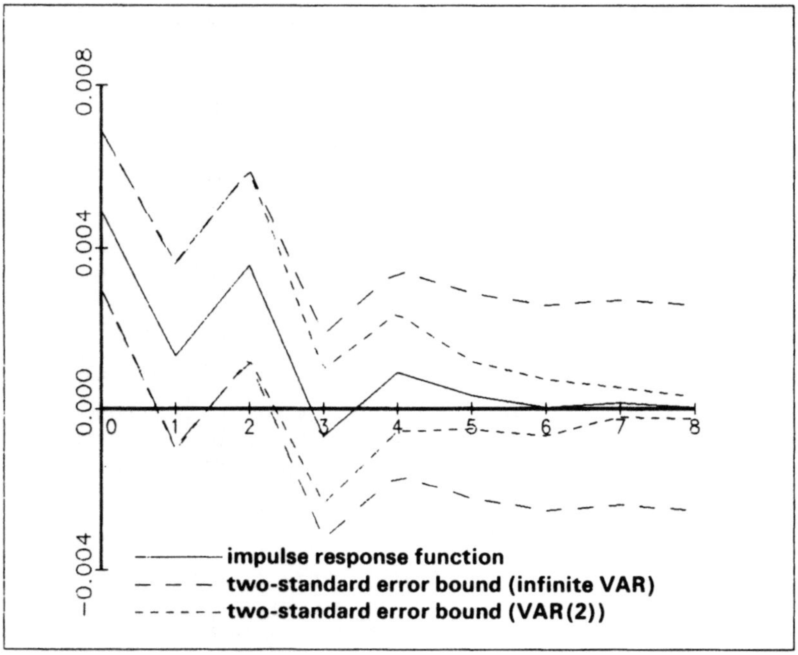

Fig. 9.2. Estimated responses of consumption to an orthogonalized impulse in income with two-standard error bounds based on finite and infinite VAR order assumptions.

osition 9.4 are seen to grow with the time lag while the standard errors from Proposition 3.6 decline almost to zero for longer lags. This reflects the additional estimation uncertainty that results from assuming that the VAR order goes to infinity with the sample size. Thereby more and more parameters are estimated as the sample size gets large.

In Table 9.2 forecast error variance decompositions of the system are shown. Again most standard errors based on the infinite VAR assumption are slightly larger than those from Chapter 3 which are also given in the table. Although it is tempting to use the estimated standard errors in checking the significance of individual forecast error variance components, we know from Remark 1 that they are not useful for that purpose as the asymptotic standard errors from Proposition 9.4 corresponding to zero forecast error variance components are zero.

9.5 Exercises

Problem 9.1

For the invertible MA(1) process $y_t = u_t + Mu_{t-1}$ and $n = 1, 2$, determine the matrix Γ_n defined in (9.2.7).

Table 9.2. Forecast error variance decomposition of the investment/income/consumption system with standard errors from two different asymptotic theories

		proportions of forecast error variance, h periods ahead, accounted for by innovations in		
forecast error in	forecast horizon h	investment $\omega_{j1,h}$	income $\omega_{j2,h}$	consumption $\omega_{j3,h}$
investment ($j = 1$)	1	1.000(.000)[a] [.000][b]	.000(.000)[.000]	.000(.000)[.000]
	2	.960(.042) [.044]	.018(.030)[.030]	.023(.031)[.033]
	3	.946(.042) [.045]	.028(.033)[.033]	.026(.029)[.032]
	4	.941(.045) [.048]	.029(.031)[.032]	.030(.032)[.036]
	8	.938(.048) [.050]	.031(.032)[.034]	.032(.035)[.039]
income ($j = 2$)	1	.018(.031) [.031]	.983(.031)[.031]	.000(.000)[.000]
	2	.060(.054) [.053]	.908(.063)[.064]	.032(.037)[.040]
	3	.070(.057) [.058]	.896(.066)[.068]	.035(.039)[.041]
	4	.068(.056) [.057]	.892(.067)[.069]	.039(.041)[.045]
	8	.069(.057) [.058]	.891(.068)[.070]	.040(.041)[.045]
consumption ($j = 3$)	1	.080(.061) [.061]	.273(.086)[.086]	.647(.090)[.090]
	2	.077(.059) [.059]	.274(.082)[.082]	.649(.088)[.088]
	3	.130(.080) [.080]	.334(.089)[.091]	.537(.091)[.092]
	4	.129(.079) [.079]	.335(.088)[.090]	.536(.089)[.091]
	8	.129(.080) [.081]	.340(.089)[.092]	.532(.091)[.093]

[a] Estimated standard error based on a finite known VAR order assumption.
[b] Estimated standard error based on an infinite VAR order assumption.

Problem 9.2

Suppose the true data generation mechanism is a univariate AR(1) process $y_t = \alpha y_{t-1} + u_t$. Assume that a univariate AR(1) is indeed fitted to the data and compare the resulting approximate forecast MSEs $\Sigma_{\hat{y}}(h)$ (given in Section 3.5) and $\Sigma_{\tilde{y}}(h)$ (given in Section 9.3.1) for $h = 1, 2, \ldots$. (Hint: see Lütkepohl (1987, pp. 76, 77).)

Problem 9.3

Suppose the true data generation process is an invertible MA(1) as in Problem 9.1. Write down explicit expressions for the asymptotic covariance matrices of $\tilde{\Phi}_i(n_T)$, $\tilde{\Theta}_i(n_T)$, $i = 1, 2$ and of $\tilde{\Psi}_m(n_T)$, $\tilde{\Xi}_m(n_T)$, $m = 1, 2$.

Problem 9.4

Let $\tilde{\Theta}_i(n_T)$ and $\hat{\Theta}_i$ be estimators of the orthogonalized impulse responses Θ_i obtained under the conditions of Propositions 9.4 and 3.6, respectively. If the true data generation mechanism is a finite order VAR(p) process and the actual process fitted to the data has order $n_T > p$, show that the asymptotic covariance matrices in (9.4.3) and (3.7.8) are identical for $i = 0, 1$.

Problem 9.5

Consider the investment/income/consumption system of Section 9.3.2 and fit a VAR(4) process to the data.

(a) Determine 95% interval forecasts for all three variables and forecast horizons $h = 1, 2, 3$ under the assumption of a known true VAR order of $p = 4$ and under the assumption of an infinite order true generation process.
(b) Determine Φ_i and Θ_i impulse responses and their asymptotic standard errors for $i = 1, 2, 3, 4$ under both the assumption of a finite and an infinite true VAR order. Compare the estimated standard errors obtained under the two alternative scenarios for all variables.

Part III
Systems with Exogenous Variables and Nonstationary Processes

In Parts I and II we have assumed that the time series of interest are generated by stationary processes. In other words, the data generating processes are assumed to have time invariant first and second moments. In this part nonstationary processes will be considered. In Chapter 10 systems with exogenous variables are discussed. These systems may be stationary if the exogenous variables are generated by stationary processes. Alternatively the exogenous variables may be nonstochastic fixed quantities. In that case the mean vectors of the time series variables of interest may be time varying. In Chapter 11 unstable nonstationary processes are considered, in Chapter 12 processes with time varying VAR structure are treated and in Chapter 13 state space models are discussed. The class of state space models is very general. It encompasses the processes considered in the previous chapters and also various kinds of nonstationary processes.

Chapter 10. Systems of Dynamic Simultaneous Equations

10.1 Background

This chapter serves to point out some possible extensions of the models considered so far and to draw attention to potential problems related to such extensions. So far we have assumed that all variables of a system are stochastic and they all have essentially the same status in that they are all determined within the system. In other words, the model describes the joint generation process of all the observable variables of interest. In practice the generation process may be affected by other observable variables which are determined outside the system of interest. Such variables are called *exogenous*. In contrast, the variables determined within the system are called *endogenous*. Exogenous variables may be stochastic or nonstochastic. Seasonal dummies fall, for instance, in the latter group whereas weather related variables such as rainfall or hours of sunshine are usually regarded as stochastic exogenous variables. As another example of the latter type of variables, if a small open economy is being studied, the price level or the output of the rest of the world may be regarded as exogenous.

A model with exogenous variables may have the form

$$\bar{A}_0 y_t = \bar{A}_1 y_{t-1} + \cdots + \bar{A}_p y_{t-p} + \bar{B}_0 x_t + \bar{B}_1 x_{t-1} + \cdots + \bar{B}_s x_{t-s} + w_t, \quad (10.1.1)$$

where $y_t = (y_{1t}, \dots, y_{Kt})'$ is a K-dimensional vector of endogenous variables, $x_t = (x_{1t}, \dots, x_{Mt})'$ is an M-dimensional vector of exogenous variables, the $\bar{A}_i$ and $\bar{B}_j$ are $(K \times K)$ and $(K \times M)$ coefficient matrices, respectively, and w_t is a K-dimensional error vector. The vector x_t may contain both stochastic and nonstochastic components. For example, it may include intercept terms, seasonal dummies and the amount of rainfall in a specific region. If the error term w_t is white noise a model of the type (10.1.1) is sometimes called a VARX(p, s) model in the following. It is referred to as a VARMAX(p, s, q) model if w_t is an MA(q) process. More generally, models of the form (10.1.1) are often called *linear systems* because they are obviously linear in all variables. In the econometrics literature the label (linear) *dynamic simultaneous equations model* (SEM) is used for such a model. Since we often have systems of economic variables in mind in the following discussion we will use this name frequently.

Other names that are occasionally found in the related literature are *transfer function models* or *distributed lag* models. These terms will become more plausible in the next section where different representations and some properties of our basic model (10.1.1) will be discussed. Estimation is briefly considered in Section 10.3 and some remarks on model specification and model checking follow in Section 10.4. Possible uses of such models, namely forecasting, multiplier analysis, and control, are treated in Sections 10.5–10.7. Concluding remarks are contained in Section 10.8. As mentioned previously, it is not the purpose of this chapter to give a detailed and complete account of all these topics. The chapter is just meant to give some guidance to possible extensions of the by now familiar models, the related problems and some further reading.

10.2 Systems with Exogenous Variables

10.2.1 Types of Variables

In the dynamic simultaneous equations model (10.1.1) we have partitioned the observables in two groups, y and x. The components of y are endogenous variables and the components of x are the exogenous variables. Although we have given some explanation of the differences between the two groups of variables we have not given a precise definition of the terms endogenous and exogenous so far. The reason is that for the present discussion it is not necessary to have such a definition. It suffices to have a partitioning into two groups. The reader should, however, have some intuition of what variables are contained in y and which ones are included in x. As mentioned previously, roughly speaking, y contains the observable *outputs* of the system, that is, the observable variables that are determined by the system. In contrast, the x variables may be regarded as *observable input variables* which are determined outside the system. In this setting the error variables w_t may be viewed as *unobservable inputs* to the system. As we have seen, nonstochastic components may be absorbed into the set of exogenous x variables. All or some of the components of x may be under full or partial control of the government or a decision or policy maker. In a control context such variables are often referred to as *instruments or instrument variables* (see Section 10.7). Sometimes the lagged endogenous variables together with the exogenous variables of a system are called *predetermined variables*. If x contains just a constant and $s = 0$ the model (10.1.1) reduces to a VAR or a VARMA model depending on the assumptions regarding w_t.

For illustrative purposes consider the following example system relating investment (INV), income (INC), and consumption (CONS) variables:

$$\begin{aligned} \text{INC}_t &= \bar{\nu}_1 + \bar{\alpha}_{11,1}\,\text{INC}_{t-1} + \bar{\alpha}_{12,1}\,\text{CONS}_{t-1} + \bar{\beta}_{12,1}\,\text{INV}_{t-1} + w_{1t}, \\ \text{CONS}_t &= \bar{\nu}_2 + \bar{\alpha}_{22,1}\,\text{CONS}_{t-1} + \bar{\alpha}_{21,0}\,\text{INC}_t + \bar{\alpha}_{21,1}\,\text{INC}_{t-1} + w_{2t}. \end{aligned} \tag{10.2.1}$$

This model is similar to those obtained for West German data in Chapter 5. An important difference is that current income appears in the consumption equation and there is no equation for investment. Thus, only income and consumption are determined within the system whereas investment is not. The fact that investment is, of course, determined within the economic system as a whole does not necessarily mean that we have to specify its generation mechanism if our main interest is with the generation mechanism of income and consumption. In terms of the representation (10.1.1) the example system can be written as

$$\begin{bmatrix} 1 & 0 \\ -\bar{\alpha}_{21,0} & 1 \end{bmatrix}\begin{bmatrix} \text{INC}_t \\ \text{CONS}_t \end{bmatrix} = \begin{bmatrix} \bar{\alpha}_{11,1} & \bar{\alpha}_{12,1} \\ \bar{\alpha}_{21,1} & \bar{\alpha}_{22,1} \end{bmatrix}\begin{bmatrix} \text{INC}_{t-1} \\ \text{CONS}_{t-1} \end{bmatrix} + \begin{bmatrix} \bar{\nu}_1 & \bar{\beta}_{12,1} \\ \bar{\nu}_2 & 0 \end{bmatrix}\begin{bmatrix} 1 \\ \text{INV}_{t-1} \end{bmatrix} + \begin{bmatrix} w_{1t} \\ w_{2t} \end{bmatrix}. \tag{10.2.2}$$

Thus $y_t = (\text{INC}_t, \text{CONS}_t)'$ and $x_t = (1, \text{INV}_t)'$ are both two-dimensional. The predetermined variables are y_{t-1} and x_{t-1}.

In this example the distinction between endogenous and exogenous variables may seem a bit arbitrary since the model represents part of an economic system in which all three variables INV, INC, and CONS are determined. It is in fact no simple matter to agree on a precise definition of exogeneity and endogeneity. A thorough discussion of the issue is given by Engle, Hendry & Richard (1983). One possibility is to call variables exogenous that are independent of the residual process w_t. In the following we will implicitly make this assumption although most results can be obtained under less restrictive conditions.

In dynamic SEMs there are sometimes identities or exact relations between some variables. For instance, the same figures may be used for supply and demand of a product. In that case an identity equating supply and demand may appear as a separate equation of a system. So far we have not excluded this possibility. However, in later sections the covariance matrix of w_t will be assumed to be nonsingular which excludes identities. Then we assume without further notice that they have been eliminated by substitution. For instance, the demand variable may be substituted for the supply variable in all instances where it appears in the system.

10.2.2 Structural Form, Reduced Form, Final Form

The representation (10.1.1) is called the *structural form* of the model if it represents the instantaneous effects of the endogenous variables properly. The instantaneous effects are reflected in the elements of $\bar{A}_0$. The idea is that the instantaneous causal links are derived from theoretical considerations and are used to place restrictions on $\bar{A}_0$. Of course, multiplication of (10.1.1) with any other nonsingular $(K \times K)$ matrix results in an equivalent representation of the process generating y_t. Such a representation is not called a structural form, however, unless it reflects the actual instantaneous effects.

The *reduced form* of the system is obtained by premultiplying (10.1.1) with $\bar{A}_0^{-1}$ which gives

$$y_t = A_1 y_{t-1} + \cdots + A_p y_{t-p} + B_0 x_t + \cdots + B_s x_{t-s} + u_t, \tag{10.2.3}$$

where $A_i := \bar{A}_0^{-1}\bar{A}_i$, $B_j := \bar{A}_0^{-1}\bar{B}_j$, and $u_t := \bar{A}_0^{-1}w_t$. We always assume without notice that the inverse of $\bar{A}_0$ exists. In Sections 10.5–10.7 we will see that the reduced form is useful for forecasting, multiplier analysis, and control purposes.

For the example model given in (10.2.2) we have

$$\bar{A}_0^{-1} = \begin{bmatrix} 1 & 0 \\ \bar{\alpha}_{21,0} & 1 \end{bmatrix}$$

and hence the reduced form is

$$\begin{bmatrix} \text{INC}_t \\ \text{CONS}_t \end{bmatrix} = A_1 \begin{bmatrix} \text{INC}_{t-1} \\ \text{CONS}_{t-1} \end{bmatrix} + B_1 \begin{bmatrix} 1 \\ \text{INV}_{t-1} \end{bmatrix} + \begin{bmatrix} u_{1t} \\ u_{2t} \end{bmatrix}, \tag{10.2.4}$$

where

$$A_1 = \begin{bmatrix} \alpha_{11,1} & \alpha_{12,1} \\ \alpha_{21,1} & \alpha_{22,1} \end{bmatrix} = \begin{bmatrix} \bar{\alpha}_{11,1} & \bar{\alpha}_{12,1} \\ \bar{\alpha}_{21,0}\bar{\alpha}_{11,1} + \bar{\alpha}_{21,1} & \bar{\alpha}_{21,0}\bar{\alpha}_{12,1} + \bar{\alpha}_{22,1} \end{bmatrix}, \tag{10.2.5}$$

$$B_1 = \begin{bmatrix} \beta_{11,1} & \beta_{12,1} \\ \beta_{21,1} & \beta_{22,1} \end{bmatrix} = \begin{bmatrix} \bar{v}_1 & \bar{\beta}_{12,1} \\ \bar{\alpha}_{21,0}\bar{v}_1 + \bar{v}_2 & \bar{\alpha}_{21,0}\bar{\beta}_{12,1} \end{bmatrix}, \tag{10.2.6}$$

and

$$\begin{bmatrix} u_{1t} \\ u_{2t} \end{bmatrix} = \begin{bmatrix} w_{1t} \\ \bar{\alpha}_{21,0} w_{1t} + w_{2t} \end{bmatrix}.$$

It is important to note that the reduced form parameters are in general nonlinear functions of the structural form parameters.

In lag operator notation the reduced form (10.2.3) can be written as

$$A(L)y_t = B(L)x_t + u_t, \tag{10.2.7}$$

where

$$A(L) := I_K - A_1 L - \cdots - A_p L^p$$

and

$$B(L) := B_0 + B_1 L + \cdots + B_s L^s.$$

If the effect of a change in an exogenous variable on the endogenous variables is of interest it is useful to solve the system (10.2.7) for the endogenous variables by multiplying with $A(L)^{-1}$. The resulting representation

$$y_t = D(L)x_t + A(L)^{-1}u_t, \tag{10.2.8}$$

where $D(L) := A(L)^{-1}B(L)$, is sometimes called the *final form* of the system. Of course, its existence requires invertibility of $A(L)$ which is guaranteed if

$$\det(A(z)) \neq 0 \qquad \text{for } |z| \leq 1. \tag{10.2.9}$$

If y_t contains just one variable, $A(L)$ is a scalar operator and the form (10.2.8) is often called a *distributed lag model* in the econometrics literature because it describes how lagged effects of changes in x_t are distributed over time. Since the lag distribution for each exogenous variable can be written as a ratio of two finite order polynomials in the lag operator the model is referred to as a *rational distributed lag model*. In the time series literature the label *rational transfer function model* is often attached to (10.2.8) in both the scalar and the vector case. The operator $D(L)$ represents the *transfer function* transferring the observable inputs into the outputs of the system.

For the example model with reduced form (10.2.4) we get a final form

$$\begin{bmatrix} \text{INC}_t \\ \text{CONS}_t \end{bmatrix} = (I_2 - A_1 L)^{-1} B_1 L \begin{bmatrix} 1 \\ \text{INV}_t \end{bmatrix} + (I_2 - A_1 L)^{-1} \begin{bmatrix} u_{1t} \\ u_{2t} \end{bmatrix}$$

$$= \left[\sum_{i=1}^{\infty} A_1^{i-1} B_1 L^i \right] \begin{bmatrix} 1 \\ \text{INV}_t \end{bmatrix} + \left[\sum_{i=0}^{\infty} A_1^i L^i \right] \begin{bmatrix} u_{1t} \\ u_{2t} \end{bmatrix}. \qquad (10.2.10)$$

Note that $B_0 = 0$ and thus, $D_0 = 0$ and $D_i = A_1^{i-1} B_1$ for $i = 1, 2, \ldots$.

The coefficient matrices $D_i = (d_{kj,i})$ of the transfer function operator

$$D(L) = \sum_{i=0}^{\infty} D_i L^i$$

contain the effects that changes in the exogenous variables have on the endogenous variables. Everything else held constant, a unit change in the j-th exogenous variable in period t induces a change of $d_{kj,i}$ units in the k-th endogenous variable in period $t + i$. The elements of the D_i matrices are therefore called *dynamic multipliers*. The accumulated effects contained in $\sum_{i=0}^{n} D_i$ are the n-th *interim multipliers* and the elements of $\sum_{i=0}^{\infty} D_i$ are the *long-run effects* or *total multipliers*. We will return to multiplier analysis in Section 10.6.

As in the example the transfer function operator $D(L)$ in general has infinite order. A finite order representation of the system is obtained by noting that $A(L)^{-1} = A(L)^*/|A(L)|$, where $A(L)^*$ denotes, as usual, the adjoint of $A(L)$. Thus, multiplying the reduced form by $A(L)^*$ gives

$$|A(L)| y_t = A(L)^* B(L) x_t + A(L)^* u_t \qquad (10.2.11)$$

which involves finite order operators only. In the econometrics literature these equations are sometimes referred to as *final equations*. Since $|A(L)|$ is a scalar operator each equation contains only one of the endogenous variables. If u_t has a finite order MA representation and there are no exogenous variables, (10.2.11) is a final equations form VARMA representation as defined in Chapter 7. This is the reason for using the name final equations form in that chapter.

Yet another relation to the VARMA processes of the previous chapters is obtained by assuming that the exogenous variables x_t are driven by a VAR(q) process, say

$$x_t = C_1 x_{t-1} + \cdots + C_q x_{t-q} + v_t,$$

where $q \le p$ and v_t is white noise. In that case the joint generation process for x_t and y_t is

$$\begin{bmatrix} I_K & -B_0 \\ 0 & I_M \end{bmatrix} \begin{bmatrix} y_t \\ x_t \end{bmatrix} = \begin{bmatrix} A_1 & B_1 \\ 0 & C_1 \end{bmatrix} \begin{bmatrix} y_{t-1} \\ x_{t-1} \end{bmatrix} + \cdots + \begin{bmatrix} A_p & B_p \\ 0 & C_p \end{bmatrix} \begin{bmatrix} y_{t-p} \\ x_{t-p} \end{bmatrix} + \begin{bmatrix} u_t \\ v_t \end{bmatrix},$$

where it is assumed without loss of generality that $s \le p$, $B_i = 0$ for $i > s$ and

$C_j = 0$ for $j > q$. If u_t is also white noise, premultiplying by

$$\begin{bmatrix} I_K & -B_0 \\ 0 & I_M \end{bmatrix}^{-1}$$

shows that the joint generation process of y_t and x_t is a VAR(p) so that y_t is a subprocess or linear transformation of a VAR(p). Therefore we know from Section 6.6.1, Corollary 6.1.1, that, as a marginal process, it has a VARMA representation. A similar result is obtained if u_t is a finite order MA process and/or x_t is generated by a VARMA(q, r) process. It is easy to extend the foregoing argument accordingly. This way we could also give a more precise definition of the stochastic process y_t described by a dynamic SEM.

10.2.3 Models with Rational Expectations

Sometimes the endogenous variables are assumed to depend not only on other endogenous and exogenous variables but also on expectations on endogenous variables. If only expectations formed in the previous period for the present period are of importance one could simply add another term involving the expectations variables to the structural form (10.1.1). Denoting the expectations variables by y_t^e may then result in a reduced form

$$y_t = A_1 y_{t-1} + \cdots + A_p y_{t-p} + F y_t^e + B_0 x_t + \cdots + B_s x_{t-s} + u_t \tag{10.2.12}$$

or

$$A(L) y_t = F y_t^e + B(L) x_t + u_t, \tag{10.2.13}$$

where F is a $(K \times K)$ matrix of parameters and $A(L)$ and $B(L)$ are the matrix polynomials in the lag operator from (10.2.7).

Following Muth (1961), the expectations y_t^e formed in period $t - 1$ are called *rational* if they are the best possible predictions given the information in period $t - 1$. In other words, y_t^e is the conditional expectation $E_{t-1}(y_t)$ given all information in period $t - 1$. In forming the predictions or expectations not only the past values of the endogenous and exogenous variables are assumed known but also the model (10.2.12) and the generation process of the exogenous variables. It is easy to see that, if the exogenous variables are generated by a VAR or VARMA process, the expectations variables can be eliminated from (10.2.12)/(10.2.13). The resulting reduced form is of a VARX or VARMAX type and consequently y_t is generated by a VARMA mechanism. To show this suppose that u_t is independent white noise and, as before, denote by E_t the conditional expectation given all information up to period t. Applying E_{t-1} to (10.2.12) then gives

$$\begin{aligned} y_t^e &= E_{t-1}(y_t) \\ &= A_1 y_{t-1} + \cdots + A_p y_{t-p} + F y_t^e + B_0 E_{t-1}(x_t) + B_1 x_{t-1} + \cdots + B_s x_{t-s} \end{aligned} \tag{10.2.14}$$

or

$$y_t^e = (A(L) - I_K) y_t + F y_t^e + B_0 E_{t-1}(x_t) + (B(L) - B_0) x_t. \tag{10.2.15}$$

Assuming that $I_K - F$ is invertible this system can be solved for y_t^e:

$$y_t^e = (I_K - F)^{-1}[(A(L) - I_K)y_t + B_0 E_{t-1}(x_t) + (B(L) - B_0)x_t]. \tag{10.2.16}$$

If x_t is generated by a VAR(q) process, say

$$x_t = C_1 x_{t-1} + \cdots + C_q x_{t-q} + v_t,$$

where v_t is independent white noise, then

$$E_{t-1}(x_t) = C_1 x_{t-1} + \cdots + C_q x_{t-q}.$$

Substituting this expression in (10.2.16) shows that y_t^e depends on lagged y_t and x_t only. Thus, substituting for y_t^e in (10.2.12) or (10.2.13) we get a standard VARX form of the model.

Thus, in theory, when the true coefficient matrices are known we can simply eliminate the term involving expectations variables and work with a standard reduced form without an expectation term. It should be clear, however, that in substituting the right-hand side of (10.2.16) for y_t^e in (10.2.12) implies nonlinear restrictions on the coefficient matrices of the reduced form without expectations terms. Taking into account such restrictions may increase the efficiency of parameter estimators. The same is true, of course, for the structural form. Therefore it is important in practice whether or not the actual relationship between the variables is partly determined by agents' expectations.

We can also go one step further and solve the reduced form for y_t. In the previous subsection we have seen that, under the present assumptions, the endogenous variables are generated by a VARMA process.

For expository purposes we have just treated a very special case where only expectations formed in period $t - 1$ for period t enter the model. Extensions can be treated in a similar way. For instance, past expectations for more than one period ahead or expectations formed in various previous periods may be of importance. If x_t is generated by a VAR(q) process they can be eliminated like in the special case considered in the foregoing. It is also easy to extend the discussion to more general generation processes of the exogenous variables. For the case of a VARMA(q, r) generation process see Problem 10.4. Furthermore, the error term u_t may have a more general structure without effecting the essential result. In summary, if past expectations only enter the model it is relatively easy to solve for the endogenous variables. They have a VARMA representation if the exogenous variables and error terms are generated by stationary VARMA processes.

A complication of the basic model that makes life a bit more difficult is the inclusion of future expectations. It is quite realistic to suppose that, for instance, the expected future price of a commodity may determine the supply in the present period. For example, if bond prices are expected to fall during the next period an investor may decide to sell now. If future expectations enter the model the solution for the endogenous variables will in general not be unique. In other words, the process that generates the endogenous variables may not be uniquely determined by the model even if the generation process of the exogenous variables is uniquely specified.

To illustrate the problem let us consider the following simple structural form:

$$\bar{A}_0 y_t = \bar{A}_1 y_{t-1} + F y_{t+1}^e + z_t, \tag{10.2.17}$$

where $y_{t+1}^e = E_t(y_{t+1})$ denotes the optimal forecast of y_{t+1} in period t. The term z_t includes all the exogenous variables and the error term, e.g.,

$$z_t = B_0 x_t + \cdots + B_s x_{t-s} + u_t. \tag{10.2.18}$$

Since the expectations make efficient use of all information available at time t,

$$e_{t+1} = y_{t+1} - E_t(y_{t+1}) \tag{10.2.19}$$

must be a zero mean white noise sequence of forecast errors uncorrelated with information available at time t. In order to solve (10.2.17) for y_t we substitute $y_{t+1} - e_{t+1}$ for y_{t+1}^e and get

$$\bar{A}_0 y_t = \bar{A}_1 y_{t-1} + F y_{t+1} - F e_{t+1} + z_t$$

or

$$G(L) y_t = -F e_t + z_{t-1},$$

where $G(L) := -F + \bar{A}_0 L - \bar{A}_1 L^2$. If this operator is invertible we get

$$y_t = -G(L)^{-1} F e_t + G(L)^{-1} z_{t-1}. \tag{10.2.20}$$

Suppose now that z_t has the structure given in (10.2.18) and that both x_t and u_t are finite order VARMA processes, then the same holds for z_t because it is just a linear transformation of VARMA processes (see Section 6.6.1, Corollary 6.1.1). Consequently, $G(L)^{-1} z_{t-1}$ is a VARMA process. Also, since e_t is white noise by assumption, $G(L)^{-1} F e_t$ is of the VARMA type. Hence the right-hand side of (10.2.20) is just the sum of VARMA processes and thus y_t must have a VARMA representation (see Section 6.6.1). However, the white noise process e_t is in general not unique. In fact, any white noise process could be used that gives rise to a y_t process satisfying the structural form relations (10.2.17). In other words, y_t is not necessarily uniquely determined by (10.2.20). Broze, Gouriéroux & Szafarz (1989) give necessary and sufficient conditions for the process e_t to be a feasible choice in (10.2.20).

In this discussion we have made simplifying assumptions that can be relaxed. For instance, a solution is also possible if $G(L)$ is not invertible. Moreover, x_t may not admit a VARMA representation. Again a solution for y_t is possible. It may not be a VARMA process, though. Further generalizations are possible by adding more lags of y_t in (10.2.17) or by including further future or past expectations terms. It can be shown that these extensions can be handled as in the framework of the simple model (10.2.17) by redefining the variables appropriately (see Broze, Gouriéroux & Szafarz (1989)). Further extensive discussions of rational expectations models can also be found in volumes by Lucas & Sargent (1981) and Pesaran (1987).

10.3 Estimation

Suppose we wish to estimate the parameters of the reduced form (10.2.3) which can be written as

$$y_t = AY_{t-1} + BX_{t-1} + B_0 x_t + u_t, \tag{10.3.1}$$

where $A := [A_1, \dots, A_p]$, $B := [B_1, \dots, B_s]$,

$$Y_t := \begin{bmatrix} y_t \\ \vdots \\ y_{t-p+1} \end{bmatrix}, \qquad X_t := \begin{bmatrix} x_t \\ \vdots \\ x_{t-s+1} \end{bmatrix},$$

and u_t is assumed to be white noise with *nonsingular* covariance matrix Σ_u. We assume that a matrix R and a vector γ exist such that

$$\boldsymbol{\beta} := \text{vec}[A, B, B_0] = R\gamma. \tag{10.3.2}$$

With these assumptions estimation of $\boldsymbol{\beta}$ and hence of A, B, and B_0 is straightforward.

For a sample of size T the system can be written compactly as

$$Y = [A, B, B_0]Z + U, \tag{10.3.3}$$

where

$$Y := [y_1, \dots, y_T], \qquad Z := \begin{bmatrix} Y_0, \dots, Y_{T-1} \\ X_0, \dots, X_{T-1} \\ x_1, \dots, x_T \end{bmatrix} \qquad \text{and} \qquad U := [u_1, \dots, u_T].$$

Vectorizing gives

$$\mathbf{y} = (Z' \otimes I_K)R\gamma + \mathbf{u},$$

where $\mathbf{y} := \text{vec}(Y)$ and $\mathbf{u} := \text{vec}(U)$. From Chapter 5 the GLS estimator is known to be

$$\hat{\gamma} = [R'(ZZ' \otimes \Sigma_u^{-1})R]^{-1}R'(Z \otimes \Sigma_u^{-1})\mathbf{y}. \tag{10.3.4}$$

This estimator is not operational because in practice Σ_u is unknown. However, as in Section 5.2.2, Σ_u may be estimated from the LS estimator

$$\gamma^{\#} = [R'(ZZ' \otimes I_K)R]^{-1}R'(Z \otimes I_K)\mathbf{y}$$

which gives residuals $\mathbf{u}^{\#} = \mathbf{y} - (Z' \otimes I_K)R\gamma^{\#}$ and an estimator

$$\Sigma_u^{\#} = U^{\#}U^{\#\prime}/T \tag{10.3.5}$$

of Σ_u, where $U^{\#}$ is such that $\text{vec}(U^{\#}) = \mathbf{u}^{\#}$. Using this estimator of the white noise covariance matrix results in the EGLS estimator

$$\hat{\hat{\gamma}} = [R'(ZZ' \otimes \Sigma_u^{\#-1})R]^{-1}R'(Z \otimes \Sigma_u^{\#-1})\mathbf{y}. \tag{10.3.6}$$

Under standard assumptions this estimator is consistent and asymptotically normal,

$$\sqrt{T}(\hat{\hat{\gamma}} - \gamma) \xrightarrow{d} N(0, \Sigma_{\hat{\hat{\gamma}}}), \tag{10.3.7}$$

where

$$\Sigma_{\hat{\hat{\gamma}}} = (R'[\text{plim}(ZZ'/T) \otimes \Sigma_u^{-1}]R)^{-1}. \tag{10.3.8}$$

One condition for this result to hold is, of course, that both $\text{plim}(ZZ'/T)$ and the inverse of the matrix in (10.3.8) exist. Further assumptions are required to guarantee the asymptotic normal distribution of the EGLS estimator. The assumptions may include the following ones: (i) u_t is standard white noise, (ii) the VAR part is stable, that is,

$$|A(z)| = |I_K - A_1 z - \cdots - A_p z^p| \neq 0 \qquad \text{for } |z| \leq 1,$$

and (iii) x_t is generated by a stationary, stable and invertible VARMA process which is independent of u_t. A precise statement of more general conditions and a proof are given, e.g., by Hannan & Deistler (1988).

The latter part of this set of assumptions requires that all the exogenous variables are stochastic. It can be modified so as to include nonstochastic variables as well. In that case the plim in (10.3.8) reduces to a nonstochastic limit in some or all components (see, e.g., Anderson (1971, Chapter 5), Harvey (1981)). The existence of $\text{plim}(ZZ'/T)$ excludes trending variables, however. More general cases that even allow for trending variables are considered by Park & Phillips (1988, 1989).

An estimator for $\boldsymbol{\beta} = R\gamma$ is obtained as $\hat{\hat{\boldsymbol{\beta}}} = R\hat{\hat{\gamma}}$. If (10.3.7) holds, this estimator also has an asymptotic normal distribution,

$$\sqrt{T}(\hat{\hat{\boldsymbol{\beta}}} - \boldsymbol{\beta}) \xrightarrow{d} N(0, \Sigma_{\hat{\hat{\boldsymbol{\beta}}}} = R\Sigma_{\hat{\hat{\gamma}}}R'). \tag{10.3.9}$$

Moreover, under general conditions, the corresponding estimator $\hat{\hat{\Sigma}}_u$ of the white noise covariance matrix is asymptotically independent of $\hat{\hat{\boldsymbol{\beta}}}$ and equivalent to the estimator UU'/T based on the unknown true residuals. For instance, for a Gaussian process,

$$\sqrt{T}\,\text{vech}(\hat{\hat{\Sigma}}_u - \Sigma_u) \xrightarrow{d} N(0, 2\mathbf{D}_K^+(\Sigma_u \otimes \Sigma_u)\mathbf{D}_K^{+\prime}), \tag{10.3.10}$$

where $\mathbf{D}_K^+ = (\mathbf{D}_K'\mathbf{D}_K)^{-1}\mathbf{D}_K'$ is the Moore-Penrose inverse of the $(K^2 \times K(K+1)/2)$ duplication matrix $\mathbf{D}_K$.

In discussing direct reduced form estimation with white noise errors we have treated the simplest possible case. The following complications are possible.

(1) Usually there will be restrictions on the structural form coefficients $\bar{A}_i$, $i = 0, \ldots, p$, and $\bar{B}_j, j = 0, \ldots, s$. Such restrictions may imply nonlinear constraints on the reduced form coefficients which are not covered by the above approach. Rational expectations assumptions may be another source of nonlinear restrictions on the reduced form parameters. Theoretically it is not difficult to handle nonlinear restrictions on the reduced form parameters. In practice numerical problems may arise in a multivariate LS or GLS estimation with nonlinear restrictions.

(2) Interest may focus on the structural rather than the reduced form. Estimation of the structural form has been discussed extensively in the econometrics literature. For recent surveys and many further references see Judge et al. (1985) or Hausman (1983). A major complication in estimating the structural form of a SEM such as (10.1.1) results from its possible nonuniqueness. Note that we have not assumed a triangular $\bar{A}_0$ matrix or a diagonal covariance matrix of w_t. Premultiplication of (10.1.1) by any nonsingular matrix results in an equivalent representation of the process. Thus, there must be restrictions on the structural form coefficients that guarantee uniqueness or identification of the structural form coefficients. Of course, in principle this identification problem is similar to that discussed in Chapter 7, Section 7.1.

(3) Identification problems may even arise in the reduced form if the error process u_t is allowed to be a finite order MA process. Clearly this identification problem is essentially the same as that discussed in Chapter 7. Hannan & Deistler (1988) provide quite general identification conditions and estimation results for this case. A further complication in this context arises if the structural form is of interest and the w_t error process is autocorrelated.

(4) So far we have just discussed models which are linear in the variables. In practice there may be nonlinear relations between the variables. Estimation of nonlinear dynamic models where the exogenous as well as the endogenous variables may enter in a nonlinear way are, for instance, discussed by Bierens (1981), Gallant (1987), and Gallant & White (1988).

10.4 Remarks on Model Specification and Model Checking

The basic principles of model specification and checking the model adequacy have been discussed in some detail in previous chapters. We will therefore make just a few remarks here. With respect to the specification there is, however, a major difference between the models considered previously and the dynamic SEMs of this chapter. While in a VAR or VARMA analysis usually relatively little prior knowledge from economic or other subject matter theory is used, such theories may well be the major building block in specifying SEMs. In that case model checking becomes of central importance in investigating the validity of the theory. Quite often theories are not available that specify the data generation process completely. For instance, the lag lengths of the endogenous and/or exogenous variables may have to be specified with statistical tools. Also, some researchers may not be prepared to rely on the available theories and therefore prefer to substitute statistical investigations for uncertain prior knowledge. Statistical specification strategies for general VARMAX models or reduced form dynamic SEMs are, for instance, proposed and discussed by Hannan & Kavalieris (1984), Hannan & Deistler (1988) and Poskitt (1989a). These strategies are based on model selection criteria of the type considered in previous chapters. An extensive literature exists on the specification of special models. For instance, distributed lag models are discussed at length in the econometrics literature (for some references see Judge et al. (1985, Chapters 9 and 10)). Specification proposals for transfer function models with one dependent variable y_t go back to the pioneering work of Box & Jenkins (1976). Other suggestions have been made by Haugh & Box (1977), Young, Jakeman & McMurtrie (1980), Liu & Hanssens (1982), Tsay (1985), and Poskitt (1989b) to name just a few.

In checking the model adequacy one may want to test various restrictions. These may range from constraints suggested by some kind of theory such as the rational expectations hypothesis to tests of the significance of extra lags. The three testing principles discussed previously, namely the LR, LM, and Wald principles (see Appendix C.5) can be used in the present context. Their asymptotic properties follow in the usual way from properties of the estimators and the model.

A residual analysis is another tool which is available in the present case. Plots of residuals may help to identify unusual values or patterns that suggest model deficiencies. Plots of residual autocorrelations may aid in checking the white noise assumption. Also a portmanteau test for overall residual autocorrelation may be developed for dynamic models with exogenous variables; see Poskitt & Tremayne (1981) for a discussion of this issue and further references.

10.5 Forecasting

10.5.1 Unconditional and Conditional Forecasts

If the future path of the exogenous variables is unknown to the forecaster then forecasts of these variables are needed in order to predict the future values of the endogenous variables on the basis of a dynamic SEM. For simplicity suppose that the exogenous variables are generated by a zero mean VAR(q) process as in Section 10.2.3,

$$x_t = C_1 x_{t-1} + \cdots + C_q x_{t-q} + v_t. \tag{10.5.1}$$

Now this process can be used to produce optimal forecasts $x_t(h)$ of x_t in the usual way. If the endogenous variables are generated by the reduced form model (10.2.3) with u_t being white noise independent of the x_t process, the optimal h-step forecast of y_{t+h} at origin t is

$$y_t(h) = A_1 y_t(h-1) + \cdots + A_p y_t(h-p) + B_0 x_t(h) + \cdots + B_s x_t(h-s), \tag{10.5.2}$$

where $y_t(j) := y_{t+j}$ and $x_t(j) := x_{t+j}$ for $j \leq 0$. This formula can be used for recursively determining forecasts for $h = 1, 2, \ldots$.

An alternative way for obtaining these predictions results from writing the generation processes of the exogenous variables in one overall model together with the reduced form SEM:

$$\begin{bmatrix} I_K & -B_0 \\ 0 & I_M \end{bmatrix} \begin{bmatrix} y_t \\ x_t \end{bmatrix} = \begin{bmatrix} A_1 & B_1 \\ 0 & C_1 \end{bmatrix} \begin{bmatrix} y_{t-1} \\ x_{t-1} \end{bmatrix} + \cdots + \begin{bmatrix} A_p & B_p \\ 0 & C_p \end{bmatrix} \begin{bmatrix} y_{t-p} \\ x_{t-p} \end{bmatrix} + \begin{bmatrix} u_t \\ v_t \end{bmatrix}, \tag{10.5.3}$$

where we assume without loss of generality that $p \geq \max(s, q)$ and set $B_i = 0$ for $i > s$ and $C_j = 0$ for $j > q$. As in Section 10.1.2, premultiplying by

$$\begin{bmatrix} I_K & -B_0 \\ 0 & I_M \end{bmatrix}^{-1} = \begin{bmatrix} I_K & B_0 \\ 0 & I_M \end{bmatrix}$$

gives a standard VAR(p) model. It is easy to see that the optimal forecasts for y_t and x_t from that model are exactly the same as those obtained by getting forecasts for x_t from (10.5.1) first and using them in the prediction formula for y_t given in (10.5.2) (see Problem 10.5). Thus, under the present assumptions, the discussion of forecasting VAR(p) processes applies. It will not be repeated here. It is not difficult to extend these ideas to sets of exogenous variables with nonstochastic components such as intercept terms or seasonal dummies. Also, some components of x_t may be generated by VARMA rather than just VAR processes.

We will refer to forecasts of y_t obtained in this way as *unconditional forecasts* because they are based on forecasts of the exogenous variables for the forecast period. Occasionally the forecaster may know some or all of the future values of the exogenous variables, for instance, because they are under the control of some decision maker. In that case he or she may be interested in *forecasts* of y_t *conditional* on a specific future path of x_t. In order to derive the optimal conditional forecasts we write the reduced form (10.2.3) in VARX(1, 0) form

$$Y_t = \mathbf{A} Y_{t-1} + \mathbf{B} x_t + U_t, \tag{10.5.4}$$

where

$$Y_t := \begin{bmatrix} y_t \\ \vdots \\ y_{t-p+1} \\ x_t \\ \vdots \\ x_{t-s+1} \end{bmatrix}, \quad U_t := \begin{bmatrix} u_t \\ 0 \\ \vdots \\ 0 \end{bmatrix} \quad ((Kp + Ms) \times 1),$$

$$\mathbf{A} := \left[\begin{array}{ccc|ccc} A_1 \dots A_{p-1} & & A_p & B_1 \dots B_{s-1} & & B_s \\ I_K & 0 & 0 & 0 & 0 & 0 \\ \ddots & \vdots & \vdots & \vdots\ \ddots & \vdots & \vdots \\ 0 \dots & I_K & 0 & 0 \dots & 0 & 0 \\ \hline & & & 0 \dots & 0 & 0 \\ & 0 & & I_M & 0 & 0 \\ & & & \ddots & \vdots & \vdots \\ & & & 0 \dots & I_M & 0 \end{array}\right], \quad \mathbf{B} := \begin{bmatrix} B_0 \\ 0 \\ \vdots \\ 0 \\ I_M \\ 0 \\ \vdots \\ 0 \end{bmatrix} \begin{matrix} \left.\vphantom{\begin{matrix} B_0 \\ 0 \\ \vdots \\ 0 \end{matrix}}\right\} (Kp \times M) \\ \left.\vphantom{\begin{matrix} I_M \\ 0 \\ \vdots \\ 0 \end{matrix}}\right\} (Ms \times M) \end{matrix}$$

$((Kp + Ms) \times (Kp + Ms))$

Successive substitution for lagged Y_t's gives

$$Y_t = \mathbf{A}^h Y_{t-h} + \sum_{i=0}^{h-1} \mathbf{A}^i \mathbf{B} x_{t-i} + \sum_{i=0}^{h-1} \mathbf{A}^i U_{t-i}. \tag{10.5.5}$$

Hence, premultiplying by the $(K \times (Kp + Ms))$ matrix $J := [I_K \quad 0 \dots 0]$ results in

$$y_{t+h} = J\mathbf{A}^h Y_t + \sum_{i=0}^{h-1} J\mathbf{A}^i \mathbf{B} x_{t+h-i} + \sum_{i=0}^{h-1} J\mathbf{A}^i J' u_{t+h-i}, \tag{10.5.6}$$

where $U_t = J'JU_t = J'u_t$ has been used. Now the optimal h-step forecast of y_t at origin t, given $x_{t+1}, \dots, x_{t+h}$ and all present and past information, is easily seen to be

$$y_t(h|x) := J\mathbf{A}^h Y_t + \sum_{i=0}^{h-1} J\mathbf{A}^i \mathbf{B} x_{t+h-i} \tag{10.5.7}$$

and the corresponding forecast error is

$$y_{t+h} - y_t(h|x) = \sum_{i=0}^{h-1} J\mathbf{A}^i J' u_{t+h-i}. \tag{10.5.8}$$

Thus, the MSE of the conditional forecast is

$$\Sigma_y(h|x) := \mathrm{MSE}(y_t(h|x)) = \sum_{i=0}^{h-1} J\mathbf{A}^i J' \Sigma_u J(\mathbf{A}^i)' J'. \tag{10.5.9}$$

Although this MSE matrix formally looks like the MSE matrix of the optimal forecast from a VAR model where $J\mathbf{A}^i J'$ is replaced by Φ_i, the MSE matrix here is in general different from the one of an unconditional forecast. This fact is easy to see by considering the different definition of the matrix $\mathbf{A}$ used in the pure VAR(p) case.

To illustrate the difference between conditional and unconditional forecasts we consider the simple reduced form

$$y_t = A_1 y_{t-1} + B_0 x_t + u_t, \tag{10.5.10}$$

where x_t is assumed to be generated by a zero mean VAR(1) process,

$$x_t = C_1 x_{t-1} + v_t.$$

Moreover we assume that u_t and v_t are independent white noise processes with covariance matrices Σ_u and Σ_v, respectively. The unconditional forecasts are obtained from the VAR process

$$\begin{bmatrix} I_K & -B_0 \\ 0 & I_M \end{bmatrix} \begin{bmatrix} y_t \\ x_t \end{bmatrix} = \begin{bmatrix} A_1 & 0 \\ 0 & C_1 \end{bmatrix} \begin{bmatrix} y_{t-1} \\ x_{t-1} \end{bmatrix} + \begin{bmatrix} u_t \\ v_t \end{bmatrix}$$

which has the standard VAR(1) form

$$\begin{bmatrix} y_t \\ x_t \end{bmatrix} = \begin{bmatrix} A_1 & B_0 C_1 \\ 0 & C_1 \end{bmatrix} \begin{bmatrix} y_{t-1} \\ x_{t-1} \end{bmatrix} + \begin{bmatrix} u_t + B_0 v_t \\ v_t \end{bmatrix}.$$

The optimal 1-step forecast from this model is

$$\begin{bmatrix} y_t(1) \\ x_t(1) \end{bmatrix} = \begin{bmatrix} A_1 & B_0 C_1 \\ 0 & C_1 \end{bmatrix} \begin{bmatrix} y_t \\ x_t \end{bmatrix}.$$

The corresponding MSE matrix is

$$\mathcal{I}_{\begin{bmatrix} y \\ x \end{bmatrix}}(1) = E\left[\begin{bmatrix} u_t + B_0 v_t \\ v_t \end{bmatrix} [(u_t + B_0 v_t)', v_t']\right] = \begin{bmatrix} \mathcal{I}_u + B_0 \mathcal{I}_v B_0' & B_0 \mathcal{I}_v \\ \mathcal{I}_v B_0' & \mathcal{I}_v \end{bmatrix}.$$

The upper left-hand corner block of this matrix is the MSE matrix of $y_t(1)$, the unconditional forecast of the endogenous variables. Thus,

$$\mathcal{I}_y(1) = \mathcal{I}_u + B_0 \mathcal{I}_v B_0'. \tag{10.5.11}$$

On the other hand, in the VARX(1, 0) representation (10.5.4) we have $\mathbf{A} = A_1$ and $\mathbf{B} = B_0$ for the present example. Hence, the conditional 1-step forecast of y_t is

$$y_t(1|x) = A_1 y_t + B_0 x_{t+1}$$

with corresponding MSE matrix

$$\mathcal{I}_y(1|x) = \mathcal{I}_u.$$

Obviously, $\mathcal{I}_y(1) - \mathcal{I}_y(1|x) = B_0 \mathcal{I}_v B_0'$ is positive semidefinite and thus the unconditional forecast is inferior to the conditional forecast. It must be kept in mind, however, that the conditional forecast is only feasible if the future values of the exogenous variables are either known or assumed. If only hypothetical values are used the conditional forecast may be quite poor if the actual values of the exogenous variables are different from the hypothetical ones. The smaller MSE of the conditional forecast is simply due to ignoring any uncertainty regarding the future path of the exogenous variables.

With the foregoing results interval forecasts and forecast regions can be set up as usual. So far we have discussed forecasting with known models. The case of estimated models will be considered next.

10.5.2 Forecasting Estimated Dynamic SEMs

In order to evaluate the consequences of using estimated instead of known processes for unconditional forecasts we can use a joint model for the endogenous and exogenous variables and then draw on results of the previous chapters. Therefore in this section we will focus on conditional forecasts only. We denote by $\hat{y}_t(h|x)$ the conditional h-step forecast (10.5.7) based on the estimated reduced form (10.2.3). The forecast error becomes

$$y_{t+h} - \hat{y}_t(h|x) = [y_{t+h} - y_t(h|x)] + [y_t(h|x) - \hat{y}_t(h|x)]. \tag{10.5.12}$$

Conditional on the exogenous variables the two terms in brackets are uncorrelated. Hence, assuming as in previous chapters that the processes used for estimation and forecasting are independent, an MSE approximation

$$\Sigma_{\hat{y}}(h|x) = \Sigma_y(h|x) + \frac{1}{T}\Omega_y(h|x) \tag{10.5.13}$$

is obtained in the by now familiar way. Here

$$\Omega_y(h|x) := E\left[\frac{\partial y_t(h|x)}{\partial \boldsymbol{\beta}'} \Sigma_{\hat{\boldsymbol{\beta}}} \frac{\partial y_t(h|x)'}{\partial \boldsymbol{\beta}}\right], \tag{10.5.14}$$

$\boldsymbol{\beta} := \mathrm{vec}[A_1, \ldots, A_p, B_1, \ldots, B_s, B_0]$ and $\Sigma_{\hat{\boldsymbol{\beta}}}$ is the covariance matrix of the asymptotic distribution of $\sqrt{T}(\hat{\boldsymbol{\beta}} - \boldsymbol{\beta})$. It is straightforward to show that

$$\begin{aligned}\frac{\partial y_t(h|x)}{\partial \boldsymbol{\beta}'} &= \frac{\partial(J\mathbf{A}^h Y_t)}{\partial \boldsymbol{\beta}'} + \sum_{i=0}^{h-1} \frac{\partial(J\mathbf{A}^i \mathbf{B} x_{t+h-i})}{\partial \boldsymbol{\beta}'} \\ &= \sum_{i=0}^{h-1} \Bigg[Y_t'(\mathbf{A}')^{h-1-i} \otimes J\mathbf{A}^i J' \\ &\quad + \sum_{j=0}^{i-1} x_{t+h-i}' \mathbf{B}'(\mathbf{A}')^{i-1-j} \otimes J\mathbf{A}^j J',\ x_{t+h-i}' \otimes J\mathbf{A}^i J' \Bigg].\end{aligned} \tag{10.5.15}$$

An estimator of $\Omega_y(h|x)$ is obtained in the usual way be replacing all unknown parameters in this expression and in $\Sigma_{\hat{\boldsymbol{\beta}}}$ by estimators and by using the average over $t = 1, \ldots, T$ for the expectation in (10.5.14).

Although we have discussed forecasting with estimated coefficients in terms of a simple VARX(p, s) model with white noise residuals it is possible to generalize these results to models with autocorrelated error process u_t. The more general case is treated, for instance, by Yamamoto (1980) and Baillie (1981).

10.6 Multiplier Analysis

In an econometric simultaneous equations analysis the marginal impact of changes in the exogenous variables is sometimes investigated. For example, if the exogenous variables are instruments for, say, the government or a central bank the consequences of changes in these instruments may be investigated. A government may, for instance, desire to know about the effects of a change in a tax rate. In that case *policy simulation* is of interest. In other cases the consequences of changes in the exogenous variables that are not under the control of any decision maker may be of interest. For instance, it may be desirable to study the future consequences of the present weather conditions.

Therefore the dynamic multipliers discussed in Section 10.2.2 are considered. They are contained in the D_i matrices of the final form operator

$$D(L) = \sum_{i=0}^{\infty} D_i L^i := A(L)^{-1} B(L),$$

where $A(L) := I_K - A_1 L - \cdots - A_p L^p$ and $B(L) := B_0 + B_1 L + \cdots + B_s L^s$ are the reduced form operators as before. The D_i matrices are conveniently obtained from the VARX(1, 0) representation (10.5.4) which implies

$$y_t = \sum_{i=0}^{\infty} J\mathbf{A}^i\mathbf{B}x_{t-i} + \sum_{i=0}^{\infty} J\mathbf{A}^i J' u_{t-i} \tag{10.6.1}$$

because $J\mathbf{A}^h Y_t \to 0$ as $h \to \infty$, if y_t is a stable, stationary process (see (10.5.6)). The D_i matrices are coefficient matrices of the exogenous variables in the final form representation. Thus,

$$D_i = J\mathbf{A}^i\mathbf{B}, \qquad i = 0, 1, \ldots, \tag{10.6.2}$$

the n-th interim multipliers are

$$M_n := D_0 + D_1 + \cdots + D_n = J(I + \mathbf{A} + \cdots + \mathbf{A}^n)\mathbf{B}, \qquad i = 0, 1, \ldots, \tag{10.6.3}$$

and the total multipliers are

$$M_\infty := \sum_{i=0}^{\infty} D_i = J(I - \mathbf{A})^{-1}\mathbf{B} = A(1)^{-1}B(1). \tag{10.6.4}$$

Having obtained the foregoing representations of the multipliers, estimation of these quantities is straightforward. Estimators of the dynamic multipliers are obtained by substituting estimators $\hat{A}_i$ and $\hat{B}_j$ of the coefficient matrices in $\mathbf{A}$ and $\mathbf{B}$. The asymptotic properties of the estimators then follow in the usual way. For completeness we mention the following result from Schmidt (1973).

In the framework of Section 10.3, suppose $\hat{\boldsymbol{\beta}}$ is a consistent estimator of $\boldsymbol{\beta} := \operatorname{vec}[A, B, B_0]$ satisfying

$$\sqrt{T}(\hat{\boldsymbol{\beta}} - \boldsymbol{\beta}) \xrightarrow{d} N(0, \Sigma_{\hat{\boldsymbol{\beta}}}).$$

Then

$$\sqrt{T}\operatorname{vec}(\hat{D}_i - D_i) \xrightarrow{d} N(0, G_i\Sigma_{\hat{\boldsymbol{\beta}}}G_i'), \tag{10.6.5}$$

where $G_0 := [0, I_{KM}]$ and

$$G_i := \frac{\partial \operatorname{vec}(D_i)}{\partial \boldsymbol{\beta}'} = \left[\sum_{j=0}^{i-1} \mathbf{B}'(\mathbf{A}')^{i-1-j} \otimes J\mathbf{A}^j J',\ I_M \otimes J\mathbf{A}^i J'\right] \qquad \text{for } i = 1, 2, \ldots.$$

The proof of this result is left as an exercise. It is also easy to find the asymptotic distribution of the interim multipliers (accumulated multipliers) and the total multipliers (see Problem 10.8).

10.7 Optimal Control

A policy or decision maker who has control over some of the exogenous variables can use a dynamic simultaneous equations model to assess interventions with a multiplier or simulation analysis as described in the previous section. However, if the decision maker has specific target values of the endogenous variables in mind he or she may wish to go a step further and determine which values of the instrument variables will produce the desired values of the endogenous variables.

Usually it will not be possible to actually achieve all targets simultaneously and sometimes the decision maker is not completely free to choose the instruments. For instance, doubling a particular tax rate or increasing the price of specific government services drastically may result in the overthrow of the government or in social unrest and is therefore not a feasible option. Therefore a loss function is usually set up in which the loss of deviations from the target values is specified. For instance, if the desired paths of the endogenous and instrument variables after period T are $y^0_{T+1}, \ldots, y^0_{T+n}$ and $x^0_{T+1}, \ldots, x^0_{T+n}$, respectively, a *quadratic* loss function has the form

$$\mathfrak{L} = \sum_{i=1}^{n} [(y_{T+i} - y^0_{T+i})' K_i (y_{T+i} - y^0_{T+i}) + (x_{T+i} - x^0_{T+i})' P_i (x_{T+i} - x^0_{T+i})], \tag{10.7.1}$$

where the K_i and P_i are symmetric positive semidefinite matrices. Since the variables are assumed to be stochastic, the loss is a random variable too. Therefore minimization of the average or expected loss, $E(\mathfrak{L})$, is usually the objective.

In a quadratic loss function the same weight is assigned to positive and negative deviations from the target values. For many situations and variables this is not quite realistic. For example, if the target is to have an unemployment rate of 2% then having less than 2% may not be a problem at all while any higher rate may be regarded as a serious problem. Nevertheless quadratic loss functions are the most common ones in applied and theoretical studies. Therefore we will also use them in the following. One reason for the popularity of this type of loss function is clearly its tractability.

In order to approach a formal solution of the optimal control problem outlined in the foregoing we assume that the economic system is described by a model like (10.1.1) with reduced form (10.2.3). However, to be able to distinguish between instrument variables and other exogenous variables we introduce a new symbol for the latter. Suppose x_t represents an $(M \times 1)$ vector of instrument variables, the $(N \times 1)$ vector z_t contains all other exogenous variables and the reduced form of the model is

$$y_t = A_1 y_{t-1} + \cdots + A_p y_{t-p} + B_0 x_t + \cdots + B_s x_{t-s} + C z_t + u_t, \tag{10.7.2}$$

where u_t is white noise. Some of the components of z_t may be lagged variables. To summarize them in a vector indexed by t is just a matter of redefining their time index.

For the present purposes it is useful to write the model in VARX(1, 0) form similar to (10.5.4),

$$Y_t = \mathbf{A} Y_{t-1} + \mathbf{B} x_t + \mathbf{C} z_t + U_t, \tag{10.7.3}$$

where Y_t, U_t, $\mathbf{A}$, and $\mathbf{B}$ are as defined in (10.5.4) and

$$\mathbf{C} := \begin{bmatrix} C \\ 0 \\ \vdots \\ 0 \end{bmatrix}$$

is a $((Kp + Ms) \times N)$ matrix. Recall that

$$Y_t := \begin{bmatrix} y_t \\ \vdots \\ y_{t-p+1} \\ x_t \\ \vdots \\ x_{t-s+1} \end{bmatrix}$$

contains current and lagged endogenous and instrument variables. Thus, the quadratic loss function specified in (10.7.1) may be rewritten in the form

$$\mathfrak{L} = \sum_{i=1}^{n} (Y_{T+i} - Y_{T+i}^0)' Q_i (Y_{T+i} - Y_{T+i}^0), \tag{10.7.4}$$

where the Q_i are symmetric positive semidefinite matrices involving the K_i and P_i.

In this framework the problem of *optimal control* may be stated as follows: Given the model (10.7.3), given the vector Y_T, given values $z_{T+1}, \ldots, z_{T+n}$ of the uncontrolled variables and given target values $y_{T+1}^0, \ldots, y_{T+n}^0$ and $x_{T+1}^0, \ldots, x_{T+n}^0$, find the values $x_{T+1}^*, \ldots, x_{T+n}^*$ that minimize the expected loss $E(\mathfrak{L})$ specified in (10.7.4). The solution to this dynamic programming problem is well documented in the control theory literature. It turns out to be

$$x_{T+i}^* = G_i Y_{T+i-1} + g_i, \qquad i = 1, \ldots, n, \tag{10.7.5}$$

where the Y_{T+i} are assumed to be obtained as

$$Y_{T+i} = \mathbf{A} Y_{T+i-1} + \mathbf{B} x_{T+i}^* + \mathbf{C} z_{T+i} + u_{T+i}.$$

Here the $(M \times (Kp + Ms))$ matrix G_i is

$$G_i := -(\mathbf{B}' H_i \mathbf{B})^{-1} \mathbf{B}' H_i \mathbf{A}$$

and the $(M \times 1)$ vector g_i is

$$g_i := -(\mathbf{B}' H_i \mathbf{B})^{-1} \mathbf{B}'(H_i \mathbf{C} z_{T+i} - h_i)$$

with

$$H_n := Q_n \qquad \text{and} \qquad H_{i-1} := Q_{i-1} + (\mathbf{A} + \mathbf{B}G_i)' H_i (\mathbf{A} + \mathbf{B}G_i), \quad \text{for } i = 1, \ldots, n-1,$$

and

$$h_n := Q_n Y_{T+n}^0 \qquad \text{and} \qquad h_{i-1} := Q_{i-1} Y_{T+i-1}^0 - \mathbf{A}' H_i (\mathbf{C} z_{T+i} + \mathbf{B} g_i) + \mathbf{A}' h_i \quad \text{for } i = 1, \ldots, n-1.$$

The actual computation of these quantities proceeds in the order $H_n, G_n, h_n, g_n, H_{n-1}, G_{n-1}, h_{n-1}, g_{n-1}, H_{n-2}, \ldots$. This solution can be found in various variations in the control theory literature (e.g., Chow (1975, 1981), Murata (1982)). Obviously, since the Y_t are random the same is true for the optimal decision rule x^*_{T+i}, $i = 1, \ldots, n$.

There are a number of problems that arise in practice in the context of optimal control as presented here. For instance, we have considered a finite planning horizon of n periods. In some situations it is of interest to find the optimal decision rule for an infinite planning period. Moreover, in practice the parameter matrices **A**, **B**, and **C** are usually unknown and have to be replaced by estimators. More generally stochastic parameter models may be considered. This, of course, introduces an additional stochastic element into the optimal decision rule. A further complication arises if the relations between the variables cannot be captured adequately by a *linear* model such as (10.7.2) but require a nonlinear specification. It is also possible to consider other types of optimization rules. In this section we have assumed that the optimal decision rule for period $T + i$ is determined on the basis of all available information in period $T + i - 1$. In particular, the realization Y_{T+i-1} is assumed given in setting up the decision rule x^*_{T+i}. Such an approach is often referred to as a *closed-loop strategy*. An alternative approach would be to determine the decision rule at the beginning of the planning period for the entire planning horizon. This approach is called an *open-loop strategy*. Although it is in general inferior to closed-loop optimization it may be of interest occasionally. These and many other topics are treated in the optimal control literature. Chow (1975, 1981) and Murata (1982) are books on the topic with emphasis on optimal decision making related to economic and econometric models. Friedmann (1981) provides the asymptotic properties of the optimal decision rule when estimators are substituted for the parameters in the control rule.

10.8 Concluding Remarks on Dynamic SEMs

In this chapter we have summarized some problems related to the estimation, specification, and analysis of dynamic models with exogenous variables. Major problem areas that were identified without giving details of possible solutions are the distinction between endogenous and exogenous variables, the identification or unique parameterization of dynamic models, the estimation, specification, and checking of structural form models as well as the treatment of nonlinear specifications. Also, we have just scratched the surface of control problems which represent one important area of applications of dynamic SEMs.

Other problems of obvious importance in the context of these models relate to the choice of the data associated with the variables. If a structural form is derived from some economic or other subject matter theory it is important that the available data represents realizations of the variables related to the theory. In particular the level of aggregation (temporal and contemporaneous) and seasonal characteristics (seasonally adjusted or unadjusted) may be of importance. The models we have considered do not allow specifically for seasonality except perhaps for seasonal dummies and other seasonal components among the exogenous variables. The seasonality aspect in the context of dynamic SEMs and models specifically designed for seasonal data are discussed, for example, by Hylleberg (1986).

So far we have essentially considered stationary processes, that is, processes with time invariant first and second moments. Mild deviations from the stationarity assumption are possible in dynamic SEMs where exogenous variables may cause changes in the mean or conditional mean of the endogenous variables. However, in discussing properties of estimators or long-run multipliers we have made assumptions that come close to assuming stationarity. For instance, if the exogenous variables are driven by a stationary VARMA process the means and second moments of the endogenous variables will be time invariant. Unfortunately, in practice most variables are nonstationary. Therefore we will discuss different types of nonstationary models in the next three chapters.

10.9 Exercises

Problem 10.1

Consider the following structural form

$$Q_t = \alpha_0 + \alpha_1 R_{t-1} + w_{1t},$$

$$P_t = \beta_0 + \beta_1 Q_t + w_{2t},$$

where R_t is a measure for the rainfall in period t, Q_t is the quantity of an agricultural product supplied in period t, and P_t is the price of the product. Derive the reduced form, the final equations and the final form of the model.

Problem 10.2

Suppose that the rainfall variable R_t is generated by a white noise process with mean μ_R. Determine the unconditional 3-step forecasts for Q_t and P_t based on the model from Problem 10.1. Determine also the conditional 3-step forecasts given $R_{t+i} = \mu_R$, $i = 1, 2, 3$. Compare the two forecasts.

Problem 10.3

Given the model of Problem 10.1, what is the marginal total or long-run effect of an additional unit of rainfall in this period?

Problem 10.4

Suppose the system y_t has the structural form

$$\bar{A}(L)y_t = \bar{F}y_t^e + \bar{B}(L)x_t + w_t,$$

where $\bar{A}(L) := \bar{A}_0 - \bar{A}_1 L - \cdots - \bar{A}_p L^p$, $\bar{B}(L) := \bar{B}_0 + \bar{B}_1 L + \cdots + \bar{B}_s L^s$ and x_t is generated by a VARMA(q, r) process

$$C(L)x_t = M(L)v_t.$$

Assume that y_t^e represents rational expectations formed in period $t-1$ and eliminate the expectations variables from the structural form.

Problem 10.5

Show that the 1-step forecast for y_t obtained from the VAR(p) model (10.5.3) is identical to the one determined from (10.5.2) if

$$x_t(1) = C_1 x_t + \cdots + C_q x_{t-q+1}$$

is used as forecast for the exogenous variables.

Problem 10.6

Show that the partial derivatives $\partial y_t(h|x)/\partial \boldsymbol{\beta}'$ have the form given in (10.5.15).

Problem 10.7

Derive a prediction test for structural change on the basis of the conditional forecasts of the endogenous variables of a dynamic SEM.

Problem 10.8

Show that the dynamic multipliers have the asymptotic distribution given in Section 10.6. Show also that the n-th interim multipliers have an asymptotic normal distribution,

$$\sqrt{T}\,\mathrm{vec}(\hat{M}_n - M_n) \xrightarrow{d} N(0, \Sigma_{\hat{\mathbf{m}}}(n)),$$

where

$$\Sigma_{\hat{\mathbf{m}}}(n) = (G_0 + \cdots + G_n)\Sigma_{\hat{\boldsymbol{\beta}}}(G_0 + \cdots + G_n)'$$

and the G_i are the $[KM \times K(Kp + M(s+1))]$ matrices defined in Section 10.6. Furthermore,

$$\sqrt{T}\,\mathrm{vec}(\hat{M}_\infty - M_\infty) \xrightarrow{d} N(0, \Sigma_{\hat{\mathbf{m}}}(\infty)),$$

where

$$\Sigma_{\hat{\mathbf{m}}}(\infty) = G_\infty \Sigma_{\hat{\boldsymbol{\beta}}} G_\infty'$$

with

$$G_\infty := [((I - \mathbf{A})^{-1}\mathbf{B})', I_M] \otimes J(I - \mathbf{A})^{-1}J'.$$

Here the notation from Section 10.6 has been used.

Problem 10.9

Derive the optimal decision rule for the control problem stated in Section 10.7. (Hint: see Chow (1975).)

Chapter 11. Nonstationary Systems with Integrated and Cointegrated Variables

11.1 Introduction

In Parts I and II we have considered stationary, stable systems. As stated in Chapter 2, a process is stationary if it has time invariant first and second moments. This implies that there are no trends (trending means) or shifts in the mean or in the covariances or specific seasonal patterns. Of course, trends and seasonal fluctuations are quite common in practice. For instance, the *original* investment, income, and consumption data used in many previous examples have trends (see Figure 3.1). Thus, if interest centers on analyzing the original variables rather than the rates of change it is necessary to have models that accommodate the nonstationary features of the data. In Chapter 2 it is demonstrated that a stable process is stationary. Unstable nonstationary processes will be considered in this chapter and other types of nonstationarities will be treated in the next chapters. We shall begin with a brief summary of important characteristics of specific unstable processes.

11.1.1 Integrated Processes

Recall that a VAR(p) process,

$$y_t = \nu + A_1 y_{t-1} + \cdots + A_p y_{t-p} + u_t, \tag{11.1.1}$$

is stable if

$$\det(I_K - A_1 z - \cdots - A_p z^p)$$

has no roots in and on the complex unit circle. For a univariate AR(1) process $y_t = \nu + \alpha y_{t-1} + u_t$ this means that

$$1 - \alpha z \neq 0 \qquad \text{for } |z| \leq 1$$

or, equivalently, $|\alpha| < 1$.

Consider the borderline case where $\alpha = 1$ and suppose that $\nu = 0$. The resulting process $y_t = y_{t-1} + u_t$ is a *random walk*. Starting the process at $t = 0$ with $y_0 = 0$ it is easy to see that

$$y_t = y_{t-1} + u_t = y_{t-2} + u_{t-1} + u_t = \cdots = \sum_{s=1}^{t} u_s. \tag{11.1.2}$$

Thus, y_t is the sum of all disturbances or innovations of the previous periods so that each disturbance has a lasting impact on the process. If u_t is white noise with variance σ_u^2,

$$E(y_t) = 0$$

and

$$\operatorname{Var}(y_t) = t \operatorname{Var}(u_t) = t\sigma_u^2.$$

Hence the variance of a random walk tends to infinity. Furthermore, the correlation

$$\operatorname{Corr}(y_t, y_{t+h}) = \frac{E\left[\left(\sum_{s=1}^{t} u_s\right)\left(\sum_{s=1}^{t+h} u_s\right)\right]}{[t\sigma_u^2(t+h)\sigma_u^2]^{1/2}}$$

$$= \frac{t}{(t^2 + th)^{1/2}} \xrightarrow[t\to\infty]{} 1$$

for any integer h. This latter property of a random walk means that y_t and y_s are strongly correlated even if they are far apart in time. It can also be shown that the expected time between two crossings of zero is infinite. These properties are often reflected in trending behavior. Examples are depicted in Figure 11.1. This kind of trend is, of course, not a deterministic but a stochastic trend.

In contrast, if ν is nonzero, $y_t = \nu + y_{t-1} + u_t$ is a *random walk with drift* and has a deterministic linear trend in the mean. To see this suppose again that it is started at $t = 0$ with $y_0 = 0$. Then

$$y_t = t\nu + \sum_{s=1}^{t} u_s$$

and $E(y_t) = t\nu$. An example of a time series generated by a random walk with drift is shown in Figure 11.2.

The previous discussion suggests that starting unstable processes at some finite time t_0 is useful. If the starting time is $t_0 = -\infty$ it is difficult to justify finite moments of the process. On the other hand, if an AR process starts at some finite time it is strictly speaking not necessarily stationary even if it is stable. To see this let $y_t = \nu + \alpha y_{t-1} + u_t$ be a univariate stable AR(1) process with $|\alpha| < 1$. Starting with y_0 at $t = 0$ gives

$$y_t = \nu \sum_{i=0}^{t-1} \alpha^i + \alpha^t y_0 + \sum_{i=0}^{t-1} \alpha^i u_{t-i}.$$

Hence,

$$E(y_t) = \nu \sum_{i=0}^{t-1} \alpha^i + \alpha^t E(y_0)$$

is not time invariant if α and $\nu \neq 0$. A similar result is obtained for the second moments. However, the first and second moments approach limit values as

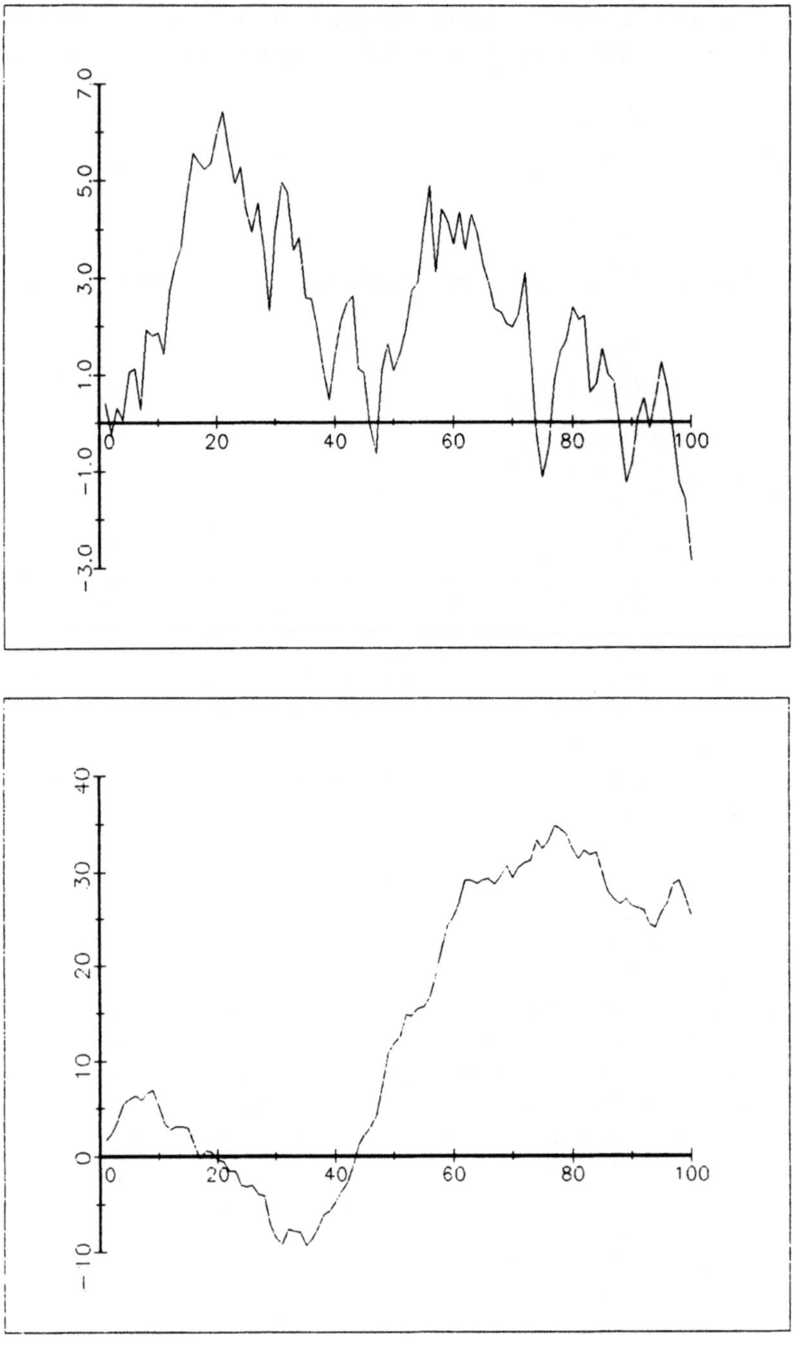

Fig. 11.1. Random walks.

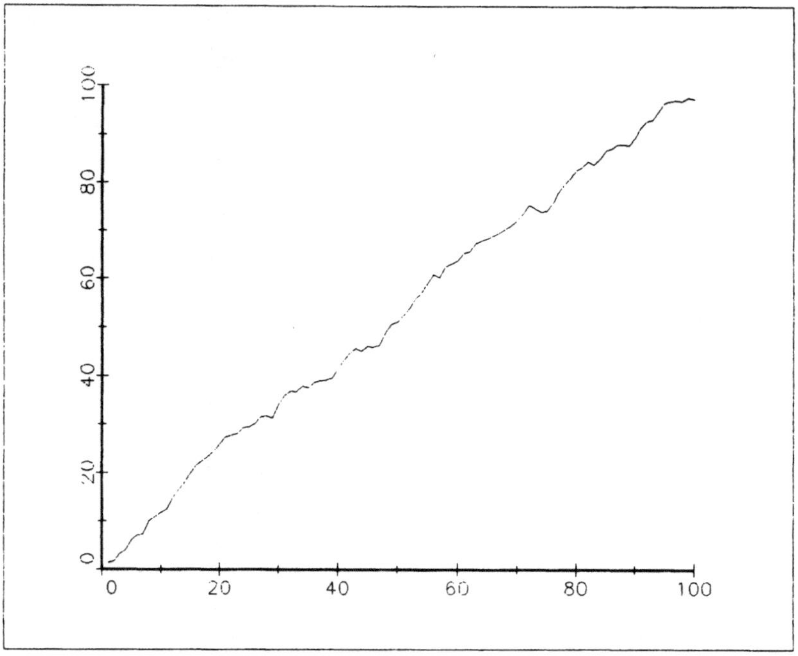

Fig. 11.2. A random walk with drift.

$t \to \infty$ and one might call such a process *asymptotically stationary.* In order to simplify matters the term "asymptotically" is sometimes dropped and such processes are then simply called stationary. For our purposes this point is of no relevance because in later sections of this chapter we will be interested in the parameters of the processes considered and possibly in their asymptotic moments. Without further warning, nonstationary, unstable processes will be assumed to begin at some finite time period.

A behavior similar to that of a random walk is also observed for higher order AR processes such as

$$y_t = \nu + \alpha_1 y_{t-1} + \cdots + \alpha_p y_{t-p} + u_t,$$

if $1 - \alpha_1 z - \cdots - \alpha_p z^p$ has a root for $z = 1$. Note that

$$1 - \alpha_1 z - \cdots - \alpha_p z^p = (1 - \lambda_1 z) \cdots (1 - \lambda_p z),$$

where $\lambda_1, \ldots, \lambda_p$ are the reciprocals of the roots of the polynomial. If the process has just one unit root (a root equal to 1) and all other roots are outside the complex unit circle its behavior is similar to that of a random walk, that is, its variances increase linearly to infinity, the correlation between variables h periods apart tends to 1 and the process has a linear trend in mean if $\nu \neq 0$. In case one of the roots is strictly inside the unit circle the process becomes explosive, that is, its variance goes to infinity at an exponential rate. Many researchers feel that such processes are unrealistic models for most economic data. Although processes with roots on the unit circle other than one are often useful we shall concentrate on the case of unit roots and all other roots outside the unit circle. This situation seems to be of considerable practical interest.

Univariate processes with d unit roots (d roots equal to 1) in their AR operator are called *integrated of order d* ($I(d)$). If there is just one unit root, i.e., the process is $I(1)$, it is quite easy to see how a stable process can be obtained: simply by taking first differences, $\Delta y_t := (1 - L)y_t = y_t - y_{t-1}$, of the original process. More generally, if the process is $I(d)$ it can be made stable by differencing d times, that is, $\Delta^d y_t = (1 - L)^d y_t$ is stable.

Consider now a K-dimensional VAR(p) process as in (11.1.1) with $\nu = 0$. It can be written as

$$A(L)y_t = u_t, \tag{11.1.3}$$

where $A(L) := I_K - A_1 L - \cdots - A_p L^p$ and L is the lag operator. Multiplying from the left by the adjoint $A(L)^*$ of $A(L)$ gives

$$|A(L)|\, y_t = A(L)^* u_t. \tag{11.1.4}$$

Thus, the VAR(p) process in (11.1.3) can be written as a VARMA process with univariate AR operator, that is, all components have the same AR operator $|A(L)|$. Each component of $A(L)^* u_t$ is a sum of finite order MA processes. Therefore it is a finite order MA process (see Chapter 6, Section 6.6.1, and Chapter 8, Section 8.2.1). Hence, the k-th component of y_t has a representation

$$|A(L)|\, y_{kt} = \bar{m}_k(L) v_{kt},$$

where v_{kt} is univariate white noise. If $|A(L)|$ has d unit roots and otherwise all roots outside the unit circle the AR operator can be written as

$$|A(L)| = \alpha(L)(1 - L)^d = \alpha(L)\Delta^d,$$

where $\alpha(L)$ is an invertible operator. Consequently, $\Delta^d y_{kt}$ is a stable process. Hence, each component becomes stable upon differencing.

Since we are considering processes which are started at some finite time t_0 we should perhaps think for a moment about the treatment of initial values when multiplying by an operator such as $A(L)^*$ in (11.1.4). One possible assumption is that the new representation is valid from period $t_0 + q$ onwards, where q is the order of the operator by which we have multiplied (in this case $A(L)^*$). Other assumptions are possible. Since the initial value problem is not of great importance in the following discussion we will not elaborate on this issue here but just mention that assumptions exist that avoid trouble.

The foregoing discussion shows that if a VAR(p) process is unstable because of unit roots only, each component can be made stable by differencing. Note, however, that, due to cancellations, it may not be necessary to difference each component as many times as there are unit roots in $|A(L)|$. To illustrate this point consider the bivariate VAR(1) process

$$\left(\begin{bmatrix} 1 & 0 \\ 0 & 1 \end{bmatrix} - \begin{bmatrix} 1 & 0 \\ 0 & 1 \end{bmatrix} L\right)\begin{bmatrix} y_{1t} \\ y_{2t} \end{bmatrix} = u_t.$$

Obviously, each component is stationary after differencing once, i.e., each component is $I(1)$, although

$$|A(L)| = \left| \begin{bmatrix} 1 - L & 0 \\ 0 & 1 - L \end{bmatrix} \right| = (1 - L)^2$$

has two unit roots. It is also possible that some components are stable and stationary as univariate processes whereas others need differencing. Examples are easy to construct.

If the VAR(p) process has a nonzero intercept term so that

$$A(L)y_t = \nu + u_t$$

and $|A(z)|$ has one or more unit roots then some of the components of y_t may have deterministic trends. Unlike the univariate case, it is also possible, however, that none of the components of y_t has a deterministic trend. This occurs if $A(L)^*\nu = 0$. For instance, if

$$A(L) = \begin{bmatrix} 1 - L & \eta L \\ 0 & 1 \end{bmatrix},$$

$|A(z)|$ obviously has a unit root and

$$A(L)^* = \begin{bmatrix} 1 & -\eta L \\ 0 & 1 - L \end{bmatrix}.$$

Hence,

$$A(L)^* \begin{bmatrix} \nu_1 \\ \nu_2 \end{bmatrix} = \begin{bmatrix} \nu_1 - \eta \nu_2 \\ \nu_2 - \nu_2 \end{bmatrix}$$

which is zero if $\nu_1 = \eta\nu_2$. Thus, in a VAR analysis an intercept term cannot be excluded a priori if none of the component series has a deterministic trend.

The following question comes to mind in this context. Suppose each component of a VAR(p) process is $I(d)$, is it possible that differencing each component individually distorts interesting features of the relationship between the original variables? If the latter were not the case a VAR or VARMA analysis could be performed after differencing as described in previous chapters. It turns out, however, that differencing may indeed distort the relationship between the original variables. Systems with cointegrated variables are examples where fitting VAR models upon differencing is inadequate. Such systems are introduced next.

11.1.2 Cointegrated Processes

Equilibrium relationships are suspected between many economic variables such as household income and expenditures or prices of the same commodity in different markets. Suppose the variables of interest are collected in the vector $y_t = (y_{1t}, \dots, y_{Kt})'$ and their long-run equilibrium relation is $cy_t = c_1 y_{1t} + \cdots + c_K y_{Kt} = 0$, where $c = (c_1, \dots, c_K)$. In any particular period this relation may not be satisfied exactly but we may have $cy_t = z_t$, where z_t is a stochastic variable representing the deviations from the equilibrium. If there really is an equilibrium it seems plausible to assume that the y variables move together and that z_t is stable. This, however, does not exclude the possibility that the y variables wander extensively as a group. Thus, it is not excluded that each variable is integrated, yet there exists a linear combination of the variables which is stationary. Integrated variables with this property are called *cointegrated*. In Figure 11.3 two artificially generated cointegrated time series are depicted.

Generally, the variables in a K-dimensional process y_t are *cointegrated of order* (d, b), briefly $y_t \sim CI(d, b)$, if all components of y_t are $I(d)$ and there exists a linear combination cy_t with $c = (c_1, \ldots, c_K) \neq 0$ which is $I(d-b)$. For instance, if all components of y_t are $I(1)$ and cy_t is stationary ($I(0)$), then $y_t \sim CI(1, 1)$. The vector c is a *cointegrating vector*. A process consisting of cointegrated variables is a *cointegrated process*. These processes were introduced by Granger (1981) and Engle & Granger (1987). Since then they have become enormously popular in theoretical and applied econometric work.

In the following we will use a slightly different definition of cointegration in order to simplify the terminology. We call a K-dimensional process y_t integrated of order d, briefly, $y_t \sim I(d)$, if $\Delta^d y_t$ is stable and $\Delta^{d-1} y_t$ is not stable. The $I(d)$ process y_t is called cointegrated if there is a linear combination cy_t which is integrated of order less than d. This definition differs from the one given by Engle & Granger (1987) in that we do not exclude components of y_t with order of

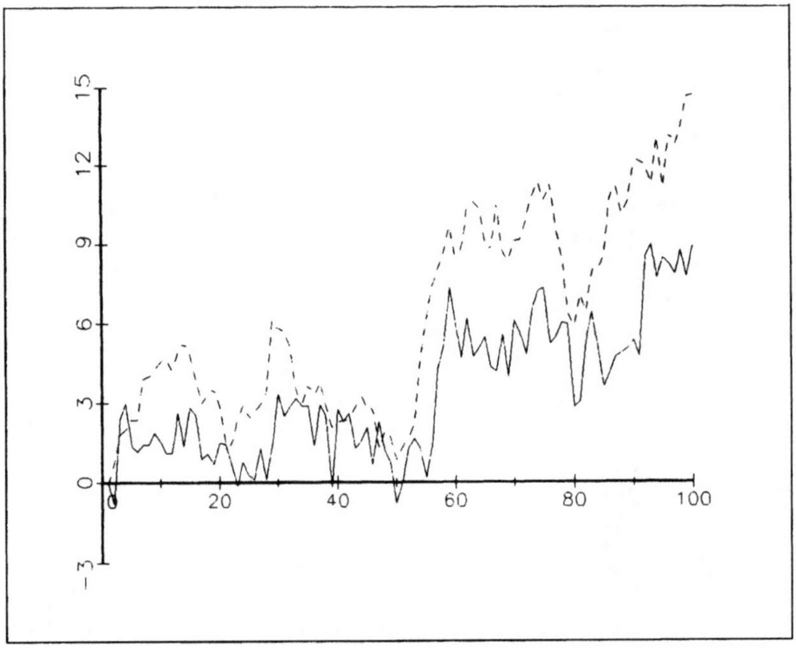

Fig. 11.3. A bivariate cointegrated time series.

integration less than d. If there is just one $I(d)$ component in y_t and all other components are stable ($I(0)$) the vector y_t is $I(d)$ according to our definition because $\Delta^d y_t$ is stable and $\Delta^{d-1} y_t$ is not. In such a case a relation cy_t that involves the stationary components only is a cointegration relation in our terms. Clearly this aspect of our definition is not in line with the original idea of cointegration as a special relation between integrated variables. However, in the following our definition simplifies the terminology as it avoids distinguishing between variables with different order of integration. The reader should keep in mind the basic ideas of cointegration when it comes to interpreting specific relationships.

Obviously, a cointegrating vector is not unique. Multiplying by a nonzero constant yields a further cointegrating vector. Also, there may be various linearly independent cointegrating vectors. For instance, if there are four variables in a system the first two may be connected by a long-run equilibrium relation and also the last two. Thus, there may be a cointegrating vector with zeroes in the last two positions and one with zeroes in the first two positions. In addition there may be a cointegration relation involving all four variables.

Before the concept of cointegration was introduced the closely related *error correction models* were discussed in the econometrics literature (see, e.g., Davidson, Hendry, Srba & Yeo (1978), Hendry & von Ungern-Sternberg (1981), Salmon (1982)). In an error correction model the changes in a variable depend on the deviation from some equilibrium relation. Suppose, for instance, that y_{1t} represents the price of a commodity in a particular market and y_{2t} is the corresponding price of the same commodity in another market. Assume furthermore that the equilibrium relation between the two variables is given by $y_{1t} = \gamma y_{2t}$ and that the changes in y_{1t} depend on the deviations from this equilibrium in period $t-1$,

$$\Delta y_{1t} = h_1(y_{1,t-1} - \gamma y_{2,t-1}) + u_{1t}.$$

A similar relation may hold for y_{2t},

$$\Delta y_{2t} = h_2(y_{1,t-1} - \gamma y_{2,t-1}) + u_{2t}.$$

In a more general error correction model the Δy_{it} may in addition depend on previous changes in both variables as, for instance, in the following model:

$$\begin{aligned} \Delta y_{1t} &= h_1(y_{1,t-1} - \gamma y_{2,t-1}) + f_{11,1}\Delta y_{1,t-1} + f_{12,1}\Delta y_{2,t-1} + u_{1t}, \\ \Delta y_{2t} &= h_2(y_{1,t-1} - \gamma y_{2,t-1}) + f_{21,1}\Delta y_{1,t-1} + f_{22,1}\Delta y_{2,t-1} + u_{2t}. \end{aligned} \tag{11.1.5}$$

Further lags of Δy_{it} may also be included.

To see the close relationship between error correction models and the concept of cointegration suppose y_{1t} and y_{2t} are both $I(1)$ variables. In that case all terms in (11.1.5) involving the Δy_{it} are stable. In addition, u_{1t} and u_{2t} are white noise errors which are also stable. Since an unstable term cannot equal a stable process,

$$h_i(y_{1,t-1} - \gamma y_{2,t-1}) = \Delta y_{it} - f_{i1,1}\Delta y_{1,t-1} - f_{i2,1}\Delta y_{2,t-1} - u_{it}$$

must be stable too. Hence, if $h_1 \neq 0$ or $h_2 \neq 0$, $y_{1t} - \gamma y_{2t}$ is stable and, thus, represents a cointegration relation.

In vector and matrix notation the model (11.1.5) can be written as

$$\Delta y_t = HCy_{t-1} + F\Delta y_{t-1} + u_t,$$

or

$$y_t - y_{t-1} = HCy_{t-1} + F(y_{t-1} - y_{t-2}) + u_t, \tag{11.1.6}$$

where $y_t := (y_{1t}, y_{2t})'$, $u_t := (u_{1t}, u_{2t})'$,

$$H := \begin{bmatrix} h_1 \\ h_2 \end{bmatrix}, \qquad C := (1, -\gamma), \qquad \text{and} \qquad F := \begin{bmatrix} f_{11,1} & f_{12,1} \\ f_{21,1} & f_{22,1} \end{bmatrix}.$$

Rearranging terms in (11.1.6) gives the VAR(2) representation

$$y_t = (I_K + F + HC)y_{t-1} - Fy_{t-2} + u_t.$$

Hence, cointegrated variables may be generated by a VAR process.

To see how cointegration can arise more generally in K-dimensional VAR models consider the VAR(2) process

$$y_t = A_1 y_{t-1} + A_2 y_{t-2} + u_t \tag{11.1.7}$$

with $y_t = (y_{1t}, \dots, y_{Kt})'$. Suppose the process is unstable with

$$|I_K - A_1 z - A_2 z^2| = (1 - \lambda_1 z)\cdots(1 - \lambda_n z) = 0 \qquad \text{for } z = 1.$$

Since the λ_i are the reciprocals of the roots of the determinantal polynomial, one or more of them must be equal to 1. All other roots are assumed to lie outside the unit circle, that is, all λ_i that are not 1 are inside the complex unit circle. Since $|I_K - A_1 - A_2| = 0$, the matrix

$$\Pi = I_K - A_1 - A_2$$

is singular. Suppose $\text{rk}(\Pi) = r < K$ so that Π can be decomposed as $\Pi = HC$, where H is $(K \times r)$ and C is $(r \times K)$. From the discussion in the previous subsection we know that each variable becomes stationary upon differencing. Let us assume that differencing once is sufficient and rewrite (11.1.7) as

$$y_t - y_{t-1} = -(I_K - A_1)y_{t-1} + (I_K - A_1)y_{t-2} - (I_K - A_1 - A_2)y_{t-2} + u_t$$

or

$$\Delta y_t = D_1 \Delta y_{t-1} - \Pi y_{t-2} + u_t, \tag{11.1.8}$$

where $D_1 = -(I_K - A_1)$, or

$$HCy_{t-2} = -\Delta y_t + D_1 \Delta y_{t-1} + u_t.$$

Since the right-hand side involves stationary terms only, HCy_{t-2} must also be stationary and it remains stationary upon multiplication by $(H'H)^{-1}H'$. In other words, Cy_t is stationary and hence each row of Cy_t represents a cointegrating relation. Note that simply taking first differences of all variables in (11.1.7) gives

$$\Delta y_t = A_1 \Delta y_{t-1} + A_2 \Delta y_{t-2} + \Delta u_t$$

which has the noninvertible MA part $\Delta u_t = u_t - u_{t-1}$. Hence, in general, the process does not admit a pure VAR representation in first differences.

In the following we will be interested in the specific case where all individual variables are $I(1)$ or $I(0)$. We call the K-dimensional VAR(p) process

$$y_t = \nu + A_1 y_{t-1} + \cdots + A_p y_{t-p} + u_t, \tag{11.1.9}$$

cointegrated of rank r if

$$\Pi = I_K - A_1 - \cdots - A_p$$

has rank r and thus Π can be written as HC with H and C' being of dimension $(K \times r)$ and of rank r. The matrix C is the *cointegrating matrix* or the *matrix of cointegrating vectors* and H is sometimes called the *loading matrix*. If $r = 0$, Δy_t has a stable VAR($p - 1$) representation and, for $r = K$, y_t is a stable VAR(p) process.

Rewriting (11.1.9) as in (11.1.8) it has a representation

$$\Delta y_t = \nu + D_1 \Delta y_{t-1} + \cdots + D_{p-1} \Delta y_{t-p+1} - \Pi y_{t-p} + u_t, \tag{11.1.10}$$

where

$$D_i = -(I_K - A_1 - \cdots - A_i), \qquad i = 1, \ldots, p - 1.$$

The process (11.1.9) has the error correction representation

$$\Delta y_t = \nu - HC y_{t-1} + F_1 \Delta y_{t-1} + \cdots + F_{p-1} \Delta y_{t-p+1} + u_t, \tag{11.1.11}$$

where

$$F_i = -(A_{i+1} + \cdots + A_p), \qquad i = 1, \ldots, p - 1.$$

Other possible representations of a cointegrated process are discussed by Hylleberg & Mizon (1989).

In the following sections we will usually work with (11.1.9) or (11.1.10). Of course, thereby we work within a much more narrow framework than that allowed in the general definition of cointegration. First, we consider $I(1)$ processes only and, second, the discussion is limited to finite order VAR processes. Also, we do not make a priori exogeneity assumptions for some of the variables as in Chapter 10.

The organization of this chapter is as follows. In the next section estimation of VAR processes with cointegrated and integrated variables is considered. Forecasting and impulse response analysis for such systems is discussed in Section 11.3 and the fourth section is devoted to model selection and model checking. Throughout we illustrate the theoretical results by an example based on U.S. money, income, and interest rate data.

11.2 Estimation of Integrated and Cointegrated VAR(p) Processes

In this section estimation of cointegrated and integrated finite order VAR(p) processes is discussed. In the first subsection Johansen's (1988, 1991) ML approach is presented. Then some other estimation methods for cointegrated processes are described. In the final subsection Bayesian estimation including the Minnesota or Litterman prior for integrated processes is discussed.

11.2.1 ML Estimation of a Gaussian Cointegrated VAR(p) Process

11.2.1a The ML Estimators and their Properties

Suppose y_t is a K-dimensional VAR(p) process with cointegration rank r, $0 < r < K$. We exclude the cases where the rank r is zero or K because they have been treated in previous chapters. For convenience the intercept term is assumed to be zero ($\nu = 0$). We will comment on the consequences of a nonzero constant term later. Thus, we desire to estimate the coefficients of the model

$$y_t = A_1 y_{t-1} + \cdots + A_p y_{t-p} + u_t, \tag{11.2.1}$$

subject to the constraint

$$\text{rk}(\Pi) = \text{rk}(I_K - A_1 - \cdots - A_p) = r. \tag{11.2.2}$$

We assume that u_t is Gaussian white noise with nonsingular covariance matrix Σ_u and the initial conditions $y_{-p+1}, \ldots, y_0$ are fixed. In order to impose the cointegration constraint it is convenient to reparameterize the model as in (11.1.8),

$$\Delta y_t = D_1 \Delta y_{t-1} + \cdots + D_{p-1} \Delta y_{t-p+1} - \Pi y_{t-p} + u_t, \tag{11.2.3}$$

and write Π as a product $\Pi = HC$, where H is $(K \times r)$ and C is $(r \times K)$ with $\text{rk}(H) = \text{rk}(C) = r$.

Defining

$$\Delta Y := [\Delta y_1, \ldots, \Delta y_T], \qquad \Delta X_t := \begin{bmatrix} \Delta y_t \\ \vdots \\ \Delta y_{t-p+2} \end{bmatrix}, \qquad \Delta X := [\Delta X_0, \ldots, \Delta X_{T-1}],$$

$$D := [D_1, \ldots, D_{p-1}], \qquad Y_{-p} := [y_{1-p}, \ldots, y_{T-p}], \tag{11.2.4}$$

the log-likelihood function for a sample of size T is easily seen to be

$$\ln l = -\frac{KT}{2} \ln 2\pi - \frac{T}{2} \ln |\Sigma_u|$$
$$- \frac{1}{2} \text{tr}[(\Delta Y - D\Delta X + HCY_{-p})' \Sigma_u^{-1} (\Delta Y - D\Delta X + HCY_{-p})]. \tag{11.2.5}$$

In the next proposition the ML estimators are given.

PROPOSITION 11.1 (*ML Estimators of a Cointegrated VAR Process*)

Let

$$M := I - \Delta X' (\Delta X \Delta X')^{-1} \Delta X,$$

$R_0 := \Delta YM,$

$R_1 := Y_{-p}M,$

$S_{ij} := R_i R_j'/T, \qquad i = 0, 1,$

G be the lower triangular matrix with positive diagonal satisfying $GS_{11}G' = I_K$,

$\lambda_1 \geq \cdots \geq \lambda_K$ be the eigenvalues of $GS_{10}S_{00}^{-1}S_{01}G'$,

and

$v_1, \ldots, v_K$ be the corresponding orthonormal eigenvectors.

The log-likelihood function in (11.2.5) is maximized for

$$C = \tilde{C} := [v_1, \ldots, v_r]'G,$$

$$H = \tilde{H} := -\Delta YMY_{-p}'\tilde{C}'[\tilde{C}Y_{-p}MY_{-p}'\tilde{C}']^{-1} = -S_{01}\tilde{C}'(\tilde{C}S_{11}\tilde{C}')^{-1},$$

$$D = \tilde{D} := (\Delta Y + \tilde{H}\tilde{C}Y_{-p})\Delta X'(\Delta X \Delta X')^{-1},$$

$$\Sigma_u = \tilde{\Sigma}_u := (\Delta Y - \tilde{D}\Delta X + \tilde{H}\tilde{C}Y_{-p})(\Delta Y - \tilde{D}\Delta X + \tilde{H}\tilde{C}Y_{-p})'/T.$$

The maximum is

$$\max \ln l = -\frac{KT}{2}\ln 2\pi - \frac{T}{2}\left[\ln|S_{00}| + \sum_{i=1}^{r}\ln(1 - \lambda_i)\right] - \frac{KT}{2}. \qquad \blacksquare$$

Proof: From Chapter 3, Section 3.4, it is known that for any fixed H and C the maximum of $\ln l$ is attained for

$$\tilde{D}(HC) = (\Delta Y + HCY_{-p})\Delta X'(\Delta X \Delta X')^{-1}.$$

Thus we replace D by $\tilde{D}(HC)$ in (11.2.5) and get

$$-\frac{KT}{2}\ln 2\pi - \frac{T}{2}\ln|\Sigma_u| - \frac{1}{2}\mathrm{tr}[(\Delta YM + HCY_{-p}M)'\Sigma_u^{-1}(\Delta YM + HCY_{-p}M)].$$

Hence, we just have to maximize this expression with respect to H, C, and Σ_u. We also know from Chapter 3 that for given H and C the maximum is attained if

$$\tilde{\Sigma}_u(HC) = (\Delta YM + HCY_{-p}M)(\Delta YM + HCY_{-p}M)'/T$$

is substituted for Σ_u. Consequently we must maximize

$$-\frac{T}{2}\ln|(\Delta YM + HCY_{-p}M)(\Delta YM + HCY_{-p}M)'/T|$$

or, equivalently, minimize the determinant with respect to H and C. Thus, all results of Proposition 11.1 follow from Proposition A.7 of Appendix A.14. $\blacksquare$

The solutions $\tilde{C}$ and $\tilde{H}$ of the optimization problem given in the proposition are not unique because, for any nonsingular $(r \times r)$ matrix F, $\tilde{H}F^{-1}$ and $F\tilde{C}$ represent another set of ML estimators for H and C. However, the proposition shows that closed form expressions for ML estimators are available. If $r = K$ the proposition still remains valid. In that case, however, ML estimation of the original model is quite simple and the detour via the reparameterization (11.2.3) becomes superfluous. The next question concerns the properties of the ML estimators of a cointegrated system. They are given in the following proposition.

PROPOSITION 11.2 (*Asymptotic Properties of the ML Estimators*)

Let y_t be a cointegrated Gaussian VAR(p) process with cointegration rank r, $0 < r < K$. The ML estimators given in Proposition 11.1 have the following asymptotic properties:

$$\operatorname{plim} \tilde{D} = D, \qquad \operatorname{plim} \tilde{H}\tilde{C} = \Pi, \qquad \operatorname{plim} \tilde{\Sigma}_u = \Sigma_u,$$

$$\sqrt{T}\operatorname{vec}([\tilde{D}, -\tilde{H}\tilde{C}] - [D, -\Pi]) \xrightarrow{d} N(0, \Sigma_{co}), \tag{11.2.6}$$

where

$$\Sigma_{co} = \left(\begin{bmatrix} I_{Kp-K} & 0 \\ 0 & C' \end{bmatrix} \Omega^{-1} \begin{bmatrix} I_{Kp-K} & 0 \\ 0 & C \end{bmatrix}\right) \otimes \Sigma_u$$

and

$$\Omega = \operatorname{plim} \frac{1}{T} \begin{bmatrix} \Delta X \Delta X' & \Delta X Y_{-p}' C' \\ C Y_{-p} \Delta X' & C Y_{-p} Y_{-p}' C' \end{bmatrix}.$$

Σ_{co} is consistently estimated by

$$\tilde{\Sigma}_{co} = \left(\begin{bmatrix} I & 0 \\ 0 & \tilde{C}' \end{bmatrix} \tilde{\Omega}^{-1} \begin{bmatrix} I & 0 \\ 0 & \tilde{C} \end{bmatrix}\right) \otimes \tilde{\Sigma}_u,$$

where

$$\tilde{\Omega} = \frac{1}{T} \begin{bmatrix} \Delta X \Delta X' & \Delta X Y_{-p}' \tilde{C}' \\ \tilde{C} Y_{-p} \Delta X' & \tilde{C} Y_{-p} Y_{-p}' \tilde{C}' \end{bmatrix}.$$

Furthermore,

$$\sqrt{T}\operatorname{vech}(\tilde{\Sigma}_u - \Sigma_u) \xrightarrow{d} N(0, 2\mathbf{D}_K^+(\Sigma_u \otimes \Sigma_u)\mathbf{D}_K^{+\prime}) \tag{11.2.7}$$

and $\tilde{\Sigma}_u$ is asymptotically independent of $\tilde{D}$ and $\tilde{H}\tilde{C}$. Here, as usual, $\mathbf{D}_K^+ = (\mathbf{D}_K'\mathbf{D}_K)^{-1}\mathbf{D}_K'$ and $\mathbf{D}_K$ is the $(K^2 \times K(K+1)/2)$ duplication matrix. ■

REMARK 1. The matrices H and C cannot be estimated consistently without further constraints. Under the assumptions of Proposition 11.2 these matrices are not identified (not unique). We have encountered a similar problem when discussing reduced rank models in Chapter 5. In that chapter we have made specific identifying assumptions in order to obtain unique parameter values and estimators. The same could be done here. Since the cointegration relations are often interpreted as long-run relations such an approach may not be helpful in the present context, however, since the parameters obtained by some arbitrary normalization may not reflect the actual long-run relations of the variables. Note, however, that the product $BC = \Pi$ can be estimated consistently without further identifying constraints. ■

REMARK 2. If C were known, ML estimators $\tilde{H}(C) := -S_{01}C'(CS_{11}C')^{-1}$ and $\tilde{D}(\tilde{H}C) := (\Delta Y + \tilde{H}(C)CY_{-p})\Delta X'(\Delta X \Delta X')^{-1}$ could be considered. The resulting estimator $[\tilde{D}(\tilde{H}C), -\tilde{H}(C)C]$ has exactly the asymptotic normal distribution given in (11.2.6) (see Lemma 11.2). In other words, whether the cointegrating matrix C is known or estimated is of no consequence for the asymptotic distribution of the ML estimators of D and Π. The reason is that, although C cannot be estimated consistently, a basis of the space spanned by its rows can be estimated "superconsistently", that is, the estimator converges more rapidly than the other estimators. This statement will become more transparent in the proof of Proposition 11.2 in Subsection 11.2.1c. ■

REMARK 3. The covariance matrix Σ_{co} is singular. This is easily seen by noting that Ω is a $[(Kp - K + r) \times (Kp - K + r)]$ matrix. Thus the rank of the $(K^2p \times K^2p)$ matrix Σ_{co} cannot be greater than $K(Kp - K + r)$ which is smaller than K^2p if $r < K$. ■

REMARK 4. The asymptotic distribution of the white noise covariance matrix estimator $\tilde{\Sigma}_u$ is the same as in the case of a stationary, stable VAR(p) model (see Chapter 3). ■

REMARK 5. The normality of the process is not essential for the asymptotic properties of the estimators $\tilde{D}$ and $\tilde{\Pi} = \tilde{H}\tilde{C}$. Much of Proposition 11.2 holds under weaker conditions. We have chosen the normality assumption for convenience. The essential conditions will become more apparent in the proof in Subsection 11.2.1c. As usual, the asymptotic distribution of $\tilde{\Sigma}_u$ may be different if u_t is not Gaussian. ■

REMARK 6. If an intercept term is included in the model it may be estimated jointly with the other coefficients by adding a 1 to ΔX_t in (11.2.4) and modifying the related quantities accordingly. In that case the estimators $\tilde{D}$, $\tilde{\Pi}$, and $\tilde{\Sigma}_u$ retain an asymptotic normal distribution. ■

In the previous chapters we have computed forecasts and impulse responses from the representation (11.2.1) of the VAR process. Therefore the asymptotic distribution of the estimators of the A_i obtained from the $\tilde{D}_i$, $\tilde{H}$, and $\tilde{C}$ is also of interest. That distribution follows easily from Proposition 11.2 because the A_i are obtained from the D_i and Π by linear transformations. Specifically

$$A_1 = I_K + D_1$$

$$A_i = D_i - D_{i-1}, \qquad i = 2, \dots, p-1,$$

and

$$A_p = -\Pi - D_{p-1}.$$

Hence,

$$A := [A_1, \dots, A_p] = [D_1, \dots, D_{p-1}, -\Pi]W + J, \tag{11.2.8}$$

where

$$J := [I_K \quad 0 \dots 0] \qquad (K \times Kp)$$

and

$$W := \begin{bmatrix} I_K & -I_K & 0 & \dots & 0 & 0 \\ 0 & I_K & -I_K & & 0 & 0 \\ \vdots & & \ddots & \ddots & & \vdots \\ 0 & 0 & \dots & & I_K & -I_K \\ 0 & 0 & \dots & & 0 & I_K \end{bmatrix} \qquad (Kp \times Kp).$$

Consequently, since

$$\operatorname{vec}([D_1, \dots, D_{p-1}, -\Pi]W) = (W' \otimes I_K) \operatorname{vec}[D_1, \dots, D_{p-1}, -\Pi],$$

we get the following implication of Proposition 11.2.

COROLLARY 11.2.1

Under the conditions of Proposition 11.2,

$$\sqrt{T} \operatorname{vec}(\tilde{A} - A) \xrightarrow{d} N(0, \Sigma_{\alpha}^{co}),$$

where $\tilde{A} := [\tilde{D}_1, \dots, \tilde{D}_{p-1}, -\tilde{H}\tilde{C}]W + J$ and

$$\Sigma_{\alpha}^{co} := \left(W' \begin{bmatrix} I & 0 \\ 0 & C' \end{bmatrix} \Omega^{-1} \begin{bmatrix} I & 0 \\ 0 & C \end{bmatrix} W \right) \otimes \Sigma_u = (W' \otimes I_K)\Sigma_{co}(W \otimes I_K),$$

which is consistently estimated by

$$\tilde{\Sigma}_{\alpha}^{co} := (W' \otimes I_K)\tilde{\Sigma}_{co}(W \otimes I_K).$$

■

If $r < K$, Σ_{α}^{co} is singular (see Remark 3), that is, $\tilde{A}$ has a singular asymptotic distribution. If an intercept term is included in the model which is estimated along with the other parameters, $\tilde{A}$ and $\tilde{\Sigma}_u$ still have asymptotic normal distributions and the asymptotic covariance matrices may be estimated analogously to the zero intercept case. Before we sketch the proof of Proposition 11.2 we give an example.

11.2.1b An Example

As an example we consider the following four-dimensional system of U.S. economic variables:

y_1 – logarithm of the real money stock M1 (ln M1),
y_2 – logarithm of GNP in billions of 1982 dollars (ln GNP),
y_3 – discount interest rate on new issues of 91-day Treasury bills (r^s),
y_4 – yield on long-term (20 years) Treasury bonds (r^l).

Quarterly data for the years 1954 to 1987 are used. The data are reproduced in Table E.3 of Appendix E. They are plotted in Figure 11.4. The GNP and M1 data are seasonally adjusted. r^s and r^l are regarded as short- and long-term interest rates, respectively. The plots in Figure 11.4 show that the series are trending together. Thus cointegration is possible.

We have fitted a VAR(2) model *with intercept* term and cointegration rank $r = 1$ to the data. A justification for the VAR order and cointegration rank will be provided in Section 11.4. Using the formulas from Proposition 11.1 gives estimates

$$\tilde{C} = [-8.05, 3.74, -116.88, 75.36]$$

and

$$\tilde{H} = \begin{bmatrix} -.0017 \\ -.0035 \\ -.0003 \\ .0006 \end{bmatrix}.$$

Hence,

$$\tilde{\Pi} = \tilde{H}\tilde{C} = \begin{bmatrix} .014 & -.007 & .203 & -.131 \\ (.005) & (.002) & (.072) & (.044) \\ .028 & -.013 & .408 & -.263 \\ (.006) & (.003) & (.085) & (.055) \\ .002 & -.001 & .031 & -.020 \\ (.006) & (.003) & (.082) & (.053) \\ -.005 & .002 & -.074 & .048 \\ (.003) & (.001) & (.041) & (.026) \end{bmatrix},$$

where the numbers in parentheses are estimated standard errors obtained from Proposition 11.2. Furthermore

$$\tilde{D} = \begin{bmatrix} .332 & .098 & -.556 & -.838 \\ (.067) & (.073) & (.104) & (.216) \\ .071 & .052 & -.169 & .549 \\ (.079) & (.086) & (.123) & (.256) \\ .179 & .080 & -.009 & .425 \\ (.076) & (.082) & (.118) & (.245) \\ .037 & .047 & .041 & .138 \\ (.038) & (.041) & (.059) & (.122) \end{bmatrix}$$

and

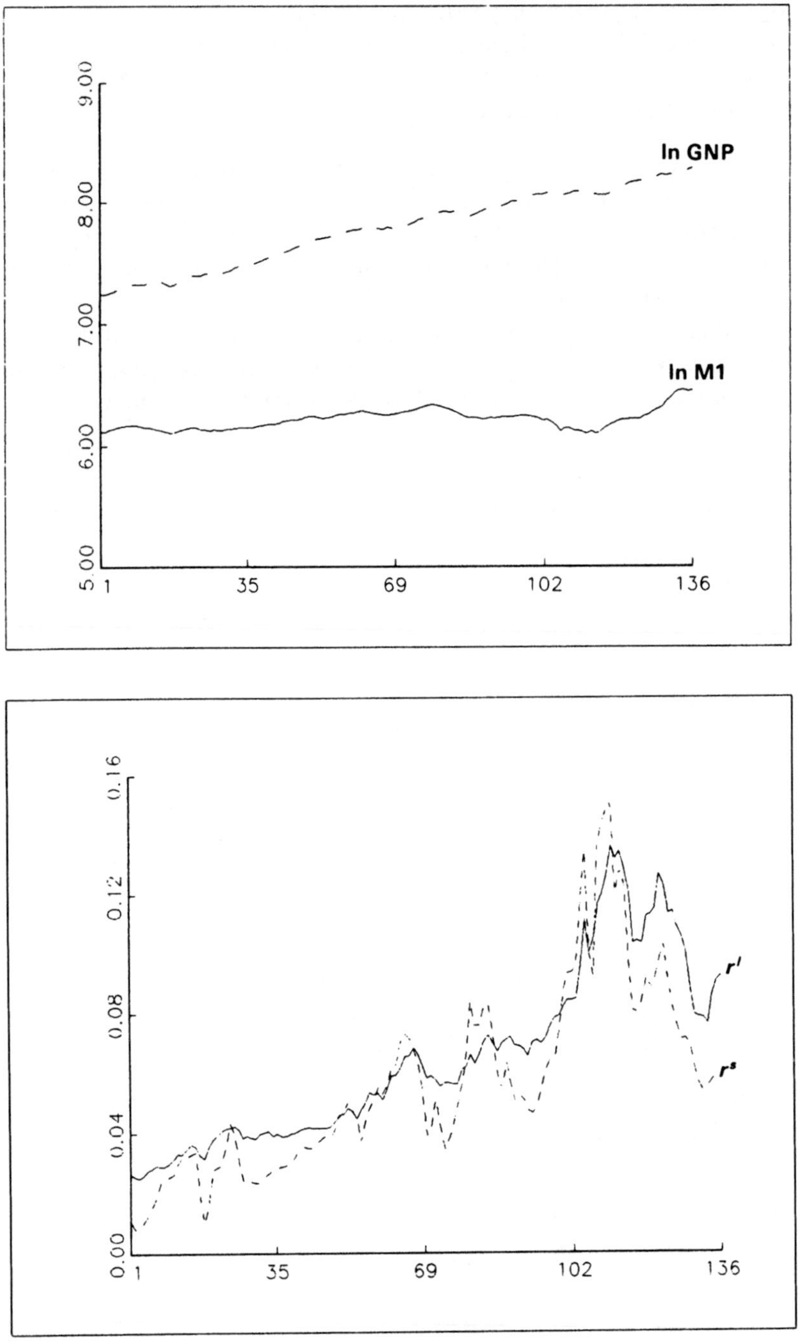

Fig. 11.4. U.S. M1, GNP, and interest rate time series.

Table 11.1. VAR(2) coefficient estimates for the U.S. system

estimation method	ν	A_1				A_2			
ML estimation	.041	1.332	.098	−.556	−.838	−.346	−.091	.354	.969
		(.067)*	(.073)	(.104)	(.216)	(.064)	(.073)	(.110)	(.207)
	.086	.071	1.052	−.169	.549	−.099	−.039	−.239	−.286
		(.079)	(.086)	(.123)	(.256)	(.076)	(.087)	(.131)	(.245)
	.005	.179	.080	.991	.425	−.181	−.079	−.022	−.405
		(.076)	(.082)	(.118)	(.245)	(.073)	(.083)	(.125)	(.235)
	−.014	.037	.047	.041	1.138	−.032	−.050	.033	−.186
		(.038)	(.041)	(.059)	(.122)	(.036)	(.042)	(.062)	(.117)
unrestricted LS estimation	.028	1.307	.106	−.554	−.814	−.318	−.101	.318	1.022
		(.070)	(.075)	(.107)	(.224)	(.070)	(.076)	(.115)	(.221)
	.129	.080	1.045	−.177	.473	−.135	−.014	−.197	−.416
		(.083)	(.088)	(.126)	(.265)	(.083)	(.090)	(.136)	(.261)
	.096	.193	.068	.978	.284	−.248	−.035	.053	−.644
		(.077)	(.081)	(.116)	(.245)	(.077)	(.083)	(.125)	(.240)
	.030	.042	.042	.034	1.065	−.064	−.027	.070	−.308
		(.038)	(.041)	(.058)	(.122)	(.038)	(.041)	(.063)	(.120)
two-step estimation	.001	1.437	.157	−.482	−1.018	−.438	−.156	.482	1.013
		(.058)	(.072)	(.104)	(.214)	(.057)	(.072)	(.104)	(.218)
	.006	.271	1.168	−.029	.151	−.286	−.156	−.032	−.259
		(.071)	(.089)	(.128)	(.266)	(.070)	(.089)	(.128)	(.269)
	−.001	.171	.086	.984	.321	−.199	−.062	.022	−.527
		(.062)	(.077)	(.111)	(.230)	(.061)	(.078)	(.111)	(.234)
	.000	−.015	.024	.004	1.160	−.001	−.011	−.001	−.274
		(.031)	(.039)	(.056)	(.116)	(.031)	(.039)	(.056)	(.118)

* Estimated standard errors in parentheses.

$$\tilde{\Sigma}_u = \begin{bmatrix} .507 & . & . & . \\ .149 & .710 & . & . \\ -.080 & .164 & .652 & . \\ -.033 & .101 & .230 & .163 \end{bmatrix} \times 10^{-4}.$$

The resulting coefficient estimates of the standard representation of the level variables are given in Table 11.1, where the standard errors are obtained from Corollary 11.2.1. The estimates of the intercept terms are obtained by adding a 1 in the first position of ΔX_t in (11.2.4) and modifying the other quantities accordingly.

11.2.1c Discussion of the Proof of Proposition 11.2

We do not give a detailed proof of Proposition 11.2 but just present some main distinctions and similarities to the stable, stationary case. This subsection may be skipped by anyone not interested in the proof.

The asymptotics developed and used in the following depend on whether or not the intercept term ν is zero in the model (11.2.1). We will therefore assume here that $\nu = 0$ as in Proposition 11.2 and refer the reader to Johansen (1991), Phillips (1988), Park & Phillips (1988, 1989), Ahn & Reinsel (1990) and the references in these papers for the case where $\nu \neq 0$. The distinction is necessary because a nonzero ν implies that the components of y_t may be integrated processes with drift which behave differently from processes without drift (see Section 11.1.1). In addition to $\nu = 0$ we assume without further notice that all other conditions of Proposition 11.2 are satisfied and we use the notation of the previous subsections. We first summarize some results from Phillips & Durlauf (1986), Johansen (1988), Ahn & Reinsel (1990), and Park & Phillips (1989) without proof. These results show some major distinctions between the stable and the presently considered unstable case.

LEMMA 11.1

$\Delta X \Delta X'/T$ has a nonstochastic probability limit; (11.2.9)

$CY_{-p}\Delta X'/T$ has a nonstochastic probability limit; (11.2.10)

$CY_{-p}U'/T \xrightarrow{p} 0$; (11.2.11)

$CY_{-p}Y_{-p}'C'/T$ has a nonstochastic, nonsingular probability limit; (11.2.12)

$CY_{-p}U'/\sqrt{T}$ converges in distribution; (11.2.13)

$CY_{-p}\Delta Y'/\sqrt{T}$ converges in distribution; (11.2.14)

$Y_{-p}\Delta X'/T$ converges in distribution; (11.2.15)

$Y_{-p}U'/T$ converges in distribution; (11.2.16)

$CY_{-p}Y_{-p}'/T$ converges in distribution; (11.2.17)

$Y_{-p}Y_{-p}'/T^2$ converges in distribution. (11.2.18)

Let $\bar{C} = (CC')(\tilde{C}C')^{-1}$, then

$T(\bar{C}\tilde{C} - C)$ converges in distribution. (11.2.19)

■

This lemma remains valid if y_t is not Gaussian but satisfies weaker conditions for its distribution (see the above references). Since the lemma implies Proposition 11.2, as will be seen in the following, that proposition also holds for many non-Gaussian processes.

Since ΔY, CY_{-p}, and ΔX contain stable variables only, essentially the same results as in the stable case hold for these quantities. This is reflected in (11.2.9)–(11.2.14). On the other hand, Y_{-p} contains unstable level variables that behave differently from stable variables. This can be seen in (11.2.15)–(11.2.18). For instance, for a stable process YX'/T was seen to have a fixed probability limit in Chapter 3. Now the corresponding quantity $Y_{-p}\Delta X'/T$ has a stochastic limit.

Similarly, for a stable process the cross-products $Y_{-p}Y'_{-p}/T$ converge in probability. In the present unstable case convergence is only obtained if we divide by T^2 instead of T and even then we just get a stochastic limit (convergence in distribution). Intuitively the reason is that integrated variables do not fluctuate around a constant mean but are trending. Thus, the sums of products and cross-products go to infinity (or minus infinity) more rapidly than for stable processes. Similar results are also obtained for other cross-products that are used in the following. They can be derived from Lemma 11.1, though, and are therefore not stated explicitly.

The different convergence results of stable and unstable variables are also the reason for the superconvergence of the transformed $\tilde{C}$ estimator in (11.2.19). For $T(\bar{C}\tilde{C} - C)$ we get T-convergence instead of the usual $\sqrt{T}$-convergence. Note, however, that we do not claim convergence of $\tilde{C}$ to C because the latter matrix is not identified under our assumptions and therefore cannot be estimated consistently. We now prove Proposition 11.2 by showing a series of lemmas.

LEMMA 11.2

For given C let $\tilde{D}(C)$ and $\tilde{H}(C)$ be the resulting ML estimators of D and H, respectively, then

$$\sqrt{T}\,\mathrm{vec}([\tilde{D}(C), -\tilde{H}(C)C] - [D, -\Pi]) \xrightarrow{d} N(0, \Sigma_{co}). \qquad \blacksquare$$

Proof: See Problem 11.3. ■

The lemma states what was already mentioned in Remark 2 of Subsection 11.2.1a, namely that exactly the limiting distribution claimed in Proposition 11.2 is obtained if C is assumed known. In a series of intermediate results we will now show that

$$\sqrt{T}([\tilde{D}, -\tilde{H}\tilde{C}] - [\tilde{D}(C), -\tilde{H}(C)C]) \xrightarrow{p} 0$$

which, using Proposition C.2(2) of Appendix C, proves (11.2.6).

LEMMA 11.3

Let $G_T := CY_{-p}MY'_{-p}C'/T$ and $\tilde{G}_T := \bar{C}\tilde{C}Y_{-p}MY'_{-p}\tilde{C}'\bar{C}'/T$. Then the following results hold:

(a) $G := \operatorname{plim} G_T = \operatorname{plim} \tilde{G}_T$ is nonstochastic and nonsingular;

(b) $\sqrt{T}(G_T - G)$ and $\sqrt{T}(\tilde{G}_T - G)$ converge in distribution to the same multivariate normal limiting distribution;

(c)' $\sqrt{T}(G_T^{-1} - G^{-1})$ and $\sqrt{T}(\tilde{G}_T^{-1} - G^{-1})$ converge in distribution to the same limiting distribution. ■

Proof: Since CY_{-p} and M contain stable variables only the existence and nonsingularity of plim G_T follow as for stable, stationary processes. With the same argument $\sqrt{T}(G_T - G)$ is seen to have a well-defined multivariate normal limiting distribution. Thus, by Proposition C.4 of Appendix C the same holds for $\sqrt{T}(G_T^{-1} - G^{-1})$ because the inverse is a differentiable function.

We will now prove that

$$\operatorname{plim} \sqrt{T}(\tilde{G}_T - G_T) = 0 \tag{11.2.20}$$

which implies all the other claims of the lemma. To see (11.2.20) note that

$$\begin{aligned}\sqrt{T}(\tilde{G}_T - G_T) &= T(\bar{C}\tilde{C} - C)Y_{-p}MY_{-p}'\tilde{C}'\bar{C}'/T^{3/2} \\ &\quad + (CY_{-p}MY_{-p}'/T^{3/2})T(\bar{C}\tilde{C} - C)'.\end{aligned}$$

Lemma 11.1 implies that

$$CY_{-p}MY_{-p}'/T^{3/2} \xrightarrow{p} 0$$

and

$$Y_{-p}MY_{-p}'\tilde{C}'\bar{C}'/T^{3/2} = \frac{1}{T^2}(Y_{-p}MY_{-p}')\sqrt{T}(\bar{C}\tilde{C} - C)' + \frac{1}{T^{3/2}}Y_{-p}MY_{-p}'C' \xrightarrow{p} 0.$$

Hence, $\sqrt{T}(\tilde{G}_T - G_T) \xrightarrow{p} 0.$ ■

LEMMA 11.4

$T[\tilde{H}\tilde{C} - \tilde{H}(C)C]$ converges in distribution. ■

Proof: From the proof of Proposition 11.1 it is easy to see that

$$\tilde{H}(C)C = -\Delta YMY_{-p}'C'[CY_{-p}MY_{-p}'C']^{-1}C$$

and

$$\tilde{H}\tilde{C} = -\Delta YMY_{-p}'\tilde{C}'\bar{C}'[\bar{C}\tilde{C}Y_{-p}MY_{-p}'\tilde{C}'\bar{C}']^{-1}\bar{C}\tilde{C}.$$

Defining

$$\begin{aligned}F_T &:= -\Delta YMY_{-p}'C'/T, \\ \tilde{F}_T &:= -\Delta YMY_{-p}'\tilde{C}'\bar{C}'/T,\end{aligned}$$

and G_T and $\tilde{G}_T$ as in Lemma 11.3, we know that $\operatorname{plim} \sqrt{T}(\tilde{G}_T^{-1} - G_T^{-1}) = 0$ (see Lemma 11.3). Lemma 11.1 implies that both $\sqrt{T}F_T$ and

$$T(\tilde{F}_T - F_T) = \frac{-\Delta YMY_{-p}'}{T}T(\bar{C}\tilde{C} - C)'$$

converge in distribution. Hence, Lemma 11.4 follows by noting that

$$\begin{aligned}T[\tilde{H}\tilde{C} - \tilde{H}(C)C] &= T[\tilde{F}_T\tilde{G}_T^{-1}\bar{C}\tilde{C} - F_TG_T^{-1}C] \\ &= T(\tilde{F}_T - F_T)\tilde{G}_T^{-1}\bar{C}\tilde{C} + F_T\tilde{G}_T^{-1}T(\bar{C}\tilde{C} - C) + \sqrt{T}F_T\sqrt{T}(\tilde{G}_T^{-1} - G_T^{-1})C\end{aligned}$$

converges in distribution. ■

LEMMA 11.5

$$\operatorname{plim} \sqrt{T}[\tilde{D} - \tilde{D}(C)] = 0. \qquad \blacksquare$$

Proof: Noting that

$$\tilde{D} = [\Delta Y + \tilde{H}\tilde{C}Y_{-p}]\Delta X'(\Delta X \Delta X')^{-1}$$

and

$$\tilde{D}(C) = [\Delta Y + \tilde{H}(C)CY_{-p}]\Delta X'(\Delta X \Delta X')^{-1},$$

Lemmas 11.4 and 11.1 imply

$$\sqrt{T}[\tilde{D} - \tilde{D}(C)] = \sqrt{T}[\tilde{H}\tilde{C} - \tilde{H}(C)C]\frac{Y_{-p}\Delta X'}{T}\left[\frac{\Delta X \Delta X'}{T}\right]^{-1} \xrightarrow{p} 0. \qquad \blacksquare$$

With this series of lemmas the result in (11.2.6) is established. The consistency of the covariance matrix estimator

$$\begin{aligned}\tilde{\Sigma}_{co} &= \left(\begin{bmatrix} I & 0 \\ 0 & \tilde{C}' \end{bmatrix}\tilde{\Omega}^{-1}\begin{bmatrix} I & 0 \\ 0 & \tilde{C} \end{bmatrix}\right) \otimes \tilde{\Sigma}_u \\ &= \left(\begin{bmatrix} I & 0 \\ 0 & \tilde{C}'\bar{C}' \end{bmatrix}\left(\frac{1}{T}\begin{bmatrix} \Delta X \Delta X' & \Delta X Y_{-p}'\tilde{C}'\bar{C}' \\ \bar{C}\tilde{C}Y_{-p}\Delta X' & \bar{C}\tilde{C}Y_{-p}Y_{-p}'\tilde{C}'\bar{C}' \end{bmatrix}\right)^{-1}\begin{bmatrix} I & 0 \\ 0 & \bar{C}\tilde{C} \end{bmatrix}\right) \otimes \tilde{\Sigma}_u\end{aligned}$$

follows from Lemma 11.1 and the consistency of $\tilde{\Sigma}_u$ which is demonstrated next.

LEMMA 11.6

$$\operatorname{plim} \sqrt{T}(\tilde{\Sigma}_u - UU'/T) = 0. \qquad \blacksquare$$

This lemma not only implies consistency of $\tilde{\Sigma}_u$ but also shows that the asymptotic distribution of

$$\sqrt{T}\operatorname{vec}(\tilde{\Sigma}_u - \Sigma_u)$$

is the same as that of

$$\sqrt{T}\operatorname{vec}(UU'/T - \Sigma_u).$$

In other words, it is independent of the other coefficients of the system and has the form given in (11.2.7) (see also Section 3.4, Proposition 3.4). Hence Proposition 11.2 is proven once Lemma 11.6 is established. We turn to the proof now.

Proof of Lemma 11.6

$$\begin{aligned}\tilde{\Sigma}_u &= (\Delta Y - \tilde{D}\Delta X + \tilde{H}\tilde{C}Y_{-p})(\Delta Y - \tilde{D}\Delta X + \tilde{H}\tilde{C}Y_{-p})'/T \\ &= [U + (D - \tilde{D})\Delta X + (\tilde{H}\tilde{C} - \Pi)Y_{-p}] \\ &\quad \times [U + (D - \tilde{D})\Delta X + (\tilde{H}\tilde{C} - \Pi)Y_{-p}]'/T\end{aligned}$$

$$= \frac{UU'}{T} + (D - \tilde{D})\frac{\Delta X U'}{T} + (\tilde{H}\tilde{C} - \Pi)\frac{Y_{-p}U'}{T}$$

$$+ \frac{U\Delta X'}{T}(D - \tilde{D})' + (D - \tilde{D})\frac{\Delta X \Delta X'}{T}(D - \tilde{D})'$$

$$+ (\tilde{H}\tilde{C} - \Pi)\frac{Y_{-p}\Delta X'}{T}(D - \tilde{D})' + \frac{UY'_{-p}}{T}(\tilde{H}\tilde{C} - \Pi)'$$

$$+ (D - \tilde{D})\frac{\Delta X Y'_{-p}}{T}(\tilde{H}\tilde{C} - \Pi)' + (\tilde{H}\tilde{C} - \Pi)\frac{Y_{-p}Y'_{-p}}{T}(\tilde{H}\tilde{C} - \Pi)'.$$

Since plim $\Delta X U'/T = 0$,

$$\sqrt{T}(D - \tilde{D})\Delta X U'/T \xrightarrow{p} 0.$$

Moreover,

$$\sqrt{T}(D - \tilde{D})\frac{\Delta X \Delta X'}{T}(D - \tilde{D})' \xrightarrow{p} 0$$

and

$$\sqrt{T}(\tilde{H}\tilde{C} - \Pi)\frac{Y_{-p}\Delta X'}{T}(D - \tilde{D})' \xrightarrow{p} 0$$

by Lemma 11.1. Thus, it remains to show that

$$\sqrt{T}(\tilde{H}\tilde{C} - \Pi)\frac{Y_{-p}U'}{T} \xrightarrow{p} 0 \tag{11.2.21}$$

and

$$\sqrt{T}(\tilde{H}\tilde{C} - \Pi)\frac{Y_{-p}Y'_{-p}}{T}(\tilde{H}\tilde{C} - \Pi)' \xrightarrow{p} 0. \tag{11.2.22}$$

In proving (11.2.21) we use that $Y_{-p}U'/T$ converges in distribution by Lemma 11.1 and

$$\sqrt{T}(\tilde{H}\tilde{C} - \Pi)\frac{Y_{-p}U'}{T} = \sqrt{T}[\tilde{H}\tilde{C} - \tilde{H}(C)C]\frac{Y_{-p}U'}{T}$$
$$+ \sqrt{T}[\tilde{H}(C) - H]\frac{CY_{-p}U'}{T}$$

which converges to zero in probability by Lemmas 11.1 and 11.4. The result in (11.2.22) follows in a similar way (see Problem 11.4). ■

11.2.2 Other Estimation Methods for Cointegrated Systems

A number of estimation methods other than ML have been proposed for cointegrated systems. We will consider two variants in this subsection.

11.2.2a Unconstrained LS Estimation

The first method ignores the cointegration restriction and estimates the process in levels of the original variables without taking differences. In other words, the system (11.2.1) is estimated by LS and the estimator of $A = [A_1, \ldots, A_p]$ is

$$\hat{A} = YX'(XX')^{-1},$$

where

$$Y := [y_1, \ldots, y_T] \qquad \text{and} \qquad X := [Y_0, \ldots, Y_{T-1}] \qquad \text{with } Y_t := \begin{bmatrix} y_t \\ \vdots \\ y_{t-p+1} \end{bmatrix}.$$

Sims, Stock & Watson (1990), and Park & Phillips (1989) have shown that the unconstrained LS estimator $\hat{A}$ has the same asymptotic properties as the ML estimator which observes the cointegration restriction. An interesting additional result is that the usual estimator of the covariance matrix is a consistent estimator of the asymptotic covariance matrix Σ_{α}^{co}. This result is summarized in the next proposition without proof.

PROPOSITION 11.3 (*Asymptotic Properties of the Unconstrained LS Estimator*)

Let y_t be a cointegrated Gaussian, K-dimensional VAR(p) process with cointegration rank r, $0 < r < K$. Then the multivariate LS estimator $\hat{A} = YX'(XX')^{-1}$ is a consistent estimator of A and

$$\sqrt{T}\,\text{vec}(\hat{A} - A) \xrightarrow{d} N(0, \Sigma_{\alpha}^{co}),$$

where Σ_{α}^{co} is the covariance matrix defined in Corollary 11.2.1. Furthermore

$$\hat{\Sigma}_{\alpha}^{co} = (XX')^{-1} \otimes ((Y - \hat{A}X)(Y - \hat{A}X)')$$

is a consistent estimator of Σ_{α}^{co}. ■

This proposition means that we may simply ignore the fact that our system is unstable and proceed as in the stationary, stable case if asymptotic properties are of interest only. In general estimated standard errors and t-statistics from usual regression programs may be used in the usual way. Yet there is an important difference to the stationary, stable case. Now the asymptotic covariance matrix Σ_{α}^{co} is singular which may cause trouble in certain inference procedures, for instance, if Wald tests are considered (see Section 11.3). In fact, in the extreme case where we have a VAR(1) process consisting of K random walks,

$$y_t = Ay_{t-1} + u_t$$

with $A = I_K$, the covariance matrix Σ_{α}^{co} will be zero. In other words, in that case

$$\text{plim}\,\sqrt{T}\,\text{vec}(\hat{A} - I_K) = 0$$

and it can be shown that $T \operatorname{vec}(\hat{A} - I_K)$ converges in distribution. The limiting distribution is nonnormal (see Park & Phillips (1988, 1989)). This result clearly shows that there is a difference between the stable and unstable cases which may be important in practice. Recall that if the asymptotic variances of some elements of $\sqrt{T}(\hat{A} - A)$ are zero the usual t-ratios will not have a limiting standard normal distribution in general. Therefore, in this case we cannot simply interpret these quantities in the usual way. Occasionally an analysis of the integration and cointegration properties of a system may reveal potential problems of this kind. Also, it must be kept in mind that there can be substantial differences between small and large sample properties. In small samples it may well be preferable to impose the cointegration restriction if it is actually satisfied. Proposition 11.3 remains essentially valid if an intercept term is included in the model to be estimated.

For the example system from Subsection 11.2.1b the unconstrained multivariate LS estimators (obtained with intercepts included) are also given in Table 11.1, where the estimated asymptotic standard errors are given in parentheses. The differences to the ML estimator $\tilde{A}$ of Subsection 11.2.1b are obvious. Given the sampling variability reflected in the estimated standard errors, the differences are not drastic, however.

11.2.2b A Two-Stage Procedure

Another popular estimator is available if there is just one cointegration relation, that is, $r = 1$. The procedure consists of two stages. In the first stage the cointegrating vector is estimated by LS and in the second stage the remaining parameters are estimated conditionally on the first stage estimator. To study this estimator in more detail we consider the model

$$\Delta y_t = D_1 \Delta y_{t-1} + \cdots + D_{p-1} \Delta y_{t-p+1} - HC y_{t-p} + u_t. \tag{11.2.23}$$

Since $r = 1$, $C = (c_1, \dots, c_K)$ is a $(1 \times K)$ vector and $Cy_t = c_1 y_{1t} + \cdots + c_K y_{Kt}$. Let us suppose that $c_1 \neq 0$. Then the cointegration relation can be written as

$$y_{1t} = \gamma_2 y_{2t} + \cdots + \gamma_K y_{Kt} + v_t,$$

where $\gamma_i := -c_i/c_1$ and v_t is a stable, stationary process. In the first stage of our estimation procedure this relation is estimated by ordinary least squares. In other words, the estimator

$$\hat{\gamma} = y_{(1)}' Y_{(2)} (Y_{(2)}' Y_{(2)})^{-1} \tag{11.2.24}$$

is used for $\gamma := (\gamma_2, \dots, \gamma_K)$. Here

$$y_{(1)} := \begin{bmatrix} y_{11} \\ \vdots \\ y_{1T} \end{bmatrix} \quad \text{and} \quad Y_{(2)} := \begin{bmatrix} y_{21} & \cdots & y_{K1} \\ \vdots & & \vdots \\ y_{2T} & \cdots & y_{KT} \end{bmatrix}.$$

In the second stage the parameters $D = [D_1, \dots, D_{p-1}]$ and H are estimated conditionally on the estimator $\hat{C}_{2s} = (1, -\hat{\gamma})$ of C. The estimators turn out to be

$$\hat{H}_{2s} = -\Delta Y M Y_{-p}' \hat{C}_{2s}' [\hat{C}_{2s} Y_{-p} M Y_{-p}' \hat{C}_{2s}']^{-1} \qquad (11.2.25)$$

and

$$\hat{D}_{2s} = (\Delta Y + \hat{H}_{2s} \hat{C}_{2s} Y_{-p}) \Delta X' (\Delta X \Delta X')^{-1}, \qquad (11.2.26)$$

where all symbols are as defined in Section 11.2.1a (see Problem 11.5).

The properties of these estimators follow from results due to Stock (1987), Phillips & Durlauf (1986), Park & Phillips (1989), Johansen (1991) and others. Specifically, Stock (1987) shows that $\hat{\gamma}$ is superconsistent. More precisely, $T(\hat{\gamma} - \gamma)$ converges in distribution and thus

$$\text{plim } T^{1-\delta}(\hat{\gamma} - \gamma) = 0$$

for any $\delta > 0$. However, there is some evidence that $\hat{\gamma}$ is biased in small samples (Phillips & Hansen (1990)). On the other hand, the superconsistency of $\hat{\gamma}$ implies that the estimators of D and Π have the same *asymptotic* properties as the ML estimators in Proposition 11.2.

PROPOSITION 11.4 (*Asymptotic Properties of the Two-Stage Estimators*)

Let y_t be a K-dimensional, cointegrated Gaussian VAR(p) process with cointegration rank $r = 1$. Then the two-stage estimator is consistent and

$$\sqrt{T}\, \text{vec}([\hat{D}_{2s}, -\hat{H}_{2s}\hat{C}_{2s}] - [D, -\Pi]) \xrightarrow{d} N(0, \Sigma_{co}),$$

where Σ_{co} is the covariance matrix given in Proposition 11.2, (11.2.6). ■

It may be worth emphasizing the difference between the ML procedure described in Section 11.2.1a and the present two-stage procedure. There is not only the difference in the way the estimators are computed but there is also a difference in assumptions. Besides the fact that we have described the two-stage procedure for cointegration rank $r = 1$ we also need to know one variable which enters the cointegration relation with nonzero coefficient. That variable may then be used as dependent variable in the first-stage cointegration regression.

The second stage in the procedure may be modified. For instance, one may just be interested in the first equation of the system. In this case the first equation may be estimated separately without taking into account the remaining ones. Thus, the two-stage procedure may be applied in a single equation modeling context.

Using the example data of Subsection 11.2.1b, the first stage of the two-stage procedure with $y_1 = \ln \text{M1}$ as dependent variable gives

$$\hat{y}_1 = .858y_2 + .207y_3 - 7.324y_4$$

or

$$\hat{\gamma} = [.858, .207, -7.324].$$

In contrast, if we normalize (divide by 8.05) the ML estimate $\tilde{C}$ given in Section 11.2.1b we get an estimate

$$\tilde{\gamma} = [.465, -14.52, 9.36],$$

which is quite different from $\hat{\gamma}$. However, the estimates for A_1 and A_2 obtained from the two-stage procedure are not all that different from the ML estimates especially taking into account the sampling variability which is reflected in the estimated standard errors. The two-stage estimates obtained with an intercept term included in the model are also given in Table 11.1.

There has been a considerable amount of research on estimation and hypothesis testing in systems with integrated and cointegrated variables. For instance, Johansen (1991), Johansen & Juselius (1990), and Lütkepohl & Reimers (1992) consider estimation with restrictions on the cointegration and loading matrices; Ahn & Reinsel (1990) discuss an ML approach which can be generalized so as to allow for restrictions on the A_i coefficients; Park & Phillips (1988, 1989) and Phillips (1988) provide general results on estimating systems with integrated and cointegrated exogenous variables; Stock (1987) considers a so-called nonlinear LS estimator, and Phillips & Hansen (1990) discuss instrumental variables estimation of models containing integrated variables.

11.2.3 Bayesian Estimation of Integrated Systems

11.2.3a Generalities

In Chapter 5, Section 5.4, we have discussed Bayesian estimation of stationary, stable VAR(p) processes. For a Gaussian process with integrated variables and a normal prior a normal posterior distribution of the VAR coefficients can be derived in a like manner. Suppose $\boldsymbol{\beta} := \mathrm{vec}[\nu, A_1, \dots, A_p]$ is the vector of VAR coefficients including an intercept vector and the prior is

$$\boldsymbol{\beta} \sim N(\boldsymbol{\beta}^*, V_{\boldsymbol{\beta}}). \tag{11.2.27}$$

Then, using the same line of reasoning as in Section 5.4, the posterior mean is

$$\bar{\boldsymbol{\beta}} = [V_{\boldsymbol{\beta}}^{-1} + (ZZ' \otimes \Sigma_u^{-1})]^{-1}[V_{\boldsymbol{\beta}}^{-1}\boldsymbol{\beta}^* + (Z \otimes \Sigma_u^{-1})\mathbf{y}]$$

and the posterior covariance matrix is

$$\bar{\Sigma}_{\boldsymbol{\beta}} = [V_{\boldsymbol{\beta}}^{-1} + (ZZ' \otimes \Sigma_u^{-1})]^{-1},$$

where

$$\mathbf{y} := \mathrm{vec}(Y) \quad \text{and} \quad Z := [Z_0, \dots, Z_{T-1}] \quad \text{with } Z_t := \begin{bmatrix} 1 \\ y_t \\ \vdots \\ y_{t-p+1} \end{bmatrix}.$$

11.2.3b The Minnesota or Litterman Prior

A possible choice of $\boldsymbol{\beta}^*$ and V_β for stable processes was discussed in Section 5.4.3. If the variables are believed to be integrated the following prior discussed by Doan, Litterman & Sims (1984) and Litterman (1986), sometimes known as Minnesota prior, could be used: (1) Set the prior mean of the first lag of each variable equal to one in its own equation and set all other coefficients at zero. In other words, if the prior means were the true parameter values each variable were a random walk. (2) Choose the prior variances of the coefficients as in Section 5.4.3. In other words, the prior variances of the intercept terms are infinite and the prior variance of $\alpha_{ij,l}$, the ij-th element of A_l, is

$$v_{ij,l} = \begin{cases} (\lambda/l)^2 & \text{if } i = j \\ (\lambda\theta\sigma_i/l\sigma_j)^2 & \text{if } i \neq j, \end{cases}$$

where λ is the prior standard deviation of $\alpha_{ii,1}$, $0 < \theta < 1$, and σ_i^2 is the i-th diagonal element of Σ_u. Thus, we get, for instance, for a bivariate VAR(2) system,

$$\begin{aligned}
y_{1t} &= \underset{(\infty)}{0} + \underset{(\lambda)}{1}\, y_{1,t-1} + \underset{(\lambda\theta\sigma_1/\sigma_2)}{0}\, y_{2,t-1} + \underset{(\lambda/2)}{0}\, y_{1,t-2} + \underset{(\lambda\theta\sigma_1/2\sigma_2)}{0}\, y_{2,t-2} + u_{1t}, \\
y_{2t} &= \underset{(\infty)}{0} + \underset{(\lambda\theta\sigma_2/\sigma_1)}{0}\, y_{1,t-1} + \underset{(\lambda)}{1}\, y_{2,t-1} + \underset{(\lambda\theta\sigma_2/2\sigma_1)}{0}\, y_{1,t-2} + \underset{(\lambda/2)}{0}\, y_{2,t-2} + u_{2t},
\end{aligned}$$

where all coefficients are set at their prior means and the numbers in parentheses are their prior standard deviations. Forgetting about the latter numbers for the moment, each of these two equations is seen to be a random walk for one of the variables. The nonzero prior standard deviations indicate that we are not sure about such a simple model. The standard deviations decline with increasing lag length because more recent lags are assumed to be more likely to belong into the model. The infinite standard deviations for the intercept terms simply reflect that we do not have any prior guess for these coefficients. Also, we do not impose covariance priors and, hence, choose V_β to be a diagonal matrix. Its inverse is

$$V_\beta^{-1} = \begin{bmatrix}
0 & & & & & & & & & \\
& 0 & & & & & & & 0 & \\
& & 1/\lambda^2 & & & & & & & \\
& & & (\sigma_1/\lambda\theta\sigma_2)^2 & & & & & & \\
& & & & (\sigma_2/\lambda\theta\sigma_1)^2 & & & & & \\
& & & & & 1/\lambda^2 & & & & \\
& & 0 & & & & (2/\lambda)^2 & & & \\
& & & & & & & (2\sigma_1/\lambda\theta\sigma_2)^2 & & \\
& & & & & & & & (2\sigma_2/\lambda\theta\sigma_1)^2 & \\
& & & & & & & & & (2/\lambda)^2
\end{bmatrix}$$

where 0 is substituted for $1/\infty^2$.

To compute $\bar{\boldsymbol{\beta}}$ requires the inversion of $V_\beta^{-1} + (ZZ' \otimes \Sigma_u^{-1})$. Since this matrix is usually quite large, Bayesian estimation is often performed separately for each of the K equations of the system. In that case

$$\bar{b}_k = [V_k^{-1} + \sigma_k^{-2} ZZ']^{-1}(V_k^{-1} b_k^* + \sigma_k^{-2} Z y_{(k)})$$

is used as an estimator for the parameters b_k of the k-th equation, that is, b_k' is the k-th row of $B := [v, A_1, \ldots, A_p]$. Here V_k is the prior covariance matrix of b_k, b_k^* is its prior mean, and $y_{(k)}'$ is the k-th row of Y. As in Chapter 5, σ_k^2 is replaced by the k-th diagonal element of the ML estimator

$$\tilde{\Sigma}_u = Y(I_T - Z'(ZZ')^{-1}Z)Y'/T$$

of the white noise covariance matrix.

11.2.3c An Example

As an example we consider again the system of U.S. variables from Section 11.2.1b. In Table 11.1 it can be seen that the last three of the four diagonal elements of A_1 are close to 1. The first diagonal element is also not drastically different from 1 although 1 is not within a two-standard error interval around the estimate. Overall a prior with mean 1 for the diagonal elements of A_1 does not appear to be unreasonable for this example. Of course, in a Bayesian analysis the prior is usually not chosen on the basis of the ML estimates.

We have estimated the system with the Minnesota prior and different values of λ and θ. Some results for a VAR(2) process are given in Table 11.2 to illustrate

Table 11.2. Bayesian estimates of the U.S. system

prior	v	A_1				A_2			
	.028	1.307	.106	−.554	−.814	−.318	−.101	.318	1.022
$\lambda = \infty, \theta = 1$	.129	.080	1.045	−.177	.473	−.135	−.014	−.197	−.416
(unrestricted)	.096	.193	.068	.978	.284	−.248	−.035	.053	−.644
	.030	.042	.042	.034	1.065	−.064	−.027	.070	−.308
	.061	1.307	.021	−.514	−.465	−.331	−.009	.212	.679
$\lambda = 1, \theta = .25$	.110	.060	1.088	−.173	.283	−.108	−.060	−.162	−.238
	.078	.119	.064	1.060	.025	−.167	−.034	−.069	−.316
	.029	.021	.029	.050	1.044	−.043	−.014	.031	−.265
	.083	1.550	.004	−.012	−.007	−.570	.002	−.000	.004
$\lambda = 1, \theta = .01$	−.015	.005	1.270	−.011	−.011	−.001	−.271	−.003	−.002
	−.032	−.003	.008	1.095	−.001	−.002	.002	−.216	−.001
	−.016	−.003	.004	.002	1.187	−.001	.001	.000	−.252
	−.045	1.009	.002	−.001	−.000	−.003	.000	−.000	.000
$\lambda = .01, \theta = .25$	.018	.001	.999	−.001	−.002	.000	−.002	−.000	−.000
	−.004	.001	−.000	.993	−.001	.000	−.000	−.002	−.000
	−.003	.000	.000	.000	.994	.000	.000	−.000	−.002

the effect of the choice of the prior variance parameters λ and θ. For this particular data set a combination $\lambda = 1$ and $\theta = .25$ leads to mild changes in the estimates only relative to unrestricted estimates ($\lambda = \infty, \theta = 1$). Decreasing θ has the effect of shrinking the off-diagonal elements towards zero. Thus, a small θ is reasonable if the variables are expected to be unrelated. The effect of a small θ is seen in Table 11.2 in the panel corresponding to $\lambda = 1$ and $\theta = .01$. On the other hand, lowering λ shrinks the diagonal elements of A_1 towards 1 and all other coefficients (except the intercept terms) towards zero. This effect is clearly observed for $\lambda = .01, \theta = .25$. Hence, if the analyst has a strong prior in favor of unrelated random walks a small λ is appropriate.

In practice one would usually choose a higher VAR order than 2 in a Bayesian analysis because chopping off the process at $p = 2$ implies a very strong prior with mean zero and variances zero for $A_3, A_4, \ldots$ which is a bit unrealistic. The above analysis is just meant to illustrate the effect of the choice of the parameters that determine the prior variance. Also, if the variables are believed to be cointegrated the Minnesota prior is not a good choice. It is more suited for a process which has a VAR representation in first differences because the basic idea underlying this prior is that the variables are roughly unrelated random walks.

11.3 Forecasting and Structural Analysis

In this section forecasting, testing for Granger-causality, and impulse response analysis based on integrated and cointegrated systems are studied. We begin with forecasting.

11.3.1 Forecasting Integrated and Cointegrated Systems

For a VAR(p) process,

$$y_t = \nu + A_1 y_{t-1} + \cdots + A_p y_{t-p} + u_t, \tag{11.3.1}$$

the optimal h-step forecast with minimal MSE is given by the conditional expectation, provided that expectation exists, even if $\det(I_K - A_1 z - \cdots - A_p z^p)$ has roots on the unit circle. In the proof of the optimality of the conditional expectation in Section 2.2.2 we have not used the stationarity and stability of the system. Thus, assuming that u_t is independent white noise, the optimal h-step forecast at origin t is

$$y_t(h) = \nu + A_1 y_t(h-1) + \cdots + A_p y_t(h-p), \tag{11.3.2}$$

where $y_t(j) := y_{t+j}$ for $j \leq 0$, just as in the stationary, stable case.

Also the forecast errors are of the same form as in the stable case. To see this we write the process (11.3.1) in VAR(1) form as

$$Y_t = \boldsymbol{\nu} + \mathbf{A} Y_{t-1} + U_t, \tag{11.3.3}$$

where

$$Y_t := \begin{bmatrix} y_t \\ \vdots \\ y_{t-p+1} \end{bmatrix}, \quad \mathbf{v} := \begin{bmatrix} v \\ 0 \\ \vdots \\ 0 \end{bmatrix}, \quad \mathbf{A} := \begin{bmatrix} A_1 & A_2 & \dots & A_{p-1} & A_p \\ I_K & 0 & \dots & 0 & 0 \\ 0 & I_K & & 0 & 0 \\ \vdots & & \ddots & \vdots & \vdots \\ 0 & 0 & \dots & I_K & 0 \end{bmatrix},$$
$$(Kp \times 1) \qquad (Kp \times 1) \qquad (Kp \times Kp)$$

$$U_t := \begin{bmatrix} u_t \\ 0 \\ \vdots \\ 0 \end{bmatrix}.$$
$$(Kp \times 1)$$

If u_t is independent white noise the optimal h-step forecast of Y_t is

$$Y_t(h) = \mathbf{v} + \mathbf{A} Y_t(h-1) = (I_{Kp} + \mathbf{A} + \cdots + \mathbf{A}^{h-1})\mathbf{v} + \mathbf{A}^h Y_t.$$

Moreover,

$$\begin{aligned} Y_{t+h} &= \mathbf{v} + \mathbf{A} Y_{t+h-1} + U_{t+h} \\ &= (I_{Kp} + \mathbf{A} + \cdots + \mathbf{A}^{h-1})\mathbf{v} + \mathbf{A}^h Y_t + U_{t+h} + \mathbf{A} U_{t+h-1} + \cdots + \mathbf{A}^{h-1} U_{t+1}. \end{aligned}$$

Hence, the forecast error for the process Y_t is

$$Y_{t+h} - Y_t(h) = U_{t+h} + \mathbf{A} U_{t+h-1} + \cdots + \mathbf{A}^{h-1} U_{t+1}.$$

Premultiplying by the $(K \times Kp)$ matrix $J := [I_K \quad 0 \dots 0]$ gives

$$\begin{aligned} y_{t+h} - y_t(h) &= J U_{t+h} + J\mathbf{A}J'JU_{t+h-1} + \cdots + J\mathbf{A}^{h-1}J'JU_{t+1} \\ &= u_{t+h} + \Phi_1 u_{t+h-1} + \cdots + \Phi_{h-1} u_{t+1}, \end{aligned} \tag{11.3.4}$$

where $J'JU_t = U_t$ and $\Phi_i = J\mathbf{A}^i J'$ have been used. Thus, the form of the forecast error is exactly the same as in the stable case and the forecast is easily seen to be unbiased, that is,

$$E[y_{t+h} - y_t(h)] = 0.$$

Furthermore, the Φ_i may be obtained from the A_i by the recursions

$$\Phi_i = \sum_{j=1}^{i} \Phi_{i-j} A_j, \qquad i = 1, 2, \dots \tag{11.3.5}$$

with $\Phi_0 = I_K$ just as in Chapter 2. Also the forecast MSE matrix becomes

$$\Sigma_y(h) = \sum_{i=0}^{h-1} \Phi_i \Sigma_u \Phi_i' \tag{11.3.6}$$

as in the stable case. Yet there is a very important difference. In the stable case the Φ_i converge to zero as $i \to \infty$ and $\Sigma_y(h)$ converges to the covariance matrix of y_t as $h \to \infty$. This result was obtained because the eigenvalues of $\mathbf{A}$ have modulus less than one in the stable case. Hence $\Phi_i = J\mathbf{A}^i J' \to 0$ as $i \to \infty$. Since the eigenvalues of $\mathbf{A}$ are just the reciprocals of the roots of the determinantal polynomial $\det(I_K - A_1 z - \cdots - A_p z^p)$ the Φ_i do not converge to zero in the presently considered unstable case where one or more of the eigenvalues of $\mathbf{A}$ are 1. Consequently some elements of the forecast MSE matrix $\Sigma_y(h)$ will approach infinity as $h \to \infty$. In other words, the forecast MSEs will be unbounded and the forecast uncertainty may become extremely large as we predict the distant future even if the structure of the process does not change. Consider, for instance, the process

$$\begin{bmatrix} y_{1t} \\ y_{2t} \end{bmatrix} = \begin{bmatrix} .5 & .3 \\ 0 & 1 \end{bmatrix} \begin{bmatrix} y_{1,t-1} \\ y_{2,t-1} \end{bmatrix} + \begin{bmatrix} u_{1t} \\ u_{2t} \end{bmatrix}$$

with $\Sigma_u = I_2$ which is easily seen to represent a cointegrated system. For this system the following MSE matrices are obtained:

$$\Sigma_y(1) = I_2,$$

$$\Sigma_y(5) = \begin{bmatrix} 1.43 & .43 \\ .43 & 5.0 \end{bmatrix},$$

$$\Sigma_y(10) = \begin{bmatrix} 1.43 & .43 \\ .43 & 10.0 \end{bmatrix},$$

$$\Sigma_y(50) = \begin{bmatrix} 1.43 & .43 \\ .43 & 50.0 \end{bmatrix}.$$

Obviously, the forecast MSE of the second variable grows with the forecast horizon, whereas the MSE of y_{1t} approaches an upper bound quickly when the forecast horizon increases.

In practice, of course, we don't know the coefficients ν, $A_1, \ldots, A_p$, and Σ_u. Replacing them by estimators creates similar problems as in the stationary, stable case considered in Chapter 3, Section 3.5. Denoting the h-step forecast based on estimated coefficients by $\hat{y}_t(h)$ and indicating estimators by carets gives

$$\hat{y}_t(h) = \hat{\nu} + \hat{A}_1 \hat{y}_t(h-1) + \cdots + \hat{A}_p \hat{y}_t(h-p), \tag{11.3.7}$$

where $\hat{y}_t(j) := y_{t+j}$ for $j \leq 0$. For this predictor the forecast error becomes

$$\begin{aligned} y_{t+h} - \hat{y}_t(h) &= [y_{t+h} - y_t(h)] + [y_t(h) - \hat{y}_t(h)] \\ &= \sum_{i=0}^{h-1} \Phi_i u_{t+h-i} + [y_t(h) - \hat{y}_t(h)], \end{aligned}$$

where the last two terms are uncorrelated. In fact, the last term has zero probability limit as in the stationary case (see Problem 11.6). Thus the forecast errors from estimated processes and processes with known coefficients are asymptotically equivalent. However, in the present case the MSE correction for estimated processes derived in Section 3.5 is difficult to justify (see Problem 11.7 and Basu & Sen Roy (1987)). This problem must be kept in mind when forecast intervals are constructed. One possible MSE estimator is

$$\hat{\Sigma}_y(h) = \sum_{i=0}^{h-1} \hat{\Phi}_i \hat{\Sigma}_u \hat{\Phi}_i', \tag{11.3.8}$$

where the $\hat{\Phi}_i$ are obtained from the estimated A_i by the recursions (11.3.5). This estimator is likely to underestimate the true forecast uncertainty on average in small samples. Therefore, there is some danger that the confidence level of corresponding forecast intervals is overstated. These statements are a bit speculative, however, because little is known about the small sample properties of forecasts based on estimated unstable processes. Engle & Yoo (1987) and Reinsel & Ahn (1988) report on simulation studies in which imposing the cointegration restriction in the estimation gave better long-range forecasts than the use of unrestricted multivariate LS estimators.

11.3.2 Testing for Granger-Causality

11.3.2a The Noncausality Restrictions

Tests for Granger-causality in integrated and cointegrated systems may be conducted just like in stable systems. From the discussion in the previous subsection it follows easily that the restrictions characterizing Granger-noncausality are exactly the same as in the stable case. More precisely, suppose that the vector y_t in (11.3.1) is partitioned in M- and $(K-M)$-dimensional subvectors z_t and x_t,

$$y_t = \begin{bmatrix} z_t \\ x_t \end{bmatrix} \quad \text{and} \quad A_i = \begin{bmatrix} A_{11,i} & A_{12,i} \\ A_{21,i} & A_{22,i} \end{bmatrix}, \qquad i = 1, \dots, p,$$

where the A_i are partitioned in accordance with the partitioning of y_t. Then x_t does not Granger-cause z_t if and only if the hypothesis

$$H_0\colon A_{12,i} = 0 \qquad \text{for } i = 1, \dots, p \tag{11.3.9}$$

is true. In turn, z_t does not Granger-cause x_t if and only if $A_{21,i} = 0$ for $i = 1, \dots, p$. In other words, in order to test for Granger-causality we just have to test a set of linear hypotheses. A Wald test for general linear restrictions is considered next.

11.3.2b A Wald Test for Linear Constraints

If the process is Gaussian and has been estimated by the ML procedure described in Section 11.2.1, the asymptotic distribution given in Corollary 11.2.1 may be used in constructing a Wald test of a general linear hypothesis

$$H_0\colon C\boldsymbol{\alpha} = c \qquad \text{against} \qquad H_1\colon C\boldsymbol{\alpha} \neq c, \tag{11.3.10}$$

where $\boldsymbol{\alpha} := \text{vec}(A_1, \dots, A_p)$, C is an $(N \times pK^2)$ matrix of rank N, and c is an $(N \times 1)$ vector. The Wald statistic and its asymptotic distribution are given in the following proposition.

PROPOSITION 11.5 (*Asymptotic Distribution of the Wald Statistic*)

Suppose $\tilde{\boldsymbol{\alpha}}$ is an estimator of the coefficient vector $\boldsymbol{\alpha}$ which has an asymptotic normal distribution,

$$\sqrt{T}(\tilde{\boldsymbol{\alpha}} - \boldsymbol{\alpha}) \xrightarrow{d} N(0, \boldsymbol{\Sigma}_{\boldsymbol{\alpha}}^{co}).$$

Moreover $\tilde{\boldsymbol{\Sigma}}_{\boldsymbol{\alpha}}^{co}$ is a consistent estimator of $\boldsymbol{\Sigma}_{\boldsymbol{\alpha}}^{co}$. Then the Wald statistic for testing (11.3.10),

$$\lambda_w = T(C\tilde{\boldsymbol{\alpha}} - c)'(C\tilde{\boldsymbol{\Sigma}}_{\boldsymbol{\alpha}}^{co}C')^{-1}(C\tilde{\boldsymbol{\alpha}} - c), \tag{11.3.11}$$

has an asymptotic $\chi^2(N)$-distribution provided

$$\text{rk}(C\tilde{\boldsymbol{\Sigma}}_{\boldsymbol{\alpha}}^{co}C') = \text{rk}(C\boldsymbol{\Sigma}_{\boldsymbol{\alpha}}^{co}C') = N. \tag{11.3.12}$$

■

Of course, this result follows from standard asymptotic theory (see Appendix C). We have chosen to state it as a proposition here because the rank condition (11.3.12) now becomes important. It is automatically satisfied for stable, full VAR processes as discussed in Chapter 3 because in that case the asymptotic covariance matrix of the coefficient estimator is nonsingular. Now, however, $\boldsymbol{\Sigma}_{\boldsymbol{\alpha}}^{co}$ is singular if the cointegration rank r is less than K (see Subsection 11.2.1a). Therefore it is possible in principle that $\text{rk}(C\boldsymbol{\Sigma}_{\boldsymbol{\alpha}}^{co}C') < N$. The proposition can be generalized by replacing the inverse in (11.3.11) by a generalized inverse. In that case the asymptotic distribution of λ_w is $\chi^2(\text{rk}(C\boldsymbol{\Sigma}_{\boldsymbol{\alpha}}^{co}C'))$ if

$$\text{rk}(C\tilde{\boldsymbol{\Sigma}}_{\boldsymbol{\alpha}}^{co}C') = \text{rk}(C\boldsymbol{\Sigma}_{\boldsymbol{\alpha}}^{co}C') \tag{11.3.13}$$

with probability one (see Andrews (1987)). Unfortunately the latter condition will not hold in general. In particular, if a cointegrated system is estimated in unconstrained form by multivariate LS and if $\boldsymbol{\Sigma}_{\boldsymbol{\alpha}}^{co}$ is estimated as in Proposition 11.3, $C\hat{\boldsymbol{\Sigma}}_{\boldsymbol{\alpha}}^{co}C'$ has rank N with probability 1 while $\text{rk}(C\boldsymbol{\Sigma}_{\boldsymbol{\alpha}}^{co}C')$ may be less than N. Andrews (1987) has shown that in such a case the asymptotic distribution of λ_w may not even be χ^2. This is another argument in favor of performing a careful analysis of the cointegration properties of a given system. It is perhaps worth pointing out that (11.3.13) may be violated, even if the cointegration rank has been specified correctly and the corresponding restrictions have been imposed in the estimation procedure (see Problem 11.8).

11.3.3 Impulse Response Analysis

11.3.3a Theoretical Considerations

Integrated and cointegrated systems must be interpreted cautiously. As mentioned in Section 11.1.2, in cointegrated systems the term Cy_t is usually thought of as representing the long-run equilibrium relations between the variables. Suppose there is just one such relation, say

$$c_1 y_{1t} + \cdots + c_K y_{Kt} = 0,$$

or, if $c_1 \neq 0$,

$$y_{1t} = -\frac{c_2}{c_1} y_{2t} - \cdots - \frac{c_K}{c_1} y_{Kt}.$$

It is tempting to argue that the long-run effect of a unit increase in y_2 will be a change of size $-c_2/c_1$ in y_1. This, however, ignores all the other relations between the variables which are summarized in a VAR(p) model. A one-time unit innovation in y_2 may affect various other variables which also have an impact on y_1. Therefore the long-run effect of a y_2-innovation on y_1 may be quite different from $-c_2/c_1$. The impulse responses may give a better picture of the relations between the variables.

In Chapter 2, Section 2.3.2, the impulse responses of stationary, stable VAR(p) processes were shown to be the coefficients of specific moving average (MA) representations. An unstable, integrated or cointegrated VAR(p) process does not possess valid MA representations of the types discussed in Chapter 2. Yet the Φ_i and Θ_i coefficient matrices can be computed as in Section 2.3.2. For the Φ_i we have seen this in Section 11.3.1, Equation (11.3.5), and from the discussion in that section it is easy to see that the elements of the $\Phi_i = (\phi_{jk,i})$ matrices may represent impulse responses just as in the stable case. More precisely, $\phi_{jk,i}$ is the response of variable j to a unit forecast error in variable k, i periods ago, if the system reflects the actual responses to forecast errors. Recall that in stable processes the responses taper off to zero as $i \to \infty$. This property does not necessarily hold in unstable systems where the effect of a one-time impulse may not die out asymptotically.

In Section 2.3 we have also considered accumulated impulse responses, responses to orthogonalized residuals and forecast error variance decompositions. These tools for structural analysis are all available for unstable systems as well using precisely the same formulas as in Chapter 2. The only quantities that cannot be computed in general are the total "long-run effects" or total multipliers Ψ_∞ and Ξ_∞ because they may not be finite.

The asymptotic properties of the other response coefficients and forecast error variance components based on estimated VAR coefficients follow from Proposition 3.6 in conjunction with Proposition 11.2 and Corollary 11.2.1. In other words, the relevant covariance matrices $\Sigma_{\hat{\alpha}}$ and $\Sigma_{\hat{\sigma}}$ have to be used in Proposition 3.6. Of course, the remarks on Proposition 3.6 regarding the estimation of standard errors etc. apply for the present case too.

11.3.3b An Example

To illustrate the impulse response analysis we use again our U.S. example system. Before we look at some impulse responses it may be worth considering the estimated cointegration relation again. In Section 11.2.1b we found the cointegration relation

$$-8.05 \ln M1_t + 3.74 \ln GNP_t - 116.88 r_t^s + 75.36 r_t^l.$$

Assuming that this relation describes a long-run money demand function and normalizing the coefficient of ln M1 gives

$$\ln M1_t = .465 \ln GNP_t - 14.52 r_t^s + 9.36 r_t^l + z_t, \tag{11.3.14}$$

where z_t is a stationary, stable process. As mentioned earlier, it is problematic to interpret the coefficients in this relation as long-run elasticities because such an interpretation ignores the dynamics of the system. For instance, a shift of ln GNP may not lead to a long-term shift of ln M1 by .465 units because a shift in ln GNP is likely to have an impact on other variables as well that may interact in the short- and in the long-run. This problem becomes apparent in the impulse responses.

In Figure 11.5 the responses of ln GNP and ln M1 to a one-time unit impulse

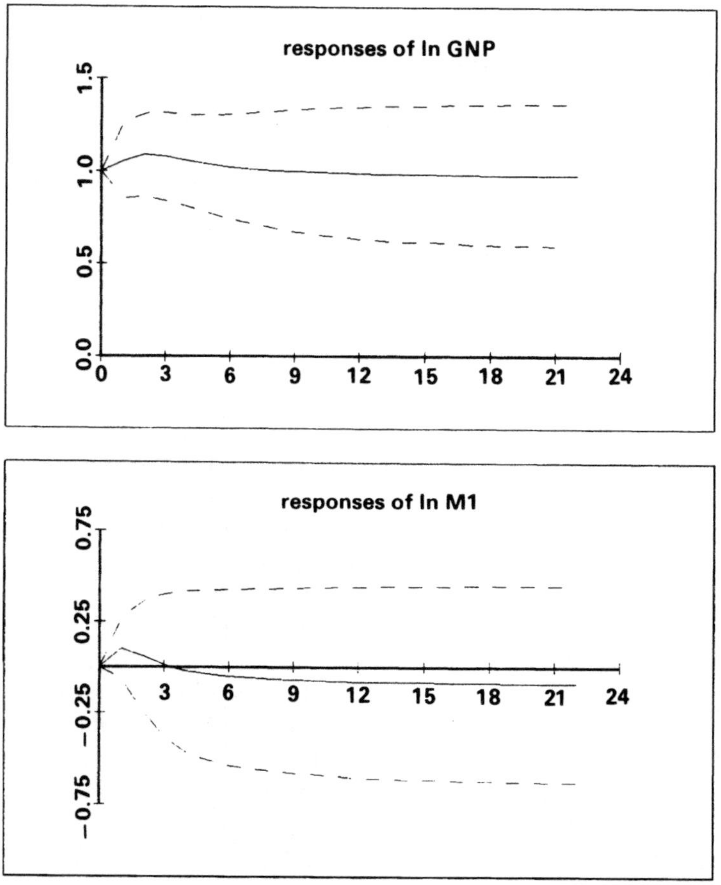

Fig. 11.5. Responses of ln GNP and ln M1 to a forecast error impulse in ln GNP.

(forecast error) in the former variable are depicted. They are computed from the ML estimates of Section 11.2.1b. Forgetting about estimation variability for the moment, the one-time impulse is seen to have a lasting effect on ln GNP. As a consequence of the unit root in the system the impulse response function does not return to zero. After 20 periods it is still close to one. In contrast, the long-run response of ln M1 is quite close to zero. It is clearly different from .465. Hence, interpreting the coefficients in (11.3.14) directly as long-run elasticities is obviously problematic in this system. In Figure 11.5 two-standard error bounds for the impulse responses are indicated with dashed lines. Taking them into account one may even argue that a one-time impulse in ln GNP has no significant effect on money demand although it has a permanent effect on income.

We emphasize again that an uncritical impulse response analysis is problematic, too. In particular, different sets of impulse responses exist and it is not clear which one properly reflects the actual reactions of the variables. The caveats of impulse response analyses are discussed in Sections 2.3 and 3.7. They are therefore not repeated here.

11.4 Model Selection and Model Checking

In model selection and checking the model adequacy some results from the stable, stationary case carry over to the unstable case. We will consider a few of them in the following. In addition tests of the cointegration rank will be treated. We begin with a discussion of VAR order selection criteria and then consider tests for the rank of cointegration. Testing for structural change is dealt with at the end of this section.

11.4.1 VAR Order Selection

It was mentioned in Section 11.3.2 that Wald tests for zero restrictions on VAR coefficients can be constructed. Hence, order selection could proceed as a sequence of tests similar to that in Section 4.2. Since the procedure and its problems are discussed in some detail in that section we will not repeat it here but focus on order selection criteria such as AIC, HQ, and SC in this subsection.

In Section 4.3 the FPE criterion was introduced for stationary, stable processes as a criterion that minimizes the forecast MSE and therefore has a justification if forecasting is the objective. We have seen in Section 11.3.1 that the forecast MSE correction used for estimated stationary processes is difficult to justify in the unstable case and, hence, the FPE criterion cannot be based on the same footing in the latter case. This, of course, does not mean that the criterion is not a useful one in some other sense for nonstationary processes. For instance, it is possible that it still provides models with excellent small sample prediction properties. It was also shown in Section 4.3 that Akaike's AIC criterion is asymptotically equivalent to the FPE criterion. Therefore, similar comments apply for AIC.

The criteria HQ and SC were justified by their ability to choose the order "correctly in large samples", that is, they are consistent criteria. It was shown by Paulsen (1984) and Tsay (1984) that the consistency property of these criteria is maintained for unstable processes. To make that statement precise we give the following result from Paulsen (1984) without proof.

PROPOSITION 11.6 (*Consistent VAR Order Estimation*)

Let

$$y_t = \nu + A_1 y_{t-1} + \cdots + A_p y_{t-p} + u_t$$

be a K-dimensional VAR(p) process with $A_p \neq 0$ and standard white noise u_t and suppose that $\det(I_K - A_1 z - \cdots - A_p z^p)$ has s roots equal to one, that is, $z = 1$ is a root with multiplicity s, and all other roots are outside the complex unit circle. Furthermore, let

$$Cr(m) = \ln|\tilde{\Sigma}_u(m)| + mc_T/T, \tag{11.4.1}$$

where $\tilde{\Sigma}_u(m)$ is the ML or pseudo ML estimator of Σ_u for a VAR(m) model based on a sample of size T and m fixed presample values as in Proposition 4.2, and c_T is a nondecreasing sequence indexed by T. Let $\hat{p}$ be such that

$$Cr(\hat{p}) = \min\{Cr(m)|m = 0, 1, \ldots, M\}$$

and suppose $M \geq p$. Then $\hat{p}$ is a consistent estimator of p if and only if $c_T \to \infty$ and $c_T/T \to 0$ as $T \to \infty$. ∎

This proposition extends Proposition 4.2 to unstable systems. It implies that AIC is not a consistent criterion while HQ and SC are both consistent. Thus, if consistent estimation is the objective we may apply HQ and SC for stable and unstable processes.

Denoting the orders chosen by AIC, HQ, and SC by $\hat{p}$(AIC), $\hat{p}$(HQ), and $\hat{p}$(SC), respectively, we also get from Proposition 4.3 that

$$\hat{p}(\text{SC}) \leq \hat{p}(\text{HQ}) \leq \hat{p}(\text{AIC}) \qquad \text{for } T \geq 16.$$

This result is obtained because Proposition 4.3 does not require any stationarity or stability assumptions. It follows as in Chapter 4 that AIC asymptotically overestimates the true order with positive probability (see Corollary 4.3.1).

Although these results are nice because they generalize the stationary case in an easy way they do not mean that AIC or FPE are order selection criteria inferior to HQ and SC. Recall that consistent order estimation may not be a relevant objective in small sample situations. In fact, the true data generating process may not admit a finite order VAR representation.

We have applied the three criteria AIC, HQ, and SC to our example data from Section 11.2.1b with a maximum order of $M = 8$. The values of the criteria are shown in Table 11.3. SC and HQ both recommend the order $\hat{p} = 2$ while $\hat{p}$(AIC) = 3. We have chosen to go with the HQ-SC selection in the previous example sections.

Table 11.3. Estimation of the VAR order of the U.S. system

VAR order m	AIC(m) × 10^3	HQ(m) × 10^3	SC(m) × 10^3
0	−2.644	−2.644	−2.644
1	−3.986	−3.972	−3.950
2	−4.052	−4.023*	−3.980*
3	−4.059*	−4.015	−3.952
4	−4.052	−3.994	−3.909
5	−4.049	−3.977	−3.871
6	−4.056	−3.969	−3.842
7	−4.048	−3.947	−3.799
8	−4.052	−3.936	−3.767

* Minimum.

In Chapter 4 we have mentioned that model selection may be based on the residual autocorrelations or portmanteau tests. Unfortunately the proofs of the asymptotic properties of these statistics given in Section 4.4 do not carry over to unstable processes.

11.4.2 Testing for the Rank of Cointegration

Based on Proposition 11.1 it is easy to derive the likelihood ratio statistic for testing a specific cointegration rank $r = r_0$ of a VAR(p) process against a larger rank of cointegration, say $r = r_1$.

Suppose we have a VAR(p) process as in (11.1.9) with intercept term and we wish to test

$$H_0: r = r_0 \qquad \text{against} \qquad H_1: r_0 < r \le r_1. \tag{11.4.2}$$

Under Gaussian assumptions the maximum of the likelihood function for a model with cointegration rank r is given in Proposition 11.1. From that result the LR statistic for testing (11.4.2) is seen to be

$$\begin{aligned}\lambda_{LR}(r_0, r_1) &= 2[\ln l(r_1) - \ln l(r_0)] \\ &= T\left[-\sum_{i=1}^{r_1} \ln(1-\lambda_i) + \sum_{i=1}^{r_0} \ln(1-\lambda_i)\right] \\ &= -T \sum_{i=r_0+1}^{r_1} \ln(1-\lambda_i),\end{aligned} \tag{11.4.3}$$

where $l(r_i)$ denotes the maximum of the Gaussian likelihood function for cointegration rank r_i. Obviously, the test value is quite easy to compute.

Unfortunately, in this case the asymptotic distribution of the LR statistic is nonstandard. In particular, it is not a χ^2-distribution. However, it just depends on the difference $K - r$ between the dimension of the process and its cointegra-

Table 11.4. Percentage points of asymptotic distributions of LR tests for the cointegration rank from Johansen & Juselius (1990)

	$K - r$	90%	95%	97.5%	99%
	1	6.69	8.08	9.66	11.58
	2	15.58	17.84	19.61	21.96
$\lambda_{LR}(r, K)$	3	28.44	31.26	34.06	37.29
	4	45.25	48.42	51.80	55.55
	5	65.96	69.98	73.03	77.91
	1	6.69	8.08	9.66	11.58
	2	12.78	14.60	16.40	18.78
$\lambda_{LR}(r, r + 1)$	3	18.96	21.28	23.36	26.15
	4	24.92	27.34	29.60	32.62
	5	30.82	33.26	35.70	38.86

tion rank as well as the alternative hypothesis. Thus, it is possible to tabulate selected percentage points of its asymptotic distribution. This has been done by Johansen & Juselius (1990) and some of their critical values are reproduced in Table 11.4. These percentage points permit tests of the hypotheses

$$H_0: r = r_0 \qquad \text{against} \qquad H_1: r_0 < r \leq K \tag{11.4.4}$$

and

$$H_0: r = r_0 \qquad \text{against} \qquad H_1: r = r_0 + 1 \tag{11.4.5}$$

Critical values for an LR test of (11.4.4) are also tabulated by Reinsel & Ahn (1988). They are very similar but not identical to those listed in Table 11.4. This, of course, is a consequence of the sampling variability in the generation process of these values.

We have applied tests of (11.4.4) and (11.4.5) in a sequential manner to the example data considered in Section 11.2.1b and give the test values in Table 11.5. In that table sequences of hypotheses are presented with r_0 ranging from $K - 1$ to 0. A testing sequence terminates when H_0 is rejected for the first time. Using a significance level of 5% in each individual test both testing sequences reject $r_0 = 0$ and, hence, support a cointegration rank of $r = 1$. This was exactly the rank used in Section 11.2.1b. Of course, for these testing sequences similar comments apply as in Section 4.2. That is, the overall Type I error will differ from the individual significance levels and the individual tests are not independent which makes it difficult to determine the actual Type I error.

It may be worth noting that the asymptotic distribution of the LR statistic is different if there is no intercept in the model. Thus, if the intercept vector v is set at zero a different set of critical values has to be used for the rank test. Relevant percentage points for this case are tabulated by Johansen (1988) and Reinsel &

Table 11.5. Tests for the rank of cointegration of the example system

hypotheses H_0	H_1	values of test statistics
$r = 3$	$r = 4$	0.01
$r = 2$	$r \geq 3$	2.65
$r = 1$	$r \geq 2$	20.65
$r = 0$	$r \geq 1$	55.97*
$r = 3$	$r = 4$	0.01
$r = 2$	$r = 3$	2.63
$r = 1$	$r = 2$	18.01
$r = 0$	$r = 1$	35.32*

* Significant at 5% level.

Ahn (1988). A different set of critical values is also needed if the intercept term is confined to the cointegration relation in estimating the system. Such a restriction arises if the process can be written in the form

$$\Delta y_t = D_1 \Delta y_{t-1} + \cdots + D_{p-1} \Delta y_{t-p+1} - HC(y_{t-p} - \eta) + u_t.$$

This case is treated in detail by Johansen (1991) and Johansen & Juselius (1990) who also provide critical values for the cointegration LR test. Confining the intercept term to the cointegration relation implies that there is no linear trend in the y_t variables (see Section 11.1). Such an assumption is sometimes reasonable in practice and therefore this particular case is important.

Based on a Monte Carlo simulation investigation Reinsel & Ahn (1988) find that the small sample critical values of the LR tests may differ slightly from the asymptotic critical values. Therefore they suggest an adjusted statistic

$$\lambda^*(r_0, K) = -(T - Kp) \sum_{i=r_0+1}^{K} \ln(1 - \lambda_i) \tag{11.4.6}$$

for testing (11.4.4). The power of likelihood ratio tests for cointegration is investigated e.g. by Reinsel & Ahn (1988) and Johansen (1989).

A number of other tests for cointegration have been proposed in the literature. Some of them are less general than Johansen's LR tests and others are based on quite different principles. For instance, Engle & Granger (1987) and Engle & Yoo (1987) propose a battery of tests that may be used for checking whether there is one cointegration relation between a set of variables. Some tests require checking for unit roots in individual univariate processes at some stage. Such tests for univariate time series have been pioneered by Fuller (1976, Chapter 8) and Dickey & Fuller (1979) (see also Dickey, Bell & Miller (1986)). A test for one unit root in the autoregressive polynomial of a VAR process is discussed by Fountis & Dickey (1989) and other tests for cointegration are proposed by Phillips & Ouliaris (1988, 1990) and Stock & Watson (1988).

These tests proceed from a given VAR order p. It is also possible to determine the VAR order and cointegration rank simultaneously using some model selection criterion. For this purpose the criteria must be modified along the lines of Section 5.3.5. The statistical properties of such a procedure are currently unknown in general.

11.4.3 Prediction Tests for Structural Change

In Chapter 4, Section 4.6, we have considered two tests for structural change that may be applied with small modifications if the data generation process is integrated or cointegrated. To see this consider a Gaussian K-dimensional VAR(p) process

$$y_t = \nu + A_1 y_{t-1} + \cdots + A_p y_{t-p} + u_t \tag{11.4.7}$$

with cointegration rank r. Denoting the optimal h-step predictor at origin T by $y_T(h)$ and its MSE matrix by $\Sigma_y(h)$ as in Section 11.3.1, the quantity

$$\tau_h = (y_{T+h} - y_T(h))'\Sigma_y(h)^{-1}(y_{T+h} - y_T(h)) \tag{11.4.8}$$

has a $\chi^2(K)$-distribution (see Sections 11.3.1 and 4.6.1). If the parameters of the process were known this statistic could be used to test whether y_{T+h} is generated by the Gaussian process given in (11.4.7).

In practice the process parameters have to be replaced by estimators and in Section 4.6.1 we have modified the forecast MSE matrix accordingly. In Section 11.3.1 we have seen that the MSE approximation used for stationary, stable processes is not appropriate in the present unstable case. Therefore we propose the statistic

$$\tau_h^{\#} = (y_{T+h} - \tilde{y}_T(h))'\tilde{\Sigma}_y(h)^{-1}(y_{T+h} - \tilde{y}_T(h))/K \tag{11.4.9}$$

which has an approximate $F(K, T - Kp - 1)$-distribution. Here

$$\tilde{y}_T(h) = \tilde{\nu} + \tilde{A}_1 \tilde{y}_T(h-1) + \cdots + \tilde{A}_p \tilde{y}_T(h-p)$$

with $\tilde{y}_T(j) := y_{T+j}$ for $j \leq 0$ and the $\tilde{A}_i$ are the ML estimators of the A_i. Moreover,

$$\tilde{\Sigma}_y(h) := \sum_{i=0}^{h-1} \tilde{\Phi}_i \tilde{\Sigma}_u \tilde{\Phi}_i',$$

where $\tilde{\Sigma}_u$ is the ML estimator of Σ_u (see Proposition 11.1) and the $\tilde{\Phi}_i$ are computed from the $\tilde{A}_i$ by the recursions in (11.3.5). The F-approximation to the distribution of $\tau_h^{\#}$ follows by noting that

$$\operatorname{plim}(\tau_h - K\tau_h^{\#}) = 0.$$

Hence, $K\tau_h^{\#}$ has an asymptotic $\chi^2(K)$-distribution and

$$\tau_h^{\#} \approx \chi^2(K)/K \approx F(K, T - Kp - 1), \tag{11.4.10}$$

where the numerator degrees of freedom are chosen in analogy with the stationary case. The quality of the F-approximation in small samples is presently unknown.

A test based on several forecasts as discussed in Section 4.6.2 may be generalized to unstable processes in a similar way. We may use

$$\bar{\lambda}_h = T \sum_{i=1}^{h} \tilde{u}'_{T+i} \tilde{\Sigma}_u^{-1} \tilde{u}_{T+i} / [(T + Kp + 1)Kh] \tag{11.4.11}$$

as a test statistic with an approximate $F(Kh, T - Kp - 1)$-distribution. Here the

$$\tilde{u}_{T+i} = y_{T+i} - \tilde{v} - \tilde{A}_1 y_{T+i-1} - \cdots - \tilde{A}_p y_{T+i-p}$$

are the ML residuals. In this case there is little theoretical justification for the term $(T + Kp + 1)$ and for the denominator degrees of freedom of the F-distribution. Nevertheless the approximate distribution follows from asymptotic theory as in the stationary, stable case (see Problem 11.9).

11.5 Exercises

11.5.1 Algebraic Exercises

Problem 11.1

(a) Determine the ML estimator for A_1 in the cointegrated VAR(1) process $y_t = A_1 y_{t-1} + u_t$ with cointegration rank 1. Note that there is no intercept term in the model.

(b) Determine the ML estimators of H, C, A_1, and v in the cointegrated VAR(1) process $y_t = v + A_1 y_{t-1} + u_t$ with cointegration rank 1.

Problem 11.2

Determine the ML estimators in a cointegrated VAR(p) process such as (11.2.3) with cointegration rank r, under the assumption that the cointegration matrix satisfies restrictions $C = RC_r$, where R and C_r are $(r \times s)$ and $(s \times K)$ matrices, respectively, with $r < s < K$. (Hint: Proceed as in the proof of Proposition 11.1.)

Problem 11.3

Prove Lemma 11.2.

(Hint: Show that $\Delta Y = D\Delta X - HCY_{-p} + U = [D, -H]\begin{bmatrix} \Delta X \\ CY_{-p} \end{bmatrix} + U$ and, hence,

$$[\tilde{D}(C), -\tilde{H}(C)] = \Delta Y[\Delta X', Y'_{-p}C'] \left[\begin{bmatrix} \Delta X \\ CY_{-p} \end{bmatrix} [\Delta X', Y'_{-p}C'] \right]^{-1}.$$

Use arguments similar to those in Section 3.2.2 to obtain the asymptotic normal distribution of this estimator and note that

$$[\tilde{D}(C), -\tilde{H}(C)C] = [\tilde{D}(C), -\tilde{H}(C)]\begin{bmatrix} I & 0 \\ 0 & C \end{bmatrix}.)$$

Problem 11.4

Show that, under the conditions of Subsection 11.2.1c,

$$\sqrt{T}(\tilde{H}\tilde{C} - \Pi)\frac{Y_{-p}Y'_{-p}}{T}(\tilde{H}\tilde{C} - \Pi)'$$

$$= \sqrt{T}[(\tilde{H}\tilde{C} - \tilde{H}(C)C) + (\tilde{H}(C) - H)C]\frac{Y_{-p}Y'_{-p}}{T}[(\tilde{H}\tilde{C} - \tilde{H}(C)C) + (\tilde{H}(C) - H)C]'$$

converges to zero in probability. (Hint: Use Lemmas 11.1 and 11.4.)

Problem 11.5

Show that the expressions in (11.2.25) and (11.2.26) are the LS estimators of H and D, respectively, conditional on $C = \hat{C}_{2s}$.

Problem 11.6

Consider a cointegrated VAR(1) process without intercept, $y_t = A_1 y_{t-1} + u_t$, and show that

$$\text{plim}[y_T(1) - \hat{y}_T(1)] = \text{plim}(A_1 - \tilde{A}_1)y_T = 0.$$

Assume that y_t is Gaussian with initial vector $y_0 = 0$ and the ML estimator $\tilde{A}_1$ is based on $y_1, \ldots, y_T$.
(Hint: Use the results from Subsection 11.2.1c and plim $y_T/T = 0$ from Phillips & Durlauf (1986).)

Problem 11.7

Consider the matrix $\Omega(h)$ used in the MSE correction in Section 3.5 and argue why its use for unstable processes is problematic. Analyze in particular the derivation in (3.5.12).

Problem 11.8

Consider a three-dimensional VAR(1) process with cointegration rank 1 and suppose the cointegrating matrix has the form $C = (c_1, c_2, 0)$. Use Corollary

11.2.1 to demonstrate that the elements in the last column of A_1 have zero asymptotic variance. Formulate a linear hypothesis on the coefficients of A_1 for which the rank condition (11.3.13) is likely to be violated if the covariance estimator of Corollary 11.2.1 is used.

Problem 11.9

Under the conditions of Section 11.4.3, show that

$$(T + Kp + 1)Kh\bar{\lambda}_h/T \xrightarrow{d} \chi^2(hK),$$

where $\bar{\lambda}_h$ is the statistic defined in (11.4.11).

11.5.2 Numerical Exercises

The following problems are based on the U.S. data given in Table E.3 of Appendix E as described in Subsection 11.2.1b. The variables are defined as in that subsection.

Problem 11.10

Assume that the data were generated by a VAR(3) process and determine the cointegration rank with the tests described in Section 11.4.2.

Problem 11.11

Modify the AIC criterion appropriately and choose the order and cointegration rank simultaneously with this criterion. Compare the result with that from Problem 11.10.

Problem 11.12

Apply the ML procedure described in Section 11.2.1 to estimate a VAR(3) process with cointegration rank $r = 1$ and intercept vector. Determine the estimates $\tilde{\nu}$, $\tilde{A}_1$, $\tilde{A}_2$, and $\tilde{A}_3$ and compare them to unrestricted LS estimates of a VAR(3) process.

Problem 11.13

Compute forecasts up to 10 periods ahead using both the unrestricted VAR(3) model and the VAR(3) model with cointegration rank 1. Compare the forecasts.

Problem 11.14

Compare the impulse responses obtained from an unrestricted and restricted VAR(3) model with cointegration rank 1.

Chapter 12. Periodic VAR Processes and Intervention Models

12.1 Introduction

In the previous chapter we have considered nonstationary VAR models with time invariant parameters. Nonstationarity, that is, time varying first and/or second moments of a process, can also be modeled in the framework of time varying parameter processes. Suppose, for instance, that the time series show a seasonal pattern. In that case a VAR(p) process with different intercept terms for each season may be a reasonable model:

$$y_t = \nu_i + A_1 y_{t-1} + \cdots + A_p y_{t-p} + u_t. \tag{12.1.1}$$

Here ν_i is a $(K \times 1)$ intercept vector associated with the i-th season, that is, in (12.1.1) the time index t is assumed to be associated with the i-th season of the year. It is easy to sec that such a process has a potentially different mean for each season of the year.

Assuming s seasons the model (12.1.1) could be written alternatively as

$$y_t = n_{1t}\nu_1 + \cdots + n_{st}\nu_s + A_1 y_{t-1} + \cdots + A_p y_{t-p} + u_t$$

where

$$n_{it} = 0 \text{ or } 1 \qquad \text{and} \qquad \sum_{i=1}^{s} n_{it} = 1. \tag{12.1.2}$$

In other words, n_{it} assumes the value of 1 if t belongs to the i-th season and is zero otherwise, that is, n_{it} is a *seasonal dummy variable.*

Of course, the model (12.1.1) is covered by the set-up of Chapter 10. In a seasonal context it is possible, however, that the other coefficients also vary for different seasons. In that case a more general model may be adequate:

$$y_t = \nu_t + A_{1t} y_{t-1} + \cdots + A_{pt} y_{t-p} + u_t \tag{12.1.3}$$

with

$$\begin{aligned} B_t &:= [\nu_t, A_{1t}, \ldots, A_{pt}] \\ &= n_{1t}[\nu_1, A_{1,1}, \ldots, A_{p,1}] + \cdots + n_{st}[\nu_s, A_{1,s}, \ldots, A_{p,s}] \\ &= n_{1t}B_1 + \cdots + n_{st}B_s \end{aligned} \tag{12.1.4a}$$

and

$$\Sigma_t := E(u_t u_t') = n_{1t}\Sigma_1 + \cdots + n_{st}\Sigma_s. \tag{12.1.4b}$$

Here the n_{it} are seasonal dummy variables as in (12.1.2), the B_i are $(K \times (Kp + 1))$ coefficient matrices and the Σ_i are $(K \times K)$ covariance matrices. The model (12.1.3) with periodically varying parameters as specified in (12.1.4) is a general *periodic VAR(p) model* with period s. Varying parameter models of this type will be discussed in Section 12.3.

The model (12.1.3) can also be used in a situation where a stationary, stable data generation process is in operation until period T_1, say, and then some outside intervention occurs after which another VAR(p) process generates the data. This case can be handled within the model class (12.1.3) by defining $s = 2$,

$$n_{1t} = \begin{cases} 1 & \text{for } t \leq T_1 \\ 0 & \text{for } t > T_1 \end{cases}$$

and

$$n_{2t} = \begin{cases} 1 & \text{for } t > T_1 \\ 0 & \text{for } t \leq T_1. \end{cases}$$

Intervention models of this type will be considered in Section 12.4. Interventions in economic systems may, for instance, be due to legislative activities or catastrophic weather conditions. Of course, there could be more than one intervention in the stretch of a time series. The general model (12.1.3) encompasses that situation when the dummy variables are chosen appropriately. In Section 12.2 some properties of the general model (12.1.3) will be given that can be derived without special assumptions regarding the movement of the parameters. These properties are valid for both the periodic and intervention models discussed in Section 12.3 and 12.4, respectively.

An important characteristic of periodic and intervention models is that only a finite number of parameter regimes exist that are associated with specific, known time periods. In other words, the parameter variations are systematic. Such a model structure is not realistic in all situations of practical interest. We will therefore discuss models with randomly varying coefficients in the next chapter.

12.2 The VAR(p) Model with Time Varying Coefficients

In this section we consider the following general form of a K-dimensional VAR(p) model with time varying coefficients:

$$y_t = \nu_t + A_{1t}y_{t-1} + \cdots + A_{pt}y_{t-p} + u_t, \tag{12.2.1}$$

where u_t is zero mean white noise with covariance matrices $E(u_t u_t') = \Sigma_t$. That is, the u_t may have time varying covariance matrices and thus may not be identically distributed. We retain the independence assumption for u_t and u_s, $s \neq t$. Of course, the constant coefficient VAR(p) model considered in previous chapters is a special case of (12.2.1). Further special cases are treated in the next sections. We will now discuss some properties of the general model.

12.2.1 General Properties

In order to derive general properties it is convenient to write the model (12.2.1) in VAR(1) form:

$$Y_t = \mathbf{v}_t + \mathbf{A}_t Y_{t-1} + U_t, \tag{12.2.2}$$

where

$$Y_t := \begin{bmatrix} y_t \\ \vdots \\ y_{t-p+1} \end{bmatrix}, \quad \mathbf{v}_t := \begin{bmatrix} v_t \\ 0 \\ \vdots \\ 0 \end{bmatrix}, \quad \mathbf{A}_t := \begin{bmatrix} A_{1,t} & \dots & A_{p-1,t} & A_{p,t} \\ I_K & & 0 & 0 \\ & \ddots & \vdots & \vdots \\ 0 & \dots & I_K & 0 \end{bmatrix}, \quad U_t := \begin{bmatrix} u_t \\ 0 \\ \vdots \\ 0 \end{bmatrix}.$$
$$(Kp \times 1) \qquad (Kp \times 1) \qquad (Kp \times Kp) \qquad (Kp \times 1)$$

By successive substitution we get

$$Y_t = \left[\prod_{j=0}^{h-1} \mathbf{A}_{t-j}\right] Y_{t-h} + \sum_{i=0}^{h-1} \left[\prod_{j=0}^{i-1} \mathbf{A}_{t-j}\right] \mathbf{v}_{t-i} + \sum_{i=0}^{h-1} \left[\prod_{j=0}^{i-1} \mathbf{A}_{t-j}\right] U_{t-i}. \tag{12.2.3}$$

Defining the $(K \times Kp)$ matrix $J := [I_K \quad 0]$ such that $y_t = J Y_t$ and premultiplying (12.2.3) by this matrix gives

$$y_t = J\left[\prod_{j=0}^{h-1} \mathbf{A}_{t-j}\right] Y_{t-h} + \sum_{i=0}^{h-1} \Phi_{it} v_t + \sum_{i=0}^{h-1} \Phi_{it} u_{t-i}, \tag{12.2.4}$$

where

$$\Phi_{it} = J\left[\prod_{j=0}^{i-1} \mathbf{A}_{t-j}\right] J' \tag{12.2.5}$$

and it has been used that $J'JU_t = U_t, JU_t = u_t$, and similar results hold for $\mathbf{v}_t$. If

$$\sum_{i=0}^{h-1} \Phi_{it} v_t$$

converges to a constant, say μ_t, for $h \to \infty$, and if the first term on the right-hand side of (12.2.4) converges to zero in mean square and the last term converges in mean square as $h \to \infty$, we get the representation

$$y_t = \mu_t + \sum_{i=0}^{\infty} \Phi_{it} u_{t-i}, \tag{12.2.6}$$

where $\mu_t = E(y_t)$. In the following it is assumed without further notice that this representation exists.

It can be used to derive the autocovariance structure of the process. For instance,

$$E[(y_t - \mu_t)(y_t - \mu_t)'] = E\left[\left[\sum_{j=0}^{\infty} \Phi_{jt} u_{t-j}\right]\left[\sum_{i=0}^{\infty} \Phi_{it} u_{t-i}\right]'\right]$$
$$= E\left[\sum_{j=0}^{\infty}\sum_{i=0}^{\infty} \Phi_{jt} u_{t-j} u'_{t-i} \Phi'_{it}\right] = \sum_{i=0}^{\infty} \Phi_{it} \Sigma_{t-i} \Phi'_{it}$$

and

$$E[(y_t - \mu_t)(y_{t-1} - \mu_{t-1})'] = E\left[\left[\sum_{j=0}^{\infty} \Phi_{jt} u_{t-j}\right]\left[\sum_{i=0}^{\infty} \Phi_{i,t-1} u_{t-1-i}\right]'\right]$$
$$= E\left[\sum_{j=-1}^{\infty}\sum_{i=0}^{\infty} \Phi_{j+1,t} u_{t-j-1} u'_{t-1-i} \Phi'_{i,t-1}\right]$$
$$= \sum_{i=0}^{\infty} \Phi_{i+1,t} \Sigma_{t-1-i} \Phi'_{i,t-1}.$$

More generally, for some integer h,

$$E[(y_t - \mu_t)(y_{t-h} - \mu_{t-h})'] = \sum_{i=0}^{\infty} \Phi_{i+h,t} \Sigma_{t-h-i} \Phi'_{i,t-h}.$$

Usually these formulas are not very useful for actually computing the autocovariances. They show, however, that the autocovariances generally depend on t and h so that the process y_t is not stationary. In addition, of course, the mean vectors μ_t may be time varying.

Optimal forecasts can be obtained either from (12.2.1) or from (12.2.6). In the former case the forecasts can be computed recursively as

$$y_t(h) = \nu_{t+h} + A_{1,t+h} y_t(h-1) + \cdots + A_{p,t+h} y_t(h-p), \tag{12.2.7}$$

where $y_t(j) := y_{t+j}$ for $j \le 0$. Using (12.2.6) gives

$$y_t(h) = \mu_{t+h} + \sum_{i=h}^{\infty} \Phi_{i,t+h} u_{t+h-i} \tag{12.2.8}$$

and the forecast error is

$$y_{t+h} - y_t(h) = \sum_{i=0}^{h-1} \Phi_{i,t+h} u_{t+h-i}. \tag{12.2.9}$$

Hence, the forecast MSE matrices turn out to be

$$\Sigma_t(h) := \mathrm{MSE}(y_t(h)) = \sum_{i=0}^{h-1} \Phi_{i,t+h} \Sigma_{t+h-i} \Phi'_{i,t+h}. \tag{12.2.10}$$

We will discuss some basics of ML estimation for the general model (12.2.1) next.

12.2.2 ML Estimation

Although specific results require specific assumptions it is useful to establish some general results related to ML estimation under Gaussian assumptions first. We write the model (12.2.1) as

$$y_t = B_t Z_{t-1} + u_t, \tag{12.2.11}$$

where $B_t := [\nu_t, A_{1t}, \ldots, A_{pt}]$, $Z_{t-1} := (1, Y'_{t-1})'$, and we assume that the $(K \times (Kp+1))$ matrices B_t depend on an $(N \times 1)$ vector $\boldsymbol{\gamma}$ of fixed time invariant parameters, that is, $B_t = B_t(\boldsymbol{\gamma})$. Furthermore, the Σ_t are assumed to depend on an $(M \times 1)$ vector $\boldsymbol{\sigma}$ of fixed parameters. The vector $\boldsymbol{\sigma}$ is disjoint of and unrelated with $\boldsymbol{\gamma}$. Examples where this situation arises will be seen in the next sections. One example, of course, is a constant coefficient model, where $B_t = B = [\nu, A_1, \ldots, A_p]$ and $\Sigma_t = \Sigma_u$ for all t and we may choose $\boldsymbol{\gamma} = \text{vec}(B)$ and $\boldsymbol{\sigma} = \text{vech}(\Sigma_u)$ if no further restrictions are imposed.

Assuming that u_t is Gaussian white noise, that is, $u_t \sim N(0, \Sigma_t)$, the log-likelihood function is

$$\ln l(\boldsymbol{\gamma}, \boldsymbol{\sigma}) = -\frac{KT}{2}\ln 2\pi - \frac{1}{2}\sum_{t=1}^{T} \ln|\Sigma_t| - \frac{1}{2}\sum_{t=1}^{T} u_t'\Sigma_t^{-1}u_t, \tag{12.2.12}$$

where any initial condition terms are ignored. The corresponding normal equations are

$$\begin{aligned} 0 = \frac{\partial \ln l}{\partial \boldsymbol{\gamma}} &= -\sum_{t=1}^{T} \frac{\partial u_t'}{\partial \boldsymbol{\gamma}} \Sigma_t^{-1} u_t = -\sum_{t=1}^{T} \frac{\partial (y_t - B_t Z_{t-1})'}{\partial \boldsymbol{\gamma}} \Sigma_t^{-1} u_t \\ &= \sum_{t=1}^{T} \frac{\partial\, \text{vec}(B_t)'}{\partial \boldsymbol{\gamma}} (Z_{t-1} \otimes I_K) \Sigma_t^{-1} u_t = \sum_t \frac{\partial\, \text{vec}(B_t)'}{\partial \boldsymbol{\gamma}} \Sigma_t^{-1} u_t Z_{t-1}' \end{aligned} \tag{12.2.13}$$

and

$$\begin{aligned} 0 = \frac{\partial \ln l}{\partial \boldsymbol{\sigma}} &= -\frac{1}{2}\sum_t \left[\frac{\partial\, \text{vec}(\Sigma_t)'}{\partial \boldsymbol{\sigma}} \frac{\partial \ln|\Sigma_t|}{\partial\, \text{vec}(\Sigma_t)}\right] - \frac{1}{2}\sum_t \left[\frac{\partial\, \text{vec}(\Sigma_t)'}{\partial \boldsymbol{\sigma}} \frac{\partial u_t'\Sigma_t^{-1}u_t}{\partial\, \text{vec}(\Sigma_t)}\right] \\ &= -\frac{1}{2}\sum_t \left[\frac{\partial\, \text{vec}(\Sigma_t)'}{\partial \boldsymbol{\sigma}} \text{vec}(\Sigma_t^{-1} - \Sigma_t^{-1}u_t u_t'\Sigma_t^{-1})\right]. \end{aligned} \tag{12.2.14}$$

Even if $\partial\, \text{vec}(B_t)'/\partial\boldsymbol{\gamma}$ is a matrix that does not depend on $\boldsymbol{\gamma}$, (12.2.13) is in general a system of equations which is nonlinear in $\boldsymbol{\gamma}$ and $\boldsymbol{\sigma}$ because $u_t = y_t - B_t Z_{t-1}$ involves $\boldsymbol{\gamma}$. However, we will see in the next sections that in many cases of interest (12.2.13) reduces to a linear system which is easy to solve. Also, a solution of (12.2.14) is easy to obtain under the conditions of the next sections.

It is furthermore possible to derive an expression for the information matrix associated with the general log-likelihood function (12.2.12). The second partial derivatives with respect to $\boldsymbol{\gamma}$ are

$$\begin{aligned} \frac{\partial^2 \ln l}{\partial\boldsymbol{\gamma}\partial\boldsymbol{\gamma}'} &= -\sum_t \left[\frac{\partial\, \text{vec}(B_t)'}{\partial \boldsymbol{\gamma}} (Z_{t-1} \otimes I_K)\Sigma_t^{-1}(Z_{t-1}' \otimes I_K)\frac{\partial\, \text{vec}(B_t)}{\partial \boldsymbol{\gamma}'}\right] \\ &\quad + \text{terms with mean zero} \\ &= -\sum_t \left[\frac{\partial\, \text{vec}(B_t)'}{\partial \boldsymbol{\gamma}} (Z_{t-1}Z_{t-1}' \otimes \Sigma_t^{-1})\frac{\partial\, \text{vec}(B_t)}{\partial \boldsymbol{\gamma}'}\right] \\ &\quad + \text{terms with mean zero.} \end{aligned} \tag{12.2.15}$$

Assuming that $\partial \operatorname{vec}(\Sigma_t)'/\partial \boldsymbol{\sigma}$ does not depend on $\boldsymbol{\sigma}$ and thus the second partial derivatives of Σ_t with respect to the elements of $\boldsymbol{\sigma}$ are zero, we get

$$\begin{aligned}\frac{\partial^2 \ln l}{\partial \boldsymbol{\sigma} \partial \boldsymbol{\sigma}'} &= -\frac{1}{2}\sum_t \left(\frac{\partial \operatorname{vec}(\Sigma_t)'}{\partial \boldsymbol{\sigma}}\left[\frac{\partial \operatorname{vec}(\Sigma_t^{-1})}{\partial \boldsymbol{\sigma}'} - (I_K \otimes \Sigma_t^{-1} u_t u_t')\frac{\partial \operatorname{vec}(\Sigma_t^{-1})}{\partial \boldsymbol{\sigma}'}\right.\right.\\ &\qquad \left.\left. - (\Sigma_t^{-1} u_t u_t' \otimes I_K)\frac{\partial \operatorname{vec}(\Sigma_t^{-1})}{\partial \boldsymbol{\sigma}'}\right]\right)\\ &= \frac{1}{2}\sum_t \left[\frac{\partial \operatorname{vec}(\Sigma_t)'}{\partial \boldsymbol{\sigma}}(\Sigma_t^{-1} \otimes \Sigma_t^{-1} - \Sigma_t^{-1} \otimes \Sigma_t^{-1} u_t u_t' \Sigma_t^{-1}\right.\\ &\qquad \left. - \Sigma_t^{-1} u_t u_t' \Sigma_t^{-1} \otimes \Sigma_t^{-1})\frac{\partial \operatorname{vec}(\Sigma_t)}{\partial \boldsymbol{\sigma}'}\right]. \end{aligned} \tag{12.2.16}$$

The assumption of zero second partial derivatives of Σ_t with respect to $\boldsymbol{\sigma}$ will be satisfied in all cases of interest in the following sections. Furthermore, it is easy to see that under the present assumptions

$$E[\partial^2 \ln l/\partial \boldsymbol{\gamma} \partial \boldsymbol{\sigma}'] = 0.$$

Consequently, the information matrix becomes

$$\begin{aligned} I(\boldsymbol{\gamma}, \boldsymbol{\sigma}) &= E\left[\frac{\partial^2 - \ln l}{\partial \begin{bmatrix} \boldsymbol{\gamma} \\ \boldsymbol{\sigma} \end{bmatrix} \partial(\boldsymbol{\gamma}', \boldsymbol{\sigma}')}\right] = -E\left[\frac{\partial^2 \ln l}{\partial \begin{bmatrix} \boldsymbol{\gamma} \\ \boldsymbol{\sigma} \end{bmatrix} \partial(\boldsymbol{\gamma}', \boldsymbol{\sigma}')}\right] \\ &= \begin{bmatrix} \sum_t \left[\frac{\partial \operatorname{vec}(B_t)'}{\partial \boldsymbol{\gamma}}[E(Z_{t-1} Z_{t-1}) \otimes \Sigma_t^{-1}]\frac{\partial \operatorname{vec}(B_t)}{\partial \boldsymbol{\gamma}'}\right] & 0 \\ 0 & \frac{1}{2}\sum_t \left[\frac{\partial \operatorname{vec}(\Sigma_t)'}{\partial \boldsymbol{\sigma}}(\Sigma_t^{-1} \otimes \Sigma_t^{-1})\frac{\partial \operatorname{vec}(\Sigma_t)}{\partial \boldsymbol{\sigma}'}\right] \end{bmatrix}. \end{aligned} \tag{12.2.17}$$

Although these expressions look a bit unwieldy in their present general form, they are quite handy if special assumptions regarding the time variation of the parameters are made. We will now turn to such special types of time varying coefficient VAR models.

12.3 Periodic Processes

As we have seen in the introduction, in periodic VAR processes the coefficients vary periodically with period s, say. In other words,

$$y_t = \nu_t + A_t Y_{t-1} + u_t, \tag{12.3.1}$$

where

$$\begin{aligned} \nu_t &= n_{1t}\nu_1 + \cdots + n_{st}\nu_s, && (K \times 1), \\ A_t &= [A_{1t}, \ldots, A_{pt}] = n_{1t}A_1 + \cdots + n_{st}A_s, && (K \times Kp), \\ \Sigma_t &= E(u_t u_t') = n_{1t}\Sigma_1 + \cdots + n_{st}\Sigma_s, && (K \times K), \end{aligned} \tag{12.3.2}$$

and the n_{it} are seasonal dummy variables assuming a value of one if t is associated with the i-th season and zero otherwise. Obviously the general framework of the previous section encompasses this model. Hence, some properties can be obtained by substituting the expressions from (12.3.2) in the general formulas of the previous section. For periodic processes, however, many properties are more easily derived via another approach which will be introduced and exploited in the next subsection.

Special models arise if only a subset of the parameters vary periodically. For instance, if $\Sigma_i = \Sigma_1$ and $A_i = A_1$ for $i = 1, \dots, s$, we have a model with seasonal means and otherwise time invariant structure. Simplifications of this kind are useful in practice because they imply a reduction in the number of free parameters to be estimated and thereby result in more efficient estimates and forecasts at least in large samples. A special case of foremost interest is, of course, a nonperiodic, constant coefficient VAR model. If the data generation process turns out to be of that type the interpretation and analysis is greatly simplified. We will consider estimation and tests of various sets of relevant hypotheses in Subsection 12.3.2.

12.3.1 A VAR Representation with Time Invariant Coefficients

Suppose we have a quarterly process with period $s = 4$ and y_1 belongs to the first quarter. Then we may define an annual process with vectors

$$\mathfrak{y}_1 := \begin{bmatrix} y_4 \\ y_3 \\ y_2 \\ y_1 \end{bmatrix}, \quad \mathfrak{y}_2 := \begin{bmatrix} y_8 \\ y_7 \\ y_6 \\ y_5 \end{bmatrix}, \dots, \mathfrak{y}_\tau := \begin{bmatrix} y_{4\tau} \\ y_{4\tau-1} \\ y_{4\tau-2} \\ y_{4\tau-3} \end{bmatrix}, \dots.$$

This process has a representation with time invariant parameter matrices. For instance, if the process for each quarter is a VAR(1),

$$\begin{aligned} y_t &= \nu_t + A_{1,t} y_{t-1} + u_t \\ &= \nu_i + A_{1,i} y_{t-1} + u_t, \qquad \text{if } t \text{ belongs to the } i\text{-th quarter,} \end{aligned}$$

then the process $\mathfrak{y}_\tau$ has the representation

$$\begin{bmatrix} I_K & -A_{1,4} & 0 & 0 \\ 0 & I_K & -A_{1,3} & 0 \\ 0 & 0 & I_K & -A_{1,2} \\ 0 & 0 & 0 & I_K \end{bmatrix} \begin{bmatrix} y_{4\tau} \\ y_{4\tau-1} \\ y_{4\tau-2} \\ y_{4\tau-3} \end{bmatrix}$$

$$= \begin{bmatrix} \nu_1 \\ \nu_2 \\ \nu_3 \\ \nu_4 \end{bmatrix} + \begin{bmatrix} 0 & 0 & 0 & 0 \\ 0 & 0 & 0 & 0 \\ 0 & 0 & 0 & 0 \\ A_{1,1} & 0 & 0 & 0 \end{bmatrix} \begin{bmatrix} y_{4\tau-4} \\ y_{4\tau-5} \\ y_{4\tau-6} \\ y_{4\tau-7} \end{bmatrix} + \begin{bmatrix} u_{4\tau} \\ u_{4\tau-1} \\ u_{4\tau-2} \\ u_{4\tau-3} \end{bmatrix}. \tag{12.3.3}$$

More generally, if we have s different regimes (seasons per year) with constant parameters within each regime and if we assume that y_1 belongs to the first season, we may define the sK-dimensional process

$$\mathfrak{y}_\tau := \underset{(sK \times 1)}{\begin{bmatrix} y_{s\tau} \\ y_{s\tau-1} \\ \vdots \\ y_{s\tau-s+1} \end{bmatrix}}, \qquad \tau = 0, \pm 1, \pm 2, \ldots.$$

This process has the following VAR(p) representation, where P is the smallest integer greater than or equal to p/s:

$$\mathfrak{A}_0 \mathfrak{y}_\tau = \mathbf{v} + \mathfrak{A}_1 \mathfrak{y}_{\tau-1} + \cdots + \mathfrak{A}_P \mathfrak{y}_{\tau-P} + \mathfrak{u}_\tau, \tag{12.3.4}$$

where

$$\mathfrak{A}_0 := \underset{(sK \times sK)}{\begin{bmatrix} I_K & -A_{1,s} & -A_{2,s} & \cdots & -A_{s-1,s} \\ 0 & I_K & -A_{1,s-1} & \cdots & -A_{s-2,s-1} \\ \vdots & & \ddots & & \vdots \\ 0 & 0 & 0 & \cdots & I_K \end{bmatrix}}, \quad \mathbf{v} := \underset{(sK \times 1)}{\begin{bmatrix} v_1 \\ v_2 \\ \vdots \\ v_s \end{bmatrix}},$$

$$\mathfrak{A}_i := \underset{(sK \times sK)}{\begin{bmatrix} A_{is,s} & A_{is+1,s} & \cdots & A_{(i+1)s-1,s} \\ A_{is-1,s-1} & A_{is,s-1} & \cdots & A_{(i+1)s-2,s-1} \\ \vdots & \vdots & & \vdots \\ A_{is-s+1,1} & A_{is-s+2,1} & \cdots & A_{is,1} \end{bmatrix}}, \qquad i = 1, \ldots, P,$$

$$\mathfrak{u}_\tau := \begin{bmatrix} u_{s\tau} \\ u_{s\tau-1} \\ \vdots \\ u_{s\tau-s+1} \end{bmatrix}.$$

All $A_{i,j}$ with $i > p$ are zero.

The process $\mathfrak{y}_\tau$ is stationary if the y_t have bounded first and second moments and the VAR operator is stable, that is,

$$\det(\mathfrak{A}_0 - \mathfrak{A}_1 z - \cdots - \mathfrak{A}_P z^P) = \det(I_{sK} - \mathfrak{A}_0^{-1}\mathfrak{A}_1 z - \cdots - \mathfrak{A}_0^{-1}\mathfrak{A}_P z^P) \neq 0 \quad \text{for } |z| \leq 1. \tag{12.3.5}$$

Note that $|\mathfrak{A}_0| = 1$.

For the example process (12.3.3) we have

$$\mathfrak{A}_0^{-1} = \begin{bmatrix} I_K & A_{1,4} & A_{1,4}A_{1,3} & A_{1,4}A_{1,3}A_{1,2} \\ 0 & I_K & A_{1,3} & A_{1,3}A_{1,2} \\ 0 & 0 & I_K & A_{1,2} \\ 0 & 0 & 0 & I_K \end{bmatrix}$$

and, thus,

$$\mathfrak{A}_0^{-1}\mathfrak{A}_1 = \begin{bmatrix} A_{1,4}A_{1,3}A_{1,2}A_{1,1} & 0 & 0 & 0 \\ A_{1,3}A_{1,2}A_{1,1} & 0 & 0 & 0 \\ A_{1,2}A_{1,1} & 0 & 0 & 0 \\ A_{1,1} & 0 & 0 & 0 \end{bmatrix}.$$

Hence,

$$\det(\mathfrak{A}_0 - \mathfrak{A}_1 z) = \det(I_K - A_{1,4}A_{1,3}A_{1,2}A_{1,1}z) \neq 0 \qquad \text{for } |z| \leq 1$$

is the stability condition for the example process. If this condition is satisfied we can, for instance, compute the autocovariances of the process $\mathfrak{y}_\tau$ in the usual way. Note, however, that stationarity of $\mathfrak{y}_\tau$ does not imply stationarity of the original process y_t. Even if $\mathfrak{y}_\tau$ has a time invariant mean vector

$$\boldsymbol{\mu} = \begin{bmatrix} \mu_4 \\ \mu_3 \\ \mu_2 \\ \mu_1 \end{bmatrix}$$

the mean vectors μ_4 and μ_3 associated with the fourth and third quarters, respectively, may be different. Similar thoughts apply for other quarters and for the autocovariances associated with different quarters.

The process $\mathfrak{y}_\tau$ corresponding to a periodic process y_t can also be used to determine an upper bound for the order p of the latter. If $\mathfrak{y}_\tau$ is stationary and its order P is selected in the usual way, we know that $p \leq sP$.

Optimal forecasts of a periodic process are easily obtained from the recursions (12.2.7). Assuming that the forecast origin t is associated with the last period of the year, we get

$$\begin{aligned} y_t(1) &= \nu_1 + A_{1,1}y_t + \cdots + A_{p,1}y_{t-p+1} \\ y_t(2) &= \nu_2 + A_{1,2}y_t(1) + \cdots + A_{p,2}y_{t-p+2} \\ &\vdots \\ y_t(s) &= \nu_s + A_{1,s}y_t(s-1) + \cdots + A_{p,s}y_t(s-p) \\ y_t(s+1) &= \nu_1 + A_{1,1}y_t(s) + \cdots + A_{p,1}y_t(s+1-p) \\ &\vdots \end{aligned}$$

12.3.2 ML Estimation and Testing for Varying Parameters

The general framework for ML estimation of the periodic VAR(p) model given in (12.3.1)/(12.3.2), under Gaussian assumptions, is laid out in Section 12.2.2. For the present case, however, a number of simplifications result and closed form expressions can be given for the estimators. In the following we discuss estimation under various types of restrictions and we consider tests of time invariance of different groups of parameters. Most of the tests are likelihood ratio (LR) tests the general form of which is discussed in Appendix C, Section C.5. Recall that the LR statistic is

$$\lambda_{LR} = 2[\ln l(\tilde{\boldsymbol{\delta}}) - \ln l(\tilde{\boldsymbol{\delta}}_r)], \tag{12.3.6}$$

where $\tilde{\boldsymbol{\delta}}$ is the unconstrained ML estimator and $\tilde{\boldsymbol{\delta}}_r$ is the restricted ML estimator obtained by maximizing the likelihood function under the null hypothesis H_0. If H_0 is true, under general conditions, the LR statistic has an asymptotic χ^2-distribution with degrees of freedom equal to the number of linearly independent restrictions. In the following we will give the maximum of the likelihood function under various sets of restrictions in addition to the ML estimators. This will enable us to set up LR tests for different sets of restrictions.

For the present case of a periodic VAR model the normal equations given in (12.2.13) reduce to

$$\begin{aligned} 0 = \frac{\partial \ln l}{\partial \gamma} &= \sum_{t=1}^{T} \frac{\partial \operatorname{vec}(n_{1t}B_1 + \cdots + n_{st}B_s)'}{\partial \gamma} \Sigma_t^{-1} u_t Z_{t-1}' \\ &= \sum_{i=1}^{s} \sum_{t=1}^{T} n_{it} \frac{\partial \operatorname{vec}(B_i)'}{\partial \gamma} \Sigma_i^{-1} u_t Z_{t-1}', \end{aligned} \tag{12.3.7}$$

where $B_i = [\nu_i, A_i]$, $i = 1, \dots, s$, and

$$\Sigma_t^{-1} = \left[\sum_{i=1}^{s} n_{it}\Sigma_i\right]^{-1} = \sum_i n_{it}\Sigma_i^{-1}$$

has been used. Moreover, (12.2.14) becomes

$$0 = \frac{\partial \ln l}{\partial \boldsymbol{\sigma}} = -\frac{1}{2}\sum_{i=1}^{s}\sum_{t=1}^{T} n_{it}\left[\frac{\partial \operatorname{vec}(\Sigma_i)'}{\partial \boldsymbol{\sigma}} \operatorname{vec}(\Sigma_i^{-1} - \Sigma_i^{-1}u_t u_t' \Sigma_i^{-1})\right]. \tag{12.3.8}$$

We will see in the following that the solution of these sets of normal equations is relatively easy in many situations. The discussion follows Lütkepohl (1992).

12.3.2a All Coefficients Time Varying

We begin with a periodic VAR(p) model for which all parameters are time varying, that is,

$$H_1: \quad B_t = [\nu_t, A_t] = \sum_{i=1}^{s} n_{it}B_i, \qquad \Sigma_t = \sum_{i=1}^{s} n_{it}\Sigma_i. \tag{12.3.9}$$

For this case

$$\gamma = \operatorname{vec}[B_1, \dots, B_s]$$

and

$$\boldsymbol{\sigma} = [\operatorname{vech}(\Sigma_1)', \dots, \operatorname{vech}(\Sigma_s)']'.$$

Using a little algebra the ML estimators can be obtained from (12.3.7) and (12.3.8):

$$\tilde{B}_i^{(1)} = \left[\sum_{t=1}^{T} n_{it} y_t Z'_{t-1}\right]\left[\sum_{t=1}^{T} n_{it} Z_{t-1} Z'_{t-1}\right]^{-1} \tag{12.3.10}$$

and

$$\tilde{\Sigma}_i^{(1)} = \sum_t n_{it}(y_t - \tilde{B}_i^{(1)} Z_{t-1})(y_t - \tilde{B}_i^{(1)} Z_{t-1})'/T\bar{n}_i \tag{12.3.11}$$

for $i = 1, \ldots, s$. Here $\bar{n}_i = \sum_{t=1}^{T} n_{it}/T$. Except for an additive constant the corresponding maximum of the log-likelihood is

$$\lambda_1 := -\frac{1}{2}\sum_t \ln|\tilde{\Sigma}_t^{(1)}| = -\frac{1}{2}T(\bar{n}_1 \ln|\tilde{\Sigma}_1^{(1)}| + \cdots + \bar{n}_s \ln|\tilde{\Sigma}_s^{(1)}|). \tag{12.3.12}$$

12.3.2b All Coefficients Time Invariant

The next case we consider is our well-known basic stationary VAR(p) model where all the coefficients are time invariant:

$$H_2: \quad B_i = B_1, \qquad \Sigma_i = \Sigma_1, \qquad i = 2, \ldots, s. \tag{12.3.13}$$

For this case we know that the ML estimators are

$$\tilde{B}_1^{(2)} = \left(\sum_t y_t Z'_{t-1}\right)\left(\sum_t Z_{t-1} Z'_{t-1}\right)^{-1} \tag{12.3.14}$$

and

$$\tilde{\Sigma}_1^{(2)} = \sum_t (y_t - \tilde{B}_1^{(2)} Z_{t-1})(y_t - \tilde{B}_1^{(2)} Z_{t-1})'/T. \tag{12.3.15}$$

The maximum log-likelihood is, except for an additive constant,

$$\lambda_2 := -\tfrac{1}{2}T \ln|\tilde{\Sigma}_1^{(2)}|. \tag{12.3.16}$$

This case is considered here because H_2 is a null hypothesis of foremost interest in the present context. Of course, if it turns out that H_2 is true we can proceed with a standard, stationary analysis. The slight change of notation relative to previous chapters is useful here to avoid confusion.

12.3.2c Time Invariant White Noise

If just the white noise covariance matrix is time invariant while the other coefficients vary, we have

$$H_3: \quad B_t = [\nu_t, A_t] = \sum_{i=1}^{s} n_{it} B_i \qquad \text{and} \qquad \Sigma_i = \Sigma_1, \quad i = 2, \ldots, s. \tag{12.3.17}$$

For this case it follows from (12.3.7) that the ML estimators of the B_i are

$$\tilde{B}_i^{(3)} = \tilde{B}_i^{(1)}, \qquad i = 1, \ldots, s, \tag{12.3.18}$$

and (12.3.8) implies

$$\tilde{\Sigma}_1^{(3)} = \sum_{i=1}^{s} \sum_{t=1}^{T} n_{it}(y_t - \tilde{B}_i^{(1)} Z_{t-1})(y_t - \tilde{B}_i^{(1)} Z_{t-1})'/T. \tag{12.3.19}$$

The resulting maximum log-likelihood turns out to be

$$\lambda_3 := -\tfrac{1}{2} T \ln |\tilde{\Sigma}_1^{(3)}|, \tag{12.3.20}$$

where again an additive constant is suppressed.

12.3.2d Time Invariant Covariance Structure

If just the intercept terms and hence the means are time varying we have the conventional case of a model with seasonal dummies and otherwise constant coefficients. In the present framework this situation may be represented as

$$H_4: \; v_t = \sum_{i=1}^{s} n_{it} v_i \quad \text{and} \quad A_i = A_1, \;\; \Sigma_i = \Sigma_1, \;\; i = 2, \ldots, s. \tag{12.3.21}$$

Under this hypothesis the ML estimators are easily obtained by defining

$$W_t = \begin{bmatrix} n_{1,t+1} \\ \vdots \\ n_{s,t+1} \\ Y_t \end{bmatrix} \quad \text{and} \quad C = [v_1, \ldots, v_s, A_1].$$

The ML estimator of C is

$$\tilde{C} = \left[\sum_t y_t W_{t-1}'\right]\left[\sum_t W_{t-1} W_{t-1}'\right]^{-1} \tag{12.3.22}$$

and that of Σ_1 is

$$\tilde{\Sigma}_1^{(4)} = \sum_t (y_t - \tilde{C} W_{t-1})(y_t - \tilde{C} W_{t-1})'/T. \tag{12.3.23}$$

Dropping again an additive constant the corresponding maximum of the log-likelihood is

$$\lambda_4 := -\tfrac{1}{2} T \ln |\tilde{\Sigma}_1^{(4)}|. \tag{12.3.24}$$

12.3.2e LR Tests

In Table 12.1 the LR tests of some hypotheses of interest are listed. The LR statistics, under general conditions, all have asymptotic χ^2-distributions with the

Table 12.1. LR tests for time varying parameters

null hypothesis	alternative hypothesis	LR statistic λ_{LR}	degrees of freedom
H_2	H_1	$2(\lambda_1 - \lambda_2)$	$(s-1)K(K(p+\frac{1}{2})+\frac{3}{2})$
H_3	H_1	$2(\lambda_1 - \lambda_3)$	$(s-1)K(K+1)/2$
H_4	H_1	$2(\lambda_1 - \lambda_4)$	$(s-1)K(Kp+(K+1)/2)$
H_2	H_3	$2(\lambda_3 - \lambda_2)$	$(s-1)K(Kp+1)$
H_2	H_4	$2(\lambda_4 - \lambda_2)$	$(s-1)K$

given degrees of freedom. For this result to hold it is important that the $\bar{n}_i$ are approximately equal for $i = 1, \ldots, s$, as assumed in periodic models. The reader is invited to check the degrees of freedom listed in Table 12.1 by counting the number of restrictions imposed under the null hypothesis.

12.3.2f *Testing a Model with Time Varying White Noise only Against one with all Coefficients Time Varying*

ML estimation of models for which all coefficients are time invariant except for the white noise covariance matrix is complicated by the implied nonlinearity of the normal equations (12.3.7). Thus, LR tests involving the hypothesis

$$H_5\colon \quad B_i = B_1, \qquad i = 2, \ldots, s \qquad \text{and} \qquad \Sigma_t = \sum_{i=1}^{s} n_{it}\Sigma_i \tag{12.3.25}$$

are computationally unattractive. If we wish to test H_5 against a model for which all parameters are time varying (H_5 against H_1), estimation under the alternative is straightforward and therefore a Wald test may be considered.

Just as a reminder, if the unrestricted estimator $\tilde{\gamma}$ of a parameter vector γ has an asymptotic normal distribution,

$$\sqrt{T}(\tilde{\gamma} - \gamma) \xrightarrow{d} N(0, \Sigma_{\tilde{\gamma}}),$$

and the restrictions under H_0 are given in the form $R\gamma = 0$, then the Wald statistic is of the form

$$\lambda_w = T\tilde{\gamma}'R'[R\tilde{\Sigma}_{\tilde{\gamma}}R']^{-1}R\tilde{\gamma}, \tag{12.3.26}$$

where $\tilde{\Sigma}_{\tilde{\gamma}}$ is a consistent estimator of $\Sigma_{\tilde{\gamma}}$. If $\text{rk}(R) = N$, $R\Sigma_{\tilde{\gamma}}R'$ is invertible and H_0 is true, the Wald statistic has an asymptotic $\chi^2(N)$-distribution (see Appendix C, Section C.5).

In the case of interest here the restrictions relate to the VAR coefficients and intercept terms only. Therefore we consider the $s(K^2p + K)$-dimensional vector $\gamma = \text{vec}[B_1, \ldots, B_s]$. The restrictions under $H_0 = H_5$ can be written as $R\gamma = 0$ with

$$R = \underset{((s-1)\times s)}{\begin{bmatrix} 1 & -1 & & 0 \\ \vdots & & \ddots & \\ 1 & 0 & & -1 \end{bmatrix}} \otimes I_{K^2p+K}. \tag{12.3.27}$$

Denoting the unrestricted ML estimator of γ by $\tilde{\gamma}$, standard asymptotic theory implies that it has an asymptotic normal distribution with

$$\Sigma_{\tilde{\gamma}} = \lim_{T\to\infty} (I(\gamma)/T)^{-1}$$

and

$$I(\gamma) = -E\left[\frac{\partial^2 \ln l}{\partial\gamma\partial\gamma'}\right]$$

is the upper left-hand block of the information matrix (12.2.17). For the present case $I(\gamma)$ is seen to be block diagonal with the i-th $((K^2p + K) \times (K^2p + K))$ block on the diagonal being

$$-E\left[\frac{\partial^2 \ln l}{\partial\gamma_i\partial\gamma_i'}\right] = E\left[\sum_t n_{it} Z_{t-1} Z_{t-1}'\right] \otimes \Sigma_i^{-1},$$

where $\gamma_i = \text{vec}(B_i)$. Thus $\Sigma_{\tilde{\gamma}}$ is also block-diagonal and, under standard assumptions, the i-th block is consistently estimated by

$$\left[\frac{1}{T}\sum_t n_{it} Z_{t-1} Z_{t-1}'\right]^{-1} \otimes \tilde{\Sigma}_i^{(1)}.$$

The resulting estimator $\tilde{\Sigma}_{\tilde{\gamma}}$ of $\Sigma_{\tilde{\gamma}}$ may be used in (12.3.26). If H_5 is true, λ_w has an asymptotic χ^2-distribution with $(s - 1)K(Kp + 1)$ degrees of freedom. A disadvantage of the test is that the computation of λ_w requires the inversion of a rather large matrix.

12.3.2g Testing a Time Invariant Model Against one with Time Varying White Noise

In order to test a stationary constant parameter model ($H_2 = H_0$) against one where the white noise covariances vary (H_5) an LM (Lagrange multiplier) test is convenient because it requires ML estimation under the null hypothesis only. In Appendix C, Section C.5, the general form of the LM statistic is given as

$$\lambda_{LM} = s(\tilde{\gamma}_r, \tilde{\sigma}_r)' I(\tilde{\gamma}_r, \tilde{\sigma}_r)^{-1} s(\tilde{\gamma}_r, \tilde{\sigma}_r), \tag{12.3.28}$$

where $I(\tilde{\gamma}_r, \tilde{\sigma}_r)$ is the information matrix of the unrestricted model evaluated at the restricted ML estimators obtained under the null hypothesis and

$$s(\boldsymbol{\gamma}, \boldsymbol{\sigma}) = \begin{bmatrix} \dfrac{\partial \ln l}{\partial \boldsymbol{\gamma}} \\ \dfrac{\partial \ln l}{\partial \boldsymbol{\sigma}} \end{bmatrix}$$

is the score vector of first partial derivatives of the log-likelihood function. In the present case $\boldsymbol{\gamma} = \text{vec}(B_1)$ is left unrestricted. Thus, $\tilde{\boldsymbol{\gamma}}_r = \tilde{\boldsymbol{\gamma}}$ and

$$\left[\frac{\partial \ln l}{\partial \boldsymbol{\gamma}}\bigg|_{\tilde{\boldsymbol{\gamma}}_r}\right] = 0.$$

Consequently, defining $\boldsymbol{\sigma} = (\text{vech}(\Sigma_1)', \ldots, \text{vech}(\Sigma_s)')'$ the LM statistic reduces to

$$\lambda_{\text{LM}} = -\frac{\partial \ln l}{\partial \boldsymbol{\sigma}'} E\left[\frac{\partial^2 \ln l}{\partial \boldsymbol{\sigma} \partial \boldsymbol{\sigma}'}\right]^{-1} \frac{\partial \ln l}{\partial \boldsymbol{\sigma}} \tag{12.3.29}$$

with all derivatives evaluated at the restricted estimator

$$\tilde{\boldsymbol{\sigma}}_r = \tilde{\boldsymbol{\sigma}}^{(2)} := \begin{bmatrix} \text{vech}(\tilde{\Sigma}_1^{(2)}) \\ \vdots \\ \text{vech}(\tilde{\Sigma}_1^{(2)}) \end{bmatrix} \qquad (sK(K+1)/2 \times 1).$$

From (12.3.8) we see that

$$\frac{\partial \ln l}{\partial \, \text{vech}(\Sigma_i)} = -\frac{1}{2} \sum_{t=1}^{T} n_{it} \mathbf{D}_K' \, \text{vec}(\Sigma_i^{-1} - \Sigma_i^{-1} u_t u_t' \Sigma_i^{-1}), \tag{12.3.30}$$

where $\mathbf{D}_K = \partial \, \text{vec}(\Sigma_i)/\partial \, \text{vech}(\Sigma_i)'$ is the $(K^2 \times K(K+1)/2)$ duplication matrix as before. Furthermore, for the present case,

$$\frac{\partial \, \text{vec}(\Sigma_t)'}{\partial \boldsymbol{\sigma}} = \begin{bmatrix} n_{1t} \mathbf{D}_K' \\ \vdots \\ n_{st} \mathbf{D}_K' \end{bmatrix} \qquad (sK(K+1)/2 \times K^2).$$

Thus, we get from (12.2.17),

$$-E\left[\frac{\partial^2 \ln l}{\partial \boldsymbol{\sigma} \partial \boldsymbol{\sigma}'}\right] = \begin{bmatrix} \frac{1}{2} T \bar{n}_1 \mathbf{D}_K'(\Sigma_1^{-1} \otimes \Sigma_1^{-1}) \mathbf{D}_K & & 0 \\ & \ddots & \\ 0 & & \frac{1}{2} T \bar{n}_s \mathbf{D}_K'(\Sigma_s^{-1} \otimes \Sigma_s^{-1}) \mathbf{D}_K \end{bmatrix}$$

which implies

$$-E\left[\frac{\partial^2 \ln l}{\partial \boldsymbol{\sigma} \partial \boldsymbol{\sigma}'}\right]^{-1} = \begin{bmatrix} 2\mathbf{D}_K^+(\Sigma_1 \otimes \Sigma_1)\mathbf{D}_K^{+\prime}/T\bar{n}_1 & & 0 \\ & \ddots & \\ 0 & & 2\mathbf{D}_K^+(\Sigma_s \otimes \Sigma_s)\mathbf{D}_K^{+\prime}/T\bar{n}_s \end{bmatrix}, \tag{12.3.31}$$

where $\mathbf{D}_K^+$ is the Moore-Penrose inverse of $\mathbf{D}_K$. Using (12.3.30) and (12.3.31) with u_t replaced by $\tilde{u}_t = y_t - \tilde{B}_1^{(2)} Z_{t-1}$ and Σ_i replaced by $\tilde{\Sigma}_1^{(2)}$ the LM statistic in (12.3.29) is easy to evaluate. Under $H_0 = H_2$ and general conditions it has an asymptotic χ^2-distribution with $(s-1)K(K+1)/2$ degrees of freedom.

12.3.3 An Example

The previously considered theoretical concepts shall now be illustrated by an example from Lütkepohl (1992). We use the first differences of logarithms of quarterly, unadjusted West German income and consumption data for the years 1960–1987 given in Table E.4 of Appendix E. The two series are plotted in Figure 12.1. Obviously they exhibit a quite strong seasonal pattern.

There are various problems that may be brought up with respect to the data. For instance, it is possible that the logarithms of the original series are cointegrated (see Chapter 11). In that case fitting a VAR process to the first differences may be inappropriate. Also, there may be structural shifts during the sample period. We ignore such problems here because we just want to provide an illustrative example for the theoretical results of the previous subsections.

Since we have quarterly data the period $s = 4$ is given naturally. Stacking the variables for each year in one long 8-dimensional vector $\mathfrak{y}_\tau$ as in Section 12.3.1 we just have 27 observations for each component of $\mathfrak{y}_\tau$. (Note that the first value of the series is lost by differencing.) Thus the largest full VAR process that can be fitted to the 8-dimensional system is a VAR(3). In such a situation application of model selection criteria is a doubtful strategy for choosing the order of $\mathfrak{y}_\tau$. Since we want to test the null hypothesis of constant coefficients it may be reasonable to choose the VAR order under H_0, that is, to assume a constant coefficient model at the VAR order selection stage. Therefore we have fitted constant coefficient VAR models to the bivariate y_t series consisting of the quarterly income and consumption variables. FPE, AIC, HQ, and SC all have chosen the order $p = 5$ when a maximum of 8 was allowed. Of course, this may not mean too much if the parameters are actually time varying. The order 5 seems to be a reasonable choice, however, because it includes lags from a whole year and the corresponding quarter of the previous year. Therefore we will work with $p = 5$ in the following.

The first test we carry out is one of H_2 against H_1, that is, a stationary model is tested against one where all the parameters are time varying. Note that we use the order $p = 5$ also for the model with time varying coefficients. The test value $\lambda_{LR} = 2(\lambda_1 - \lambda_2) = 223.79$ is clearly significant at the 1% level since in this case the number of degrees of freedom of the asymptotic χ^2-distribution is 75. Thus, we conclude that at least some parameters are not time invariant. To see whether the white noise series may be regarded as stationary we also test H_3 against H_1. The resulting test value is $\lambda_{LR} = 2(\lambda_1 - \lambda_3) = 35.95$ which is also significant at the 1% level since we now have 9 degrees of freedom. Next we use the Wald test described in Section 12.3.2f to see whether the VAR coefficients and intercept terms may be assumed constant through time. In other words we test H_5 against H_1. The test value becomes $\lambda_w = 347$. Comparing this with critical values from the $\chi^2(66)$-distribution we again reject the null hypothesis H_5 at the 1% level. The reader is invited to perform further tests on these data. The tests performed so far support a full periodic model.

Of course, it is possible that a periodic model does not adequately capture the

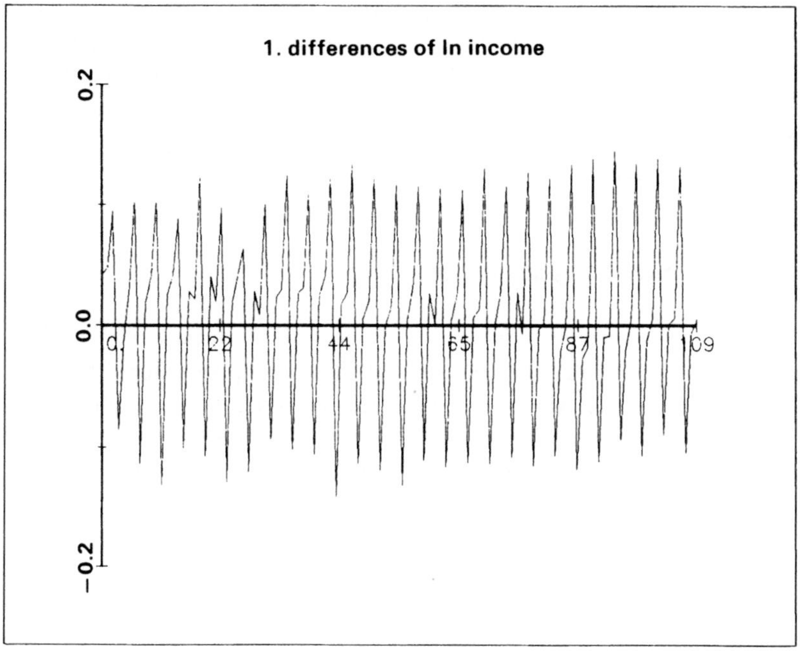

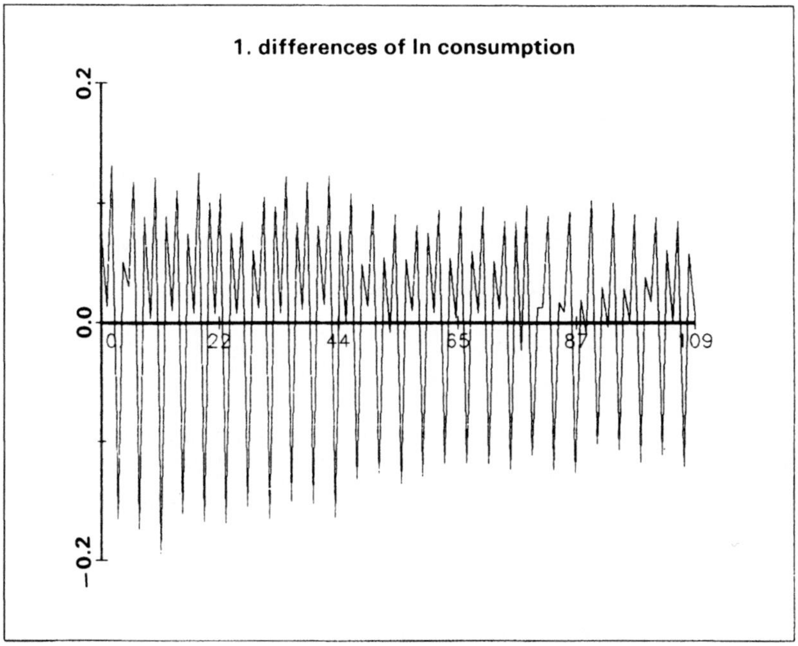

Fig. 12.1. Unadjusted West German income and consumption series.

characteristics of the data generating process. In that case the tests may not have much relevance. To check the adequacy of a periodic model similar tools may be used as in the stationary nonperiodic case. For instance, a residual analysis could be performed in a similar fashion as for nonperiodic VAR models. The properties of tests for model adequacy may be derived from the stationary representation of the annual process $\mathfrak{y}_\tau$.

12.3.4 Bibliographical Notes and Extensions

Early discussions of periodic time series models include those by Gladyshev (1961) and Jones & Brelsford (1967). Pagano (1978) studies properties of periodic autoregressions while Cleveland & Tiao (1979) consider periodic univariate ARMA models and Tiao & Grupe (1980) explore the consequences of fitting nonperiodic models to data generated by a periodic model. Cipra (1985) discusses inference for periodic moving average processes and Li & Hui (1988) develop an algorithm for ML estimation of periodic ARMA models. A Bayesian analysis of periodic autoregressions is given by Anděl (1983, 1987) and an application of periodic modeling can be found, for instance, in Osborn & Smith (1989). This list of references is by no means complete but serves to show that quite some work has been done in the past on periodic processes and that extensions of the presently considered models are possible.

12.4 Intervention Models

In the introduction an intervention model was described as one where a particular stationary data generation mechanism is in operation until period T_1, say, and another process generates the data after period T_1. For instance,

$$y_t = \nu_1 + A_1 Y_{t-1} + u_t, \qquad E(u_t u_t') = \Sigma_1, \quad t \leq T_1 \tag{12.4.1}$$

and

$$y_t = \nu_2 + A_2 Y_{t-1} + u_t, \qquad E(u_t u_t') = \Sigma_2, \quad t > T_1. \tag{12.4.2}$$

In the present case it makes a difference whether the intervention is modeled within the intercept form of the process like in (12.4.1)/(12.4.2) or within a mean-adjusted representation. We will consider both cases in turn following Lütkepohl (1992).

12.4.1 Interventions in the Intercept Model

Before we consider more general situations it may be useful to study the consequences of (12.4.1) and (12.4.2) in a little more detail. For simplicity suppose that $A_2 = A_1$ and $\Sigma_2 = \Sigma_1$ so that there is just a shift in intercept terms. In this case the mean of y_t is

$$E(y_t) = \begin{cases} \sum_{i=0}^{\infty} \Phi_i v_1, & t \le T_1 \\ \sum_{i=0}^{t-T_1} \Phi_i v_2 + \sum_{i=t-T_1+1}^{\infty} \Phi_i v_1, & t > T_1 \end{cases}$$

where the Φ_i are the coefficient matrices of the moving average representation of the mean adjusted process, i.e.,

$$\sum_{i=0}^{\infty} \Phi_i z^i = (I_K - A_{11} z - \cdots - A_{p1} z^p)^{-1}.$$

Hence, after the intervention the process mean does not reach a fixed new level immediately but only asymptotically,

$$E(y_t) \xrightarrow[t\to\infty]{} \sum_{i=0}^{\infty} \Phi_i v_2.$$

In the more general situation where all coefficients change due to the intervention, similar results also hold for the autocovariance structure. Of course, such a behaviour may be quite plausible in practice because a system may react slowly to an intervention. On the other hand it is also conceivable that an abrupt change occurs. For the case of a change in the mean we will discuss this situation in Subsection 12.4.2.

Before that we note that a model of the form considered in Section 12.3 may be used for intervention models as well with properly specified n_{it} as mentioned in the introduction (Section 12.1). The hypotheses considered in Section 12.3.2 are also of interest in the present context. The test statistics may be computed with the same formulas as in Section 12.3.2. However, the statistics do not necessarily have the indicated asymptotic distributions in the present case. The problem is that the ML estimators given in the previous section may not be consistent anymore. To see this, consider, for instance, the hypothesis H_1 (all parameters time varying) from Section 12.3.2a and the model in (12.4.1)/(12.4.2). If T_1 is some fixed finite point and $T > T_1$,

$$\tilde{B}_1 = [\tilde{v}_1, \tilde{A}_1] = \left[\sum_{t=1}^{T_1} y_t Z'_{t-1}\right]\left[\sum_{t=1}^{T_1} Z_{t-1} Z'_{t-1}\right]^{-1}$$

will not be consistent because the sample information regarding $B_1 = [v_1, A_1]$ does not increase when T goes to infinity. As a way out of this problem it may be assumed that T_1 increases with T. For instance, T_1 may be a fixed proportion of T. Then, under common assumptions,

$$\text{plim}\ \tilde{B}_1 = \text{plim}\left[\frac{1}{T_1}\sum_{t=1}^{T_1} y_t Z'_{t-1}\right]\text{plim}\left[\frac{1}{T_1}\sum_{t=1}^{T_1} Z_{t-1} Z'_{t-1}\right]^{-1} = B_1.$$

Also asymptotic normality is easy to obtain in this case and the test statistics have the limiting χ^2-distributions given in Section 12.3.2.

A logical problem may arise if more than one intervention is present. In that case it may not be easy to justify the assumption that all subperiods approach infinity with the sample period T. Whether or not this is a problem of practical relevance must be decided on the basis of the as yet unknown small sample properties of the tests. In any event, the large sample χ^2-distribution is just meant to be a guide for the small sample performance of the tests and as such it may be used if the periods between the interventions are reasonably large. Since no small sample results are available it is not clear, however, how large is large enough to obtain a good approximation to the asymptotic χ^2-distributions of the test statistics.

12.4.2 A Discrete Change in the Mean

We have seen that in an intercept model like (12.4.1)/(12.4.2) the mean smoothly approaches a new level after the intervention. Occasionally it may be more plausible to assume that there is a one time jump in the process mean after time T_1. In such a situation a model in mean-adjusted form,

$$y_t - \mu_t = A_1(y_{t-1} - \mu_{t-1}) + \cdots + A_p(y_{t-p} - \mu_{t-p}) + u_t, \tag{12.4.3}$$

is easier to work with. Here $\mu_t = E(y_t)$ and, for simplicity, it is assumed that all other coefficients are time invariant. Therefore the second subscript is dropped from the VAR coefficient matrices. We also assume Gaussian white noise u_t with time invariant covariance, $u_t \sim N(0, \Sigma_u)$. Suppose

$$\mu_t = n_{1t}\mu_1 + \cdots + n_{st}\mu_s, \qquad n_{it} = 0 \text{ or } 1, \qquad \sum_{i=1}^{s} n_{it} = 1. \tag{12.4.4}$$

In other words, there are s interventions so that for each i, the n_{it}, $t = 1, \dots, T$, are a sequence of zeroes and ones, the latter appearing in consecutive positions.

In general the exact ML estimation of the model (12.4.3) results in nonlinear normal equations and is therefore unattractive. However, the μ_i may be estimated as

$$\tilde{\mu}_i = \frac{1}{T\bar{n}_i} \sum_{t=1}^{T} n_{it} y_t, \qquad i = 1, \dots, s. \tag{12.4.5}$$

Providing $T\bar{n}_i = \sum_t n_{it}$ approaches infinity with T, it can be shown that under general assumptions, $\tilde{\mu}_i$ is consistent and

$$\sqrt{T\bar{n}_i}(\tilde{\mu}_i - \mu_i) \xrightarrow{d} N(0, \Sigma_{\tilde{\mu}}), \tag{12.4.6}$$

where

$$\Sigma_{\tilde{\mu}} = (I_K - A_1 - \cdots - A_p)^{-1}\Sigma_u(I_K - A_1 - \cdots - A_p)'^{-1}$$

(see Chapter 3, Section 3.3, and Problem 12.6). Note that the asymptotic covariance matrix does not depend on i. Furthermore, the $\tilde{\mu}_i$ are asymptotically independent. Hence, it is quite easy to perform a Wald test of the hypothesis

$$H_6: \quad \mu_i = \mu_1, \quad i = 2, \dots, s \qquad \text{or} \qquad R\begin{bmatrix} \mu_1 \\ \vdots \\ \mu_s \end{bmatrix} = 0, \tag{12.4.7}$$

where R has a similar structure as in (12.3.27). The corresponding Wald statistic is

$$\lambda_w = T[\sqrt{\bar{n}_1}\tilde{\mu}_1', \dots, \sqrt{\bar{n}_s}\tilde{\mu}_s']R'[R(I_s \otimes \tilde{\Sigma}_{\tilde{\mu}})R']^{-1}R\begin{bmatrix} \sqrt{\bar{n}_1}\tilde{\mu}_1 \\ \vdots \\ \sqrt{\bar{n}_s}\tilde{\mu}_s \end{bmatrix}, \tag{12.4.8}$$

where $[R(I_s \otimes \tilde{\Sigma}_{\tilde{\mu}})R']^{-1}$ reduces to

$$\begin{bmatrix} 2 & 1 & \dots & 1 \\ 1 & 2 & & 1 \\ \vdots & & \ddots & \vdots \\ 1 & 1 & \dots & 2 \end{bmatrix}^{-1} \otimes \tilde{\Sigma}_{\tilde{\mu}}^{-1}$$

and $\tilde{\Sigma}_{\tilde{\mu}}$ is estimated in the usual way. In other words,

$$\tilde{A} = \left[\sum_t \tilde{y}_t \tilde{Y}_{t-1}'\right]\left[\sum_t \tilde{Y}_{t-1}\tilde{Y}_{t-1}'\right]^{-1}$$

and

$$\tilde{\Sigma}_u = \sum_t (\tilde{y}_t - \tilde{A}\tilde{Y}_{t-1})(\tilde{y}_t - \tilde{A}\tilde{Y}_{t-1})'/T,$$

where $\tilde{y}_t := y_t - \tilde{\mu}_t$ and

$$\tilde{Y}_{t-1} := \begin{bmatrix} y_{t-1} - \tilde{\mu}_{t-1} \\ \vdots \\ y_{t-p} - \tilde{\mu}_{t-p} \end{bmatrix}.$$

Under H_6, λ_w has an asymptotic χ^2-distribution with $(s-1)K$ degrees of freedom.

12.4.3 An Illustrative Example

As an example of testing for structural change in the present framework we consider again the seasonally adjusted quarterly West German investment, income, and consumption data given in Table E.1 of Appendix E. The data were first used in Chapter 3. As in that and some other chapters we perform the analysis for the first differences of logarithms (rates of change) of the data. In Chapter 4 a possible structural break was found after year 1978 when the second oil price crisis occurred. We will now investigate whether this result can be confirmed with the tools of the present chapter.

For an event like a drastic oil price increase a smooth adjustment of the general economic conditions seems more plausible than a discrete change. Therefore the intercept version of an intervention model is chosen with $n_{1t} = 1$, $n_{2t} = 0$, for $t \le 1978.4$ and $n_{1t} = 0$, $n_{2t} = 1$ for $t \ge 1979.1$. Since a VAR(2) model performed reasonably well in Chapter 4 for the period 1960–1978 we use VAR(2) processes for both subperiods. This choice is legitimate under the null hypothesis of no structural change after 1978.

We first test a stationary model (H_2) against one in which all parameters are allowed to vary (H_1). The resulting value of the LR statistic is $\lambda_{LR} = 64.11$. From Table 12.1 we have

$$(s-1)K(K(p+\tfrac{1}{2})+\tfrac{3}{2}) = 27$$

degrees of freedom since $s = 2$, $K = 3$, and $p = 2$. Hence, we can reject the null hypothesis of time invariance at the 1% level of significance ($\chi^2(27)_{.99} = 46.96$). This, of course, does not necessarily mean that all coefficients are really time varying. For instance, the white noise covariance matrix may be time invariant while the other coefficients vary. To check this possibility we test H_3 against H_1 (all coefficients time varying). The value of the LR statistic becomes $\lambda_{LR} = 33.46$ and the number of degrees of freedom for this test is 6. Thus, the test value exceeds the critical value of the χ^2-distribution for a 1% significance level ($\chi^2(6)_{.99} = 16.81$) and we reject the null hypothesis. Further tests on the data are possible and the reader is invited to carry them out.

It is perhaps worth pointing out that we have only four years of data or 16 observations for each variable after the potential structural change. The quality of the χ^2-approximations to the distributions of the LR statistics may therefore be doubtful.

12.4.4 Extensions and References

Although we have used the label "intervention" for the type of change that occurs in the models considered in the previous subsections they could also be regarded as outliers if, for instance, a change in the process mean occurs for a small number of periods only. Tsay (1988) discusses univariate time series models with outliers and structural changes and lists a number of further references. By appropriate choice of the dummy variables n_{it} it is possible to combine periodic and intervention or outlier models. Extensions of the present framework to VARMA or restricted VAR models are possible in principle. The related practical problems may be considerable, however.

More general forms of interventions in the process mean are discussed by Box & Tiao (1975) and Abraham (1980). They assume that interventions have occurred at $t = T_1, \dots, T_k$ and they define a vector $I_t = (I_t(T_1), \dots, I_t(T_k))'$ of dummy variables that may be of the type

$$I_t(T_i) = \begin{cases} 0 & \text{for } t < T_i \\ 1 & \text{for } t \ge T_i \end{cases}$$

or of the type

$$I_t(T_i) = \begin{cases} 0 & \text{for } t \neq T_i \\ 1 & \text{for } t = T_i. \end{cases}$$

They model the interventions as $R(L)I_t$, where $R(L)$ is a matrix of rational functions in the lag operator.

12.5 Exercises

Problem 12.1

Suppose y_t is a periodic K-dimensional VAR(1) process,

$$y_t = \nu_1 + A_{11}y_{t-1} + u_t, \qquad E(u_t u_t') = \Sigma_1, \qquad \text{if } t \text{ is even}$$

and

$$y_t = \nu_2 + A_{12}y_{t-1} + u_t, \qquad E(u_t u_t') = \Sigma_2, \qquad \text{if } t \text{ is odd.}$$

a) Derive explicit expressions for the means μ_t and the matrices Φ_{it}.
b) Derive the autocovariances $E[(y_t - \mu_t)(y_{t-h} - \mu_{t-h})']$ for $h = 1, 2, 3$ and for both cases, t even and t odd. Write down explicitly the assumptions used in deriving the autocovariance matrices.

Problem 12.2

Assume that the process y_t given in Problem 12.1 is bivariate with

$$A_{11} = \begin{bmatrix} .5 & .3 \\ .8 & 1.2 \end{bmatrix}$$

and

$$A_{12} = \begin{bmatrix} .6 & .4 \\ .8 & .5 \end{bmatrix}.$$

Is the corresponding process $\mathfrak{y}_\tau = (y_{2\tau}', y_{2\tau-1}')'$ stable?

Problem 12.3

Give the forecasts $y_t(h)$, $h = 1, 2, 3$, t odd, for the process from Problem 12.1 and give explicit expressions for the forecast MSE matrices.

Problem 12.4

For the process given in Problem 12.1 construct an LM test of the hypotheses

$$H_0\colon \nu_1 = \nu_2, \qquad A_{11} = A_{12}, \qquad \Sigma_1 = \Sigma_2$$

against

$$H_1: \nu_1 = \nu_2, \qquad A_{11} = A_{12}, \qquad \Sigma_1 \neq \Sigma_2.$$

Give an explicit expression for the LM statistic.

Problem 12.5

Suppose the process from Problem 12.1 is in operation until period T_1 and after that another periodic VAR(1) process of the same type but with different coefficients generates a set of variables. Define dummy variables in such a way that the complete process can be written in the form (12.1.3)/(12.1.4).

Problem 12.6

Show that (12.4.6) holds. (Hint: See Chapter 3, Section 3.3.)

Problem 12.7

In 1974 the Deutsche Bundesbank officially changed from an interest rate target to a money supply target. Use the two interest rate series given in Table E.5 and test for an intervention in 1974 within the framework discussed in Section 12.4. Use tests for different types of interventions related to shifts in the intercept terms, the VAR coefficients, and the white noise covariances.

Chapter 13. State Space Models

13.1 Background

State space models may be regarded as generalizations of the models considered so far. They have been used extensively in system theory, the physical sciences, and engineering. The terminology is therefore largely from these fields. The general idea behind these models is that an observed multiple time series $y_1, \dots, y_T$ depends upon a possibly unobserved state z_t which is driven by a stochastic process. The relation between y_t and z_t is described by the *observation* or *measurement equation*

$$y_t = \mathbf{H}_t z_t + v_t, \tag{13.1.1}$$

where $\mathbf{H}_t$ is a matrix that may also depend on the period of time t and v_t is the *observation error* which is typically assumed to be white noise. The *state vector* or state of nature is generated by the dynamic process

$$z_t = \mathbf{B}_{t-1} z_{t-1} + w_{t-1} \tag{13.1.2}$$

which is often called the *transition equation* because it describes the transition of the state of nature from period $t-1$ to period t. $\mathbf{B}_t$ is a coefficient matrix that may depend on t and w_t is white noise. The system (13.1.1)/(13.1.2) is one form of a *linear state space system*.

The following example from Meinhold & Singpurwalla (1983) may illustrate the related concepts. Suppose we wish to trace a satellite's orbit. The state vector z_t may then consist of the position and the speed of the satellite in period t with respect to the center of the earth. The state cannot be measured directly but, for example, the distance from a certain observatory may be observed. These measurements constitute the observed vectors y_t. As another example consider the income of an individual which may depend on unobserved factors such as intelligence, special abilities, special interests and so on. In this case the state vector consists of the variables that describe the abilities of the person and y_t is his or her observed income.

The reader may recall that all the models considered so far have been written in a form similar to (13.1.1)/(13.1.2) at some stage. For instance, our standard (zero mean) VAR(p) model can be written in VAR(1) form as

$$Y_t = \mathbf{A} Y_{t-1} + U_t, \tag{13.1.3}$$

where

$$Y_t := \begin{bmatrix} y_t \\ \vdots \\ y_{t-p+1} \end{bmatrix}, \qquad U_t := \begin{bmatrix} u_t \\ 0 \\ \vdots \\ 0 \end{bmatrix}, \quad \text{etc.}$$

With $z_t := Y_t$, $\mathbf{B}_t := \mathbf{A}$, and $w_{t-1} := U_t$ equation (13.1.3) may be regarded as the transition equation of a state space model. The corresponding measurement equation is

$$y_t = [I_K \quad 0 \ldots 0] Y_t$$

with $\mathbf{H}_t := [I_K \quad 0 \ldots 0]$ and $v_t := 0$.

In the next section we will introduce a slightly more general version of a linear state space model, we will review all the previous models and we will cast them into state space form. As we have seen, the representations of the models used in the previous chapters are useful for many purposes. There are occasions, however, where a state space representation makes life easier. We have actually used state space representations of some models without explicitly mentioning this fact. We will also consider some further models that have been discussed in the literature and are special cases of the general state space class. This way we will give an overview of a number of important models that have been considered in the multiple time series literature.

In Section 13.3 we will discuss the Kalman filter which is an extremely useful tool in the analysis of state space models. Given the observable vectors y_t it provides estimates of the state vectors and measures of the precision of these estimates. In a situation where the state vector consists of unobservable variables such estimates may be of interest. In a system such as (13.1.1)/(13.1.2) the matrices $\mathbf{B}_t$ and $\mathbf{H}_t$ and the covariance matrices of v_t and w_t will often depend on unknown parameters. The Kalman filter is also helpful in estimating these parameters. This issue will be discussed in Section 13.4.

In this chapter we will just give a brief introduction to some basic concepts related to state space models and the Kalman filter. Various textbooks exist that provide broader introductions to the topic. Examples are Jazwinski (1970) and Anderson & Moore (1979).

13.2 State Space Models

13.2.1 The General Linear State Space Model

As mentioned in the previous section a *linear state space model* consists of a *transition* or *system equation*

$$z_{t+1} = \mathbf{B}_t z_t + \mathbf{F}_t x_t + w_t, \qquad t = 0, 1, 2, \ldots \tag{13.2.1a}$$

or equivalently

$$z_t = \mathbf{B}_{t-1} z_{t-1} + \mathbf{F}_{t-1} x_{t-1} + w_{t-1}, \qquad t = 1, 2, \ldots, \tag{13.2.1b}$$

and a *measurement* or *observation equation*

$$y_t = \mathbf{H}_t z_t + \mathbf{G}_t x_t + v_t, \qquad t = 1, 2, \ldots \tag{13.2.2}$$

Here

y_t is a $(K \times 1)$ vector of observable *output* or *endogenous variables*,
z_t is an $(N \times 1)$ *state vector* or the *state of nature*,
x_t is an $(M \times 1)$ vector of observable *inputs* or *instruments* or *policy variables*,
v_t is a $(K \times 1)$ vector of *observation* or *measurement errors* or *noise*,
w_t is an $(N \times 1)$ vector of *system* or *transition equation errors* or *noise*,
$\mathbf{H}_t$ is a $(K \times N)$ *measurement matrix*,
$\mathbf{G}_t$ is a $(K \times M)$ *input matrix* of the observation equation,
$\mathbf{B}_t$ is an $(N \times N)$ *transition* or *system matrix*,

and

$\mathbf{F}_t$ is an $(N \times M)$ *input matrix* of the transition equation.

The matrices $\mathbf{H}_t$, $\mathbf{G}_t$, $\mathbf{B}_t$, and $\mathbf{F}_t$ are assumed to be known at time t. Although they are in general allowed to vary at least some of them will often be time invariant. In practice at least some of the elements of these matrices are usually unknown and have to be estimated. This issue is deferred to Section 13.4. It is perhaps noteworthy that the process is assumed to be started from an *initial state* z_0 and a given *initial input* x_0.

To complete the description of the model we make the following stochastic assumptions for the noise processes and the initial state:

The joint process

$$\begin{bmatrix} w_t \\ v_t \end{bmatrix}$$

is a zero mean white noise process with possibly time dependent covariance matrices

$$\begin{bmatrix} \Sigma_{w_t} & \Sigma_{w_t v_t} \\ \Sigma_{v_t w_t} & \Sigma_{v_t} \end{bmatrix}.$$

The initial state z_0 is uncorrelated with w_t, v_t for all t and has a distribution with mean μ_{z_0} and covariance matrix Σ_{z_0}. The input sequence $x_0, x_1, \ldots$ is nonstochastic. If the observed inputs are actually stochastic the analysis is assumed to be conditional on a given sequence of inputs.

With these assumptions we can derive stochastic properties of the states and the system outputs. Successive substitution in (13.2.1b) implies

$$z_t = \mathbf{\Phi}_{t,t} z_0 + \sum_{i=1}^{t} \mathbf{\Phi}_{i-1,t}(\mathbf{F}_{t-i} x_{t-i} + w_{t-i}), \tag{13.2.3}$$

where

$$\mathbf{\Phi}_{0,t} = I_N \qquad \text{and} \qquad \mathbf{\Phi}_{i,t} = \prod_{j=1}^{i} \mathbf{B}_{t-j}, \qquad i = 1, 2, \ldots$$

(see also Section 12.2.1). Hence,

$$\mu_{z_t} := E(z_t) = \mathbf{\Phi}_{t,t}\mu_{z_0} + \sum_{i=1}^{t} \mathbf{\Phi}_{i-1,t}\mathbf{F}_{t-i}x_{t-i} \tag{13.2.4}$$

and

$$\begin{aligned} \mathrm{Cov}(z_t, z_{t+h}) &= E[(z_t - \mu_{z_t})(z_{t+h} - \mu_{z_{t+h}})'] \\ &= \mathbf{\Phi}_{t,t}\Sigma_{z_0}\mathbf{\Phi}'_{t+h,t+h} + \sum_{i=1}^{t} \mathbf{\Phi}_{i-1,t}\Sigma_{w_{t-i}}\mathbf{\Phi}'_{h+i-1,t+h}. \end{aligned} \tag{13.2.5}$$

Under the aforementioned stochastic assumptions it is also easy to obtain the mean and covariance matrix of the output process:

$$\mu_{y_t} := E(y_t) = \mathbf{H}_t E(z_t) + \mathbf{G}_t x_t$$

and

$$\mathrm{Cov}(y_t, y_{t+h}) = \mathbf{H}_t \,\mathrm{Cov}(z_t, z_{t+h})\mathbf{H}'_t.$$

Generally the means and autocovariances of the y_t are obviously not time invariant. Thus, in general y_t is a nonstationary process.

We will now consider various special cases of state space models which are obtained by specific definitions of the state vector, the inputs, the noise processes, and the matrices $\mathbf{H}_t$, $\mathbf{G}_t$, $\mathbf{B}_t$, and $\mathbf{F}_t$. These matrices and the white noise covariance matrices will often not depend on t in which case we will suppress the subscript for notational simplicity.

13.2.1a A Finite Order VAR Process

Although we have mentioned earlier how to cast a VAR(p) process

$$y_t = \nu + A_1 y_{t-1} + \cdots + A_p y_{t-p} + u_t \tag{13.2.6}$$

in state space form it may be useful to consider this model again because it illustrates that often different state space models can represent a particular process. One possible state space representation is obtained by defining

$$Y_t := \begin{bmatrix} y_t \\ \vdots \\ y_{t-p+1} \end{bmatrix}, \quad \boldsymbol{\nu} := \begin{bmatrix} \nu \\ 0 \\ \vdots \\ 0 \end{bmatrix}, \quad \mathbf{A} := \begin{bmatrix} A_1 & \ldots & A_{p-1} & A_p \\ I_K & & 0 & 0 \\ & \ddots & \vdots & \vdots \\ 0 & \ldots & I_K & 0 \end{bmatrix}, \quad U_t := \begin{bmatrix} u_t \\ 0 \\ \vdots \\ 0 \end{bmatrix}. \tag{13.2.7}$$

Hence,

$$Y_t = \mathbf{A}Y_{t-1} + \mathbf{v} + U_t, \tag{13.2.8}$$

$$y_t = [I_K \quad 0\dots 0]\,Y_t \tag{13.2.9}$$

is a state space model with state vector $z_t := Y_t$, $\mathbf{B} := \mathbf{A}$, $\mathbf{F} := \mathbf{v}$, $x_t := 1$, $w_t := U_{t+1}$, $\mathbf{H} := [I_K \quad 0\dots 0]$, $\mathbf{G} := 0$, $v_t := 0$.

An alternative possibility is to define the state vector as

$$z_t := \begin{bmatrix} 1 \\ y_t \\ \vdots \\ y_{t-p+1} \end{bmatrix}$$

and choose

$$\mathbf{B} := \begin{bmatrix} 1 & 0 & \dots & 0 & 0 \\ \nu & A_1 & \dots & A_{p-1} & A_p \\ 0 & I_K & & 0 & 0 \\ \vdots & & \ddots & \vdots & \vdots \\ 0 & 0 & \dots & I_K & 0 \end{bmatrix} \quad \text{and} \quad w_t := \begin{bmatrix} 0 \\ u_{t+1} \\ 0 \\ \vdots \\ 0 \end{bmatrix}$$

so that

$$z_{t+1} = \mathbf{B}z_t + w_t$$

and

$$y_t = [0 \quad I_K \quad 0\dots 0]\,z_t$$

which describes the same process as (13.2.8)/(13.2.9). It may be worth pointing out that in the present framework the process is assumed to be started at time $t = 0$ while we have assumed an infinite past of the process in some previous chapters.

13.2.1b A VARMA(p, q) Process

A state space representation of the VARMA(p, q) process

$$y_t = \nu + A_1 y_{t-1} + \cdots + A_p y_{t-p} + u_t + M_1 u_{t-1} + \cdots + M_q u_{t-q} \tag{13.2.10}$$

is known from Chapter 6, Section 6.3.2. It is obtained by choosing a state vector

$$z_t := \begin{bmatrix} y_t \\ \vdots \\ y_{t-p+1} \\ u_t \\ \vdots \\ u_{t-q+1} \end{bmatrix}, \quad \text{transition noise } w_t := \begin{bmatrix} u_{t+1} \\ 0 \\ \vdots \\ 0 \\ u_{t+1} \\ 0 \\ \vdots \\ 0 \end{bmatrix},$$

an input sequence $x_t := 1$ as in (13.2.8), $\mathbf{B} := \mathbf{A}$ from Chapter 6, Equation (6.3.8), $\mathbf{F} := \boldsymbol{\nu}$ defined similarly as in (13.2.7), $\mathbf{H} := [I_K \quad 0 \dots 0]$, $\mathbf{G} := 0$, and $v_t := 0$. For many purposes this is not the most useful state space representation of a VARMA model. Other state space representations are given by Aoki (1987), Hannan & Deistler (1988), and Wei (1990).

13.2.1c *The VARX Model*

The VARX model

$$y_t = A_1 y_{t-1} + \cdots + A_p y_{t-p} + B_0 x_t + \cdots + B_s x_{t-s} + u_t \tag{13.2.11}$$

considered in Chapter 10 is easily cast in state space form by choosing the state vector

$$z_t := \begin{bmatrix} y_t \\ \vdots \\ y_{t-p+1} \\ x_t \\ \vdots \\ x_{t-s+1} \end{bmatrix},$$

and the transition equation

$$z_t = \left[\begin{array}{cccc|cccc} A_1 & \dots & A_{p-1} & A_p & B_1 & \dots & B_{s-1} & B_s \\ I & & 0 & 0 & 0 & \dots & 0 & 0 \\ & \ddots & \vdots & \vdots & \vdots & & \vdots & \vdots \\ 0 & \dots & I & 0 & 0 & \dots & 0 & 0 \\ \hline & & & & 0 & \dots & 0 & 0 \\ & & & & I & & 0 & 0 \\ & & 0 & & & \ddots & \vdots & \vdots \\ & & & & 0 & \dots & I & 0 \end{array}\right] z_{t-1} + \begin{bmatrix} B_0 \\ 0 \\ \vdots \\ 0 \\ I \\ 0 \\ \vdots \\ 0 \end{bmatrix} x_t + \begin{bmatrix} u_t \\ 0 \\ \vdots \\ 0 \end{bmatrix}. \tag{13.2.12}$$

The corresponding observation equation becomes

$$y_t = [I_K \quad 0 \dots 0] z_t.$$

It is also possible to extend the model so as to allow for a finite order MA(q) error process in (13.2.11) (see Problem 13.1).

13.2.1d *Systematic Sampling and Aggregation*

Suppose that annual data is available whereas a decision maker is interested in, say, quarterly figures. Let η_{it} be an $(M \times 1)$ vector of variables associated with the i-th quarter of year t and suppose the vector of all quarterly variables associated with year t,

$$\eta_t := \begin{bmatrix} \eta_{1t} \\ \eta_{2t} \\ \eta_{3t} \\ \eta_{4t} \end{bmatrix},$$

is generated by the VAR(p) process

$$\eta_t = A_1\eta_{t-1} + \cdots + A_p\eta_{t-p} + u_t.$$

Then we may define a state vector

$$z_t := \begin{bmatrix} \eta_t \\ \vdots \\ \eta_{t-p+1} \end{bmatrix}$$

and a transition equation

$$z_t = \mathbf{A}z_{t-1} + U_t,$$

where $\mathbf{A}$ and U_t are the same quantities as in (13.2.7). If the yearly values are obtained by adding (aggregating) the quarterly figures the observation equation becomes

$$y_t = [I_M \quad I_M \quad I_M \quad I_M \quad 0\ldots0]z_t,$$

where M is the dimension of η_{it}. Alternatively, if the annual figures are obtained by systematic sampling, that is, by taking, say, the fourth quarter values as the annual figures, the observation equation becomes

$$y_t = [0 \quad 0 \quad 0 \quad I_M \quad 0\ldots0]z_t.$$

Extensions of this framework to the case where η_t is generated by a VARMA or VARMAX process are straightforward. For applications of state space models in aggregation and systematic sampling problems see Nijman (1985), Harvey (1984), Harvey & Pierse (1984), Jones (1980), Ansley & Kohn (1983).

The examples considered so far have in common that the system matrices **H**, **G**, **B**, and **F** are all time invariant and the state vector contains at least some observed or observable variables. In contrast the state vector is unobservable in the next two examples while the system matrices remain time invariant.

13.2.1e Structural Time Series Models

In a structural time series model the observed time series is viewed as a sum of unobserved components such as a trend, a seasonal component, and an irregular component (see, e.g., Kitagawa (1981), Harvey & Todd (1983), Harvey (1989)). For instance, for a univariate time series $y_1, \ldots, y_T$ the structural model may have the form

$$y_t = \mu_t + \gamma_t + u_t, \tag{13.2.13}$$

where μ_t is a trend component and γ_t is a seasonal component. Harvey & Todd (1983) assume a local approximation to a linear trend function for which both the level and the slope are shifting. They postulate a process

$$\mu_t = \mu_{t-1} + \beta_{t-1} + \eta_t \qquad \text{with } \beta_t = \beta_{t-1} + \xi_t \tag{13.2.14}$$

as the trend generation mechanism. Here η_t and ξ_t are assumed to be white noise processes. This trend model is a mixture of two random walks which are discussed in Chapter 11 (see Section 11.1.1). For the seasonal component it is assumed that the sum over the seasonal factors of a full year is approximately zero,

$$\gamma_t = -\sum_{j=1}^{s-1} \gamma_{t-j} + \omega_t, \tag{13.2.15}$$

where s is the number of seasons and ω_t is white noise. The three white noise processes η_t, ξ_t, and ω_t are assumed to be independent.

This model can be set up in state space form by defining the state vector to be

$$z_t := \begin{bmatrix} \mu_t \\ \beta_t \\ \gamma_t \\ \vdots \\ \gamma_{t-s+2} \end{bmatrix}$$

and, hence, the transition equation is

$$z_t = \left[\begin{array}{cc|cccc} 1 & 1 & & & 0 & \\ 0 & 1 & & & & \\ \hline & & -1 & \dots & -1 & -1 \\ & & 1 & & 0 & 0 \\ & 0 & & \ddots & \vdots & \vdots \\ & & 0 & \dots & 1 & 0 \end{array}\right] z_{t-1} + \begin{bmatrix} \eta_t \\ \xi_t \\ \omega_t \\ 0 \\ \vdots \\ 0 \end{bmatrix}. \tag{13.2.16}$$

The corresponding measurement equation is

$$y_t = [1 \quad 0 \quad 1 \quad 0 \dots 0] z_t + u_t. \tag{13.2.17}$$

Multivariate generalizations of this model are possible (see Harvey (1987)).

13.2.1f Factor Analytic Models

In a classical factor analytic setting it is assumed that a set of K observed variables y_t depends linearly on $N < K$ unobserved common factors f_t and on individual factors u_t. In other words,

$$y_t = \mathbf{L} f_t + u_t, \tag{13.2.18}$$

where $\mathbf{L}$ is a $(K \times N)$ matrix of *factor loadings* and the components of u_t are typically assumed to be uncorrelated, that is, Σ_u is a diagonal matrix (Anderson (1984), Morrison (1976)). One objective of a factor analysis is the construction of the unobserved factors f_t. We may view (13.2.18) as the measurement equation of a state space model and, if the factors f_t and f_s are independent for $t \neq s$, we may specify a trivial transition equation $f_t = w_{t-1}$.

However, if y_t consists of time series variables it may be more reasonable to assume that the factors are autocorrelated. For example, they may be generated by a VAR or VARMA process. Also the individual factors u_t may be autocorrelated. *Dynamic factor analytic models* of this type were considered by Geweke (1977) and Engle & Watson (1981). Assuming that

$$f_t = A_1 f_{t-1} + \cdots + A_p f_{t-p} + \eta_t$$

and

$$u_t = C_1 u_{t-1} + \cdots + C_q u_{t-q} + \varepsilon_t,$$

where η_t and ε_t are white noise processes, a state space model can be obtained by specifying a state vector

$$z_t := \begin{bmatrix} f_t \\ \vdots \\ f_{t-p+1} \\ u_t \\ \vdots \\ u_{t-q+1} \end{bmatrix},$$

and a transition equation

$$z_t = \left[\begin{array}{cccc|cccc} A_1 & \dots & A_{p-1} & A_p & & & & \\ I & & 0 & 0 & & 0 & & \\ & \ddots & \vdots & \vdots & & & & \\ 0 & \dots & I & 0 & & & & \\ \hline & & & & C_1 & \dots & C_{q-1} & C_q \\ & & & & I & & 0 & 0 \\ & 0 & & & & \ddots & \vdots & \vdots \\ & & & & 0 & \dots & I & 0 \end{array}\right] z_{t-1} + \begin{bmatrix} \eta_t \\ 0 \\ \vdots \\ 0 \\ \varepsilon_t \\ 0 \\ \vdots \\ 0 \end{bmatrix}. \tag{13.2.19}$$

The corresponding measurement equation is

$$y_t = \mathbf{L} f_t + u_t = [\mathbf{L} \quad 0 \dots 0 \quad I_K \quad 0 \dots 0] z_t. \tag{13.2.20}$$

An extension to the case where f_t and u_t are generated by VARMA processes is left to the reader (see Problem 13.2). If exogenous variables are added to the original model (13.2.18) and in addition the factors are dynamic processes we obtain the dynamic MIMIC models of Engle & Watson (1981).

In all the previous examples the system matrices $\mathbf{H}_t$, $\mathbf{G}_t$, $\mathbf{B}_t$, and $\mathbf{F}_t$ are time invariant. We will now consider models where at least some elements of these matrices vary through time.

13.2.1g VARX Models with Systematically Varying Coefficients

We extend the varying coefficient VAR models of Chapter 12 slightly by adding exogenous variables and assuming that a given multiple time series is generated according to

$$y_t = A_{1t}y_{t-1} + \cdots + A_{pt}y_{t-p} + F_t x_t + u_t. \tag{13.2.21}$$

The vector x_t may simply include an intercept term or seasonal dummies. It may also include other exogenous variables and even lags of exogenous variables. Using

$$Y_t := \begin{bmatrix} y_t \\ \vdots \\ y_{t-p+1} \end{bmatrix}, \quad \mathbf{A}_t := \begin{bmatrix} A_{1,t} & \dots & A_{p-1,t} & A_{p,t} \\ I & & 0 & 0 \\ & \ddots & \vdots & \vdots \\ 0 & \dots & I & 0 \end{bmatrix}, \quad \mathbf{F}_t := \begin{bmatrix} F_t \\ 0 \\ \vdots \\ 0 \end{bmatrix}, \quad \text{and}$$

$$w_{t-1} := \begin{bmatrix} u_t \\ 0 \\ \vdots \\ 0 \end{bmatrix}$$

gives a transition equation

$$Y_t = \mathbf{A}_t Y_{t-1} + \mathbf{F}_t x_t + w_{t-1} \tag{13.2.22}$$

and a measurement equation

$$y_t = [I_K \quad 0 \dots 0]\, Y_t. \tag{13.2.23}$$

Obviously, the transition matrix $\mathbf{B}_{t-1} := \mathbf{A}_t$ and the input matrix $\mathbf{F}_t$ of the transition equation are time dependent in this state space model.

13.2.1h Random Coefficient VARX Models

So far all the original models either have time invariant, constant coefficients or, as in the previous subsection, systematically varying coefficients. We will now consider models with random coefficients and demonstrate how they can be cast in state space form. Let us begin with a simple multivariate regression model of the form

$$y_t = C_t x_t + v_t = (x_t' \otimes I)\,\mathrm{vec}(C_t) + v_t. \tag{13.2.24}$$

Assuming that the parameter vector $\gamma_t := \mathrm{vec}(C_t)$ is generated by a VAR(q) model,

$$\gamma_t = \nu + B_1 \gamma_{t-1} + \cdots + B_q \gamma_{t-q} + u_t, \tag{13.2.25}$$

we may define the state vector as

$$z_t := \begin{bmatrix} \gamma_t \\ \vdots \\ \gamma_{t-q+1} \end{bmatrix}$$

and get a state space model with the following transition and measurement equations, respectively,

$$z_t = \begin{bmatrix} B_1 & \dots & B_{q-1} & B_q \\ I & & 0 & 0 \\ & \ddots & \vdots & \vdots \\ 0 & \dots & I & 0 \end{bmatrix} z_{t-1} + \begin{bmatrix} \nu \\ 0 \\ \vdots \\ 0 \end{bmatrix} + \begin{bmatrix} u_t \\ 0 \\ \vdots \\ 0 \end{bmatrix},$$

$$y_t = [x_t' \otimes I, 0, \dots, 0] z_t + v_t.$$

Obviously the measurement matrix

$$\mathbf{H}_t := [x_t' \otimes I, 0, \dots, 0]$$

is time varying. It may, in fact, be random if the exogenous variables are regarded as stochastic variables. Such an assumption is mandatory if x_t contains lagged y_t variables. To see this point more clearly let us explicitly introduce lagged y_t's in (13.2.24):

$$\begin{aligned} y_t &= \mathbf{A}_t Y_{t-1} + C_t x_t + v_t \\ &= (Y_{t-1}' \otimes I)\,\mathrm{vec}(\mathbf{A}_t) + (x_t' \otimes I)\,\mathrm{vec}(C_t) + v_t, \end{aligned} \tag{13.2.26}$$

where $\mathbf{A}_t := [A_{1t}, \dots, A_{pt}]$ and $Y_{t-1}' := [y_{t-1}', \dots, y_{t-p}']$. Now suppose that $\gamma_t = \mathrm{vec}(C_t)$ is generated by the VAR(q) process (13.2.25) and $\boldsymbol{\alpha}_t = \mathrm{vec}(\mathbf{A}_t)$ is driven by a VAR(r) process

$$\boldsymbol{\alpha}_t = D_1 \boldsymbol{\alpha}_{t-1} + \cdots + D_r \boldsymbol{\alpha}_{t-r} + \eta_t \tag{13.2.27}$$

which is independent of γ_t. Defining the state vector as

$$z_t := \begin{bmatrix} \gamma_t \\ \vdots \\ \gamma_{t-q+1} \\ \boldsymbol{\alpha}_t \\ \vdots \\ \boldsymbol{\alpha}_{t-r+1} \end{bmatrix}$$

the following state space model is obtained:

$$z_t = \left[\begin{array}{cccc|cccc} B_1 & \dots & B_{q-1} & B_q & & & & \\ I & & 0 & 0 & & 0 & & \\ & \ddots & \vdots & \vdots & & & & \\ 0 & \dots & I & 0 & & & & \\ \hline & & & & D_1 & \dots & D_{r-1} & D_r \\ & 0 & & & I & & 0 & 0 \\ & & & & & \ddots & \vdots & \vdots \\ & & & & 0 & \dots & I & 0 \end{array}\right] z_{t-1} + \begin{bmatrix} v \\ 0 \\ \vdots \\ 0 \\ \vdots \\ 0 \end{bmatrix} + \begin{bmatrix} u_t \\ 0 \\ \vdots \\ 0 \\ \eta_t \\ 0 \\ \vdots \\ 0 \end{bmatrix}, \tag{13.2.28}$$

$$y_t = [x_t' \otimes I, 0, \dots, 0, Y_{t-1}' \otimes I, 0, \dots, 0]z_t + v_t. \tag{13.2.29}$$

Further extensions of this model are possible. For instance, $\boldsymbol{\alpha}_t$ and $\boldsymbol{\gamma}_t$ may be individually or jointly generated by a VARMA rather than a finite order VAR process. Moreover, exogenous input variables with constant coefficients could appear in (13.2.26). These extensions are left to the reader (see Problem 13.3).

The number of publications on random coefficient models is vast in both the econometrics and the time series literature. Famous examples from the earlier econometrics literature on the topic are Hildreth & Houck (1968), Swamy (1971), and Cooley & Prescott (1973, 1976). Recent surveys of the related literature are Chow (1984) and Nicholls & Pagan (1985). Both include extensive reference lists. On the time series side a number of references can be found in the monograph by Nicholls & Quinn (1982). Other important work on the topic includes the seminal article by Doan, Litterman & Sims (1984) who investigate the potential of random coefficient VAR models with Bayesian restrictions for econometric time series analysis.

13.2.2 Nonlinear State Space Models

There are also time series models that do not fall into the linear state space framework considered so far. Therefore it may be worth pointing out that more general nonlinear state space models have been studied in recent publications. A very general formulation has the form

$$z_{t+1} = \mathbf{b}_t(z_t, x_t, w_t, \boldsymbol{\delta}_1) \tag{13.2.30}$$

for the transition equations and

$$y_t = \mathbf{h}_t(z_t, x_t, v_t, \boldsymbol{\delta}_2) \tag{13.2.31}$$

for the measurement equations. Here $\boldsymbol{\delta}_1$ and $\boldsymbol{\delta}_2$ are vectors of parameters.

Bilinear time series models are examples for which the linear state space framework is too narrow. A very simple univariate bilinear time series model has the form

$$y_t = \alpha y_{t-1} + u_t + \beta y_{t-1} u_{t-1},$$

where u_t is univariate white noise. The product term $\beta y_{t-1} u_{t-1}$ distinguishes this model from a linear specification. Bilinear models have been found useful in modeling nonnormal phenomena (see, e.g., Granger & Andersen (1978)).

A general multivariate bilinear time series model may be specified as follows:

$$y_t = A_1 y_{t-1} + \cdots + A_p y_{t-p} + u_t + M_1 u_{t-1} + \cdots + M_q u_{t-q}$$
$$+ \sum_{i=1}^{r} \sum_{j=1}^{s} C_{ij} \operatorname{vec}(y_{t-i} u'_{t-j}). \tag{13.2.32}$$

Assuming without loss of generality that $p \geq r$ and $q \geq s$ and defining

$$z_t := \begin{bmatrix} y_t \\ \vdots \\ y_{t-p+1} \\ u_t \\ \vdots \\ u_{t-q+1} \end{bmatrix}, \quad \mathbf{B} := \left[\begin{array}{cccc|cccc} A_1 & \dots & A_{p-1} & A_p & M_1 & \dots & M_{q-1} & M_q \\ I & & 0 & 0 & 0 & \dots & 0 & 0 \\ & \ddots & \vdots & \vdots & \vdots & & \vdots & \vdots \\ 0 & \dots & I & 0 & 0 & \dots & 0 & 0 \\ \hline & & & & 0 & \dots & 0 & 0 \\ & & & & I & & 0 & 0 \\ & & 0 & & & \ddots & \vdots & \vdots \\ & & & & 0 & \dots & I & 0 \end{array}\right],$$

$$w_t := \begin{bmatrix} u_{t+1} \\ 0 \\ \vdots \\ 0 \\ u_{t+1} \\ 0 \\ \vdots \\ 0 \end{bmatrix},$$

and a matrix $\mathbf{C}$ which contains the elements of the C_{ij} matrices in a suitable arrangement, we get a bilinear state space model of the form

$$z_{t+1} = \mathbf{B} z_t + w_t + \mathbf{C} \operatorname{vec}(z_t z'_t), \tag{13.2.33}$$

$$y_t = [I_K \quad 0 \dots 0] z_t. \tag{13.2.34}$$

Obviously, the transition equation involves a nonlinear term, namely $\operatorname{vec}(z_t z'_t)$. Hence, (13.2.33)/(13.2.34) is an example of a nonlinear state space system.

The work of Granger & Andersen (1978) and others on univariate bilinear models has stimulated investigations in this area. Much of this work is documented in a monograph by Subba Rao & Gabr (1984). Recent work on multivariate bilinear models includes Stensholt & Tjøstheim (1987) and Liu (1989).

With all these examples we have not nearly exhausted the range of models that have been used and studied in the recent time series literature. Important omissions are threshold autoregressive models analyzed by Tong (1983) and exponential autoregressive models introduced by Ozaki (1980) and Haggan & Ozaki (1980). A general nonlinear model class is considered by Priestley (1980) and reviews of many nonlinear models and extensive lists of references are given by Priestley (1988) and by Anděl (1989). So far applications of these models in analyzing multiple time series are rare, however.

In contrast, the ARCH (autoregressive conditional heteroscedasticity) models of Engle (1982), Engle & Bollerslev (1986) appear to be promissing in a multiple time series context. Since in many time series periods with large volatility and periods with reduced variation cluster together Engle (1982) suggests a model where the conditional variance depends on the history of the series. A simple version of an ARCH model is of the following type:

$$y_t = (\alpha_0 + \alpha_1 y_{t-1}^2)^{1/2} u_t, \tag{13.2.35}$$

where u_t is a standard normal white noise process, that is, $u_t \sim N(0, 1)$. For this process

$$E(y_t|y_{t-1}) = 0,$$

$$\operatorname{Var}(y_t|y_{t-1}) = E(y_t^2|y_{t-1}) = \alpha_0 + \alpha_1 y_{t-1}^2.$$

Moreover, the unconditional mean and variance are

$$E(y_t) = 0 \qquad \text{and} \qquad \operatorname{Var}(y_t) = \alpha_0/(1 - \alpha_1).$$

Thus, although the unconditional moments are time invariant, the conditional variance varies. It is not difficult to write the model (13.2.35) in state space form. ARCH models have been extended to the multivariate case by Engle, Granger & Kraft (1984), Diebold & Nerlove (1987), Bollerslev, Engle & Wooldridge (1988) and others. We will not study these models here but turn to the Kalman filter and its use in analyzing state space models now.

13.3 The Kalman Filter

The Kalman filter was originally developed by Kalman (1960) and Kalman & Bucy (1961). It is a tool to recursively estimate the state z_t given observations $y_1, \ldots, y_T$ of the output variables. Under normality assumptions the optimal estimator produced by the Kalman filter is the conditional expectation $E[z_t|y_1, \ldots, y_T]$. The Kalman filter also provides the conditional covariance matrix $\operatorname{Cov}(z_t|y_1, \ldots, y_T)$ which may serve as a measure for estimation or prediction uncertainty. Of course, for $t > T$ the estimator is actually a forecast or prediction at origin T in the terminology of the previous chapters. The computation of the estimators $E[z_t|y_1, \ldots, y_t]$, $t = 1, \ldots, T$, is called *filtering* to distinguish it from the forecasting problem.

In some of the examples of Section 13.2 estimation of the state vectors is of obvious interest, for instance, if the state vector consists of time varying coefficients as in Section 13.2.1h or if the state vector contains the unobserved factors of a dynamic factor analytic model as in Section 13.2.1f. In other cases where the state vector is not of foremost interest or where it consists of observable variables the conditional means and covariance matrices can still be useful in evaluating the likelihood function. We will return to this point in Section 13.4. Now the Kalman filter recursions will be given.

13.3.1 The Kalman Filter Recursions

13.3.1a Assumptions for the State Space Model

We assume a linear state space model with transition equation

$$z_t = \mathbf{B}z_{t-1} + \mathbf{F}x_{t-1} + w_{t-1} \tag{13.3.1}$$

and with measurement equation

$$y_t = \mathbf{H}_t z_t + \mathbf{G}x_t + v_t \tag{13.3.2}$$

for $t = 1, 2, \ldots$. Note that both input matrices and the transition matrix are time invariant and known. This condition is satisfied in most of the example models of Section 13.2. The measurement matrices $\mathbf{H}_t$ are assumed to be known and nonstochastic at time t. This assumption does not exclude lagged output variables from $\mathbf{H}_t$ because the past output variables are given at time t. The input sequence x_t, $t = 0, 1, \ldots$, is assumed to be nonstochastic. The white noise processes w_t and v_t are independent. They are both Gaussian with time invariant covariances,

$$w_t \sim N(0, \Sigma_w), \qquad t = 0, 1, \ldots,$$

$$v_t \sim N(0, \Sigma_v), \qquad t = 1, \ldots.$$

Also the initial state is Gaussian, $z_0 \sim N(\mu_0, \Sigma_0)$, and it is independent of v_t, w_{t-1}, $t = 1, \ldots$. The initial state may be a constant, nonstochastic vector in which case $\Sigma_0 = 0$.

With the exception of the normality assumption the foregoing conditions are satisfied for most of the example models of Section 13.2 under the usual assumptions entertained for these models. It is possible to derive recursions similar to those given below under more general conditions. If the normality assumption is dropped the recursions given below can still be justified.

13.3.1b The Recursions

We will use the following additional notation in stating the Kalman filter recursions:

$$z_{t|s} := E(z_t | y_1, \ldots, y_s),$$
$$\Sigma_z(t|s) := \mathrm{Cov}(z_t | y_1, \ldots, y_s),$$
$$y_{t|s} := E(y_t | y_1, \ldots, y_s), \tag{13.3.3}$$
$$\Sigma_y(t|s) := \mathrm{Cov}(y_t | y_1, \ldots, y_s),$$

$(z|y) \sim N(\mu, \Sigma)$ symbolizes that the conditional distribution of z given y is a multivariate normal with mean μ and covariance Σ.

Under the conditions of Section 13.3.1a,

$$(z_t | y_1, \ldots, y_{t-1}) \sim N(z_{t|t-1}, \Sigma_z(t|t-1)), \qquad \text{for } t = 2, \ldots, T, \tag{13.3.4}$$

$$(z_t | y_1, \ldots, y_t) \sim N(z_{t|t}, \Sigma_z(t|t)), \qquad \text{for } t = 1, \ldots, T, \tag{13.3.5}$$

$$(y_t | y_1, \ldots, y_{t-1}) \sim N(y_{t|t-1}, \Sigma_y(t|t-1)), \qquad \text{for } t = 2, \ldots, T, \tag{13.3.6}$$

and

$$(z_t | y_1, \ldots, y_T) \sim N(z_{t|T}, \Sigma_z(t|T)), \tag{13.3.7}$$

$$(y_t | y_1, \ldots, y_T) \sim N(y_{t|T}, \Sigma_y(t|T)), \qquad \text{for } t > T. \tag{13.3.8}$$

The conditional means and covariance matrices can be obtained by the following *Kalman filter recursions* which are graphically depicted in Figure 13.1:

Initialization: $z_{0|0} := \mu_0$, $\Sigma_z(0|0) := \Sigma_0$.

Prediction step $(1 \leq t \leq T)$:

$$z_{t|t-1} = \mathbf{B} z_{t-1|t-1} + \mathbf{F} x_{t-1},$$
$$\Sigma_z(t|t-1) = \mathbf{B}\Sigma_z(t-1|t-1)\mathbf{B}' + \Sigma_w,$$
$$y_{t|t-1} = \mathbf{H}_t z_{t|t-1} + \mathbf{G} x_t,$$
$$\Sigma_y(t|t-1) = \mathbf{H}_t \Sigma_z(t|t-1)\mathbf{H}_t' + \Sigma_v.$$

Correction step $(1 \leq t \leq T)$:

$$z_{t|t} = z_{t|t-1} + \mathbf{P}_t(y_t - y_{t|t-1}),$$
$$\Sigma_z(t|t) = \Sigma_z(t|t-1) - \mathbf{P}_t \Sigma_y(t|t-1)\mathbf{P}_t',$$

where

$$\mathbf{P}_t := \Sigma_z(t|t-1)\mathbf{H}_t' \Sigma_y(t|t-1)^{-1} \qquad \textit{(Kalman filter gain)}.$$

If the inverse of $\Sigma_y(t|t-1)$ does not exist it may be replaced by a generalized inverse. The recursions proceed by performing the prediction step for $t = 1$. Then the correction step is carried out for $t = 1$. Then the prediction and correction steps are repeated for $t = 2$, and so on.

Forecasting step $(t > T)$:

$$z_{t|T} = \mathbf{B} z_{t-1|T} + \mathbf{F} x_{t-1},$$

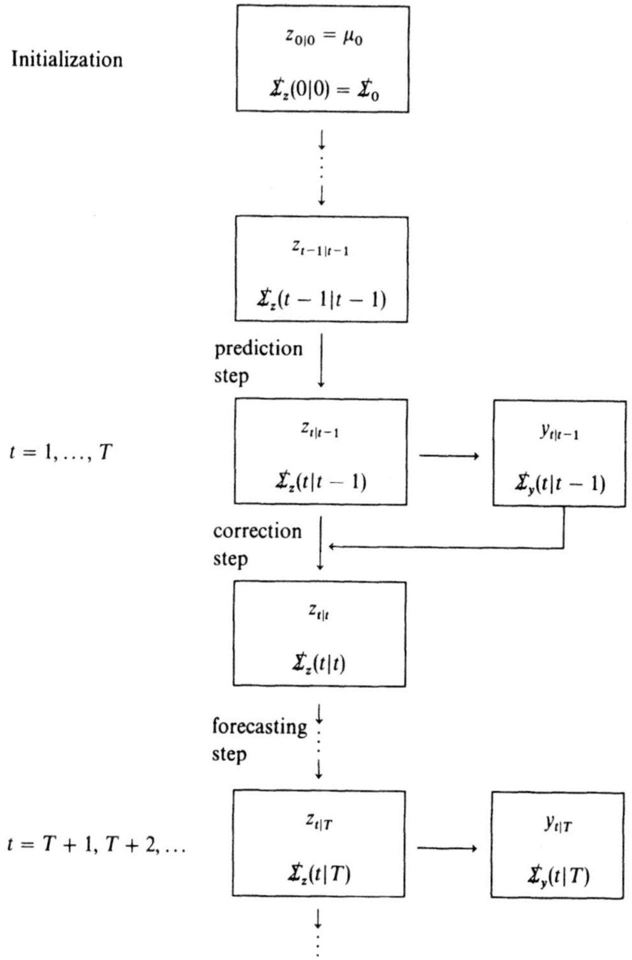

Figure 13.1. Kalman Filter Recursions

$$\Sigma_z(t|T) = \mathbf{B}\Sigma_z(t-1|T)\mathbf{B}' + \Sigma_w,$$
$$y_{t|T} = \mathbf{H}_t z_{t|T} + \mathbf{G}x_t,$$
$$\Sigma_y(t|T) = \mathbf{H}_t\Sigma_z(t|T)\mathbf{H}_t' + \Sigma_v.$$

The forecasting step may be carried out recursively for $t = T + 1, T + 2, \ldots$.

13.3.1c Computational Aspects and Extensions

In practice, in running through the Kalman filter recursions, computational inaccuracies may accumulate in such a way that the actually computed covariance matrices are not positive semidefinite. These and other computational issues are discussed in Anderson & Moore (1979, Chapter 6) and numerical modifications of the recursions are suggested that may help to overcome the possible difficulties (see also Schneider (1990a)).

As mentioned previously, it is possible to justify the Kalman filter recursions even if the initial state and the white noise processes are not Gaussian. In that case the quantities obtained by the recursions are no longer moments of conditional normal distributions, however. For other interpretations of the quantities see, for example, Schneider (1988).

Sometimes reconstruction of the state vectors given all the information $y_1, \ldots, y_T$ is of interest. For instance, in the random coefficient models of Subsection 13.2.1h, where the state vector z_t contains the coefficients associated with period t, one may want to estimate the states and hence the coefficients given all the sample information $y_1, \ldots, y_T$. We will see a detailed example in Section 13.5. Recursions are also available to compute $z_{t|T}$ and $\Sigma_z(t|T)$ for $t < T$. The evaluation of $z_{t|T}$ for $t < T$ is known as *smoothing*. Under the assumptions of the previous subsection (including normality),

$$(z_t | y_1, \ldots, y_T) \sim N(z_{t|T}, \Sigma_z(t|T))$$

for $t = 0, 1, \ldots, T$. The conditional moments may be obtained recursively for $t = T-1, T-2, \ldots, 0$ as follows.

Smoothing step ($t < T$):

$$z_{t|T} = z_{t|t} + \mathbf{S}_t(z_{t+1|T} - z_{t+1|t}),$$

$$\Sigma_z(t|T) = \Sigma_z(t|t) - \mathbf{S}_t[\Sigma_z(t+1|t) - \Sigma_z(t+1|T)]\mathbf{S}_t',$$

where

$$\mathbf{S}_t := \Sigma_z(t|t)\mathbf{B}'\Sigma_z(t+1|t)^{-1} \qquad (\textit{Kalman smoothing matrix}),$$

(see Anderson & Moore (1979)).

13.3.2 Proof of the Kalman Filter Recursions

The proof follows Anderson & Moore (1979, pp. 39–41) and Meinhold & Singpurwalla (1983). It may be skipped without loss of continuity. We proceed inductively and we use the following property of multivariate normal distributions (see Propositions B.1 and B.2 of Appendix B):

$y \sim N(\mu_y, \Sigma_y)$ and $z \sim N(\mu_z, \Sigma_z)$ are independent $(K \times 1)$ random vectors

$$\Rightarrow y + z \sim N(\mu_y + \mu_z, \Sigma_y + \Sigma_z), \qquad (13.3.9)$$

$$y \sim N(\mu_y, \Sigma_y) \Rightarrow Ay + c \sim N(A\mu_y + c, A\Sigma_y A'), \qquad (13.3.10)$$

and

$$\begin{bmatrix} z \\ y \end{bmatrix} \sim N\left(\begin{bmatrix} \mu_z \\ \mu_y \end{bmatrix}, \begin{bmatrix} \Sigma_z & \Sigma_{zy} \\ \Sigma_{yz} & \Sigma_y \end{bmatrix}\right)$$

$$\Rightarrow (z|y) \sim N(\mu_z + \Sigma_{zy}\Sigma_y^{-1}(y - \mu_y), \quad \Sigma_z - \Sigma_{zy}\Sigma_y^{-1}\Sigma_{yz}). \tag{13.3.11}$$

Here Σ_y^{-1} may be replaced by a generalized inverse if Σ_y is singular.

We will now demonstrate the prediction and correction step for $t = 1$. With that goal in mind we note that by (13.3.9) and (13.3.10) and the joint normality of w_0 and v_1, z_1 and y_1 are jointly normally distributed,

$$\begin{aligned}
\begin{bmatrix} z_1 \\ y_1 \end{bmatrix} &= \begin{bmatrix} I \\ \mathbf{H}_1 \end{bmatrix} z_1 + \begin{bmatrix} 0 \\ \mathbf{G}x_1 \end{bmatrix} + \begin{bmatrix} 0 \\ I \end{bmatrix} v_1 \\
&= \begin{bmatrix} I \\ \mathbf{H}_1 \end{bmatrix} \mathbf{B} z_0 + \begin{bmatrix} \mathbf{F}x_0 \\ \mathbf{G}x_1 + \mathbf{H}_1 \mathbf{F} x_0 \end{bmatrix} + \begin{bmatrix} I & 0 \\ \mathbf{H}_1 & I \end{bmatrix} \begin{bmatrix} w_0 \\ v_1 \end{bmatrix} \\
&\sim N\left(\begin{bmatrix} \mathbf{B}\mu_0 + \mathbf{F}x_0 \\ \mathbf{H}_1(\mathbf{B}\mu_0 + \mathbf{F}x_0) + \mathbf{G}x_1 \end{bmatrix}, \right. \\
&\qquad \left. \begin{bmatrix} I & 0 \\ \mathbf{H}_1 & I \end{bmatrix} \begin{bmatrix} \Sigma_w & 0 \\ 0 & \Sigma_v \end{bmatrix} \begin{bmatrix} I & \mathbf{H}_1' \\ 0 & I \end{bmatrix} + \begin{bmatrix} \mathbf{B} \\ \mathbf{H}_1 \mathbf{B} \end{bmatrix} \Sigma_0 [\mathbf{B}', \mathbf{B}'\mathbf{H}_1'] \right).
\end{aligned}$$

Hence,

$$\begin{aligned}
z_{1|0} &= E(z_1) = \mathbf{B}\mu_0 + \mathbf{F}x_0 = \mathbf{B}z_{0|0} + \mathbf{F}x_0, \\
\Sigma_z(1|0) &= \mathrm{Cov}(z_1) = \Sigma_w + \mathbf{B}\Sigma_0\mathbf{B}' = \mathbf{B}\Sigma_z(0|0)\mathbf{B}' + \Sigma_w, \\
y_{1|0} &= E(y_1) = \mathbf{H}_1 z_{1|0} + \mathbf{G}x_1, \\
\Sigma_y(1|0) &= \mathrm{Cov}(y_1) = \mathbf{H}_1\Sigma_w\mathbf{H}_1' + \Sigma_v + \mathbf{H}_1\mathbf{B}\Sigma_0\mathbf{B}'\mathbf{H}_1' = \mathbf{H}_1\Sigma_z(1|0)\mathbf{H}_1' + \Sigma_v,
\end{aligned}$$

which proves the prediction step for $t = 1$. By (13.3.11), the conditional distribution of z_1 given y_1 is

$$\begin{aligned}
(z_1|y_1) \sim N[&z_{1|0} + \Sigma_z(1|0)\mathbf{H}_1'\Sigma_y(1|0)^{-1}(y_1 - y_{1|0}), \\
&\Sigma_z(1|0) - \Sigma_z(1|0)\mathbf{H}_1'\Sigma_y(1|0)^{-1}\mathbf{H}_1\Sigma_z(1|0)]
\end{aligned}$$

which proves the correction step for $t = 1$.

Now suppose the normal distributions in (13.3.4)–(13.3.6) and the prediction and correction steps are correct for $t - 1$. Then, using the transition and measurement equations, z_t and y_t have a joint normal distribution

$$\begin{aligned}
\begin{bmatrix} z_t \\ y_t \end{bmatrix} &= \begin{bmatrix} I \\ \mathbf{H}_t \end{bmatrix} z_t + \begin{bmatrix} 0 \\ \mathbf{G} \end{bmatrix} x_t + \begin{bmatrix} 0 \\ I \end{bmatrix} v_t \\
&= \begin{bmatrix} I \\ \mathbf{H}_t \end{bmatrix} (\mathbf{B}z_{t-1} + \mathbf{F}x_{t-1} + w_{t-1}) + \begin{bmatrix} 0 \\ \mathbf{G} \end{bmatrix} x_t + \begin{bmatrix} 0 \\ I \end{bmatrix} v_t.
\end{aligned}$$

By the induction assumption and (13.3.9)/(13.3.10) this term has the following conditional normal distribution given $y_1, \ldots, y_{t-1}$:

$$N\left(\begin{bmatrix} \mathbf{B}z_{t-1|t-1}+\mathbf{F}x_{t-1} \\ \mathbf{H}_t(\mathbf{B}z_{t-1|t-1}+\mathbf{F}x_{t-1})+\mathbf{G}x_t \end{bmatrix}, \begin{bmatrix} \Sigma_z(t|t-1) & \\ \mathbf{H}_t\Sigma_z(t|t-1) & \mathbf{H}_t\Sigma_z(t|t-1)\mathbf{H}_t' + \Sigma_v \end{bmatrix}\right), \tag{13.3.12}$$

where $\Sigma_z(t|t-1) = \mathbf{B}\Sigma_z(t-1|t-1)\mathbf{B}' + \Sigma_w$. This proves the prediction step. Application of (13.3.11) to (13.3.12) gives the conditional distribution of z_t given $y_1, \ldots, y_t$ and proves the correction step.

It remains to prove the forecasting step. Again by induction $(z_t|y_1,\ldots,y_T)$ and $(y_t|y_1,\ldots,y_T)$ both have normal distributions with the first and second moments as stated in the forecasting step.

13.4 Maximum Likelihood Estimation of State Space Models

In this section we consider ML estimation of the state space system given in Section 13.3.1. We collect the time invariant unknown parameters from the matrices $\mathbf{B}$, $\mathbf{F}$, $\mathbf{H}_t$, $\mathbf{G}$, Σ_w, Σ_v, Σ_0, and the vector μ_0 in the parameter vector $\boldsymbol{\delta}$. For a given $\boldsymbol{\delta}$ the matrices are assumed to be uniquely determined and at least twice continuously differentiable with respect to the elements of $\boldsymbol{\delta}$. For instance, in the state space model (13.2.8)/(13.2.9) which represents the finite order VAR process (13.2.6),

$$\boldsymbol{\delta} = \begin{bmatrix} \mathrm{vec}[v, A_1,\ldots,A_p] \\ \mathrm{vech}(\Sigma_u) \end{bmatrix}$$

if no constraints are placed on the VAR coefficients or Σ_u and if the initial conditions $y_{-p+1}, \ldots, y_0$ are assumed to be known and fixed. The objective in this section is to estimate $\boldsymbol{\delta}$. We will set up the log-likelihood function first. Then we discuss its maximization and finally the asymptotic properties of the ML estimators are considered.

13.4.1 The Log-Likelihood Function

By Bayes' Theorem the sample density function can be written as

$$\begin{aligned} f(y_1,\ldots,y_T;\boldsymbol{\delta}) &= f(y_1;\boldsymbol{\delta})f(y_2,\ldots,y_T|y_1;\boldsymbol{\delta}) \\ &\;\;\vdots \\ &= f(y_1;\boldsymbol{\delta})f(y_2|y_1;\boldsymbol{\delta})\cdots f(y_T|y_1,\ldots,y_{T-1};\boldsymbol{\delta}). \end{aligned}$$

Thus, using the notation of the previous section and assuming that y_t has dimension K the Gaussian log-likelihood for the present case becomes

$$\begin{aligned} \ln l(\boldsymbol{\delta}|y_1,\ldots,y_T) &= \ln f(y_1,\ldots,y_T;\boldsymbol{\delta}) \\ &= \ln f(y_1;\boldsymbol{\delta}) + \sum_{t=2}^{T} \ln f(y_t|y_1,\ldots,y_{t-1};\boldsymbol{\delta}) \end{aligned}$$

$$= -\frac{KT}{2}\ln(2\pi) - \frac{1}{2}\sum_{t=1}^{T}\ln|\Sigma_y(t|t-1)|$$

$$-\frac{1}{2}\sum_{t=1}^{T}(y_t - y_{t|t-1})'\Sigma_y(t|t-1)^{-1}(y_t - y_{t|t-1}), \qquad (13.4.1)$$

where we have used the result

$$(y_t|y_1,\ldots,y_{t-1}) \sim N(y_{t|t-1}, \Sigma_y(t|t-1))$$

from Section 13.3.1b. Here both $y_{t|t-1}$ and $\Sigma_y(t|t-1)$ depend in general on the parameter vector $\boldsymbol{\delta}$. If a specific vector $\boldsymbol{\delta}$ is given all the quantities in the log-likelihood function can be computed with the Kalman filter recursions. Thus, the Kalman filter is seen to be a useful tool for evaluating the log-likelihood function of a wide range of models. Note also that we have struggled with likelihood approximations for VARMA processes in Chapter 7. In the present framework the exact likelihood may be obtained (see also Solo (1984)).

To simplify the expression for the log-likelihood given in (13.4.1) we use the following notation:

$$e_t(\boldsymbol{\delta}) := y_t - y_{t|t-1} \quad \text{and } \Sigma_t(\boldsymbol{\delta}) := \Sigma_y(t|t-1). \qquad (13.4.2)$$

This notation makes the dependence on $\boldsymbol{\delta}$ explicit. Occasionally we will, however, drop $\boldsymbol{\delta}$. With this notation the log-likelihood function becomes

$$\ln l(\boldsymbol{\delta}) = -\frac{KT}{2}\ln(2\pi) - \frac{1}{2}\sum_{t=1}^{T}[\ln|\Sigma_t(\boldsymbol{\delta})| + e_t'(\boldsymbol{\delta})\Sigma_t(\boldsymbol{\delta})^{-1}e_t(\boldsymbol{\delta})]. \qquad (13.4.3)$$

13.4.2 The Identification Problem

Recall from the discussion in Chapter 7 that unique maximization of the likelihood function and asymptotic inference require an identified or unique parameterization. Identification is not automatic in the present context because, for instance, VARMA models are not identified without specific restrictions and VARMA processes are just special cases of the presently considered models. Hence, the identification or uniqueness problem is inherent in the general linear state space model, too. We will state the problem here again in sufficient generality to cover the present case.

Let $\mathbf{y} := \text{vec}(y_1,\ldots,y_T)$ be the vector of observed random variables and denote its distribution by $F(\mathbf{y}; \boldsymbol{\delta}_0)$, where $\boldsymbol{\delta}_0$ is the *true parameter vector*. We assume that the true distribution of $\mathbf{y}$ is a member of the parametric family

$$\{F(\mathbf{y}; \boldsymbol{\delta})|\boldsymbol{\delta} \in \Delta\},$$

where $\Delta \subset \mathbb{R}^n$ is the parameter space. The vector $\boldsymbol{\delta}_0$ is *identified* or *identifiable* if it is the only vector in Δ which gives rise to the distribution of $\mathbf{y}$. In other words, for any $\boldsymbol{\delta}_1 \in \Delta$,

$$\boldsymbol{\delta}_1 \neq \boldsymbol{\delta}_0 \Rightarrow F(\mathbf{y}; \boldsymbol{\delta}_1) \neq F(\mathbf{y}; \boldsymbol{\delta}_0) \quad \text{(for at least one } \mathbf{y}). \tag{13.4.4}$$

To compute ML estimators and to derive asymptotic properties it is actually sufficient if $\boldsymbol{\delta}_0$ has a neighborhood in which it is uniquely determined by the true distribution of $\mathbf{y}$. To distinguish this case from one where uniqueness follows for the whole parameter space the vector $\boldsymbol{\delta}_0$ or the model is often called *locally identified* or *locally identifiable* if there exists a neighborhood $U(\boldsymbol{\delta}_0)$ such that (13.4.4) holds for any $\boldsymbol{\delta}_1 \in U(\boldsymbol{\delta}_0)$. In contrast, the model or parameter vector is *globally identifiable* or *globally identified* if (13.4.4) holds for all $\boldsymbol{\delta}_1 \in \Delta$.

Since the negative log-likelihood function has a locally unique minimum if its Hessian matrix is positive definite, identification conditions for state space models may be formulated via the information matrix. If we are interested in *asymptotic* properties of estimators it is sufficient to obtain identification in large samples. Hence, under some regularity conditions, the identification assumption may be disguised in the requirement of a positive definite asymptotic information matrix. In a later proposition giving asymptotic properties of the ML estimators, to ensure identification, we will include the condition that the sequence of normalized information matrices, $I(\boldsymbol{\delta}_0)/T$, is bounded from below by a positive definite matrix as T goes to infinity. In special case models other identification conditions are often easier to deal with and are therefore preferred. For example, for VARMA processes the identification conditions given in Section 7.1.2 may be used.

13.4.3 Maximization of the Log-Likelihood Function

From Chapter 7 we know that maximization of the log-likelihood function is in general a nonlinear maximization problem. Therefore numerical optimization methods are required for its solution. One possibility is a gradient algorithm as described in Section 7.3.2 for iteratively minimizing $-\ln l$. Recall that the general form of the i-th iteration step is

$$\boldsymbol{\delta}_{i+1} = \boldsymbol{\delta}_i - s_i D_i \left[\frac{\partial - \ln l}{\partial \boldsymbol{\delta}} \bigg|_{\boldsymbol{\delta}_i} \right], \tag{13.4.5}$$

where s_i is the step length and D_i is a positive definite direction matrix. The inverse information matrix is one possible choice for this matrix. In that case the method is called *scoring algorithm.* We will provide the ingredients for this algorithm in the following, that is, we will give expressions for the gradient of $\ln l$ and an estimator of the information matrix. There are various ways to choose the step length s_i. For instance, it could be chosen so as to optimize the progress towards the minimum. Another alternative would be to simply set $s_i = 1$. We will not discuss the step length selection in further detail here.

13.4.3a The Gradient of the Log-Likelihood

From (13.4.3) we get

$$\frac{\partial \ln l}{\partial \boldsymbol{\delta}'} = -\frac{1}{2}\sum_{t=1}^{T}\left[\operatorname{vec}\left(\frac{\partial \ln|\Sigma_t|}{\partial \Sigma_t}\right)'\frac{\partial \operatorname{vec}(\Sigma_t)}{\partial \boldsymbol{\delta}'} + \frac{\partial \operatorname{tr}(e_t'\Sigma_t^{-1}e_t)}{\partial \boldsymbol{\delta}'}\right]$$

$$= -\frac{1}{2}\sum_t\left[\operatorname{vec}(\Sigma_t^{-1})'\frac{\partial \operatorname{vec}(\Sigma_t)}{\partial \boldsymbol{\delta}'} + 2e_t'\Sigma_t^{-1}\frac{\partial e_t}{\partial \boldsymbol{\delta}'}\right.$$

$$\left. - \operatorname{vec}(\Sigma_t^{-1}e_te_t'\Sigma_t^{-1})'\frac{\partial \operatorname{vec}(\Sigma_t)}{\partial \boldsymbol{\delta}'}\right]$$

$$= -\frac{1}{2}\sum_t\left[\operatorname{vec}[\Sigma_t^{-1}(I_K - e_te_t'\Sigma_t^{-1})]'\frac{\partial \operatorname{vec}(\Sigma_t)}{\partial \boldsymbol{\delta}'} + 2e_t'\Sigma_t^{-1}\frac{\partial e_t}{\partial \boldsymbol{\delta}'}\right], \tag{13.4.6}$$

where $\partial e_t/\partial \boldsymbol{\delta}'$ may be replaced by $-\partial y_{t|t-1}/\partial \boldsymbol{\delta}'$.

13.4.3b The Information Matrix

Using $E(e_te_t') = \Sigma_t(= \Sigma_y(t|t-1))$ and $E(e_t) = 0$, straightforward application of the rules for matrix and vector differentiation yields the information matrix

$$I(\boldsymbol{\delta}_0) = E\left[\frac{\partial^2 - \ln l}{\partial \boldsymbol{\delta}\partial \boldsymbol{\delta}'}\bigg|_{\boldsymbol{\delta}_0}\right]$$

$$= \frac{1}{2}\sum_{t=1}^{T}\left[\frac{\partial \operatorname{vec}(\Sigma_t)'}{\partial \boldsymbol{\delta}}(\Sigma_t^{-1}\otimes \Sigma_t^{-1})\frac{\partial \operatorname{vec}(\Sigma_t)}{\partial \boldsymbol{\delta}'}\right.$$

$$\left. + 2E\left(\frac{\partial e_t'}{\partial \boldsymbol{\delta}}\Sigma_t^{-1}\frac{\partial e_t}{\partial \boldsymbol{\delta}'}\right)\right]. \tag{13.4.7}$$

Since the true parameter values involved in this expression are unknown they are replaced by estimators and the expectation is simply dropped. For instance, in the i-th iteration of the scoring algorithm $\boldsymbol{\delta}_i$ is used as an estimator for $\boldsymbol{\delta}_0$.

13.4.3c Discussion of the Scoring Algorithm

The scoring algorithm may have poor convergence properties far away from the maximum of the log-likelihood function. On the other hand it has very good convergence properties close to the maximum. Unfortunately it is quite expensive in terms of computation time because it requires (possibly numerical) evaluation of derivatives in each iteration. Therefore other maximization methods have been proposed in the literature. Notably the EM (expectation step – maximization step) algorithm of Dempster, Laird & Rubin (1977) has been found useful in practice (see Watson & Engle (1983), Schneider (1990a)). The EM algorithm is an iterative algorithm which has the advantage of involving much cheaper computations in each iteration step than the scoring algorithm. On the other hand, convergence of the former is slower than that of the latter algorithm. Nicholls & Pagan (1985) and Schneider (1990a,b) suggest combining the EM and the scoring algorithms. This may be especially useful if no good initial estimator $\boldsymbol{\delta}_1$ is available from where to start the scoring algorithm.

13.4.4 Asymptotic Properties of the ML Estimators

We consider the state space model from Section 13.3.1 with transition equation

$$z_t = \mathbf{B}z_{t-1} + \mathbf{F}x_{t-1} + w_{t-1} \tag{13.4.8}$$

and measurement equation

$$y_t = \mathbf{H}_t z_t + \mathbf{G}x_t + v_t. \tag{13.4.9}$$

All assumptions of Section 13.3.1a are assumed to be satisfied. In addition we assume that

(i) the true parameter vector $\boldsymbol{\delta}_0$ is in the interior of the parameter space;
(ii) $\mathbf{H}_t = (x_t \otimes I)J$, where J is a known selection matrix such as $J = [I_K \quad 0 \dots 0]$ or $\mathbf{H}_t = \mathbf{H}$ is a time invariant nonstochastic matrix;
(iii) the inputs x_t are nonstochastic and uniformly bounded, that is, there exist real numbers c_1 and c_2 such that $c_1 \le x_t'x_t \le c_2$ for all $t = 0, 1, 2, \dots$;
(iv) the sequence of normalized information matrices is bounded from below by a positive definite matrix, that is, there exists a constant c such that

$$I(\boldsymbol{\delta}_0)/T > cI_n \qquad \text{as } T \to \infty;$$

(v) the eigenvalues of $\mathbf{B}$ have all modulus less than 1.

As we have seen in Section 13.4.2, (iv) is an identification condition. The last assumption is a stability condition and (iii) guarantees that the input variables have no trend. We have seen in Chapter 11 that the standard asymptotic theory does not apply for trending variables. Therefore they are excluded here. With these assumptions the following proposition can be established.

PROPOSITION 13.1 (*Asymptotic Properties of the ML Estimators*)

With all the assumptions stated in the foregoing the ML estimator $\tilde{\boldsymbol{\delta}}$ of $\boldsymbol{\delta}_0$ is consistent and asymptotically normally distributed,

$$\sqrt{T}(\tilde{\boldsymbol{\delta}} - \boldsymbol{\delta}_0) \xrightarrow{d} N(0, \Sigma_{\tilde{\boldsymbol{\delta}}}), \tag{13.4.10}$$

where

$$\Sigma_{\tilde{\boldsymbol{\delta}}} = \lim (I(\boldsymbol{\delta}_0)/T)^{-1}$$

is the inverse asymptotic information matrix. It is consistently estimated by substituting the ML estimators for unknown parameters in (13.4.7), dropping the expectation operator and dividing by T. ■

Pagan (1980) gives a proof of this proposition based on Crowder (1976) (see also Schneider (1988)). Other sets of conditions are possible to accommodate the situation where the inputs x_t are stochastic. They may, in fact, contain lagged y_t's. Moreover, $\mathbf{B}$ may have eigenvalues on the unit circle if it does not contain unknown parameters. The reader is referred to the articles by Pagan (1980), Nicholls & Pagan (1985), Schneider (1988), and to a book by Caines (1988) for details. When particular models are considered different sets of assumptions are often preferable for two reasons. First, other sets of conditions may be easier to verify or to understand for special models. Second, the conditions of Proposition 13.1 or the modifications mentioned in the foregoing may not be satisfied. We will see an example of the latter case shortly.

A number of alternatives to ML estimation have been suggested, see, e.g., Anderson & Moore (1979), Nicholls & Pagan (1985), and Schneider (1988) for more details and references.

It may be worth noting that application of the Kalman filter to systems with estimated parameters produces state estimates and precision matrices that do not take into account the estimation variability. Watanabe (1985) and Hamilton (1986) consider the properties of state estimators obtained with estimated parameter Kalman filter recursions.

13.5 A Real Data Example

As an example we consider a time varying parameter dynamic consumption function

$$\begin{aligned} y_t &= \gamma_{0t} + \gamma_{1t}x_t + \gamma_{2t}x_{t-1} + \gamma_{3t}y_{t-1} + \gamma_{4t}x_{t-2} + \gamma_{5t}y_{t-2} + v_t \\ &= X_t'\boldsymbol{\gamma}_t + v_t, \end{aligned} \tag{13.5.1}$$

where

$$X_t := \begin{bmatrix} 1 \\ x_t \\ x_{t-1} \\ y_{t-1} \\ x_{t-2} \\ y_{t-2} \end{bmatrix} \quad \text{and} \quad \boldsymbol{\gamma}_t := \begin{bmatrix} \gamma_{0t} \\ \gamma_{1t} \\ \gamma_{2t} \\ \gamma_{3t} \\ \gamma_{4t} \\ \gamma_{5t} \end{bmatrix}.$$

Here y_t and x_t represent rates of change (first differences of logarithms) of consumption and income, respectively. Suppose that the coefficient vector $\boldsymbol{\gamma}_t$ differs from $\boldsymbol{\gamma}_{t-1}$ by an additive random disturbance, that is,

$$\boldsymbol{\gamma}_t = \boldsymbol{\gamma}_{t-1} + w_{t-1}. \tag{13.5.2}$$

In other words, γ_t is driven by a (multivariate) random walk. Clearly, (13.5.1) and (13.5.2) represent the measurement and transition equations of a state space model. We complete the model by assuming that v_t and w_t are independent Gaussian white noise processes, $v_t \sim N(0, \sigma_v^2)$ and $w_t \sim N(0, \Sigma_w)$, where

$$\Sigma_w = \begin{bmatrix} \sigma_{w_0}^2 & & 0 \\ & \ddots & \\ 0 & & \sigma_{w_5}^2 \end{bmatrix} \tag{13.5.3}$$

is a diagonal matrix. Furthermore the initial state γ_0 is also assumed to be normally distributed, $\gamma_0 \sim N(\bar{\gamma}_0, \Sigma_0)$, and independent of v_t and w_t. Admittedly our assumed model is quite simple. Still, it is useful to illustrate some concepts considered in the previous sections.

Assuming that a sample $\mathbf{y} = (y_1, \dots, y_T)'$ is available, the log-likelihood function of our model is

$$\ln l(\sigma_v^2, \Sigma_w, \bar{\gamma}_0, \Sigma_0 | \mathbf{y})$$
$$= -\frac{T}{2}\ln(2\pi) - \frac{1}{2}\sum_{t=1}^{T} \ln|\Sigma_y(t|t-1)| - \frac{1}{2}\sum_{t=1}^{T} (y_t - y_{t|t-1})^2/\Sigma_y(t|t-1), \tag{13.5.4}$$

where $\Sigma_y(t|t-1)$ is a scalar ((1 × 1) matrix) as y_t is a univariate variable. The log-likelihood function may be evaluated with the Kalman filter recursions for given parameters σ_v^2, Σ_w, $\bar{\gamma}_0$, and Σ_0. The maximization problem may be solved with an iterative algorithm. Once estimates of the parameters σ_v^2 and Σ_w are available, estimates $\gamma_{t|T}$ of the coefficients of the consumption function (13.5.1) may be obtained with the smoothing recursions given in Section 13.3.1c.

Using first differences of logarithms of the quarterly consumption and income data for the years 1960 to 1982 given in Appendix E, Table E.1, we have estimated the parameters of the state space model (13.5.1)/(13.5.2). The ML estimates of the parameters of interest, namely the variances σ_v^2 and $\sigma_{w_i}^2$, $i = 0, 1, \dots, 5$, together with estimated standard errors (square roots of the diagonal elements of the estimated inverse information matrix) and corresponding t-ratios are given in Table 13.1.

The interpretation of the standard errors and t-ratios needs caution for various reasons. In Proposition 13.1, where the asymptotic distribution of the ML

Table 13.1. ML estimates for the example

parameter	estimate	standard error	t-ratio
σ_v^2	3.91×10^{-5}	1.99×10^{-5}	1.97
$\sigma_{w_0}^2$	2.04×10^{-5}	2.33×10^{-5}	.88
$\sigma_{w_1}^2$	$.14 \times 10^{-2}$	1.08×10^{-2}	.13
$\sigma_{w_2}^2$	$.46 \times 10^{-2}$	$.92 \times 10^{-2}$	.50
$\sigma_{w_3}^2$	$.45 \times 10^{-2}$	1.11×10^{-2}	.41
$\sigma_{w_4}^2$	$.51 \times 10^{-2}$	$.94 \times 10^{-2}$	.54
$\sigma_{w_5}^2$	$.62 \times 10^{-2}$	1.16×10^{-2}	.54

estimators is given, we have assumed that all eigenvalues of the transition matrix **B** have modulus less than 1. This condition is clearly not satisfied in the present example, where $\mathbf{B} = I_6$ and thus all six eigenvalues are equal to 1. However, as mentioned in Section 13.4.4, the condition on the eigenvalues of **B** is not crucial if **B** is a known matrix which does not contain unknown parameters. Of course, setting **B** at I_6 is just an assumption which may or may not be adequate.

A further deviation from the assumptions of Proposition 13.1 is that the inputs X_t contain lagged endogenous variables and hence are stochastic. Again we have mentioned in Section 13.4.4 that this assumption is not necessarily critical. The conditions of Proposition 13.1 could be modified so as to allow for lagged dependent variables.

Another assumption that may be problematic is the normality of the white noise sequences and the initial state. The normality assumption may be checked by computing the skewness and kurtosis of the standardized quantities $(y_t - y_{t|t-1})/\Sigma_y(t|t-1)^{1/2}$. A test for nonnormality may then be based on the χ^2-statistic involving both skewness and kurtosis as described in Chapter 4, Section 4.5.1. For the present example the statistic assumes the value 3.00 and has a $\chi^2(2)$-distribution under the null hypothesis of normality. Thus, it is not significant at any conventional level.

Finally, we have assumed in Proposition 13.1 that the true parameter values lie in the interior of the parameter space. Given that the variance estimates are quite small compared to their estimated standard errors it is possible that at least the $\sigma^2_{w_i}$ are in fact zero and thus lie on the boundary of the feasible parameter space. If the $\sigma^2_{w_i}$ are actually zero the γ_t are time-invariant in our model which would be a hypothesis of considerable interest. It would permit us to work with a constant coefficient specification. Unfortunately, if $\sigma^2_{w_i} = 0$ the corresponding t-ratio does not have an asymptotic standard normal distribution in general. Thus, we cannot use the t-ratios given in Table 13.1 for testing the null hypotheses $\sigma^2_{w_i} = 0$, $i = 0, 1, \ldots, 5$.

In the present context we may ignore the problems related to the asymptotic theory for the moment and simply regard the model as a descriptive tool. Using the estimated values of the parameters of the model we may consider the smoothing estimates $\boldsymbol{\gamma}_{t|T}$ of the states (the coefficients of the consumption function). They are plotted in Figure 13.2. The two-standard error bounds which are also shown in the figure are computed from the $\Sigma_\gamma(t|T)$. These quantities are obtained with the smoothing recursions given in Section 13.3.1c. From the plots in Figure 13.2 it can be seen that the intercept term γ_{0t} is the only coefficient that exhibits substantial variation through time. For instance, a considerable downturn is observed in 1966/1967 where the West German economy was in a recession. All the other coefficients show relatively little variation through time although γ_{3t}, γ_{4t}, and γ_{5t} (the coefficients of y_{t-1}, x_{t-2}, and y_{t-2}, respectively) have a tendency to decline in the second half of the 1970s. However, given the estimated two-standard error bounds, overall the results support a specification with constant coefficients of current and lagged income and lagged consumption.

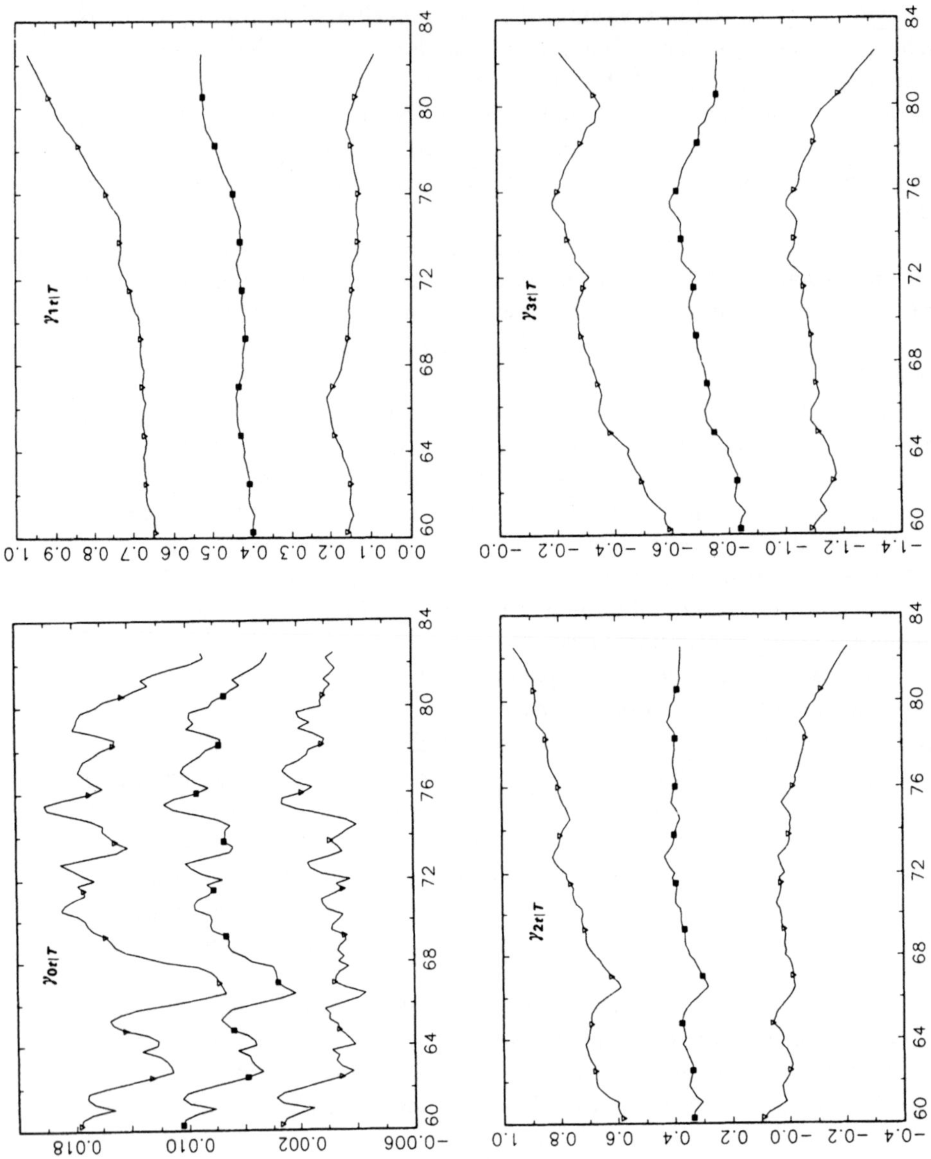

Fig. 13.2. Smoothing estimates of the consumption function coefficients (–■– coefficient estimate, –▽– estimated two-standard error bound).

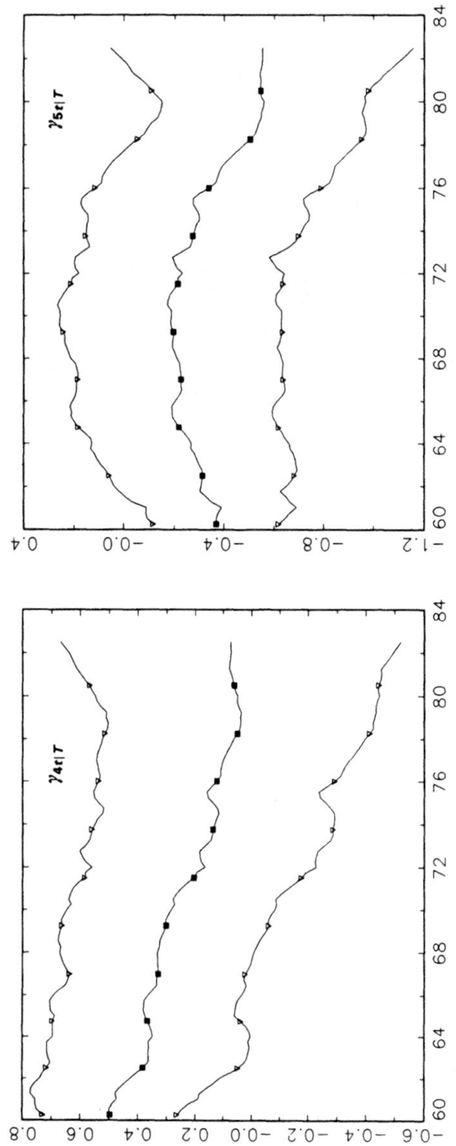

Fig. 13.2. (continued)

As mentioned previously, this example is quite simplistic. It is just meant to illustrate some of the concepts discussed in this chapter. In a more general model other lags of income and/or consumption could appear on the right-hand side of the consumption function, the coefficients could be generated by a more general VAR model and the covariance matrix of w_t could be nondiagonal. Moreover, the consumption function may just be a part of a system of equations.

Given that we have discussed different models for the same data in previous chapters the example also illustrates that there is not just one possible model or model class for the generation process of a multiple time series. The reader may wonder which of the models we have considered in this and the previous chapters is "best". That, however, depends on the questions of interest. In other words, the time series analyst has to decide on the model with the objective of his analysis in mind. In this book we have just tried to introduce some of the possible tools in this venture. With these tools in hand the analyst is hoped to be able to approach his or her problems of interest in a superior way, with an improved sense of the available possibilities and the potential pitfalls.

13.6 Exercises

Problem 13.1

Write the VARMAX model

$$\begin{aligned} y_t = {} & A_1 y_{t-1} + \cdots + A_p y_{t-p} + B_0 x_t + \cdots + B_s x_{t-s} \\ & + u_t + M_1 u_{t-1} + \cdots + M_q u_{t-q} \end{aligned}$$

in state space form.

Problem 13.2

Suppose in the dynamic factor analytic model $y_t = \mathbf{L} f_t + u_t$ the common factors f_t are generated by the VARMA(p, q) process

$$f_t = A_1 f_{t-1} + \cdots + A_p f_{t-p} + \eta_t + M_1 \eta_{t-1} + \cdots + M_q \eta_{t-q}$$

and the individual factors u_t are generated by the VARMA(r, s) process

$$u_t = C_1 u_{t-1} + \cdots + C_r u_{t-r} + \varepsilon_t + D_1 \varepsilon_{t-1} + \cdots + D_s \varepsilon_{t-s}.$$

Write the model in state space form.

Problem 13.3

Assume that y_t is generated according to

$$y_t = A_t Y_{t-1} + C_t x_t + v_t,$$

where $A_t := [A_{1t}, \ldots, A_{pt}]$ and $Y_{t-1} := [y'_{t-1}, \ldots, y'_{t-p}]'$. Suppose that $\boldsymbol{\alpha}_t :=$

vec$[A_t \quad C_t]$ is driven by the VARMA(r, s) process

$$\boldsymbol{\alpha}_t = D_1\boldsymbol{\alpha}_{t-1} + \cdots + D_r\boldsymbol{\alpha}_{t-r} + \boldsymbol{\eta}_t + M_1\boldsymbol{\eta}_{t-1} + \cdots + M_s\boldsymbol{\eta}_{t-s}.$$

Write the model in state space form.

Problem 13.4

Write down explicitly the first two steps of the Kalman filter recursions.

Problem 13.5

Suppose the scalar observable variable y_t is generated by the random coefficient regression model

$$y_t = \nu + x_t\beta_t + v_t, \qquad t = 1, \dots, T,$$

where $\beta_t = \alpha\beta_{t-1} + w_t$ is driven by an AR(1) process. Suppose further that v_t and w_t are independent zero mean Gaussian white noise processes with variance 1 and let β_0 be a standard normal random variable.

a) Determine the conditional distribution of β_t given $y_1, \dots, y_{t-1}$.
b) Write down the log-likelihood function of the model and derive its gradient. Find an expression for the information matrix.

Problem 13.6

Consider the K-dimensional Gaussian stable VAR(1) process $y_t = Ay_{t-1} + u_t$ with $y_0 \sim N(0, 0)$ and $u_t \sim N(0, \Sigma_u)$ for $t = 1, 2, \dots$. Use the Kalman filter recursions to determine $y_{t|t-1}$.

a) Show that $y_{t|t-1} = Ay_{t-1}$.
b) Show that the conditions of Proposition 13.1 are satisfied if

$$\boldsymbol{\delta} = \begin{bmatrix} \text{vec}(A) \\ \text{vech}(\Sigma_u) \end{bmatrix}.$$

Problem 13.7

Repeat the analysis of Section 13.5 with the same data and the state space model consisting of the measurement equation

$$y_t = \gamma_{0t} + \gamma_{1t}x_t + \gamma_{2t}y_{t-1} + v_t$$

and the transition equation

$$\begin{bmatrix} \gamma_{0,t} \\ \gamma_{1,t} \\ \gamma_{2,t} \end{bmatrix} = \begin{bmatrix} \gamma_{0,t-1} \\ \gamma_{1,t-1} \\ \gamma_{2,t-1} \end{bmatrix} + w_{t-1}.$$

Appendices

Appendix A. Vectors and Matrices

The following summary of matrix and vector algebra is not meant to be an introduction to the subject but is just a brief review of terms and rules used in the text. Most of them can be found in books such as Graybill (1969), Searle (1982), Anderson (1984, Appendix), Magnus & Neudecker (1988), Magnus (1988). Therefore proofs or further references are only provided in exceptional cases.

A.1 Basic Definitions

A *matrix* is a rectangular array of numbers. For instance,

$$\begin{bmatrix} 3 & -5 \\ .3 & 0 \end{bmatrix}, \qquad (0 \quad 1 \quad 0), \qquad \begin{bmatrix} 3 & 5 & .3 & .3 \\ 2 & 2 & 2 & 2 \end{bmatrix}$$

are matrices. More generally,

$$A = (a_{ij}) = \begin{bmatrix} a_{11} & \dots & a_{1n} \\ \vdots & & \vdots \\ a_{m1} & \dots & a_{mn} \end{bmatrix} \tag{A.1.1}$$

is a matrix with m rows and n columns. Such a matrix is briefly called $(m \times n)$ matrix, m being the *row dimension* and n being the *column dimension*. The numbers a_{ij} are the *elements* or *components* of A. In the following it is assumed that the elements of all matrices considered are real numbers unless otherwise stated. In other words, we will be concerned with real rather than complex matrices. If the dimensions m and n are clear from the context or if they are of no importance, the notation $A = (a_{ij})$ means that a_{ij} is a *typical element* of A, that is, A consists of elements a_{ij}, $i = 1, \dots, m$, $j = 1, \dots, n$.

A $(1 \times n)$ matrix is a *row vector* and an $(m \times 1)$ matrix is a *column vector* which is often denoted by a lower case letter in the text. If not otherwise noted, all vectors will be column vectors in the following. Instead of $(m \times 1)$ matrix we sometimes say $(m \times 1)$ vector or simply *m-vector* or *m-dimensional* vector.

An $(m \times m)$ matrix with the number of rows equal to the number of columns is a *square matrix*. An $(m \times m)$ square matrix

$$\begin{bmatrix} a_{11} & 0 & \dots & 0 \\ 0 & a_{22} & & 0 \\ \vdots & & \ddots & \vdots \\ 0 & 0 & \dots & a_{mm} \end{bmatrix}$$

with zeroes off the main diagonal is a *diagonal matrix*. If all the diagonal elements of a diagonal matrix are one, it is an *identity* or *unit matrix*. An $(m \times m)$ identity matrix is denoted by I_m or simply by I if the dimension is unimportant or obvious from the context. A square matrix with all elements below (above) the main diagonal being zero is called *upper (lower) triangular* or simply *triangular matrix*. A matrix consisting of zeroes only is a *null matrix* or *zero matrix*. Usually in this text such a matrix is simply denoted by 0 and its dimensions have to be figured out from the context.

The *transpose* of the $(m \times n)$ matrix A given in (A.1.1) is the $(n \times m)$ matrix

$$A' = \begin{bmatrix} a_{11} & \dots & a_{m1} \\ \vdots & & \vdots \\ a_{1n} & \dots & a_{mn} \end{bmatrix},$$

the n rows of A' being the n columns of A. The matrix A is *symmetric* if $A' = A$. For instance,

$$\begin{bmatrix} 3 & 3 & 1 \\ 0 & 1 & 0 \end{bmatrix} \quad \text{is the transpose of} \quad \begin{bmatrix} 3 & 0 \\ 3 & 1 \\ 1 & 0 \end{bmatrix}$$

and

$$\begin{bmatrix} 2 & -1 \\ -1 & 0 \end{bmatrix}$$

is a symmetric matrix.

A.2 Basic Matrix Operations

Let $A = (a_{ij})$ and $B = (b_{ij})$ be $(m \times n)$ matrices. The two matrices are *equal*, $A = B$, if $a_{ij} = b_{ij}$ for all i, j. The following matrix operations are basic:

$$A + B := (a_{ij} + b_{ij}). \qquad \textit{(addition)}$$

$$A - B := (a_{ij} - b_{ij}). \qquad \textit{(subtraction)}$$

For a real constant c,

$$cA = Ac := (ca_{ij}). \qquad \textit{(multiplication by a scalar)}$$

Let $C = (c_{ij})$ be an $(n \times r)$ matrix, then the product

$$AC := \left(\sum_{j=1}^{n} a_{ij} c_{jk} \right) \qquad \textit{(multiplication)}$$

is an $(m \times r)$ matrix. For instance,

$$\begin{bmatrix} 3 & 3 \\ 2 & 1 \end{bmatrix} \begin{bmatrix} 2 & 3 & 0 \\ 2 & 4 & -1 \end{bmatrix} = \begin{bmatrix} 3\cdot 2 + 3\cdot 2 & 3\cdot 3 + 3\cdot 4 & 3\cdot 0 - 3\cdot 1 \\ 2\cdot 2 + 1\cdot 2 & 2\cdot 3 + 1\cdot 4 & 2\cdot 0 - 1\cdot 1 \end{bmatrix}$$

$$= \begin{bmatrix} 12 & 21 & -3 \\ 6 & 10 & -1 \end{bmatrix}.$$

If the column dimension of A is the same as the row dimension of C so that A and C can be multiplied, the two matrices are *conformable*. In the product AC the matrix C is *premultiplied* by A and A is *postmultiplied* by C.

Rules: Suppose A, B, and C are matrices with suitable dimensions so that the following operations are defined and c is a scalar.

(1) $A + B = B + A$.
(2) $(A + B) + C = A + (B + C)$.
(3) $A(B + C) = AB + AC$.
(4) $c(A + B) = cA + cB$.
(5) $AB \neq BA$ in general.
(6) $(AB)C = A(BC)$.
(7) $(AB)' = B'A'$.
(8) $AI = IA = A$.
(9) AA' and $A'A$ are symmetric matrices.

A.3 The Determinant

The *determinant* of an $(m \times m)$ square matrix $A = (a_{ij})$ is the sum of all products

$$(-1)^p a_{1i_1} a_{2i_2} \cdots a_{mi_m}$$

consisting of precisely one element from each row and each column multiplied by -1 or 1, depending on the permutation $i_1, \ldots, i_m$ of the subscripts. The -1 is used if the number of inversions of $i_1, \ldots, i_m$ to obtain the order $1, 2, \ldots, m$ is odd and 1 is used otherwise. The sum is taken over all $m!$ permutations of the column subscripts.

For a (1×1) matrix the determinant equals the value of the single element and for $m > 1$ the determinant may be defined recursively as follows. Suppose

$$A = \begin{bmatrix} a_{11} & a_{12} \\ a_{21} & a_{22} \end{bmatrix}$$

is a (2×2) matrix. Then the determinant is

$$\det(A) = |A| = a_{11}a_{22} - a_{12}a_{21}. \tag{A.3.1}$$

For instance,

$$\det\begin{bmatrix} 3 & 1 \\ 2 & 2 \end{bmatrix} = 4.$$

To specify the determinant of a general $(m \times m)$ matrix $A = (a_{ij})$ we define the *minor* of the ij-th element a_{ij} as the determinant of the $((m-1) \times (m-1))$ matrix that is obtained by deleting the i-th row and j-th column from A. The *cofactor* of a_{ij}, denoted by A_{ij}, is the minor multiplied by $(-1)^{i+j}$. Now

$$\det(A) = |A| = a_{i1}A_{i1} + \cdots + a_{im}A_{im} = a_{1j}A_{1j} + \cdots + a_{mj}A_{mj} \tag{A.3.2}$$

for some i or $j \in \{1, \ldots, m\}$. Of course, it does not matter which row or column is chosen in (A.3.2) because the determinant of a matrix is a unique number.

For example, for the (3×3) matrix

$$A = \begin{bmatrix} 2 & 1 & 3 \\ 0 & 2 & 1 \\ 1 & -1 & 4 \end{bmatrix} \tag{A.3.3}$$

the minor of the upper right-hand corner element is

$$\det\begin{bmatrix} 0 & 2 \\ 1 & -1 \end{bmatrix} = -2.$$

The cofactor is also -2 since $(-1)^{1+3} = 1$. Developing by the first row gives

$$|A| = 2\det\begin{bmatrix} 2 & 1 \\ -1 & 4 \end{bmatrix} - 1\det\begin{bmatrix} 0 & 1 \\ 1 & 4 \end{bmatrix} + 3\det\begin{bmatrix} 0 & 2 \\ 1 & -1 \end{bmatrix} = 13.$$

The same result is obtained by developing by any other row or column, e.g., using the first column gives

$$|A| = 2\det\begin{bmatrix} 2 & 1 \\ -1 & 4 \end{bmatrix} - 0\det\begin{bmatrix} 1 & 3 \\ -1 & 4 \end{bmatrix} + 1\det\begin{bmatrix} 1 & 3 \\ 2 & 1 \end{bmatrix} = 13.$$

Rules: In the following rules $A = (a_{ij})$ and $B = (b_{ij})$ are $(m \times m)$ matrices and c is a scalar.

(1) $\det(I_m) = 1$.
(2) If A is a diagonal matrix, $\det(A) = a_{11}a_{22}\cdots a_{mm}$.
(3) If A is a lower or upper triangular matrix, $|A| = a_{11}\cdots a_{mm}$.
(4) If A contains a row or column of zeroes, $|A| = 0$.
(5) If B is obtained from A by adding to one row (column) a scalar multiple of another row (column), then $|A| = |B|$.
(6) If A has two identical rows or columns, $|A| = 0$.
(7) $\det(cA) = c^m \det(A)$.
(8) $|AB| = |A||B|$.
(9) If C is an $(m \times n)$ matrix, $\det(I_m + CC') = \det(I_n + C'C)$.

A.4 The Inverse, the Adjoint, and Generalized Inverses

A.4.1 Inverse and Adjoint of a Square Matrix

An $(m \times m)$ square matrix A is *nonsingular* or *regular* or *invertible* if there exists a unique $(m \times m)$ matrix B such that $AB = I_m$. The matrix B is denoted by A^{-1}. It is the *inverse* of A,

$$AA^{-1} = A^{-1}A = I_m.$$

For $m > 1$, the $(m \times m)$ matrix of cofactors,

$$A^* = \begin{bmatrix} A_{11} & \dots & A_{1m} \\ \vdots & \ddots & \vdots \\ A_{m1} & \dots & A_{mm} \end{bmatrix}'$$

is the *adjoint* of A. For a (1×1) matrix A we define the adjoint to be 1, that is, $A^* = 1$. To compute the inverse of the $(m \times m)$ matrix A, the relation

$$A^{-1} = |A|^{-1}A^* \tag{A.4.1}$$

is sometimes useful. For this expression to be meaningful, $|A|$ has to be nonzero. Indeed A is nonsingular if and only if $\det(A) \neq 0$.

As an example consider the matrix given in (A.3.3). Its adjoint is

$$A^* = \begin{bmatrix} \begin{vmatrix} 2 & 1 \\ -1 & 4 \end{vmatrix} & -\begin{vmatrix} 0 & 1 \\ 1 & 4 \end{vmatrix} & \begin{vmatrix} 0 & 2 \\ 1 & -1 \end{vmatrix} \\ -\begin{vmatrix} 1 & 3 \\ -1 & 4 \end{vmatrix} & \begin{vmatrix} 2 & 3 \\ 1 & 4 \end{vmatrix} & -\begin{vmatrix} 2 & 1 \\ 1 & -1 \end{vmatrix} \\ \begin{vmatrix} 1 & 3 \\ 2 & 1 \end{vmatrix} & -\begin{vmatrix} 2 & 3 \\ 0 & 1 \end{vmatrix} & \begin{vmatrix} 2 & 1 \\ 0 & 2 \end{vmatrix} \end{bmatrix}' = \begin{bmatrix} 9 & -7 & -5 \\ 1 & 5 & -2 \\ -2 & 3 & 4 \end{bmatrix}.$$

Consequently,

$$A^{-1} = \frac{1}{13}\begin{bmatrix} 9 & -7 & -5 \\ 1 & 5 & -2 \\ -2 & 3 & 4 \end{bmatrix}.$$

Multiplying this matrix by A is easily seen to result in the (3×3) identity matrix.

Rules:

(1) For an $(m \times m)$ square matrix A, $AA^* = A^*A = |A|I_m$.
(2) An $(m \times m)$ matrix A is nonsingular if and only if $\det(A) \neq 0$.

In the following $A = (a_{ij})$ and B are nonsingular $(m \times m)$ matrices and $c \neq 0$ is a scalar constant.

(3) $A^{-1} = A^*/|A|$.
(4) $(A')^{-1} = (A^{-1})'$.
(5) $(AB)^{-1} = B^{-1}A^{-1}$.
(6) $(cA)^{-1} = \frac{1}{c}A^{-1}$.
(7) $I_m^{-1} = I_m$.
(8) If A is a diagonal matrix, then A^{-1} is also diagonal with diagonal elements $1/a_{ii}$.
(9) For an $(m \times n)$ matrix C, $(I_m + CC')^{-1} = I_m - C(I_n + C'C)^{-1}C'$.

A.4.2 Generalized Inverses

Let A be an $(m \times n)$ matrix. Any matrix B satisfying $ABA = A$ is a *generalized inverse* of A. For example, if

$$A = \begin{bmatrix} 1 & 0 \\ 0 & 0 \end{bmatrix},$$

the following matrices are generalized inverses of A:

$$\begin{bmatrix} 1 & 0 \\ 0 & 1 \end{bmatrix}, \quad \begin{bmatrix} 1 & 0 \\ 0 & 0 \end{bmatrix}, \quad \begin{bmatrix} 1 & 0 \\ 0 & \frac{1}{2} \end{bmatrix}.$$

Obviously a generalized inverse is not unique in general. An $(n \times m)$ matrix B is called *Moore-Penrose (generalized) inverse* of A if it satisfies the following four conditions:

$$\begin{aligned} ABA &= A, \\ BAB &= B, \\ (AB)' &= AB, \\ (BA)' &= BA. \end{aligned} \tag{A.4.2}$$

The Moore-Penrose inverse of A is denoted by A^+, it exists for any $(m \times n)$ matrix and is unique.

Rules: (See Magnus & Neudecker (1988, p. 33, Theorem 5).)

(1) $A^+ = A^{-1}$ if A is nonsingular.
(2) $(A^+)^+ = A$.
(3) $(A')^+ = (A^+)'$.
(4) $A'AA^+ = A^+AA' = A'$.
(5) $A'A^{+\prime}A^+ = A^+A^{+\prime}A' = A^+$.
(6) $(A'A)^+ = A^+A^{+\prime}$, $(AA')^+ = A^{+\prime}A^+$.
(7) $A^+ = (A'A)^+A' = A'(AA')^+$.

A.5 The Rank

Let $x_1, \ldots, x_n$ be $(m \times 1)$ vectors. They are *linearly independent* if for the constants $c_1, \ldots, c_n$,

$$c_1 x_1 + \cdots + c_n x_n = 0$$

implies $c_1 = \cdots = c_n = 0$. Equivalently, defining the $(n \times 1)$ vector $c = (c_1, \ldots, c_n)'$ and the $(m \times n)$ matrix $X = (x_1, \ldots, x_n)$, the columns of X are linearly independent if $Xc = 0$ implies $c = 0$. The columns of X are *linearly dependent* if $c_1 x_1 + \cdots + c_n x_n = 0$ holds with at least one $c_i \neq 0$. In that case

$$x_i = d_1 x_1 + \cdots + d_{i-1} x_{i-1} + d_{i+1} x_{i+1} + \cdots + d_n x_n,$$

where $d_j = -c_j/c_i$. In other words, $x_1, \ldots, x_n$ are linearly dependent if at least one of the vectors is a linear combination of the other vectors.

If $n > m$, the columns of X are linearly dependent. Consequently if $x_1, \ldots, x_n$ are linearly independent, then $n \leq m$.

Let $a_1, \ldots, a_n$ be the columns of the $(m \times n)$ matrix $A = (a_1, \ldots, a_n)$. That is, the a_i are $(m \times 1)$ vectors. The *rank* of A, briefly $\mathrm{rk}(A)$, is the maximum number of linearly independent columns of A. Thus, if $n \leq m$ and the $a_1, \ldots, a_n$ are linearly independent, $\mathrm{rk}(A) = n$. The maximum number of linearly independent columns of A equals the maximum number of independent rows. Hence the rank may be defined equivalently as the maximum number of linearly independent rows. If $m \geq n$ $(m \leq n)$ then we say that A has *full rank* if $\mathrm{rk}(A) = n$ $(\mathrm{rk}(A) = m)$.

Rules: Let A be an $(m \times n)$ matrix.

(1) $\mathrm{rk}(A) \leq \min(m, n)$.
(2) $\mathrm{rk}(A) = \mathrm{rk}(A')$.
(3) $\mathrm{rk}(AA') = \mathrm{rk}(A'A) = \mathrm{rk}(A)$.
(4) If B is a nonsingular $(n \times n)$ matrix, then $\mathrm{rk}(AB) = \mathrm{rk}(A)$.
(5) If $\mathrm{rk}(A) = m$, then $A^+ = A'(AA')^{-1}$.
(6) If $\mathrm{rk}(A) = n$, then $A^+ = (A'A)^{-1}A'$.
(7) If B is an $(n \times r)$ matrix, $\mathrm{rk}(AB) \leq \min\{\mathrm{rk}(A), \mathrm{rk}(B)\}$.
(8) If A is $(m \times m)$, then $\mathrm{rk}(A) = m$ if and only if $|A| \neq 0$.

A.6 Eigenvalues and -vectors – Characteristic Values and Vectors

The *eigenvalues* or *characteristic values* or *characteristic roots* of an $(m \times m)$ square matrix A are the roots of the polynomial in λ given by $\det(A - \lambda I_m)$ or $\det(\lambda I_m - A)$. The determinant is sometimes called the *characteristic determinant* and the polynomial is the *characteristic polynomial* of A. Since the roots of a polynomial are complex numbers the eigenvalues are also complex in general. A number λ_i is an eigenvalue of A, if the columns of $(A - \lambda_i I_m)$ are linearly dependent. Consequently, there exists an $(m \times 1)$ vector $v_i \neq 0$ such that

$$(A - \lambda_i I_m)v_i = 0 \qquad \text{or} \qquad Av_i = \lambda_i v_i.$$

A vector with this property is an *eigenvector* or *characteristic vector* corresponding to the eigenvalue λ_i. Of course, any nonzero scalar multiple of v_i is also an eigenvector corresponding to λ_i.

As an example consider the matrix

$$A = \begin{bmatrix} 1 & 0 \\ 1 & 3 \end{bmatrix}.$$

Its eigenvalues are the roots of

$$|A - \lambda I_2| = \det\begin{bmatrix} 1-\lambda & 0 \\ 1 & 3-\lambda \end{bmatrix} = (1-\lambda)(3-\lambda).$$

Hence $\lambda_1 = 1$ and $\lambda_2 = 3$ are the eigenvalues of A. Corresponding eigenvectors are obtained by solving

$$\begin{bmatrix} 1 & 0 \\ 1 & 3 \end{bmatrix}\begin{bmatrix} v_{11} \\ v_{21} \end{bmatrix} = \begin{bmatrix} v_{11} \\ v_{21} \end{bmatrix} \qquad \text{and} \qquad \begin{bmatrix} 1 & 0 \\ 1 & 3 \end{bmatrix}\begin{bmatrix} v_{12} \\ v_{22} \end{bmatrix} = 3\begin{bmatrix} v_{12} \\ v_{22} \end{bmatrix}.$$

Thus,

$$\begin{bmatrix} v_{11} \\ v_{21} \end{bmatrix} = \begin{bmatrix} 1 \\ -\frac{1}{2} \end{bmatrix} \qquad \text{and} \qquad \begin{bmatrix} v_{12} \\ v_{22} \end{bmatrix} = \begin{bmatrix} 0 \\ 1 \end{bmatrix}$$

are eigenvectors corresponding to λ_1 and λ_2, respectively.

In the following rules the *modulus* of a complex number $z = z_1 + iz_2$ is used. Here z_1 and z_2 are the real and imaginary parts of z, respectively. The modulus $|z|$ of z is defined as

$$|z| := \sqrt{z_1^2 + z_2^2}.$$

If $z_2 = 0$ so that z is a real number the modulus is just the absolute value of z which justifies the notation.

Rules:

(1) If A is symmetric, then all its eigenvalues are real numbers.
(2) The eigenvalues of a diagonal matrix are its diagonal elements.
(3) The eigenvalues of a triangular matrix are its diagonal elements.
(4) An $(m \times m)$ matrix has at most m eigenvalues.
(5) Let $\lambda_1, \ldots, \lambda_m$ be the eigenvalues of the $(m \times m)$ matrix A, then $|A| = \lambda_1 \cdots \lambda_m$, that is, the determinant is the product of the eigenvalues.
(6) Let λ_i and λ_j be *distinct* eigenvalues of A with corresponding eigenvectors v_i and v_j. Then v_i and v_j are linearly independent.
(7) All eigenvalues of the $(m \times m)$ matrix A have modulus less than 1 if and only if $\det(I_m - Az) \neq 0$ for $|z| \leq 1$, that is, the polynomial $\det(I_m - Az)$ has no roots in and on the complex unit circle.

A.7 The Trace

The *trace* of an $(m \times m)$ square matrix $A = (a_{ij})$ is the sum of its diagonal elements,

$$\operatorname{tr} A := a_{11} + \cdots + a_{mm}.$$

For example,

$$\operatorname{tr}\begin{bmatrix} 1 & 0 \\ 1 & 3 \end{bmatrix} = 4.$$

Rules: A and B are $(m \times m)$ matrices and $\lambda_1, \ldots, \lambda_m$ are the eigenvalues of A.

(1) $\operatorname{tr}(A + B) = \operatorname{tr} A + \operatorname{tr} B$.
(2) $\operatorname{tr} A = \operatorname{tr} A'$.
(3) If C is $(m \times n)$ and D is $(n \times m)$, $\operatorname{tr}(CD) = \operatorname{tr}(DC)$.
(4) $\operatorname{tr} A = \lambda_1 + \cdots + \lambda_m$.

A.8 Some Special Matrices and Vectors

A.8.1 Idempotent and Nilpotent Matrices

An $(m \times m)$ matrix A is *idempotent* if $AA = A^2 = A$. Examples of idempotent matrices are $A = I_m$, $A = 0$, and

$$A = \begin{bmatrix} \frac{1}{3} & \frac{1}{3} & \frac{1}{3} \\ \frac{1}{3} & \frac{1}{3} & \frac{1}{3} \\ \frac{1}{3} & \frac{1}{3} & \frac{1}{3} \end{bmatrix}.$$

An $(m \times m)$ matrix A is *nilpotent* if there exists a positive integer i such that $A^i = 0$. For instance, the (2×2) matrices

$$A = \begin{bmatrix} 0 & 3 \\ 0 & 0 \end{bmatrix} \quad \text{and} \quad B = \begin{bmatrix} 1 & -1 \\ 1 & -1 \end{bmatrix}$$

are nilpotent because $A^2 = B^2 = 0$.

Rules: In the following rules A is an $(m \times m)$ matrix.

(1) If A is a diagonal matrix it is idempotent if and only if all the diagonal elements are either zero or one.
(2) If A is symmetric and idempotent, $\operatorname{rk}(A) = \operatorname{tr}(A)$.
(3) If A is idempotent and $\operatorname{rk}(A) = m$, then $A = I_m$.
(4) If A is idempotent, then $I_m - A$ is idempotent.

(5) If A is symmetric and idempotent, then $A^+ = A$.
(6) If B is an $(m \times n)$ matrix, then BB^+ and B^+B are idempotent.
(7) If A is idempotent all its eigenvalues are zero or one.
(8) If A is nilpotent all its eigenvalues are zero.

A.8.2 Orthogonal Matrices and Vectors

Two $(m \times 1)$ *vectors* x and y are *orthogonal* if $x'y = 0$. They are *orthonormal* if they are orthogonal and have unit length, where the length of a vector x is $\|x\| := \sqrt{x'x}$.

An $(m \times m)$ square *matrix* A is *orthogonal* if its transpose is its inverse, $A'A = AA' = I_m$. In other words, A is orthogonal if its rows and columns are orthonormal vectors.

Examples of orthogonal vectors are

$$x = \begin{bmatrix} 5 \\ 0 \\ 0 \end{bmatrix} \quad \text{and} \quad y = \begin{bmatrix} 0 \\ 2 \\ 0 \end{bmatrix}.$$

The following four matrices are orthogonal

$$\begin{bmatrix} 0 & 1 \\ 1 & 0 \end{bmatrix}, \quad \begin{bmatrix} \cos\varphi & \sin\varphi \\ -\sin\varphi & \cos\varphi \end{bmatrix},$$

$$\begin{bmatrix} 0 & 1 & 0 \\ 1 & 0 & 0 \\ 0 & 0 & 1 \end{bmatrix}, \quad \begin{bmatrix} 1/\sqrt{3} & 1/\sqrt{3} & 1/\sqrt{3} \\ 1/\sqrt{2} & -1/\sqrt{2} & 0 \\ 1/\sqrt{6} & 1/\sqrt{6} & -2/\sqrt{6} \end{bmatrix}.$$

Rules:

(1) I_m is orthogonal.
(2) If A is an orthogonal matrix, $\det(A) = 1$ or -1.
(3) If A and B are orthogonal and conformable matrices, then AB is orthogonal.
(4) If λ_i, λ_j are distinct eigenvalues of a *symmetric* matrix A, then the corresponding eigenvectors v_i and v_j are orthogonal.

A.8.3 Definite Matrices and Quadratic Forms

Let A be a *symmetric* $(m \times m)$ matrix and x an $(m \times 1)$ vector. The function $x'Ax$ is called a *quadratic form* in x. The symmetric matrix A or the corresponding quadratic form is

(i) *positive definite* if $x'Ax > 0$ for all m-vectors $x \neq 0$;
(ii) *positive semidefinite* if $x'Ax \geq 0$ for all m-vectors x;

(iii) *negative definite* if $x'Ax < 0$ for all m-vectors $x \neq 0$;
(iv) *negative semidefinite* if $x'Ax \leq 0$ for all m-vectors x;
(v) *indefinite* if $x'Ax > 0$ for some x and $x'Ax < 0$ for other x.

Rules: In the following rules A is a symmetric $(m \times m)$ matrix.

(1) $A = (a_{ij})$ is positive definite if and only if all its principle minors are positive, where

$$\det \begin{bmatrix} a_{11} & \cdots & a_{1i} \\ \vdots & \ddots & \vdots \\ a_{i1} & \cdots & a_{ii} \end{bmatrix}$$

is the i-th principle minor of A.
(2) A is negative definite (semidefinite) if and only if $-A$ is positive definite (semidefinite).
(3) If A is positive or negative definite, it is nonsingular.
(4) All eigenvalues of a positive (negative) definite matrix are greater (smaller) than zero.
(5) A diagonal matrix is positive (negative) definite if and only if all its diagonal elements are positive (negative).
(6) If A is positive definite and B an $(m \times n)$ matrix, then $B'AB$ is positive semidefinite.
(7) If A is positive definite and B an $(m \times n)$ matrix with $\text{rk}(B) = n$, then $B'AB$ is positive definite.
(8) If A is positive definite, then A^{-1} is positive definite.
(9) If A is idempotent, then it is positive semidefinite.

With these rules it is easy to check that

$$\begin{bmatrix} 2 & 1 \\ 1 & 1 \end{bmatrix} \quad \text{and} \quad \begin{bmatrix} 3 & 1 & 0 \\ 1 & 1 & 0 \\ 0 & 0 & 4 \end{bmatrix}$$

are positive definite matrices and

$$\begin{bmatrix} 1 & 1 \\ 1 & 1 \end{bmatrix} \quad \text{and} \quad \begin{bmatrix} 1 & 0 \\ 0 & 0 \end{bmatrix}$$

are positive semidefinite matrices.

A.9 Decomposition and Diagonalization of Matrices

A.9.1 The Jordan Canonical Form

Let A be an $(m \times m)$ matrix with eigenvalues $\lambda_1, \ldots, \lambda_n$. Then there exists a nonsingular matrix P such that

$$P^{-1}AP = \begin{bmatrix} \Lambda_1 & & 0 \\ & \ddots & \\ 0 & & \Lambda_n \end{bmatrix} =: \Lambda \qquad \text{or} \qquad A = P\Lambda P^{-1}, \tag{A.9.1}$$

where

$$\Lambda_i = \begin{bmatrix} \lambda_i & 1 & 0 & \dots & 0 \\ 0 & \lambda_i & 1 & & 0 \\ \vdots & & \ddots & \ddots & \vdots \\ 0 & 0 & & \ddots & 1 \\ 0 & 0 & & \dots & \lambda_i \end{bmatrix}.$$

This decomposition of A is the *Jordan canonical form*. Since the eigenvalues of A may be complex numbers, Λ and P may be complex matrices. If multiple roots of the characteristic polynomial exist, they may have to appear more than once in the list $\lambda_1, \dots, \lambda_n$.

The Jordan canonical form has some important implications. For instance, it implies that

$$A^j = (P\Lambda P^{-1})^j = P\Lambda^j P^{-1}$$

and it can be shown that

$$\Lambda_i^j = \begin{bmatrix} \lambda_i^j & \binom{j}{1}\lambda_i^{j-1} & \dots & \binom{j}{r_i - 1}\lambda_i^{j-r_i+1} \\ 0 & \lambda_i^j & \dots & \binom{j}{r_i - 2}\lambda_i^{j-r_i+2} \\ \vdots & & \ddots & \vdots \\ 0 & 0 & \dots & \lambda_i^j \end{bmatrix}.$$

Therefore we have the following rules.

Rules: Suppose A is a real $(m \times m)$ matrix with eigenvalues $\lambda_1, \dots, \lambda_n$ which have all modulus less than 1, that is, $|\lambda_i| < 1$ for $i = 1, \dots, n$. Furthermore, let Λ and P be the matrices given in (A.9.1).

(1) $A^j = P\Lambda^j P^{-1} \xrightarrow[j\to\infty]{} 0.$
(2) $\sum_{j=0}^{\infty} A^j = (I_m - A)^{-1}$ exists.
(3) The sequence A^j, $j = 0, 1, 2, \dots$, is absolutely summable, that is, $\sum_{j=0}^{\infty} |\alpha_{kl,j}|$ is finite for all $k, l = 1, \dots, m$, where $\alpha_{kl,j}$ is a typical element of A^j. (See Section C.3 regarding the concept of absolute summability.)

A.9.2 Decomposition of Symmetric Matrices

If A is a symmetric $(m \times m)$ matrix then there exists an orthogonal matrix P such that

$$P'AP = \Lambda = \begin{bmatrix} \lambda_1 & & 0 \\ & \ddots & \\ 0 & & \lambda_m \end{bmatrix} \quad \text{or} \quad A = P\Lambda P', \tag{A.9.2}$$

where the λ_i are the eigenvalues of A and the columns of P are the corresponding eigenvectors. Here all matrices are real again because the eigenvalues of a symmetric matrix are real numbers. Denoting the i-th column of P by p_i and using that $p_i'p_j = 0$ for $i \neq j$, we get

$$A = P\Lambda P' = \sum_{i=1}^{m} \lambda_i p_i p_i'. \tag{A.9.3}$$

Moreover,

$$A^2 = P\Lambda P'P\Lambda P' = P\Lambda^2 P'$$

and, more generally,

$$A^k = P\Lambda^k P'.$$

If A is a positive definite symmetric $(m \times m)$ matrix all eigenvalues are positive so that the notation

$$\Lambda^{1/2} := \begin{bmatrix} \sqrt{\lambda_1} & & 0 \\ & \ddots & \\ 0 & & \sqrt{\lambda_m} \end{bmatrix}$$

makes sense. Defining $Q = P\Lambda^{1/2}P'$ we get $QQ = A$. In generalization of the terminology for positive real numbers, Q may be called a *square root of* A and may be denoted by $A^{1/2}$.

A.9.3 The Choleski Decomposition of a Positive Definite Matrix

If A is a positive definite $(m \times m)$ matrix there exists an upper triangular and a lower triangular matrix P with positive main diagonal such that

$$P^{-1}AP'^{-1} = I_m \quad \text{or} \quad A = PP'. \tag{A.9.4}$$

Similarly, if A is positive semidefinite with $\text{rk}(A) = n < m$, then there exists a nonsingular matrix P such that

$$P^{-1}AP'^{-1} = \begin{bmatrix} I_n & 0 \\ 0 & 0 \end{bmatrix}. \tag{A.9.5}$$

Alternatively, $A = QQ'$, where

$$Q = P\begin{bmatrix} I_n & 0 \\ 0 & 0 \end{bmatrix}.$$

For instance,

$$\begin{bmatrix} 26 & 3 & 0 \\ 3 & 9 & 0 \\ 0 & 0 & 81 \end{bmatrix} = \begin{bmatrix} 5 & 1 & 0 \\ 0 & 3 & 0 \\ 0 & 0 & 9 \end{bmatrix} \begin{bmatrix} 5 & 0 & 0 \\ 1 & 3 & 0 \\ 0 & 0 & 9 \end{bmatrix}$$

$$= \begin{bmatrix} \sqrt{26} & 0 & 0 \\ 3/\sqrt{26} & 15/\sqrt{26} & 0 \\ 0 & 0 & 9 \end{bmatrix} \begin{bmatrix} \sqrt{26} & 3/\sqrt{26} & 0 \\ 0 & 15/\sqrt{26} & 0 \\ 0 & 0 & 9 \end{bmatrix}.$$

The decomposition $A = PP'$, where P is lower triangular with positive main diagonal, is sometimes called *Choleski decomposition*. Computer programs are available to determine the matrix P for a given positive definite matrix A. If a lower triangular matrix P is supplied by the program an upper triangular matrix Q can be obtained as follows: Define an $(m \times m)$ matrix

$$G = \begin{bmatrix} 0 & \dots & 0 & 1 \\ 0 & \dots & 1 & 0 \\ \vdots & \cdot^{\cdot^{\cdot}} & & \vdots \\ 1 & & 0 & 0 \end{bmatrix}$$

with ones on the diagonal from the upper right-hand corner to the lower left-hand corner and zeroes elsewhere. Note that $G' = G$ and $G^{-1} = G$. Suppose a decomposition of the $(m \times m)$ matrix A is desired. Then decompose $B = GAG$ as $B = PP'$, where P is lower triangular. Hence

$$A = GBG = GPGGP'G = QQ',$$

where $Q = GPG$ is upper triangular.

A.10 Partitioned Matrices

Let the $(m \times n)$ matrix A be *partitioned* into *submatrices* $A_{11}, A_{12}, A_{21}, A_{22}$ with dimensions $(p \times q)$, $(p \times (n - q))$, $((m - p) \times q)$, and $((m - p) \times (n - q))$, respectively, so that

$$A = \begin{bmatrix} A_{11} & A_{12} \\ A_{21} & A_{22} \end{bmatrix}. \tag{A.10.1}$$

For such a partitioned matrix a number of useful results hold.

Rules:

(1) $$A' = \begin{bmatrix} A'_{11} & A'_{21} \\ A'_{12} & A'_{22} \end{bmatrix}.$$

(2) If $n = m$ and $q = p$ and A, A_{11}, and A_{22} are nonsingular, then

$$A^{-1} = \begin{bmatrix} D & -DA_{12}A_{22}^{-1} \\ -A_{22}^{-1}A_{21}D & A_{22}^{-1} + A_{22}^{-1}A_{21}DA_{12}A_{22}^{-1} \end{bmatrix}$$

$$= \begin{bmatrix} A_{11}^{-1} + A_{11}^{-1}A_{12}GA_{21}A_{11}^{-1} & -A_{11}^{-1}A_{12}G \\ -GA_{21}A_{11}^{-1} & G \end{bmatrix},$$

where $D = (A_{11} - A_{12}A_{22}^{-1}A_{21})^{-1}$ and $G = (A_{22} - A_{21}A_{11}^{-1}A_{12})^{-1}$.

(3) Under the conditions of (2),

$$(A_{11} - A_{12}A_{22}^{-1}A_{21})^{-1} = A_{11}^{-1} + A_{11}^{-1}A_{12}(A_{22} - A_{21}A_{11}^{-1}A_{12})^{-1}A_{21}A_{11}^{-1}.$$

(4) Under the conditions of (2), if A_{12} and A_{21} are null matrices,

$$A^{-1} = \begin{bmatrix} A_{11}^{-1} & 0 \\ 0 & A_{22}^{-1} \end{bmatrix}.$$

(5) If A is a square matrix ($n = m$) and A_{11} is square and nonsingular, then $|A| = |A_{11}| \cdot |A_{22} - A_{21}A_{11}^{-1}A_{12}|$.

(6) If A is a square matrix and A_{22} is square and nonsingular, then $|A| = |A_{22}| \cdot |A_{11} - A_{12}A_{22}^{-1}A_{21}|$.

A.11 The Kronecker Product

Let $A = (a_{ij})$ and $B = (b_{ij})$ be $(m \times n)$ and $(p \times q)$ matrices, respectively. The $(mp \times nq)$ matrix

$$A \otimes B := \begin{bmatrix} a_{11}B & \dots & a_{1n}B \\ \vdots & & \vdots \\ a_{m1}B & \dots & a_{mn}B \end{bmatrix} \tag{A.11.1}$$

is the *Kronecker product* or *direct product* of A and B. For example, the Kronecker product of

$$A = \begin{bmatrix} 3 & 4 & -1 \\ 2 & 0 & 0 \end{bmatrix} \quad \text{and} \quad B = \begin{bmatrix} 5 & -1 \\ 3 & 3 \end{bmatrix} \tag{A.11.2}$$

is

$$A \otimes B = \begin{bmatrix} 15 & -3 & 20 & -4 & -5 & 1 \\ 9 & 9 & 12 & 12 & -3 & -3 \\ 10 & -2 & 0 & 0 & 0 & 0 \\ 6 & 6 & 0 & 0 & 0 & 0 \end{bmatrix}$$

and

$$B \otimes A = \begin{bmatrix} 15 & 20 & -5 & -3 & -4 & 1 \\ 10 & 0 & 0 & -2 & 0 & 0 \\ 9 & 12 & -3 & 9 & 12 & -3 \\ 6 & 0 & 0 & 6 & 0 & 0 \end{bmatrix}.$$

Rules: In the following rules suitable dimensions are assumed.

(1) $A \otimes B \neq B \otimes A$ in general.
(2) $(A \otimes B)' = A' \otimes B'$.
(3) $A \otimes (B + C) = A \otimes B + A \otimes C$.
(4) $(A \otimes B)(C \otimes D) = AC \otimes BD$.
(5) If A and B are invertible, $(A \otimes B)^{-1} = A^{-1} \otimes B^{-1}$.
(6) If A and B are square matrices with eigenvalues λ_A, λ_B, respectively, and corresponding eigenvectors v_A, v_B, then $\lambda_A \lambda_B$ is an eigenvalue of $A \otimes B$ with eigenvector $v_A \otimes v_B$.
(7) If A and B are $(m \times m)$ and $(n \times n)$ square matrices, respectively, then $|A \otimes B| = |A|^n |B|^m$.
(8) If A and B are square matrices,

$$\operatorname{tr}(A \otimes B) = \operatorname{tr}(A)\operatorname{tr}(B).$$

(9) $(A \otimes B)^+ = A^+ \otimes B^+$.

A.12 The vec and vech Operators and Related Matrices

A.12.1 The Operators

Let $A = (a_1, \ldots, a_n)$ be an $(m \times n)$ matrix with $(m \times 1)$ columns a_i. The *vec operator* transforms A into an $(mn \times 1)$ vector by stacking the columns, that is,

$$\operatorname{vec}(A) = \begin{bmatrix} a_1 \\ \vdots \\ a_n \end{bmatrix}.$$

For instance, if A and B are as in (A.11.2),

$$\operatorname{vec}(A) = \begin{bmatrix} 3 \\ 2 \\ 4 \\ 0 \\ -1 \\ 0 \end{bmatrix} \quad \text{and} \quad \operatorname{vec}(B) = \begin{bmatrix} 5 \\ 3 \\ -1 \\ 3 \end{bmatrix}.$$

Rules: Let A, B, C be matrices with appropriate dimensions.

(1) $\operatorname{vec}(A + B) = \operatorname{vec}(A) + \operatorname{vec}(B)$.
(2) $\operatorname{vec}(ABC) = (C' \otimes A)\operatorname{vec}(B)$.
(3) $\operatorname{vec}(AB) = (I \otimes A)\operatorname{vec}(B) = (B' \otimes I)\operatorname{vec}(A)$.
(4) $\operatorname{vec}(ABC) = (I \otimes AB)\operatorname{vec}(C) = (C'B' \otimes I)\operatorname{vec}(A)$.
(5) $\operatorname{vec}(B')'\operatorname{vec}(A) = \operatorname{tr}(AB) = \operatorname{tr}(BA) = \operatorname{vec}(A')'\operatorname{vec}(B)$.

(6) $\text{tr}(ABC) = \text{vec}(A')'(C' \otimes I)\,\text{vec}(B)$
$= \text{vec}(A')'(I \otimes B)\,\text{vec}(C)$
$= \text{vec}(B')'(A' \otimes I)\,\text{vec}(C)$
$= \text{vec}(B')'(I \otimes C)\,\text{vec}(A)$
$= \text{vec}(C')'(B' \otimes I)\,\text{vec}(A)$
$= \text{vec}(C')'(I \otimes A)\,\text{vec}(B).$

The vech operator is closely related to vec. It stacks the elements on and below the main diagonal of a square matrix only. For instance,

$$\text{vech}\begin{bmatrix} \alpha_{11} & \alpha_{12} & \alpha_{13} \\ \alpha_{21} & \alpha_{22} & \alpha_{23} \\ \alpha_{31} & \alpha_{32} & \alpha_{33} \end{bmatrix} = \begin{bmatrix} \alpha_{11} \\ \alpha_{21} \\ \alpha_{31} \\ \alpha_{22} \\ \alpha_{32} \\ \alpha_{33} \end{bmatrix}.$$

In general, if A is an $(m \times m)$ matrix, vech(A) is an $m(m+1)/2$-dimensional vector. The vech operator is usually applied to symmetric matrices in order to collect the separate elements only.

A.12.2 The Elimination, Duplication, and Commutation Matrices

The vec and vech operators are related by the *elimination matrix* $\mathbf{L}_m$ and the *duplication matrix* $\mathbf{D}_m$. The former is an $(m(m+1)/2 \times m^2)$ matrix such that for an $(m \times m)$ square matrix A

$$\text{vech}(A) = \mathbf{L}_m \,\text{vec}(A). \tag{A.12.1}$$

Thus, e.g., for $m = 3$,

$$\mathbf{L}_3 = \begin{bmatrix} 1 & 0 & 0 & 0 & 0 & 0 & 0 & 0 & 0 \\ 0 & 1 & 0 & 0 & 0 & 0 & 0 & 0 & 0 \\ 0 & 0 & 1 & 0 & 0 & 0 & 0 & 0 & 0 \\ 0 & 0 & 0 & 0 & 1 & 0 & 0 & 0 & 0 \\ 0 & 0 & 0 & 0 & 0 & 1 & 0 & 0 & 0 \\ 0 & 0 & 0 & 0 & 0 & 0 & 0 & 0 & 1 \end{bmatrix}.$$

The duplication matrix $\mathbf{D}_m$ is $(m^2 \times m(m+1)/2)$ and is defined so that for any symmetric $(m \times m)$ matrix A

$$\text{vec}(A) = \mathbf{D}_m \,\text{vech}(A). \tag{A.12.2}$$

For instance, for $m = 3$,

$$\mathbf{D}_3 = \begin{bmatrix} 1 & 0 & 0 & 0 & 0 & 0 \\ 0 & 1 & 0 & 0 & 0 & 0 \\ 0 & 0 & 1 & 0 & 0 & 0 \\ 0 & 1 & 0 & 0 & 0 & 0 \\ 0 & 0 & 0 & 1 & 0 & 0 \\ 0 & 0 & 0 & 0 & 1 & 0 \\ 0 & 0 & 1 & 0 & 0 & 0 \\ 0 & 0 & 0 & 0 & 1 & 0 \\ 0 & 0 & 0 & 0 & 0 & 1 \end{bmatrix}.$$

Since the rank of $\mathbf{D}_m$ is easily seen to be $m(m + 1)/2$, the matrix $\mathbf{D}'_m\mathbf{D}_m$ is invertible. Thus, left-multiplication of (A.12.2) by $(\mathbf{D}'_m\mathbf{D}_m)^{-1}\mathbf{D}'_m$ gives

$$(\mathbf{D}'_m\mathbf{D}_m)^{-1}\mathbf{D}'_m \operatorname{vec}(A) = \operatorname{vech}(A). \tag{A.12.3}$$

Note, however, that $(\mathbf{D}'_m\mathbf{D}_m)^{-1}\mathbf{D}'_m \neq \mathbf{L}_m$ in general because (A.12.3) holds for symmetric matrices A only while (A.12.1) holds for arbitrary square matrices A.

The *commutation matrix* $\mathbf{K}_{mn}$ is another matrix that is occasionally useful in dealing with the vec operator. $\mathbf{K}_{mn}$ is an $(mn \times mn)$ matrix defined such that for any $(m \times n)$ matrix A,

$$\operatorname{vec}(A') = \mathbf{K}_{mn} \operatorname{vec}(A)$$

or, equivalently,

$$\operatorname{vec}(A) = \mathbf{K}_{nm} \operatorname{vec}(A').$$

For example,

$$\mathbf{K}_{32} = \begin{bmatrix} 1 & 0 & 0 & 0 & 0 & 0 \\ 0 & 0 & 0 & 1 & 0 & 0 \\ 0 & 1 & 0 & 0 & 0 & 0 \\ 0 & 0 & 0 & 0 & 1 & 0 \\ 0 & 0 & 1 & 0 & 0 & 0 \\ 0 & 0 & 0 & 0 & 0 & 1 \end{bmatrix}$$

because for $A = \begin{bmatrix} \alpha_{11} & \alpha_{12} \\ \alpha_{21} & \alpha_{22} \\ \alpha_{31} & \alpha_{32} \end{bmatrix}$

$$\operatorname{vec}(A') = \begin{bmatrix} \alpha_{11} \\ \alpha_{12} \\ \alpha_{21} \\ \alpha_{22} \\ \alpha_{31} \\ \alpha_{32} \end{bmatrix} = \mathbf{K}_{32} \begin{bmatrix} \alpha_{11} \\ \alpha_{21} \\ \alpha_{31} \\ \alpha_{12} \\ \alpha_{22} \\ \alpha_{32} \end{bmatrix} = \mathbf{K}_{32} \operatorname{vec}(A).$$

Rules:

(7) $\mathbf{L}_m \mathbf{D}_m = I_{m(m+1)/2}$.
(8) $\mathbf{K}_{mm} \mathbf{D}_m = \mathbf{D}_m$.
(9) $\mathbf{K}_{m1} = \mathbf{K}_{1m} = I_m$.
(10) $\mathbf{K}'_{mn} = \mathbf{K}_{mn}^{-1} = \mathbf{K}_{nm}$.
(11) $\operatorname{tr} \mathbf{K}_{mm} = m$.
(12) $\det(\mathbf{K}_{mn}) = (-1)^{mn(m-1)(n-1)/4}$.
(13) $\operatorname{tr}(\mathbf{D}'_m \mathbf{D}_m) = m^2$, $\operatorname{tr}(\mathbf{D}'_m \mathbf{D}_m)^{-1} = m(m+3)/4$.
(14) $\det(\mathbf{D}'_m \mathbf{D}_m) = 2^{m(m-1)/2}$.
(15) $\operatorname{tr}(\mathbf{D}_m \mathbf{D}'_m) = m^2$.
(16) $|\mathbf{D}'_m(A \otimes A)\mathbf{D}_m| = 2^{m(m-1)/2}|A|^{m+1}$, where A is an $(m \times m)$ matrix.
(17) $(\mathbf{D}'_m(A \otimes A)\mathbf{D}_m)^{-1} = (\mathbf{D}'_m\mathbf{D}_m)^{-1}\mathbf{D}'_m(A^{-1} \otimes A^{-1})\mathbf{D}_m(\mathbf{D}'_m\mathbf{D}_m)^{-1}$, if A is a nonsingular $(m \times m)$ matrix.
(18) $\mathbf{L}_m \mathbf{L}'_m = I_{m(m+1)/2}$.
(19) $\mathbf{L}_m \mathbf{L}'_m$ and $\mathbf{L}_m \mathbf{K}_{mm} \mathbf{L}'_m$ are idempotent.

Let A and B be lower triangular $(m \times m)$ matrices. Then we have:

(20) $\mathbf{L}_m(A \otimes B)\mathbf{L}'_m$ is lower triangular.
(21) $\mathbf{L}'_m \mathbf{L}_m(A' \otimes B)\mathbf{L}'_m = (A' \otimes B)\mathbf{L}'_m$.
(22) $[\mathbf{L}_m(A' \otimes B)\mathbf{L}'_m]^s = \mathbf{L}_m((A')^s \otimes B^s)\mathbf{L}'_m$ for $s = 0, 1, \ldots$ and for $s = \ldots, -2, -1$, if A^{-1} and B^{-1} exist.

Let G be $(m \times n)$, F $(p \times q)$, and $\mathbf{b}$ $(p \times 1)$. Then the following holds:

(23) $\mathbf{K}_{pm}(G \otimes F) = (F \otimes G)\mathbf{K}_{qn}$.
(24) $\mathbf{K}_{pm}(G \otimes F)\mathbf{K}_{nq} = F \otimes G$.
(25) $\mathbf{K}_{pm}(G \otimes \mathbf{b}) = \mathbf{b} \otimes G$.
(26) $\mathbf{K}_{pm}(\mathbf{b} \otimes G) = G \otimes \mathbf{b}$.
(27) $\operatorname{vec}(G \otimes F) = (I_n \otimes \mathbf{K}_{qm} \otimes I_p)(\operatorname{vec}(G) \otimes \operatorname{vec}(F))$.

A.13 Vector and Matrix Differentiation

In the following it will be assumed that all derivatives exist and are continuous. Let $f(\beta)$ be a scalar function that depends on the $(n \times 1)$ vector $\beta = (\beta_1, \ldots, \beta_n)'$.

$$\frac{\partial f}{\partial \beta} := \begin{bmatrix} \frac{\partial f}{\partial \beta_1} \\ \vdots \\ \frac{\partial f}{\partial \beta_n} \end{bmatrix}, \qquad \frac{\partial f}{\partial \beta'} := \left[\frac{\partial f}{\partial \beta_1}, \ldots, \frac{\partial f}{\partial \beta_n}\right]$$

are $(n \times 1)$ and $(1 \times n)$ vectors of first partial derivatives, respectively, and

$$\frac{\partial^2 f}{\partial\beta\partial\beta'} := \left[\frac{\partial^2 f}{\partial\beta_i\partial\beta_j}\right] = \begin{bmatrix} \dfrac{\partial^2 f}{\partial\beta_1\partial\beta_1} & \cdots & \dfrac{\partial^2 f}{\partial\beta_1\partial\beta_n} \\ \vdots & & \vdots \\ \dfrac{\partial^2 f}{\partial\beta_n\partial\beta_1} & \cdots & \dfrac{\partial^2 f}{\partial\beta_n\partial\beta_n} \end{bmatrix}$$

is the $(n \times n)$ *Hessian matrix* of second order partial derivatives. If $f(A)$ is a scalar function of an $(m \times n)$ matrix $A = (a_{ij})$, then

$$\frac{\partial f}{\partial A} := \left[\frac{\partial f}{\partial a_{ij}}\right]$$

is an $(m \times n)$ matrix of partial derivatives. If the $(m \times n)$ matrix $A = (a_{ij})$ depends on the scalar β, then

$$\frac{\partial A}{\partial \beta} := \left[\frac{\partial a_{ij}}{\partial \beta}\right]$$

is an $(m \times n)$ matrix. If $y(\beta) = (y_1(\beta), \ldots, y_m(\beta))'$ is an $(m \times 1)$ vector that depends on the $(n \times 1)$ vector β, then

$$\frac{\partial y}{\partial \beta'} := \begin{bmatrix} \dfrac{\partial y_1}{\partial\beta_1} & \cdots & \dfrac{\partial y_1}{\partial\beta_n} \\ \vdots & & \vdots \\ \dfrac{\partial y_m}{\partial\beta_1} & \cdots & \dfrac{\partial y_m}{\partial\beta_n} \end{bmatrix}$$

is an $(m \times n)$ matrix and

$$\frac{\partial y'}{\partial \beta} := \left(\frac{\partial y}{\partial \beta'}\right)'.$$

For example, if $\beta = (\beta_1, \beta_2)'$ and $f(\beta) = \beta_1^2 - 2\beta_1\beta_2$, then

$$\frac{\partial f}{\partial \beta} = \begin{bmatrix} \dfrac{\partial f}{\partial\beta_1} \\ \dfrac{\partial f}{\partial\beta_2} \end{bmatrix} = \begin{bmatrix} 2\beta_1 - 2\beta_2 \\ -2\beta_1 \end{bmatrix}.$$

If

$$y(\beta) = \begin{bmatrix} \beta_1^3 + \beta_2 \\ e^{\beta_1} \end{bmatrix}, \qquad \text{then} \qquad \frac{\partial y}{\partial \beta'} = \begin{bmatrix} 3\beta_1^2 & 1 \\ e^{\beta_1} & 0 \end{bmatrix}.$$

The following two propositions are useful for deriving rules for vector and matrix differentiation.

PROPOSITION A.1 (*Chain Rule for Vector Differentiation*)

Let α and β be $(m \times 1)$ and $(n \times 1)$ vectors, respectively, and suppose $h(\alpha)$ is $(p \times 1)$ and $g(\beta)$ is $(m \times 1)$. Then, with $\alpha = g(\beta)$,

$$\underset{(p \times n)}{\frac{\partial h(g(\beta))}{\partial \beta'}} = \frac{\partial h(\alpha)}{\partial \alpha'}\frac{\partial g(\beta)}{\partial \beta'}.$$

■

PROPOSITION A.2 (*Product Rules for Vector Differentiation*)

(a) Suppose β is $(m \times 1)$, $a(\beta) = (a_1(\beta), \ldots, a_n(\beta))'$ is $(n \times 1)$, $c(\beta) = (c_1(\beta), \ldots, c_p(\beta))'$ is $(p \times 1)$ and $A = (a_{ij})$ is $(n \times p)$ and does not depend on β. Then

$$\frac{\partial[a(\beta)'Ac(\beta)]}{\partial \beta'} = c(\beta)'A'\frac{\partial a(\beta)}{\partial \beta'} + a(\beta)'A\frac{\partial c(\beta)}{\partial \beta'}.$$

(b) If β is a (1×1) scalar, $A(\beta)$ is $(m \times n)$ and $B(\beta)$ is $(n \times p)$, then

$$\frac{\partial AB}{\partial \beta} = \frac{\partial A}{\partial \beta}B + A\frac{\partial B}{\partial \beta}.$$

(c) If β is an $(m \times 1)$ vector, $A(\beta)$ is $(n \times p)$ and $B(\beta)$ is $(p \times q)$, then

$$\frac{\partial \operatorname{vec}(AB)}{\partial \beta'} = (I_q \otimes A)\frac{\partial \operatorname{vec}(B)}{\partial \beta'} + (B' \otimes I_n)\frac{\partial \operatorname{vec}(A)}{\partial \beta'}.$$

■

Proof:

(a)
$$\frac{\partial(a'Ac)}{\partial \beta'} = \frac{\partial\left(\sum_{i,j} a_i a_{ij} c_j\right)}{\partial \beta'}$$
$$= \sum_{i,j}\left[\frac{\partial a_i}{\partial \beta'}a_{ij}c_j + a_i a_{ij}\frac{\partial c_j}{\partial \beta'}\right]$$
$$= c'A'\frac{\partial a}{\partial \beta'} + a'A\frac{\partial c}{\partial \beta'}.$$

(b) $AB = \left[\sum_j a_{ij}b_{jk}\right]$ and

$$\frac{\partial\left(\sum_j a_{ij}b_{jk}\right)}{\partial \beta} = \sum_j\left[\frac{\partial a_{ij}}{\partial \beta}b_{jk} + a_{ij}\frac{\partial b_{jk}}{\partial \beta}\right].$$

(c) Follows from (b) by stacking the columns of AB and writing the resulting columns $\partial \operatorname{vec}(AB)/\partial \beta_i$ for $i = 1, \ldots, m$ in one matrix. ■

The following rules are now easy to verify.

Rules:

(1) Let A be an $(m \times n)$ matrix and β be an $(n \times 1)$ vector. Then

$$\frac{\partial A\beta}{\partial \beta'} = A \qquad \text{and} \qquad \frac{\partial \beta' A'}{\partial \beta} = A'.$$

Proof: This result is a special case of Proposition A.2(a). ■

(2) Let A be $(m \times m)$ and β be $(m \times 1)$. Then

$$\frac{\partial \beta' A\beta}{\partial \beta} = (A + A')\beta \qquad \text{and} \qquad \frac{\partial \beta' A\beta}{\partial \beta'} = \beta'(A' + A).$$

Proof: See Proposition A.2(a). ■

(3) If A is $(m \times m)$ and β is $(m \times 1)$, then

$$\frac{\partial^2 \beta' A\beta}{\partial \beta \partial \beta'} = A + A'.$$

Proof: Follows from (1) and (2). ■

(4) If A is a symmetric $(m \times m)$ matrix and β an $(m \times 1)$ vector, then

$$\frac{\partial^2 \beta' A\beta}{\partial \beta \partial \beta'} = 2A.$$

Proof: See (3). ■

(5) Let Ω be a symmetric $(n \times n)$ matrix and $c(\beta)$ an $(n \times 1)$ vector that depends on the $(m \times 1)$ vector β. Then

$$\frac{\partial c(\beta)'\Omega c(\beta)}{\partial \beta'} = 2c(\beta)'\Omega \frac{\partial c(\beta)}{\partial \beta'}$$

and

$$\frac{\partial^2 c(\beta)'\Omega c(\beta)}{\partial \beta \partial \beta'} = 2\left[\frac{\partial c(\beta)'}{\partial \beta}\Omega\frac{\partial c(\beta)}{\partial \beta'} + (c(\beta)'\Omega \otimes I_m)\frac{\partial\, \mathrm{vec}(\partial c(\beta)'/\partial \beta)}{\partial \beta'}\right].$$

In particular, if y is an $(n \times 1)$ vector and X an $(n \times m)$ matrix,

$$\frac{\partial (y - X\beta)'\Omega(y - X\beta)}{\partial \beta'} = -2(y - X\beta)'\Omega X$$

and

$$\frac{\partial^2 (y - X\beta)'\Omega(y - X\beta)}{\partial \beta \partial \beta'} = 2X'\Omega X.$$

Proof: Follows from Proposition A.2(a). ■

(6) Suppose β is $(m \times 1)$, $B(\beta)$ is $(n \times p)$, A is $(k \times n)$, and C is $(p \times q)$ and the latter two matrices do not depend on β. Then

$$\frac{\partial \operatorname{vec}(ABC)}{\partial \beta'} = (C' \otimes A)\frac{\partial \operatorname{vec}(B)}{\partial \beta'}.$$

Proof: Follows from Rule 2, Section A.12, and Proposition A.1. ■

(7) Suppose β is $(m \times 1)$, $A(\beta)$ is $(n \times p)$, $D(\beta)$ is $(q \times r)$, and C is $(p \times q)$ and does not depend on β. Then

$$\frac{\partial \operatorname{vec}(ACD)}{\partial \beta'} = (I_r \otimes AC)\frac{\partial \operatorname{vec}(D)}{\partial \beta'} + (D'C' \otimes I_n)\frac{\partial \operatorname{vec}(A)}{\partial \beta'}.$$

Proof: Follows from Proposition A.2(c) by setting $B = CD$ and noting that $\partial \operatorname{vec}(CD)/\partial \beta' = (I_r \otimes C)\partial \operatorname{vec}(D)/\partial \beta'$. ■

(8) If β is $(m \times 1)$ and $A(\beta)$ is $(n \times n)$, then, for any positive integer h,

$$\frac{\partial \operatorname{vec}(A^h)}{\partial \beta'} = \left[\sum_{i=0}^{h-1} (A')^{h-1-i} \otimes A^i\right]\frac{\partial \operatorname{vec}(A)}{\partial \beta'}.$$

Proof: Follows inductively from Proposition A.2(c). The result is evident for $h = 1$. Assuming it holds for $h - 1$ gives

$$\frac{\partial \operatorname{vec}(AA^{h-1})}{\partial \beta'} = (I_n \otimes A)\left[\sum_{i=0}^{h-2} (A')^{h-2-i} \otimes A^i\right]\frac{\partial \operatorname{vec}(A)}{\partial \beta'} + ((A')^{h-1} \otimes I_n)\frac{\partial \operatorname{vec}(A)}{\partial \beta'}.$$ ■

(9) If A is a nonsingular $(m \times m)$ matrix, then

$$\frac{\partial \operatorname{vec}(A^{-1})}{\partial \operatorname{vec}(A)'} = -(A^{-1})' \otimes A^{-1}.$$

Proof: Using Proposition A.2(c),

$$0 = \frac{\partial \operatorname{vec}(I_m)}{\partial \operatorname{vec}(A)'} = \frac{\partial \operatorname{vec}(A^{-1}A)}{\partial \operatorname{vec}(A)'} = (I_m \otimes A^{-1})\frac{\partial \operatorname{vec}(A)}{\partial \operatorname{vec}(A)'} + (A' \otimes I_m)\frac{\partial \operatorname{vec}(A^{-1})}{\partial \operatorname{vec}(A)'}.$$ ■

(10) Let A be a symmetric positive definite $(m \times m)$ matrix and let P be a lower triangular $(m \times m)$ matrix with positive elements on the main diagonal such that $A = PP'$. Moreover, let $\mathbf{L}_m$ be an $(m(m+1)/2 \times m^2)$ elimination matrix such that $\mathbf{L}_m \operatorname{vec}(A) = \operatorname{vech}(A)$ consists of the elements on and

below the diagonal of A only. Then

$$\frac{\partial\, \text{vech}(P)}{\partial\, \text{vech}(A)'} = \{\mathbf{L}_m[(I_m \otimes P)\mathbf{K}_{mm} + (P \otimes I_m)]\mathbf{L}_m'\}^{-1}$$

$$= \{\mathbf{L}_m(I_{m^2} + \mathbf{K}_{mm})(P \otimes I_m)\mathbf{L}_m'\}^{-1},$$

where $\mathbf{K}_{mm}$ is an $(m^2 \times m^2)$ commutation matrix such that $\mathbf{K}_{mm}\text{vec}(P) = \text{vec}(P')$.

Proof: See Lütkepohl (1989b). ■

(11) If $A = (a_{ij})$ is an $(m \times m)$ matrix, then

$$\frac{\partial\, \text{tr}(A)}{\partial A} = I_m.$$

Proof: $\text{tr}\, A = a_{11} + \cdots + a_{mm}$. Hence,

$$\frac{\partial\, \text{tr}(A)}{\partial a_{ij}} = \begin{cases} 0 & \text{if } i \neq j \\ 1 & \text{if } i = j. \end{cases}$$ ■

(12) If $A = (a_{ij})$ is $(m \times n)$ and $B = (b_{ij})$ is $(n \times m)$, then

$$\frac{\partial\, \text{tr}(AB)}{\partial A} = B'.$$

Proof: Follows because $\text{tr}(AB) = \sum_{j=1}^n a_{1j}b_{j1} + \cdots + \sum_{j=1}^n a_{mj}b_{jm}$. ■

(13) Suppose A is an $(m \times n)$ matrix and B, C are $(m \times m)$ and $(n \times m)$, respectively. Then

$$\frac{\partial\, \text{tr}(BAC)}{\partial A} = B'C'.$$

Proof: Follows from Rule 12 since $\text{tr}(BAC) = \text{tr}(ACB)$. ■

(14) Let A, B, C, D be $(m \times n)$, $(n \times n)$, $(m \times n)$, and $(n \times m)$ matrices, respectively. Then

$$\frac{\partial\, \text{tr}(DABA'C)}{\partial A} = CDAB + D'C'AB'.$$

Proof: See Murata (1982, Appendix, Theorem 6a). ■

(15) Let A, B, and C be $(m \times m)$ matrices and suppose A is nonsingular. Then

$$\frac{\partial\, \text{tr}(BA^{-1}C)}{\partial A} = -(A^{-1}CBA^{-1})'.$$

Proof: By Rule 6 of Section A.12,

$$\operatorname{tr}(BA^{-1}C) = \operatorname{vec}(B')'(C' \otimes I_m)\operatorname{vec}(A^{-1}).$$

Hence, using (9),

$$\begin{aligned}\frac{\partial \operatorname{tr}(BA^{-1}C)}{\partial \operatorname{vec}(A)'} &= -\operatorname{vec}(B')'(C' \otimes I_m)((A^{-1})' \otimes A^{-1}) \\ &= -[(A^{-1}C \otimes A^{-1\prime})\operatorname{vec}(B')]' \\ &= -[\operatorname{vec}(A^{-1\prime}B'C'A^{-1\prime})]'\end{aligned}$$

by Rule 2 of Section A.12. ■

(16) Let $A = (a_{ij})$ be an $(m \times m)$ matrix. Then

$$\frac{\partial |A|}{\partial A} = A^{*\prime},$$

where A^* is the adjoint of A.

Proof: Developing by the i-th row gives

$$|A| = a_{i1}A_{i1} + \cdots + a_{im}A_{im},$$

where A_{ij} is the cofactor of a_{ij}. Hence,

$$\frac{\partial |A|}{\partial a_{ij}} = A_{ij}$$

because A_{ij} does not contain a_{ij}. ■

(17) If A is a nonsingular $(m \times m)$ matrix with $|A| > 0$, then

$$\frac{\partial \ln|A|}{\partial A} = (A')^{-1}.$$

Proof: Using Proposition A.1 (chain rule),

$$\frac{\partial \ln|A|}{\partial A} = \frac{\partial \ln|A|}{\partial |A|} \cdot \frac{\partial |A|}{\partial A} = \frac{1}{|A|}A^{*\prime} = (A')^{-1}.$$ ■

PROPOSITION A.3 (*Taylor's Theorem*)

Let $f(\beta)$ be a scalar valued function of the $(m \times 1)$ vector β. Suppose $f(\beta)$ is at least twice continuously differentiable on an open set S that contains β_0, β, and the entire line segment between β_0 and β. Then there exists a point $\bar{\beta}$ on the line segment such that

$$f(\beta) = f(\beta_0) + \frac{\partial f(\beta_0)}{\partial \beta'}(\beta - \beta_0) + \frac{1}{2}(\beta - \beta_0)' \frac{\partial^2 f(\bar{\beta})}{\partial \beta \partial \beta'}(\beta - \beta_0), \qquad \text{(A.13.1)}$$

where $\partial f(\beta_0)/\partial \beta' := (\partial f/\partial \beta'|_{\beta_0})$. ■

The expansion of f given in (A.13.1) is a *second order Taylor expansion* at or around β_0.

A.14 Optimization of Vector Functions

Suppose $f(\beta)$ is a real valued (scalar) differentiable function of the $(m \times 1)$ vector β. A necessary condition for a local optimum (minimum or maximum) at $\tilde{\beta}$ is that

$$\frac{\partial f}{\partial \beta} = 0 \qquad \text{for } \beta = \tilde{\beta}, \qquad \text{that is,} \qquad \frac{\partial f(\tilde{\beta})}{\partial \beta} := \left[\left. \frac{\partial f}{\partial \beta} \right|_{\tilde{\beta}} \right] = 0.$$

In other words, f has a *stationary point* at $\tilde{\beta}$. If this condition is satisfied and the *Hessian matrix* of second partial derivatives

$$\frac{\partial^2 f}{\partial \beta \partial \beta'}$$

is negative (positive) definite for $\beta = \tilde{\beta}$, then $\tilde{\beta}$ is a local maximum (minimum).

If a set of constraints is given in the form

$$\varphi(\beta) = (\varphi_1(\beta), \ldots, \varphi_n(\beta))' = 0,$$

that is, $\varphi(\beta)$ is an $(n \times 1)$ vector, then a local optimum subject to these constraints is assumed at a stationary point of the *Lagrange function*

$$LG(\beta, \lambda) = f(\beta) - \lambda' \varphi(\beta),$$

where λ is an $(n \times 1)$ vector of *Lagrange multipliers*. In other words, a necessary condition for a constrained local optimum is that

$$\frac{\partial LG}{\partial \beta} = 0 \qquad \text{and} \qquad \frac{\partial LG}{\partial \lambda} = 0$$

hold simultaneously.

The following optimization result is the basis for results given in Chapter 5.

PROPOSITION A.4 (*Maximum of* $\mathrm{tr}(B'\Omega B)$)

Let Ω be a positive semidefinite symmetric $(K \times K)$ matrix with eigenvalues $\lambda_1 \geq \lambda_2 \geq \ldots \geq \lambda_K$ and corresponding orthonormal $(K \times 1)$ eigenvectors $v_1, v_2, \ldots, v_K$. Moreover, let B be a $(K \times r)$ matrix with $B'B = I_r$. Then the maximum of $\mathrm{tr}(B'\Omega B)$ with respect to B is obtained for

$$B = \hat{B} = [v_1, \ldots, v_r]$$

and

$$\max_B \mathrm{tr}(B'\Omega B) = \lambda_1 + \cdots + \lambda_r.$$

■

Proof: The proposition follows from Theorem 6, p. 205, of Magnus & Neudecker (1988) by induction. For $r = 1$ our result is just a special case of that theorem. For $r > 1$, assuming that the proposition holds for $r - 1$ and denoting the columns of B by $b_1, \dots, b_r$,

$$\begin{aligned}
\operatorname{tr}(B'\Omega B) &= \operatorname{tr}\begin{bmatrix} b_1' \\ \vdots \\ b_r' \end{bmatrix} \Omega [b_1, \dots, b_r] = \operatorname{tr}\begin{bmatrix} b_1'\Omega b_1 & & * \\ & \ddots & \\ * & & b_r'\Omega b_r \end{bmatrix} \\
&= b_1'\Omega b_1 + \cdots + b_{r-1}'\Omega b_{r-1} + b_r'\Omega b_r \\
&= \lambda_1 + \cdots + \lambda_{r-1} + b_r'\Omega b_r
\end{aligned}$$

and $\max b_r'\Omega b_r = v_r'\Omega v_r = \lambda_r$, under the conditions of the proposition, by the aforementioned theorem from Magnus & Neudecker (1988). ■

The next proposition may be regarded as a corollary of Proposition A.4.

PROPOSITION A.5 (*Minimum of* $\operatorname{tr}(Y - BCX)'\Sigma_u^{-1}(Y - BCX)$)

Let Y, X, Σ_u, B, and C be matrices of dimensions $(K \times T), (Kp \times T), (K \times K), (K \times r)$, and $(r \times Kp)$, respectively, with Σ_u positive definite, $\operatorname{rk}(B) = \operatorname{rk}(C) = r$, $\operatorname{rk}(X) = Kp$, and $\operatorname{rk}(Y) = K$. Then a minimum of

$$\operatorname{tr}(Y - BCX)'\Sigma_u^{-1}(Y - BCX) \tag{A.14.1}$$

with respect to B and C is obtained for

$$B = \hat{B} = \Sigma_u^{1/2}\hat{V} \quad \text{and} \quad C = \hat{C} = \hat{V}'\Sigma_u^{-1/2}YX'(XX')^{-1}, \tag{A.14.2}$$

where $\hat{V} = [\hat{v}_1, \dots, \hat{v}_r]$ is the $(K \times r)$ matrix of the orthonormal eigenvectors corresponding to the r largest eigenvalues of

$$\frac{1}{T}\Sigma_u^{-1/2}YX'(XX')^{-1}XY'\Sigma_u^{-1/2}$$

in nonincreasing order. ■

Proof: We first assume $\Sigma_u = I_K$.

$$\begin{aligned}
\operatorname{tr}(Y - BCX)'(Y - BCX) &= \operatorname{tr}(Y - BCX)(Y - BCX)' \\
&= (\operatorname{vec}(Y) - \operatorname{vec}(BCX))'(\operatorname{vec}(Y) - \operatorname{vec}(BCX)) \\
&= [\operatorname{vec}(Y) - (X' \otimes B)\operatorname{vec}(C)]' \\
&\quad \times [\operatorname{vec}(Y) - (X' \otimes B)\operatorname{vec}(C)].
\end{aligned} \tag{A.14.3}$$

A derivation similar to that in Section 3.2.1 shows that this sum of squares is minimized with respect to vec(C) when this vector is chosen to be

$$\begin{aligned}\text{vec}(\hat{C}) &= [(X \otimes B')(X' \otimes B)]^{-1}(X \otimes B')\,\text{vec}(Y)\\ &= (XX' \otimes B'B)^{-1}\,\text{vec}(B'YX')\\ &= \text{vec}[(B'B)^{-1}B'YX'(XX')^{-1}].\end{aligned}$$

Since we may normalize the columns of B we choose $B'B = I_r$ without loss of generality. Hence,

$$\hat{C} = B'YX'(XX')^{-1}. \tag{A.14.4}$$

Substituting for C in (A.14.3) gives

$$\begin{aligned}&\text{tr}(Y - BB'YX'(XX')^{-1}X)(Y - BB'YX'(XX')^{-1}X)'\\ &\quad = \text{tr}(YY') - \text{tr}(BB'YX'(XX')^{-1}XY') - \text{tr}(YX'(XX')^{-1}XY'BB')\\ &\qquad + \text{tr}(BB'YX'(XX')^{-1}XX'(XX')^{-1}XY'BB')\\ &\quad = \text{tr}(YY') - \text{tr}(B'YX'(XX')^{-1}XY'B),\end{aligned}$$

where again $B'B = I_r$ has been used. This expression is minimized with respect to B, where

$$\frac{1}{T}\,\text{tr}\,B'YX'(XX')^{-1}XY'B$$

assumes its maximum. By Proposition A.4 the maximum is attained if B consists of the eigenvectors corresponding to the r largest eigenvalues of

$$\frac{1}{T}YX'(XX')^{-1}XY'$$

which proves the proposition for $\Sigma_u = I_K$.

If $\Sigma_u \neq I_K$,

$$\text{tr}(Y - BCX)'\Sigma_u^{-1}(Y - BCX) = \text{tr}(Y^\# - B^\# CX)'(Y^\# - B^\# CX)$$

has to be minimized with respect to $B^\#$ and C. Here $Y^\# = \Sigma_u^{-1/2}Y$ and $B^\# = \Sigma_u^{-1/2}B$. From the above derivation the solution is $\hat{B}^\# = \hat{V}$ and

$$\hat{C} = \hat{B}^{\#\prime}Y^\# X'(XX')^{-1} = \hat{V}'\Sigma_u^{-1/2}YX'(XX')^{-1},$$

where the columns of $\hat{V}$ are the eigenvectors corresponding to the r largest eigenvalues of

$$\frac{1}{T}Y^\# X'(XX')^{-1}XY^{\#\prime} = \frac{1}{T}\Sigma_u^{-1/2}YX'(XX')^{-1}XY'\Sigma_u^{-1/2}.$$

Hence, $\hat{B} = \Sigma_u^{1/2}\hat{B}^\# = \Sigma_u^{1/2}\hat{V}$. ■

A result similar to that in Proposition A.4 also holds for the maximum and minimum of a determinant. The following proposition is a slight modification of Theorem 15 of Magnus & Neudecker (1988, Chapter 11).

PROPOSITION A.6 (*Maximum and Minimum of* $|C\Omega C'|$)

Let Ω be a positive definite symmetric $(K \times K)$ matrix with eigenvalues $\lambda_1 \geq \lambda_2 \geq \ldots \geq \lambda_K$ and corresponding orthonormal $(K \times 1)$ eigenvectors $v_1, \ldots, v_K$. Furthermore, let C be an $(r \times K)$ matrix with $CC' = I_r$. Then

$$\max_C |C\Omega C'| = \lambda_1 \cdots \lambda_r$$

and the maximum is attained for

$$C = \hat{C} = [v_1, \ldots, v_r]'.$$

Moreover,

$$\min_C |C\Omega C'| = \lambda_K \lambda_{K-1} \cdots \lambda_{K-r+1}$$

and the minimum is attained for

$$C = \hat{C} = [v_K, \ldots, v_{K-r+1}]'. \qquad \blacksquare$$

An important implication of this proposition is used in Chapter 11 and is stated next.

PROPOSITION A.7 (*Minimum of* $|(Y - BCX)(Y - BCX)'/T|$)

Let Y and X be $(K \times T)$ matrices of rank K and let B and C be of rank r and dimensions $(K \times r)$ and $(r \times K)$, respectively. Furthermore, let $\lambda_1 \geq \ldots \geq \lambda_K$ be the eigenvalues of

$$(XX')^{-1/2} XY'(YY')^{-1} YX'(XX')^{-1/2'}$$

and the corresponding orthonormal eigenvectors are $v_1, \ldots, v_K$. Here $(XX')^{-1/2}$ is some matrix satisfying

$$(XX')^{-1/2}(XX')(XX')^{-1/2'} = I_K.$$

Then

$$\min_{B,C} |(Y - BCX)(Y - BCX)'/T| = |YY'/T|(1 - \lambda_1)\cdots(1 - \lambda_r)$$

and the minimum is attained for

$$C = \hat{C} = [v_1, \ldots, v_r]'(XX')^{-1/2}$$

and

$$B = \hat{B} = YX'\hat{C}'(\hat{C}XX'\hat{C}')^{-1}. \qquad \blacksquare$$

A proof of this proposition can be found in Tso (1981). It should be noted that the minimizing matrices $\hat{B}$ and $\hat{C}$ are not unique. Any nonsingular $(r \times r)$ matrix F leads to another set of minimizing matrices $F\hat{C}$, $\hat{B}F^{-1}$.

A.15 Problems

The following problems refer to the matrices

$$A = \begin{bmatrix} 5 & 2 \\ -1 & 1 \end{bmatrix}, \quad B = \begin{bmatrix} 6 & 0 & 0 \\ -6 & 1 & 0 \end{bmatrix}, \quad C = \begin{bmatrix} 1 & 4 & 0 \\ 2 & 2 & 2 \\ 1 & 2 & 0 \end{bmatrix},$$

$$D = \begin{bmatrix} 5 & 2 \\ 2 & 1 \end{bmatrix}, \quad H(\beta) = \begin{bmatrix} 4\beta_1 & 2\beta_1 + \beta_2 \\ 1 + \beta_2 & 3 \end{bmatrix}$$

Problem A.1

Determine $A + D$, $A - 2D$, A', AB, BC, $B'A$, $B'A'$, $A \otimes D$, $B \otimes D$, $D \otimes B$, $B' \otimes D'$, $B + BC$, $\operatorname{tr} A$, $\operatorname{tr} D$, $\det A$, $|D|$, $|C|$, $\operatorname{vec}(B)$, $\operatorname{vec}(B')$, $\operatorname{vech}(C)$, $\mathbf{K}_{33}$, A^{-1}, D^{-1}, $(A \otimes D)^{-1}$, $\operatorname{rk}(C)$, $\operatorname{rk}(B)$, $\det(A \otimes D)$, $\operatorname{tr}(A \otimes D)$, C^{-1} (use the rules for the partitioned inverse).

Problem A.2

Determine the eigenvalues of A, D, and $A \otimes D$.

Problem A.3

Find an upper triangular matrix Q such that $D = QQ'$ and find an orthogonal matrix P such that $D = P\Lambda P'$, where Λ is a diagonal matrix with the eigenvalues of D on the main diagonal. Compute D^5.

Problem A.4

Is $F = I_2 - BB'$ idempotent? Is BB' positive definite?

Problem A.5

Determine the following derivatives:

(a) $\dfrac{\partial \det(H)}{\partial \beta}$, (b) $\dfrac{\partial^2 \det(H)}{\partial \beta \partial \beta'}$, (c) $\dfrac{\partial \operatorname{tr} H}{\partial \beta}$, (d) $\dfrac{\partial \operatorname{vec}(H)}{\partial \beta'}$,

(e) $\dfrac{\partial \operatorname{vec}(H^2)}{\partial \beta'}$, (f) $\dfrac{\partial H(\beta)\beta}{\partial \beta'}$,

where $\beta = (\beta_1, \beta_2)'$.

Problem A.6

Determine the stationary points of $|H|$ with respect to β. Are they local extrema?

Problem A.7

Give a second order Taylor expansion of $\det(H)$ around $\beta = (0, 0)'$.

Appendix B. Multivariate Normal and Related Distributions

B.1 Multivariate Normal Distributions

A K-dimensional vector of continuous random variables $y = (y_1, \ldots, y_K)'$ has a *multivariate normal distribution* with mean vector $\mu = (\mu_1, \ldots, \mu_K)'$ and covariance matrix Σ, briefly

$$y \sim N(\mu, \Sigma),$$

if its distribution has the probability density function (p.d.f.)

$$f(y) = \frac{1}{(2\pi)^{K/2}} |\Sigma|^{-1/2} \exp\left[-\frac{1}{2}(y - \mu)'\Sigma^{-1}(y - \mu) \right]. \tag{B.1.1}$$

Alternatively, $y \sim N(\mu, \Sigma)$ if for any K-vector c for which $c'\Sigma c \neq 0$ the linear combination $c'y$ has a univariate normal distribution, that is, $c'y \sim N(c'\mu, c'\Sigma c)$ (see Rao (1973, Chapter 8)). This definition of a multivariate normal distribution is useful because it carries over to the case where Σ is positive semidefinite and singular, while the multivariate density in (B.1.1) is only meaningful if Σ is positive definite and hence nonsingular. It must be emphasized, however, that the two definitions are equivalent if Σ is positive definite rather than just positive semidefinite. Another possibility to define a multivariate normal distribution with singular covariance matrix may be found in Anderson (1984).

The following results regarding the multivariate normal and related distributions are useful. Many of them are stated in Judge et al. (1985, Appendix A). Proofs can be found in Rao (1973, Chapter 8) and Hogg & Craig (1978, Chapter 12).

PROPOSITION B.1 (*Marginal and Conditional Distributions of a Multivariate Normal*)

Let y_1 and y_2 be two random vectors such that

$$\begin{bmatrix} y_1 \\ y_2 \end{bmatrix} \sim N\left(\begin{bmatrix} \mu_1 \\ \mu_2 \end{bmatrix}, \begin{bmatrix} \Sigma_{11} & \Sigma_{12} \\ \Sigma_{21} & \Sigma_{22} \end{bmatrix} \right),$$

where the partitioning of the mean vector and covariance matrix corresponds to that of the vector $(y_1', y_2')'$. Then

$$y_1 \sim N(\mu_1, \Sigma_{11})$$

and the conditional distribution of y_1 given $y_2 = c$ is also multivariate normal,

$$(y_1|y_2 = c) \sim N(\mu_1 + \Sigma_{12}\Sigma_{22}^{-1}(c - \mu_2), \Sigma_{11} - \Sigma_{12}\Sigma_{22}^{-1}\Sigma_{21}).$$

If Σ_{22} is singular, the inverse may be replaced by a generalized inverse. Moreover, y_1 and y_2 are independent if and only if $\Sigma_{12} = \Sigma_{21}' = 0$. ■

PROPOSITION B.2 (*Linear Transformation of a Multivariate Normal*)

Suppose $y \sim N(\mu, \Sigma)$ is $(K \times 1)$, A is an $(M \times K)$ matrix and c an $(M \times 1)$ vector. Then

$$x = Ay + c \sim N(A\mu + c, A\Sigma A').$$ ■

B.2 Related Distributions

Suppose $y \sim N(0, I_K)$. The distribution of $z = y'y$ is a (central) *chi-square distribution* with K degrees of freedom,

$$z \sim \chi^2(K).$$

PROPOSITION B.3 (*Distributions of Quadratic Forms*)

(1) Suppose $y \sim N(0, I_K)$ and A is a symmetric idempotent $(K \times K)$ matrix with $\mathrm{rk}(A) = n$. Then

$$y'Ay \sim \chi^2(n).$$

(2) If $y \sim N(0, \Sigma)$, where Σ is a positive definite $(K \times K)$ matrix, then

$$y'\Sigma^{-1}y \sim \chi^2(K).$$

(3) Let $y \sim N(0, QA)$, where Q is a symmetric, idempotent $(K \times K)$ matrix with $\mathrm{rk}(Q) = n$ and A is a positive definite $(K \times K)$ matrix. Then

$$y'A^{-1}y \sim \chi^2(n).$$

(4) Suppose $y \sim N(0, \Sigma)$, where Σ is a nonsingular $(K \times K)$ covariance matrix. Then

$$y'Ay \text{ has a } \chi^2\text{-distribution} \quad \Leftrightarrow \quad A\Sigma A = A.$$

In this case $y'Ay \sim \chi^2(n)$ if $\mathrm{rk}(A) = n$. ■

PROPOSITION B.4 (*Independence of a Normal Vector and a Quadratic Form*)

Suppose $y \sim N(\mu, \sigma^2 I_K)$, A is a symmetric, idempotent $(K \times K)$ matrix, B is an $(M \times K)$ matrix and $BA = 0$. Then By is stochastically independent of the random variable $y'Ay$. ■

PROPOSITION B.5 (*Independence of Quadratic Forms*)

If $y \sim N(0, \sigma^2 I_K)$ and A and B are symmetric, idempotent $(K \times K)$ matrices with $AB = 0$, then $y'Ay$ and $y'By$ are stochastically independent. ■

If $z \sim N(0, 1)$ and $u \sim \chi^2(m)$ are independent, then

$$T = \frac{z}{\sqrt{u/m}}$$

has a *t-distribution* with m degrees of freedom, $T \sim t(m)$. If $u \sim \chi^2(m)$ and $v \sim \chi^2(n)$ are independent, then

$$\frac{u/m}{v/n} \sim F(m, n),$$

that is, the ratio of two independent χ^2 random variables, each divided by its degrees of freedom, has an *F-distribution* with m and n degrees of freedom. The numbers m and n indicate the numerator and denominator degrees of freedom, respectively.

PROPOSITION B.6 (*Distributions of Ratios of Quadratic Forms*)

(1) Suppose $x \sim N(0, I_m)$ and $y \sim N(0, I_n)$ are independent. Then

$$\frac{x'x/m}{y'y/n} \sim F(m, n).$$

(2) If $y \sim N(0, I_K)$ and A and B are symmetric, idempotent $(K \times K)$ matrices with $\text{rk}(A) = m$, $\text{rk}(B) = n$ and $AB = 0$, then

$$\frac{y'Ay/m}{y'By/n} \sim F(m, n).$$

(3) $z \sim F(m, n) \Rightarrow 1/z \sim F(n, m)$. ■

If $y \sim N(\mu, I_K)$, then $y'y$ has a *noncentral χ^2-distribution* with K degrees of freedom and *noncentrality parameter* (or simply *noncentrality*) $\tau = \mu'\mu$. Briefly,

$$y'y \sim \chi^2(K; \tau).$$

The noncentrality parameter is sometimes defined differently in the literature. For instance, $\lambda = (1/2)\mu'\mu$ is sometimes called noncentrality parameter. Let $w \sim \chi^2(m; \tau)$ and $v \sim \chi^2(n)$ be independent random variables, then

$$\frac{w/m}{v/n} \sim F(m, n; \tau),$$

that is, the ratio has a *noncentral F-distribution* with m and n degrees of freedom and noncentrality parameter τ.

PROPOSITION B.7 (*Quadratic Form with Noncentral χ^2-Distribution*)

If $y \sim N(\mu, \Sigma)$ with positive definite $(K \times K)$ covariance matrix Σ, then $y'\Sigma^{-1}y \sim \chi^2(K; \mu'\Sigma^{-1}\mu)$. ■

Appendix C. Convergence of Sequences of Random Variables and Asymptotic Distributions

It is often difficult to derive the exact distribution of estimators. In that case their asymptotic or limiting properties, when the sample size gets large, are of interest. The limiting properties are then regarded as approximations to the properties for the sample size available. In order to study the limiting properties some concepts of convergence of sequences of random variables and vectors are useful. They are discussed in Section C.1 and asymptotic properties of estimators are considered in Section C.2. Infinite sums of random variables are treated in Section C.3. Maximum likelihood estimators and their asymptotic properties are considered in Section C.4 and some common testing principles are treated in Section C.5.

This appendix contains a brief summary of results used in the text. Many of these results can be found in Judge et al. (1985, Section 5.8). A more complete discussion and proofs are provided in Roussas (1973), Serfling (1980) and other more advanced books on statistics. Further references will be given in the following.

C.1 Concepts of Stochastic Convergence

Let $x_1, x_2, \dots$ or $\{x_T\}$, $T = 1, 2, \dots$, be a sequence of scalar random variables which are all defined on a common probability space $(\Omega, \mathcal{M}, P)$. The sequence $\{x_T\}$ *converges in probability* to the random variable x (which is also defined on $(\Omega, \mathcal{M}, P)$) if for every $\varepsilon > 0$,

$$\lim_{T\to\infty} P(|x_T - x| > \varepsilon) = 0$$

or, equivalently,

$$\lim P(|x_T - x| < \varepsilon) = 1.$$

This type of stochastic convergence is abbreviated as

$$\text{plim}\, x_T = x \qquad \text{or} \qquad x_T \xrightarrow{p} x.$$

The limit x may be a fixed, nonstochastic real number which is then regarded as a degenerate random variable that takes on one particular value with probability one.

The sequence $\{x_T\}$ *converges almost surely* (a.s.) or *with probability one* to the random variable x if for every $\varepsilon > 0$

$$P\left(\lim_{T\to\infty} |x_T - x| < \varepsilon\right) = 1.$$

This type of convergence is often written as $x_T \xrightarrow{\text{a.s.}} x$ and is sometimes called *strong convergence*.

The sequence $\{x_T\}$ *converges in quadratic mean* or *mean square error* to x, briefly $x_T \xrightarrow{\text{q.m.}} x$, if

$$\lim_{T\to\infty} E(x_T - x)^2 = 0.$$

This type of convergence requires that the mean and variance of the x_T and x exist.

Finally, denoting the distribution function of x_T and x by F_T and F, respectively, the sequence $\{x_T\}$ is said to *converge in distribution* or *weakly* or *in law* to x, if for all real numbers c for which F is continuous,

$$\lim_{T\to\infty} F_T(c) = F(c).$$

This type of convergence is abbreviated as $x_T \xrightarrow{d} x$. It must be emphasized that we do not require the convergence of the sequence of p.d.f.s of the x_T to the p.d.f. of x. In fact, we do not even require that the distributions of the x_T have p.d.f.s. Even if they do have p.d.f.s, convergence in distribution does not imply their convergence to the p.d.f. of x.

All these concepts of stochastic convergence can be extended to sequences of random vectors (multivariate random variables). Suppose $\{x_T = (x_{1T}, \ldots, x_{KT})'\}$, $T = 1, 2, \ldots$, is a sequence of K-dimensional random vectors and $x = (x_1, \ldots, x_K)'$ is a K-dimensional random vector.

$$\operatorname{plim} x_T = x \quad \text{or} \quad x_T \xrightarrow{p} x \quad \text{if} \quad \operatorname{plim} x_{kT} = x_k \text{ for } k = 1, \ldots, K.$$

$$x_T \xrightarrow{\text{a.s.}} x \quad \text{if} \quad x_{kT} \xrightarrow{\text{a.s.}} x_k \quad \text{for } k = 1, \ldots, K.$$

$$x_T \xrightarrow{\text{q.m.}} x \quad \text{if} \quad \lim E(x_T - x)'(x_T - x) = 0.$$

$$x_T \xrightarrow{d} x \quad \text{if} \quad \lim F_T(c) = F(c) \text{ for all continuity points of } F.$$

Here F_T and F are the joint distribution functions of x_T and x respectively. Almost sure convergence and convergence in probability can be defined for matrices in the same way in terms of convergence of the individual elements. Convergence in quadratic mean and in distribution is easily extended to sequences of random matrices by vectorizing them. In the following proposition the relationships between the different modes of convergence are stated and useful rules for establishing convergence are given.

PROPOSITION C.1 (*Convergence Properties of Sequences of Random Variables*)

Suppose $\{x_T\}$ is a sequence of K-dimensional random variables and x is a K-dimensional random variable. Then the following relations hold:

(1) $x_T \xrightarrow{\text{a.s.}} x \quad \Rightarrow \quad x_T \xrightarrow{p} x \quad \Rightarrow \quad x_T \xrightarrow{d} x.$
(2) $x_T \xrightarrow{\text{q.m.}} x \quad \Rightarrow \quad x_T \xrightarrow{p} x \quad \Rightarrow \quad x_T \xrightarrow{d} x.$
(3) If x is a fixed, nonstochastic vector, then $x_T \xrightarrow{\text{q.m.}} x \Leftrightarrow [\lim E(x_T) = x$ and $\lim E(x_T - Ex_T)'(x_T - Ex_T) = 0]$.
(4) If x is a fixed, nonstochastic vector, then

$$x_T \xrightarrow{p} x \quad \Leftrightarrow \quad x_T \xrightarrow{d} x.$$

(5) (Slutsky's Theorem) If $g: \mathbb{R}^K \to \mathbb{R}^m$ is a continuous function, then

$$x_T \xrightarrow{p} x \quad \Rightarrow \quad g(x_T) \xrightarrow{p} g(x) \quad [\text{plim } g(x_T) = g(\text{plim } x_T)],$$

$$x_T \xrightarrow{d} x \quad \Rightarrow \quad g(x_T) \xrightarrow{d} g(x),$$

and

$$x_T \xrightarrow{\text{a.s.}} x \quad \Rightarrow \quad g(x_T) \xrightarrow{\text{a.s.}} g(x).$$

■

PROPOSITION C.2 (*Properties of Convergence in Probability and in Distribution*)

Suppose $\{x_T\}$ and $\{y_T\}$ are sequences of $(K \times 1)$ random vectors, $\{A_T\}$ is a sequence of $(K \times K)$ random matrices, x is a $(K \times 1)$ random vector, c is a fixed $(K \times 1)$ vector, and A is a fixed $(K \times K)$ matrix.

(1) If plim x_T, plim y_T, and plim A_T exist, then
  (a) $\text{plim}(x_T \pm y_T) = \text{plim } x_T \pm \text{plim } y_T$;
  (b) $\text{plim}(c'x_T) = c' \text{ plim } x_T$;
  (c) $\text{plim } x_T'y_T = (\text{plim } x_T)'(\text{plim } y_T)$;
  (d) $\text{plim } A_T x_T = \text{plim}(A_T) \text{ plim}(x_T)$.
(2) If $x_T \xrightarrow{d} x$ and $\text{plim}(x_T - y_T) = 0$, then $y_T \xrightarrow{d} x$.
(3) If $x_T \xrightarrow{d} x$ and $\text{plim } y_T = c$, then
  (a) $x_T \pm y_T \xrightarrow{d} x \pm c$;
  (b) $y_T'x_T \xrightarrow{d} c'x$.
(4) If $x_T \xrightarrow{d} x$ and $\text{plim } A_T = A$, then $A_T x_T \xrightarrow{d} Ax$.
(5) If $x_T \xrightarrow{d} x$ and $\text{plim } A_T = 0$, then $\text{plim } A_T x_T = 0$. ■

PROPOSITION C.3 (*Limits of Sequences of t and F Random Variables*)

(1) $t(T) \xrightarrow[T\to\infty]{d} N(0, 1)$

(that is, a sequence of random variables with t-distributions with T degrees of freedom converges to a standard normal distribution as the degrees of freedom go to infinity).

(2) $JF(J, T) \xrightarrow[T \to \infty]{d} \chi^2(J)$. ■

C.2 Asymptotic Properties of Estimators and Test Statistics

Suppose we have a sequence of $(m \times n)$ estimators $\{\hat{B}_T\}$ for an $(m \times n)$ parameter matrix B, where T denotes the sizes of the sample (length of the time series) on which the estimator is based. For simplicity we will delete the subscript T in the following and we will mean the sequence of estimators when we use the term "estimator".

The estimator $\hat{B}$ is *consistent* if $\operatorname{plim} \hat{B} = B$. In the literature this type of consistency is sometimes called *weak consistency*. However, in this text we simply use the term consistency instead. The estimator is *strongly consistent* if $\hat{B} \xrightarrow{\text{a.s.}} B$, and the estimator is *mean square consistent* if $\hat{B} \xrightarrow{\text{q.m.}} B$. By Proposition C.1 both strong consistency and mean square consistency imply consistency.

Let $\hat{\beta}$ be an estimator (a sequence of estimators) of the $(K \times 1)$ vector β. The estimator is said to have an *asymptotic normal distribution* if $\sqrt{T}(\hat{\beta} - \beta)$ converges in distribution to a random vector with multivariate normal distribution $N(0, \mathbf{\Sigma})$, that is,

$$\sqrt{T}(\hat{\beta} - \beta) \xrightarrow{d} N(0, \mathbf{\Sigma}). \tag{C.2.1}$$

In that case, for large T, $N(\beta, \mathbf{\Sigma}/T)$ is usually used as an approximation to the distribution of $\hat{\beta}$. Equivalently (C.2.1) may be defined by requiring that

$$\frac{\sqrt{T}c'(\hat{\beta} - \beta)}{(c'\mathbf{\Sigma}c)^{1/2}} \xrightarrow{d} N(0, 1),$$

for any $(K \times 1)$ vector c for which $c'\mathbf{\Sigma}c \neq 0$. The following proposition provides some useful rules for determining the asymptotic distributions of estimators and test statistics.

PROPOSITION C.4 (*Asymptotic Properties of Estimators*)

Suppose $\hat{\beta}$ is an estimator of the $(K \times 1)$ vector β with

$$\sqrt{T}(\hat{\beta} - \beta) \xrightarrow{d} N(0, \mathbf{\Sigma}).$$

Then the following rules hold:

(1) If $\operatorname{plim} \hat{A} = A$, then

$$\sqrt{T}\hat{A}(\hat{\beta} - \beta) \xrightarrow{d} N(0, A\mathbf{\Sigma}A').$$

(See Schmidt (1976, p. 251)).

(2) If $R \neq 0$ is an $(M \times K)$ matrix, then

$$\sqrt{T}(R\hat{\beta} - R\beta) \xrightarrow{d} N(0, R\mathbf{\Sigma}R').$$

(3) If $g(\beta) = (g_1(\beta), \ldots, g_m(\beta))'$ is a vector-valued continuously differentiable function with $\partial g/\partial\beta' \neq 0$ at β, then

$$\sqrt{T}[g(\hat{\beta}) - g(\beta)] \xrightarrow{d} N\left[0, \frac{\partial g(\beta)}{\partial \beta'} \Sigma \frac{\partial g(\beta)'}{\partial \beta}\right].$$

If $\partial g/\partial\beta' = 0$ at β, $\sqrt{T}[g(\hat{\beta}) - g(\beta)] \xrightarrow{p} 0$. (See Serfling (1980, p. 122–124)).

(4) If Σ is nonsingular,

$$T(\hat{\beta} - \beta)'\Sigma^{-1}(\hat{\beta} - \beta) \xrightarrow{d} \chi^2(K).$$

(5) If Σ is nonsingular and plim $\hat{\Sigma} = \Sigma$, then

$$T(\hat{\beta} - \beta)'\hat{\Sigma}^{-1}(\hat{\beta} - \beta) \xrightarrow{d} \chi^2(K).$$

(6) If $\Sigma = QA$, where Q is symmetric, idempotent of rank n and A is positive definite, then

$$T(\hat{\beta} - \beta)'A^{-1}(\hat{\beta} - \beta) \xrightarrow{d} \chi^2(n).$$ ■

C.3 Infinite Sums of Random Variables

The MA representation of a VAR process is often an infinite sum of random vectors. As in the study of infinite sums of real numbers we must specify precisely what we mean by such an infinite sum. The concept of absolute convergence is basic in the following. A doubly infinite squence of real numbers $\{a_i\}$, $i = 0, \pm 1, \pm 2, \ldots$, is *absolutely summable* if

$$\lim_{n\to\infty} \sum_{i=-n}^{n} |a_i|$$

exists and is finite. The limit is usually denoted by

$$\sum_{i=-\infty}^{\infty} |a_i|.$$

The following theorem provides a justification for working with infinite sums of random variables. A proof may be found in Fuller (1976, pp. 29–31).

PROPOSITION C.5 (*Existence of Infinite Sums of Random Variables*)

Suppose $\{a_i\}$ is an absolutely summable sequence of real numbers and $\{z_t\}$, $t = 0, \pm 1, \pm 2, \ldots$, is a sequence of random variables satisfying

$$E(z_t^2) \leq c, \quad t = 0, \pm 1, \pm 2, \ldots,$$

for some finite constant c. Then there exists a sequence of random variables $\{y_t\}$, $t = 0, \pm 1, \ldots$, such that

$$\sum_{i=-n}^{n} a_i z_{t-i} \xrightarrow[n\to\infty]{\text{q.m.}} y_t$$

and thus

$$\operatorname{plim} \sum_{i=-n}^{n} a_i z_{t-i} = y_t.$$

The random variables y_t are uniquely determined except on a set of probability zero. If, in addition, the z_t are independent random variables, then

$$\sum_{i=-n}^{n} a_i z_{t-i} \xrightarrow{\text{a.s.}} y_t. \qquad \blacksquare$$

This theorem makes precise what we mean by a (univariate) infinite MA

$$y_t = \sum_{i=0}^{\infty} \Phi_i u_{t-i},$$

where u_t is univariate zero mean white noise with variance $\sigma_u^2 < \infty$. Defining $a_i = 0$ for $i < 0$ and $a_i = \Phi_i$ for $i \geq 0$ and assuming that $\{a_i\}$ is absolutely summable, the proposition guarantees that the process y_t is uniquely defined as a limit in mean square, except on a set of probability zero. The latter qualification may be ignored for practical purposes because we may always change a random variable on a set of probability zero without changing its probability characteristics. The requirement for the MA coefficients to be absolutely summable is satisfied if y_t is a stable AR process. For instance, if $y_t = \alpha y_{t-1} + u_t$ is an AR(1) process, $\Phi_i = \alpha^i$ which is an absolutely summable sequence for $|\alpha| < 1$. With respect to the moments of an infinite sum of random variables the following result holds:

PROPOSITION C.6 (*Moments of Infinite Sums of Random Variables*)

Suppose $\{z_t\}$ satisfies the conditions of Proposition C.5, $\{a_i\}$ and $\{b_i\}$ are absolutely summable sequences and

$$y_t = \sum_{i=-\infty}^{\infty} a_i z_{t-i}, \qquad x_t = \sum_{i=-\infty}^{\infty} b_i z_{t-i}.$$

Then

$$E(y_t) = \lim_{n\to\infty} \sum_{i=-n}^{n} a_i E(z_{t-i})$$

and

$$E(y_t x_t) = \lim_{n\to\infty} \sum_{i=-n}^{n} \sum_{j=-n}^{n} a_i b_j E(z_{t-i} z_{t-j})$$

and, in particular,

$$E(y_t^2) = \lim_{n\to\infty} \sum_{i=-n}^{n} \sum_{j=-n}^{n} a_i a_j E(z_{t-i} z_{t-j}). \qquad \blacksquare$$

Proof: Fuller (1976, pp. 32–33). ■

All these concepts and results may be extended to vector processes. A *sequence of* $(K \times K)$ *matrices* $\{A_i = (a_{mn,i})\}$, $i = 0, \pm 1, \ldots$, is *absolutely summable* if each sequence $\{a_{mn,i}\}$, $m, n = 1, \ldots, K$, is absolutely summable. Equivalently, $\{A_i\}$ may be defined to be absolutely summable if the sequence $\{\|A_i\|\}$ is summable, where

$$\|A_i\| = [\operatorname{tr}(A_i A_i')]^{1/2} = \left(\sum_m \sum_n a_{mn,i}^2\right)^{1/2}.$$

To see the equivalence of the two definitions note that

$$|a_{mn,i}| \le \|A_i\| \le \sum_m \sum_n |a_{mn,i}|.$$

Hence,

$$\sum_{i=-\infty}^{\infty} |a_{mn,i}| \tag{C.3.1}$$

exists and is finite if

$$\sum_{i=-\infty}^{\infty} \|A_i\| \tag{C.3.2}$$

is finite. In turn, if (C.3.1) is finite for all m, n, then, for all h,

$$\sum_{i=-h}^{h} \|A_i\| \le \sum_{i=-h}^{h} \sum_m \sum_n |a_{mn,i}|$$

so that (C.3.2) is finite. Thus, the two definitions are indeed equivalent.

PROPOSITION C.7 (*Existence of Infinite Sums of Random Vectors*)

Suppose $\{A_i\}$ is an absolutely summable sequence of real $(K \times K)$ matrices and $\{z_t\}$ is a sequence of K-dimensional random variables satisfying

$$E(z_t' z_t) \le c, \qquad t = 0, \pm 1, \pm 2, \ldots,$$

for some finite constant c. Then there exists a sequence of K-dimensional random variables $\{y_t\}$ such that

$$\sum_{i=-n}^{n} A_i z_{t-i} \xrightarrow[n\to\infty]{\text{q.m.}} y_t.$$

The sequence is uniquely determined except on a set of probability zero. ■

Proof: Analogous to Fuller (1976, pp. 29–31); replace the absolute value by $\|\cdot\|$. ■

This proposition ensures that the infinite MA representations of the VAR processes considered in this text are well-defined as it can be shown that the MA coefficient matrices of a stable VAR process form an absolutely summable sequence. With respect to the moments of infinite sums we have the following result.

PROPOSITION C.8 (*Moments of Infinite Sums of Random Vectors*)

Suppose $\{z_t\}$ satisfies the conditions of Proposition C.7, $\{A_i\}$ and $\{B_i\}$ are absolutely summable sequences of $(K \times K)$ matrices and

$$y_t = \sum_{i=-\infty}^{\infty} A_i z_{t-i}, \qquad x_t = \sum_{i=-\infty}^{\infty} B_i z_{t-i}.$$

Then

$$E(y_t) = \lim_{n\to\infty} \sum_{i=-n}^{n} A_i E(z_{t-i})$$

and

$$E(y_t x_t') = \lim_{n\to\infty} \sum_{i=-n}^{n} \sum_{j=-n}^{n} A_i E(z_{t-i} z_{t-j}') B_j',$$

where the limit of a sequence of matrices is the matrix of limits of the sequences of individual elements. ■

Proof: Along similar lines as the proof of Fuller (1976, Theorem 2.2.2, pp. 32–33). ■

While we have restricted the discussion to absolutely summable sequences of coefficients, it may be worth mentioning that infinite sums of random variables and vectors can be defined in more general terms.

C.4 Maximum Likelihood Estimation

Suppose $y_1, y_2, \ldots$ is a sequence of K-dimensional random vectors the first T of which have joint probability density function $f_T(y_1, \ldots, y_T; \boldsymbol{\delta}_0)$, where $\boldsymbol{\delta}_0$ is an unknown $(M \times 1)$ vector of parameters that does not depend on T. It is assumed to be from a subset Δ of the M-dimensional Euclidean space $\mathbb{R}^M$. Suppose further that $f_T(\cdot; \boldsymbol{\delta})$ has a known functional form and one wishes to estimate $\boldsymbol{\delta}_0$.

For a fixed realization $y_1, \ldots, y_T$ the function

$$l(\boldsymbol{\delta}) = l(\boldsymbol{\delta} | y_1, \ldots, y_T) = f_T(y_1, \ldots, y_T; \boldsymbol{\delta}),$$

viewed as a function of $\boldsymbol{\delta}$, is the *likelihood function*. Its natural logarithm $\ln l(\boldsymbol{\delta}|\cdot)$ is the *log-likelihood function*. A vector $\tilde{\boldsymbol{\delta}}$, maximizing the likelihood function or log-likelihood function, is called a *maximum likelihood* (ML) *estimate*, that is, if

$$l(\tilde{\boldsymbol{\delta}}) = \sup_{\boldsymbol{\delta} \in \Delta} l(\boldsymbol{\delta}),$$

then $\tilde{\boldsymbol{\delta}}$ is an ML estimate. Here sup denotes the supremum, that is, the least upper bound, which may exist even if the maximum does not. In general, $\tilde{\boldsymbol{\delta}}$ depends on $y_1, \dots, y_T$, that is $\tilde{\boldsymbol{\delta}} = \tilde{\boldsymbol{\delta}}(y_1, \dots, y_T)$. Replacing the fixed values $y_1, \dots, y_T$ by their corresponding random vectors, $\tilde{\boldsymbol{\delta}}$ is an *ML estimator* of $\boldsymbol{\delta}_0$ if the functional dependence on $y_1, \dots, y_T$ is such that $\tilde{\boldsymbol{\delta}}$ is a random vector.

If $l(\boldsymbol{\delta})$ is a differentiable function of $\boldsymbol{\delta}$, the vector of first partial derivatives of $\ln l(\boldsymbol{\delta})$, that is,

$$s(\boldsymbol{\delta}) = \partial \ln l(\boldsymbol{\delta})/\partial \boldsymbol{\delta},$$

regarded as a random vector (a function of the random vectors $y_1, \dots, y_T$) is the *score vector*. It vanishes at $\boldsymbol{\delta} = \tilde{\boldsymbol{\delta}}$ if the maximum of $\ln l(\boldsymbol{\delta})$ is attained at an interior point of the parameter space Δ. Minus the expectation of the matrix of second partial derivatives of $\ln l$, evaluated at the true parameter vector $\boldsymbol{\delta}_0$,

$$I(\boldsymbol{\delta}_0) = -E\left[\left.\frac{\partial^2 \ln l}{\partial \boldsymbol{\delta} \partial \boldsymbol{\delta}'}\right|_{\boldsymbol{\delta}_0}\right],$$

is the *information matrix* for $\boldsymbol{\delta}_0$. The matrix

$$I_a(\boldsymbol{\delta}_0) = \lim_{T \to \infty} I(\boldsymbol{\delta}_0)/T,$$

if it exists, is the *asymptotic information matrix* for $\boldsymbol{\delta}_0$. If it is nonsingular its inverse is a lower bound for the covariance matrix of the asymptotic distribution of any consistent estimator with asymptotic normal distribution. In other words, if $\hat{\boldsymbol{\delta}}$ is a consistent estimator of $\boldsymbol{\delta}_0$ with

$$\sqrt{T}(\hat{\boldsymbol{\delta}} - \boldsymbol{\delta}_0) \xrightarrow{d} N(0, \Sigma_{\hat{\boldsymbol{\delta}}}),$$

then $I_a(\boldsymbol{\delta}_0)^{-1} \le \Sigma_{\hat{\boldsymbol{\delta}}}$, that is, $\Sigma_{\hat{\boldsymbol{\delta}}} - I_a(\boldsymbol{\delta}_0)^{-1}$ is positive semidefinite. Under quite general regularity conditions an ML estimator $\tilde{\boldsymbol{\delta}}$ for $\boldsymbol{\delta}_0$ is consistent and

$$\sqrt{T}(\tilde{\boldsymbol{\delta}} - \boldsymbol{\delta}_0) \xrightarrow{d} N(0, I_a(\boldsymbol{\delta}_0)^{-1}).$$

Thus, in large samples, $\tilde{\boldsymbol{\delta}}$ is approximately distributed as $N(\boldsymbol{\delta}_0, I_a(\boldsymbol{\delta}_0)^{-1}/T)$.

C.5 Likelihood Ratio, Lagrange Multiplier, and Wald Tests

Three principles for constructing tests of statistical hypotheses are employed frequently in the text. We consider testing of

$$H_0\colon \varphi(\boldsymbol{\delta}_0) = 0 \quad \text{against} \quad H_1\colon \varphi(\boldsymbol{\delta}_0) \neq 0, \tag{C.5.1}$$

where $\boldsymbol{\delta}_0$ is the true $(M \times 1)$ parameter vector as in the previous subsection and $\varphi\colon \mathbb{R}^M \to \mathbb{R}^N$ is a continuously differentiable function so that $\varphi(\boldsymbol{\delta})$ is of dimension $(N \times 1)$. We assume that $[\partial\varphi/\partial\boldsymbol{\delta}'|_{\boldsymbol{\delta}_0}]$ has rank N. This implies that $N \le M$ and the N restrictions for the parameter vector are distinguishable in a neighborhood of $\boldsymbol{\delta}_0$. Often the hypotheses can be written alternatively as

$$H_0\colon \boldsymbol{\delta}_0 = g(\boldsymbol{\gamma}_0) \quad \text{against} \quad H_1\colon \boldsymbol{\delta}_0 \neq g(\boldsymbol{\gamma}_0), \tag{C.5.2}$$

where γ_0 is an $(M-N)$-dimensional vector and $g: \mathbb{R}^{M-N} \to \mathbb{R}^M$ is a continuously differentiable function in a neighborhood of γ_0 (see Gallant (1987, pp. 57–58)).

The *likelihood ratio* (LR) *test* of (C.5.1) or (C.5.2) is based on the statistic

$$\lambda_{LR} = 2(\ln l(\tilde{\boldsymbol{\delta}}) - \ln l(\tilde{\boldsymbol{\delta}}_r)),$$

where $\tilde{\boldsymbol{\delta}}$ denotes the unconstrained ML estimator and $\tilde{\boldsymbol{\delta}}_r$ is the restricted ML estimator of $\boldsymbol{\delta}_0$ under H_0, that is, $\tilde{\boldsymbol{\delta}}_r$ is obtained by maximizing $\ln l$ over the parameter space restricted by the conditions stated in H_0. Under suitable regularity conditions we have

$$\lambda_{LR} \xrightarrow{d} \chi^2(N). \tag{C.5.3}$$

The *Lagrange multiplier* (LM) *statistic* for testing (C.5.1) or (C.5.2) is of the form

$$\lambda_{LM} = s(\tilde{\boldsymbol{\delta}}_r)' I(\tilde{\boldsymbol{\delta}}_r)^{-1} s(\tilde{\boldsymbol{\delta}}_r), \tag{C.5.4}$$

where $s(\boldsymbol{\delta})$ denotes the score vector and $I(\boldsymbol{\delta})$ the information matrix as before. In the LM statistic both functions are evaluated at the restricted estimator of $\boldsymbol{\delta}_0$. Under H_0, λ_{LM} has an asymptotic $\chi^2(N)$-distribution if weak regularity conditions are satisfied. The name derives from the fact that it can be written as

$$\lambda_{LM} = \tilde{\lambda}' \left[\frac{\partial \varphi}{\partial \boldsymbol{\delta}'}\bigg|_{\tilde{\boldsymbol{\delta}}_r} \right] I(\tilde{\boldsymbol{\delta}}_r)^{-1} \left[\frac{\partial \varphi'}{\partial \boldsymbol{\delta}}\bigg|_{\tilde{\boldsymbol{\delta}}_r} \right] \tilde{\lambda}, \tag{C.5.5}$$

where $\tilde{\lambda}$ is the vector of Lagrange multipliers for which the Lagrange function has a stationary point (see Appendix A, A.14).

The equivalence of (C.5.4) and (C.5.5) can be seen by recalling that the constrained minimum of $-\ln l$ is attained at a stationary point of the Lagrange function

$$\mathrm{LG}(\boldsymbol{\delta}, \lambda) = -\ln l(\boldsymbol{\delta}) + \lambda' \varphi(\boldsymbol{\delta}).$$

In other words, $\tilde{\boldsymbol{\delta}}_r$ satisfies

$$0 = \left[\frac{\partial \mathrm{LG}}{\partial \boldsymbol{\delta}'}\bigg|_{(\tilde{\boldsymbol{\delta}}_r, \tilde{\lambda})} \right] = -\left[\frac{\partial \ln l}{\partial \boldsymbol{\delta}'}\bigg|_{\tilde{\boldsymbol{\delta}}_r} \right] + \tilde{\lambda}' \left[\frac{\partial \varphi}{\partial \boldsymbol{\delta}'}\bigg|_{\tilde{\boldsymbol{\delta}}_r} \right] = -s(\tilde{\boldsymbol{\delta}}_r)' + \tilde{\lambda}' \left[\frac{\partial \varphi}{\partial \boldsymbol{\delta}'}\bigg|_{\tilde{\boldsymbol{\delta}}_r} \right].$$

The *Wald statistic* is based on an unconstrained estimator which is asymptotically normal,

$$\sqrt{T}(\tilde{\boldsymbol{\delta}} - \boldsymbol{\delta}_0) \xrightarrow{d} N(0, \Sigma_{\tilde{\boldsymbol{\delta}}}).$$

By Proposition C.4(3) it follows that

$$\sqrt{T}[\varphi(\tilde{\boldsymbol{\delta}}) - \varphi(\boldsymbol{\delta}_0)] \xrightarrow{d} N\left[0, \left(\frac{\partial \varphi}{\partial \boldsymbol{\delta}'}\bigg|_{\boldsymbol{\delta}_0} \right) \Sigma_{\tilde{\boldsymbol{\delta}}} \left(\frac{\partial \varphi'}{\partial \boldsymbol{\delta}}\bigg|_{\boldsymbol{\delta}_0} \right) \right].$$

Thus, if H_0 is true and the covariance matrix is invertible,

$$\lambda_w = T\varphi(\tilde{\boldsymbol{\delta}})' \left[\left(\frac{\partial \varphi}{\partial \boldsymbol{\delta}'}\bigg|_{\tilde{\boldsymbol{\delta}}} \right) \tilde{\Sigma}_{\tilde{\boldsymbol{\delta}}} \left(\frac{\partial \varphi'}{\partial \boldsymbol{\delta}}\bigg|_{\tilde{\boldsymbol{\delta}}} \right) \right]^{-1} \varphi(\tilde{\boldsymbol{\delta}}) \xrightarrow{d} \chi^2(N), \tag{C.5.6}$$

where $\tilde{\Sigma}_{\tilde{\boldsymbol{\delta}}}$ is a consistent estimator of $\Sigma_{\tilde{\boldsymbol{\delta}}}$. The statistic λ_w is the Wald statistic.

In summary we have three test statistics with equivalent asymptotic distributions under the null hypothesis. The LR statistic involves both the restricted and the unrestricted ML estimators, the LM statistic is based on the restricted estimator only, and the Wald statistic requires just the unrestricted estimator. The choice among the three statistics is often based on computational convenience. Wald tests have the disadvantage that they are not invariant under transformations of the restrictions. In other words, if the restrictions can be written in two equivalent ways (e.g., $\delta_i = 0$ and $\delta_i^2 = 0$) the corresponding Wald tests may have different small sample properties. Their small sample power may be low (see Gregory & Veall (1985), Breusch & Schmidt (1988)).

Appendix D. Evaluating Properties of Estimators and Test Statistics by Simulation and Resampling Techniques

If asymptotic theory is difficult or only small samples are available properties of estimators and test statistics are sometimes investigated by heavy use of the computer. The idea is to simulate the distribution (or some of its properties) of the random variables of interest by artificially sampling from some known distribution. Generally, if the random variable or vector of interest, say $q = q(z)$, is a function of a random vector z with a known distribution F_z, then samples $z_1, \dots, z_N$ are drawn from F_z and the empirical distribution of q given by $q_n = q(z_n)$, $n = 1, \dots, N$, is determined. The characteristics of the actual distribution of q are then inferred from the empirical distribution.

Often the statistics of interest in this book are functions of multiple time series generated by VAR(p) processes. Therefore we will briefly describe in the next section how to simulate such time series. Afterwards some more details are given on simulation and resampling techniques for evaluating estimators and test statistics.

D.1 Simulating a Multiple Time Series with VAR Generation Process

To simulate a multiple time series of dimension K and length T we first generate a series of (often independent) disturbance vectors $u_{-s}, \dots, u_0, u_1, \dots, u_T$. If a series of Gaussian disturbances is desired, i.e. $u_t \sim N(0, \Sigma_u)$, we may choose K independent univariate standard normal variates $v_1, \dots, v_K$ and multiply by a $(K \times K)$ matrix P for which $PP' = \Sigma_u$, that is,

$$u_t = P\begin{bmatrix} v_1 \\ \vdots \\ v_K \end{bmatrix}.$$

This process is repeated $T + s + 1$ times until we have the desired series of disturbances. Programs for generating (pseudo) standard normal variates are available on most computers. Also facilities for generating random numbers from other distributions are usually available and may be used in a like manner to obtain disturbances from other distributions of interest.

For a given set of parameters $\nu, A_1, \ldots, A_p$, where ν is $(K \times 1)$ and the A_i are $(K \times K)$, and a given set of starting values $y_{-p+1}, \ldots, y_0$, the u_t may be used to simulate a time series $y_1, \ldots, y_T$ with VAR(p) generation process recursively as

$$y_t = \nu + A_1 y_{t-1} + \cdots + A_p y_{t-p} + u_t$$

starting with $t = 1, t = 2$, etc. until $t = T$. There are different ways to obtain the initial values. Assuming the desired process is stable they may be set at zero or at the process mean $\mu = (I_K - A_1 - \cdots - A_p)^{-1}\nu$. Since the choice of initial values has some impact on the generated time series, a number of presample values $y_t, t = -s, \ldots, 0$, is often generated and then discarded in the subsequent analysis.

A possibility to ensure the same correlation structure for the initial values and the rest of the time series is to determine the covariance matrix of p consecutive y_t vectors, say Σ_Y. Using the results of Chapter 2, Section 2.1, that matrix may be obtained from

$$\operatorname{vec}(\Sigma_Y) = (I_{(Kp)^2} - \mathbf{A} \otimes \mathbf{A})^{-1} \operatorname{vec}(\Sigma_U),$$

where

$$\mathbf{A} = \underset{(Kp \times Kp)}{\begin{bmatrix} A_1 & A_2 & \ldots & A_{p-1} & A_p \\ I_K & 0 & \ldots & 0 & 0 \\ 0 & I_K & & 0 & 0 \\ \vdots & & \ddots & \vdots & \vdots \\ 0 & 0 & \ldots & I_K & 0 \end{bmatrix}} \quad \text{and} \quad \Sigma_U = \underset{(Kp \times Kp)}{\begin{bmatrix} \Sigma_u & 0 & \ldots & 0 \\ 0 & 0 & \ldots & 0 \\ \vdots & \vdots & \ddots & \vdots \\ 0 & 0 & \ldots & 0 \end{bmatrix}}.$$

Then a $(Kp \times Kp)$ matrix Q is chosen such that $QQ' = \Sigma_Y$ and p initial starting vectors are obtained as

$$\begin{bmatrix} y_0 \\ \vdots \\ y_{-p+1} \end{bmatrix} = Q \begin{bmatrix} v_1 \\ \vdots \\ v_{Kp} \end{bmatrix} + \begin{bmatrix} \mu \\ \vdots \\ \mu \end{bmatrix},$$

where the v_i are independent variates with mean zero and unit variance.

D.2 Evaluating Distributions of Functions of Multiple Time Series by Simulation

Suppose we are interested in the function $q_T = q(y_1, \ldots, y_T)$ of some VAR(p) process y_t, where q_T is of dimension $(M \times 1)$. q_T may be some estimator or test statistic. To investigate the distribution F_T of q_T we generate a large number, say N, of independent multiple time series of length T and compute the corresponding values of q_T, say $q_T(n), n = 1, \ldots, N$. The properties of F_T are then estimated from the empirical distribution of the $q_T(n)$. For instance, the mean vector of q_T is estimated by

$$\frac{1}{N}\sum_{n=1}^{N} q_T(n).$$

Analogously we may estimate the variances, standard deviations, quantiles or other characteristics of F_T. Obviously such a procedure may be quite costly in terms of computer time.

D.3 Evaluating Distributions of Functions of Multiple Time Series by Resampling

If the distribution of the disturbances in unknown so-called *bootstrap* or *resampling* methods may be applied to investigate the distributions of functions of stochastic processes or multiple time series. Suppose a time series $y_1, \ldots, y_T$ and the presample values required for estimation are available and assume that it is generated by a VAR(p) process. Fitting a VAR(p) process to this time series we get coefficient estimates $\hat{\nu}, \hat{A}_1, \ldots, \hat{A}_p$ and a series of residuals $\hat{u}_1, \ldots, \hat{u}_T$. Now new series of disturbances may be generated by sampling with replacement from these vectors. Thereby we obtain series

$$\hat{u}_{-s}(n), \ldots, \hat{u}_1(n), \ldots, \hat{u}_T(n), \qquad n = 1, \ldots, N,$$

which may be used to generate N time series of length T. In this procedure the true coefficients are replaced by $\hat{\nu}, \hat{A}_1, \ldots, \hat{A}_p$ and the original presample values may be applied for each generated time series. Alternatively new presample values are generated for each of the N time series. The empirical distributions of the functions of interest are then obtained from these N time series as described previously.

Appendix E. Data Used for Examples and Exercises

Table E.1. Quarterly, Seasonally Adjusted West German Fixed Investment, Disposable Income, and Consumption Expenditures in Billions of DM

year	quarter	investment	income	consumption	year	quarter	investment	income	consumption
1960	I	180	451	415	1969	I	315	922	798
	II	179	465	421		II	339	949	816
	III	185	485	434		III	364	979	837
	IV	192	493	448		IV	371	988	858
1961	I	211	509	459	1970	I	375	1025	881
	II	202	520	458		II	432	1063	905
	III	207	521	479		III	453	1104	934
	IV	214	540	487		IV	460	1131	968
1962	I	231	548	497	1971	I	475	1137	983
	II	229	558	510		II	496	1178	1013
	III	234	574	516		III	494	1211	1034
	IV	237	583	525		IV	498	1256	1064
1963	I	206	591	529	1972	I	526	1290	1101
	II	250	599	538		II	519	1314	1102
	III	259	610	546		III	516	1346	1145
	IV	263	627	555		IV	531	1385	1173
1964	I	264	642	574	1973	I	573	1416	1216
	II	280	653	574		II	551	1436	1229
	III	282	660	586		III	538	1462	1242
	IV	292	694	602		IV	532	1493	1267
1965	I	286	709	617	1974	I	558	1516	1295
	II	302	734	639		II	524	1557	1317
	III	304	751	653		III	525	1613	1355
	IV	307	763	668		IV	519	1642	1371
1966	I	317	766	679	1975	I	526	1690	1402
	II	314	779	686		II	510	1759	1452
	III	306	808	697		III	519	1756	1485
	IV	304	785	688		IV	538	1780	1516
1967	I	292	794	704	1976	I	549	1807	1549
	II	275	799	699		II	570	1831	1567
	III	273	799	709		III	559	1873	1588
	IV	301	812	715		IV	584	1897	1631
1968	I	280	837	724	1977	I	611	1910	1650
	II	289	853	746		II	597	1943	1685
	III	303	876	758		III	603	1976	1722
	IV	322	897	779		IV	619	2018	1752

Table E.1. (continued)

year	quarter	invest-ment	income	consump-tion	year	quarter	invest-ment	income	consump-tion
1978	I	635	2040	1774	1981	I	833	2521	2145
	II	658	2070	1807		II	860	2545	2164
	III	675	2121	1831		III	870	2580	2206
	IV	700	2132	1842		IV	830	2620	2225
1979	I	692	2199	1890	1982	I	801	2639	2235
	II	759	2253	1958		II	824	2618	2237
	III	782	2276	1948		III	831	2628	2250
	IV	816	2318	1994		IV	830	2651	2271
1980	I	844	2369	2061					
	II	830	2423	2056					
	III	853	2457	2102					
	IV	852	2470	2121					

Source: Deutsche Bundesbank.

Table E.2. Quarterly, Seasonally Adjusted U.S. Fixed Investment (y_1) and Change in Business Inventories (y_2)

year	quarter	y_1	y_2	year	quarter	y_1	y_2
1947	I	69.6	0.1	1954	I	82.8	−3.4
	II	67.6	−0.9		II	84.1	−4.1
	III	69.5	−2.9		III	87.0	−2.7
	IV	74.7	2.7		IV	88.5	1.5
1948	I	77.1	4.1	1955	I	92.1	5.9
	II	77.4	5.6		II	96.1	8.0
	III	76.6	6.9		III	98.3	7.8
	IV	76.1	5.3		IV	98.8	9.2
1949	I	71.8	−0.3	1956	I	96.6	7.5
	II	68.9	−7.1		II	97.4	5.5
	III	68.5	−2.5		III	97.6	4.9
	IV	70.6	−7.7		IV	96.6	5.4
1950	I	75.4	4.4	1957	I	96.2	2.5
	II	82.3	7.7		II	95.3	2.9
	III	88.2	8.0		III	96.4	3.7
	IV	86.9	22.1		IV	94.9	−3.0
1951	I	83.4	13.4	1958	I	90.0	−6.8
	II	80.3	19.9		II	87.2	−6.2
	III	79.4	14.6		III	88.0	0.3
	IV	78.6	7.0		IV	93.0	5.3
1952	I	79.3	7.3	1959	I	98.3	5.0
	II	80.3	−2.7		II	101.6	13.0
	III	75.3	5.4		III	102.6	−0.4
	IV	80.6	7.2		IV	101.4	8.2
1953	I	83.9	3.9	1960	I	104.9	13.5
	II	84.2	5.1		II	101.8	4.9
	III	84.4	1.9		III	98.8	3.0
	IV	83.8	−5.0		IV	98.6	−3.9

Table E.2. (continued)

year	quarter	y_1	y_2	year	quarter	y_1	y_2
1961	I	97.7	−3.8	1967	I	136.4	14.6
	II	99.2	1.9		II	139.6	7.5
	III	101.3	6.6		III	141.1	12.2
	IV	104.6	6.7		IV	145.5	13.8
1962	I	106.1	10.6	1968	I	148.9	6.3
	II	109.9	9.2		II	148.9	11.8
	III	111.1	8.0		III	150.7	9.2
	IV	110.1	4.7		IV	155.0	7.6
1963	I	110.7	7.6	1969	I	159.1	9.8
	II	116.0	7.0		II	158.4	12.2
	III	118.5	9.3		III	158.1	13.4
	IV	122.0	7.1		IV	154.3	6.8
1964	I	124.0	6.1	1970	I	151.8	2.9
	II	124.0	8.0		II	150.0	4.8
	III	124.9	7.3		III	150.4	6.3
	IV	126.4	7.9		IV	149.5	3.3
1965	I	133.4	13.4	1971	I	154.3	7.9
	II	137.9	10.6		II	158.4	10.0
	III	140.1	12.4		III	162.1	5.0
	IV	143.8	8.8		IV	166.0	3.7
1966	I	147.5	13.5	1972	I	174.3	4.8
	II	146.2	17.8		II	176.1	10.1
	III	145.0	15.1		III	178.2	12.1
	IV	139.7	20.5		IV	186.7	10.8

Source: U.S. Department of Commerce, Bureau of Economic Analysis, *The National Income and Product Accounts of the United States, 1929–1974.*

Table E.3. Seasonally Adjusted Real U.S. Money ($M1$) and GNP in 1982 Dollars; Discount Rate on 91-Day Treasury Bills (r^{short}) and Yield on Long-Term Treasury Bonds (r^{long})

year	quarter	$M1$	GNP	r^{short}	r^{long}
1954	I	450.9	1406.8	0.010800	0.026133
	II	453.0	1401.2	0.008133	0.025233
	III	459.1	1418.0	0.008700	0.024900
	IV	464.6	1438.8	0.010367	0.025667
1955	I	469.6	1469.6	0.012600	0.027467
	II	473.1	1485.7	0.015133	0.028167
	III	474.6	1505.5	0.018633	0.029267
	IV	474.3	1518.7	0.023467	0.028900
1956	I	475.4	1515.0	0.023800	0.028867
	II	472.9	1522.6	0.025967	0.029900
	III	468.7	1523.7	0.025967	0.031267
	IV	467.5	1540.6	0.030633	0.033000

Table E.3. (continued)

year	quarter	$M1$	GNP	r^{short}	r^{long}
1957	I	464.7	1553.3	0.031700	0.032733
	II	461.2	1552.4	0.031567	0.034333
	III	457.1	1561.5	0.033800	0.036300
	IV	453.0	1537.3	0.033433	0.035333
1958	I	447.5	1506.1	0.018367	0.032567
	II	449.6	1514.2	0.010200	0.031533
	III	454.2	1550.0	0.017100	0.035700
	IV	458.5	1586.7	0.027867	0.037533
1959	I	464.1	1606.4	0.028000	0.039167
	II	466.3	1637.0	0.030200	0.040600
	III	468.1	1629.5	0.035333	0.041567
	IV	460.0	1643.4	0.043000	0.041667
1960	I	459.2	1671.6	0.039433	0.042233
	II	455.7	1666.8	0.030900	0.041067
	III	459.5	1668.4	0.023933	0.038300
	IV	455.9	1654.1	0.023600	0.039067
1961	I	458.0	1671.3	0.023767	0.038267
	II	461.5	1692.1	0.023267	0.038033
	III	462.2	1716.3	0.023233	0.039733
	IV	465.4	1754.9	0.024767	0.040067
1962	I	467.4	1777.9	0.027400	0.040600
	II	468.5	1796.4	0.027167	0.038900
	III	466.5	1813.1	0.028567	0.039800
	IV	467.7	1810.1	0.028033	0.038767
1963	I	471.9	1834.6	0.029100	0.039133
	II	474.8	1860.0	0.029433	0.039800
	III	477.7	1892.5	0.032800	0.040133
	IV	479.9	1906.1	0.034967	0.041067
1964	I	481.9	1948.7	0.035367	0.041567
	II	484.8	1965.4	0.034800	0.041633
	III	491.3	1985.2	0.035067	0.041433
	IV	495.6	1993.7	0.036867	0.041400
1965	I	498.3	2036.9	0.039000	0.041500
	II	497.6	2066.4	0.038800	0.041433
	III	501.7	2099.3	0.038600	0.041967
	IV	507.8	2147.6	0.041567	0.043500
1966	I	511.8	2190.1	0.046333	0.045567
	II	511.8	2195.8	0.045967	0.045833
	III	506.2	2218.3	0.050500	0.047800
	IV	503.1	2229.2	0.045800	0.046967
1967	I	507.1	2241.8	0.045267	0.044400
	II	510.8	2255.2	0.036567	0.047100
	III	518.0	2287.7	0.043467	0.049333
	IV	521.3	2300.6	0.047867	0.053300
1968	I	521.9	2327.3	0.050633	0.052433
	II	525.4	2366.9	0.055067	0.053033
	III	528.3	2385.3	0.052267	0.050733
	IV	533.8	2383.0	0.055800	0.054200

Table E.3. (continued)

year	quarter	$M1$	GNP	r^{short}	r^{long}
1969	I	536.5	2416.5	0.061400	0.058833
	II	532.8	2419.8	0.062400	0.059133
	III	527.6	2433.2	0.070467	0.061367
	IV	523.2	2423.5	0.073167	0.065333
1970	I	521.4	2408.6	0.072600	0.065633
	II	518.1	2406.5	0.067533	0.068200
	III	519.4	2435.8	0.063833	0.066500
	IV	521.2	2413.8	0.053600	0.062667
1971	I	524.7	2478.6	0.038600	0.058233
	II	530.8	2478.4	0.042067	0.058833
	III	534.1	2491.1	0.050500	0.057500
	IV	536.5	2491.0	0.042333	0.055200
1972	I	542.6	2545.6	0.034333	0.056500
	II	547.8	2595.1	0.037467	0.056567
	III	554.4	2622.1	0.042400	0.056267
	IV	562.5	2671.3	0.048500	0.056100
1973	I	565.2	2734.0	0.056400	0.061000
	II	560.1	2741.0	0.066100	0.062267
	III	556.1	2738.3	0.083900	0.065967
	IV	548.5	2762.8	0.074633	0.063000
1974	I	542.1	2747.4	0.076033	0.066367
	II	532.8	2755.2	0.082667	0.070500
	III	522.8	2719.3	0.082833	0.072700
	IV	511.9	2695.4	0.073333	0.069733
1975	I	505.3	2642.7	0.058700	0.067033
	II	506.8	2669.6	0.054000	0.069733
	III	505.5	2714.9	0.063333	0.070933
	IV	501.0	2752.7	0.056833	0.072233
1976	I	502.2	2804.4	0.049533	0.069100
	II	505.5	2816.9	0.051667	0.068867
	III	502.6	2828.6	0.051700	0.067900
	IV	505.3	2856.8	0.046967	0.065500
1977	I	508.1	2896.0	0.046233	0.070133
	II	507.7	2942.7	0.048267	0.070967
	III	509.4	3001.8	0.054733	0.069767
	IV	513.0	2994.1	0.061367	0.071600
1978	I	513.7	3020.5	0.064100	0.075800
	II	514.1	3115.9	0.064833	0.078500
	III	512.5	3142.6	0.073167	0.079333
	IV	509.7	3181.6	0.086800	0.081967
1979	I	503.3	3181.7	0.093600	0.084367
	II	496.1	3178.9	0.093733	0.084367
	III	497.2	3207.4	0.096300	0.084833
	IV	487.6	3201.3	0.118033	0.096067
1980	I	474.0	3233.4	0.134600	0.111500
	II	451.2	3157.0	0.100500	0.100167
	III	464.9	3159.1	0.092367	0.104333
	IV	465.3	3199.2	0.137100	0.116400

Table E.3. (continued)

year	quarter	$M1$	GNP	r^{short}	r^{long}
1981	I	455.3	3261.1	0.143667	0.120100
	II	453.7	3250.2	0.148300	0.126567
	III	448.9	3264.6	0.150867	0.136000
	IV	447.1	3219.3	0.120233	0.132300
1982	I	451.2	3170.4	0.128933	0.134467
	II	447.1	3179.9	0.123600	0.129433
	III	449.1	3154.5	0.097067	0.122000
	IV	464.9	3159.3	0.079333	0.103400
1983	I	475.8	3186.6	0.080800	0.104367
	II	484.3	3258.3	0.084200	0.103467
	III	493.6	3306.4	0.091867	0.112600
	IV	496.4	3365.1	0.087933	0.113233
1984	I	497.5	3451.7	0.091333	0.115433
	II	500.4	3498.0	0.098433	0.126867
	III	501.5	3520.6	0.103433	0.123400
	IV	502.2	3535.2	0.089733	0.113733
1985	I	511.0	3577.5	0.081833	0.114267
	II	518.2	3599.2	0.075233	0.109133
	III	533.9	3635.8	0.071033	0.105900
	IV	543.2	3662.4	0.071467	0.100800
1986	I	553.4	3719.3	0.068867	0.089033
	II	576.8	3711.6	0.061300	0.079467
	III	598.0	3721.3	0.055333	0.078867
	IV	620.0	3734.7	0.053400	0.078400
1987	I	631.9	3776.7	0.055333	0.076367
	II	634.8	3823.0	0.057333	0.085767
	III	630.1	3865.3	0.060333	0.090833
	IV	630.5	3923.0	0.060033	0.092400

Source: Business Conditions Digest.

Table E.4. Unadjusted West German Real per Capita Personal Disposable Income and Personal Consumption Expenditures

year	quarter	income	consumption	year	quarter	income	consumption
1960	I	1684.6080	1505.3198	1963	I	1931.2619	1673.9302
	II	1757.2068	1632.8338		II	1980.3270	1829.8065
	III	1842.7211	1655.8021		III	2065.5097	1848.9135
	IV	2027.3177	1889.8641		IV	2256.8667	2065.9135
1961	I	1859.0561	1600.8018	1964	I	2040.4667	1758.5866
	II	1832.6429	1686.0382		II	2098.8673	1894.7205
	III	1893.5360	1739.6939		III	2145.8642	1910.8631
	IV	2098.9234	1958.8067		IV	2427.6995	2165.6707
1962	I	1873.2676	1645.0461	1965	I	2178.3079	1830.4957
	II	1908.6426	1798.0200		II	2268.4690	2024.4865
	III	1989.5797	1805.1480		III	2313.5771	2040.4966
	IV	2203.5812	2038.5067		IV	2548.2016	2274.2020

Table E.4. (continued)

year	quarter	income	consumption
1966	I	2239.8707	1919.7148
	II	2286.0157	2070.1321
	III	2379.5459	2085.6752
	IV	2535.5248	2269.4856
1967	I	2247.8021	1942.0016
	II	2311.8334	2063.5617
	III	2332.9532	2088.5606
	IV	2579.1389	2320.2063
1968	I	2348.2887	1965.2127
	II	2404.7787	2164.5546
	III	2478.4108	2182.9295
	IV	2808.6086	2467.3345
1969	I	2535.4286	2121.2628
	II	2605.4529	2307.8719
	III	2688.1167	2333.5107
	IV	2998.7831	2626.8251
1970	I	2695.4440	2252.5326
	II	2764.9544	2445.3018
	III	2879.4832	2484.5415
	IV	3255.6348	2810.1885
1971	I	2825.4761	2383.4010
	II	2877.2233	2575.9185
	III	2955.7203	2578.9525
	IV	3378.4279	2872.9650
1972	I	3014.1083	2517.0234
	II	3030.1896	2645.5609
	III	3092.0762	2682.6305
	IV	3491.3989	2962.6573
1973	I	3098.3107	2607.4582
	II	3091.6898	2755.3575
	III	3144.9089	2732.3144
	IV	3535.0880	2988.6982
1974	I	3096.3842	2606.4549
	II	3110.8586	2748.4852
	III	3222.0775	2775.8346
	IV	3613.9943	3012.7843
1975	I	3233.5315	2647.7614
	II	3319.4594	2854.9956
	III	3323.1825	2880.3328
	IV	3723.1654	3164.0112
1976	I	3312.5138	2808.3615
	II	3324.6223	2965.6652
	III	3418.3893	2978.0314
	IV	3824.4625	3280.6247
1977	I	3414.6557	2914.1164
	II	3438.2362	3095.7771
	III	3487.2942	3122.1189
	IV	3973.8991	3438.6965
1978	I	3546.9684	3053.1464
	II	3553.1513	3216.7262
	III	3655.2346	3254.8354
	IV	4102.7937	3543.8848
1979	I	3680.9675	3130.1370
	II	3783.8147	3406.7708
	III	3755.9600	3327.0867
	IV	4261.4059	3669.2334
1980	I	3797.3271	3279.6486
	II	3786.5219	3324.2316
	III	3790.9261	3369.4241
	IV	4282.9525	3682.5376
1981	I	3848.1635	3250.4014
	II	3764.6489	3308.7353
	III	3759.0830	3339.1146
	IV	4294.2124	3661.0499
1982	I	3811.5833	3224.8444
	II	3709.8919	3291.6504
	III	3641.5356	3260.4145
	IV	4180.7126	3611.8419
1983	I	3734.1328	3259.6856
	II	3698.3154	3361.7589
	III	3664.9417	3347.9925
	IV	4237.3911	3700.2983
1984	I	3847.0869	3323.8678
	II	3766.3393	3424.4785
	III	3783.3845	3430.6044
	IV	4327.6089	3753.7867
1985	I	3887.8617	3335.6273
	II	3839.1482	3469.0466
	III	3859.0148	3533.5180
	IV	4431.0216	3859.7411
1986	I	4049.0191	3452.2034
	II	4053.1962	3669.5392
	III	4077.4384	3673.5696
	IV	4653.9432	3999.3784
1987	I	4191.4281	3540.2152
	II	4163.4328	3750.5110
	III	4177.7770	3775.5819
	IV	4811.1361	4150.3215

Source: Deutsches Institut für Wirtschaftsforschung, Berlin.

Table E.5. West German Interest Rate on 3-Months Loans in the Money Market (Frankfurt/Main) (i^{short}) and Yields on Bonds Outstanding for Total Fixed Interest Securities (i^{long})

year	month	i^{short}	i^{long}	year	month	i^{short}	i^{long}	year	month	i^{short}	i^{long}
1960	1	4.36	6.20	1964	1	3.33	6.00	1968	1	3.32	6.90
	2	4.47	6.20		2	3.33	6.00		2	3.45	7.00
	3	4.71	6.20		3	3.46	6.00		3	3.52	7.00
	4	4.59	6.20		4	3.55	6.10		4	3.64	6.90
	5	4.64	6.20		5	3.69	6.20		5	3.68	6.70
	6	5.25	6.40		6	3.75	6.20		6	3.72	6.70
	7	5.58	6.60		7	3.69	6.30		7	3.59	6.70
	8	5.44	6.50		8	3.91	6.30		8	3.55	6.60
	9	5.61	6.40		9	4.08	6.30		9	3.54	6.50
	10	6.10	6.40		10	5.42	6.30		10	4.75	6.50
	11	5.38	6.20		11	5.50	6.30		11	4.50	6.60
	12	5.06	6.20		12	5.36	6.30		12	4.44	6.50
1961	1	4.66	6.10	1965	1	3.88	6.30	1969	1	3.87	6.50
	2	4.13	6.10		2	3.95	6.40		2	3.91	6.50
	3	3.73	6.00		3	4.28	6.40		3	4.21	6.60
	4	3.34	5.80		4	4.52	6.50		4	4.40	6.70
	5	3.23	5.70		5	4.69	6.70		5	4.38	6.80
	6	3.13	5.70		6	4.80	6.80		6	5.50	6.90
	7	3.16	5.80		7	5.08	6.90		7	5.78	7.10
	8	3.09	5.90		8	5.27	7.00		8	6.50	7.10
	9	3.06	6.00		9	5.36	7.10		9	6.94	7.20
	10	4.05	6.00		10	6.66	7.20		10	7.42	7.30
	11	3.71	6.00		11	6.58	7.30		11	7.69	7.30
	12	3.83	6.00		12	6.55	7.40		12	8.83	7.40
1962	1	2.93	5.90	1966	1	5.23	7.30	1970	1	9.29	7.50
	2	2.94	5.90		2	5.36	7.30		2	9.51	7.60
	3	3.02	5.80		3	5.69	7.40		3	9.81	7.90
	4	3.08	5.80		4	6.21	7.60		4	9.86	8.00
	5	3.08	5.90		5	6.35	7.70		5	9.93	8.20
	6	3.13	6.00		6	6.81	7.90		6	9.88	8.60
	7	3.29	6.00		7	6.89	8.10		7	9.59	8.60
	8	3.26	6.10		8	7.00	8.10		8	9.16	8.40
	9	3.20	6.10		9	6.80	8.10		9	9.36	8.40
	10	4.45	6.20		10	7.88	8.00		10	9.53	8.50
	11	4.33	6.20		11	7.73	7.90		11	8.84	8.60
	12	4.35	6.20		12	7.57	7.60		12	8.12	8.30
1963	1	3.29	6.10	1967	1	5.69	7.40	1971	1	7.50	7.90
	2	3.31	6.10		2	5.56	7.30		2	7.47	7.90
	3	3.53	6.10		3	5.04	7.20		3	7.46	8.00
	4	3.56	6.10		4	4.48	6.90		4	6.36	8.00
	5	3.60	6.10		5	3.69	6.80		5	6.16	8.10
	6	3.83	6.10		6	3.98	6.90		6	6.80	8.30
	7	3.98	6.10		7	3.51	6.90		7	7.66	8.50
	8	3.81	6.10		8	3.56	6.90		8	7.56	8.50
	9	3.75	6.10		9	3.43	6.90		9	7.59	8.40
	10	5.13	6.10		10	4.20	6.90		10	7.80	8.30
	11	5.00	6.10		11	4.00	7.00		11	6.79	8.20
	12	4.98	6.10		12	4.07	7.00		12	6.63	8.10

Table E.5. (continued)

year	month	i^{short}	i^{long}	year	month	i^{short}	i^{long}	year	month	i^{short}	i^{long}
1972	1	5.19	7.90	1976	1	3.93	8.40	1980	1	8.86	8.10
	2	4.88	7.70		2	3.72	8.20		2	8.97	8.50
	3	4.80	7.80		3	3.74	7.80		3	9.64	9.50
	4	4.78	8.00		4	3.62	7.80		4	10.22	9.60
	5	4.71	8.20		5	3.77	8.00		5	10.26	8.80
	6	4.65	8.30		6	4.14	8.30		6	10.11	8.30
	7	4.65	8.40		7	4.47	8.40		7	9.70	8.00
	8	4.80	8.30		8	4.56	8.30		8	8.98	7.90
	9	5.32	8.30		9	4.56	8.10		9	8.97	8.30
	10	6.88	8.40		10	4.85	8.00		10	9.08	8.50
	11	8.07	8.60		11	4.69	7.60		11	9.45	9.00
	12	8.60	8.70		12	4.93	7.40		12	10.20	9.10
1973	1	7.89	8.60	1977	1	4.78	7.20	1981	1	9.47	9.20
	2	7.96	8.60		2	4.71	7.10		2	10.67	9.90
	3	8.77	8.70		3	4.73	7.00		3	13.60	10.40
	4	10.62	8.90		4	4.62	6.60		4	13.19	10.40
	5	12.42	9.40		5	4.44	6.40		5	13.20	11.00
	6	13.62	10.20		6	4.28	6.40		6	13.09	11.10
	7	14.30	10.30		7	4.29	6.30		7	12.96	11.20
	8	14.57	10.10		8	4.12	6.10		8	12.90	11.50
	9	14.25	9.80		9	4.15	6.00		9	12.50	11.30
	10	14.49	9.90		10	4.13	6.00		10	11.78	10.60
	11	13.62	9.60		11	4.15	6.00		11	11.08	10.20
	12	13.20	9.70		12	3.98	6.00		12	10.82	9.90
1974	1	12.09	9.70	1978	1	3.58	5.80	1982	1	10.46	10.00
	2	10.67	10.00		2	3.46	5.70		2	10.27	9.90
	3	11.20	10.70		3	3.51	5.60		3	9.87	9.60
	4	10.07	10.80		4	3.56	5.60		4	9.33	9.10
	5	9.10	10.80		5	3.60	5.80		5	9.18	8.90
	6	9.46	10.90		6	3.68	6.00		6	9.28	9.20
	7	9.48	10.90		7	3.75	6.30		7	9.46	9.50
	8	9.65	10.90		8	3.70	6.60		8	9.00	9.20
	9	9.69	10.80		9	3.70	6.40		9	8.18	8.80
	10	9.78	10.90		10	3.95	6.30		10	7.58	8.40
	11	9.04	10.60		11	3.85	6.60		11	7.31	8.20
	12	8.60	9.90		12	4.06	6.60		12	6.62	8.00
1975	1	7.74	9.40	1979	1	3.89	6.70	1983	1	5.82	7.70
	2	6.43	9.00		2	4.15	7.00		2	5.83	7.70
	3	5.71	8.90		3	4.47	7.10		3	5.45	7.40
	4	4.89	8.80		4	5.54	7.20		4	5.20	7.40
	5	4.99	8.50		5	5.92	7.60		5	5.33	7.70
	6	4.88	8.40		6	6.46	8.00		6	5.57	8.10
	7	4.66	8.40		7	6.84	7.90		7	5.57	8.20
	8	3.88	8.60		8	7.09	7.70		8	5.71	8.30
	9	3.93	8.70		9	7.89	7.80		9	5.88	8.40
	10	4.07	8.70		10	8.76	7.90		10	6.18	8.20
	11	4.12	8.70		11	9.65	8.30		11	6.30	8.20
	12	4.21	8.60		12	9.58	8.00		12	6.48	8.30

Table E.5. (continued)

year	month	i^{short}	i^{long}	year	month	i^{short}	i^{long}
1984	1	6.12	8.20	1986	1	4.67	6.40
	2	5.95	8.10		2	4.49	6.30
	3	5.86	7.90		3	4.54	6.00
	4	5.84	7.90		4	4.49	5.60
	5	6.10	8.00		5	4.60	5.90
	6	6.13	8.10		6	4.60	6.00
	7	6.13	8.10		7	4.63	6.00
	8	6.02	7.90		8	4.57	5.80
	9	5.82	7.70		9	4.50	5.80
	10	6.07	7.40		10	4.59	6.00
	11	5.96	7.20		11	4.69	6.10
	12	5.83	7.00		12	4.81	6.00
1985	1	5.87	7.10	1987	1	4.49	5.90
	2	6.16	7.50		2	3.97	5.70
	3	6.39	7.70		3	3.99	5.60
	4	6.02	7.30		4	3.89	5.50
	5	5.84	7.10		5	3.76	5.40
	6	5.68	7.00		6	3.70	5.50
	7	5.34	6.80		7	3.83	5.80
	8	4.79	6.50		8	3.95	6.00
	9	4.69	6.40		9	3.99	6.20
	10	4.81	6.60		10	4.70	6.50
	11	4.84	6.70		11	3.94	6.00
	12	4.83	6.60		12	3.65	5.80

Source: Deutsche Bundesbank.

References

Abraham, B. (1980), "Intervention Analysis and Multiple Time Series," *Biometrika*, 67, 73–78.
Ahn, S.K. (1988), "Distribution for Residual Autocovariances in Multivariate Autoregressive Models with Structured Parameterization," *Biometrika*, 75, 590–593.
Ahn, S.K. & G.C. Reinsel (1988), "Nested Reduced-Rank Autoregressive Models for Multiple Time Series," *Journal of the American Statistical Association*, 83, 849–856.
Ahn, S.K. & G.C. Reinsel (1990), "Estimation of Partially Nonstationary Multivariate Autoregressive Models," *Journal of the American Statistical Association*, 85, 813–823.
Akaike, H. (1969), "Fitting Autoregressive Models for Prediction," *Annals of the Institute of Statistical Mathematics*, 21, 243–247.
Akaike, H. (1971), "Autoregressive Model Fitting for Control," *Annals of the Institute of Statistical Mathematics*, 23, 163–180.
Akaike, H. (1973), "Information Theory and an Extension of the Maximum Likelihood Principle," in B.N. Petrov & F. Csáki (eds.), *2nd International Symposium on Information Theory*, Budapest: Académiai Kiadó, 267–281.
Akaike, H. (1974), "A New Look at the Statistical Model Identification," *IEEE Transactions on Automatic Control*, AC-19, 716–723.
Akaike, H. (1976), "Canonical Correlation Analysis of Time Series and the Use of an Information Criterion," in R.K. Mehra & D.G. Lainiotis (eds.), *Systems Identification: Advances and Case Studies*, New York: Academic Press, pp. 27–96.
Anděl, J. (1983), "Statistical Analysis of Periodic Autoregression," *Aplikace Mathematiky*, 28, 364–385.
Anděl, J. (1987), "On Multiple Periodic Autoregression," *Aplikace Mathematiky*, 32, 63–80.
Anděl, J. (1989), "On Nonlinear Models for Time Series," *Statistics*, 20, 615–632.
Anderson, B.D.O. & J.B. Moore (1979), *Optimal Filtering*, Englewood Cliffs, NJ: Prentice-Hall.
Anderson, T.W. (1971), *The Statistical Analysis of Time Series*, New York: John Wiley.
Anderson, T.W. (1984), *An Introduction to Multivariate Statistical Analysis, 2nd ed.*, New York: John Wiley.
Andrews, D.W.K. (1987), "Asymptotic Results for Generalized Wald Tests," *Econometric Theory*, 3, 348–358.
Ansley, C.F. & R. Kohn (1983), "Exact Likelihood of Vector Autoregressive-Moving Average Process with Missing or Aggregated Data," *Biometrika*, 70, 275–278.
Aoki, M. (1987), *State Space Modeling of Time Series*, Berlin: Springer-Verlag.
Baillie, R.T. (1981), "Prediction from the Dynamic Simultaneous Equation Model with Vector Autoregressive Errors," *Econometrica*, 49, 1331–1337.
Baringhaus, L. & N. Henze (1988), "A Consistent Test for Multivariate Normality Based on the Empirical Characteristic Function," *Metrika*, 35, 339–348.
Barone, P. (1987), "A Method for Generating Independent Realizations of a Multivariate Normal Stationary and Invertible ARMA(p, q) Process," *Journal of Time Series Analysis*, 8, 125–130.
Basu, A.K. & S. Sen Roy (1986), "On some Asymptotic Results for Multivariate Autoregressive Models with Estimated Parameters," *Calcutta Statistical Association Bulletin*, 35, 123–132.
Basu, A.K. & S. Sen Roy (1987), "On Asymptotic Prediction Problems for Multivariate Autoregres-

sive Models in the Unstable Nonexplosive Case," *Calcutta Statistical Association Bulletin*, 36, 29–37.

Berk, K.N. (1974), "Consistent Autoregressive Spectral Estimates," *Annals of Statistics*, 2, 489–502.

Bhansali, R.J. (1978), "Linear Prediction by Autoregressive Model Fitting in the Time Domain," *Annals of Statistics*, 6, 224–231.

Bierens, H.J. (1981), *Robust Methods and Asymptotic Theory in Nonlinear Econometrics*, Berlin: Springer-Verlag.

Bollerslev, T., R.F. Engle & J.M. Wooldridge (1988), "A Capital Asset Pricing Model with Time-Varying Covariates," *Journal of Political Economy*, 96, 116–131.

Box, G.E.P. & G.M. Jenkins (1976), *Time Series Analysis: Forecasting and Control*, San Francisco: Holden-Day.

Box, G.E.P. & G.C. Tiao (1975), "Intervention Analysis with Applications to Economic and Environmental Problems," *Journal of the American Statistical Association*, 70, 70–79.

Breusch, T.S. & P. Schmidt (1988), "Alternative Forms of the Wald Test: How Long is a Piece of String," *Communications in Statistics, Theory and Methods*, 17, 2789–2795.

Brockwell, P.J. & R.A. Davis (1987), *Time Series: Theory and Methods*, New York: Springer-Verlag.

Broze, L., C. Gouriéroux & A. Szafarz (1989), *Reduced Forms of Rational Expectations Models*, Paris: CEPREMAP publication.

Caines, P.E. (1988), *Linear Stochastic Systems*, New York: John Wiley.

Chitturi, R.V. (1974), "Distribution of Residual Autocorrelations in Multiple Autoregressive Schemes," *Journal of the American Statistical Association*, 69, 928–934.

Chow, G.C. (1975), *Analysis and Control of Dynamic Economic Systems*, New York: John Wiley.

Chow, G.C. (1981), *Econometric Analysis by Control Methods*, New York: John Wiley.

Chow, G.C. (1984), "Random and Changing Coefficient Models," in Z. Griliches & M.D. Intriligator (eds.), *Handbook of Econometrics, Vol. II*, Amsterdam: North-Holland, pp. 1213–1245.

Cipra, T. (1985), "Periodic Moving Average Process," *Aplikace Matematiky*, 30, 218–229.

Cleveland, W.P. & G.C. Tiao (1979), "Modeling Seasonal Time Series," *Economie Appliquée*, 32, 107–129.

Cooley, T.F. & E. Prescott (1973), "Varying Parameter Regression: A Theory and Some Applications," *Annals of Economic and Social Measurement*, 2, 463–474.

Cooley, T.F. & E. Prescott (1976), "Estimation in the Presence of Stochastic Parameter Variation," *Econometrica*, 44, 167–184.

Cooper, D.M. & E.F. Wood (1982), "Identifying Multivariate Time Series Models," *Journal of Time Series Analysis*, 3, 153–164.

Crowder, M.J. (1976), "Maximum-Likelihood Estimation for Dependent Observations," *Journal of the Royal Statistical Society*, B38, 45–53.

Davidson, J.E.H., D.F. Hendry, F. Srba & S. Yeo (1978), "Econometric Modelling of the Aggregate Time-series Relationship between Consumer's Expenditure and Income in the United Kingdom," *Economic Journal*, 88, 661–692.

Davies, N., C.M. Triggs & P. Newbold (1977), "Significance Levels of the Box-Pierce Portmanteau Statistic in Finite Samples," *Biometrika*, 64, 517–522.

Deistler, M. & E.J. Hannan (1981), "Some Properties of the Parameterization of ARMA Systems with Unknown Order," *Journal of Multivariate Analysis*, 11, 474–484.

Deistler, M. & B.M. Pötscher (1984), "The Behaviour of the Likelihood Function for ARMA Models," *Advances in Applied Probability*, 16, 843–865.

Dempster, A.P., N.M. Laird & D.B. Rubin (1977), "Maximum-Likelihood From Incomplete Data via the EM-Algorithm," *Journal of the Royal Statistical Society*, B39, 1–38.

Dickey, D.A., W.A. Bell & R.B. Miller (1986), "Unit Roots in Time Series Models: Tests and Implications," *American Statistician*, 40, 12–26.

Dickey, D.A. & W.A. Fuller (1979), "Distribution of the Estimators for Autoregressive Time Series with Unit Root," *Journal of the American Statistical Association*, 74, 427–431.

Diebold, F.X. & M. Nerlove (1987), "Factor Structure in a Multivariate ARCH Model of Exchange Rate Fluctuations," in T. Pukkila & S. Puntanen (eds.) *Proceedings of the Second International*

Tampere Conference in Statistics, Department of Mathematical Sciences/Statistics, University of Tampere, Tampere, Finland, pp. 429–438.

Doan, T., R.B. Litterman & C.A. Sims (1984), "Forecasting and Conditional Projection Using Realistic Prior Distributions," *Econometric Reviews*, 3, 1–144.

Dufour, J.-M. (1985), "Unbiasedness of Predictions from Estimated Vector Autoregressions," *Econometric Theory*, 1, 387–402.

Dufour, J.-M. & R. Roy (1985), "Some Robust Exact Results on Sample Autocorrelations and Tests of Randomness," Journal of Econometrics, 29, 257–273.

Dunsmuir, W.T.M. & E.J. Hannan (1976), "Vector Linear Time Series Models," *Advances in Applied Probability*, 8, 339–364.

Engle, R.F. (1982), "Autoregressive Conditional Heteroscedasticity with Estimates of the Variance of United Kingdom Inflation," *Econometrica*, 50, 987–1007.

Engle, R.F. & T. Bollerslev (1986), "Modelling the Persistence in Conditional Variances," *Econometric Reviews*, 5, 1–50.

Engle, R.F. & C.W.J. Granger (1987), "Co-Integration and Error Correction: Representation, Estimation and Testing," *Econometrica*, 55, 251–276.

Engle, R.F., C.W.J. Granger & D. Kraft (1984), "Combining Competing Forecasts of Inflation Using a Bivariate ARCH Model," *Journal of Economic Dynamics and Control*, 8, 151–165.

Engle, R.F., D.F. Hendry & J.F. Richard (1983), "Exogeneity," *Econometrica*, 51, 277–304.

Engle, R.F. & M. Watson (1981), "A One-Factor Multivariate Time Series Model of Metropolitan Wage Rates," *Journal of the American Statistical Association*, 76, 774–781.

Engle, R.F. & B.S. Yoo (1987), "Forecasting and Testing in Co-Integrated Systems," *Journal of Econometrics*, 35, 143–159.

Fountis, N.G. & D.A. Dickey (1989), "Testing for a Unit Root Nonstationarity in Multivariate Autoregressive Time Series," *Annals of Statistics*, 17, 419–428.

Friedmann, R. (1981), "The Reliability of Policy Recommendations and Forecasts from Linear Econometric Models," *International Economic Review*, 22, 415–428.

Fuller, W.A. (1976), *Introduction to Statistical Time Series*, New York: John Wiley.

Gallant, A.R. (1987), *Nonlinear Statistical Models*, New York: John Wiley.

Gallant, A.R. & H. White (1988), *A Unified Theory of Estimation and Inference for Nonlinear Dynamic Models*, Oxford: Basil Blackwell.

Gasser, T. (1975), "Goodness-of-fit Tests for Correlated Data," *Biometrika*, 62, 563–570.

Geweke, J. (1977), "The Dynamic Factor Analysis of Economic Time-Series Models," in D.J. Aigner & A.S. Goldberger (eds.), *Latent Variables in Socio-Economic Models*, New York: North-Holland, pp. 365–383.

Geweke, J. (1984), "Inference and Causality in Economic Time Series Models," Chapter 19 in Z. Griliches & M.D. Intriligator (eds.), *Handbook of Econometrics, Vol. II*, Amsterdam: North-Holland, pp. 1101–1144.

Geweke, J., R. Meese & W. Dent (1983), "Comparing Alternative Tests of Causality in Temporal Systems: Analytic Results and Experimental Evidence," *Journal of Econometrics*, 21, 161–194.

Gladyshev, E.G. (1961), "Periodically Correlated Random Sequences," *Soviet Mathematics*, 2, 385–388.

Granger, C.W.J. (1969a), "Prediction with a Generalized Cost of Error Function," *Operations Research Quarterly*, 20, 199–207.

Granger, C.W.J. (1969b), "Investigating Causal Relations by Econometric Models and Cross-Spectral Methods," *Econometrica*, 37, 424–438.

Granger, C.W.J. (1981), "Some Properties of Time Series Data and Their Use in Econometric Model Specification," *Journal of Econometrics*, 16, 121–130.

Granger, C.W.J. (1982), "Generating Mechanisms, Models, and Causality," Chapter 8 in W. Hildenbrand (ed.), *Advances in Econometrics*, Cambridge: Cambridge University Press, pp. 237–253.

Granger, C.W.J. & A.P. Andersen (1978), *An Introduction to Bilinear Time Series Models*, Göttingen: Vandenhoeck & Ruprecht.

Granger, C.W.J. & P. Newbold (1986), *Forecasting Economic Time Series, 2nd Ed.*, New York: Academic Press.

Graybill, F.A. (1969), *Introduction to Matrices with Applications in Statistics*, Belmont, Ca.: Wadsworth.

Gregory, A.W. & M.R. Veall (1985), "Formulating Wald Tests of Nonlinear Restrictions," *Econometrica*, 53, 1465–1468.

Haggan, V. & T. Ozaki (1980), "Amplitude-Dependent Exponential AR Model Fitting for Non-linear Random Vibrations," in O.D. Anderson (ed.), *Time Series*, Amsterdam: North-Holland, pp. 57–71.

Hamilton, J.D. (1986), "A Standard Error for the Estimated State Vector of a State Space Model," *Journal of Econometrics*, 33, 387–397.

Hannan, E.J. (1969), "The Identification of Vector Mixed Autoregressive Moving Average Systems," *Biometrika*, 56, 223–225.

Hannan, E.J. (1970), *Multiple Time Series*, New York: John Wiley.

Hannan, E.J. (1976), "The Identification and Parameterization of ARMAX and State Space Forms," *Econometrica*, 44, 713–723.

Hannan, E.J. (1979), "The Statistical Theory of Linear Systems," in P.R. Krishnaiah (ed.), *Developments in Statistics*, New York: Academic Press, pp. 83–121.

Hannan, E.J. (1981), "Estimating the Dimension of a Linear System," *Journal of Multivariate Analysis*, 11, 459–473.

Hannan, E.J. & M. Deistler (1988), *The Statistical Theory of Linear Systems*, New York: John Wiley.

Hannan, E.J. & L. Kavalieris (1984), "Multivariate Linear Time Series Models," *Advances in Applied Probability*, 16, 492–561.

Hannan, E.J. & B.G. Quinn (1979), "The Determination of the Order of an Autoregression," *Journal of the Royal Statistical Society*, B41, 190–195.

Hannan, E.J. & J. Rissanen (1982), "Recursive Estimation of Mixed Autoregressive Moving Average Order," *Biometrika*, 69, 81–94.

Harvey, A.C. (1981), *The Econometric Analysis of Time Series*, Oxford: Allan.

Harvey, A.C. (1984), "A Unified View of Statistical Forecasting Procedures," *Journal of Forecasting*, 3, 245–275.

Harvey, A.C. (1987), "Applications of the Kalman Filter in Econometrics," in T.F. Bewley (ed.), *Advances in Econometrics-Fifth World Congress, Vol. I*, Cambridge: Cambridge University Press, pp. 285–313.

Harvey, A.C. (1989), *Forecasting, Structural Time Series Models, and the Kalman Filter*, Cambridge: Cambridge University Press.

Harvey, A.C. & R.G. Pierse (1984), "Estimating Missing Observations in Economic Time Series," *Journal of the American Statistical Association*, 79, 125–131.

Harvey, A.C. & P.H.J. Todd (1983), "Forecasting Economic Time Series with Structural and Box-Jenkins Models: A Case Study," *Journal of Business & Economic Statistics*, 1, 299–315.

Haugh, L.D. & G.E.P. Box (1977), "Identification of Dynamic Regression (Distributed Lag) Models Connecting two Time Series," *Journal of the American Statistical Association*, 72, 121–130.

Hausman, J.A. (1983), "Specification and Estimation of Simultaneous Equation Models," in Z. Griliches & M.D. Intriligator (eds.), *Handbook of Econometrics, Vol. I*, Amsterdam: North-Holland, pp. 391–448.

Hendry, D.F. & T. von Ungern-Sternberg (1981), "Liquidity and Inflation Effects on Consumer's Expenditure," in A.S. Deaton (ed.), *Essays in the Theory and Measurement of Consumer's Behavior*, Cambridge: Cambridge University Press.

Hildreth, C. & J.P. Houck (1968), "Some Estimators for a Linear Model with Random Coefficients," *Journal of the American Statistical Association*, 63, 584–595.

Hillmer, S.C. & G.C. Tiao (1979), "Likelihood Function of Stationary Multiple Autoregressive Moving Average Models," *Journal of the American Statistical Association*, 74, 652–660.

Hogg, R.V. & A.T. Craig (1978), *Introduction to Mathematical Statistics*, New York: Macmillan.

Hosking, J.R.M. (1980), "The Multivariate Portmanteau Statistic," *Journal of the American Statistical Association*, 75, 602–608.

Hosking, J.R.M. (1981a), "Equivalent Forms of the Multivariate Portmanteau Statistic," *Journal of the Royal Statistical Society*, B43, 261–262.

Hosking, J.R.M. (1981b), "Lagrange-multiplier Tests of Multivariate Time Series Models," *Journal of the Royal Statistical Society*, B43, 219–230.

Hsiao, C. (1979), "Autoregressive Modeling of Canadian Money and Income Data," *Journal of the American Statistical Association*, 74, 553–560.

Hsiao, C. (1982), "Time Series Modelling and Causal Ordering of Canadian Money, Income and Interest Rates," in O.D. Anderson (ed.), *Time Series Analysis: Theory and Practice 1*, Amsterdam: North-Holland, 671–699.

Hylleberg, S. (1986), *Seasonality in Regression*, Orlando: Academic Press.

Hylleberg, S. & G.E. Mizon (1989), "Cointegration and Error Correction Mechanisms," *Economic Journal*, 99, 113–125.

Jarque, C.M. & A.K. Bera (1987), "A Test for Normality of Observations and Regression Residuals," *International Statistical Review*, 55, 163–172.

Jazwinski, A.H. (1970), *Stochastic Processes and Filtering Theory*, New York: Academic Press.

Jenkins, G.M. & A.S. Alavi (1981), "Some Aspects of Modelling and Forecasting Multivariate Time Series," *Journal of Time Series Analysis*, 2, 1–47.

Johansen, S. (1988), "Statistical Analysis of Cointegration Vectors," *Journal of Economic Dynamics and Control*, 12, 231–254.

Johansen, S. (1989), "The Power Function of the Likelihood Ratio Test for Cointegration," *Institute of Mathematical Statistics, University of Copenhagen*, working paper.

Johansen, S. (1991), "Estimation and Hypothesis Testing of Cointegration Vectors in Gaussian Vector Autoregressive Models." *Econometrica*, 59, 1551–1580.

Johansen, S. & K. Juselius (1990), "Maximum Likelihood Estimation and Inference on Cointegration – With Applications to the Demand for Money," *Oxford Bulletin of Economics and Statistics*, 52, 169–210.

Jones, R.H. (1980), "Maximum Likelihood Fitting of ARMA Models to Time Series With Missing Observations," *Technometrics*, 22, 389–395.

Jones, R.H. & W.M. Brelsford (1967), "Time Series with Periodic Structure," *Biometrika*, 54, 403–408.

Judge, G.G., W.E. Griffiths, R.C. Hill, H. Lütkepohl & T.-C. Lee (1985), *The Theory and Practice of Econometrics, 2nd ed.*, New York: John Wiley.

Kalman, R.E. (1960), "A New Approach to Linear Filtering and Prediction Problems," *Journal of Basic Engineering*, 82, 35–45.

Kalman, R.E. & R.S. Bucy (1961), "New Results in Linear Filtering and Prediction Theory," *Journal of Basic Engineering*, 83, 95–108.

Kitagawa, G. (1981), "A Nonstationary Time Series Model and its Fitting by a Recursive Filter," *Journal of Time Series Analysis*, 2, 103–116.

Kohn, R. (1979), "Asymptotic Estimation and Hypothesis Testing Results for Vector Linear Time Series Models," *Econometrica*, 47, 1005–1030.

Kohn, R. (1981), "A Note on an Alternative Derivation of the Likelihood of an Autoregressive Moving Average Process," *Economics Letters*, 7, 233–236.

Lewis, R. & G.C. Reinsel (1985), "Prediction of Multivariate Time Series by Autoregressive Model Fitting," *Journal of Multivariate Analysis*, 16, 393–411.

Li, W.K. & Y.V. Hui (1988), "An Algorithm for the Exact Likelihood of Periodic Autoregressive Moving Average Models," *Communications in Statistics – Simulation and Computation*, 17, 1483–1494.

Li, W.K. & A.I. McLeod (1981), "Distribution of the Residual Autocorrelations in Multivariate ARMA Time Series Models," *Journal of the Royal Statistical Society*, B43, 231–239.

Litterman, R.B. (1986), "Forecasting With Bayesian Vector Autoregressions – Five Years of Experience," *Journal of Business & Economic Statistics*, 4, 25–38.

Liu, J. (1989), "On the Existence of a General Multiple Bilinear Time Series," *Journal of Time Series Analysis*, 10, 341–355.

Liu, L.-M. & D.M. Hanssens (1982), "Identification of Multiple-Input Transfer Function Models," *Communications in Statistics-Theory and Methods*, 11, 297–314.

Ljung, G.M. & G.E.P. Box (1978), "On a Measure of Lack of Fit in Time Series Models," *Biometrika*, 65, 297–303.

Lucas, R.E. Jr. & T.J. Sargent (eds.) (1981), *Rational Expectations and Econometric Practice, Vol. 1 und 2*, Minneapolis: The University of Minnesota Press.

Lütkepohl, H. (1984), "Linear Transformations of Vector ARMA Processes," *Journal of Econometrics*, 26, 283–293.

Lütkepohl, H. (1985), "Comparison of Criteria for Estimating the Order of a Vector Autoregressive Process," *Journal of Time Series Analysis*, 6, 35–52, "Correction," 8 (1987), 373.

Lütkepohl, H. (1986), *Prognose aggregierter Zeitreihen*, Göttingen: Vandenhoeck & Ruprecht.

Lütkepohl, H. (1987), *Forecasting Aggregated Vector ARMA Processes*, Berlin: Springer-Verlag.

Lütkepohl, H. (1988a), "Prediction Tests for Structural Stability," *Journal of Econometrics*, 39, 267–296.

Lütkepohl, H. (1988b), "Asymptotic Distribution of the Moving Average Coefficients of an Estimated Vector Autoregressive Process," *Econometric Theory*, 4, 77–85.

Lütkepohl, H. (1989a), "Prediction Tests for Structural Stability of Multiple Time Series," *Journal of Business & Economic Statistics*, 7, 129–135.

Lütkepohl, H. (1989b), "A Note on the Asymptotic Distribution of Impulse Response Functions of Estimated VAR Models with Orthogonal Residuals," *Journal of Econometrics*, 42, 371–376.

Lütkepohl, H. (1990a), "Asymptotic Distributions of Impulse Response Functions and Forecast Error Variance Decompositions of Vector Autoregressive Models," *Review of Economics and Statistics*, 72, 116–125.

Lütkepohl, H. (1990b), "Testing for Causation Between two Variables in Higher Dimensional VAR Models," paper presented at the World Congress of the Econometric Society, Barcelona.

Lütkepohl, H. (1992), "Testing for Time Varying Parameters in Vector Autoregressive Models," in W.E. Griffiths, H. Lütkepohl & M.E. Bock (eds.), *Readings in Econometric Theory and Practice*, Amsterdam: North-Holland, pp. 243–264.

Lütkepohl, H. & D.S. Poskitt (1991), "Estimating Orthogonal Impulse Responses via Vector Autoregressive Models," *Econometric Theory*, 7, 487–496.

Lütkepohl, H. & H.-E. Reimers (1992), "Impulse Response Analysis of Cointegrated Systems," *Journal of Economic Dynamics and Control*, 16, 53–78.

Lütkepohl, H. & W. Schneider (1989), "Testing for Nonnormality of Autoregressive Time Series," *Computational Statistics Quarterly*, 5, 151–168.

Magnus, J.R. (1988), *Linear Structures*, London: Charles Griffin.

Magnus, J.R. & H. Neudecker (1986), "Symmetry, 0-1 Matrices and Jacobians: A Review," *Econometric Theory*, 2, 157–190.

Magnus, J.R. & H. Neudecker (1988), *Matrix Differential Calculus with Applications in Statistics and Econometrics*, Chichester: John Wiley.

Mann, H.B. & A. Wald (1943), "On the Statistical Treatment of Linear Stochastic Difference Equations," *Econometrica*, 11, 173–220.

Mardia, K.V. (1980), "Tests for Univariate and Multivariate Normality," in: *Handbook of Statistics, Vol. 1*, Amsterdam: North-Holland, pp. 279–320.

Meinhold, R.J. & N.D. Singpurwalla (1983), "Understanding the Kalman-Filter," *American Statistician*, 37, 123–127.

Mittnik, S. (1990), "Computation of Theoretical Autocovariance Matrices of Multivariate Autoregressive Moving Average Time Series," *Journal of the Royal Statistical Society*, B52, 151–155.

Mood, A.M., F.A. Graybill & D.C. Boes (1974), *Introduction to the Theory of Statistics, 3rd ed.*, Auckland: McGraw-Hill.

Morrison, D.F. (1976), *Multivariate Statistical Methods, 2nd ed.*, New York: McGraw-Hill.

Murata, Y. (1982), *Optimal Control Methods for Linear Discrete-Time Economic Systems*, New York: Springer-Verlag.

Muth, J.F. (1961), "Rational Expectations and the Theory of Price Movements," *Econometrica*, 29, 315–335.

Nankervis, J.C. & N.E. Savin (1988), "The Student's t Approximation in a Stationary First Order Autoregressive Model," *Econometrica*, 56, 119–145.

Nicholls, D.F. & A.R. Pagan (1985), "Varying Coefficient Regression," in E.J. Hannan, P.R. Krishnaiah & M.M. Rao (eds.), *Handbook of Statistics, Vol. 5*, Amsterdam: North-Holland, pp. 413–449.

Nicholls, D.F. & A.L. Pope (1988), "Bias in the Estimation of Multivariate Autoregressions," *Australian Journal of Statistics*, 30A, 296–309.

Nicholls, D.F. & B.G. Quinn (1982), *Random Coefficient Autoregressive Models: An Introduction*, New York: Springer-Verlag.

Nijman, T.E. (1985), *Missing Observations in Dynamic Macroeconomic Modeling*, Amsterdam: Free University Press.

Osborn, D.R. & J.P. Smith (1989), "The Performance of Periodic Autoregressive Models in Forecasting Seasonal U.K. Consumption," *Journal of Business & Economic Statistics*, 7, 117–127.

Ozaki, T. (1980), "Non-linear Time Series Models for Nonlinear Random Vibrations," *Journal of Applied Probability*, 17, 84–93.

Pagan, A. (1980), "Some Identification and Estimation Results for Regression Models with Stochastically Varying Coefficients," *Journal of Econometrics*, 13, 341–363.

Pagano, M. (1978), "On Periodic and Multiple Autoregressions," *Annals of Statistics*, 6, 1310–1317.

Pankratz, A. (1983), *Forecasting With Univariate Box-Jenkins Models: Concepts and Cases*, New York: John Wiley.

Park, J.Y. & P.C.B. Phillips (1988), "Statistical Inference in Regressions with Integrated Processes: Part 1," *Econometric Theory*, 4, 468–497.

Park, J.Y. & P.C.B. Phillips (1989). "Statistical Inference in Regressions with Integrated Processes: Part 2," *Econometric Theory*, 5, 95–131.

Paulsen, J. (1984), "Order Determination of Multivariate Autoregressive Time Series With Unit Roots," *Journal of Time Series Analysis*, 5, 115–127.

Paulsen, J. & D. Tjøstheim (1985), "On the Estimation of Residual Variance and Order in Autoregressive Time Series," *Journal of the Royal Statistical Society*, B47, 216–228.

Penm, J.H.W. & R.D. Terrell (1982), "On the Recursive Fitting of Subset Autoregressions," *Journal of Time Series Analysis*, 3, 43–59.

Penm, J.H.W. & R.D. Terrell (1984), "Multivariate Subset Autoregressive Modelling With Zero Constraints for Detecting 'Overall Causality'," *Journal of Econometrics*, 24, 311–330.

Penm, J.H.W. & R.D. Terrell (1986), "The 'Derived' Moving Average Model and its Role in Causality," in J. Gani & M.B. Priestley (eds.), *Essays in Time Series and Allied Processes*, Sheffield: Applied Probability Trust, pp. 99–111.

Pesaran, M.H. (1987), *The Limits to Rational Expectations*, Oxford: Basil Blackwell.

Phillips, P.C.B. (1988), "Multiple Regression with Integrated Time Series," *Contemporary Mathematics*, 80, 79–105.

Phillips, P.C.B. & S.N. Durlauf (1986), "Multiple Time Series Regression with Integrated Processes," *Review of Economic Studies*, 53, 473–495.

Phillips, P.C.B. & B.E. Hansen (1990), "Statistical Inference in Instrumental Variables Regression with $I(1)$ Processes," *Review of Economic Studies*, 57, 99–125.

Phillips, P.C.B. & S. Ouliaris (1988), "Testing for Cointegration Using Principle Components Methods," *Journal of Economic Dynamics and Control*, 12, 205–230.

Phillips, P.C.B. & S. Ouliaris (1990), "Asymptotic Properties of Residual Based Tests for Cointegration," *Econometrica*, 58, 165–193.

Poskitt, D.S. (1987), "A Modified Hannan-Rissanen Strategy for Mixed Autoregressive-Moving Average Order Determination," *Biometrika*, 74, 781–790.

Poskitt, D.S. (1989a), "A Two Stage Least Squares Procedure for the Identification of ARMAX Models," working paper.

Poskitt, D.S. (1989b), "A Method for the Estimation of Transfer Function Models," *Journal of the Royal Statistical Society*, B51, 29–46.

Poskitt, D.S. & A.R Tremayne (1981), "An Approach to Testing Linear Time Series Models," *Annals of Statistics*, 9, 974–986.

Poskitt, D.S. & A.R. Tremayne (1982), "Diagnostic Tests for Multiple Time Series Models," *Annals of Statistics*, 10, 114–120.

Priestley, M.B. (1980), "State-Dependent Models: A General Approach to Non-Linear Time Series Analysis," *Journal of Time Series Analysis*, 1, 47–71.

Priestley, M.B. (1981), *Spectral Analysis and Time Series*, London: Academic Press.

Priestley, M.B. (1988), *Non-Linear and Non-Stationary Time Series Analysis*, London: Academic Press.

Quenouille, M.H. (1957), *The Analysis of Multiple Time-Series*, London: Griffin.

Quinn, B.G. (1980), "Order Determination for a Multivariate Autoregression," *Journal of the Royal Statistical Society*, B42, 182–185.

Rao, C.R. (1973), *Linear Statistical Inference and Its Applications, 2nd ed.*, New York: John Wiley.

Reinsel, G. (1983), "Some Results on Multivariate Autoregressive Index Models," *Biometrika*, 70, 145–156.

Reinsel, G.C. & S.K. Ahn (1988), "Asymptotic Distribution of the Likelihood Ratio Test for Cointegration in the Nonstationary Vector AR Model," Technical Report, *University of Wisconsin, Madison, Department of Statistics.*

Rohatgi, V.K. (1976), *An Introduction to Probability Theory and Mathematical Statistics*, New York: John Wiley.

Roussas, G.G. (1973), *A First Course in Mathematical Statistics*, Reading, Ma.: Addison-Wesley.

Salmon, M. (1982), "Error Correction Mechanisms," *Economic Journal*, 92, 615–629.

Samaranayake, V.A. & D.P. Hasza (1988), "Properties of Predictors for Multivariate Autoregressive Models with Estimated Parameters," *Journal of Time Series Analysis*, 9, 361–383.

Schlittgen, R. & B.H.J. Streitberg (1984), *Zeitreihenanalyse*, München: R. Oldenbourg.

Schmidt, P. (1973), "The Asymptotic Distribution of Dynamic Multipliers," *Econometrica*, 41, 161–164.

Schmidt, P. (1976), *Econometrics*, New York: Marcel Dekker.

Schneider, W. (1988), "Analytical Uses of Kalman Filtering in Econometrics – A Survey," *Statistical Papers*, 29, 3–33.

Schneider, W. (1990a), "Maximum Likelihood Estimation of Seemingly Unrelated Regression Equations with Time-Varying Coefficients by Kalman Filtering – A Comparison of Scoring, EM- and an Adaptive EM-Method," *Computers and Mathematics with Applications*, forthcoming.

Schneider, W. (1990b), "Implementing a Stability Test of the West German Money Demand Function," P. Hackl & A. Westlund (eds.), *Economic Structural Change, Analysis and Forecasting*, Berlin: Springer, forthcoming.

Schwarz, G. (1978), "Estimating the Dimension of a Model," *Annals of Statistics*, 6, 461–464.

Searle, S.R. (1982), *Matrix Algebra Useful for Statistics*, New York: John Wiley.

Serfling, R.J. (1980), *Approximation Theorems of Mathematical Statistics*, New York: John Wiley.

Shibata R. (1980), "Asymptotically Efficient Selection of the Order of the Model for Estimating Parameters of a Linear Process," *Annals of Statistics*, 8, 147–164.

Sims, C.A. (1980), "Macroeconomics and Reality," *Econometrica*, 48, 1–48.

Sims, C.A. (1981), "An Autoregressive Index Model for the U.S. 1948–1975," in J. Kmenta & J.B. Ramsey (eds.), *Large-Scale Macro-Econometric Models*, Amsterdam: North-Holland, pp. 283–327.

Sims, C.A., J.H. Stock & M.W. Watson (1990), "Inference in Linear Time Series Models with Some Unit Roots," *Econometrica*, 58, 113–144.

Solo, V. (1984), "The Exact Likelihood for a Multivariate ARMA Model," *Journal of Multivariate Analysis*, 15, 164–173.

Stensholt, B.K. & D. Tjøstheim (1987), "Multiple Bilinear Time Series Models," *Journal of Time Series Analysis*, 8, 221–233.

Stock, J.H. (1987), "Asymptotic Properties of Least Squares Estimators of Cointegrating Vectors," *Econometrica*, 55, 1035–1056.

Stock, J.H. & M.W. Watson (1988), "Testing for Common Trends," *Journal of the American Statistical Association*, 83, 1097–1107.

Subba Rao, T. & M.M. Gabr (1984), *An Introduction to Bispectral Analysis and Bilinear Time Series Models*, New York: Springer-Verlag.

Swamy, P.A.V.B. (1971), *Statistical Inference in Random Coefficient Regression Models*, Berlin: Springer-Verlag.

Theil, H. (1971), *Principles of Econometrics*, Santa Barbara: John Wiley.

Tiao, G.C. & G.E.P. Box (1981), "Modeling Multiple Time Series With Applications," *Journal of the American Statistical Association*, 76, 802–816.

Tiao, G.C. & M.R. Grupe (1980), "Hidden Periodic Autoregressive-Moving Average Models in Time Series Data," *Biometrika*, 67, 365–373.

Tiao, G.C. & R.S. Tsay (1989), "Model Specification in Multivariate Time Series (with discussion)," *Journal of the Royal Statistical Society*, B51, 157–213.

Tjøstheim, D. & J. Paulsen (1983), "Bias of Some Commonly-Used Time Series Estimates," *Biometrika*, 70, 389–399.

Tong, H. (1983), *Threshold Models in Non-linear Time Series Analysis*, New York: Springer-Verlag.

Tsay, R.S. (1984), "Order Selection in Nonstationary Autoregressive Models," *Annals of Statistics*, 12, 1425–1433.

Tsay, R.S. (1985), "Model Identification in Dynamic Regression (Distributed Lag) Models," *Journal of Business & Economic Statistics*, 3, 228–237.

Tsay, R.S. (1988), "Outliers, Level Shifts and Variance Changes in Time Series," *Journal of Forecasting*, 7, 1–20.

Tsay, R.S. (1989a), "Parsimonious Parameterization of Vector Autoregressive Moving Average Models," *Journal of Business & Economic Statistics*, 7, 327–341.

Tsay, R.S. (1989b), "Identifying Multivariate Time Series Models," *Journal of Time Series Analysis*, 10, 357–372.

Tso, M.K.-S. (1981), "Reduced-rank Regression and Canonical Analysis," *Journal of the Royal Statistical Society*, B43, 183–189.

Velu, R.P., G.C. Reinsel & D.W. Wichern (1986), "Reduced Rank Models for Multiple Time Series," *Biometrika*, 73, 105–118.

Wallis, K.F. (1977), "Multiple Time Series Analysis and the Final Form of Econometric Models," *Econometrica*, 45, 1481–1497.

Watanabe, N. (1985), "Note on the Kalman Filter with Estimated Parameters," *Journal of Time Series Analysis*, 6, 269–278.

Watson, M.W. & R.F. Engle (1983), "Alternative Algorithms for the Estimation of Dynamic Factor, MIMIC and Varying Coefficient Regression Models," *Journal of Econometrics*, 23, 385–400.

Wei, W.W.S. (1990), *Time Series Analysis: Univariate and Multivariate Methods*, Redwood City, Ca.: Addison-Wesley.

White, H. & G.M. MacDonald (1980), "Some Large-Sample Tests for Nonnormality in the Linear Regression Model," *Journal of the American Statistical Association*, 75, 16–28.

Wold, H. (1938), *A Study in the Analysis of Stationary Time-Series*, Uppsala: Almqvist and Wiksells.

Yamamoto, T. (1980), "On the Treatment of Autocorrelated Errors in the Multiperiod Prediction of Dynamic Simultaneous Equation Models," *International Economic Review*, 21, 735–748.

Young, P.C., A.J. Jakeman & R. McMurtrie (1980), "An Instrumental Variable Method for Model Order Identification," *Automatica*, 16, 281–294.

Zellner, A. & F. Palm (1974), "Time Series Analysis and Simultaneous Equation Econometric Models," *Journal of Econometrics*, 2, 17–54.

List of Propositions and Definitions

Chapter 2. Stable Vector Autoregressive Processes

PROPOSITION 2.1: Stationarity Condition 20
PROPOSITION 2.2: Characterization of Granger-Noncausality 38
PROPOSITION 2.3: Characterization of Instantaneous Causality ... 41
PROPOSITION 2.4: Zero Impulse Responses 45
PROPOSITION 2.5: Zero Orthogonalized Impulse Responses 55

Chapter 3. Estimation of Vector Autoregressive Processes

DEFINITION 3.1: Standard White Noise 66
PROPOSITION 3.1: Asymptotic Properties of the LS Estimator 66
PROPOSITION 3.2: Asymptotic Properties of the White Noise Covariance Matrix Estimators 68
PROPOSITION 3.3: Asymptotic Properties of the Sample Mean 77
PROPOSITION 3.4: Asymptotic Properties of ML Estimators 85
PROPOSITION 3.5: Asymptotic Distribution of the Wald Statistic .. 94
PROPOSITION 3.6: Asymptotic Distributions of Impulse Responses . 98

Chapter 4. VAR Order Selection and Checking the Model Adequacy

PROPOSITION 4.1: Asymptotic Distribution of the LR Statistic 123
PROPOSITION 4.2: Consistency of VAR Order Estimators 131
PROPOSITION 4.3: Small Sample Comparison of AIC, HQ, and SC . 133
PROPOSITION 4.4: Asymptotic Distributions of White Noise Autocovariances and Autocorrelations 140
PROPOSITION 4.5: Asymptotic Distributions of Residual Autocovariances 146
PROPOSITION 4.6: Asymptotic Distributions of Residual Autocorrelations 147
PROPOSITION 4.7: Approximate Distribution of the Portmanteau Statistic 150
PROPOSITION 4.8: Asymptotic Distribution of Skewness and Kurtosis 153
PROPOSITION 4.9: Asymptotic Distribution of Residual Skewness and Kurtosis 156

Chapter 5. VAR Processes with Parameter Constraints

PROPOSITION 5.1:	Asymptotic Properties of the GLS Estimator ..	170
PROPOSITION 5.2:	Asymptotic Properties of the EGLS Estimator .	170
PROPOSITION 5.3:	Asymptotic Properties of the Implied Restricted EGLS Estimator	171
PROPOSITION 5.4:	Asymptotic Properties of the White Noise Covariance Estimator	172
PROPOSITION 5.5:	Asymptotic Properties of the Restricted ML Estimators	174
PROPOSITION 5.6:	EGLS Estimator of Parameters Arranged Equationwise	175
PROPOSITION 5.7:	Asymptotic Distributions of Residual Autocovariances and Autocorrelations	187
PROPOSITION 5.8:	Approximate Distribution of the Portmanteau Statistic	188
PROPOSITION 5.9:	LS Estimator of the Reduced Rank VAR Model	197
PROPOSITION 5.10:	Asymptotic Properties of Reduced Rank LS Estimators	199

Chapter 6. Vector Autoregressive Moving Average Processes

PROPOSITION 6.1:	Linear Transformation of an MA(q) Process ...	231
PROPOSITION 6.2:	Forecast Efficiency of Linearly Transformed VARMA Processes	234
PROPOSITION 6.3:	Characterization of Noncausality	237

Chapter 7. Estimation of VARMA Models

DEFINITION 7.1:	Final Equations Form	246
DEFINITION 7.2:	Echelon Form	246
PROPOSITION 7.1:	Asymptotic Properties of ML Estimators	271

Chapter 9. Fitting Finite Order VAR Models to Infinite Order Processes

PROPOSITION 9.1:	Properties of the LS Estimator of an Approximating VAR Model	306
PROPOSITION 9.2:	Asymptotic Properties of the White Noise Covariance Matrix Estimator	308
PROPOSITION 9.3:	Asymptotic Distributions of Estimated Forecasts	310
PROPOSITION 9.4:	Asymptotic Distributions of Impulse Responses .	314

Chapter 11. Nonstationary Systems with Integrated and Cointegrated Variables

PROPOSITION 11.1:	ML Estimators of a Cointegrated VAR Process	356
PROPOSITION 11.2:	Asymptotic Properties of the ML Estimators ..	358

PROPOSITION 11.3: Asymptotic Properties of the Unconstrained LS Estimator 369
PROPOSITION 11.4: Asymptotic Properties of the Two-Stage Estimators 371
PROPOSITION 11.5: Asymptotic Distribution of the Wald Statistic .. 379
PROPOSITION 11.6: Consistent VAR Order Estimation 383

Chapter 13. State Space Models

PROPOSITION 13.1: Asymptotic Properties of the ML Estimators .. 438

Appendix A. Vectors and Matrices

PROPOSITION A.1: Chain Rule for Vector Differentiation 469
PROPOSITION A.2: Product Rules for Vector Differentiation 469
PROPOSITION A.3: Taylor's Theorem 473
PROPOSITION A.4: Maximum of $\mathrm{tr}(B'\Omega B)$ 474
PROPOSITION A.5: Minimum of $\mathrm{tr}(Y - BCX)'\Sigma_u^{-1}(Y - BCX)$ 475
PROPOSITION A.6: Maximum and Minimum of $|C\Omega C'|$ 477
PROPOSITION A.7: Minimum of $|(Y - BCX)(Y - BCX)'/T|$ 477

Appendix B. Multivariate Normal and Related Distributions

PROPOSITION B.1: Marginal and Conditional Distributions of a Multivariate Normal 480
PROPOSITION B.2: Linear Transformation of a Multivariate Normal 481
PROPOSITION B.3: Distributions of Quadratic Forms 481
PROPOSITION B.4: Independence of a Normal Vector and a Quadratic Form 481
PROPOSITION B.5: Independence of Quadratic Forms 482
PROPOSITION B.6: Distributions of Ratios of Quadratic Forms 482
PROPOSITION B.7: Quadratic Form with Noncentral χ^2-Distribution 483

Appendix C. Convergence of Sequences of Random Variables and Asymptotic Distributions

PROPOSITION C.1: Convergence Properties of Sequences of Random Variables 486
PROPOSITION C.2: Properties of Convergence in Probability and in Distribution 486
PROPOSITION C.3: Limits of Sequences of t and F Random Variables 486
PROPOSITION C.4: Asymptotic Properties of Estimators 487
PROPOSITION C.5: Existence of Infinite Sums of Random Variables . 488
PROPOSITION C.6: Moments of Infinite Sums of Random Variables . 489
PROPOSITION C.7: Existence of Infinite Sums of Random Vectors . 490
PROPOSITION C.8: Moments of Infinite Sums of Random Vectors . 491

Index of Notation

Most of the notation is clearly defined in the text where it is used. The following list is meant to provide some general guidelines. Occasionally, in the text a symbol has a meaning which differs from the one specified in this list when confusion is unlikely. For instance, A usually stands for a VAR coefficient matrix whereas in the Appendix it is often a general matrix.

General Symbols

$=$	equals
$:=$	equals by definition
$\Rightarrow$	implies
$\Leftrightarrow$	is equivalent to
$\sim$	is distributed as
$\in$	element of
$\subset$	subset of
$\cup$	union
$\cap$	intersection
$\sum$	summation sign
$\prod$	product sign
$\rightarrow$	converges to, approaches
$\xrightarrow{p}$	converges in probability to
$\xrightarrow{\text{a.s.}}$	converges almost surely to
$\xrightarrow{\text{q.m.}}$	converges in quadratic mean to
$\xrightarrow{d}$	converges in distribution to
lim	limit
plim	probability limit
max	maximum
min	minimum
sup	supremum, least upper bound
ln	natural logarithm
exp	exponential function
$\|z\|$	absolute value or modulus of z
$\mathbb{R}$	real numbers
$\mathbb{R}^m$	m-dimensional Euclidean space
$\mathbb{C}$	complex numbers

L	lag operator
E	expectation
Var	variance
Cov	covariance, covariance matrix
MSE	mean squared error (matrix)
Pr	probability
$l(\cdot)$	likelihood function
$\ln l$	log-likelihood function
$l_0(\cdot)$	approximate likelihood function
$\ln l_0$	approximate log-likelihood function
λ_{LM}	Lagrange multiplier statistic
λ_{LR}	likelihood ratio statistic
λ_w	Wald statistic
P_h	portmanteau statistic
$\bar{P}_h$	modified portmanteau statistic
d.f.	degrees of freedom
K	dimension of a stochastic process or time series
T	sample size, time series length
AIC	Akaike information criterion
FPE	final prediction error (criterion)
HQ	Hannan-Quinn (criterion)
SC	Schwarz criterion

Distributions and Stochastic Processes

$N(\mu, \Sigma)$	(multivariate) normal distribution with mean (vector) μ and variance (covariance matrix) Σ
$\chi^2(m)$	χ^2-distribution with m degrees of freedom
$F(m, n)$	F-distribution with m numerator and n denominator degrees of freedom
$t(m)$	t-distribution with m degrees of freedom
AR	autoregressive (process)
AR(p)	autoregressive process of order p
ARMA	autoregressive moving average (process)
ARMA(p, q)	autoregressive moving average process of order (p, q)
MA	moving average (process)
MA(q)	moving average process of order q
VAR	vector autoregressive (process)
VAR(p)	vector autoregressive process of order p
VARMA	vector autoregressive moving average (process)
VARMA(p, q)	vector autoregressive moving average process of order (p, q)

Vector and Matrix Operations

M'	transpose of M
M^*	adjoint of M

M^{-1}	inverse of M
M^{+}	Moore-Penrose generalized inverse of M
$M^{1/2}$	square root of M
M^{k}	k-th power of M
MN	matrix product of M and N
$+$	plus
$-$	minus
$\otimes$	Kronecker product
$\det(M)$, $\det M$	determinant of M
$\|M\|$	determinant of M
$\\|M\\|$	norm of M
$\mathrm{rk}(M)$, $\mathrm{rk}\, M$	rank of M
$\mathrm{tr}(M)$, $\mathrm{tr}\, M$	trace of M
vec	column stacking operator
vech	column stacking operator for symmetric matrices (stacks the elements on and below the main diagonal only)
$\dfrac{\partial \varphi}{\partial \beta'}$	vector or matrix of first order partial derivatives of φ with respect to β
$\dfrac{\partial^2 \varphi}{\partial \beta \partial \beta'}$	Hessian matrix of φ, matrix of second order partial derivatives of φ with respect to β

General Matrices

$\mathbf{D}_m$	$(m^2 \times m(m+1)/2)$ duplication matrix
I_m	$(m \times m)$ unit or identity matrix
$I(\cdot)$	information matrix
$I_a(\cdot)$	asymptotic information matrix
J	$:= [I_K \quad 0 \ldots 0]$
$\mathbf{K}_{mn}$	$(mn \times mn)$ commutation matrix
$\mathbf{L}_m$	$(m(m+1)/2 \times m^2)$ elimination matrix
0	zero or null matrix or vector

Vectors and Matrices Related to Stochastic Processes and Multiple Time Series

u_t	K-dimensional white noise process
u_{kt}	k-th element of u_t
$u_{(k)}$	$:= \begin{bmatrix} u_{k1} \\ \vdots \\ u_{kT} \end{bmatrix}$
U	$:= [u_1, \ldots, u_T]$
$\mathbf{u}$	$:= \mathrm{vec}(U)$

U_t $:= \begin{bmatrix} u_t \\ 0 \\ \vdots \\ 0 \end{bmatrix}$ or $\begin{bmatrix} u_t \\ 0 \\ \vdots \\ 0 \\ 0 \\ u_t \\ 0 \\ \vdots \\ 0 \end{bmatrix}$

v_t white noise process
w_t white noise process
y_t K-dimensional stochastic process
y_{kt} k-th element of y_t

$y_{(k)}$ $:= \begin{bmatrix} y_{k1} \\ \vdots \\ y_{kT} \end{bmatrix}$

$\bar{y}$ $:= \sum_{t=1}^{T} y_t/T$, sample mean (vector)
$y_t(h)$ h-step forecast of y_{t+h} at origin t
Y $:= [y_1, \dots, y_T]$
$\mathbf{y}$ $:= \mathrm{vec}(Y)$

Y_t $:= \begin{bmatrix} y_t \\ \vdots \\ y_{t-p+1} \end{bmatrix}$ or $\begin{bmatrix} y_t \\ \vdots \\ y_{t-p-1} \\ u_t \\ \vdots \\ u_{t-q+1} \end{bmatrix}$ or $\begin{bmatrix} y_t \\ \vdots \\ y_{t-p+1} \\ x_t \\ \vdots \\ x_{t-s+1} \end{bmatrix}$

Z_t $:= \begin{bmatrix} 1 \\ y_t \\ \vdots \\ y_{t-p+1} \end{bmatrix}$

Matrices and Vectors Related to VAR and VARMA Representations (Parts I and II)

A_i VAR coefficient matrix
A $:= [A_1, \dots, A_p]$
$\boldsymbol{\alpha}$ $:= \mathrm{vec}(A)$

$\mathbf{A}$ $:= \begin{bmatrix} A_1 & \dots & A_{p-1} & A_p \\ I_K & & 0 & 0 \\ & \ddots & \vdots & \vdots \\ 0 & \dots & I_K & 0 \end{bmatrix}$ or $\begin{bmatrix} \mathbf{A}_{11} & \mathbf{A}_{12} \\ \mathbf{A}_{21} & \mathbf{A}_{22} \end{bmatrix}$

$\mathbf{A}_{11}$ $:= \begin{bmatrix} A_1 & \dots & A_{p-1} & A_p \\ I_K & & 0 & 0 \\ & \ddots & \vdots & \vdots \\ 0 & \dots & I_K & 0 \end{bmatrix}$ $(Kp \times Kp)$

$\mathbf{A}_{12}$ $:= \begin{bmatrix} M_1 & \dots & M_{q-1} & M_q \\ 0 & \dots & 0 & 0 \\ \vdots & & \vdots & \vdots \\ 0 & \dots & 0 & 0 \end{bmatrix}$ $(Kq \times Kq)$

$\mathbf{A}_{21}$ $:= 0 \quad (Kq \times Kp)$

$\mathbf{A}_{22}$ $:= \begin{bmatrix} 0 & 0 \\ I_{K(q-1)} & 0 \end{bmatrix}$ $(Kq \times Kq)$

M_i MA coefficient matrix

$\mathbf{m}$ $:= \text{vec}[M_1, \dots, M_q]$

$\mathbf{M}$ $:= \begin{bmatrix} \mathbf{M}_{11} & \mathbf{M}_{12} \\ \mathbf{M}_{21} & \mathbf{M}_{22} \end{bmatrix}$ $(K(p+q) \times K(p+q))$

$\mathbf{M}_{11}$ $:= \begin{bmatrix} -M_1 & \dots & -M_{q-1} & -M_q \\ I_K & & 0 & 0 \\ & \ddots & \vdots & \vdots \\ 0 & \dots & I_K & 0 \end{bmatrix}$ $(Kq \times Kq)$

$\mathbf{M}_{12}$ $:= \begin{bmatrix} -A_1 & \dots & -A_{p-1} & -A_p \\ 0 & \dots & 0 & 0 \\ \vdots & & \vdots & \vdots \\ 0 & \dots & 0 & 0 \end{bmatrix}$ $(Kq \times Kp)$

$\mathbf{M}_{21}$ $:= 0 \quad (Kp \times Kq)$

$\mathbf{M}_{22}$ $:= \begin{bmatrix} 0 & 0 \\ I_{K(p-1)} & 0 \end{bmatrix}$ $(Kp \times Kp)$

Φ_i coefficient matrix of canonical MA representation

Π_i coefficient matrix of pure VAR representation

Impulse Responses and Related Quantities

Φ_i matrix of forecast error impulse responses

Ψ_m $:= \sum_{i=0}^{m} \Phi_i$, matrix of accumulated forecast error impulse responses

Ψ_∞ $:= \sum_{i=0}^{\infty} \Phi_i$, matrix of total or long-run forecast error impulse responses

Θ_i matrix of orthogonalized impulse responses

Ξ_m $:= \sum_{i=0}^{m} \Theta_i$, matrix of accumulated orthogonalized impulse responses

Ξ_∞ $:= \sum_{i=0}^{\infty} \Theta_i$, matrix of total or long-run orthogonalized impulse responses
$\omega_{jk,h}$ proportion of h-step forecast error variance of variable j accounted for by innovations in variable k

Moment Matrices

Γ $:= \text{plim } ZZ'/T$
$\Gamma_y(h)$ $:= \text{Cov}(y_t, y_{t-h})$ for a stationary process y_t
$R_y(h)$ correlation matrix corresponding to $\Gamma_y(h)$
Σ_u $:= E(u_t u_t') = \text{Cov}(u_t)$, white noise covariance matrix
Σ_y $:= E(y_t - \mu)(y_t - \mu)' = \text{Cov}(y_t)$, covariance matrix of a stationary process y_t
P lower triangular Choleski decomposition of Σ_u
$\Sigma_{\hat{\alpha}}$ covariance matrix of the asymptotic distribution of $\sqrt{T}(\hat{\alpha} - \alpha)$
$\Omega(h)$ correction term for MSE matrix of h-step forecast
$\Sigma_y(h)$ MSE or forecast error covariance matrix of h-step forecast of y_t
$\Sigma_{\hat{y}}(h)$ approximate MSE matrix of h-step forecast of estimated process y_t

Author Index

Abraham, B., 412, 509
Ahn, S.K., 139, 146, 152, 364, 372, 378, 385, 386, 509, 516
Aigner, D.J., 511
Akaike, H., 128, 129, 298, 509
Alavi, A . S., 298, 513
Anděl, J., 408, 428, 509
Andersen, A. P., 427, 511
Anderson, B.D.O., 416, 432, 439, 509
Anderson, O.D., 512, 513
Anderson, T.W., 95, 161, 332, 423, 449, 480, 509
Andrews, D.W.K., 379, 509
Ansley, C.F., 421, 509
Aoki, M., 298, 420, 509

Baillie, R.T., 338, 509
Baringhaus, L., 155, 509
Barone, P., 24, 509
Basu, A.K., 87, 377, 509
Bell, W.A., 386, 510
Bera, A.K., 155, 513
Berk, K.N., 306, 510
Bewley, T.F., 512
Bhansali, R.J., 306, 510
Bierens, H.J., 333, 510
Bock, M.E., 514
Boes, D.C., 514
Bollerslev, T., 428, 510, 511
Box, G.E.P., 152, 284, 286, 298, 334, 412, 510, 512, 514, 517
Brelsford, W.M., 408, 513
Breusch, T.S., 494, 510
Brockwell, P.J., 230, 510
Broze, L., 330, 510
Bucy, R.S., 428, 513

Caines, P.E., 439, 510
Chitturi, R.V., 139, 510
Chow, G.C., 341, 342, 345, 426, 510
Cipra, T., 408, 510
Cleveland, W.P., 408, 510
Cooley, T.F., 426, 510
Cooper, D.M., 298, 510
Craig, A.T., 480, 512
Crowder, M.J., 439, 510
Csáki, F., 509

Davidson, J.E.H., 353, 510
Davies, N., 152, 510
Davis, R.A., 230, 510
Deaton, A.S., 512
Deistler, M., 219, 248, 259, 267, 268, 271, 289, 290, 291, 292, 294, 295, 332, 333, 420, 510, 512
Dempster, A.P., 437, 510
Dent, W., 95, 511
Dickey, D.A., 386, 510, 511
Diebold, F.X., 428, 510
Doan, T., 208, 373, 426, 511
Dufour, J.-M., 86, 142, 511
Dunsmuir, W.T.M., 271, 511
Durlauf, S.N., 364, 371, 389, 515

Engle, R.F., 325, 352, 378, 386, 423, 424, 428, 437, 510, 511, 517

Fountis, N.G., 386, 511
Friedmann, R., 342, 511
Fuller, W.A., 66, 77, 141, 386, 488, 490, 491, 510, 511

Gabr, M.M., 427, 516
Gallant, A.R., 333, 493, 511
Gani, J., 515
Gasser, T., 154, 511
Geweke, J., 43, 95, 423, 511
Gladyshev, E.G., 408, 511
Goldberger, A.S., 511
Gouriéroux, C., 330, 510
Granger, C.W.J., 28, 35, 43, 298, 352, 386, 427, 428, 511, 512
Graybill, F.A., 449, 512, 514
Gregory, A.W., 494, 512

Griffiths, W.E., 513, 514
Griliches, Z., 510, 511, 512
Grupe, M.R., 408, 517

Hackl, P., 516
Haggan, V., 428, 512
Hamilton, J.D., 439, 512
Hannan, E.J., 20, 124, 130, 132, 141, 219, 248, 267, 268, 271, 286, 289, 290, 291, 292, 294, 295, 332, 333, 420, 510, 511, 512, 515
Hansen, B.E., 371, 372, 515
Hanssens, D.M., 334, 514
Harvey, A.C., 332, 421, 422, 512
Hasza, D.P., 87, 516
Haugh, L.D., 334, 512
Hausman, J.A., 333, 512
Hendry, D.F., 325, 353, 510, 511, 512
Henze, N., 155, 509
Hildenbrand, W., 511
Hildreth, C., 426, 512
Hill, R.C., 513
Hillmer, S.C., 255, 259, 512
Hogg, R.V., 480, 512
Hosking, J.R.M., 139, 152, 298, 300, 512, 513
Houck, J.P., 426, 512
Hsiao, C., 183, 513
Hui, Y.V., 408, 513
Hylleberg, S., 343, 355, 513

Intriligator, M.D., 510, 511, 512

Jakeman, A.J., 334, 517
Jarque, C.M., 155, 513
Jazwinski, A.H., 416, 513
Jenkins, G.M., 284, 286, 298, 334, 510, 513
Johansen, S., 355, 364, 371, 372, 385, 386, 513
Jones, R.H., 408, 421, 513
Judge, G.G., 211, 262, 263, 264, 333, 334, 480, 484, 513
Juselius, K., 372, 385, 386, 513

Kalman, R.E., 428, 513
Kavalieris, L., 267, 268, 289, 294, 333, 512
Kitagawa, G., 421, 513
Kmenta, J., 516
Kohn, R., 255, 271, 298, 421, 509, 513
Kraft, D., 428, 511
Krishnaiah, P.R., 512, 515

Lainiotis, D.G., 509
Laird, N.M., 437, 510
Lee, T.-C., 513
Lewis, R., 306, 310, 513
Li, W.K., 139, 300, 408, 513
Litterman, R.B., 208, 373, 426, 511, 513
Liu, J., 427, 513
Liu, L.-M., 334, 514
Ljung, G.M., 152, 514
Lucas, R.E., Jr., 330, 514
Lütkepohl, H., 56, 98, 103, 120, 138, 156, 161, 162, 163, 164, 183, 231, 233, 234, 235, 236, 308, 310, 314, 315, 316, 318, 372, 400, 406, 408, 472, 513, 514

MacDonald, G.M., 155, 517
Magnus, J.R., 198, 449, 454, 475, 476, 514
Mann, H.B., 66, 514
Mardia, K.V., 155, 514
McLeod, A.I., 139, 300, 513
McMurtrie, R., 334, 517
Meese, R., 95, 511
Mehra, R.K., 509
Meinhold, R.J., 415, 432, 514
Miller, R.B., 386, 510
Mittnik, S., 227, 514
Mizon, G.E., 355, 513
Mood, A.M., 514
Moore, J.B., 416, 432, 439, 509
Morrison, D.F., 423, 514
Murata, Y., 341, 342, 472, 514
Muth, J.F., 328, 514

Nankervis, J.C., 75, 515
Nerlove, M., 428, 510
Neudecker, H., 198, 449, 454, 475, 476, 514
Newbold, P., 28, 152, 298, 510, 511
Nicholls, D.F., 75, 426, 438, 439, 515
Nijman, T.E., 421, 515

Osborn, D.R., 408, 515
Ouliaris, S., 386, 515
Ozaki, T., 428, 512, 515

Pagan, A., 426, 438, 439, 515
Pagano, M., 408, 515
Palm, F., 285, 287, 517
Pankratz, A., 515
Park, J.Y., 332, 364, 369, 370, 371, 372, 515
Paulsen, J., 75, 79, 126, 130, 131, 383, 515, 517
Penm, J.H.W., 180, 183, 515
Pesaran, M.H., 330, 515
Petrov, B.N., 509
Phillips, P.C.B., 332, 364, 369, 370, 371, 372, 386, 389, 515
Pierse, R.G., 421, 512
Pope, A. L., 75, 515

Poskitt, D.S., 268, 286, 289, 295, 297, 298, 300, 308, 314, 315, 333, 334, 514, 515, 516
Pötscher, B.M., 259, 510
Prescott, E., 426, 510
Priestley, M.B., 428, 515, 516
Pukkila, T., 510
Puntanen, S., 510

Quenouille, M.H., 297, 516
Quinn, B.G., 130, 131, 132, 134, 426, 512, 515, 516

Ramsey, J.B., 516
Rao, C.R., 480, 516
Rao, M.M., 515
Reimers, H.-E., 372, 514
Reinsel, G.C., 194, 198, 199, 201, 306, 310, 364, 372, 378, 385, 386, 509, 513, 516, 517
Richard, J.F., 325, 511
Rissanen, J., 286, 512
Rohatgi, V.K., 516
Roussas, G.G., 484, 516
Roy, R., 142, 511
Rubin, D.B., 437, 510

Salmon, M., 353, 516
Samaranayake, V.A., 87, 516
Sargent, T.J., 330, 514
Savin, N.E., 75, 515
Schlittgen, R., 516
Schmidt, P., 339, 487, 494, 510, 516
Schneider, W., 156, 432, 437, 438, 439, 514, 516
Schwarz, G., 132, 516
Searle, S.R., 449, 516
Sen Roy, S., 87, 377, 509
Serfling, R.J., 484, 488, 516
Shibata, R., 133, 516
Sims, C.A., 58, 59, 208, 369, 373, 426, 511, 516
Singpurwalla, N.D., 415, 432, 514
Smith, J.P., 408, 515
Solo, V., 435, 516
Srba, F., 353, 510
Stensholt, B.K., 427, 516
Stock, J.H., 369, 371, 372, 386, 516
Streitberg, B.H.J., 516
Subba Rao, T., 427, 516
Swamy, P.A.V.B., 426, 517
Szafarz, A., 330, 510

Terrell, R.D., 180, 183, 515
Theil, H., 52, 517
Tiao, G.C., 255, 259, 297, 298, 408, 412, 510, 512, 517
Tjøstheim, D., 75, 79, 126, 131, 427, 515, 516, 517
Todd, P.H.J., 421, 422, 512
Tong, H., 428, 517
Tremayne, A.R., 298, 300, 334, 515, 516
Triggs, C.M., 152, 510
Tsay, R.S., 298, 334, 383, 412, 517
Tso, M.K.-S., 477, 517

Ungern-Sternberg, T. von, 353, 512

Veall, M.R., 494, 512
Velu, R.P., 198, 199, 201, 517

Wald, A., 66, 514
Wallis, K.F., 285, 517
Watanabe, N., 439, 517
Watson, M.W., 369, 386, 423, 424, 437, 511, 516, 517
Wei, W.W.S., 420, 517
Westlund, A., 516
White, H., 155, 333, 511, 517
Wichern, D.W., 198, 199, 201, 517
Wold, H., 20, 517
Wood, E.F., 298, 510
Wooldridge, J.M., 428, 510

Yamamoto, T., 338, 517
Yeo, S., 353, 510
Yoo, B.S., 378, 386, 511
Young, P.C., 334, 517

Zellner, A., 285, 287, 517

Subject Index

Absolutely summable sequence, 488, 490
Accumulated forecast error impulse responses, 48, 97
Accumulated impulse responses, 48, 97
Accumulated orthogonalized impulse responses, 97
Adjoint of a matrix, 453
Aggregation
  contemporaneous —, 230–231, 235
  — of MA processes, 231–232
  — of VARMA processes, 235–236
  temporal —, 230–231, 235–236, 420–421
AIC, 129, 181, 202
Akaike's information criterion (*see* AIC)
Almost sure convergence, 485
Analysis
  Bayesian —, 206–212, 372–375
  causality —, 35–43, 236–238, 378–379
  impulse response —, 43–56, 178–179, 238, 281, 379–382
  multiplier —, 43–44, 48, 327, 338–339
  — of multiple time series, 1–2
  structural —, 35–59
AR process (*see* autoregressive process)
ARCH model, 428
Asymptotic distribution
  — of autocorrelations, 140
  — of autocovariances, 140
  — of EGLS estimator, 170–171
  — of forecast error variance decomposition, 98–100, 314–315
  — of GLS estimator, 170
  — of impulse responses, 98–100, 178–179, 314–315, 379–382
  — of kurtosis, 153, 156
  — of LM statistic, 300, 493
  — of LR statistic, 123, 493
  — of LS estimator, 66, 306–307
  — of ML estimator, 85, 174, 271–278, 358, 438–439
  — of multivariate LS estimator, 66, 306–307
  — of residual autocorrelations, 147, 187
  — of residual autocovariances, 146, 187
  — of skewness, 153, 156
  — of Wald statistic, 94, 379, 493
  — of Yule-Walker estimator, 79
Asymptotic information matrix, 492
Asymptotic normal distribution, 487
Asymptotic normality, 487
Asymptotic properties
  — of estimated forecasts, 85–93, 309–311, 337–338
  — of estimators, 487–488
  — of impulse responses, 98–100, 281, 379–380
  — of LM statistic, 121–125, 493
  — of LR statistic, 299–300, 493
  — of LS estimator, 65–70, 305–309, 369
  — of ML estimator, 82–85, 174, 271–278, 358, 438–439
  — of sample mean, 77
  — of Wald statistic, 94, 379, 493
  — of white noise covariance matrix estimator, 68–69, 174, 271–272, 358
  — of Yule-Walker estimator, 79
Asymptotically stationary process, 349
Autocorrelation function
  asymptotic distribution of —, 138–152
  computation of —, 25–27
  definition of —, 25
  estimation of —, 138–152
  — of residuals, 142–150, 186–188
  — of a stochastic process, 25
  — of a VAR process, 25–27
  — of a VARMA process, 227
Autocorrelation matrix
  computation of —, 25–27
  definition of —, 25
  estimation of —, 138–152
  — of residuals, 142–150, 186–188
  — of a VAR process, 25–27
  — of a VARMA process, 227
Autocovariance function
  asymptotic distribution of —, 138–152

Autocovariance function (*continued*)
computation of —, 21–25
definition of —, 10
estimation of —, 138–150
— of MA process, 219
— of residuals, 142–150, 186–188
— of a VAR process, 11, 21–25
— of a VAR(1) process, 10, 21–23
— of a VARMA process, 226–228
Autoregressive conditional heteroscedasticity, 428
Autoregressive process, 4
(*see also* VAR process)
Autoregressive representation of VARMA process
infinite order —, 222
pure —, 222

Backshift operator (*see* lag operator)
Backward shift operator (*see* lag operator)
Bayesian estimation
basics of —, 206
— of Gaussian VAR process, 206–212
— of integrated systems, 372–375
— with normal priors, 206–208, 372–374
Bilinear state space model, 427
Bilinear time series model, 426
Bonferroni's inequality, 34
Bonferroni's method, 34–35
Bootstrap, 497
Bottom-up specification of subset VAR model, 182–183
Box-Jenkins methodology, 284, 286

Canonical MA representation, 16
Causality, 35–43, 45, 93–97, 236–238, 378–379
definition of —, 35–37
Granger —, 36
instantaneous —, 36
Wold —, 52
Chain rule for vector differentiation, 469
Characteristic
— determinant, 455
— polynomial
— — of a matrix, 455
— — of an operator, 12
reverse —, 12
— value of a matrix, 455
— vector of a matrix, 456
Checking the adequacy
— of cointegrated systems, 387–388
— of dynamic SEMs, 333–334
— of reduced rank VAR models, 201–202
— of subset VAR models, 186–189
— of VAR models, 138–164
— of VARMA models, 298–301
Checking the whiteness
— of VAR residuals, 138–152
— of VARMA residuals, 300–301
Chi-square distribution, 481
noncentral —, 482
Choleski decomposition of a positive definite matrix, 462
Closed-loop control strategy, 324
Cofactor of an element of a square matrix, 452
Cointegrated process, 351–355
— of order (d, b), 352
Cointegrated system, 351–355
Cointegrated VAR process, 355
checking the adequacy of —, 387–388
estimation of —, 355–372
forecasting of —, 375–378
Granger-causality in —, 378
impulse response analysis of —, 379–382
least squares estimation of —, 369–370
ML estimation of —, 356–368
structural analysis of —, 378–382
two-stage estimation of —, 370–372
Cointegrating
— matrix, 355
— vector, 352
Cointegration matrix, 355
Cointegration rank, 355
LR test for —, 384–386
testing for —, 384–387
Column vector, 449
Commutation matrix, 466
Complex matrix, 460
Complex number
modulus of —, 456
Conditional expectation, 28–30
Conditional forecast, 335
Conditional likelihood function, 256
Conditional normal distribution, 480–481
Confidence interval
— for forecast error variance components, 101–102
— for forecasts, 33–35
— for impulse responses, 100
Confidence region for forecasts, 33–35
Consistency, 487
— in quadratic mean, 487
mean square —, 487
strong —, 487
super —, 359, 371
weak —, 487

Consistent estimation
— of Kronecker indices, 292
— of VAR order, 130–132, 383
— of white noise covariance matrix, 69
Constrained VAR models
linear constraints, 168–192
nonlinear constraints, 192–205
Constraints (*see* restrictions)
Contemporaneous aggregation of VARMA process, 235
Control
optimal — (*see* optimal control)
Convergence
almost sure —, 485
— in distribution, 485
— in law, 485
— in mean square, 485
— in probability, 484
— in quadratic mean, 485
stochastic —, 484–487
strong —, 485
weak —, 485
— with probability one, 485
Correlation
— function (*see* autocorrelation function)
— matrix, 25
Covariance
— function (*see* autocovariance function)
— matrix, 10

Data generating process, 3
Data generation process, 3
Decomposition of matrices, 459–462
Choleski —, 462
Jordan —, 459–460
Definite
— matrix, 458–459
— quadratic form, 458–459
Degree
McMillan —, 247
Determinant of a matrix, 451
Determinantal polynomial of a VAR process, 12
Deterministic
— component of a stochastic process, 20
— trend, 347
Diagnostic checking
— of cointegrated systems, 387–388
— of dynamic SEMs, 333–334
— of reduced rank VAR models, 201–202
— of restricted VAR models, 186–188
— of VAR models, 138–164
— of VARMA models, 298–301
Diagonal matrix, 450
Diagonalization of a matrix, 460–461
Difference operator, 350
Differencing, 350
Differentiation of vectors and matrices, 467–474
Direct product (*see* Kronecker product)
Direction matrix, 263
Discrete stochastic process, 3
Distributed lag model, 323, 326–327
rational —, 326–327
Distribution
asymptotic —, 487
χ^2 —, 481
conditional —, 481
F —, 482
multivariate normal —, 480–481
normal —, 480–481
posterior —, 206
prior —, 206
t —, 482
Distribution function, 2
— of a random variable, 2
— of a random vector, 2
Drift of a random walk, 347
Dummy variable, 391
seasonal —, 391
Duplication matrix, 465
Dynamic
— factor analytic model, 423
— MIMIC model, 424
— multipliers, 44, 327
Dynamic SEM
checking the adequacy of —, 333–334
estimation of —, 331–333
final equations of —, 327
final form of —, 326
forecasting of —, 334–338
conditional —, 335
unconditional —, 335
multipliers of —, 338–339
optimal control of —, 339–342
rational expectations in —, 328–330
reduced form of —, 325
specification of —, 333–334
structural form of —, 325
Dynamic simultaneous equations model (*see* dynamic SEM)

Echelon form
— of a VARMA process, 246–247
specification of —, 289–297
— VARMA representation, 246–247
Econometric dynamic simultaneous equations model (*see* dynamic SEM)

Econometric model (*see* dynamic SEM)
Effect of linear transformations
— on forecast efficiency, 234
— on MA process, 231
— on VARMA orders, 232
Efficiency
— of estimators, 172–173
— of forecasts, 27–28, 234
EGLS estimation
asymptotic properties of —, 170–173
implied restricted —, 171
— of parameters arranged equationwise, 174–175
restricted —, 170–175
Eigenvalue of a matrix, 455
Eigenvector of a matrix, 456
Elimination matrix, 465
EM algorithm, 437
Empirical distribution, 495
generation of —, 495
Endogenous variable, 323–325, 417
Equilibrium, 351–352
Equilibrium relation, 351–352
Error correction model, 353
Error process,
— of the measurement equation of a state space model, 417
— of the transition equation of a state space model, 417
— of a VAR process, 9
Estimated GLS estimation (*see* EGLS estimation)
Estimated generalized least squares estimation (*see* EGLS estimation)
Estimation
Bayesian —, 206–212
EGLS —, 170
GLS —, 169
least squares — (see LS —)
LS —, 62–75, 171
maximum likelihood —, 80–85, 173–174, 356–368, 394–396, 491–492
multivariate least squares —, 62–75, 305–309
— of autocorrelations, 139–150
— of autocovariances, 139–150
— of cointegrated systems, 355–372
— of dynamic SEMs, 331–333
— of integrated VAR processes, 372–375
— of periodic models, 399–402
— of process mean, 76–78
— of reduced rank VAR model, 197–201
— of state space models, 434–439
— of time varying coefficient models, 394–396
— of VAR models, 62–85, 305–309
— of VARMA models, 241–283
— of white noise covariance matrix, 67–70, 171–172
preliminary —, 265–268
restricted —, 169–177, 194, 197–201
EGLS —, 169–173
GLS —, 169–173
two-stage —, 370–372
— with linear restrictions, 169–177
— with nonlinear restrictions, 194, 197–201
— with unknown process mean, 78
Yule-Walker —, 78–79
Euclidean norm, 490
Exact likelihood function, 252–254
Exogenous variable, 323–325
systems with —, 324–325
Expectation
conditional —, 28–30
rational —, 328
Expectation step-maximization step algorithm (*see* EM algorithm)

F-distribution, 482
noncentral —, 482
Factor analytic model
dynamic —, 423
Factor loadings, 423
Feedback system, 36
Filter
Kalman —, 428–434
Filtering, 428
Final equations form
— of a dynamic simultaneous equations model, 327
specification of —, 285–289
— VARMA representation, 246
Final form of a dynamic simultaneous equations system, 326
Final prediction error criterion (*see* FPE criterion)
Finite order MA process, 217–220
Finite order VAR process, 9
Forecast
conditional —, 335
estimated —, 85–93, 278–281, 309–313
h-step —, 27
interval —, 33–35
minimum MSE —, 28–32
— of VAR process, 29–35, 85–93, 309–313
— of VARMA process, 228–230, 278–281
optimal —, 27–28

point —, 28–33
unconditional —, 335
Forecast efficiency, 27–32
Forecast error, 28–32, 85–86
— covariance matrix, 31–32
— impulse responses, 43–48
— MA representation, 16
— variance, 56–58
— — component, 57, 97
proportion of —, 57
Forecast error variance decomposition, 56–58, 313–317
asymptotic distribution of —, 97–104, 178–179, 314–315
critique of —, 58
interpretation of —, 56–58
Forecast horizon, 27
Forecast interval, 33–35, 89
Forecast MSE matrix, 28, 31–32, 230
approximate —, 87–89, 279–281, 310
Forecast origin, 27
Forecast region, 33–35, 89
Forecasting, 27–35
— of cointegrated systems, 375–378
— of dynamic SEMs, 334–338
— of estimated VAR process, 85–93, 177–178
— of estimated VARMA process, 278–281
— of infinite order VAR process, 309–313
— of integrated systems, 375–378
— of MA process, 32
— of restricted VAR process, 177–178
— of VAR process, 27–35
— of VARMA process, 228–230
FPE criterion, 128
Fundamental MA representation, 16

Gain (*see* Kalman gain)
Gaussian likelihood function
— of cointegrated process, 356
— of MA process, 252–255
— of state space model, 434–435
— of VAR process, 80–81
— of VARMA process, 255–259
Gaussian process, 12, 33
VAR, 33
VARMA, 221
white noise, 12, 67
General linear state space model (*see* state space model)
Generalized inverse of a matrix, 454
Generalized least squares estimation (*see* GLS estimation)
Generating process of a time series, 3
Generation process of a time series, 3
Global identification, 435–436
Globally identified model, 435–436
GLS estimation, 169–173
asymptotic properties of —, 170
Gradient
— algorithm, 262
— of log-likelihood function, 437
— of vector function, 262
Granger-causality, 35–43
characterization of —, 38, 378
critique of —, 41–43
definition of —, 36
— in cointegrated systems, 378–379
— in VAR models, 37–39, 93–95
— in VARMA models, 236–238
interpretation of —, 41–43
test for —, 93–95, 378–379
Wald test for —, 93–95, 378–379

Hannan-Kavalieris procedure, 294–295
Hannan-Quinn criterion (*see* HQ criterion)
Hessian matrix, 468
HQ criterion, 132, 181, 202

Idempotent matrix, 457
Identifiable model (*see* identified model)
Identification
global —, 436
local —, 436
— of dynamic simultaneous equations system, 333
— of state space model, 435–436
— of VARMA model, 241–252
— of VARMAX model, 333
Identification problem, 241–252
Identified model
globally —, 436
locally —, 436
state space, 435–436
VARMA, 241–252
Identity matrix, 450
Impact multiplier, 54
Impulse response, 43–56
accumulated —, 48
asymptotic distribution of —, 97–104, 178–179, 314–315
estimation of —, 97, 178–179
forecast error —, 43–48
long-run —, 48
— of cointegrated systems, 379–382
— of VAR models, 43–56
— of VARMA models, 238, 281
orthogonal —, 48–55
orthogonalized —, 48–55

Impulse response (*continued*)
total —, 48
zero —, 45, 55
Impulse response analysis, 43–56
critique of —, 55–56
— of cointegrated systems, 379–382
— of VAR models, 43–56
— of VARMA models, 238, 281
Indefinite
— matrix, 459
— quadratic form, 459
Independent white noise, 29
Index model, 194
Infinite order MA representation
— of a stationary process, 20
— of a time varying coefficient process, 393
— of a VAR process, 13–19
— of a VARMA process, 221
Infinite order VAR representation
— of an MA process, 218
— of a VARMA process, 222
Infinite sum of random variables
convergence of —, 488–491
definition of —, 488–491
Information criterion (*see* AIC)
Information matrix, 492
asymptotic —, 492
— of state space model, 437
— of time varying coefficient VAR model, 396
— of VAR process, 82–85
— of VARMA process, 264–265
Initial
— estimator (*see* preliminary estimator of VARMA process)
— input, 417
— state, 417
Innovation
— accounting, 58
forecast error —, 43–48
orthogonalized —, 48–55
— process, 9
Input
— matrix of a state space model, 417
observable —, 324, 417
unobservable —, 324
— variables, 324
Inputs of a state space model, 417
Instantaneous causality, 35–43
characterization of —, 40–41
critique of —, 41–43
definition of —, 36
interpretation of —, 41–43
tests for —, 95–97
Instantaneous effect, 52
Instrument
observable —, 324, 417
— variable, 324, 417
Instrumental variable estimation, 372
Integrated
— of order d, 350
— process, 346–351
— time series, 346–351
— variable, 346–351
Integration
order of —, 350
Intercept VAR model, 9
Interim multipliers, 48, 327
Interpretation
classical versus Bayesian —, 211–212
— of forecast error variance decomposition, 56–58
— of Granger-causality, 41–43
— of impulse responses, 55–56
— of instantaneous causality, 41–43
— of VAR model, 58–59
— of VARMA model, 236–238
Interval forecast, 33–35, 89
Intervention
— in intercept model, 408–410
testing for —, 409–410
Intervention model, 392
estimation of —, 408–411
specification of —, 408–411
Inverse
— of a matrix, 453
— of an operator, 17–18
Invertible
— MA operator, 219
— MA process, 218–219
— matrix, 453
— operator, 17
— VARMA process, 222
Iterative optimization algorithm
EM algorithm, 437
Newton algorithm, 263
scoring algorithm, 264, 436–438

Jordan canonical form, 460

Kalman filter, 428–434
— correction step, 430
— forecasting step, 430
— gain, 430
— initialization, 430
— prediction step, 430
— recursions, 429–432
— smoothing step, 432

Kalman gain, 430
Kalman smoothing matrix, 432
Kronecker indices, 247
determination of —, 289–297
estimation of —, 289–297
— of echelon form, 247
— of VARMA process, 247
specification of —, 289–297
Kronecker product, 463
Kurtosis
asymptotic distribution of —, 153, 156
measure of multivariate —, 152–158

Lag operator, 17
Lagrange function, 474
Lagrange multipliers, 474
Lagrange multiplier statistic, 299–300, 404–405, 493
asymptotic distribution of —, 300, 405, 493
Lagrange multiplier test, 298–300, 404–405, 493
Least squares estimation
asymptotic properties of —, 65–70, 171–173, 198–201, 306–307
multivariate —, 62–75, 197–201, 305–309
— of cointegrated VAR process, 369–370
— of reduced rank VAR model, 197–201
— of VAR process, 62–75, 305–309
restricted —, 171–173
small sample properties of —, 73–75
— with mean-adjusted data, 75–78
Least squares estimator of white noise covariance matrix, 67–70, 308–309
asymptotic properties of —, 68, 308–309
Left-coprime operator, 245
Likelihood function, 491
conditional —, 256
— of cointegrated process, 356
— of MA process, 252–255
— of state space model, 434–435
— of time varying coefficient VAR model, 395
— of VAR process, 80–81
— of VARMA process, 255–259
Likelihood ratio statistic
asymptotic distribution of —, 123, 493
definition of —, 121, 493
Likelihood ratio test, 493
— for cointegration rank, 384–387
— of linear restrictions, 121–125
— of nonlinear restrictions, 492–494
— of periodicity, 402–403
— of varying parameters, 400–403
— of zero restrictions, 121–125
Limiting distribution (*see* asymptotic distribution)
Linear constraints
— for VAR coefficients, 168–169
Linear minimum MSE predictor, 30–33
Linear state space system, 415–418
Linear system, 323
Linear transformation
— of MA process, 231–232
— of multivariate normal distribution, 481
— of VARMA process, 232–235
Linearly dependent vectors, 455
Linearly independent vectors, 455
Litterman prior
— for nonstationary process, 373–375
— for stationary process, 208–210
LM statistic (*see* Lagrange multiplier statistic)
LM test (*see* Lagrange multiplier test)
Loading matrix, 355
Locally identified model, 436
Log-likelihood function, 491
Long-run
— effect, 48, 327
— multiplier, 48, 327
Loss function, 27–28
— of forecast, 27–28
quadratic —, 340
LR statistic (*see* likelihood ratio statistic)
LR test (*see* likelihood ratio test)
LS estimation of VAR process (*see* least squares estimation of VAR process)

MA operator, 219
MA process,
autocovariances of —, 219
finite order —, 217–220
infinite order —, 13–19
invertible —, 218–219
likelihood function of —, 252–255
MA representation
canonical —, 16, 223
forecast error —, 16, 223
fundamental —, 16
— of a stationary process, 20
— of a VAR process, 13–19
— of a VARMA process, 221
prediction error —, 16, 223
Marginal distribution, 480–481
Matrix, 449
— addition, 450
adjoint of —, 453
characteristic determinant of —, 455
characteristic polynomial of —, 455

Matrix (*continued*)
characteristic root of —, 455
characteristic value of —, 455
characteristic vector of —, 456
Choleski decomposition of —, 462
cofactor of an element of —, 452
column dimension of —, 449
commutation —, 466
conformable —, 451
decomposition of —, 459–462
determinant of —, 451
diagonal —, 450
diagonalization of —, 460–461
— differentiation, 467–474
duplication —, 465
eigenvalue of —, 455
eigenvector of —, 456
element of —, 449
elimination —, 465
full rank —, 455
generalized inverse of —, 454
Hessian —, 468
idempotent —, 457
identity —, 450
indefinite —, 459
inverse of —, 453
invertible —, 453
Jordan canonical form of —, 460
lower triangular —, 450
minor of an element of —, 452
Moore-Penrose inverse of —, 454
— multiplication, 450
— multiplication by a scalar, 450
negative definite —, 459
negative semidefinite —, 459
nilpotent —, 457
nonsingular —, 453
null —, 450
— operations, 450–451
— operator
left-coprime —, 245
unimodular —, 245
orthogonal —, 458
partitioned —, 462
positive definite —, 458
positive semidefinite —, 458
rank of —, 455
regular —, 453
row dimension of —, 449
— rules, 449–477
square —, 449
square root of —, 461
— subtraction, 450
symmetric —, 450
trace of —, 457
transpose of —, 450
triangular —, 450
typical element of —, 449
unit —, 450
upper triangular —, 450
zero —, 450
Maximum likelihood estimation (*see* ML estimation)
McMillan degree
— of echelon form, 247
— of VARMA process, 247
Mean-adjusted
— process, 75
— VAR process, 75
Mean square convergence, 485
Mean squared error of forecast (*see* forecast MSE)
Mean vector of a VAR process, 11, 75
Measurement
— errors, 417
— equation of state space model, 415, 417
— matrix, 417
MIMIC models, 424
Minimization
— algorithms, 262–264
iterative —, 262–264
numerical —, 262–264
Minnesota prior (*see* Litterman prior)
Minor of an element of a square matrix, 452
ML estimates
computation of —, 81–82, 259–270, 434–439
ML estimation, 491–492
— of cointegrated system, 356–368
— of periodic VAR process, 399–402
— of restricted VAR process, 173–174
— of state space model, 434–439
— of VAR process, 80–85
— of VAR process with time varying coefficients, 394–396
— of VARMA process, 252–278
pseudo —, 123
Model checking
— of cointegrated systems, 387–388
— of dynamic SEMs, 333–334
— of reduced rank VAR models, 201–202
— of restricted VAR models, 186–189
— of state space models, 441
— of subset VAR models, 186–189
— of VAR models, 138–164
— of VARMA models, 298–301
Model selection
— of cointegrated processes, 382–387

— of reduced rank VAR models, 201–202
— of subset VAR models, 179–185
— of VAR models, 118–138
— of VARMA models, 284–298
Model specification
— of cointegrated processes, 382–387
— of dynamic SEMs, 333–334
— of periodic VAR models, 399–408
— of reduced rank VAR models, 201–202
— of subset VAR models, 179–185
— of VAR models, 118–138
— of VARMA models, 284–298
Model specification criteria
AIC, 129, 181, 202
FPE, 128
HQ, 132, 181, 202
SC, 132, 181, 202
Modified portmanteau statistic, 152, 188
approximate distribution of —, 152, 188
Modified portmanteau test, 152, 188, 300
Modulus of a complex number, 456
Moore-Penrose (generalized) inverse, 454
Moving average
— process (*see* MA process)
— representation (*see* MA representation)
MSE matrix, 28, 31–32, 230
approximate —, 87–89, 279–281, 310
MSE of forecast, 28, 31–32, 87–89, 230, 279–281, 310
Multiple time series, 2
Multiplicative operator, 193–194
Multiplier
— analysis, 43, 327, 338–339
dynamic —, 44, 327
impact —, 54
interim —, 48, 327
long-run —, 48, 327
total —, 48, 327
Multivariate
— autoregressive process (*see* VAR process)
— least squares estimation, 62–79
— of infinite order VAR process, 305–309
— of reduced rank VAR process, 197–201
— of VAR process, 62–75
— normal distribution, 480–481
— stochastic process, 3
— time series (*see* multiple time series)
— white noise process (*see* vector white noise)

Negative definite
— matrix, 459
— quadratic form, 459
Negative semidefinite matrix, 459
Newton algorithm, 263
Newton-Raphson algorithm, 263
Nilpotent matrix, 457
Noise (*see* white noise)
Noncentral
— χ^2-distribution, 482
— F-distribution, 482
Noncentrality, 482
Noncentrality parameter
— of a χ^2-distribution, 482
— of an F-distribution, 482
Nondeterministic
— process, 32
purely —, 32
Nonlinear
— optimization algorithm (*see* iterative optimization algorithm)
— parameter restrictions, 192–197
— state space model, 426–428
Nonnegative definite matrix (*see* positive semidefinite matrix)
Nonnormality
tests for —, 152–158
Nonsingular matrix, 453
Nonstationary
— process, 346, 391–392, 418, 424–426
— time series, 346
— VAR process, 350, 355–356, 391–392
Normal distribution
distributions related to —, 481–483
multivariate —, 480–481
properties of —, 480–481
Normal equations
— for VAR coefficient estimates, 64
— for VAR process with time varying coefficients, 395
— for VARMA estimation, 260–262
Normal prior for Gaussian VAR process, 206–208, 372

Observable
— input, 324
— output, 324
— variables, 324
Observation
— equation of state space model, 415, 417
— error, 415, 417
— noise, 417
OLS estimation (*see* least squares estimation)
Open-loop strategy, 342
Operator
left-coprime —, 245
MA —, 219
unimodular —, 245
VAR —, 17

Optimal
— forecast, 28–33
— linear predictor, 30–33
— predictor, 28–33
Optimal control, 339–342
closed-loop —, 342
open-loop —, 342
problem of —, 341
Optimization
— algorithms, 262–264
— of vector functions, 474–477
Order of
— MA process, 218
— VAR process, 9, 119
— VARMA process, 220
Order determination
criteria for —, 128–138
— for cointegrated process, 382–384
— for VAR process, 118–138
tests for —, 119–127
Order estimation
consistent —, 130–132
criteria for —, 128–138
— for cointegrated processes, 382–384
— of VAR process, 128–138
Ordinary least squares estimation (*see* least squares estimation)
Orthogonal
— innovations, 40
— matrix, 458
— residuals, 40
— vectors, 458
— white noise, 40
Orthogonalized
— impulses, 48–55
— innovations, 48–55
Orthogonalized impulse responses, 48–55
accumulated —, 97
Orthonormal vectors, 458
Outlier, 412
Output
observable —, 324
— of a state space system, 417

Parameter constraints (*see* restrictions)
Partitioned matrix, 462
rules for —, 462–463
Period of a stochastic process, 396
Periodic process (*see* periodic VAR process)
Periodic VAR process
— definition of, 396–399
— estimation of, 399–402
— specification of, 399–408
Point forecast, 28–33
Policy
— simulation, 338
— variable, 417
Portmanteau statistic, 150–152, 188, 300
approximate distribution of —, 150–151, 188, 301
modified —, 152, 188, 300
Portmanteau test, 150–152, 188, 300–301
modified —, 152, 188, 300–301
Positive definite
— matrix, 458
— quadratic form, 458
Positive semidefinite matrix, 458
Poskitt's procedure, 295–297
Posterior
— density, 206
— mean, 206
— p.d.f., 206
Postmultiplication, 451
Predetermined variable, 324
Prediction (*see* forecasting)
Prediction error representation, 16
Prediction tests for structural change
— based on one forecast, 159–161
— based on several forecasts, 161–163
— for cointegrated systems, 387–388
— of VAR processes, 159–164
— of VARMA processes, 301
Predictor (*see* forecast)
Preliminary estimation of VARMA process, 265–268
Preliminary estimator of VARMA process, 266–268
Premultiplication, 451
Presample values, 62
Prior
normal —, 206–208
— p.d.f., 206
Process
ARCH —, 428
cointegrated —, 351–355
Gaussian —, 12, 33
invertible MA —, 218–219
invertible VARMA —, 222
periodic —, 392
stable VAR —, 11–12
stable VARMA —, 221
stationary stochastic —, 19
stochastic —, 3
VAR —, 4, 9
VARMA —, 220–223
white noise —, 9
Product rule for vector differentiation, 469
Pseudo ML estimator, 123

Pure MA representation of a VARMA process, 221
Pure VAR representation of a VARMA process, 222
Purely nondeterministic process, 32

Quadratic form, 458
  distribution of —, 481
  indefinite —, 459
  negative definite —, 459
  negative semidefinite —, 459
  nonnegative definite — (*see* positive semidefinite)
  positive definite —, 458
  positive semidefinite —, 458

Random coefficient VARX model, 424–426
Random variable, 2
  distribution function of —, 2
Random vector, 2
  distribution function of —, 2
Random walk, 346
Random walk with drift, 347
Rank of cointegration
  LR test for —, 384–387
  testing for —, 384–387
Rank of a matrix, 455
Ratio of quadratic forms, 482
  distribution of —, 482
Rational
— distributed lag model, 326–327
— expectations, 328
— transfer function, 327
— transfer function model, 327
Real matrices, 449
Recursions
  Kalman filter —, 429–432
Recursive computation
— of derivatives, 260–261
— of forecasts, 29, 31
— of residuals, 269
Recursive model, 52
Reduced form of a dynamic SEM, 54, 325
Reduced rank VAR model, 195–205
  asymptotic properties of estimators of —, 198–201
  checking of —, 201–202
  estimation of —, 197–201
  forecasting of —, 201
  LS estimator of —, 197–201
  multivariate LS estimation of —, 197–201
  specification of —, 201–202
  structural analysis of —, 201
Regular matrix, 453
Resampling, 497
Resampling technique, 497
Residual autocorrelation
  asymptotic properties of —, 147, 186–187
  estimation of —, 142–150, 186–187
— of VAR process, 142–150, 186–187
— of VARMA process, 300–301
Residual autocovariance
  asymptotic properties of —, 146, 186–187
  estimation of —, 142–150, 186–187
— of VAR process, 142–150, 186–187
— of VARMA process, 300–301
Residuals of VAR process
  checking the whiteness of —, 138–152, 188
Residuals of VARMA process
  checking the whiteness of —, 300–301
  estimation of —, 266
Restricted estimation of VAR models, 169–177
  asymptotic properties of —, 170–175
  EGLS, 170–173
  GLS, 169–170
  LS, 171
  ML, 173–174
Restrictions for VAR coefficients
— for individual equations, 174–175
  Granger-causality —, 37–39
  linear —, 168–169
  nonlinear —, 192–197
  tests of —, 95–96, 121–125
  Wald test of —, 95–96
  zero —, 179–185
Restrictions for VARMA coefficients
  Granger-causality —, 236–238
  identifying —, 246–248
  linear —, 257
  LM test of —, 298–300
  tests of —, 298–300
Restrictions on white noise covariance, 40–41, 175–177
Reverse characteristic polynomial of a VAR process, 12
Row vector, 449
RRVAR model (*see* reduced rank VAR model)

Sample
— autocorrelations, 140
— autocovariances, 139
— mean, 76–78
— size (*see* time series length)
Sampling
  systematic — (*see* systematic sampling)
SC, 132, 181, 202

Schwarz criterion (*see* SC)
Score vector, 492
Scoring algorithm, 264, 436–438
Seasonal dummies, 391
Seasonal model, 391
Seasonal operator, 192–193
Seasonal process, 391
Seasonal time series, 391
Second order partial derivatives
  matrix of — (*see* Hessian matrix)
Second order Taylor expansion, 474
SEM (*see also* dynamic SEM), 323
Shift operator (*see* lag operator)
Simulation techniques
  evaluating properties of estimators by —, 495–497
  evaluating properties of test statistics by —, 495–497
Simultaneous equations model (*see* dynamic SEM)
Simultaneous equations system (*see* dynamic SEM)
Skewness
  asymptotic distribution of —, 153, 156
  measure of multivariate —, 152–158
Small sample properties
  investigation of —, 73, 495–497
— of estimated forecasts, 91–93
— of estimators, 495–497
— of LS estimator, 73–75
— of test statistics, 495–497
— of VAR order selection criteria, 133–138
Smoothing, 432
Smoothing matrix
  Kalman —, 432
Smoothing step, 432
Specification of
— cointegrated systems, 382–387
— dynamic SEMs, 333–334
— echelon form, 289–297
— final equations form, 285–289
— reduced rank VAR models, 201–202
— subset VAR models, 179–185
— VAR models, 118–138
— VARMA models, 284–298
Square root of a matrix, 461
Stability
— of a VAR process, 11
— of a VARMA process, 221
Stability condition, 11–12
Stable
— VAR operator, 11–12
— VAR process, 11–12
— VARMA process, 221
Standard form of VAR model, 9
Standard VARMA representation, 242
Standard white noise, 66
State of nature (*see* state vector)
State space model
  estimation of —, 434–439
  identification of —, 435–436
    global —, 436
    local —, 436
  linear —, 415–418
  log-likelihood function of —, 434–435
  ML estimation of —, 434–439
  nonlinear —, 426–428
State space representation
— of factor analytic model, 422–424
— of random coefficient VARX model, 424–426
— of VAR process, 418–419
— of VARMA process, 419–420
— of VARX process, 420
— of VARX process with systematically varying coefficients, 424
State space system (*see* state space model)
State vector, 415, 417
Stationarity, 19
  asymptotic —, 349
— of a stochastic process, 19
Stationarity condition, 20
Stationary point of a function, 474
Stationary stochastic process, 19
  MA representation of —, 20
  VAR process, 19–20
  Wold representation of —, 20
Step direction, 262
Stochastic convergence
  almost surely, 485
  concepts of —, 484–487
— in distribution, 485
— in law, 485
— in probability, 484
— in quadratic mean, 485
  strong —, 485
  weak —, 485
— with probability one, 485
Stochastic process, 2
  cointegrated —, 351–355
  MA, 217–220
  multivariate —, 3
  nonstationary —, 346, 391–392, 418, 424–426
  periodic —, 396–399
  stationary —, 19
  VAR, 4, 9
  VARMA, 220–223

VARMAX, 323
VARX, 323, 420, 424–426
vector —, 3
white noise, 9
Stochastic trend, 347
Strong consistency, 487
Structural analysis
— of cointegrated systems, 378–382
— of dynamic SEMs, 338–339
— of reduced rank VAR models, 204–205
— of subset VAR models, 178–179, 192–193
— of VAR models, 35–59
— of VARMA models, 236–238
Structural change, 159
prediction tests for — (*see* prediction tests for structural change)
testing for —, 159–164, 301, 387–388, 402–405, 411
Structural form of a dynamic SEM, 54, 325
Structural time series model, 421–422
Submatrix, 462
Subprocess, 21
Subset VAR model, 179–192
checking of —, 186–189
specification of —, 179–185
bottom-up strategy, 122–183
top-down strategy, 180–182
structural analysis of —, 192–193
Superconsistent, 359, 371
Superconvergence, 365
Symmetric matrix, 450
System equation, 416
System matrix, 417
System of dynamic simultaneous equations (*see* dynamic SEM)
System with exogenous variables, 324–343
Systematic sampling, 420–421
Systematically varying coefficients
— of VAR models, 391–394
— of VARX models, 424

Taylor expansion, 474
second order —, 474
Taylor's theorem, 473
Temporal aggregation, 230–231, 235–236, 420–421
Testing for
— causal relations, 93–97, 378–379
— Granger-causality, 93–95, 378–379
— instantaneous causality, 95–97
— nonnormality, 152–158
— — of white noise process, 152–155
— — of VAR process, 155–158
— periodicity, 402–408
— rank of cointegration, 384–387
— rank of reduced rank VAR model, 201–202
— structural change, 159–164, 301, 387–388, 402–405, 411
— — based on one forecast period, 159–161
— — based on several forecast periods, 161–163
— whiteness of residuals, 150–152, 188, 300–301
— zero autocorrelation
— — of subset VAR model, 188
— — of VAR process, 150–152
— — of VARMA process, 300–301
— — of white noise process, 139–142
Tests of parameter restrictions, 492–494
linear restrictions, 93–94, 121–125
nonlinear restrictions, 298–300
Threshold models, 428
Time invariant
— autocovariances, 19, 402
— coefficients, 401
— mean, 19
— moments, 19
Time series
analysis of —, 1
— length, 3
multiple —, 2
nonstationary —, 346
seasonal —, 391
stable —, 13–14
univariate —, 1
Time varying
— coefficients, 391–396
— parameters, 391–396
randomly —, 424–426
systematically —, 391–396
Top-down strategy for subset VAR specification, 180–182
Total
— effect of an impulse, 48
— impulse response, 48
— multiplier, 48, 327
Trace of a matrix, 457
Transfer function, 327
Transfer function model, 323, 327
rational —, 327
Transformation
linear —, 231–235, 481
— of MA process, 231–232
— of VARMA process, 232–235
Transition equation
— errors, 417

Transition equation (*continued*)
— noise, 417
— of a state space model, 415, 416
Transition matrix, 417
Transpose of a matrix, 450
Trend
  deterministic —, 347
  stochastic —, 347
Trend component (*see* trend)
Triangular matrix
  lower —, 450
  upper —, 450
Two-stage estimation
  asymptotic properties of —, 371
— of cointegrated systems, 370–372

Unconditional forecast, 335
Unimodular operator, 245
Univariate time series, 1

VAR model (*see* VAR process)
VAR order estimator
  consistent —, 130
  small sample properties of —, 133–138
  strongly consistent —, 130
VAR order selection
  AIC criterion for —, 129
  comparison of criteria for —, 132–138
  consistent —, 130–132
  criteria for —, 128–138
  FPE criterion for —, 128
  HQ criterion for —, 132
  SC criterion for —, 132
  sequence of tests for —, 119–127
  testing scheme for —, 125–126
VAR process, 4, 9
  autocovariances of —, 10–11, 21–25
  autocorrelations of —, 25–27
  checking the adequacy of —, 138–164
  estimation of —, 62–85, 305–309
  finite order —, 9
  forecasting of —, 28–35, 85–93, 309–313
  Granger-causality in —, 35–43
  impulse response analysis of —, 43–56, 97–114, 313–317
  infinite order —, 215, 305–319
  intercept form of —, 9
  interpretation of —, 58–59
  LS estimation of —, 62–79, 305–309
  MA representation of —, 13–19
  mean-adjusted —, 75
  nonstationary —, 355, 392–399
  order estimation of —, 128–138
  order determination of —, 118–138
  reduced rank —, 195–205
  specification of —, 118–138
  stable —, 11–12
  standard form of —, 9
  state space representation of —, 418–419
  stationary —, 19–21
  subset —, 179–192
  unstable —, 355
  VAR(1) representation of —, 11
— with linear parameter restrictions, 168–192
— with nonlinear parameter restrictions, 192–205
— with parameter constraints, 167–214
— with time varying coefficients, 392–396
  Yule-Walker estimation of —, 78–79
VAR representation of a VARMA process
  infinite order —, 222
  pure —, 222
Variance decomposition, 56–58
VARMA model (*see* VARMA process)
VARMA process,
  aggregation of —, 235–236
  autocorrelations of —, 227
  autocovariances of —, 226–228
  checking the adequacy of —, 298–301
  definition of —, 220–223
  echelon form of —, 246–247
  estimation of —, 241–283
  final equations form of —, 246
  forecasting of —, 228–230, 278–281
  Granger-causality in —, 236–238
  identifiability of —, 241–252
  identification of —, 241–252
  impulse response analysis of —, 238, 281
— in standard form, 242
  interpretation of —, 236–238
  invertible —, 222
  linear transformation of —, 231–235
  MA representation of —, 221
  ML estimation of —, 252–278
  nonuniqueness of —, 241–246
  preliminary estimation of —, 265–268
— representation in standard form, 242
  specification of —, 284–298
  stable —, 221
  standard form —, 242
  state space representation of —, 419–420
  transformation of —, 230–236
  VAR representation of —, 222
  VAR(1) representation of —, 223–226
VARMAX model, 323
VARX model, 323, 420, 424 (*see also* dynamic SEM)

random coefficient —, 424–426
Vec operator, 464
Vech operator, 465
Vector autoregressive moving average process (*see* VARMA process)
Vector autoregressive process (*see* VAR process)
Vector differentiation, 467–474
Vector process (*see* vector stochastic process)
Vector stochastic process, 3
periodic process (*see* periodic VAR process)
VAR process (*see* VAR process)
VARMA process (*see* VARMA process)
VARMAX process (*see* VARMAX process)
VARX process (*see* VARX process)
white noise process (*see* white noise)
Vector time series (*see* multiple time series)
Vector white noise, 9

Wald statistic, 93–94, 379, 403, 493
asymptotic distribution of —, 94, 379, 403, 493
Wald test, 494
— for Granger-causality, 93–94, 378–379
— for instantaneous causality, 95–97
— of linear constraints, 93–94, 378–379
— of nonlinear constraints, 493–494
— of zero constraints, 95–96, 403–404
Weak consistency, 487
White noise
covariance matrix of —, 9
Gaussian —, 12, 67
independent —, 29
— process, 9
standard —, 66
testing for —, 139–142, 150–152, 188, 300–301
vector —, 9
zero mean —, 9
White noise assumption
checking of —, 139–142
testing of —, 150–152, 188, 300–301
White noise covariance matrix estimator
asymptotic properties of —, 68–69, 174, 271–272, 358
White noise process (*see* white noise)
Whiteness of residuals
checking the —, 138–152, 188, 300–301
testing for —, 150–152, 188, 300–301
Wold-causality, 52
Wold decomposition, 20
Wold decomposition theorem, 20
Wold representation of a stationary stochastic process, 20

Yule-Walker
— equations, 21, 23
— estimation of VAR process, 78–79
— estimator, 78–79

Zero impulse responses, 45, 55
Zero mean
— VAR process, 30–31
— VARMA process, 226
— white noise, 9
Zero orthogonalized impulse responses, 55

Lightning Source UK Ltd.
Milton Keynes UK
15 December 2009

147564UK00013B/2/P

9 783540 569404